Springer Series in Statistics

For further volumes:
http://www.springer.com/series/692

Jon Wakefield

Bayesian and Frequentist Regression Methods

Springer

Jon Wakefield
Departments of Statistics and Biostatistics
University of Washington
Seattle, Washington
USA

ISSN 0172-7397
ISBN 978-1-4419-0924-4   ISBN 978-1-4419-0925-1 (eBook)
DOI 10.1007/978-1-4419-0925-1
Springer New York Heidelberg Dordrecht London

Library of Congress Control Number: 2012952935

© Springer Science+Business Media New York 2013
This work is subject to copyright. All rights are reserved by the Publisher, whether the whole or part of the material is concerned, specifically the rights of translation, reprinting, reuse of illustrations, recitation, broadcasting, reproduction on microfilms or in any other physical way, and transmission or information storage and retrieval, electronic adaptation, computer software, or by similar or dissimilar methodology now known or hereafter developed. Exempted from this legal reservation are brief excerpts in connection with reviews or scholarly analysis or material supplied specifically for the purpose of being entered and executed on a computer system, for exclusive use by the purchaser of the work. Duplication of this publication or parts thereof is permitted only under the provisions of the Copyright Law of the Publisher's location, in its current version, and permission for use must always be obtained from Springer. Permissions for use may be obtained through RightsLink at the Copyright Clearance Center. Violations are liable to prosecution under the respective Copyright Law.
The use of general descriptive names, registered names, trademarks, service marks, etc. in this publication does not imply, even in the absence of a specific statement, that such names are exempt from the relevant protective laws and regulations and therefore free for general use.
While the advice and information in this book are believed to be true and accurate at the date of publication, neither the authors nor the editors nor the publisher can accept any legal responsibility for any errors or omissions that may be made. The publisher makes no warranty, express or implied, with respect to the material contained herein.

Printed on acid-free paper

Springer is part of Springer Science+Business Media (www.springer.com)

In the order of my meeting them, this book is dedicated to:
   Norma Maureen Wakefield
   Eric Louis Wakefield
   Samantha Louise Wakefield
   Felicity Zoe Moodie
   Eleanor Anna Wakefield
   Eric Stephen Wakefield

Preface

The past 25 years have seen great advances in both Bayesian and frequentist methods for data analysis. The most significant advance for the Bayesian approach has been the development of Markov chain Monte Carlo methods for estimating expectations with respect to the posterior, hence allowing flexible inference and routine implementation for a wide range of models. In particular, this development has led to the more widespread use of hierarchical models for dependent data. With respect to frequentist methods, estimating functions have emerged as a unifying approach for determining the properties of estimators. Generalized estimating equations provide a particularly important example of this methodology that allows inference for dependent data.

The aim of this book is to provide a modern description of Bayesian and frequentist methods of regression analysis and to illustrate the use of these methods on real data. Many books describe one or the other of the Bayesian or frequentist approaches to regression modeling in different contexts, and many mathematical statistics texts describe the theory behind Bayesian and frequentist approaches without providing a detailed description of specific methods. References to such texts are given at the end of Chaps. 2 and 3. Bayesian and frequentist methods are not viewed here as competitive, but rather as complementary techniques, and in this respect this book has some uniqueness.

In embarking on the writing of this book, I have been influenced by many current and former colleagues. My early training was in the Mathematics Department at the University of Nottingham and my first permanent academic teaching position was in the Mathematics Department at Imperial College of Science, Technology and Medicine in London. During this period I was introduced to the Bayesian paradigm and was greatly influenced by Adrian Smith, both as a lecturer and as a Ph.D. adviser. I have also benefited, and continue to benefit, from numerous conversations with Dave Stephens who I have known for over 25 years. Following my move to the University of Washington in Seattle I was exposed to a very modern view of frequentist methods in the Department of Biostatistics. In particular, Scott Emerson, Patrick Heagerty and Thomas Lumley have provided constant stimulation. These interactions, among many others, have influenced the way I now think about statistics, and it is this exposure which I hope has allowed me to write a balanced account of Bayesian and frequentist methods. There is some theory in this book and some data analysis, but the focus is on material that lies between these endeavors and concerns methods. At the University of Washington there is an advanced three-course regression methods sequence and this book arose out of my teaching of the three courses in the sequence.

If modern computers had been available a 100 years ago, the discipline of statistics would have developed in a dramatically different fashion to the way in which it actually evolved. In particular, there would probably be less dependence on linear and generalized linear models, which are mathematically and computationally convenient. While these model classes are still useful and do possess a number of convenient mathematical and computational properties, I believe they should be viewed as just two choices within a far wider range of models that are now available. The approach to modeling that is encouraged in this book is to first specify the model suggested by the background science and to then proceed to examining the mathematical and computational aspects of the model.

As a preparation for this book, the reader is assumed to have a grasp of calculus and linear algebra and have taken first courses in probability and statistical theory. The content of this book is as follows. An introductory chapter describes a number of motivating examples and discusses general issues that need consideration before a regression analysis is carried out. This book is then broken into five parts: I, Inferential Approaches; II, Independent Data; III, Dependent Data; IV, Nonparametric Modeling; V, Appendices. The first two chapters of Part I provide descriptions of the frequentist and Bayesian approaches to inference, with a particular emphasis on the rationale of each approach and a delineation of situations in which one or the other approach is preferable. The third chapter in Part I discusses model selection and hypothesis testing. Part II considers independent data and contains three chapters on the linear model, general regression models (including generalized linear models), and binary data models. The two chapters of Part III consider dependent data with linear models and general regression models. Mixed models and generalized estimating equations are the approaches to inference that are emphasized. Part IV contains three chapters on nonparametric modeling with an emphasis on spline and kernel methods. The examples and simulation studies of this book were almost exclusively carried out within the freely available R programming environment. The code for the examples and figures may be found at:

> http://faculty.washington.edu/jonno/regression-methods.html

along with the inevitable errata and links to datasets. Exercises are included at the end of all chapters but the first. Many of these exercises concern analyses of real data. In my own experience, a full understanding of methods requires their implementation and application to data.

In my own teaching I have based three one-quarter courses on the following. *Regression Methods for Independent Data* is based on Part II, dipping into topics in Part I as needed and using motivating examples from Chap. 1. *Regression Methods for Dependent Data* centers on Part II, again using examples from Chap. 1, and building on the independent data material. Finally, *Nonparametric Regression and Classification* is based on the material in Part IV. The latter course is stand-alone in the sense of not requiring the independent and dependent data courses though extra material on a number of topics, including linear and generalized linear models and mixed models, will need to be included if not previously encountered.

In the 2003–2004 academic year I was the Genentech Professor and received funding specifically to work on this book. The staff at Springer have been very helpful at all stages. John Kimmel was the editor during most of the writing of this book and I am appreciative of his gentle prodding and advice. About 18 months from the completion of this book, Marc Strauss stepped in and has also been very supportive. Many of my colleagues have given comments on various chapters, but I would like to specifically thank Lurdes Inoue, Katie Kerr, Erica Moodie, Zoe Moodie, Ken Rice, Dave Stephens, Jon Wellner, Daniela Witten, and Simon Wood for feedback on different parts of this book. Finally, lest we forget, I would like to thank all of those students who suffered through initial presentations of this material—I hope your sacrifices were not in vain...

Seattle, WA   Jon Wakefield
June 2012

Contents

1 **Introduction and Motivating Examples** 1
   1.1   Introduction 1
   1.2   Model Formulation 1
   1.3   Motivating Examples 5
         1.3.1   Prostate Cancer 5
         1.3.2   Outcome After Head Injury 9
         1.3.3   Lung Cancer and Radon 10
         1.3.4   Pharmacokinetic Data 12
         1.3.5   Dental Growth 16
         1.3.6   Spinal Bone Mineral Density 18
   1.4   Nature of Randomness... 20
   1.5   Bayesian and Frequentist Inference 22
   1.6   The Executive Summary 23
   1.7   Bibliographic Notes.. 24

Part I Inferential Approaches

2 **Frequentist Inference** 27
   2.1   Introduction 27
   2.2   Frequentist Criteria 29
   2.3   Estimating Functions 32
   2.4   Likelihood 36
         2.4.1   Maximum Likelihood Estimation 36
         2.4.2   Variants on Likelihood 44
         2.4.3   Model Misspecification 46
   2.5   Quasi-likelihood 49
         2.5.1   Maximum Quasi-likelihood Estimation 49
         2.5.2   A More Complex Mean–Variance Model 53
   2.6   Sandwich Estimation 56
   2.7   Bootstrap Methods 63
         2.7.1   The Bootstrap for a Univariate Parameter............ 64

		2.7.2	The Bootstrap for Regression	66
		2.7.3	Sandwich Estimation and the Bootstrap	66
	2.8	Choice of Estimating Function		70
	2.9	Hypothesis Testing		72
		2.9.1	Motivation	72
		2.9.2	Preliminaries	73
		2.9.3	Score Tests	74
		2.9.4	Wald Tests	75
		2.9.5	Likelihood Ratio Tests	75
		2.9.6	Quasi-likelihood	76
		2.9.7	Comparison of Test Statistics	77
	2.10	Concluding Remarks		79
	2.11	Bibliographic Notes		80
	2.12	Exercises		80
3	**Bayesian Inference**			**85**
	3.1	Introduction		85
	3.2	The Posterior Distribution and Its Summarization		86
	3.3	Asymptotic Properties of Bayesian Estimators		89
	3.4	Prior Choice		90
		3.4.1	Baseline Priors	90
		3.4.2	Substantive Priors	93
		3.4.3	Priors on Meaningful Scales	95
		3.4.4	Frequentist Considerations	96
	3.5	Model Misspecification		99
	3.6	Bayesian Model Averaging		100
	3.7	Implementation		102
		3.7.1	Conjugacy	102
		3.7.2	Laplace Approximation	106
		3.7.3	Quadrature	107
		3.7.4	Integrated Nested Laplace Approximations	109
		3.7.5	Importance Sampling Monte Carlo	110
		3.7.6	Direct Sampling Using Conjugacy	112
		3.7.7	Direct Sampling Using the Rejection Algorithm	114
	3.8	Markov Chain Monte Carlo		121
		3.8.1	Markov Chains for Exploring Posterior Distributions	121
		3.8.2	The Metropolis–Hastings Algorithm	122
		3.8.3	The Metropolis Algorithm	123
		3.8.4	The Gibbs Sampler	123
		3.8.5	Combining Markov Kernels: Hybrid Schemes	125
		3.8.6	Implementation Details	125
		3.8.7	Implementation Summary	133
	3.9	Exchangeability		134
	3.10	Hypothesis Testing with Bayes Factors		137
	3.11	Bayesian Inference Based on a Sampling Distribution		140
	3.12	Concluding Remarks		143

	3.13	Bibliographic Notes	145
	3.14	Exercises	145

4 Hypothesis Testing and Variable Selection … 153
 4.1 Introduction … 153
 4.2 Frequentist Hypothesis Testing … 153
  4.2.1 Fisherian Approach … 154
  4.2.2 Neyman–Pearson Approach … 154
  4.2.3 Critique of the Fisherian Approach … 154
  4.2.4 Critique of the Neyman–Pearson Approach … 155
 4.3 Bayesian Hypothesis Testing with Bayes Factors … 156
  4.3.1 Overview of Approaches … 156
  4.3.2 Critique of the Bayes Factor Approach … 158
  4.3.3 A Bayesian View of Frequentist Hypothesis Testing … 159
 4.4 The Jeffreys–Lindley Paradox … 161
 4.5 Testing Multiple Hypotheses: General Considerations … 164
 4.6 Testing Multiple Hypotheses: Fixed Number of Tests … 165
  4.6.1 Frequentist Analysis … 166
  4.6.2 Bayesian Analysis … 171
 4.7 Testing Multiple Hypotheses: Variable Selection … 178
 4.8 Approaches to Variable Selection and Modeling … 179
  4.8.1 Stepwise Methods … 181
  4.8.2 All Possible Subsets … 183
  4.8.3 Bayesian Model Averaging … 185
  4.8.4 Shrinkage Methods … 185
 4.9 Model Building Uncertainty … 185
 4.10 A Pragmatic Compromise to Variable Selection … 188
 4.11 Concluding Comments … 189
 4.12 Bibliographic Notes … 190
 4.13 Exercises … 190

Part II Independent Data

5 Linear Models … 195
 5.1 Introduction … 195
 5.2 Motivating Example: Prostate Cancer … 195
 5.3 Model Specification … 196
 5.4 A Justification for Linear Modeling … 198
 5.5 Parameter Interpretation … 199
  5.5.1 Causation Versus Association … 199
  5.5.2 Multiple Parameters … 201
  5.5.3 Data Transformations … 205
 5.6 Frequentist Inference … 209
  5.6.1 Likelihood … 209
  5.6.2 Least Squares Estimation … 214

	5.6.3	The Gauss–Markov Theorem	215
	5.6.4	Sandwich Estimation	216
5.7	Bayesian Inference	221
5.8	Analysis of Variance	224
	5.8.1	One-Way ANOVA	224
	5.8.2	Crossed Designs	227
	5.8.3	Nested Designs	229
	5.8.4	Random and Mixed Effects Models	230
5.9	Bias-Variance Trade-Off	231
5.10	Robustness to Assumptions	236
	5.10.1	Distribution of Errors	237
	5.10.2	Nonconstant Variance	237
	5.10.3	Correlated Errors	238
5.11	Assessment of Assumptions	239
	5.11.1	Review of Assumptions	239
	5.11.2	Residuals and Influence	240
	5.11.3	Using the Residuals	243
5.12	Example: Prostate Cancer	245
5.13	Concluding Remarks	247
5.14	Bibliographic Notes	248
5.15	Exercises	249

6 General Regression Models ... 253

6.1	Introduction	253
6.2	Motivating Example: Pharmacokinetics of Theophylline	254
6.3	Generalized Linear Models	256
6.4	Parameter Interpretation	259
6.5	Likelihood Inference for GLMs	260
	6.5.1	Estimation	260
	6.5.2	Computation	263
	6.5.3	Hypothesis Testing	267
6.6	Quasi-likelihood Inference for GLMs	270
6.7	Sandwich Estimation for GLMs	272
6.8	Bayesian Inference for GLMs	273
	6.8.1	Prior Specification	273
	6.8.2	Computation	274
	6.8.3	Hypothesis Testing	275
	6.8.4	Overdispersed GLMs	276
6.9	Assessment of Assumptions for GLMs	278
6.10	Nonlinear Regression Models	283
6.11	Identifiability	284
6.12	Likelihood Inference for Nonlinear Models	285
	6.12.1	Estimation	285
	6.12.2	Hypothesis Testing	287
6.13	Least Squares Inference	289
6.14	Sandwich Estimation for Nonlinear Models	290

	6.15	The Geometry of Least Squares	291
	6.16	Bayesian Inference for Nonlinear Models	294
		6.16.1 Prior Specification	294
		6.16.2 Computation	294
		6.16.3 Hypothesis Testing	295
	6.17	Assessment of Assumptions for Nonlinear Models	298
	6.18	Concluding Remarks	299
	6.19	Bibliographic Notes	299
	6.20	Exercises	300

7 Binary Data Models 305
- 7.1 Introduction 305
- 7.2 Motivating Examples 306
  - 7.2.1 Outcome After Head Injury 306
  - 7.2.2 Aircraft Fasteners 306
  - 7.2.3 Bronchopulmonary Dysplasia 307
- 7.3 The Binomial Distribution 308
  - 7.3.1 Genesis 308
  - 7.3.2 Rare Events 309
- 7.4 Generalized Linear Models for Binary Data 310
  - 7.4.1 Formulation 310
  - 7.4.2 Link Functions 312
- 7.5 Overdispersion 313
- 7.6 Logistic Regression Models 316
  - 7.6.1 Parameter Interpretation 316
  - 7.6.2 Likelihood Inference for Logistic Regression Models 318
  - 7.6.3 Quasi-likelihood Inference for Logistic Regression Models 321
  - 7.6.4 Bayesian Inference for Logistic Regression Models 321
- 7.7 Conditional Likelihood Inference 327
- 7.8 Assessment of Assumptions 331
- 7.9 Bias, Variance, and Collapsibility 334
- 7.10 Case-Control Studies 337
  - 7.10.1 The Epidemiological Context 337
  - 7.10.2 Estimation for a Case-Control Study 338
  - 7.10.3 Estimation for a Matched Case-Control Study 341
- 7.11 Concluding Remarks 343
- 7.12 Bibliographic Notes 344
- 7.13 Exercises 345

Part III Dependent Data

8 Linear Models 353
- 8.1 Introduction 353
- 8.2 Motivating Example: Dental Growth Curves 354

8.3		The Efficiency of Longitudinal Designs	356
8.4		Linear Mixed Models	359
	8.4.1	The General Framework	359
	8.4.2	Covariance Models for Clustered Data	360
	8.4.3	Parameter Interpretation for Linear Mixed Models	363
8.5		Likelihood Inference for Linear Mixed Models	364
	8.5.1	Inference for Fixed Effects	365
	8.5.2	Inference for Variance Components via Maximum Likelihood	367
	8.5.3	Inference for Variance Components via Restricted Maximum Likelihood	368
	8.5.4	Inference for Random Effects	376
8.6		Bayesian Inference for Linear Mixed Models	381
	8.6.1	A Three-Stage Hierarchical Model	381
	8.6.2	Hyperpriors	382
	8.6.3	Implementation	386
	8.6.4	Extensions	388
8.7		Generalized Estimating Equations	391
	8.7.1	Motivation	391
	8.7.2	The GEE Algorithm	392
	8.7.3	Estimation of Variance Parameters	395
8.8		Assessment of Assumptions	400
	8.8.1	Review of Assumptions	400
	8.8.2	Approaches to Assessment	402
8.9		Cohort and Longitudinal Effects	413
8.10		Concluding Remarks	416
8.11		Bibliographic Notes	416
8.12		Exercises	417

9 General Regression Models 425

9.1		Introduction	425
9.2		Motivating Examples	426
	9.2.1	Contraception Data	426
	9.2.2	Seizure Data	427
	9.2.3	Pharmacokinetics of Theophylline	428
9.3		Generalized Linear Mixed Models	430
9.4		Likelihood Inference for Generalized Linear Mixed Models	432
9.5		Conditional Likelihood Inference for Generalized Linear Mixed Models	437
9.6		Bayesian Inference for Generalized Linear Mixed Models	441
	9.6.1	Model Formulation	441
	9.6.2	Hyperpriors	441
9.7		Generalized Linear Mixed Models with Spatial Dependence	445
	9.7.1	A Markov Random Field Prior	445
	9.7.2	Hyperpriors	447

9.8	Conjugate Random Effects Models		450
9.9	Generalized Estimating Equations for Generalized Linear Models		451
9.10	GEE2: Connected Estimating Equations		452
9.11	Interpretation of Marginal and Conditional Regression Coefficients		455
9.12	Introduction to Modeling Dependent Binary Data		457
9.13	Mixed Models for Binary Data		458
	9.13.1	Generalized Linear Mixed Models for Binary Data	458
	9.13.2	Likelihood Inference for the Binary Mixed Model	462
	9.13.3	Bayesian Inference for the Binary Mixed Model	462
	9.13.4	Conditional Likelihood Inference for Binary Mixed Models	465
9.14	Marginal Models for Dependent Binary Data		467
	9.14.1	Generalized Estimating Equations	467
	9.14.2	Loglinear Models	468
	9.14.3	Further Multivariate Binary Models	471
9.15	Nonlinear Mixed Models		475
9.16	Parameterization of the Nonlinear Model		477
9.17	Likelihood Inference for the Nonlinear Mixed Model		479
9.18	Bayesian Inference for the Nonlinear Mixed Model		482
	9.18.1	Hyperpriors	482
	9.18.2	Inference for Functions of Interest	484
9.19	Generalized Estimating Equations		487
9.20	Assessment of Assumptions for General Regression Models		489
9.21	Concluding Remarks		492
9.22	Bibliographic Notes		495
9.23	Exercises		496

Part IV Nonparametric Modeling

10 Preliminaries for Nonparametric Regression 503

10.1	Introduction		503
10.2	Motivating Examples		504
	10.2.1	Light Detection and Ranging	505
	10.2.2	Ethanol Data	505
10.3	The Optimal Prediction		506
	10.3.1	Continuous Responses	507
	10.3.2	Discrete Responses with K Categories	508
	10.3.3	General Responses	510
	10.3.4	In Practice	511
10.4	Measures of Predictive Accuracy		511
	10.4.1	Continuous Responses	512
	10.4.2	Discrete Responses with K Categories	515
	10.4.3	General Responses	517

	10.5	A First Look at Shrinkage Methods	517
		10.5.1 Ridge Regression	517
		10.5.2 The Lasso	523
	10.6	Smoothing Parameter Selection	526
		10.6.1 Mallows C_P	527
		10.6.2 K-Fold Cross-Validation	529
		10.6.3 Generalized Cross-Validation	532
		10.6.4 AIC for General Models	534
		10.6.5 Cross-Validation for Generalized Linear Models	538
	10.7	Concluding Comments	542
	10.8	Bibliographic Notes	543
	10.9	Exercises	543
11	**Spline and Kernel Methods**		**547**
	11.1	Introduction	547
	11.2	Spline Methods	547
		11.2.1 Piecewise Polynomials and Splines	547
		11.2.2 Natural Cubic Splines	552
		11.2.3 Cubic Smoothing Splines	553
		11.2.4 B-Splines	556
		11.2.5 Penalized Regression Splines	557
		11.2.6 A Brief Spline Summary	560
		11.2.7 Inference for Linear Smoothers	560
		11.2.8 Linear Mixed Model Spline Representation: Likelihood Inference	563
		11.2.9 Linear Mixed Model Spline Representation: Bayesian Inference	567
	11.3	Kernel Methods	572
		11.3.1 Kernels	574
		11.3.2 Kernel Density Estimation	575
		11.3.3 The Nadaraya–Watson Kernel Estimator	578
		11.3.4 Local Polynomial Regression	580
	11.4	Variance Estimation	584
	11.5	Spline and Kernel Methods for Generalized Linear Models	587
		11.5.1 Generalized Linear Models with Penalized Regression Splines	587
		11.5.2 A Generalized Linear Mixed Model Spline Representation	591
		11.5.3 Generalized Linear Models with Local Polynomials	592
	11.6	Concluding Comments	593
	11.7	Bibliographic Notes	593
	11.8	Exercises	594

12	**Nonparametric Regression with Multiple Predictors**		597
	12.1	Introduction	597
	12.2	Generalized Additive Models	598
		12.2.1 Model Formulation	598
		12.2.2 Computation via Backfitting	599
	12.3	Spline Methods with Multiple Predictors	601
		12.3.1 Natural Thin Plate Splines	602
		12.3.2 Thin Plate Regression Splines	603
		12.3.3 Tensor Product Splines	604
	12.4	Kernel Methods with Multiple Predictors	607
	12.5	Smoothing Parameter Estimation	608
		12.5.1 Conventional Approaches	608
		12.5.2 Mixed Model Formulation	608
	12.6	Varying-Coefficient Models	610
	12.7	Regression Trees	614
		12.7.1 Hierarchical Partitioning	614
		12.7.2 Multiple Adaptive Regression Splines	622
	12.8	Classification	624
		12.8.1 Logistic Models with K Classes	625
		12.8.2 Linear and Quadratic Discriminant Analysis	626
		12.8.3 Kernel Density Estimation and Classification	630
		12.8.4 Classification Trees	634
		12.8.5 Bagging	636
		12.8.6 Random Forests	639
	12.9	Concluding Comments	643
	12.10	Bibliographic Notes	644
	12.11	Exercises	644

Part V Appendices

A	**Differentiation of Matrix Expressions**	649
B	**Matrix Results**	653
C	**Some Linear Algebra**	655
D	**Probability Distributions and Generating Functions**	657
E	**Functions of Normal Random Variables**	667
F	**Some Results from Classical Statistics**	669
G	**Basic Large Sample Theory**	673

References	675
Index	689

Chapter 1
Introduction and Motivating Examples

1.1 Introduction

This book examines how a response is related to covariates using mathematical models whose unknown parameters we wish to estimate using available information—this endeavor is known as *regression analysis*. In this first chapter, we will begin in Sect. 1.2 by making some general comments about model formulation. In Sect. 1.3, a number of examples will be described in order to motivate the material to follow in the remainder of this book. In Sect. 1.4, we examine, in simple idealized scenarios, how "randomness" is induced by not controlling for covariates in a model. Section 1.5 briefly contrasts the Bayesian and frequentist approaches to inference, and Sect. 1.7 gives references that expand on the material of this chapter. Finally, Sect. 1.6 summarizes the overall message of this book which is that in many instances, carefully thought out Bayesian and frequentist analyses will provide similar conclusions; however, situations in which one or the other approach may be preferred are also described.

1.2 Model Formulation

In a regression analysis, the following steps may be followed:

1. Formulate a model based on the nature of the data, the subject matter context, and the aims of the data analysis.
2. Examine the mathematical properties of the initial model with respect to candidate inference procedures. This examination will focus on whether specific methods are suited to both the particular context under consideration and the specific questions of interest in the analysis.
3. Consider the computational aspects of the model.

The examination in steps 2 and 3 may suggest that we need to change the model.[1] Historically, the range of model forms that were available for regression modeling was severely limited by computational and, to a lesser extent, mathematical considerations. For example, though *generalized linear models* contain a flexible range of alternatives to the linear model, a primary motivation for their formulation was ease of fitting and mathematical tractability. Hence, step 3 in particular took precedent over step 1.

Specific aspects of the initial model formulation will now be discussed in more detail. When carrying out a regression analysis, careful consideration of the following issues is vital and in many instances will outweigh in importance the particular model chosen or estimation method used. The interpretation of parameters also depends vitally on the following issues.

Observational Versus Experimental Data

An important first step in data analysis is to determine whether the data are experimental or observational in nature. In an experimental study, the experimenter has control over at least some aspects of the study. For example, units (e.g., patients) may be randomly assigned to covariate groups of interest (e.g., treatment groups). If this randomization is successfully implemented, any differences in response will (in expectation) be due to group assignment only, allowing a causal interpretation of the estimated parameters. The beauty of randomization is that the groups are balanced with respect to all covariates, crucially including those that are *unobserved*.

In an observational study, we never know whether observed differences between the responses of groups of interest are due, at least partially, to other "confounding" variables related to group membership. If the confounders are measured, then there is some hope for controlling for the variability in response that is not due to group membership, but if the confounders are unobserved variables, then such control is not possible. In the epidemiology and biostatistics literature, this type of discrepancy between the estimate and the "true" quantity of interest is often described as bias due to confounding. In later chapters, this issue will be examined in detail, since it is a primary motivation for regression modeling. In observational studies, estimated coefficients are traditionally described as *associations*, and causality is only alluded to more informally via consideration of the combined evidence of different studies and scientific plausibility. We expand upon this discussion in Sect. 1.4.

Predictive models are more straightforward to build than causal models. To quote Freedman (1997), "For description and prediction, the numerical values of the individual coefficients fade into the background; it is the whole linear combination on the right-hand side of the equation that matters. For causal inference, it is the individual coefficients that do the trick."

---

[1] To make clear, we are not suggesting refining the model based on inadequacies of fit; this is a dangerous enterprise, as we discuss in Chap. 4.

Study Population

Another important step is to determine the population from which the data were collected so that the individuals to whom inferential conclusions apply may be determined. Extrapolation of inference beyond the population providing the data is a risky enterprise.

Throughout this book, we will take a superpopulation view in which probability models are assumed to describe variability with respect to a hypothetical, infinite population. The study population that exists in practice consists of N units, of which n are sampled. To summarize:

$$\text{Superpopulation} (\infty) \quad \rightarrow \quad \text{Study Population} (N) \quad \rightarrow \quad \text{Sample} (n)$$

Inference for the parameters of a superpopulation may be contrasted with a survey sampling perspective in which the focus is upon characteristics of the responses of the N units; in the latter case, a full census ($n = N$) will obviate the need for statistical analysis.

The Sampling Scheme

The data collection procedure has implications for the analysis, in terms of the models that are appropriate, the questions that may be asked, and the inferential approach that may be adopted. In the most straightforward case, the data arise through random sampling from a well-defined population. In other situations, the random samples may be drawn from within covariate-defined groups, which may improve efficiency of estimation by concentrating the sampling in informative groups but may limit the range of questions that can be answered by the data due to the restrictions on the sampling scheme. In more complex situations, the data may result from outcome-dependent sampling. For example, a case-control study is an outcome-dependent sampling scheme in which the binary response of interest is fixed by design, and the random variables are the covariates sampled within each of the outcome categories (cases and controls). For such data, care is required because the majority of conventional approaches will not produce valid inference, and analysis is carried out most easily using logistic regression models. Similar issues are encountered in the analysis of matched case-control studies, in which cases and controls are matched upon additional (confounder) variables. Bias in parameters of interest will occur if such data are analyzed using methods for unmatched studies, again because the sampling scheme has not been acknowledged. In the case of individually matched cases and controls (in which, for example, for each case a control is picked with the same gender, age, and race), conventional likelihood-based methods are flawed because the number of parameters (including one parameter for each case-control pair) increases with the sample size (providing an example of the importance of paying attention to the regularity conditions required for valid inference)—*conditional* likelihood provides a valid inferential approach in this case. The analysis of data from case-control studies is described in Chap. 7.

Missing Data

Measurements may be missing on the responses which can lead to bias in estimation, depending on the reasons for the absence. It is clear that bias will arise when the probability of missingness depends on the size of the response that would have been observed. An extreme example is when the result of a chemical assay is reported as "below the lower limit of detection"; such a variable may be reported as the (known) lower limit, or as a zero, and analyzing the data using these values can lead to substantial bias. Removing these observations will also lead to bias. In the analysis of individual-level data over time (to give so-called longitudinal data) another common mechanism for missing observations is when individuals drop out of the study.

Aim of the Analysis

The primary aim of the analysis should always be kept in mind; in particular, is the purpose descriptive, exploratory (e.g., for hypothesis generation), confirmatory (with respect to an a priori hypothesis), or predictive? Regression models can be used for each of these endeavors, but the manner of their use will vary. Large data sets can often be succinctly described using parsimonious[2] regression models. Exploratory studies are often informal in nature, and many different models may be fitted in order to gain insights into the structure of the data. In general, however, great care must be taken with data dredging since spurious associations may be discovered due to chance alone.

The level of sophistication of the analysis, and the assumptions required, will vary as the aims and abundance of data differ. For example, if one has a million observations independently sampled from a population, and one requires inference for the mean of the population, then inference may be based on the sample mean and sample standard deviation alone, without recourse to more sophisticated models and approaches—we would expect such inference to be reliable, being based on few assumptions. Similarly, inference is straightforward if we are interested in the average response at an observed covariate value for which abundant data were recorded.

---

[2]The Oxford English Dictionary describes *parsimony* as "...that no more causes or forces should be assumed than are necessary to account for the facts," which serves our purposes, though care is required in the use of the words "causes," "forces," and "facts."

However, if such data are not available (e.g., when the number of covariates becomes large or the sample size is small), or if interpolation is required, regression models are beneficial, as they allow the totality of the data to estimate global parameters and smooth across unstructured variability. To answer many statistical questions, very simple approaches will often suffice; the *art* of statistical analysis is deciding upon when a more sophisticated approach is necessary/warranted, since dependence on assumptions usually increases with increasing sophistication.

1.3 Motivating Examples

We now introduce a number of examples to illustrate different data collection procedures, types of data, and study aims. We highlight the distinguishing features of the data in each example and provide a signpost to the chapter in which appropriate methods of analysis may be found.

In general, data $\{Y_i, x_i, i = 1, \ldots, n\}$ will be available on n units, with Y_i representing the univariate response variable and $x_i = [1, x_{i1}, \ldots, x_{ik}]$ the row vector of explanatory variables on unit i. Variables written as uppercase letters will represent random variables, and those in lowercase fixed quantities, with boldface representing vectors and matrices.

1.3.1 Prostate Cancer

We describe a dataset analyzed by Tibshirani (1996) and originally presented by Stamey et al. (1989). The data were collected on $n = 97$ men before radical prostatectomy, which is a major surgical operation that removes the entire prostate gland along with some surrounding tissue. We take as response, Y, the log of prostate specific antigen (PSA); PSA is a concentration and is measured in ng/ml. In Stamey et al. (1989), PSA was proposed as a preoperative marker to predict the clinical stage of cancer. As well as modeling the stage of cancer as a function of PSA, the authors also examined PSA as a function of age and seven other histological and morphometric covariates. We take as our aim the building of a predictive model for PSA, using the eight covariates:

- log(can vol): The log of cancer volume, measured in milliliters (cc). The area of cancer was measured from digitized images and multiplied by a thickness to produce a volume.
- log(weight): The log of the prostate weight, measured in grams.
- Age: The age of the patient, in years.
- log(BPH): The log of the amount of benign prostatic hyperplasia (BPH), a noncancerous enlargement of the prostate gland, as an area in a digitized image and reported in cm^2.

- SVI: The seminal vesicle invasion, a 0/1 indicator of whether prostate cancer cells have invaded the seminal vesicle.
- log(cap pen): The log of the capsular penetration, which represents the level of extension of cancer into the capsule (the fibrous tissue which acts as an outer lining of the prostate gland). Measured as the linear extent of penetration, in cm.
- Gleason: The Gleason score, a measure of the degree of aggressiveness of the tumor. The Gleason grading system assigns a grade (1–5) to each of the two largest areas of cancer in the tissue samples with 1 being the least aggressive and 5 the most aggressive; the two grades are then added together to produce the Gleason score.
- PGS45: The percentage of Gleason scores that are 4 or 5.

The BPH and capsular penetration variables originally contained zeros, and a small number was substituted before the log transform was taken. It is not clear from the original paper why the log transform was taken though PSA varies over a wide range, and so linearity of the mean model may be aided by the log transform. It is also not clear why the variable PGS45 was constructed. If initial analyses were carried out to find variables that were associated with PSA, then significance levels of hypothesis tests will not be accurate (since they are not based on an a priori hypotheses but rather are the result of data dredging).

Carrying out exploratory data analysis (EDA) is a vital step in any data analysis. Such an enterprise includes the graphical and tabular examination of variables, the checking of the data for errors (for example, to see if variables are within their admissible ranges), and the identification of outlying (unusual) observations or influential observations that when perturbed lead to large changes in inference. This book is primarily concerned with methods, and the level of EDA that is performed will be less than would be desirable in a serious data analysis.

Figure 1.1 displays the response plotted against each of the covariates and indicates a number of associations. The association between Y and log(can vol) appears particularly strong. In observational settings such as this, there are often strong dependencies between the covariates. We may investigate these dependencies using scatterplots (or tables, if both variables are discrete). Figure 1.2 gives an indication of the dependencies between those variables that exhibit the strongest associations; log(can vol) is strongly associated with a number of other covariates. Consequently, we might expect that adding log(can vol) to a model for log(PSA) that contains other covariates will change the estimated associations between log(PSA) and the other variables.

We define Y_i as the log of PSA and $x_i = [1, x_{i1}, \ldots, x_{i8}]$ as the 1×9 row vector associated with patient i, $i = 1, \ldots, n = 97$. We may write a general mean model as $E[Y_i \mid x_i] = f(x_i, \beta)$ where $f(\cdot, \cdot)$ represents the functional form and β unknown regression parameters. The most straightforward form is the multiple linear regression

$$f(x_i, \beta) = \beta_0 + \sum_{j \in C} x_{ij} \beta_j, \tag{1.1}$$

1.3 Motivating Examples

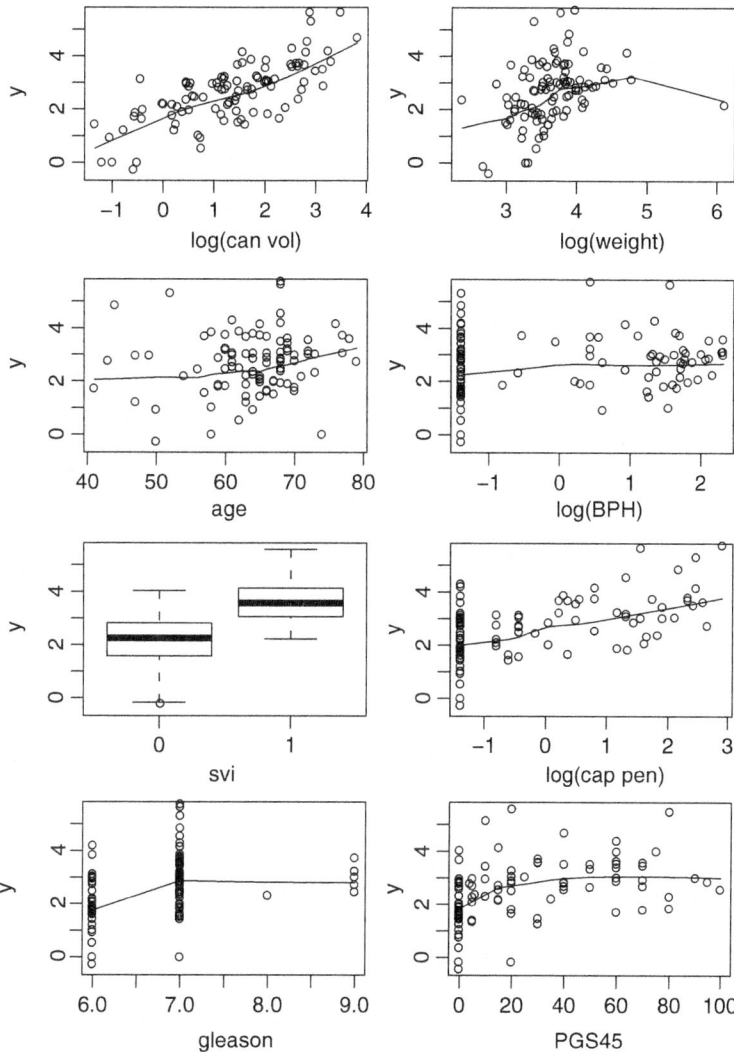

Fig. 1.1 The response $y = \log(\text{PSA})$ plotted versus each of the eight explanatory variables, x, in the prostate cancer study, with local smoothers superimposed for continuous covariates

where C corresponds to the subset of elements of $\{1, 2, \ldots, 8\}$ whose associated covariates we wish to include in the model and $\beta = [\beta_0, \{\beta_j, j \in C\}]^\mathrm{T}$. The interpretation of each of the coefficients β_j depends crucially on knowing the scaling and units of measurement of the associated variables x_j.

Most of the x variables in this study are measured with error (as is clear from their derivation, e.g., log(BPH) is derived from a digitized image), and if we are interested in estimating causal effects, then this aspect needs to be acknowledged

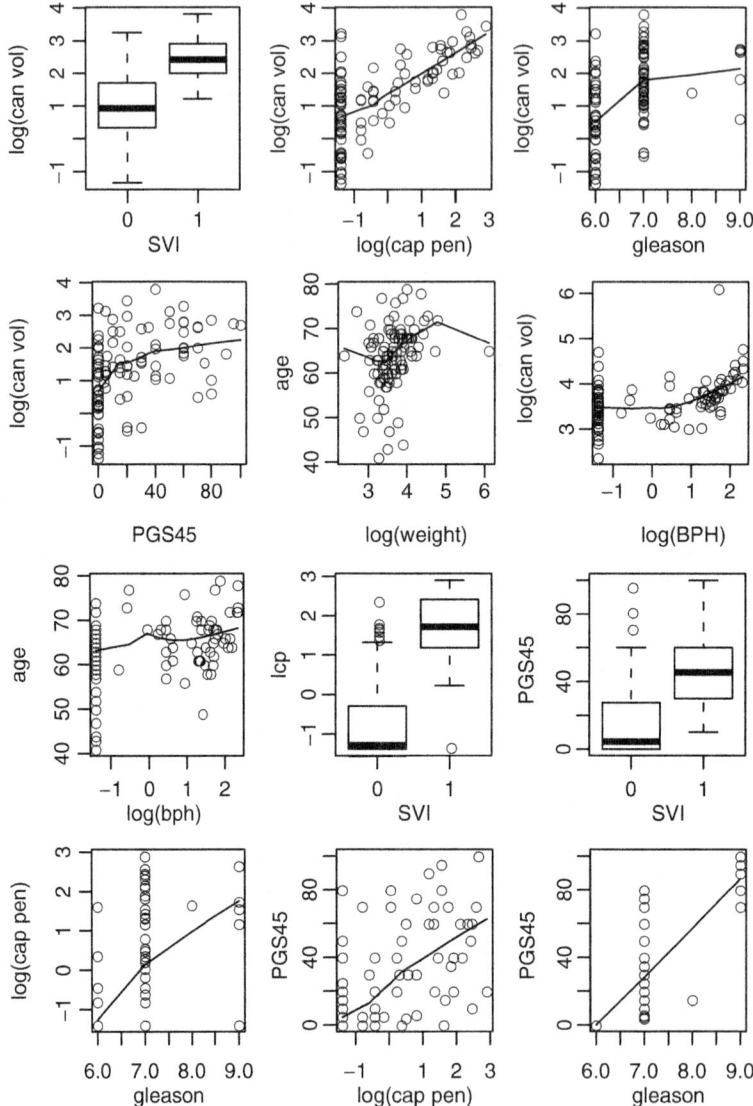

Fig. 1.2 Associations between selected explanatory variables in the prostate cancer study, with local smoothers superimposed for continuous covariates

in the models that are fitted, since inference is affected in this situation, which is known as *errors-in-variables*.

Distinguishing Features. Inference for multiple linear regression models is described in Chap. 5, including a discussion of parameter interpretation. Chapter 4 discusses the difficult but important topics of model formulation and selection.

1.3 Motivating Examples

Table 1.1 Outcome after head injury as a function of four covariates: pupils, hematoma present, coma score, and age

	Pupils	Good				Poor			
	Hematoma present	No		Yes		No		Yes	
	Coma score	Low	High	Low	High	Low	High	Low	High
	1–25 Dead	9	5	5	7	58	11	32	12
	Alive	47	77	11	24	29	24	13	16
Age	26–54 Dead	19	6	21	14	45	7	61	15
(years)	Alive	15	44	18	38	11	16	11	21
	≥55 Dead	7	12	19	25	20	7	42	17
	Alive	1	6	2	15	0	2	7	7

1.3.2 Outcome After Head Injury

Table 1.1 reports data presented by Titterington et al. (1981) in a study initiated by the Institute of Neurological Sciences in Glasgow. These data were collected prospectively by neurosurgeons between 1968 and 1976. The original aim was to predict recovery for individual patients on the basis of data collected shortly after the injury. The data that we consider contain information on a binary outcome, $Y = 0/1$, corresponding to dead/alive after head injury, and the covariates: pupils (with good corresponding to a reaction to light and poor to no reaction), coma score (representing depth of coma, low or high), hematoma present (no/yes), and age (categorized as 1–25, 26–54, ≥55).

The response of interest here is $p(\boldsymbol{x}) = \Pr(Y = 1 \mid \boldsymbol{x})$; the probability that a patient with covariates $\boldsymbol{x}$ is alive. This quantity must lie in the range $[0,1]$, and so, at least in this respect, linear models are unappealing. To illustrate, suppose we have a univariate continuous covariate x and the model

$$p(x) = \beta_0 + \beta_1 x.$$

While probabilities not close to zero or one may change at least approximately linearly with x, it is extremely unlikely that this behavior will extend to the extremes, where the probability–covariate relationship must flatten out in order to remain in the correct range. An additional, important, consideration is that linear models commonly assume that the variance is constant and, in particular, does not depend on the mean. For a binary outcome with probability of response $p(x)$, the Bernoulli variance is $p(x)[1 - p(x)]$ and so depends on the mean. As we will see, accurate inference depends crucially on having modeled the mean–variance relationship appropriately.

A common model for binary data is the logistic regression model, in which the odds of death, $p(x)/[1 - p(x)]$, is modeled as a function of x. For example, the linear logistic regression model is

$$\frac{p(x)}{1 - p(x)} = \exp(\beta_0 + \beta_1 x).$$

This form is mathematically appealing, since the modeled probabilities are constrained to lie within [0,1], though the interpretation of the parameters β_0 and β_1 is not straightforward.

Distinguishing Features. Chapter 7 is dedicated to the modeling of binary data. In this chapter, logistic regression models are covered in detail, along with alternatives. Formulating predictive models and assessing the predictive power of such models is considered in Chaps. 10–12.

1.3.3 Lung Cancer and Radon

We now describe an example in which the data arise from a spatial ecological study. In an ecological study, the unit of analysis is the group rather than the individual. In spatial epidemiological studies, due primarily to reasons of confidentiality, data on disease, population, and exposure are often available as aggregates across area. It is these areas that constitute the (ecological) group level at which the data are analyzed. In this example, we examine the association between lung cancer incidence (over the years 1998–2002) and residential radon at the level of the county, in Minnesota. Radon is a naturally occurring radioactive gas produced by the breakdown of uranium in soil, rock, and water and is a known carcinogen for lung cancer (Darby et al. 2001). However, in many ecological studies, when the association between lung cancer incidence and residential radon is estimated, radon appears protective. *Ecological bias* is an umbrella term that refers to the distortion of individual-level associations due to the process of aggregation. There are many facets to ecological bias (Wakefield 2008), but an important issue in the lung cancer/radon context is the lack of control for confounding, a primary source being smoking.

Let Y_i denote the lung cancer incidence count and x_i the average radon in county $i = 1, \ldots, n = 87$. Age and gender are strongly associated with lung cancer incidence, and a standard approach to controlling these factors is to form *expected counts* $E_i = \sum_{j=1}^{J} N_{ij} q_j$ in which we multiply the population in stratum j and county i, N_{ij}, by a "reference" probability of lung cancer in stratum j, q_j, to obtain the expected count in stratum j. Summing over all J stratum gives the total expected count. Intuitively, these counts are what we would expect if the disease rates in county i conform with the reference. A summary response measure in county i is the standardized morbidity ratio (SMR), given by Y_i/E_i. Counties with SMRs greater than 1 have an excess of cases, when compared to that expected.

Figure 1.3 maps the SMRs in counties of Minnesota, and we observe more than twofold variability with areas of high incidence in the northeast of the state. Figure 1.4 maps the average radon by county, with low radon in the counties to the northeast. This negative association is confirmed in Fig. 1.5 in which we plot the SMRs versus average radon, with a smoother indicating the local trend.

1.3 Motivating Examples

Fig. 1.3 Standardized morbidity ratios for lung cancer in the period 1998–2002 by county in Minnesota

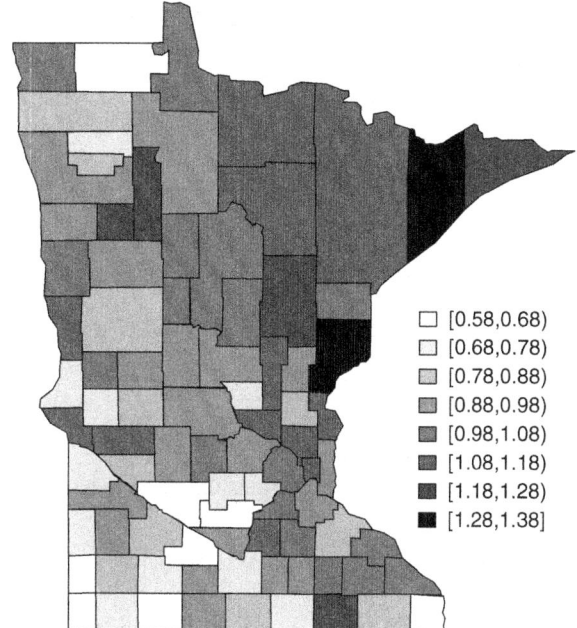

Fig. 1.4 Average radon (pCi/liter) by county in Minnesota

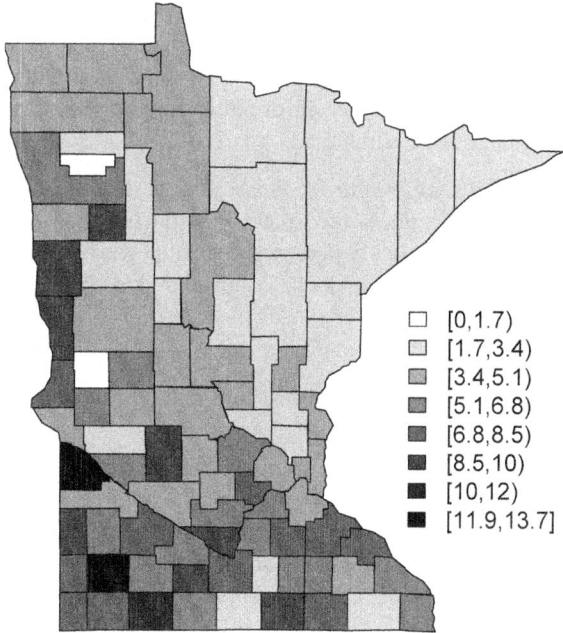

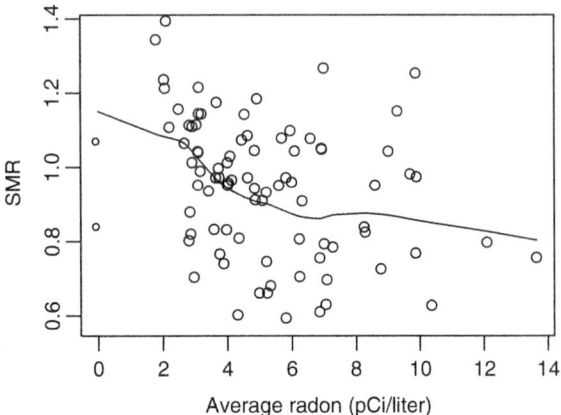

Fig. 1.5 Standardized morbidity ratios versus average radon (pCi/liter) by county in Minnesota

A simple model that constrains the mean to be positive is the *loglinear regression*

$$\log \mathrm{E}\left[\frac{Y_i}{E_i}\bigg|x_i\right] = \beta_0 + \beta_1 x_i$$

$i = 1, \ldots, n$. We might combine this form with a Poisson model for the counts. However, in a Poisson model, the variance is constrained to equal the mean, which is often too restrictive in practice, since excess-Poisson variability is often encountered. Hence, we would prefer to fit a more flexible model. We might also be concerned with residual spatial dependence between disease counts in counties that are close to each other. Information on confounder variables, especially smoking, would also be desirable.

Distinguishing Features. Poisson regression models for independent data, and extensions to allow for excess-Poisson variation, are described in Chap. 6. Such models are explicitly designed for nonnegative response variables. Accounting for residual spatial dependence is considered in Chap. 9.

1.3.4 Pharmacokinetic Data

Pharmacokinetics is the study of the time course of a drug and its metabolites after introduction into the body. A typical experiment consists of a known dose of drug being administered via a particular route (e.g., orally or via an injection) at a known time. Subsequently, blood samples are taken, and the concentration of the drug is measured. The data are in the form of n pairs of points $[x_i, y_i]$, where x_i denotes the sampling time at which the ith blood sample is taken and y_i denotes the ith measured concentration, $i = 1, \ldots, n$. We describe in some detail some of the contextual scientific background in order to motivate a particular regression model.

A typical dataset, taken from Upton et al. (1982), is tabulated in Table 1.2 and plotted in Fig. 1.6. These data were collected after a subject was given an oral dose

1.3 Motivating Examples

Table 1.2 Concentration (y) of the drug theophylline as a function of time (x), obtained from a subject who was administered an oral dose of size 4.53 mg/kg

Observation number i	Time (hours) x_i	Concentration (mg/liter) y_i
1	0.27	4.40
2	0.58	6.90
3	1.02	8.20
4	2.02	7.80
5	3.62	7.50
6	5.08	6.20
7	7.07	5.30
8	9.00	4.90
9	12.15	3.70
10	24.17	1.05

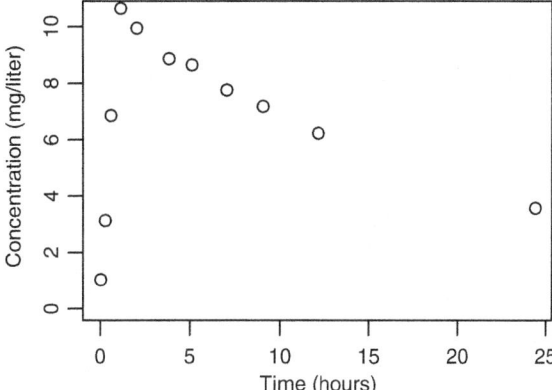

Fig. 1.6 Concentration of theophylline plotted versus time for the data of Table 1.2

of 4.53 mg/kg of the antiasthmatic agent theophylline. The concentration of drug was determined in subsequent blood samples using a chemical assay (a method for determining the amount of a specific substance in a sample). Data were collected over a period slightly greater than 24 h following drug administration.

Pharmacokinetic experiments are important as they help in understanding the absorption, distribution, and elimination processes of drugs. Such an understanding provides information that may be used to decide upon the sizes and timings of doses that should be administered in order to achieve concentrations falling within a desired therapeutic window. Often the concentration of drug acts as a surrogate for the therapeutic response. The aim of a pharmacokinetic trial may be dose recommendation for a specific population, for example, to determine a dose size for the packaging, or recommendations for a particular patient based on covariates, which is known as *individualization*. A typical question is, for the patient who produced the data in Table 1.2, what dose could we give at 25 h to achieve a concentration of 10 mg/l at 37 h?

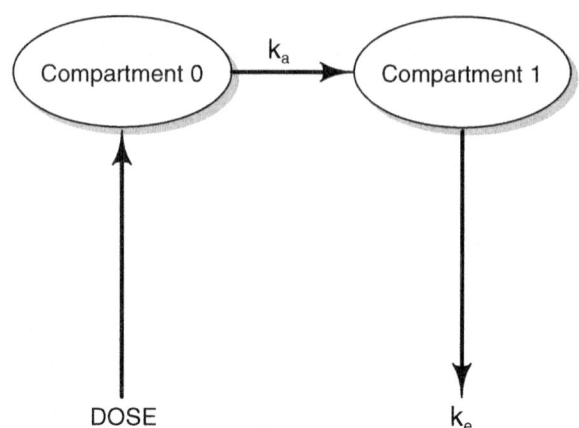

Fig. 1.7 Representation of a one-compartment system with oral dosing. Concentrations are measured in compartment 1

The processes determining drug concentrations are very complicated, but simple compartmental models (e.g., Godfrey 1983) have been found to mimic the concentrations observed in patients. The basic idea is to model the body as a system of compartments within each of which the kinetics of the drug flow is assumed to be similar. We consider the simplest possible model for modeling drug concentrations following the administration of an oral dose. The model is represented in Fig. 1.7 and assumes that the body consists of a compartment into which the drug is introduced and from which absorption occurs into a second "blood compartment." The compartments are labeled retrospectively as 0 and 1 in Fig. 1.7. Subsequently, elimination from compartment 1 occurs with blood samples taken from this compartment.

We now describe in some detail the one-compartment model with first-order absorption and elimination. Let $w_k(t)$ represent the amount of drug in compartment k at time t, $k = 0, 1$. The drug flow between the compartments is described by the differential equations

$$\frac{dw_0}{dt} = -k_a w_0, \tag{1.2}$$

$$\frac{dw_1}{dt} = k_a w_0 - k_e w_1, \tag{1.3}$$

where $k_a > 0$ is the absorption rate constant associated with the flow from compartment 0 to compartment 1 and $k_e > 0$ is the elimination rate constant (see Fig. 1.7). At time zero, the initial dose is $w_0(0) = D$, and solving the pair of differential equations (1.2) and (1.3), subject to this condition, gives the amount of drug in the body at time x as

$$w_1(x) = \frac{Dk_a}{k_a - k_e} \left[\exp(-k_e x) - \exp(-k_a x) \right]. \tag{1.4}$$

1.3 Motivating Examples

We do not measure the amount of total drug but drug concentration, and so we need to normalize (1.4) by dividing $w_1(x)$ by the volume $V > 0$ of the blood compartment to give

$$\mu(x) = \frac{Dk_a}{V(k_a - k_e)}\left[\exp(-k_e x) - \exp(-k_a x)\right]. \quad (1.5)$$

so that $\mu(x)$ is the drug concentration in the blood compartment at time x. Equation (1.5) describes a model that is nonlinear in the parameters V, k_a and k_e; for reasons that will be examined in detail in Chap. 6, inference for such models is more difficult than for their linear counterparts.

We have so far ignored the stochastic element of the model. An obvious error model is

$$y_i = \mu(x_i) + \epsilon_i,$$

with $E[\epsilon_i] = 0$, $\text{var}(\epsilon_i) = \sigma_\epsilon^2$, $i = 1,\ldots,n$, and $\text{cov}(\epsilon_i, \epsilon_j) = 0$, $i \neq j$. We may go one stage further and assume $\epsilon_i \mid \sigma_\epsilon^2 \sim_{iid} N(0, \sigma_\epsilon^2)$ where $\sim_{iid}$ is shorthand for "is independent and identically distributed as." There are a number of potential difficulties with this error model, beyond the distributional choice of normality. Concentrations must be nonnegative, and so we might expect the magnitude of errors to decrease with decreasing "true" concentration $\mu(x)$, a phenomenon that is often confirmed by examination of assay validation data. The error terms are likely to reflect not only assay precision, however, but also model misspecification, and given the simple one-compartment system we have assumed, this could be substantial. We might therefore expect the error terms to display correlation across time. In this example, the scientific context therefore provides not only a mean function but also information on how the variance of the data changes with the mean.

One simple solution, to at least some of these difficulties, is to take the logarithm of (1.5) and fit the model:

$$\log y_i = \log \mu(x_i) + \delta_i.$$

We may further assume $E[\delta_i] = 0$, $\text{var}(\delta_i) = \sigma_\delta^2$, $i = 1,\ldots,n$, and $\text{cov}(\delta_i, \delta_j) = 0$, $i \neq j$, multiplicative errors on the original scale and additive errors on the log scale give

$$\text{var}(Y) = \mu(x)^2 \text{var}(e^\delta) \approx \mu(x)^2 \sigma_\delta^2$$

for small δ.

There are two other issues that are relevant to modeling in this example. The first is that in pharmacokinetic analyses, interest often focuses on *derived* parameters of interest, which are functions of $[V, k_a, k_e]$. In particular, we may wish to make inference for the time to maximum concentration, the maximum concentration, the clearance (initial dose divided by the area under the concentration curve), and the elimination half-life, which are given by

$$x_{\max} = \frac{1}{k_a - k_e} \log\left(\frac{k_a}{k_e}\right)$$

$$c_{\max} = \mu(x_{\max}) = \frac{D}{V}\left(\frac{k_e}{k_a}\right)^{k_e/(k_a-k_e)}$$

$$\text{Cl} = V \times k_e$$

$$t_{1/2} = \frac{\log 2}{k_e}.$$

A second issue is that model (1.5) is unidentifiable in the sense that the parameters $[V, k_a, k_e]$ give the same curve as the parameters $[V k_e/k_a, k_e, k_a]$. This identifiability problem can be overcome via a restriction such as constraining the absorption rate to exceed the elimination rate, $k_a > k_e > 0$, though this complicates inference.

Often the data available for individualization will be sparse. For example, suppose we only observed the first two observations in Table 1.2. In this situation, inference is impossible without additional information (since there are more parameters than data points), which suggests a Bayesian approach in which prior information on the unknown parameters is incorporated into the analysis.

Distinguishing Features. Model (1.5) is nonlinear in the parameters. Such models will be considered in Chap. 6, including their use in situations in which additional information on the parameters is incorporated via the specification of a prior distribution. The data in Table 1.2 are from a single subject. In the original study, data were available for 12 subjects, and ideally we would like to analyze the totality of data; hierarchical models provide one framework for such an analysis. Hierarchical nonlinear models are considered in Chap. 9.

1.3.5 Dental Growth

Table 1.3 gives dental measurements of the distance in millimeters from the center of the pituitary gland to the pteryo-maxillary fissure in 11 girls and 16 boys recorded at the ages of 8, 10, 12, and 14 years. These data were originally analyzed in Potthoff and Roy (1964).

Figure 1.8 plots these data, and we see that dental growth for each child increases in an approximately linear fashion. Three inferential situations are:

1. *Summarization.* For each of the boy and girl populations, estimate the mean and standard deviation of pituitary gland measurements at each of the four ages.
2. *Population inference.* For each of the populations of boys and girls from which these data were sampled, estimate the average linear growth over the age range 8–14 years. Additionally, estimate the average dental distance, with an associated interval estimate, at an age of 9 years.
3. *Individual inference.* For a specific boy or girl in the study, estimate the rate of growth over the age range 8–14 years and predict the growth at 15 years. Additionally, for an unobserved girl, from the same population that produced the sampled girls, obtain a predictive growth curve, along with an interval envelope.

1.3 Motivating Examples

Table 1.3 Dental growth data for boys and girls

Girl	Age (years)				Boy	Age (years)			
	8	10	12	14		8	10	12	14
1	21.0	20.0	21.5	23.0	1	26.0	25.0	29.0	31.0
2	21.0	21.5	24.0	25.5	2	21.5	22.5	23.0	26.5
3	20.5	24.0	24.5	26.0	3	23.0	22.5	24.0	27.5
4	23.5	24.5	25.0	26.5	4	25.5	27.5	26.5	27.0
5	21.5	23.0	22.5	23.5	5	20.0	23.5	22.5	26.0
6	20.0	21.0	21.0	22.5	6	24.5	25.5	27.0	28.5
7	21.5	22.5	23.0	25.0	7	22.0	22.0	24.5	26.5
8	23.0	23.0	23.5	24.0	8	24.0	21.5	24.5	25.5
9	20.0	21.0	22.0	21.5	9	23.0	20.5	31.0	26.0
10	16.5	19.0	19.0	19.5	10	27.5	28.0	31.0	31.5
11	24.5	25.0	28.0	28.0	11	23.0	23.0	23.5	25.0
					12	21.5	23.5	24.0	28.0
					13	17.0	24.5	26.0	29.5
					14	22.5	25.5	25.5	26.0
					15	23.0	24.5	26.0	30.0
					16	22.0	21.5	23.5	25.0

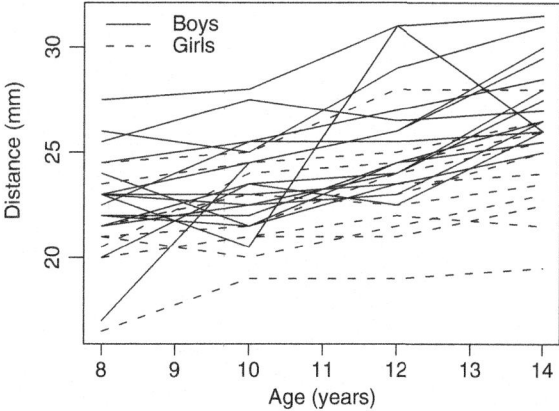

Fig. 1.8 Dental growth data for boys and girls: distance plotted versus age

With 16 boys and 11 girls, inference for situation 1 can be achieved by simply evaluating the sample mean and standard deviation at each time point; these quantities are given in Table 1.4. These simple summaries are straightforward to construct and are based on independence of individuals. To obtain interval estimates for the means and standard deviations, one must be prepared to make assumptions (such as approximate normality of the measurements), since for these data the sample sizes are not large and we might be wary of appealing to large sample (asymptotic) arguments.

Table 1.4 Sample means and standard deviations (SDs) for girls and boys, by age group

Age (years)	Girls Mean (mm)	SD (mm)	Boys Mean (mm)	SD (mm)
8	21.2	2.1	22.9	2.5
10	22.2	1.9	23.8	2.1
12	23.1	2.4	25.7	2.7
14	24.1	2.4	27.5	2.1

For situation 2, we may fit a linear model relating distance to age. Since there are no data at 9 years, to obtain an estimate of the dental distance, we again require a model relating distance to age. In situation 3, we may wish to use the totality of data as an aid to providing inference for a specific child. For a new girl from the same population, we clearly need to use the existing data and a model describing between-girl differences.

For longitudinal (repeated measures) data such as these, we cannot simply fit models to the totality of the data on boys or girls and assume independence of measurements; we need to adjust for the correlation between measurements on the same child. There is clearly dependence between such measurements. For example, boy 10 has consistently higher measurements than the majority of boys. There are two distinct approaches to modeling longitudinal data. In the *marginal* approach, the average response is modeled as a function of covariates (including time), and standard errors are empirically adjusted for dependence. In the *conditional* approach, the response of each individual is modeled as a function of individual-specific parameters that are assumed to arise from a distribution, so that the overall variability is partitioned into within- and between-child components. The marginal approach is designed for estimating population-level questions (as posed in situation 2) based on minimal assumptions. Conditional approaches can answer a greater number of inferential questions but require an increased number of assumptions which decreases their robustness to model misspecification.

Distinguishing Features. Chapter 8 describes linear models for dependent data such as these.

1.3.6 Spinal Bone Mineral Density

Bachrach et al. (1999) analyze longitudinal data on spinal bone mineral density (SBMD) measurements on 230 women aged between 8 and 27 years and of one of four ethnic groups: Asian, Black, Hispanic, and White. The aim of this study was to examine ethnic differences in SBMD.

Figure 1.9 displays the SBMD measurements by individual, with one panel for each of the four races. The relationship between SBMD and age is clearly nonlinear, and there are also woman-specific differences in overall level so that observations

1.3 Motivating Examples

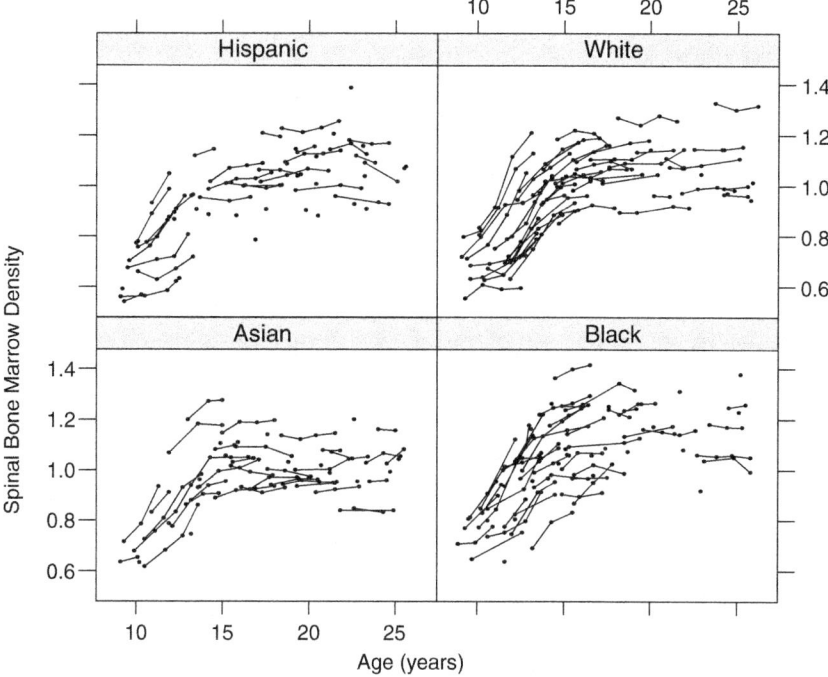

Fig. 1.9 Spinal bone mineral density measurements as a function of age and ethnicity. Points that are connected represent measurements from the same woman

on the same woman are correlated. Letting Y_{ij} represent the SBMD measurement on woman i at age age_{ij}, we might propose a mean model of the form

$$\mathrm{E}[Y_{ij} \mid \text{age}_{ij}] = \boldsymbol{x}_i \boldsymbol{\beta} + f(\text{age}_{ij}) + b_i$$

where $\boldsymbol{x}_i$ is a 1×4 row vector with a single one and three zeroes that represents the ethnicity of woman i (coded in the order Hispanic, White, Asian, Black), with $\boldsymbol{\beta} = [\beta_H, \beta_W, \beta_A, \beta_B]^{\mathrm{T}}$ the 4×1 vector of associated regression coefficients, $f(\text{age}_{ij})$ is a function that varies smoothly with age, and b_i is a woman-specific intercept which is included to account for dependencies of measurements on the same individual. The relationship between SBMD and age is not linear and not of primary interest. Consequently, we would like to use a flexible model form, and we may not be concerned if this model does not contain easily interpretable parameters. Nonparametric regression is the term we use to refer to flexible mean modeling.

Distinguishing Features. The analysis of these data requires both a flexible mean model for the age effect and acknowledgement of the dependence of measurements on the same woman. Chapters 10–12 describe models that allow for these possibilities.

1.4 Nature of Randomness

Regression models consist of both deterministic and stochastic (random) components, and a consideration of the sources of the randomness is worthwhile, both to interpret parameters contained in the deterministic component and to model the stochastic component. We initially consider an idealized situation in which a response is completely deterministic, given sufficient information, and randomness is only induced by missing information.[3] Let y denote a variable with values $y_1, \ldots, y_N$ within a population. We begin with a very simple deterministic model

$$y_i = \beta_0 + \beta_1 x_i + \gamma z_i \qquad (1.6)$$

for $i = 1, \ldots, N$, so that, given x_i and z_i (and knowing β_0, β_1 and γ), y_i is completely determined. Suppose we only measure y_i and x_i and assume the model

$$Y_i = \beta_0^\star + \beta_1^\star x_i + \epsilon_i.$$

To interpret $\beta_0^\star$ and $\beta_1^\star$, we need to understand the relationship between x_i and z_i, $i = 1, \ldots, N$. To this end, write

$$z_i = a + b x_i + \delta_i, \qquad (1.7)$$

$i = 1, \ldots, N$. This form does not in any sense assume that a linear association is appropriate or "correct", rather it is the linear approximation to $\mathrm{E}[Z \,|\, x]$. In (1.7), we may take a and b as the least squares estimates from fitting a linear model to the data $[x_i, z_i]$, $i = 1, \ldots, N$. Substitution of (1.7) into (1.6) yields

$$y_i = \beta_0 + \beta_1 x_i + \gamma(a + b x_i + \delta_i)$$
$$= \beta_0^\star + \beta_1^\star x_i + \epsilon_i$$

where

$$\begin{aligned}\beta_0^\star &= \beta_0 + a\gamma \\ \beta_1^\star &= \beta_1 + b\gamma \\ \epsilon_i &= \gamma \delta_i, \quad i = 1, \ldots, N,\end{aligned} \qquad (1.8)$$

---

[3] When simulations are performed, pseudorandom numbers are generated via deterministic sequences. For example, consider the sequence generated by the *congruential generator*

$$X_i = a X_{i-1}, \ \mathrm{mod}(m)$$

along with initial value (or "seed") X_0. Then X_i takes values in $0, 1, \ldots, m-1$, and pseudorandom numbers are obtained as $U_i = X_i/m$, where X_0, a, and m are chosen so that the U_i's have (approximately) the properties of uniform $\mathrm{U}(0, 1)$ random variables. However, if X_0, a, and m are known, the randomness disappears! Ripley (1987, Chap. 2) provides a discussion of pseudorandom variable generation and specifically "good" choices of a and m.

1.4 Nature of Randomness

so that $\beta_1^\star$ is a combination of the direct effect of x_i on y_i, *and* the effect of z_i, through the linear association between z_i and x_i. This development illustrates the problems in nonrandomized situations of estimating the causal effect of x_i on y_i, that is, β_1. Turning to the stochastic component (1.8) illustrates that properties of ϵ_i are inherited from δ_i. Hence, assumptions such as constancy of variance of ϵ_i depend on the nature of z_i and, in particular, on the joint distribution of x_i and z_i.

Increasing slightly the realism, we extend the original deterministic model to

$$y_i = \beta_0 + \sum_{j=1}^{p} \beta_j x_{ij} + \sum_{k=1}^{q} \gamma_k z_{ik}. \tag{1.9}$$

Suppose we only measure $x_{i1}, \ldots, x_{ip}$ and assume the simple model

$$Y_i = \beta_0^\star + \sum_{j=1}^{p} \beta_j^\star x_{ij} + \epsilon_i, \tag{1.10}$$

where the errors, ϵ_i, now correspond to the totality of scaled versions of the z_{ik}'s that remain after extracting the linear associations with the x_{ij}'s by analogy with (1.7) and (1.8).

Viewing the error terms as sums of random variables and considering the central limit theorem (Appendix G) naturally leads to the normal distribution as a plausible error distribution. There is no compelling reason to believe that the variance of this normal distribution will be constant across the space of the x variables, however.

We have distinguished between the regression coefficients in the assumed model (1.10), denoted by $\beta_j^\star$, and those in the original model (1.9), denoted β_j. In general, $\beta_j \neq \beta_j^\star$, because of the possible effects of *confounding* which occurs due to dependencies between x_{ij} and elements of $\mathbf{z}_i = [z_{i1}, \ldots, z_{iq}]$. In the example just considered, only if x_{ij} is linearly independent of the z_{ik} will the coefficients β_j and $\beta_j^\star$ coincide. For nonlinear models, the relationship between the two sets of coefficients is even more complex.

This development illustrates that an aim of regression modeling is often to "explain" the error terms using observed covariates. In general, error terms represent not only unmeasured variables but also data anomalies, such as inaccurate recording of responses and covariates, and model misspecification. Clearly the nature of the randomness, and the probabilities we attach to different events, is conditional upon the information that we have available and, specifically, the variables we measure.

Similar considerations can be given to other types of random variables. For example, suppose we wish to model a binary random variable Y taking values coded as 0 and 1. Sometimes it will be possible to link Y to an underlying continuous *latent* variable and use similar arguments to that above. To illustrate, Y could be an indicator of low birth weight and is a simple function of the true birth weight, U, which is itself associated with many covariates. We may then model the probability of low birth weight as a function of covariates $\mathbf{x}$, via

$$p(\mathbf{x}) = \Pr(Y = 1 \mid \mathbf{x}) = \Pr(U \leq u_0 \mid \mathbf{x}) = \mathrm{E}[Y \mid \mathbf{x}],$$

where u_0 is the threshold value that determines whether a child is classified as low birth weight or not. This development is taken further in Sects. 7.6.1 and 9.13.

The above gives one a way of thinking about where the random terms in models arise from, namely as unmeasured covariates. In terms of distributional assumptions, some distributions arise naturally as a consequence of simple physical models. For example, suppose we are interested in modeling the number of events occurring over time. The process we now describe has been found empirically to model a number of phenomena, for example the arrival of calls at a telephone exchange or the emission of particles from a radioactive source. Let the rate of occurrences be denoted by $\rho > 0$ and $N(t, t + \Delta t)$ be the number of events in the interval $(t, t + \Delta t]$. Suppose that, informally speaking, Δt tends to zero from above and that

$$\Pr[N(t, t + \Delta t) = 0] = 1 - \rho \Delta t + o(\Delta t),$$
$$\Pr[N(t, t + \Delta t) = 1] = \rho \Delta t + o(\Delta t),$$

so that $\Pr[N(t, t + \Delta t) > 1] = o(\Delta t)$. The notation $o(\Delta t)$ represents a function that tends to zero more rapidly than Δt. Finally, suppose that $N(t, t + \Delta t)$ is independent of occurrences in $(0, t]$. Then we have a *Poisson process*, and the number of events occurring in the fixed interval $(t, t + h]$ is a Poisson random variable with mean ρh.

Other distributions are "artificial." For example, a number of distributions arise as functions of normal random variables (such as Student's t, Snedecor's F, and chi-squared random variables) or may be dreamt up for flexible and convenient modeling (as is the case for the so-called Pearson family of distributions).

Models can arise from idealized views of the phenomenon under study, but then we might ask: "If we could measure absolutely everything we wanted to, would there be any randomness left?" In all but the simplest experiments, this question is probably not that practically interesting, but the central idea of quantum mechanics tells us that probability is still needed, because some experimental outcomes are fundamentally unpredictable (e.g., Feynman 1951).

1.5 Bayesian and Frequentist Inference

What distinguishes the field of statistics from the use of statistical techniques in a particular discipline is a principled approach to inference in the face of uncertainty. There are two dominant approaches to inference, which we label as Bayesian and frequentist, and each produces inferential procedures that are optimal with respect to different criteria.

In Chaps. 2 and 3, we describe, respectively, the frequentist and Bayesian approaches to statistical inference. Central to the philosophy of each approach is the interpretation of probability that is taken. In the frequentist approach, as the name suggests, probabilities are viewed as limiting frequencies under infinite hypothetical replications of the situation under consideration. Inferential recipes, such as specific estimators, are assessed with respect to their performance under repeated sampling of the data, with model parameters viewed as fixed, albeit unknown, constants. By contrast, in the Bayesian approach that is described in this book, probabilities are viewed as subjective and are interpreted conditional on the available information. As a consequence, assigned probabilities concerning the same event may differ between individuals. In this sense probabilities do not exist as they vary as a function of the available information. All unknown parameters in a model are treated as random variables, and inference is based upon the (posterior) probability distribution of these parameters, given the data and other available information. Practically speaking, the interpretation of probability is less relevant than the number of assumptions that are required for valid inference (which has implications for the robustness of analysis) and the breadth of inferential questions that can be answered using a particular approach.

It should be stressed that many issues arising in the analysis of regression data (such as the nature of the sampling scheme, parameter interpretation, and misspecification of the mean model) are independent of philosophy and in practice are usually of far greater importance than the inferential approach taken to analysis.

Each of the frequentist and Bayesian approaches have their merits and can often be used in tandem, an approach we follow and advocate throughout this book. If substantive conclusions differ between different approaches, then discovering the reasons for the discrepancies can be informative as it may reveal that a particular analysis is leaning on inappropriate assumptions or that relevant information is being ignored by one of the approaches. Those situations in which one of the approaches is more or less suitable will also be distinguished throughout this book, with a short summary being given in the next section.

1.6 The Executive Summary

I would like to briefly summarize my view on when to take Bayesian or frequentist approaches to estimation. As the examples throughout this book show, on many occasions, if one is careful in execution, both approaches to analysis will yield essentially equivalent inference. For small samples, the Bayesian approach with thoughtfully specified priors is often the only way to go because of the difficulty in obtaining well-calibrated frequentist intervals. An example of such a sparse data occasion is given at the end of Sect. 6.16. For medium to large samples, unless there is strong prior information that one wishes to incorporate, a robust frequentist approach using sandwich estimation (or quasi-likelihood if one has faith in the variance model) is very appealing since consistency is guaranteed under relatively mild conditions. For highly complex models (e.g., with many random effects), a Bayesian approach is often the most convenient way to formulate the model, and computation under the Bayesian approach is the most straightforward. The modeling of spatial dependence in Sect. 9.7 provides one such example in which the Bayesian approach is the simplest to implement. The caveat to complex modeling is that in most cases consistency of inference is only available if all stages of the model are correctly specified. Consequently, if one really cares about interval estimates, then extensive model checking will be necessary. If formal inference is not required but rather one is in an exploratory phase, then there is far greater freedom to experiment with the approaches that one is most familiar with, including nonparametric regression. In this setting, using procedures that are less well-developed statistically is less dangerous.

In contrast to estimation, hypothesis testing using frequentist and Bayesian methods can often produce starkly differing results, even in large samples. As discussed in Chap. 4, I think that hypothesis testing is a very difficult endeavor, and tests applied using the frequentist approach, as currently practiced (with α levels being fixed regardless of sample size), can be very difficult to interpret. In general, I prefer estimation to hypothesis testing.

As a final comment, as noted, in many instances carefully conducted frequentist and Bayesian approaches will lead to similar substantive conclusions; hence, the choice between these approaches can often be based on that which is most natural (i.e., based on training and experience) to the analyst. Consequently, throughout this book, methods are discussed in terms of their advantages and shortcomings, but a strong recommendation of one method over another is usually not given as there is often no reason for stating a preference.

1.7 Bibliographic Notes

Rosenbaum (2002) provides an in-depth discussion of the analysis of data from observational studies, and an in-depth treatment of causality is the subject of Pearl (2009). A classic text on survey sampling is Cochran (1977) with Korn and Graubard (1999) and Lumley (2010) providing more recent presentations. Regression from a survey sampling viewpoint is discussed in the edited volume of Chambers and Skinner (2003). Errors-in-variables is discussed in detail by Carroll et al. (2006) and missing data by Little and Rubin (2002). Johnson et al. (1994, 1995, 1997); Kotz et al. (2000), and Johnson et al. (2005) provide a thorough discussion of the genesis of univariate and multivariate discrete and continuous probability distributions and, in particular, their relationships to naturally occurring phenomena. Barnett (2009) provides a discussion of the mechanics and relative merits of Bayesian and frequentist approaches to inference; see also Cox (2006).

Part I
Inferential Approaches

Chapter 2
Frequentist Inference

2.1 Introduction

Inference from data can take many forms, but primary inferential aims will often be *point estimation*, to provide a "best guess" of an unknown parameter, and *interval estimation*, to produce ranges for unknown parameters that are supported by the data. Under the frequentist approach, parameters and hypotheses are viewed as unknown but fixed (nonrandom) quantities, and consequently there is no possibility of making probability statements about these unknowns.[1] As the name suggests, the frequentist approach is characterized by a frequency view of probability, and the behavior of inferential procedures is evaluated under hypothetical repeated sampling of the data.

Frequentist procedures are not typically universally applicable to all models/sample sizes and often require "fixes." For example, a number of variants of likelihood have been developed for use in particular situations (Sect. 2.4.2). In contrast, the Bayesian approach, described in Chap. 3, is completely prescriptive, though there are significant practical hurdles to overcome (such as likelihood and prior specification) in pursuing that prescription. In addition, in situations in which frequentist procedures encounter difficulties, Bayesian approaches typically require very careful prior specification to avoid posterior distributions that exhibit anomalous behavior.

The outline of this chapter is as follows. We begin our discussion in Sect. 2.2 with an overview of criteria by which frequentist procedures may be evaluated. In Sect. 2.3 we present a general development of *estimating functions* which provide a unifying framework for defining and establishing the properties of commonly used frequentist procedures. Two important classes of estimating functions are then

---

[1] *Random effects* models provide one example in which parameters are viewed as random from a frequentist perspective and are regarded as arising from a population of such effects. Frequentist inference for such models is described in Part III of this book.

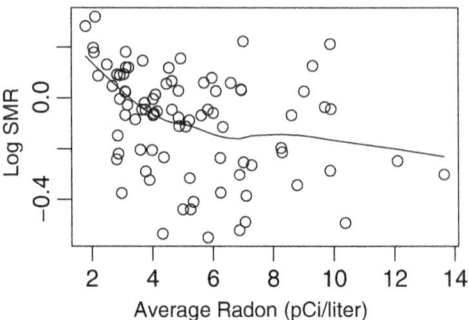

Fig. 2.1 Exploratory plot of log SMR for lung cancer versus average residential radon, with a local smoother superimposed, for 85 counties in Minnesota

introduced: those arising from the specification of a likelihood function, in Sect. 2.4, and those from a quasi-likelihood function, in Sect. 2.5. A recurring theme is the assessment of frequentist procedures under model misspecification. In Sect. 2.6 we discuss the *sandwich estimation* technique which provides estimation of the standard error of estimators in more general circumstances than were assumed in deriving the estimator. Section 2.7 introduces the bootstrap, which is a simulation-based method for making inference with reduced assumptions. Section 2.8 discusses the choice of an estimating function. Hypothesis testing is considered in Sect. 2.9, and the chapter ends with concluding remarks in Sect. 2.10. To provide some numerical relief to the mostly methodological development of this chapter, we provide one running example.

Example: Lung Cancer and Radon

We consider the data introduced in Sect. 1.3.3 and examine the association between counts of lung cancer incidence, Y_i, and the average residential radon, x_i, in county i with $i = 1, \ldots, 85$, indexing the counties within which radon measurements were available (in two counties no radon data were reported). We examine the association using the loglinear model

$$\log \mathrm{E}[\mathrm{SMR}_i \mid x_i] = \beta_0 + \beta_1 x_i. \tag{2.1}$$

where $\mathrm{SMR}_i = Y_i/E_i$ (with E_i the expected count) is the standardized mortality ratio in county i (Sect. 1.3.3) and is a summary measure that controls for the differing age and gender populations across counties. We take as our parameter of interest $\exp(\beta_1)$ which is the multiplicative change in risk associated with a 1 pCi/l increase in radon. In the epidemiological literature this parameter is referred to as the *relative risk*; here it corresponds to the risk ratio for two areas whose radon exposures x differ by one unit.

To first order, $\mathrm{E}[\log \mathrm{SMR} \mid x] \approx \log \mathrm{E}[\mathrm{SMR} \mid x]$, and so if (2.1) is an appropriate model, a plot of $\log \mathrm{SMR}_i$ versus x_i should display an approximately linear trend; Fig. 2.1 shows this plot with a local smoother superimposed and indicates a negative association. This example is illustrative, and so distracting issues, such as the effect of additional covariates (including smoking, the major confounder) and residual spatial dependence in the counts, will be conveniently ignored.

2.2 Frequentist Criteria

In this section we describe frequentist criteria by which competing estimators may be compared and discuss conditions under which optimal estimators exist under these criteria. Under the frequentist approach to inference, the fundamental outlook is that statistical procedures are assessed with respect to their performance under hypothetical, repeated sampling of the data, under fixed values of the parameters. In this section, for simplicity, we consider the estimation of a *univariate* parameter θ and let $\boldsymbol{Y} = [Y_1, \ldots, Y_n]^\mathrm{T}$ represent a vector of n random variables and $\boldsymbol{y} = [y_1, \ldots, y_n]^\mathrm{T}$ a realization. Often inference will be summarized via a $100(1-\alpha)\%$ *confidence interval* for θ, which is an interval $[\,a(\boldsymbol{Y}), b(\boldsymbol{Y})\,]$ such that

$$\Pr\{\theta \in [\,a(\boldsymbol{Y}), b(\boldsymbol{Y})\,]\} = 1 - \alpha, \tag{2.2}$$

for all θ, where the probability statement is with respect to the distribution of $\boldsymbol{Y}$ and $1 - \alpha$ is known as the *coverage* probability. For interpretation it is crucial to recognize that the random quantities in (2.2) are the endpoints of the interval $[\,a(\boldsymbol{Y}), b(\boldsymbol{Y})\,]$, so that we are not assigning a probability statement to θ. The correct interpretation of a confidence interval is that, under hypothetical repeated sampling, a proportion $1 - \alpha$ of the intervals created will contain the true value θ. We emphasize that we cannot say that the specific interval $[\,a(\boldsymbol{y}), b(\boldsymbol{y})\,]$ contains θ with probability $1 - \alpha$.

Ideally, we would like to determine the shortest possible confidence interval for a given α. The search for such intervals is closely linked to the determination of optimal point estimators of θ. The point *estimator* $\widehat{\theta}(\boldsymbol{Y})$ of θ represents a random variable, with an associated *sampling* distribution, while the point *estimate* $\widehat{\theta}(\boldsymbol{y})$ is a specific value. In any given situation a host of potential estimators are available, and we require criteria by which to judge competing choices. Heuristically speaking, a good estimator will have a sampling distribution that is concentrated "close" to the true value θ, where "close" depends on the distance measure that we apply to the distribution of $\widehat{\theta}(\boldsymbol{Y})$.

One natural measure of closeness is the *mean squared error* (MSE) of $\widehat{\theta}(\boldsymbol{Y})$ which arises from a quadratic loss function for estimation and is defined as

$$\begin{aligned}\mathrm{MSE}\left[\widehat{\theta}(\boldsymbol{Y})\right] &= \mathrm{E}_{\boldsymbol{Y}|\theta}\left[\left(\widehat{\theta}(\boldsymbol{Y}) - \theta\right)^2\right] \\ &= \mathrm{var}_{\boldsymbol{Y}|\theta}\left[\widehat{\theta}(\boldsymbol{Y})\right] + \mathrm{bias}\left[\widehat{\theta}(\boldsymbol{Y})\right]^2\end{aligned}$$

where the *bias* of the estimator is

$$\text{bias}\left[\widehat{\theta}(\boldsymbol{Y})\right] = \text{E}_{\boldsymbol{Y}|\theta}\left[\widehat{\theta}(\boldsymbol{Y})\right] - \theta.$$

This notation stresses that all expectations are with respect to the sampling distribution of the estimator, given the true value of the parameter; this is a crucial aspect but the notation is cumbersome and so will be suppressed. Finding estimators with minimum MSE for all values of θ is not possible. For example, $\widehat{\theta}(\boldsymbol{Y}) = 3$ has zero MSE for $\theta = 3$ (and so is optimal for this θ!) but is, in general, a disastrous estimator.

An elegant theory, which is briefly summarized in Appendix G, has been developed to characterize uniformly minimum-variance *unbiased* estimators (UMVUEs). The theory depends first on writing down a full probability model for the data, $p(\boldsymbol{y} \mid \theta)$. We assume conditional independence so that $p(\boldsymbol{y} \mid \theta) = \prod_{i=1}^{n} p(y_i \mid \theta)$. The Cramér–Rao lower bound for any unbiased estimator $\widehat{\phi}$ of a scalar function of interest $\phi = \phi(\theta)$ is

$$\text{var}(\widehat{\phi}) \geq -\frac{[\phi'(\theta)]^2}{\text{E}\left[\frac{\partial^2 l}{\partial \theta^2}\right]}, \tag{2.3}$$

where $l(\theta) = \sum_{i=1}^{n} \log p(y_i \mid \theta)$ is the log of the joint distribution of the data, viewed as a function of θ. If $T(\boldsymbol{Y})$ is a sufficient statistic of dimension 1, then, under suitable regularity conditions, there is a unique function $\phi(\theta)$ for which a UMVUE exists and its variance attains the Cramér–Rao lower bound. Further, a UMVUE only exists when the data are independently sampled from a one-parameter exponential family. Specifically, suppose that $p(y_i \mid \theta)$ is of one-parameter exponential family form, so that its distribution may be written, for suitably defined functions, as

$$p(y \mid \theta) = \exp\left[\theta T(y) - b(\theta) + c(y)\right]. \tag{2.4}$$

In this situation, there is a unique function of θ for which a UMVUE exists. Unfortunately, this theory only covers a narrow range of circumstances. There are methods available for constructing estimators with the minimal attainable variance in additional situations but even this wider class of models does not come close to covering the range of models that we would like to consider for practical application. UMVUEs are also not always sensible; see Exercise 2.2.

As discussed in Sect. 1.2, model formulation should begin with a model that we would like to fit, before proceeding to examine its mathematical properties. As we will see, exponential family models can provide robust inference, in the sense of performing well even if certain aspects of the assumed model are wrong, but to only consider these models is unnecessarily restrictive.

We now discuss how estimators may be compared in general circumstances *asymptotically*, that is, as $n \to \infty$. There are two hypothetical situations that are being considered here. The first is the repeated sampling aspect for fixed n, and the second is allowing $n \to \infty$. The asymptotic properties of frequentist procedures

2.2 Frequentist Criteria

may be used in two respects. The first is to justify particular procedures, and the second is to carry out inference, for example, to construct confidence intervals. We might question the relevance of asymptotic criteria, since in any practical situation n is finite, and an inconsistent or asymptotically inefficient estimator may have better finite sample properties (a reduced MSE for instance) than a consistent alternative. On the other hand, for many commonly used models, asymptotic inference is often accurate for relatively small sample sizes (as we will see in later chapters).

While unbiasedness of estimators, per se, is of debatable value, a fundamentally important frequentist criterion for assessing an estimator is *consistency*. *Weak consistency* states that as $n \to \infty$, $\widehat{\theta}_n \to_p \theta$ (Appendix F), that is,

$$\Pr(|\widehat{\theta}_n - \theta| > \epsilon) \to 0 \quad \text{as} \quad n \to \infty \quad \text{for any} \quad \epsilon > 0.$$

Intuitively, the distribution of a consistent estimator concentrates more and more around the true value as the sample size increases. In all but pathological cases, a consistent estimator is asymptotically unbiased, though the contrary is not true. For example, consider the model with $E[Y_i \mid \theta] = \theta$, $i = 1, \ldots, n$, and the estimator $\widehat{\theta} = Y_1$, this estimator is unbiased but inconsistent.

When assessing an estimator, once consistency has been established, asymptotic normality of the estimator is then typically sought, and interest focuses on the variance of the estimator. In particular, the *asymptotic relative efficiency*, or more simply the *efficiency*, allows an estimator $\widetilde{\theta}_n$ to be compared to the estimator with the smallest variance $\widehat{\theta}_n$ via

$$\frac{\text{var}(\widetilde{\theta}_n)}{\text{var}(\widehat{\theta}_n)}.$$

The $100(1-\alpha)\%$ asymptotic confidence interval associated with an estimator $\widehat{\theta}_n$ is

$$\widehat{\theta}_n \pm z_{1-\alpha/2} \times \sqrt{\text{var}(\widehat{\theta}_n)} \qquad (2.5)$$

where $Z \sim N(0,1)$ and $\Pr(Z < z_{1-\alpha/2}) = 1 - \alpha/2$. If $\widehat{\theta}_n$ is asymptotically efficient, then interval (2.5) is (asymptotically) the shortest available. Maximum likelihood estimation (Sect. 2.4) provides a method for finding efficient estimators.

A difficulty with the interpretation of frequentist inferential summaries is that all probability statements refer to hypothetical data replications and to the estimator, and not to the *estimate* from a specific realization of data. This can lead to intervals with poor properties. Exercise 2.1 describes an instance in which the confidence coverage is correct on average, but for some realizations of the data, the interval has 100% coverage.

We summarize this section and provide a road map to the remainder of the chapter. A fundamental, desirable criterion is to produce confidence intervals that are the shortest possible. Only in stylized situations may estimators with minimum variance be found in non-asymptotic situations. Asymptotically, the picture is rosier, however. In the next section we describe a general class of estimators and give results concerning consistency and asymptotic normality. Subsequently, we show that maximum likelihood estimators attain the smallest asymptotic variance (subject to regularity conditions) *if* the model is correctly specified. We then consider quasi-likelihood, sandwich estimation, and the bootstrap, each of which is designed to reduce the reliance of inference on a full probability model specification.

2.3 Estimating Functions

In the last section we saw that optimal estimators can be found when a full probability model is assumed. The need to specify a full probability model for the data is undesirable. While a practical context may suggest a mean model and perhaps an appropriate mean–variance relationship, it is rare to have faith in a choice for the *distribution* of the data. In this section we give a framework within which the asymptotic properties of a broad range of estimation recipes may be evaluated.

Let $Y = [Y_1, \ldots, Y_n]$ represent n observations from a distribution indexed by a p-dimensional parameter θ, with $\text{cov}(Y_i, Y_j \mid \theta) = 0$, $i \neq j$. In the following we will not rigorously derive asymptotic results and only informally discuss regularity conditions under which the results hold. The models discussed subsequently will, unless otherwise stated, obey the necessary conditions.

In the following, for ease of presentation, we assume that Y_i, $i = 1, \ldots, n$, are independent and identically distributed (iid).[2] An *estimating function* is a function,

$$G_n(\theta) = \frac{1}{n} \sum_{i=1}^{n} G(\theta, Y_i), \tag{2.6}$$

of the same dimension as θ for which

$$\mathrm{E}[G_n(\theta)] = 0 \tag{2.7}$$

for all θ. The estimating function $G_n(\theta)$ is a random variable because it is a function of Y. The corresponding *estimating equation* that defines the estimator $\widehat{\theta}_n$ has the form

$$G_n(\widehat{\theta}_n) = \frac{1}{n} \sum_{i=1}^{n} G(\widehat{\theta}_n, Y_i) = 0. \tag{2.8}$$

For inference the asymptotic properties of the estimating function are derived (which is why we index the estimating function by n), and these are transferred to the resultant estimator. The estimator $\widehat{\theta}_n$ that solves (2.8) will often be unavailable in closed form and so deriving its distribution from that of the estimating function

---

[2] In a regression setting we have *independently distributed* observations only, because the distribution of the outcome changes as a function of covariates.

2.3 Estimating Functions

is an ingenious step, because the estimating function may be constructed to be a simple (e.g., linear) function of the data. The estimating function defined in (2.6) is a sum of random variables, which provides the opportunity to evaluate its asymptotic properties via a central limit theorem since the first two moments will often be straightforward to calculate. The art of constructing estimating functions is to make them dependent on distribution-free quantities, for example, the first two moments of the data; robustness of inference to misspecification of higher moments often follows.

We now state an important result that will be used repeatedly in the context of frequentist inference.

Result 2.1. Suppose that $\widehat{\boldsymbol{\theta}}_n$ is a solution to the estimating equation

$$\boldsymbol{G}_n(\boldsymbol{\theta}) = \frac{1}{n}\sum_{i=1}^{n} \boldsymbol{G}(\boldsymbol{\theta}, Y_i) = \boldsymbol{0},$$

that is, $\boldsymbol{G}_n(\widehat{\boldsymbol{\theta}}_n) = \boldsymbol{0}$. Then $\widehat{\boldsymbol{\theta}}_n \to_p \boldsymbol{\theta}$ (consistency) and

$$\sqrt{n}\,(\widehat{\boldsymbol{\theta}}_n - \boldsymbol{\theta}) \to_d \mathrm{N}_p\left[\boldsymbol{0},\, \boldsymbol{A}^{-1}\boldsymbol{B}(\boldsymbol{A}^{\mathrm{T}})^{-1}\right] \qquad (2.9)$$

(asymptotic normality), where

$$\boldsymbol{A} = \boldsymbol{A}(\boldsymbol{\theta}) = \mathrm{E}\left[\frac{\partial}{\partial \boldsymbol{\theta}^{\mathrm{T}}} \boldsymbol{G}(\boldsymbol{\theta}, Y)\right]$$

and

$$\boldsymbol{B} = \boldsymbol{B}(\boldsymbol{\theta}) = \mathrm{E}[\boldsymbol{G}(\boldsymbol{\theta}, Y)\boldsymbol{G}(\boldsymbol{\theta}, Y)^{\mathrm{T}}] = \mathrm{var}\left[\boldsymbol{G}(\boldsymbol{\theta}, Y)\right].$$

Outline Derivation

We refer the interested reader to van der Vaart (1998, Sect. 5.2) for a proof of consistency and present an outline derivation of asymptotic normality, based on van der Vaart (1998, Sect. 5.3). For simplicity we assume that θ is univariate.

We expand $G_n(\theta)$ in a Taylor series around the true value θ:

$$0 = G_n(\widehat{\theta}_n) = G_n(\theta) + (\widehat{\theta}_n - \theta)\left.\frac{dG_n}{d\theta}\right|_{\theta} + \frac{1}{2}(\widehat{\theta}_n - \theta)^2 \left.\frac{d^2 G_n}{d\theta^2}\right|_{\widetilde{\theta}}, \qquad (2.10)$$

where $\widetilde{\theta}$ is a point between $\widehat{\theta}_n$ and θ. We rewrite (2.10) as

$$\sqrt{n}\,(\widehat{\theta}_n - \theta) = \frac{-\sqrt{n}\, G_n(\theta)}{\left.\frac{dG_n}{d\theta}\right|_{\theta} + \frac{1}{2}(\widehat{\theta}_n - \theta) \left.\frac{d^2 G_n}{d\theta^2}\right|_{\widetilde{\theta}}} \qquad (2.11)$$

and determine the asymptotic distribution of the right-hand side, beginning with the distribution of $G_n(\theta)$. To apply a central limit theorem, note that $\mathrm{E}[G_n(\theta)] = 0$ and

$$n \times \mathrm{var}\,[G_n(\theta)] = \mathrm{var}\,[G(\theta,Y)] = \mathrm{E}\,[G(\theta,Y)^2] = B$$

(which we assume is finite). Consequently, by the central limit theorem (Appendix G),

$$\sqrt{n}\,G_n(\theta) \to_d \mathrm{N}\,[0, B(\theta)]. \qquad (2.12)$$

We now transfer the properties of the estimating function to the estimator $\widehat{\theta}_n$ via (2.11). The first term of the denominator of (2.11),

$$\left.\frac{dG_n}{d\theta}\right|_\theta = \frac{1}{n}\sum_{i=1}^n \left.\frac{d}{d\theta}G(\theta,Y_i)\right|_\theta,$$

is an average and so converges to its expectation, provided this expectation exists, by the weak law of large numbers (Appendix G)

$$\left.\frac{dG_n}{d\theta}\right|_\theta \to_p \mathrm{E}\left[\frac{d}{d\theta}G(\theta,Y)\right] = A(\theta).$$

Due to consistency, $\widehat{\theta}_n \to_p \theta$, and the second term in the denominator of (2.11) includes the average

$$\left.\frac{d^2G_n}{d\theta^2}\right|_{\tilde\theta} = \frac{1}{n}\sum_{i=1}^n \frac{d^2}{d\theta^2}G(\theta,Y_i),$$

which, by the law of large numbers, tends to its expectation, that is,

$$\left.\frac{d^2G_n}{d\theta^2}\right|_{\tilde\theta} \to_p \mathrm{E}\left[\frac{d^2}{d\theta^2}G(\theta,Y)\right],$$

provided this average exists. Hence, the second term in the denominator of (2.11) converges in probability to zero and so, by Slutsky's theorem (Appendix G)

$$\sqrt{n}\,(\widehat{\theta}_n - \theta) \to_d \mathrm{N}\left(0, \frac{B}{A^2}\right),$$

as required, where we have suppressed the dependence of $A(\theta)$ and $B(\theta)$ on θ.

$\square$

In practice, $\boldsymbol{A} = \boldsymbol{A}(\boldsymbol{\theta})$ and $\boldsymbol{B} = \boldsymbol{B}(\boldsymbol{\theta})$ are replaced by $\boldsymbol{A}_n(\widehat{\boldsymbol{\theta}}_n)$ and $\boldsymbol{B}_n(\widehat{\boldsymbol{\theta}}_n)$, respectively, with asymptotic normality continuing to hold due to Slutsky's theorem.

In the sections that follow we describe a number of approaches for constructing and using estimating functions. These approaches differ in the number of assumptions that are required for both specifying the estimating function and making inference. At one extreme, in a fully *model-based* approach, a full probability

2.3 Estimating Functions

distribution is specified for the data and is used to both specify the estimating function and to evaluate the expectations required in the calculation of A and B. At the other extreme, minimal assumptions are made on the data to construct the estimating function, and the expectations required to evaluate $\text{var}(\widehat{\boldsymbol{\theta}}_n)$ are calculated empirically from the observed data (see Sect. 2.6).

In the independent but not identically distributed case

$$[A_n^{-1} B_n (A_n^{\text{T}})^{-1}]^{-1/2} (\widehat{\boldsymbol{\theta}}_n - \boldsymbol{\theta}) \to_d N_p(\mathbf{0}, I_p), \tag{2.13}$$

where

$$A_n = \text{E}\left[\frac{\partial}{\partial \boldsymbol{\theta}^{\text{T}}} G_n(\boldsymbol{\theta})\right]$$

$$B_n = \text{E}\left[G_n(\boldsymbol{\theta}) G_n(\boldsymbol{\theta})^{\text{T}}\right] = \text{var}\left[G_n(\boldsymbol{\theta})\right].$$

The previous independent *and* identically distributed situation is a special case, with $A_n = nA$ and $B_n = nB$, in which case (2.13) simplifies to (2.9).

The *sandwich* form of the variance of $\widehat{\boldsymbol{\theta}}_n$ in (2.9) and (2.13)—the covariance of the estimating function, flanked by the expectation of the inverse of the Jacobian matrix of the transformation from the estimating function to the parameter—is one that will appear repeatedly.

Estimators derived from an estimating function are invariant in the sense that if we are interested in a function, $\phi = g(\boldsymbol{\theta})$, then the estimator is $\widehat{\phi}_n = g(\widehat{\boldsymbol{\theta}}_n)$. The delta method (Appendix G) allows the transfer of inference from the parameters of the model to quantities of interest. Specifically, suppose

$$\sqrt{n}\,(\widehat{\boldsymbol{\theta}}_n - \boldsymbol{\theta}) \to_d N_p\left[\mathbf{0}, V(\boldsymbol{\theta})\right].$$

Then, by the delta method,

$$\sqrt{n}\,\left[g(\widehat{\boldsymbol{\theta}}_n) - g(\boldsymbol{\theta})\right] \to_d N\left[0, g'(\boldsymbol{\theta}) V(\boldsymbol{\theta}) g'(\boldsymbol{\theta})^{\text{T}}\right],$$

where $g'(\boldsymbol{\theta})$ is the $1 \times p$ vector of derivatives of $g(\cdot)$ with respect to elements of $\boldsymbol{\theta}$. For example, for $p = 2$

$$\text{var}\left[g(\boldsymbol{\theta})\right] = V_{11} \left(\left.\frac{\partial g}{\partial \theta_1}\right|_{\boldsymbol{\theta}}\right)^2 + 2 V_{12} \left(\left.\frac{\partial g}{\partial \theta_1}\right|_{\boldsymbol{\theta}}\right)\left(\left.\frac{\partial g}{\partial \theta_2}\right|_{\boldsymbol{\theta}}\right) + V_{22} \left(\left.\frac{\partial g}{\partial \theta_2}\right|_{\boldsymbol{\theta}}\right)^2,$$

where V_{jk} denotes the (j, k)th element of V, $j, k = 1, 2$. Again in practice, $\widehat{\boldsymbol{\theta}}_n$ replaces $\boldsymbol{\theta}$ in $\text{var}[g(\boldsymbol{\theta})]$. The accuracy of the asymptotic distribution depends on the parameterization adopted. A rule of thumb is to obtain the asymptotic distribution for a reparameterized parameter defined on the real line; one may then transform back to the parameter of interest, to construct confidence intervals, for example.

The implementation of a frequentist approach usually requires a maximization or root-finding algorithm, but most statistical software packages now contain reliable routines for such endeavors in the majority of situations encountered in practice; hence, we will rarely discuss computational details (in contrast to the Bayesian approach for which computation is typically more challenging).

2.4 Likelihood

For reasons that will become evident, likelihood provides a popular approach to statistical inference and our coverage reflects this. Let $p(y \mid \theta)$ be a full probability model for the observed data given a p dimensional vector of parameters, θ. The probability model for the full data is based upon the context and all relevant accumulated knowledge. The level of belief in this model will clearly be context specific, and in many situations, there will be insufficient information available to confidently specify all components of the model. Depending on the confidence in the likelihood, which in turn depends on the sample size (since large n allows more reliable examination of the assumptions of the model), the likelihood may be effectively viewed as approximately "correct," in which case inference proceeds as if the true model were known. Alternatively the likelihood may be seen as an initial working model from which an estimating function is derived; the properties of the subsequent estimator may then be determined under a more general model.

Definition. Viewing $p(y \mid \theta)$ as a function of θ gives the *likelihood function*, denoted $L(\theta)$.

A key point is that $L(\theta)$ is *not* a probability distribution in θ, hence the name likelihood.[3]

2.4.1 Maximum Likelihood Estimation

The value of θ that maximizes $L(\theta)$ and hence gives the highest probability (density) to the observed data, denoted $\widehat{\theta}$, is known as the maximum likelihood estimator (MLE).

In Part II of this book, we consider models that are appropriate when the data are conditionally independent given θ so that

$$p(y \mid \theta) = \prod_{i=1}^{n} p(y_i \mid \theta).$$

---

[3] We use the label "likelihood" in this section, but strictly speaking we are considering *frequentist* likelihood, since we will evaluate the frequentist properties of an estimator derived from the likelihood. This contrasts with a pure likelihood view, as described in Royall (1997), in which properties are derived from the likelihood function alone, without resorting to frequentist arguments.

2.4 Likelihood

For the remainder of this chapter, we assume such conditional independence holds. For both computation and analysis, it is convenient to consider the *log-likelihood* function

$$l(\boldsymbol{\theta}) = \log L(\boldsymbol{\theta}) = \sum_{i=1}^{n} \log p(Y_i \mid \boldsymbol{\theta})$$

and the *score* function

$$\boldsymbol{S}(\boldsymbol{\theta}) = \frac{\partial l(\boldsymbol{\theta})}{\partial \boldsymbol{\theta}} = \left[\frac{\partial l(\boldsymbol{\theta})}{\partial \theta_1}, \ldots, \frac{\partial l(\boldsymbol{\theta})}{\partial \theta_p}\right]^{\mathrm{T}}$$
$$= [S_1(\boldsymbol{\theta}), \ldots, S_p(\boldsymbol{\theta})]^{\mathrm{T}},$$

which is the $p \times 1$ vector of derivatives of the log-likelihood. As we now illustrate, the score satisfies the requirements of an estimating function.

Definition. Fisher's expected information in a sample of size n is the $p \times p$ matrix

$$\boldsymbol{I}_n(\boldsymbol{\theta}) = -\mathrm{E}\left[\frac{\partial^2}{\partial \boldsymbol{\theta} \partial \boldsymbol{\theta}^{\mathrm{T}}} l(\boldsymbol{\theta})\right] = -\mathrm{E}\left[\frac{\partial \boldsymbol{S}(\boldsymbol{\theta})}{\partial \boldsymbol{\theta}^{\mathrm{T}}}\right].$$

Result. Under suitable regularity conditions,

$$\mathrm{E}[\boldsymbol{S}(\boldsymbol{\theta})] = \mathrm{E}\left[\frac{\partial l}{\partial \boldsymbol{\theta}}\right] = \boldsymbol{0}, \tag{2.14}$$

and

$$\boldsymbol{I}_n(\boldsymbol{\theta}) = -\mathrm{E}\left[\frac{\partial \boldsymbol{S}(\boldsymbol{\theta})}{\partial \boldsymbol{\theta}^{\mathrm{T}}}\right] = \mathrm{E}\left[\boldsymbol{S}(\boldsymbol{\theta})\boldsymbol{S}(\boldsymbol{\theta})^{\mathrm{T}}\right]. \tag{2.15}$$

Proof. For simplicity we give a prove for the situation in which θ is univariate, and the observations are independent and identically distributed. Under these circumstances

$$I_n(\theta) = nI_1(\theta),$$

where

$$I_1(\theta) = -\mathrm{E}\left[\frac{d^2}{d\theta^2} \log p(Y \mid \theta)\right].$$

The expectation of the score is

$$\mathrm{E}[S(\theta)] = \sum_{i=1}^{n} \mathrm{E}\left[\frac{d}{d\theta} \log p(Y_i \mid \theta)\right] = n\mathrm{E}\left[\frac{d}{d\theta} \log p(Y \mid \boldsymbol{\theta})\right]$$

and, under regularity conditions that allow the interchange of differentiation and integration,

$$\mathrm{E}\left[\frac{d}{d\theta}\log p(Y \mid \theta)\right] = \int \left(\frac{d}{d\theta}\log p(y \mid \theta)\right) p(y \mid \theta) dy$$
$$= \int \frac{d}{d\theta} p(y \mid \theta) \frac{p(y \mid \theta)}{p(y \mid \theta)} dy = \frac{d}{d\theta} \int p(y \mid \theta) dy = 0,$$
(2.16)

which proves (2.14).
From (2.16),

$$0 = \frac{d}{d\theta}\left[\int \left(\frac{d}{d\theta}\log p(y \mid \theta)\right) p(y \mid \theta) dy\right]$$
$$= \int \frac{d}{d\theta} \left(\frac{d}{d\theta}\log p(y \mid \theta) p(y \mid \theta)\right) dy$$
$$= \int \left(\frac{d^2}{d\theta^2}\log p(y \mid \theta)\right) p(y \mid \theta) dy + \int \left(\frac{d}{d\theta}\log p(y \mid \theta)\right) \left(\frac{d}{d\theta} p(y \mid \theta)\right) dy$$
$$= \int \left(\frac{d^2}{d\theta^2}\log p(y \mid \theta)\right) p(y \mid \theta) dy + \int \left(\frac{d}{d\theta}\log p(y \mid \theta)\right)^2 p(y \mid \theta) dy$$
$$= \mathrm{E}\left[\frac{d^2}{d\theta^2}\log p(Y \mid \theta)\right] + \mathrm{E}\left[\left(\frac{d}{d\theta}\log p(Y \mid \theta)\right)^2\right],$$

which proves (2.15). □

Viewing the score as an estimating function,

$$G_n(\boldsymbol{\theta}) = \frac{1}{n}S(\boldsymbol{\theta}) = \frac{1}{n}\sum_{i=1}^{n}\frac{d}{d\theta}\log p(Y_i \mid \theta),$$

shows that the MLE satisfies $G_n(\widehat{\boldsymbol{\theta}}_n) = \mathbf{0}$. We have already seen that

$$\mathrm{E}[G_n(\boldsymbol{\theta})] = \frac{1}{n}\mathrm{E}[S(\boldsymbol{\theta})] = \mathbf{0},$$

and to apply Result 2.1 of Sect. 2.3, we require

$$\boldsymbol{A}(\boldsymbol{\theta}) = \mathrm{E}\left[\frac{\partial}{\partial \boldsymbol{\theta}^{\mathrm{T}}} \boldsymbol{G}(\boldsymbol{\theta}, Y)\right] = \mathrm{E}\left[\frac{\partial^2}{\partial \boldsymbol{\theta} \partial \boldsymbol{\theta}^{\mathrm{T}}}\log p(Y \mid \boldsymbol{\theta})\right]$$

and

$$\boldsymbol{B}(\boldsymbol{\theta}) = \mathrm{E}\left[\boldsymbol{G}(\boldsymbol{\theta}, Y)\boldsymbol{G}(\boldsymbol{\theta}, Y)^{\mathrm{T}}\right] = \mathrm{E}\left[\left(\frac{\partial}{\partial \boldsymbol{\theta}}\log p(Y \mid \boldsymbol{\theta})\right)\left(\frac{\partial}{\partial \boldsymbol{\theta}}\log p(Y \mid \boldsymbol{\theta})\right)^{\mathrm{T}}\right].$$

2.4 Likelihood

Equation (2.15) shows that

$$I_1(\boldsymbol{\theta}) = -A(\boldsymbol{\theta}) = B(\boldsymbol{\theta})$$

and, from (2.12)

$$n^{-1/2}S(\boldsymbol{\theta}) \to_d N[\mathbf{0}, I_1(\boldsymbol{\theta})]. \qquad (2.17)$$

From Result 2.1, the asymptotic distribution of the MLE is therefore

$$\sqrt{n}\,(\widehat{\boldsymbol{\theta}}_n - \boldsymbol{\theta}) \to_d N_p\left[\mathbf{0}, I_1(\boldsymbol{\theta})^{-1}\right]. \qquad (2.18)$$

For independent, but not necessarily identically distributed, random variables $Y_1, \ldots, Y_n$,

$$I_n(\boldsymbol{\theta}) = -A_n(\boldsymbol{\theta}) = B_n(\boldsymbol{\theta}),$$

and

$$I_n(\boldsymbol{\theta})^{1/2}(\widehat{\boldsymbol{\theta}}_n - \boldsymbol{\theta}) \to_d N_p(\mathbf{0}, I_p), \qquad (2.19)$$

The information is scaling the statistic and should be growing with n for the asymptotic distribution to be appropriate. Intuitively, the *curvature* of the log-likelihood, as measured by the second derivative, determines the variability of the estimator; the greater the curvature, the smaller the variance of the estimator. The distribution of $\widehat{\boldsymbol{\theta}}_n$ is sometimes written as

$$\widehat{\boldsymbol{\theta}}_n \to_d N_p\left[\boldsymbol{\theta}, I_n(\boldsymbol{\theta})^{-1}\right],$$

but this is a little sloppy since the limiting distribution should be independent of n. The variance of the score-based estimating function has the property that $A = A^T$ because the matrix of second derivatives is symmetric, that is,

$$\frac{\partial^2 l}{\partial \theta_j \partial \theta_k} = \frac{\partial^2 l}{\partial \theta_k \partial \theta_j}$$

for $j, k = 1, \ldots, p$.

If there is a unique maximum, then the MLE is consistent and asymptotically normal. The Cramér–Rao bound was given in (2.3). In the present terminology, for any unbiased estimator, $\widetilde{\boldsymbol{\theta}}$, the bound is $\text{var}(\widetilde{\boldsymbol{\theta}}) \geq I_n(\boldsymbol{\theta})^{-1}$ so that the MLE is asymptotically efficient. Asymptotic efficiency under correct model specification is a primary motivation for the widespread use of MLEs.

For inference via (2.18), we may also replace the *expected* information by the *observed* information,

$$I_n^\star = -\frac{\partial^2}{\partial \boldsymbol{\theta} \partial \boldsymbol{\theta}^T} l(\boldsymbol{\theta}).$$

Asymptotically, their use is equivalent since $I_n^\star \to_p I_n$ as $n \to \infty$ by the weak law of large numbers (Appendix G).

The regularity conditions required to derive the asymptotic distribution of the MLE include identifiability so that each element of the parameter space θ should correspond to a different model $p(y \mid \theta)$, otherwise there would be no unique value of θ to which $\widehat{\theta}$ would converge. We require the interchange of differentiation and integration, and so the range of the data cannot depend on an unknown parameter. Additionally, the true parameter value must lie in the interior of the parameter space, and the Taylor series expansion that was used to determine the asymptotic distribution of $\widehat{\theta}$ requires a well-behaved derivative and so the amount of information must increase with sample size. One situation in which one must be wary is when the number of parameters increases with sample size—this number cannot increase too quickly—see Exercise 2.6 for a model in which this condition is violated.

In Sect. 2.4.3, we examine the effects on inference based on the MLE of model misspecification and, in Sects. 2.6 and 2.7, describe methods for determining properties of the estimator that do not depend on correct specification of the full probability model.

Example: Binomial Likelihood

For a single observation from a binomial distribution, $Y \mid p \sim \text{Binomial}(n, p)$, the log-likelihood is

$$l(p) = Y \log p + (n - Y) \log(1 - p),$$

where we omit the term $\log \binom{n}{Y}$ because it is constant with respect to p. The score is

$$S(p) = \frac{dl}{dp} = \frac{Y}{p} - \frac{n - Y}{1 - p},$$

and setting $S(\widehat{p}) = 0$ gives $\widehat{p} = Y/n$. In addition

$$\frac{d^2 l}{dp^2} = -\frac{Y}{p^2} - \frac{n - Y}{(1 - p)^2},$$

and

$$I(p) = -\mathrm{E}\left[\frac{d^2 l}{dp^2}\right] = \frac{n}{p(1 - p)}.$$

We therefore see that the amount of information in the data for p is greater if p is closer to 0 or 1. This is intuitively reasonable since the variance of Y is $np(1 - p)$ and so there is less variability in the data (and hence less uncertainty) if p is close to 0 or 1. The asymptotic distribution of the MLE is

$$\sqrt{n}(\widehat{p} - p) \to_d \mathrm{N}\left[p, p(1 - p)\right],$$

2.4 Likelihood

so that an asymptotic 95% confidence interval for p is

$$\left[\widehat{p} - 1.96 \times \sqrt{\frac{\widehat{p}(1-\widehat{p})}{n}}, \widehat{p} + 1.96 \times \sqrt{\frac{\widehat{p}(1-\widehat{p})}{n}}\right].$$

Unfortunately, the endpoints of this interval are not guaranteed to lie in $(0,1)$. To rectify this shortcoming, we may parameterize in terms of the logit of p, $\theta = \log[p/(1-p)]$. We could derive the asymptotic distribution using the delta method, but instead we reparameterize the model to give

$$l(\theta) = Y\theta - n\log\left[1 + \exp(\theta)\right],$$

and, proceeding as in the previous parameterization,

$$\widehat{\theta} = \log\left(\frac{Y}{n-Y}\right)$$

and

$$I(\theta) = \frac{n[1+\exp(\theta)]^2}{\exp(\theta)}$$

to give

$$\sqrt{n}(\widehat{\theta} - \theta) \to_d N\left(0, \frac{\exp(\theta)}{[1+\exp(\theta)]^2}\right).$$

An asymptotic 95% confidence interval for p follows from transforming the endpoints of the interval for θ:

$$\left[\frac{\exp\left(\widehat{\theta} - 1.96 \times \sqrt{\text{var}(\widehat{\theta})/n}\right)}{1 + \exp\left(\widehat{\theta} - 1.96 \times \sqrt{\text{var}(\widehat{\theta})/n}\right)}, \frac{\exp\left(\widehat{\theta} + 1.96 \times \sqrt{\text{var}(\widehat{\theta})/n}\right)}{1 + \exp\left(\widehat{\theta} + 1.96 \times \sqrt{\text{var}(\widehat{\theta})/n}\right)}\right].$$

The endpoints will be contained in $(0,1)$, though $\widehat{\theta}$ is undefined if $Y = 0$ or $Y = n$.

Example: Lung Cancer and Radon

Consider the model

$$Y_i \mid \beta \sim_{ind} \text{Poisson}(\mu_i),$$

with $\mu_i = E_i \exp(x_i \beta)$, $x_i = [1, x_i]$, $i = 1, \ldots, n$, and $\beta = [\beta_0, \beta_1]^T$. The probability distribution of y is

$$p(y \mid \beta) = \exp\left(\sum_{i=1}^{n} y_i \log \mu_i - \sum_{i=1}^{n} \mu_i - \sum_{i=1}^{n} \log y_i!\right)$$

to give log-likelihood

$$l(\boldsymbol{\beta}) = \boldsymbol{\beta}^{\mathrm{T}} \sum_{i=1}^{n} \boldsymbol{x}_i^{\mathrm{T}} Y_i - \sum_{i=1}^{n} E_i \exp(\boldsymbol{x}_i \boldsymbol{\beta})$$

and 2×1 score vector (estimating function)

$$\begin{aligned} \boldsymbol{S}(\boldsymbol{\beta}) = \frac{\partial l}{\partial \boldsymbol{\beta}} &= \sum_{i=1}^{n} \boldsymbol{x}_i^{\mathrm{T}} [Y_i - E_i \exp(\boldsymbol{x}_i \boldsymbol{\beta})] \\ &= \boldsymbol{x}^{\mathrm{T}} [\boldsymbol{Y} - \boldsymbol{\mu}(\boldsymbol{\beta})] , \end{aligned} \qquad (2.20)$$

where $\boldsymbol{x} = [\boldsymbol{x}_1^{\mathrm{T}}, \ldots, \boldsymbol{x}_n^{\mathrm{T}}]^{\mathrm{T}}$, $\boldsymbol{Y} = [Y_1, \ldots, Y_n]^{\mathrm{T}}$, and $\boldsymbol{\mu} = [\mu_1, \ldots, \mu_n]^{\mathrm{T}}$. The equation $\boldsymbol{S}(\widehat{\boldsymbol{\beta}}) = \boldsymbol{0}$ does not, in general, have a closed-form solution, but, pathological datasets aside, numerical solution is straightforward. Asymptotic inference is based on

$$\boldsymbol{I}_n(\widehat{\boldsymbol{\beta}}_n)^{1/2}(\widehat{\boldsymbol{\beta}}_n - \boldsymbol{\beta}) \to_d \mathrm{N}_2(\boldsymbol{0}, \mathrm{I}_2),$$

where the information matrix is

$$\boldsymbol{I}_n(\widehat{\boldsymbol{\beta}}_n) = \mathrm{var}(\boldsymbol{S}) = \sum_{i=1}^{n} \boldsymbol{x}_i^{\mathrm{T}} \mathrm{var}(Y_i) \boldsymbol{x}_i = \boldsymbol{x}^{\mathrm{T}} \boldsymbol{V} \boldsymbol{x},$$

with $\boldsymbol{V}$ the diagonal matrix with elements $\mathrm{var}(Y_i) = E_i \exp(\boldsymbol{x}_i \boldsymbol{\beta})$, $i = 1, \ldots, n$. In this case, the expected and observed information coincide. In practice, the information is estimated by replacing $\boldsymbol{\beta}$ by $\widehat{\boldsymbol{\beta}}_n$. An important observation is that if the mean is correctly specified the score, (2.20) is a consistent estimator of zero, and $\widehat{\boldsymbol{\beta}}_n$ is a consistent estimator of $\boldsymbol{\beta}$. In particular, if the data do not conform to $\mathrm{var}(Y_i) = \mu_i$, we still have a consistent estimator, but the standard errors will be incorrect.

For the lung cancer data, we have $n = 85$, and the MLE is $\widehat{\boldsymbol{\beta}} = [0.17, -0.036]^{\mathrm{T}}$ with

$$\boldsymbol{I}(\widehat{\boldsymbol{\beta}})^{-1} = \begin{bmatrix} 0.027^2 & -0.95 \times 0.027 \times 0.0054 \\ -0.95 \times 0.027 \times 0.0054 & 0.0054^2 \end{bmatrix}.$$

The estimated standard errors of $\widehat{\beta}_0$ and $\widehat{\beta}_1$ are 0.027 and 0.0054, respectively, and an asymptotic 95% confidence interval for β_1 is $[-0.047, -0.026]$. Leaning on asymptotic normality is appropriate with the large sample size here. A useful inferential summary is an asymptotic 95% confidence interval for the area-level relative risk associated with a one-unit increase in residential radon, which is

$$\exp(-0.036 \pm 1.96 \times 0.0054) = [0.954, 0.975].$$

This interval suggests that the decrease in lung cancer incidence associated with a one-unit increase in residential radon is between 2.5% and 4.6%, though we stress

Example: Weibull Model

The Weibull distribution is useful for the modeling of survival and reliability data and is of the form

$$p(y \mid \boldsymbol{\theta}) = \theta_1 \theta_2^{\theta_1} y^{\theta_1 - 1} \exp\left[-(\theta_2 y)^{\theta_1}\right],$$

where $y > 0$, $\boldsymbol{\theta} = [\theta_1, \theta_2]^\mathsf{T}$ and $\theta_1, \theta_2 > 0$. The mean and variance of the Weibull distribution are

$$E[Y \mid \boldsymbol{\theta}] = \Gamma(1/\theta_1 + 1)/\theta_2$$
$$\mathrm{var}(Y \mid \boldsymbol{\theta}) = [\Gamma(2/\theta_1 + 1) - \Gamma(1/\theta_1 + 1)^2]/\theta_2^2,$$

where

$$\Gamma(\alpha) = \int_0^\infty x^{\alpha - 1} \exp(-x) dx$$

is the gamma function. Therefore, the first two moments are not simple functions of θ_1 and θ_2. With independent and identically distributed observations Y_i, $i = 1, \ldots, n$, from a Weibull distribution the log-likelihood is

$$l(\boldsymbol{\theta}) = n \log \theta_1 + n \theta_1 \log \theta_2 + (\theta_1 - 1) \sum_{i=1}^{n} \log Y_i - \theta_2^{\theta_1} \sum_{i=1}^{n} Y_i^{\theta_1},$$

with score equations

$$S_1(\boldsymbol{\theta}) = \frac{\partial l}{\partial \theta_1} = \frac{n}{\theta_1} + n \log \theta_2 + \sum_{i=1}^{n} \log Y_i - \theta_2^{\theta_1} \sum_i^{n} Y_i^{\theta_1} \log(\theta_2 Y_i)$$

$$S_2(\boldsymbol{\theta}) = \frac{\partial l}{\partial \theta_2} = \frac{n \theta_1}{\theta_2} - \theta_1 \theta_2^{\theta_1 - 1} \sum_{i=1}^{n} Y_i^{\theta_1},$$

which have no closed-form solution and are not a function of a sufficient statistic of dimension less than n. Hence, consistency of $\widehat{\boldsymbol{\theta}}_n$, where $\boldsymbol{S}(\widehat{\boldsymbol{\theta}}_n) = \boldsymbol{0}$, cannot be determined from consideration of the first moment (or even the first two moments) of the data only, unlike the Poisson example. In particular, consistency under model misspecification cannot easily be determined.

2.4.2 Variants on Likelihood

Estimation via the likelihood, as defined by $L(\boldsymbol{\theta}) = p(\boldsymbol{y} \mid \boldsymbol{\theta})$, is not always universally applied. In some situations, such as when regularity conditions are violated, alternative versions are required to provide procedures that produce estimators with desirable properties. In other situations, alternative likelihoods provide estimators with better small sample properties, perhaps because nuisance parameters are dealt with more efficiently. Unfortunately, the construction of these likelihoods is not prescriptive and can require a great deal of ingenuity. We describe conditional, marginal, and profile likelihoods.

Conditional Likelihood

Suppose $\boldsymbol{\lambda}$ represent parameters of interest, with $\boldsymbol{\phi}$ being nuisance parameters. Suppose the distribution for $\boldsymbol{y}$ can be factorized as

$$p(\boldsymbol{y} \mid \boldsymbol{\lambda}, \boldsymbol{\phi}) \propto p(\boldsymbol{t}_1 \mid \boldsymbol{t}_2, \boldsymbol{\lambda}) p(\boldsymbol{t}_2 \mid \boldsymbol{\lambda}, \boldsymbol{\phi}), \qquad (2.21)$$

where $\boldsymbol{t}_1$ and $\boldsymbol{t}_2$ are statistics, that is, functions of $\boldsymbol{y}$. Then inference for $\boldsymbol{\lambda}$ may be based on the *conditional likelihood*

$$L_c(\boldsymbol{\lambda}) = p(\boldsymbol{t}_1 \mid \boldsymbol{t}_2, \boldsymbol{\lambda}). \qquad (2.22)$$

The conditional likelihood has similar properties to a regular likelihood. Conditional likelihoods may be used in situations in which we wish to eliminate nuisance parameters. The conditioning statistic, $\boldsymbol{t}_2$, is not ancillary (Appendix F), so that it does depend on $\boldsymbol{\lambda}$, and so some information may be lost in the act of conditioning, but the benefits of elimination are assumed to outweigh this loss. Conditional likelihoods will be used in Sect. 7.7 in the context of Fisher's exact test and individually matched case-control studies (in which the number of parameters increases with sample size) and in Sects. 9.5 and 9.13.4 to eliminate random effects in mixed effects models.

Marginal Likelihood

Let $\boldsymbol{S}_1, \boldsymbol{S}_2, \boldsymbol{A}$ be a minimal sufficient statistic where $\boldsymbol{A}$ is ancillary (Appendix F), and suppose we have the factorization

$$p(\boldsymbol{y} \mid \boldsymbol{\lambda}, \boldsymbol{\phi}) \propto p(\boldsymbol{s}_1, \boldsymbol{s}_2, \boldsymbol{a} \mid \boldsymbol{\lambda}, \boldsymbol{\phi})$$
$$= p(\boldsymbol{a}) p(\boldsymbol{s}_1 \mid \boldsymbol{a}, \boldsymbol{\lambda}) p(\boldsymbol{s}_2 \mid \boldsymbol{s}_1, \boldsymbol{a}, \boldsymbol{\lambda}, \boldsymbol{\phi})$$

2.4 Likelihood

where λ are parameters of interest and ϕ are the remaining (nuisance) parameters. In contrast to conditional likelihood, marginal likelihoods are based on *averaging* over parts of the data to obtain $p(s_1 \mid a, \lambda)$, though operationally marginal likelihoods are often derived without the need for explicit averaging.

Inference for λ may be based on the *marginal* likelihood

$$L_m(\lambda) = p(s_1 \mid a, \lambda)$$

and is desirable if inference is simplified or if problems with standard likelihood methods are to be avoided.

These advantages may outweigh the loss of efficiency in ignoring the term $p(s_2 \mid s_1, a, \lambda, \phi)$. If there is no ancillary statistic, then the marginal likelihood is

$$L_m(\lambda) = p(s_1 \mid \lambda).$$

The marginal likelihood has similar properties to a regular likelihood. We will make use of marginal likelihoods for variance component estimation in mixed effects models in Sect. 8.5.3.

Example: Normal Linear Model

Assume $Y \mid \beta, \sigma^2 \sim N_n(x\beta, \sigma^2 I_n)$ where x is the $n \times (k+1)$ design matrix and $\dim(\beta) = k+1$. Suppose the parameter of interest is $\lambda = \sigma^2$, with remaining parameters $\phi = \beta$. The MLE for σ^2 is

$$\widetilde{\sigma}^2 = \frac{1}{n}(y - x\widehat{\beta})^\mathsf{T}(y - x\widehat{\beta}) = \frac{\text{RSS}}{n}$$

with $\widehat{\beta} = (x^\mathsf{T} x)^{-1} x^\mathsf{T} Y$. It is well known that $\widetilde{\sigma}^2$ has finite sample bias, because the estimation of β is not taken into account. The minimal sufficient statistics are $s_1 = S^2 = \text{RSS}/(n-k-1)$ and $s_2 = \widehat{\beta}$. We write the probability density for y in terms of s_1 and s_2:

$$p(y \mid \sigma^2, \beta) = (2\pi\sigma^2)^{-n/2} \exp\left[-\frac{1}{2\sigma^2}(y - x\beta)^\mathsf{T}(y - x\beta)\right]$$

$$\propto \sigma^{-n} \exp\left[-\frac{1}{2\sigma^2}(n-k-1)s^2\right] \exp\left[-\frac{1}{2\sigma^2}(\widehat{\beta} - \beta)^\mathsf{T} x^\mathsf{T} x (\widehat{\beta} - \beta)\right]$$

$$= p(s_1 \mid \sigma^2) p(s_2 \mid \beta, \sigma^2)$$

where going between the first and second line is straightforward if we recognize that

$$(y - x\beta)^\mathsf{T}(y - x\beta) = (y - x\widehat{\beta} + x\widehat{\beta} - x\beta)^\mathsf{T}(y - x\widehat{\beta} + x\widehat{\beta} - x\beta)$$

$$= (y - x\widehat{\beta})^\mathsf{T}(y - x\widehat{\beta}) + (\widehat{\beta} - \beta)^\mathsf{T} x^\mathsf{T} x (\widehat{\beta} - \beta), \quad (2.23)$$

with the cross term disappearing because of independence between $\widehat{\beta}$ and the vector of residuals $y - x\widehat{\beta}$. Consequently, the marginal likelihood is

$$L_m(\sigma^2) = p(s^2 \mid \sigma^2).$$

Since the data are normal

$$\frac{(n-k-1)s^2}{\sigma^2} \sim \chi^2_{n-k-1} = \text{Ga}\left(\frac{n-k-1}{2}, \frac{1}{2}\right),$$

and so

$$p(s^2 \mid \sigma^2) = \left(\frac{n-k-1}{2\sigma^2}\right)^{(n-k-1)/2} \frac{(s^2)^{(n-k-1)/2-1}}{\Gamma\left(\frac{n-k-1}{2}\right)} \exp\left[-\frac{(n-k-1)s^2}{2\sigma^2}\right],$$

to give

$$l_m = \log L_m = -(n-k-1)\log\sigma - \frac{(n-k-1)s^2}{2\sigma^2},$$

and marginal likelihood estimator $\widehat{\sigma}^2 = s^2$, the usual unbiased estimator.

Profile Likelihood

Profile likelihood provides a method of examining the behavior of a subset of the parameters. If $\theta = [\lambda, \phi]$, where λ again represents a vector of parameters of interest and ϕ the remaining parameters, then the profile likelihood $L_p(\lambda)$ for λ is defined as

$$L_p(\lambda) = \max_\phi L(\lambda, \phi). \tag{2.24}$$

If $\widetilde{\lambda}$ denotes the maximum of $L_p(\lambda)$ and $\widehat{\theta} = \left[\widehat{\lambda}, \widehat{\phi}\right]$ is the MLE, then $\widetilde{\lambda} = \widehat{\lambda}$. Profile likelihoods will be encountered in Sect. 8.5, in the context of the estimation of variance components in linear mixed effects models.

2.4.3 Model Misspecification

In the following, we begin by assuming independent observations. We have seen that if the assumed model is correct then the MLE, $\widehat{\theta}$, has asymptotic distribution

$$\sqrt{n}\left(\widehat{\theta}_n - \theta\right) \to_d N_p\left[0, I_1(\theta)^{-1}\right].$$

2.4 Likelihood

In this section we examine the effects of model misspecification. We first determine exactly what quantity the MLE is estimating under misspecification and then examine the asymptotic distribution of the MLE. Let $p(y \mid \boldsymbol{\theta})$ and $p_{\text{T}}(y)$ denote the *assumed* and *true* densities, respectively.

The average of the log-likelihood is such that

$$\frac{1}{n} \sum_{i=1}^{n} \log p(Y_i \mid \boldsymbol{\theta}) \to_{a.s.} \mathrm{E}_{\text{T}}[\log p(Y \mid \boldsymbol{\theta})], \tag{2.25}$$

by the strong law of large numbers. Hence, asymptotically the MLE maximizes the expectation of the assumed log-likelihood under the true model and $\widehat{\boldsymbol{\theta}}_n \to_p \boldsymbol{\theta}_{\text{T}}$. We now investigate what $\boldsymbol{\theta}_{\text{T}}$ represents when we have assumed an incorrect model. We write

$$\mathrm{E}_{\text{T}}[\log p(Y \mid \boldsymbol{\theta})] = \mathrm{E}_{\text{T}}\left[\log p_{\text{T}}(Y) - \log p_{\text{T}}(Y) + \log p(Y \mid \boldsymbol{\theta})\right]$$
$$= \mathrm{E}_{\text{T}}[\log p_{\text{T}}(Y)] - \mathrm{KL}(p_{\text{T}}, p), \tag{2.26}$$

where

$$\mathrm{KL}(f, g) = \int \log \frac{f(y)}{g(y)} f(y) \, dy \geq 0,$$

is the Kullback–Leibler measure of the "distance" between the densities f and g (the measure is not symmetric so is not a conventional distance measure). The first term of (2.26) does not depend on $\boldsymbol{\theta}$, and so the MLE minimizes $\mathrm{KL}(p_{\text{T}}, p)$, and is therefore that value of $\boldsymbol{\theta}$ which makes the assumed model closest, in a Kullback–Leibler sense, to the true model.

We let $\boldsymbol{S}_n(\boldsymbol{\theta})$ denote the score under the assumed model and state the following result, along with a heuristic derivation.

Result. Suppose $\widehat{\boldsymbol{\theta}}_n$ is a solution to the estimating equation $\boldsymbol{S}_n(\boldsymbol{\theta}) = \boldsymbol{0}$, that is, $\boldsymbol{S}_n(\widehat{\boldsymbol{\theta}}_n) = \boldsymbol{0}$. Then

$$\sqrt{n}\,(\widehat{\boldsymbol{\theta}}_n - \boldsymbol{\theta}_{\text{T}}) \to_d \mathrm{N}_p\left[\boldsymbol{0}, \boldsymbol{J}^{-1} \boldsymbol{K}(\boldsymbol{J}^{\text{T}})^{-1}\right] \tag{2.27}$$

where

$$\boldsymbol{J} = \boldsymbol{J}(\boldsymbol{\theta}_{\text{T}}) = \mathrm{E}_{\text{T}}\left[\frac{\partial^2}{\partial \boldsymbol{\theta} \partial \boldsymbol{\theta}^{\text{T}}} \log p(Y \mid \boldsymbol{\theta}_{\text{T}})\right],$$

and

$$\boldsymbol{K} = \boldsymbol{K}(\boldsymbol{\theta}_{\text{T}}) = \mathrm{E}_{\text{T}}\left[\left(\frac{\partial}{\partial \boldsymbol{\theta}} \log p(Y \mid \boldsymbol{\theta}_{\text{T}})\right) \left(\frac{\partial}{\partial \boldsymbol{\theta}} \log p(Y \mid \boldsymbol{\theta}_{\text{T}})\right)^{\text{T}}\right].$$

Outline Derivation

The derivation closely follows that of Result 2.1, and for simplicity we again assume θ is one-dimensional. We first obtain the expectation and variance of

$$\frac{1}{n}S_n(\theta) = \frac{1}{n}\sum_{i=1}^{n}\frac{d}{d\theta}\log p(y_i \mid \theta),$$

in order to derive the asymptotic distribution of $S_n(\theta)$. Subsequently, we obtain the distribution of $\widehat{\theta}_n$.

Recall that θ_T is that value which minimizes the Kullback–Leibler distance, that is,

$$\begin{aligned}
0 &= \frac{d}{d\theta}\text{KL}(\theta)\bigg|_{\theta_T} = \left[\frac{d}{d\theta}\int \log \frac{p_T(y)}{p(y \mid \theta)}p_T(y)dy\right]\bigg|_{\theta_T} \\
&= \left[\int \frac{d}{d\theta}\log p_T(y)p_T(y)dy - \int \frac{d}{d\theta}\log p(y \mid \theta)p_T(y)dy\right]\bigg|_{\theta_T} \\
&= 0 - \left[\int \left(\frac{d}{d\theta}\log p(y \mid \theta)\right)p_T(y)dy\right]\bigg|_{\theta_T},
\end{aligned}$$

and so $E_T[S(\theta_T)] = 0$ (and we have assumed that we can interchange the order of differentiation and integration).

For the second moment,

$$\frac{1}{n}\sum_{i=1}^{n}\left(\frac{d}{d\theta}\log p(y_i \mid \theta)\right)^2 \to_p E_T\left[\left(\frac{d}{d\theta}\log p(Y \mid \theta_T)\right)^2\right] = K,$$

which we assume exists. Hence, by the central limit theorem

$$\frac{1}{n}S(\theta_T) \to_d N(0, K).$$

Expanding $S_n(\theta)$ in a Taylor series around θ_T:

$$0 = \frac{1}{n}S_n(\widehat{\theta}_n) = \frac{1}{n}S_n(\theta_T) + (\widehat{\theta}_n - \theta_T)\frac{1}{n}\frac{dS_n}{d\theta}\bigg|_{\theta_T} + \frac{1}{2}(\widehat{\theta}_n - \theta_T)^2\frac{1}{n}\frac{d^2 S_n}{d\theta^2}\bigg|_{\widetilde{\theta}},$$

where $\widetilde{\theta}$ is between $\widehat{\theta}_n$ and θ_T and

$$\frac{1}{n}\frac{dS_n(\theta)}{d\theta}\bigg|_{\theta_T} = \frac{1}{n}\sum_{i=1}^{n}\frac{d^2}{d\theta^2}\log p(y \mid \theta)\bigg|_{\theta_T} \to_p E_T\left[\frac{d^2}{d\theta^2}\log p(Y \mid \theta_T)\right] = J.$$

2.5 Quasi-likelihood

Following the outline derivation of Result 2.1 gives

$$\sqrt{n}\,(\widehat{\theta}_n - \theta_{\text{T}}) \to_d \text{N}\left(0, \frac{K}{J^2}\right),$$

as required.

Example: Exponential Assumed Model, Gamma True Model

Suppose that the assumed model is exponential with mean θ but that the true model is gamma $\text{Ga}(\alpha, \beta)$. Minimizing the Kullback–Leibler distance with respect to θ corresponds to maximizing (2.25), that is

$$\text{E}_{\text{T}}\left[-\log\theta - \frac{Y}{\theta}\right] = \log\theta - \frac{\alpha/\beta}{\theta},$$

so that $\theta_{\text{T}} = \alpha/\beta$ is the quantity that is being estimated by the MLE. Hence, the closest exponential distribution to the gamma distribution, in a Kullback–Leibler sense, is the one that possesses the same mean.

2.5 Quasi-likelihood

2.5.1 Maximum Quasi-likelihood Estimation

In this section we describe an estimating function that is based upon the mean and variance of the data only. Specifically, we assume that the first two moments are of the form

$$\text{E}[\boldsymbol{Y} \mid \boldsymbol{\beta}] = \boldsymbol{\mu}(\boldsymbol{\beta})$$
$$\text{var}(\boldsymbol{Y} \mid \boldsymbol{\beta}) = \alpha \boldsymbol{V}\,[\boldsymbol{\mu}(\boldsymbol{\beta})]$$

where $\boldsymbol{\mu}(\boldsymbol{\beta}) = [\mu_1(\boldsymbol{\beta}), \ldots, \mu_n(\boldsymbol{\beta})]^{\text{T}}$ represents the regression function, $\boldsymbol{V}$ is a diagonal matrix (so the observations are assumed uncorrelated), with

$$\text{var}(Y_i \mid \boldsymbol{\beta}) = \alpha V\,[\mu_i(\boldsymbol{\beta})],$$

and $\alpha > 0$ is a scalar that does not depend upon $\boldsymbol{\beta}$. We assume $\boldsymbol{\beta} = [\beta_0, \ldots, \beta_k]^{\text{T}}$ so that the dimension of $\boldsymbol{\beta}$ is $k+1$. The aim is to obtain the asymptotic properties of an estimator of $\boldsymbol{\beta}$ based on these first two moments only. The specification of the mean function in a parametric regression setting is unavoidable, and efficiency will clearly depend on the form of the variance model.

To motivate an estimating function, consider the sum of squares

$$(Y - \mu)^T V^{-1}(Y - \mu)/\alpha, \qquad (2.28)$$

where $\mu = \mu(\beta)$ and $V = V(\beta)$. To minimize this sum of squares, there are two ways to proceed. Perhaps the more obvious route is to acknowledge that both μ and V are functions of β and differentiate with respect to β to give

$$-2D^T V^{-1}(Y - \mu)/\alpha + (Y - \mu)^T \frac{\partial V^{-1}}{\partial \beta}(Y - \mu)/\alpha, \qquad (2.29)$$

where D is the $n \times p$ matrix of derivatives with elements $\partial \mu_i / \partial \beta_j$, $i = 1, \ldots, n$, $j = 1, \ldots, p$. Unfortunately, (2.29) is not ideal as an estimating function because it does not necessarily have expectation zero when we only assume $E[Y \mid \beta] = \mu$, because of the presence of the second term. If the expectation of the estimating function is not zero, then an inconsistent estimator of β results.

Alternatively, we may temporarily forget that V is a function of β when we differentiate (2.28) and solve the estimating equation

$$D(\widehat{\beta})^T V(\widehat{\beta})^{-1} \left[Y - \mu(\widehat{\beta}) \right]/\alpha = 0.$$

As shorthand we write this estimating function as

$$U(\beta) = D^T V^{-1}(Y - \mu)/\alpha. \qquad (2.30)$$

This estimating function is linear in the data and so its properties are straightforward to evaluate. In particular,

1. $E[U(\beta)] = 0$, assuming $E[Y \mid \beta] = \mu(\beta)$.
2. $\text{var}[U(\beta)] = D^T V^{-1} D/\alpha$, assuming $\text{var}(Y \mid \beta) = V$.
3. $-E\left[\frac{\partial U}{\partial \beta}\right] = D^T V^{-1} D/\alpha = \text{var}[U(\beta)]$, assuming $E[Y \mid \beta] = \mu(\beta)$.

The similarity of these properties with those of the score function (Sect. 2.4.1) is apparent and has led to (2.30) being referred to as a *quasi-score* function. Let $\widehat{\beta}_n$ represent the root of (2.30), that is, $U(\widehat{\beta}_n) = 0$. We can apply Result 2.1 directly to obtain the asymptotic distribution of the maximum quasi-likelihood estimator (MQLE) as

$$(D^T V^{-1} D)^{1/2} (\widehat{\beta}_n - \beta) \to_d N_{k+1}(0, \alpha I_{k+1}),$$

where we have assumed that α is known. Using (B.4) in Appendix B

$$E[(Y - \mu)^T V^{-1}(\mu)(Y - \mu)]/\alpha = n,$$

and so if μ were known, an unbiased estimator of α would be

$$\widehat{\alpha}_n = (Y - \mu)^T V^{-1}(\mu)(Y - \mu)/n.$$

2.5 Quasi-likelihood

A degree of freedom corrected (but not in general unbiased) estimate is given by the Pearson statistic divided by its degrees of freedom:

$$\widehat{\alpha}_n = \frac{1}{n-k-1} \sum_{i=1}^{n} \frac{(Y_i - \widehat{\mu}_i)^2}{V(\widehat{\mu}_i)}, \quad (2.31)$$

where $\widehat{\mu}_i = \widehat{\mu}_i(\widehat{\boldsymbol{\beta}})$. This estimator of the scale parameter is consistent so long as the assumed variance model is correct. The asymptotic distribution that is used in practice is therefore

$$(\widehat{\boldsymbol{D}}^\mathrm{T}\widehat{\boldsymbol{V}}^{-1}\widehat{\boldsymbol{D}}/\widehat{\alpha}_n)^{1/2}(\widehat{\boldsymbol{\beta}}_n - \boldsymbol{\beta}) \to_d \mathrm{N}_{k+1}(\boldsymbol{0}, \boldsymbol{I}_{k+1}).$$

The inclusion of an estimate for α is justified by applying Slutsky's theorem (Appendix G) to $\widehat{\alpha}_n \times \boldsymbol{U}(\widehat{\boldsymbol{\beta}}_n)$. As usual in such asymptotic calculations, the uncertainty in $\widehat{\alpha}_n$ is not reflected in the variance for $\widehat{\boldsymbol{\beta}}_n$. This development reveals a mixing of inferential approaches with $\widehat{\boldsymbol{\beta}}_n$ a MQLE and $\widehat{\alpha}_n$ a method of moments estimator. A justification for the latter estimator is that it is likely to be consistent in a wider range of circumstances than a likelihood-based estimator. A crucial observation is that if the mean function is correctly specified, the estimator $\widehat{\boldsymbol{\beta}}_n$ is consistent also. Asymptotically appropriate standard errors result if the mean–variance relationship is correctly specified. McCullagh (1983) and Godambe and Heyde (1987) discuss the close links between consistency, the quasi-score function (2.30), and membership of the exponential family; see also Chap. 6.

As an aside, in the above, the mean model does not need to be "correct" since we are simply estimating a specified form of association, and estimation will be performed regardless of whether this model is appropriate. Of course, the usefulness of inference does depend on an appropriate mean model.

As a function of μ, we have the quasi-score

$$\frac{Y - \mu}{\alpha V(\mu)}, \quad (2.32)$$

and integration of this quantity gives

$$l(\mu, \alpha) = \int_y^\mu \frac{y - t}{\alpha V(t)} dt,$$

which, if it exists, behaves like a log-likelihood. As an example, for the model $\mathrm{E}[Y] = \mu$ and $\mathrm{var}(Y) = \alpha\mu$

$$l(\mu, \alpha) = \int_y^\mu \frac{y - t}{\alpha t} dt = \frac{1}{\alpha}[y \log \mu - \mu + c],$$

where $c = -y \log y - y$ and $y \log \mu - \mu$ is the log-likelihood of a Poisson random variable. Table 2.1 lists some distributions that correspond to particular choices of variance function.

Table 2.1 Variance functions and quasi log-likelihoods

Variance $V(\mu)$	Quasi log likelihood	Distribution
1	$-\frac{1}{\alpha}\left[\frac{1}{2}(y-\mu)^2\right]$	$N(\mu,\alpha)$
μ	$\frac{1}{\alpha}(y\log\mu - \mu)$	Poisson(μ)
μ^2	$\frac{1}{\alpha}\left(-\frac{y}{\mu} - \log\mu\right)$	Ga$(1/\alpha, \mu/\alpha)$
$n\mu(1-\mu)$	$\frac{1}{\alpha}\left[y\log\left(\frac{\mu}{1-\mu}\right) + n\log(1-\mu)\right]$	Binomial(n, μ)
$\mu + \mu^2/b$	$\frac{1}{\alpha}\left[y\log\left(\frac{\mu}{b+\mu}\right) + b\log\left(\frac{b}{b+\mu}\right)\right]$	NegBin(μ, b), b known
$\mu^2(1-\mu)^2$	$\frac{1}{\alpha}\left[(2y-1)\log\left(\frac{\mu}{1-\mu}\right) - \frac{y}{\mu} - \frac{1-y}{1-\mu}\right]$	No distribution

In all cases $E[Y] = \mu$. The parameterizations of the distributional forms are as in Appendix D. For the Poisson, binomial, and negative binomial distributions, these are the forms that the quasi-score corresponds to when $\alpha = 1$

The word "quasi" refers to the fact that the score may or may not correspond to a probability function. For example, in Table 2.1, the variance function $\mu^2(1-\mu)^2$ does not correspond to a probability distribution. In most cases, there is an implied distributional kernel, but the addition of the variance multiplier α often produces a mean–variance relationship that is not present in the implied distribution.

We emphasize that the first two moments do not uniquely define a distribution. For example, the negative binomial distribution may be derived as the marginal distribution of

$$Y \mid \mu, \theta \sim \text{Poisson}(\mu\theta) \tag{2.33}$$

$$\theta \sim \text{Ga}(b, b) \tag{2.34}$$

so that $E[Y] = \mu$ and

$$\text{var}(Y) = E[\text{var}(Y \mid \theta)] + \text{var}(E[Y \mid \theta]) = \mu + \frac{\mu^2}{b}. \tag{2.35}$$

These latter two moments are also recovered if we replace the gamma distribution with a lognormal distribution. Specifically, assume the model

$$Y \mid \theta^\star \sim \text{Poisson}(\theta^\star)$$

$$\theta^\star \sim \text{LogNorm}(\eta, \sigma^2)$$

and let $\mu = E[\theta] = \exp(\eta + \sigma^2/2)$. Then,

$$\text{var}(\theta^\star) = E[\theta^\star]^2 \left[\exp(\sigma^2) - 1\right] = \mu^2 \left[\exp(\sigma^2) - 1\right].$$

Under this model, $E[Y] = \mu$ and

$$\text{var}(Y) = E[\text{var}(Y \mid \theta^\star)] + \text{var}[E(Y \mid \theta^\star)] = \mu + \mu^2 \left[\exp(\sigma^2) - 1\right]$$

2.5 Quasi-likelihood

which, on writing $b^\star = [\exp(\sigma^2) - 1]^{-1}$, gives the same form of quadratic variance function, (2.35), as with the gamma model.

If the estimating function (2.30) corresponds to the score function for a particular probability distribution, then the subsequent estimator corresponds to the MLE (because α does not influence the estimation of β), though the variance of the estimator will usually differ. A great advantage of the use of quasi-likelihood is its computational simplicity.

A prediction interval for an observable, Y, is not possible with quasi-likelihood since there is no probabilistic mechanism with which to reflect the stochastic component of the prediction.

Example: Lung Cancer and Radon

We return to the lung cancer example and now assume the quasi-likelihood model

$$\mathrm{E}[Y_i \mid \beta] = E_i \exp(x_i\beta), \quad \mathrm{var}(Y_i \mid \beta) = \alpha \mathrm{E}[Y_i \mid \beta].$$

Fitting this model yields identical point estimates to the MLEs and $\widehat{\alpha} = 2.81$ so that the quasi-likelihood standard errors are $\sqrt{\widehat{\alpha}} = 1.68$ times larger than the Poisson model-based standard errors. The variance–covariance matrix is

$$(\widehat{D}^\mathrm{T}\widehat{V}^{-1}\widehat{D})^{-1}\widehat{\alpha} = \begin{bmatrix} 0.045^2 & -0.95 \times 0.045 \times 0.0090 \\ -0.95 \times 0.045 \times 0.0090 & 0.0090^2 \end{bmatrix}.$$

An asymptotic 95% confidence interval for the relative risk associated with a one-unit increase in radon is $[0.947, 0.982]$ which is $\sqrt{\widehat{\alpha}} = 1.68$ wider than the Poisson interval evaluated previously.

2.5.2 A More Complex Mean–Variance Model

For comparison, we now describe a more general model than considered under the quasi-likelihood approach. Suppose we specify the first two moments of the data as

$$\mathrm{E}[Y_i \mid \beta] = \mu_i(\beta) \tag{2.36}$$

$$\mathrm{var}(Y_i \mid \beta) = V_i(\alpha, \beta), \tag{2.37}$$

where α is an $r \times 1$ vector of parameters that appear only in the variance model. Let $\widehat{\alpha}_n$ be a consistent estimator of α. We state without proof the following result. The estimator $\widehat{\beta}_n$ that satisfies the estimating equation

$$G(\widehat{\boldsymbol{\beta}}_n, \widehat{\boldsymbol{\alpha}}_n) = \boldsymbol{D}(\widehat{\boldsymbol{\beta}}_n)\boldsymbol{V}^{-1}(\widehat{\boldsymbol{\alpha}}_n, \widehat{\boldsymbol{\beta}}_n)\left[\boldsymbol{Y} - \boldsymbol{\mu}(\widehat{\boldsymbol{\beta}}_n)\right] \qquad (2.38)$$

has asymptotic distribution

$$(\widehat{\boldsymbol{D}}^{\mathrm{T}}\widehat{\boldsymbol{V}}^{-1}\widehat{\boldsymbol{D}})^{1/2}(\widehat{\boldsymbol{\beta}}_n - \boldsymbol{\beta}) \to_d \mathrm{N}_{k+1}(\mathbf{0}, \mathrm{I}_{k+1})$$

where $\widehat{\boldsymbol{D}} = \boldsymbol{D}(\widehat{\boldsymbol{\beta}}_n)$ and $\widehat{\boldsymbol{V}} = \boldsymbol{V}(\widehat{\boldsymbol{\alpha}}_n, \widehat{\boldsymbol{\beta}}_n)$.

The difference between this model and that in the quasi-likelihood approach is that $\boldsymbol{V}$ may now depend on additional variance–covariance parameters $\boldsymbol{\alpha}$ in a more complex way. Under quasi-likelihood it is assumed that $\mathrm{var}(Y_i) = \alpha V_i(\mu_i)$, so that the estimating function does not depend on α. Consequently, $\widehat{\boldsymbol{\beta}}$ also does not depend on α, though the standard errors are proportional to $\sqrt{\alpha}$. This is a motivating factor in the development of quasi-likelihood, since standard software may be used for implementation and, perhaps more importantly, consistency of $\boldsymbol{\beta}$ is guaranteed if the mean model is correctly specified.

The form of the mean–variance relationship given by (2.36) and (2.36) suggests an iterative scheme for estimation of $\boldsymbol{\beta}$ and $\boldsymbol{\alpha}$. Set $t = 0$ and let $\widehat{\boldsymbol{\alpha}}^{(0)}$ be an initial estimate for $\boldsymbol{\alpha}$. Now iterate between

1. Solve $G(\boldsymbol{\beta}, \widehat{\boldsymbol{\alpha}}^{(t)}) = \mathbf{0}$ to give $\widehat{\boldsymbol{\beta}}^{(t+1)}$,
2. Estimate $\widehat{\boldsymbol{\alpha}}^{(t+1)}$ with $\widehat{\mu}_i = \mu_i\left(\widehat{\boldsymbol{\beta}}^{(t+1)}\right)$. Set $t \to t+1$ and return to 1.

The model given by (2.36) and (2.36) is more flexible than that provided by quasi-likelihood but requires the correct specification of mean and variance for a consistent estimator of $\boldsymbol{\beta}$.

Example: Lung Cancer and Radon

As an example of the mean–variance model discussed in the previous section, we fit a negative binomial model to the lung cancer data. This model is motivated via the random effects formulation given by (2.33) and (2.34) with loglinear model $\mu_i = \mu_i(\boldsymbol{\beta}) = E_i \exp(\beta_0 + \beta_1 x_i)$, $i = 1, \ldots, n$. In the lung cancer context, the random effects are area-specific perturbations from the mean μ_i. The introduction of the random effects may be seen as a device for inducing overdispersion. Integrating over θ_i, we obtain the negative binomial distribution

$$\Pr(y_i \mid \boldsymbol{\beta}, b) = \frac{\Gamma(y_i + b)}{\Gamma(b) y_i!} \frac{\mu_i^{y_i} b^b}{(\mu_i + b)^{y_i + b}},$$

for $y_i = 0, 1, 2, \ldots$, with

$$\mathrm{E}[Y_i \mid \boldsymbol{\beta}] = \mu_i(\boldsymbol{\beta})$$

$$\mathrm{var}(Y_i \mid \boldsymbol{\beta}, b) = \mu_i(\boldsymbol{\beta})\left[1 + \frac{\mu_i(\boldsymbol{\beta})}{b}\right], \qquad (2.39)$$

2.5 Quasi-likelihood

so that smaller values of b correspond to greater degrees of overdispersion and as $b \to \infty$ we recover the Poisson model. For consistency with later chapters we use b rather than α for the parameter occurring in the variance model. Care is required with the negative binomial distribution since a number of different parameterizations are available; see Exercise 2.4. The log-likelihood is

$$l(\boldsymbol{\beta}, b) = \sum_{i=1}^{n} \log \frac{\Gamma(y_i + b)}{\Gamma(b) y_i!} + y_i \log \mu_i + b \log b - (y_i + b) \log(\mu_i + b) \quad (2.40)$$

giving the score function for $\boldsymbol{\beta}$ as

$$\boldsymbol{S}(\boldsymbol{\beta}) = \frac{\partial l}{\partial \boldsymbol{\beta}} = \sum_{i=1}^{n} \left(\frac{\partial \mu_i}{\partial \boldsymbol{\beta}} \right)^{\mathrm{T}} \frac{y_i - \mu_i}{\mu_i + \mu_i^2 / b}$$

$$= \sum_{i=1}^{n} \boldsymbol{D}(\boldsymbol{\beta})_i^{\mathrm{T}} \boldsymbol{V}_i^{-1}(b) \left[Y_i - \mu_i(\boldsymbol{\beta}) \right]$$

which corresponds to (2.38). Hence, for fixed b, we can solve this estimating equation to obtain an estimator $\widehat{\boldsymbol{\beta}}$. Usually we will also wish to estimate b (as opposed to assuming a fixed value). One possibility is maximum likelihood though a quick glance at (2.40) reveals that no closed-form estimator will be available and numerical maximization will be required (which is not a great impediment). We describe an alternative method of moments estimator which may be more robust.

For the *quadratic* variance model (2.39), the variance is

$$\mathrm{var}(Y_i \mid \boldsymbol{\beta}, b) = \mathrm{E}[(Y_i - \mu_i)^2] = \mu_i (1 + \mu_i / b),$$

so that

$$b^{-1} = \mathrm{E} \left[\frac{(Y_i - \mu_i)^2 - \mu_i}{\mu_i^2} \right],$$

for $i = 1, \ldots, n$, leading to the method of moments estimator

$$\widehat{b} = \left[\frac{1}{n - k - 1} \sum_{i=1}^{n} \frac{(Y_i - \widehat{\mu}_i)^2 - \widehat{\mu}_i}{\widehat{\mu}_i^2} \right]^{-1}, \quad (2.41)$$

with $k = 1$ in the lung cancer example. If we have a consistent estimator $\widehat{b}$ (which follows if the quadratic variance model is correct) and the mean correctly specified, then valid inference follows from

$$(\widehat{\boldsymbol{D}}^{\mathrm{T}} \widehat{\boldsymbol{V}}(\widehat{b})^{-1} \widehat{\boldsymbol{D}})^{1/2} (\widehat{\boldsymbol{\beta}} - \boldsymbol{\beta}) \to_d \mathrm{N}_2(\boldsymbol{0}, \boldsymbol{I}_2).$$

We fit this model to the lung cancer data. The estimates (standard errors) are $\widehat{\beta}_0 = 0.090\ (0.047)$ and $\widehat{\beta}_1 = -0.030\ (0.0085)$. The latter point estimate differs a little

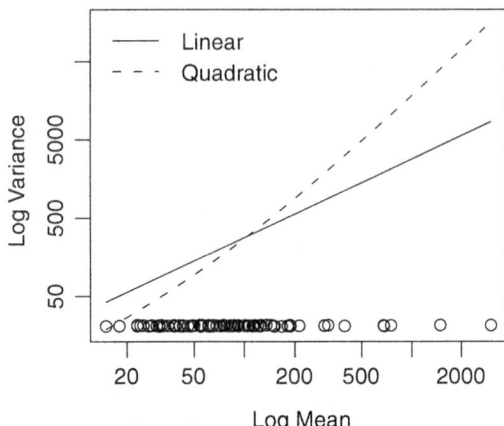

Fig. 2.2 Linear and quadratic variance functions for the lung cancer data

from the MLE (and MQLE) of -0.036, reflecting the different variance weighting in the estimating function. The moment-based estimator was $\widehat{b} = 57.8$ (the MLE is 61.3 and so close to this value). An asymptotic 95% confidence interval for the relative risk $\exp(\beta_1)$ is [0.955,0.987], so that the upper limit is closer to unity than the intervals we have seen previously.

In terms of the first two moments, the difference between quasi-likelihood and the negative binomial model is that the variances are, respectively, linear and quadratic functions of the mean. In Fig. 2.2, we plot the estimated linear and quadratic variance functions over the range of the mean for these data. To produce a clearer plot, the log of the variance is plotted against the log of the mean, and the log of the observed counts, y_i, $i = 1, \ldots, 85$, is added to the plot (with a small amount of jitter). Over the majority of the data, the two variance functions are similar, but for large values of the mean in particular, the variance functions are considerably different which leads to the differences in inference, since large observations are being weighted very differently by the two variance functions. Based on this plot, we might expect even greater differences. However, closer examination of the data reveals that the x's associated with the large y values are all in the midrange, and consequently, these points are not influential.

Examination of the residuals gives some indication that the quadratic mean–variance model is more appropriate for these data (see Sect. 6.9). It is typically very difficult to distinguish between the two models, unless there are sufficient points across a large spread of mean values.

2.6 Sandwich Estimation

A general method of avoiding stringent modeling conditions when the variance of an estimator is calculated is provided by *sandwich estimation*. Recall from Sect. 2.3 the estimating function

2.6 Sandwich Estimation

$$G_n(\boldsymbol{\theta}) = \frac{1}{n}\sum_{i=1}^{n} G(\boldsymbol{\theta}, Y_i).$$

Based on independent and identically distributed observations, we have the sandwich form for the variance

$$\text{var}(\widehat{\boldsymbol{\theta}}_n) = \frac{\boldsymbol{A}^{-1}\boldsymbol{B}(\boldsymbol{A}^{\text{T}})^{-1}}{n} \tag{2.42}$$

where

$$\boldsymbol{A} = \text{E}\left[\frac{\partial}{\partial \boldsymbol{\theta}}\boldsymbol{G}(\boldsymbol{\theta}, Y)\right]$$

and

$$\boldsymbol{B} = \text{E}[\boldsymbol{G}(\boldsymbol{\theta}, Y)\boldsymbol{G}(\boldsymbol{\theta}, Y)^{\text{T}}].$$

For (2.42) to be asymptotically appropriate, the expectations need to be evaluated under the true model (as discussed in Sect. 2.4.3).

So far we have used an assumed model to calculate the expectations. An alternative is to evaluate $\boldsymbol{A}$ and $\boldsymbol{B}$ *empirically* via

$$\widehat{\boldsymbol{A}}_n = \frac{1}{n}\sum_{i=1}^{n}\frac{\partial}{\partial \boldsymbol{\theta}}\boldsymbol{G}(\widehat{\boldsymbol{\theta}}, Y_i),$$

and

$$\widehat{\boldsymbol{B}}_n = \frac{1}{n}\sum_{i=1}^{n}\boldsymbol{G}(\widehat{\boldsymbol{\theta}}, Y_i)\boldsymbol{G}(\widehat{\boldsymbol{\theta}}, Y_i)^{\text{T}}.$$

By the weak law of large numbers, $\widehat{\boldsymbol{A}}_n \to_p \boldsymbol{A}$ and $\widehat{\boldsymbol{B}}_n \to_p \boldsymbol{B}$, and

$$\text{var}(\widehat{\boldsymbol{\theta}}_n) = \frac{\widehat{\boldsymbol{A}}^{-1}\widehat{\boldsymbol{B}}(\widehat{\boldsymbol{A}}^{\text{T}})^{-1}}{n} \tag{2.43}$$

is a consistent estimator of the variance. The great advantage of sandwich estimation is that it provides a consistent estimator of the variance in very broad situations. An important assumption is that the observations are uncorrelated (this will be relaxed in Part III of the book when generalized estimating equations are described).

We now consider the situation in which the estimating function arises from the score and suppose we have independent and identically distributed data. In this situation

$$\boldsymbol{G}_n(\boldsymbol{\theta}) = \frac{1}{n}\sum_{i=1}^{n}\frac{\partial}{\partial \boldsymbol{\theta}}l_i(\boldsymbol{\theta}),$$

with $l_i(\boldsymbol{\theta}) = \log p(Y_i \mid \boldsymbol{\theta})$, to give

$$\boldsymbol{A} = \frac{1}{n}\sum_{i=1}^{n}\text{E}\left[\frac{\partial^2}{\partial \boldsymbol{\theta}\partial \boldsymbol{\theta}^{\text{T}}}l(\boldsymbol{\theta})\right]$$

and
$$B = \frac{1}{n} \sum_{i=1}^{n} \mathrm{E}\left[\left(\frac{\partial}{\partial \boldsymbol{\theta}} l(\boldsymbol{\theta})\right)\left(\frac{\partial}{\partial \boldsymbol{\theta}} l(\boldsymbol{\theta})\right)^{\mathrm{T}}\right]$$

where $l(\boldsymbol{\theta}) = \log p(Y \mid \boldsymbol{\theta})$. Then, *under the model*,

$$\boldsymbol{I}_1(\boldsymbol{\theta}) = -\boldsymbol{A}(\boldsymbol{\theta}) = \boldsymbol{B}(\boldsymbol{\theta}), \tag{2.44}$$

so that
$$\mathrm{var}(\widehat{\boldsymbol{\theta}}_n) = \frac{\boldsymbol{A}^{-1}\boldsymbol{B}(\boldsymbol{A}^{\mathrm{T}})^{-1}}{n} = \frac{\boldsymbol{I}_1(\boldsymbol{\theta})^{-1}}{n}.$$

The sandwich estimator (2.43) is based on

$$\boldsymbol{A} = \frac{1}{n} \sum_{i=1}^{n} \frac{\partial^2}{\partial \boldsymbol{\theta} \partial \boldsymbol{\theta}^{\mathrm{T}}} l_i(\boldsymbol{\theta})\bigg|_{\widehat{\boldsymbol{\theta}}}$$

and
$$\boldsymbol{B} = \frac{1}{n} \sum_{i=1}^{n} \left(\frac{\partial}{\partial \boldsymbol{\theta}} l_i(\boldsymbol{\theta})\right)\left(\frac{\partial}{\partial \boldsymbol{\theta}} l_i(\boldsymbol{\theta})\right)^{\mathrm{T}}\bigg|_{\widehat{\boldsymbol{\theta}}}.$$

The sandwich method can be applied to general estimating functions, not just those arising from a score equation (in Sect. 2.4.3, we considered the latter in the context of model misspecification).

Suppose we assume $\mathrm{E}[Y_i] = \mu_i$ and $\mathrm{var}(Y_i) = \alpha V(\mu_i)$, and $\mathrm{cov}(Y_i, Y_j) = 0$, $i, j = 1, \ldots, n, i \neq j$, as a *working* covariance model. Under this specification, it is natural to take the quasi-score function (2.30) as an estimating function, and in this case, the variance of the resultant estimator is

$$\mathrm{var}_{\mathrm{S}}(\widehat{\boldsymbol{\beta}}_n) = (\boldsymbol{D}^{\mathrm{T}}\boldsymbol{V}^{-1}\boldsymbol{D})^{-1}\boldsymbol{D}^{\mathrm{T}}\boldsymbol{V}^{-1}\mathrm{var}(\boldsymbol{Y})\boldsymbol{V}^{-1}\boldsymbol{D}(\boldsymbol{D}^{\mathrm{T}}\boldsymbol{V}^{-1}\boldsymbol{D})^{-1}.$$

The appropriate variance is obtained by substituting in the correct form for $\mathrm{var}(\boldsymbol{Y})$. The latter is, of course, unknown but a simple "sandwich" estimator of the variance is given by

$$\widehat{\mathrm{var}}(\widehat{\boldsymbol{\beta}}_n) = (\boldsymbol{D}^{\mathrm{T}}\boldsymbol{V}^{-1}\boldsymbol{D})^{-1}\boldsymbol{D}^{\mathrm{T}}\boldsymbol{V}^{-1}\mathrm{diag}(\boldsymbol{R}\boldsymbol{R}^{\mathrm{T}})\boldsymbol{V}^{-1}\boldsymbol{D}(\boldsymbol{D}^{\mathrm{T}}\boldsymbol{V}^{-1}\boldsymbol{D})^{-1},$$

where $\boldsymbol{R} = [R_1, \ldots, R_n]^{\mathrm{T}}$ is the $n \times 1$ vector of (unstandardized) residuals

$$R_i = Y_i - \mu_i(\widehat{\boldsymbol{\beta}}),$$

so that $\mathrm{diag}(\boldsymbol{R}\boldsymbol{R}^{\mathrm{T}})$ is the $n \times n$ diagonal matrix with diagonal elements $\left[Y_i - \mu_i(\widehat{\boldsymbol{\beta}})\right]^2$ for $i = 1, \ldots, n$. This estimator is consistent for the variance of $\widehat{\boldsymbol{\beta}}$, under correct

2.6 Sandwich Estimation

Table 2.2 Components of estimation under the assumption of independent outcomes and for one-dimensional β

	Likelihood	Quasi-likelihood
$G(\beta) = \sum_i G_i(\beta)$	$\sum_i \frac{\partial}{\partial \beta} \log L_i$	$\frac{1}{\alpha} \sum_i \left(\frac{\partial \mu_i}{\partial \beta}\right) \frac{Y_i - \mu_i}{V_i}$
$A = \sum_i \mathrm{E}\left[\frac{\partial G_i}{\partial \beta}\right]$	$\sum_i \mathrm{E}\left[\frac{\partial^2}{\partial \beta^2} \log L_i\right]$	$-\frac{1}{\alpha} \sum_i \left(\frac{\partial \mu_i}{\partial \beta}\right)^2 \frac{1}{V_i}$
$\widehat{A} = \sum_i \frac{\partial G_i}{\partial \beta}\big\|_{\widehat{\beta}}$	$\sum_i \frac{\partial^2}{\partial \beta^2} \log L_i$	$-\frac{1}{\alpha} \sum_i \left(\frac{\partial \mu_i}{\partial \beta}\right)^2 \frac{1}{V_i}$
$B = \sum_i \mathrm{E}[G_i(\beta)^2]$	$\sum_i \mathrm{E}\left[\left(\frac{\partial}{\partial \beta} \log L_i\right)^2\right]$	$\frac{1}{\alpha} \sum_i \left(\frac{\partial \mu_i}{\partial \beta}\right)^2 \frac{1}{V_i}$
$\widehat{B} = \sum_i G_i(\widehat{\beta})^2$	$\sum_i \left(\frac{\partial}{\partial \beta} \log L_i\right)^2$	$\frac{1}{\alpha^2} \sum_i \left(\frac{\partial \mu_i}{\partial \beta}\right)^2 \frac{(Y_i - \widehat{\mu}_i)^2}{V_i^2}$
Model-based variance	$\left\{\sum_i \mathrm{E}\left[\frac{\partial^2}{\partial \beta^2} \log L_i\right]\right\}^{-1}$	$\alpha \left\{\sum_i \left(\frac{\partial \mu_i}{\partial \beta}\right) \frac{1}{V_i}\right\}^{-1}$
Sandwich variance	$\dfrac{\sum_i \left(\frac{\partial}{\partial \beta} \log L_i\right)^2}{\left[\sum_i \frac{\partial^2}{\partial \beta^2} \log L_i\right]^2}$	$\dfrac{\sum_i \left(\frac{\partial \mu_i}{\partial \beta}\right)^2 \frac{(Y_i - \widehat{\mu}_i)^2}{V_i^2}}{\left[\sum_i \left(\frac{\partial \mu_i}{\partial \beta}\right)^2 \frac{1}{V_i}\right]^2}$

The likelihood model is $p(\boldsymbol{y} \mid \beta) = \prod_i L_i(\beta)$, and the quasi-likelihood model has $\mathrm{E}[Y_i \mid \beta] = \mu_i(\beta)$, $\mathrm{var}(Y_i \mid \beta) = \alpha V_i(\beta)$, $i = 1, \ldots, n$, and $\mathrm{cov}(Y_i, Y_j \mid \beta) = 0$, $i \neq j$. The expected information is $-\sum_i \mathrm{E}\left[\frac{\partial^2}{\partial \beta^2} \log L_i\right]$, and the observed information is $-\sum_i \frac{\partial^2}{\partial \beta^2} \log L_i$. The sandwich estimator is $\widehat{A}^{-1} \widehat{B} \widehat{A}^{-1}$ which simplifies to $-\widehat{A}^{-1}$ under the model

specification of the mean, and with uncorrelated data. There is finite sample bias in R_i as an estimate of $Y_i - \mu_i(\beta)$ and versions that adjust for the estimation of the parameters β are available; see Kauermann and Carroll (2001).

The great advantage of sandwich estimation is that it provides a consistent estimator of the variance in very broad situations and the use of the empirical residuals is very appealing. There are two things to bear in mind when one considers the use of the sandwich technique, however. The first is that, unless the sample size is sufficiently large, the sandwich estimator may be highly unstable; in terms of mean squared error, model-based estimators may be preferable for small- to medium-sized n (for small samples one would want to avoid the reliance on the asymptotic distribution anyway). Consequently, *empirical* is a better description of the estimator than *robust*. The second consideration is that if the assumed mean–variance model is correct, then a model-based estimator is more efficient.

In many cases, quasi-likelihood with a model-based variance estimate may be viewed as an intermediary between the full model specification and sandwich estimation, in that the form of the variance function separates estimation of β and α, to give consistency of β in broad circumstances, though the standard error will not be consistently estimated unless the variance function is correct. Table 2.2 provides a summary and comparison of the various elements of the likelihood and quasi-likelihood methods, with sandwich estimators for each.

Example: Poisson Mean

We report the results of a small simulation study to illustrate the efficiency-robustness trade-off of variance estimation. Data were simulated from the model $Y_i \mid \delta \sim \text{Poisson}(\delta)$, $i = 1, \ldots, n$, where $\delta \sim_{iid} \text{Gamma}(\theta b, b)$. This setup gives marginal moments

$$\text{E}[Y_i] = \theta$$

$$\text{var}(Y_i) = \text{E}[Y_i] \times \left(1 + \frac{1}{b}\right) = \text{E}[Y_i] \times \alpha.$$

We take $\theta = 10$ and $\alpha = 1, 2, 3$ corresponding to no excess-Poisson variability, and variability that is two and three times the mean. We estimate θ and then form an asymptotic confidence interval based on a Poisson likelihood, quasi-likelihood, and sandwich estimation.

For a univariate estimator $\widehat{\theta}$ arising from a generic estimating function $G(\theta, Y)$:

$$\sqrt{n}(\widehat{\theta} - \theta) \to_d \text{N}\left(0, \frac{B}{A^2}\right).$$

where

$$A = \text{E}\left[\frac{d^2}{d\theta^2} G(\theta)\right], \quad B = \text{E}\left[\left(\frac{d}{d\theta} G(\theta)\right)^2\right].$$

Under the Poisson model

$$l_i(\theta) = -\theta + Y_i \log \theta$$

and

$$G(\theta, Y_i) = S_i(\theta) = \frac{dl_i}{d\theta} = \frac{Y_i - \theta}{\theta}$$

$$\frac{d^2 l_i}{d\theta^2} = -\frac{Y_i}{\theta^2},$$

to give the familiar MLE, $\widehat{\theta} = \overline{Y}$. As we already know

$$I_1(\theta) = -A = -\text{E}\left[\frac{d^2 l}{d\theta^2}\right] = B = \text{var}\left(\frac{(Y - \theta)^2}{\theta^2}\right) = \frac{\text{var}(Y)}{\theta^2} = \frac{1}{\theta},$$

under the assumption that $\text{var}(Y) = \theta$. The Poisson model-based variance estimator is therefore

$$\widehat{\text{var}}(\widehat{\theta}) = \frac{1}{nI_1(\widehat{\theta})} = \frac{\overline{Y}}{n}.$$

Under the Poisson model, the variance equals the mean, and given the efficiency of the latter, it makes sense to estimate the variance by the sample mean.

2.6 Sandwich Estimation

The quasi-likelihood estimator is derived from the quasi-score

$$G(\theta, Y_i) = U_i(\theta) = \frac{Y_i - \theta}{\alpha\theta},$$

and

$$\text{var}(\widehat{\theta}) = (\widehat{\boldsymbol{D}}^{\mathsf{T}}\widehat{\boldsymbol{V}}^{-1}\widehat{\boldsymbol{D}})^{-1}\widehat{\alpha}$$

where the scale parameter is estimated using the method of moments

$$\widehat{\alpha} = \frac{1}{n-1}\sum_{i=1}^{n}\frac{(Y_i - \widehat{\theta})^2}{\widehat{\theta}}.$$

The quasi-likelihood estimator of the variance is

$$\widehat{\text{var}}(\widehat{\theta}) = \frac{s^2}{n},$$

where

$$s^2 = \frac{1}{n-1}\sum_{i=1}^{n}(Y_i - \widehat{\theta})^2.$$

For sandwich estimation based on the score

$$\widehat{A} = -\frac{1}{n}\sum_{i=1}^{n}\frac{Y_i}{\widehat{\theta}^2} = -\frac{1}{\overline{Y}},$$

and

$$\widehat{B} = \frac{1}{n}\sum_{i=1}^{n}\frac{(Y_i - \widehat{\theta})^2}{\widehat{\theta}^2} = \frac{(n-1)s^2}{n\widehat{\theta}^2}.$$

Hence,

$$\widehat{\text{var}}(\widehat{\theta}) = \frac{s^2(n-1)/n}{n}. \tag{2.45}$$

Estimation of $\text{var}(Y_i)$ by $(Y_i - \overline{Y})^2$ produces the variance estimator (2.45). Estimating $\text{var}(Y_i)$ by $n(Y_i - \overline{Y})^2/(n-1)$ would reproduce the degrees of freedom adjusted quasi-likelihood estimator.

Table 2.3 gives the 95% confidence interval coverage for the model-based, quasi-likelihood, and sandwich estimator variance estimates as a function of the sample size n and overdispersion/scalar parameter α. We see that when the Poisson model is correct ($\alpha = 1$), the model-based standard errors produce accurate coverage for all values of n. For small n, the quasi-likelihood and sandwich estimators have low coverage, due to the instability in variance estimation, with sandwich estimation being slightly poorer in performance. As the level of overdispersion increases, the performance of the model-based approach starts to deteriorate as the standard error is underestimated, resulting in low coverage. For $\alpha = 2, 3$, the quasi-likelihood and

Table 2.3 Percent confidence interval coverage for the Poisson mean example, based on 100,000 simulations

	Overdispersion								
	$\alpha = 1$			$\alpha = 2$			$\alpha = 3$		
n	Model	Quasi	Sand	Model	Quasi	Sand	Model	Quasi	Sand
5	95	87	84	83	87	84	74	86	83
10	94	92	90	83	91	90	73	91	89
15	95	93	92	84	92	92	75	92	91
20	95	93	93	83	93	93	73	93	92
25	95	94	93	83	94	93	74	93	93
50	95	94	94	83	94	94	74	94	94
100	95	95	94	83	95	94	74	95	94

The nominal coverage is 95%. The overdispersion is given by $\alpha = \text{var}(Y)/E[Y]$

sandwich estimators again give low coverage for small values of n, due to instability, but for larger values, the coverage quickly improves. The adjusted degrees of freedom used by quasi-likelihood give slightly improved estimation over the naive sandwich estimator.

This example shows the efficiency-robustness trade-off. If the model is correct (which corresponds here to $\alpha = 1$), then the model-based approach performs well. The sandwich and quasi-likelihood approaches are more robust to variance misspecification, but can be unstable when the sample size is small. The choice of which variance model to use depends crucially on our faith in the model. The use of a Poisson model is a risky enterprise, however, since it does not contain an additional variance parameter.

Example: Lung Cancer and Radon

Returning to the lung cancer and radon example, we calculate sandwich standard errors, assuming that counts in different areas are uncorrelated. We take as "working model" a Poisson likelihood, with maximum likelihood estimation of β. The estimating function is

$$S(\beta) = D^\mathrm{T} V^{-1}(Y - \mu) = x^\mathrm{T}(Y - \mu),$$

as derived previously, (2.20). Under this model

$$(A^{-1} B A^\mathrm{T})^{1/2}(\widehat{\beta}_n - \beta) \to_d N_2(0, I_2),$$

with sandwich ingredients

$$A = D^\mathrm{T} V^{-1} D$$
$$B = D^\mathrm{T} V^{-1} \text{var}(Y) V^{-1} D,$$

estimators

$$\widehat{A} = \widehat{D}^{\mathrm{T}}\widehat{V}^{-1}\widehat{D}$$

$$\widehat{B} = \widehat{D}^{\mathrm{T}}\widehat{V}^{-1} \begin{bmatrix} \widehat{\sigma}_1^2 & 0 & \cdots & 0 \\ 0 & \widehat{\sigma}_2^2 & \cdots & 0 \\ \vdots & \vdots & \ddots & \vdots \\ \cdots & \cdots & \cdots & \widehat{\sigma}_n^2 \end{bmatrix} \widehat{V}^{-1}\widehat{D}$$

and with $\widehat{\sigma}_i^2 = (Y_i - \widehat{\mu}_i)^2$, for $i = 1, \ldots, n$. Substitution of the required data quantities yields the variance–covariance matrix

$$\begin{bmatrix} 0.043^2 & -0.87 \times 0.043 \times 0.0080 \\ -0.87 \times 0.043 \times 0.0080 & 0.0080^2 \end{bmatrix}.$$

The estimated standard errors of $\widehat{\beta}_0$ and $\widehat{\beta}_1$ are 0.043 and 0.0080, respectively, and are 60% and 49% larger than their likelihood counterparts, though slightly smaller than the quasi-likelihood versions. An asymptotic 95% confidence interval for the relative risk associated with a one-unit increase in radon is $[0.949, 0.980]$.

We have a linear exponential family likelihood and so a consistent estimator of the loglinear association between lung cancer incidence and radon, as is clear from (2.20). If the outcomes are independent, then a consistent sandwich variance estimator is obtained and the large sample size indicates asymptotic inference is appropriate. However, in the context of these data, independence is a little dubious as we may have residual spatial dependence, particularly since we have not controlled for confounders such as smoking which may have spatial structure (and hence will induce spatial dependence). Sandwich standard errors do not account for such dependence (unless we can lean on replication across time). In Sect. 9.7, we describe a model that allows for residual spatial dependence in the counts. Although the loglinear association is consistently estimated, this of course says nothing about causality or about the appropriateness of the mean model.

2.7 Bootstrap Methods

With respect to estimation and hypothesis testing, the fundamental frequentist inferential summary is the distribution of an estimator under hypothetical repeated sampling from the distribution of the data. So far we have concentrated on the use of the asymptotic distribution of the estimator under an assumed model, though sandwich estimation (and to a lesser extent quasi-likelihood) provided one method by which we could relax the reliance on the assumed model. The bootstrap is a computational technique for alleviating some forms of model misspecification. The bootstrap may also be used, to some extent, to account for a "non-asymptotic" sample size. We first describe its use in single parameter settings before moving to a regression context.

2.7.1 The Bootstrap for a Univariate Parameter

Suppose $Y_1, \ldots, Y_n$, are an independent and identically distributed sample from a distribution function F that depends on a univariate parameter θ. Let $\widehat{\theta}(\boldsymbol{Y})$ represent an estimator of θ. We may be interested in estimation of

(i) $\text{var}_F[\widehat{\theta}(\boldsymbol{Y})]$
(ii) $\text{Pr}_F[a < \widehat{\theta}(\boldsymbol{Y}) < b]$

where we have emphasized that these summaries are evaluated under the sampling distribution of the data F. Estimation of (i) is of particular interest if the sampling distribution of $\widehat{\theta}$ is approximately normal, in which case a $100(1-\alpha)\%$ confidence interval is

$$\widehat{\theta}(\boldsymbol{Y}) + \text{bias}_F\left[\widehat{\theta}(\boldsymbol{Y})\right] \pm z_{1-\alpha/2}\sqrt{\text{var}_F(\widehat{\theta})} \qquad (2.46)$$

where $\text{bias}_F\left[\widehat{\theta}(\boldsymbol{Y})\right]$ is the bias of the estimator, and $z_{1-\alpha/2}$ the $(1-\alpha/2)$ quantile of an $N(0,1)$ random variable. More generally, interest may focus on a function of interest $T(F)$.

The bootstrap is an idea that is so simple it seems, at first sight, like cheating but it turns out to be statistically valid in many circumstances, so long as care is taken in its implementation. The idea is to first draw B *bootstrap samples* of size n, $\boldsymbol{Y}_b^\star = [Y_{b1}^\star, \ldots, Y_{bn}^\star]$, $b = 1, \ldots, B$, from an estimate of F, $\widehat{F}$. In the *nonparametric* bootstrap, the estimate of F is F_n, the empirical estimate of the distribution function that places a mass of $1/n$ at each of the observed Y_i, $i = 1, \ldots, n$. Bootstrap samples are obtained by sampling a new dataset $Y_{bi}^\star$, $i = 1, \ldots, n$, from $\widehat{F}_n$, *with replacement*. If one has some faith in the assumed model, then $\widehat{F}$ may be based upon this model, which we call $F_{\widehat{\theta}}$ where $\widehat{\theta} = \widehat{\theta}(\boldsymbol{y})$, to give a second implementation. In this case, bootstrap samples are obtained by sampling $Y_{bi}^\star$, $i = 1, \ldots, n$, as independent and identically distributed samples from $F_{\widehat{\theta}}$, to give a *parametric* bootstrap estimator.

Intuitively, we are replacing the distribution of

$$\widehat{\theta}_n - \theta$$

with

$$\widehat{\theta}_n^\star - \widehat{\theta}_n.$$

Much theory is available to support the use of the bootstrap; early references are Bickel and Freedman (1981) and Singh (1981); see also van der Vaart (1998). Further references to the bootstrap are given in Sect. 2.11. As a simple example of the sort of results that are available, we quote the following, a proof of which may be found in Bickel and Freedman (1981).

2.7 Bootstrap Methods

Result. Consider a bootstrap estimator of the sample mean, μ, of the distribution F and assume $E[Y^2] < \infty$ and let the variance of F be σ^2. Then we know that $\sqrt{n}(\overline{Y}_n - \mu) \to_d N(0, \sigma^2)$, and for almost every sequence $Y_1, Y_2, \ldots$,

$$\sqrt{n}(\overline{Y}_n^\star - \overline{Y}_n) \to_d N(0, \sigma^2).$$

The distribution of other functions of interest can be obtained via the delta method; see van der Vaart (1998). There are two approximations that are being used in the bootstrap. First, we are estimating F by $\widehat{F}$, and second, we are estimating the quantity of interest, for example, (i) or (ii), using B samples from $\widehat{F}$. For example, if (i) is of interest, an obvious estimator of $\text{var}_F(\widehat{\theta})$ is

$$\widehat{\text{var}}_F(\widehat{\theta}) = \frac{1}{B} \sum_{b=1}^{B} \left[\widehat{\theta}(\boldsymbol{Y}_b^\star) - \frac{1}{B} \sum_{b=1}^{B} \widehat{\theta}(\boldsymbol{Y}_b^\star) \right]^2. \tag{2.47}$$

In this case, the two approximations are

$$\text{var}_F\left(\widehat{\theta}\right) \approx \text{var}_{\widehat{F}}\left(\widehat{\theta}^\star\right) \approx \widehat{\text{var}}_{\widehat{F}}\left(\widehat{\theta}^\star\right)$$

and the first approximation may be poor if the estimate $\widehat{F}$ is not close to $\widehat{F}$, but we can control the second approximation by choosing large B. For the nonparametric bootstrap, we could, in principle, enumerate all possible samples, but there are n^n of these, of which $\binom{2n-1}{n}$ are distinct, which is far too large a number to evaluate in practice.

There are many possibilities for computation of confidence limits, as required in (ii). If normality of $\widehat{\theta}$ is reasonable, then (2.46) is straightforward to use with the variance estimated by (2.47) and the bias by

$$\widehat{\text{bias}}_F\left[\widehat{\theta}(\boldsymbol{Y})\right] = \widehat{\theta}(\boldsymbol{y}) - \frac{1}{B} \sum_{b=1}^{B} \widehat{\theta}(\boldsymbol{Y}_b^\star).$$

As a simple alternative, the *bootstrap percentile interval* for a confidence interval of coverage $1 - \alpha$ is

$$\left[\widehat{\theta}_{\alpha/2}^\star, \widehat{\theta}_{1-\alpha/2}^\star \right]$$

where $\widehat{\theta}_{\alpha/2}^\star$ and $\widehat{\theta}_{1-\alpha/2}^\star$ are the $\alpha/2$ and $1 - \alpha/2$ quantiles of the bootstrap estimates $\widehat{\theta}(\boldsymbol{Y}_b^\star)$, $b = 1, \ldots, B$. More refined bootstrap confidence interval procedures are described in Davison and Hinkley (1997). For example, Exercise 2.9 outlines the derivation of a confidence interval based on a pivot. In Sect. 2.7.3, we illustrate the close links between bootstrap variance estimation and sandwich estimation.

The bootstrap method does not work for all functions of interest. In particular, it fails in situations when the tail behavior is not well behaved, for example, a bootstrap for the maximum $Y_{(n)}$ will be disastrous.

2.7.2 The Bootstrap for Regression

The parametric and nonparametric methods provide two distinct versions of the bootstrap, and in a regression context, another important distinction is between *resampling residuals* and *resampling cases*. We illustrate the difference by considering the model

$$y_i = f(\boldsymbol{x}_i, \boldsymbol{\beta}) + \epsilon_i, \qquad (2.48)$$

where the residuals ϵ_i are such that $\mathrm{E}[\epsilon_i] = 0$, $i = 1, \ldots, n$ and are assumed uncorrelated. The two methods are characterized according to whether we take F to be the distribution of Y only or of $\{Y, \boldsymbol{X}\}$. In the resampling residuals approach, the covariates $\boldsymbol{x}_i$ are considered as fixed, and bootstrap datasets are formed as

$$Y_i^{(b)} = f(\boldsymbol{x}_i, \widehat{\boldsymbol{\beta}}) + \epsilon_{bi},$$

where a number of options are available for sampling ϵ_{bi}, $b = 1, \ldots, B$, $i = 1, \ldots, n$. The simplest, nonparametric, version is to sample ϵ_{bi} with replacement from

$$e_i = y_i - f(\boldsymbol{x}_i, \widehat{\boldsymbol{\beta}}) - \frac{1}{n}\sum_{i=1}^{n}\left[y_i - f(\boldsymbol{x}_i, \widehat{\boldsymbol{\beta}})\right].$$

Various refinements of this simple approach are possible. If we are willing to assume (say) that $\epsilon_i \mid \sigma^2 \sim_{iid} \mathrm{N}(0, \sigma^2)$, then a parametric resampling residuals method samples $\epsilon_{bi} \sim \mathrm{N}(0, \widehat{\sigma}^2)$ based on an estimate $\widehat{\sigma}^2$. In a model such as (2.48), the meaning of residuals is clear, but in generalized linear models (Chap. 6), for example, this is not the case and many alternative definitions exist.

The resampling residuals method has the advantage of respecting the "design," that is, $\boldsymbol{x}_1, \ldots, \boldsymbol{x}_n$. A major disadvantage, however, is that we are leaning heavily on the assumed mean–variance relationship, and we would often prefer to protect ourselves against an assumed model. The resampling case method forms bootstrap datasets by sampling with replacement from $\{Y_i, \boldsymbol{X}_i, i = 1, \ldots, n\}$ and does not assume a mean–variance model. Again parametric and nonparametric versions are available, but the latter is preferred since the former requires a model for the joint distribution of the response and covariates which is likely to be difficult to specify. When cases are resampled, the design in each bootstrap sample will not in general correspond to that in the original dataset which, though not ideal (since it leads to wider confidence intervals than necessary), will have little impact on inference, except when there are outliers in the data; if the outliers are sampled multiple times, then instability may result.

2.7.3 Sandwich Estimation and the Bootstrap

In this section we heuristically show why we would often expect sandwich and bootstrap variance estimates to be in close correspondence. For simplicity, we

2.7 Bootstrap Methods

consider a univariate parameter θ, and let $\widehat{\theta}_n$ denote the MLE arising from a sample of size n. In a change of notation, we denote the score by $\boldsymbol{S}(\theta) = [S_1(\theta), \ldots, S_n(\theta)]^\mathrm{T}$, where $S_i(\theta) = dl_i/d\theta$ is the contribution to the score from observation Y_i, $i = 1, \ldots, n$. Hence,

$$S(\theta) = \sum_{i=1}^n S_i(\theta) = \boldsymbol{S}(\theta)^\mathrm{T}\boldsymbol{1}$$

where $\boldsymbol{1}$ is an $n \times 1$ vector of 1's. The sandwich form of the asymptotic variance of $\widehat{\theta}_n$ is

$$\mathrm{var}(\widehat{\theta}_n) = \frac{1}{n}\frac{B}{A^2}$$

where

$$A(\theta) = \mathrm{E}\left[\frac{dS}{d\theta}\right], \quad B(\theta) = \mathrm{E}\left[S(\theta)^2\right].$$

These quantities may be empirically estimated via

$$\widehat{A}_n = \frac{1}{n}\frac{dS}{d\theta}\bigg|_{\widehat{\theta}_n} = \frac{1}{n}\sum_{i=1}^n \frac{dS_i}{d\theta}\bigg|_{\widehat{\theta}_n}$$

$$\widehat{B}_n = \frac{1}{n}\boldsymbol{S}(\theta)^\mathrm{T}\boldsymbol{S}(\theta)\bigg|_{\widehat{\theta}_n} = \frac{1}{n}\sum_{i=1}^n S_i(\theta)^2\bigg|_{\widehat{\theta}_n}.$$

A convenient representation of a bootstrap sample is $\boldsymbol{Y}^\star = \boldsymbol{Y} \times \boldsymbol{D}$ where $\boldsymbol{D} = \mathrm{diag}(D_1, \ldots, D_n)$ is a diagonal matrix consisting of a multinomial random variable

$$\begin{bmatrix} D_1 \\ \vdots \\ D_n \end{bmatrix} \sim \mathrm{Multinomial}\left[n, \left(\frac{1}{n}, \ldots, \frac{1}{n}\right)\right]$$

with

$$\mathrm{E}\left([D_1, \ldots, D_n]^\mathrm{T}\right) = \boldsymbol{1}$$

$$\mathrm{var}\left([D_1, \ldots, D_n]^\mathrm{T}\right) = \mathrm{I}_n - \frac{1}{n}\boldsymbol{1}\boldsymbol{1}^\mathrm{T} \to \mathrm{I}_n$$

as $n \to \infty$. The MLE of θ in the bootstrap sample is denoted $\widehat{\theta}_n^\star$ and satisfies $S^\star(\widehat{\theta}_n^\star) = 0$, where $S^\star(\theta)$ is the score corresponding to $\boldsymbol{Y}^\star$. Note that

$$S^\star(\theta) = \sum_{i=1}^n S_i^\star(\theta) = \sum_{i=1}^n S_i(\theta)D_i.$$

We consider a one-step Newton–Raphson approximation (see Sect. 6.5.2 for a more detailed description of this method) to $\widehat{\theta}_n^\star$ and show that this leads to a bootstrap variance estimate that is approximately equal to the sandwich variance estimate. The following informal derivation is carried out without stating regularity conditions. It is important to emphasize that throughout we are conditioning on Y and therefore on $\widehat{\theta}_n$. A first-order Taylor series approximation

$$0 = S^\star(\widehat{\theta}_n^\star) \approx S^\star(\widehat{\theta}_n) + (\widehat{\theta}_n^\star - \widehat{\theta}_n) \left. \frac{dS^\star}{d\theta} \right|_{\widehat{\theta}_n}$$

leads to the one-step approximation

$$\widehat{\theta}_n^\star \approx \widehat{\theta}_n - \frac{S^\star(\widehat{\theta}_n)}{\frac{d}{d\theta} S^\star(\theta)\big|_{\widehat{\theta}_n}}.$$

The bootstrap score evaluated at $\widehat{\theta}_n$ is

$$\sum_{i=1}^n S_i^\star(\widehat{\theta}_n) = \sum_{i=1}^n S_i(\widehat{\theta}_n) D_i \neq 0,$$

unless the bootstrap sample coincides with the original sample, that is, unless $D = I_n$. We replace $\left[\frac{d}{d\theta} S^\star(\theta)\big|_{\widehat{\theta}_n} \right]$ by its limit

$$E\left[\frac{d}{d\theta} S^\star(\theta) \Big|_{\widehat{\theta}_n} \right] = E\left[\sum_{i=1}^n \frac{d}{d\theta} S_i(\theta) D_i \Big|_{\widehat{\theta}_n} \right] = \sum_{i=1}^n \frac{d}{d\theta} S_i(\theta)\Big|_{\widehat{\theta}_n} E[D_i] = n \times \widehat{A}_n$$

where $\widehat{A}_n = \frac{1}{n} \frac{d}{d\theta} S(\theta)\big|_{\widehat{\theta}_n}$. Therefore, the one-step bootstrap estimator is approximated by

$$\widehat{\theta}_n^\star \approx \widehat{\theta}_n - \frac{S(\widehat{\theta}_n)^\mathrm{T} D}{n \widehat{A}_n}$$

and is approximately unbiased as an estimator since

$$E[\widehat{\theta}_n^\star - \widehat{\theta}_n] \approx -\frac{S(\widehat{\theta}_n)^\mathrm{T} E[D]}{n \widehat{A}_n} = -\frac{S(\widehat{\theta}_n)^\mathrm{T} \mathbf{1}}{n \widehat{A}_n} = 0$$

and, recall, $\widehat{\theta}_n$ is being held constant. The variance is

$$\mathrm{var}(\widehat{\theta}_n^\star - \widehat{\theta}_n) \approx \frac{S(\widehat{\theta}_n)^\mathrm{T} \mathrm{var}([D_1,\ldots,D_n]^\mathrm{T}) S(\widehat{\theta}_n)}{(n\widehat{A}_n)^2} = \frac{S(\widehat{\theta}_n)^\mathrm{T} \left(I - \frac{1}{n}\mathbf{1}\mathbf{1}^\mathrm{T}\right) S(\widehat{\theta}_n)}{(n\widehat{A}_n)^2}$$

$$\approx \frac{S(\widehat{\theta}_n)^\mathrm{T} I S(\widehat{\theta}_n)}{(n\widehat{A}_n)^2} = \frac{n\widehat{B}_n}{(n\widehat{A}_n)^2} = \frac{\widehat{B}_n}{n\widehat{A}_n^2},$$

2.7 Bootstrap Methods

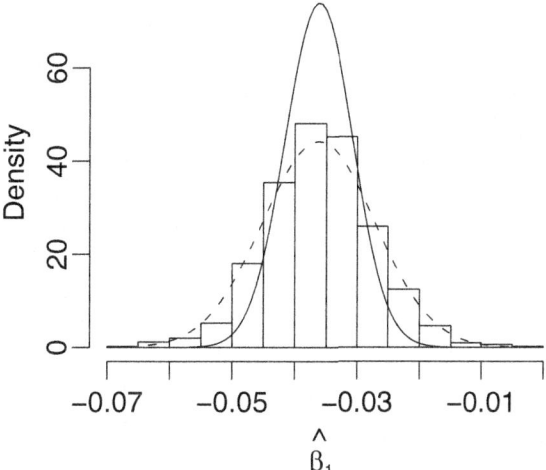

Fig. 2.3 Sampling distribution of $\widehat{\beta}_1$ arising from the nonparametric bootstrap samples. The *solid curve* is the asymptotic distribution of the MLE under the Poisson model, and the *dashed line* is the asymptotic distribution under the quasi-Poisson model

which is the sandwich estimator. Hence, $\text{var}(\widehat{\theta}_n^\star - \widehat{\theta}_n)$ approximates $\text{var}(\widehat{\theta}_n - \theta)$, which is a fundamental link in the bootstrap. For a more theoretical treatment, see Arcones and Giné (1992) and Sect. 10.3 of Kosorok (2008).

Example: Lung Cancer and Radon

For the lung cancer and radon example, we implement the nonparametric bootstrap resampling $B = 1{,}000$ sets of n case triples $[Y_{bi}^\star, E_{bi}^\star, x_{bi}^\star]$, $b = 1, \ldots, B$, $i = 1, \ldots, n$. Figure 2.3 displays the histogram of estimates arising from the bootstrap samples, along with the asymptotic normal approximations to the sampling distribution of the estimator under the Poisson and quasi-Poisson models. We see that the distribution under the quasi-likelihood model is much wider than that under the Poisson model. This is not surprising since we have already seen that the lung cancer data are overdispersed relative to a Poisson distribution. The bootstrap histogram and quasi-Poisson sampling distribution are very similar, however.

Table 2.4 summarizes inference for β_1 under a number of different methods and again confirms the similarity of asymptotic inference under the quasi-Poisson model and nonparametric bootstrap. In this example the similarity in the intervals from quasi-likelihood, sandwich estimation, and the nonparametric bootstrap is reassuring. The point estimates from the Poisson, quasi-likelihood, and sandwich approaches are identical. The point estimate from the quadratic variance model (that arises from a negative binomial model) is slightly closer to zero for these data, due to the difference in the variance models over the large range of counts in these data.

Table 2.4 Comparison of inferential summaries over various approaches, for the lung cancer and radon example

Inferential method	$\widehat{\beta}_1$	s.e.$(\widehat{\beta}_1)$	95% CI for $\exp(\beta_1)$
Poisson	−0.036	0.0054	0.954, 0.975
Quasi-likelihood	−0.036	0.0090	0.947, 0.982
Quadratic variance	−0.030	0.0085	0.955, 0.987
Sandwich estimation	−0.036	0.0080	0.949, 0.980
Bootstrap normal	−0.036	0.0087	0.948, 0.981
Bootstrap percentile	−0.036	0.0087	0.949, 0.981

The last two lines refer to nonparametric bootstrap approaches, with intervals based on normality of the sampling distribution of the estimator ("Normal") and on taking the 2.5% and 97.5% points of this distribution ("Percentile")

2.8 Choice of Estimating Function

The choice of estimating function is driven by the conflicting aims of *efficiency* and *robustness to model misspecification*. If the likelihood corresponds to the true model, then MLEs are asymptotically efficient so that asymptotic confidence intervals have minimum length. However, if the assumed model is incorrect, then there are no guarantees of even consistency of estimation.

Basing estimating functions on simple model-free functions of the data often provides robustness. As we discuss in Sect. 5.6.3, the classic Gauss–Markov theorem states, informally, that among estimators that are linear in the data, the least squares estimator has smallest variance, and this result is true for fixed sample sizes. There is also a Gauss–Markov theorem for estimating functions. Suppose $E[Y_i \mid \beta] = \mu_i(\beta)$, $\text{var}(Y_i) = \sigma_i^2$ and $\text{cov}(Y_i, Y_j) = 0$, $i \neq j$, and consider the class of *linear unbiased estimating functions* (of zero) that are of the form

$$G(\beta) = \sum_{i=1}^{n} a_i(\beta) [Y_i - \mu_i(\beta)], \qquad (2.49)$$

where $a_i(\beta)$ are specified nonrandom functions, subject to $\sum_{i=1}^{n} a_i(\beta) = c$, a constant (this is to avoid obtaining an arbitrarily small variance by multiplying the estimating function by a constant). The estimating function (2.49) provides a consistent estimator $\widehat{\beta}$ so long as the mean $\mu_i(\beta)$ is correctly specified. It can be shown, for example, Godambe and Heyde (1987), that

$$E[UU^{\mathsf{T}}] \leq E[GG^{\mathsf{T}}], \qquad (2.50)$$

where

$$U(\beta) = D^{\mathsf{T}} V^{-1} (Y - \mu)/\alpha,$$

so that this estimating function has the smallest variance. Quasi-likelihood estimators are therefore asymptotically optimal in the class of linear estimating functions

2.8 Choice of Estimating Function

and will be asymptotically efficient if the quasi-score functions correspond to the score of the likelihood of the true data-generating model. Of course a superior estimator (in terms of efficiency) may result from an estimating function that is not linear in the data, if the data arise from a model for which the score function is not linear. The consideration of *quadratic* estimating functions illustrates the efficiency-robustness trade-off.

Result (2.50) is true for an *estimating function* based on a finite sample size n, though there is no such result for the derived *estimator*. However, the estimator derived from the estimating function is asymptotically efficient (e.g., McCullagh 1983). The optimal estimating equation is that which has minimum expected distance from the score equation corresponding to the true model. We reemphasize that a consistent estimator of the parameters in the assumed regression model is obtained from the quasi-score (2.50), and the variance of the estimator will be appropriate so long as the second moment of the data has been specified correctly.

To motivate the class of quadratic estimating functions suppose

$$Y_i \mid \boldsymbol{\beta} \sim_{ind} N\left[\mu_i(\boldsymbol{\beta}), \sigma_i^2(\boldsymbol{\beta})\right],$$

$i = 1, \ldots, n$. The log-likelihood is

$$l(\boldsymbol{\beta}) = -\sum_{i=1}^{n} \log \sigma_i(\boldsymbol{\beta}) - \frac{1}{2}\sum_{i=1}^{n} \frac{[Y_i - \mu_i(\boldsymbol{\beta})]^2}{\sigma_i(\boldsymbol{\beta})^2},$$

which gives the quadratic score equations

$$\boldsymbol{S}(\boldsymbol{\beta}) = \frac{\partial l}{\partial \boldsymbol{\beta}}$$
$$= \sum_{i=1}^{n} \frac{\{Y_i - \mu_i(\boldsymbol{\beta})\}}{\sigma_i(\boldsymbol{\beta})^2} \frac{\partial \mu_i}{\partial \boldsymbol{\beta}} + \sum_{i=1}^{n} \frac{\{[Y_i - \mu_i(\boldsymbol{\beta})]^2 - \sigma_i(\boldsymbol{\beta})^2\}}{\sigma_i(\boldsymbol{\beta})^3} \frac{\partial \sigma_i}{\partial \boldsymbol{\beta}}. \quad (2.51)$$

If the first two moments are correctly specified, then $E[\boldsymbol{S}(\boldsymbol{\beta})] = \boldsymbol{0}$, so that a consistent estimator is obtained.

In general, we may consider

$$\sum_{i=1}^{n} a_i(\boldsymbol{\beta})[Y_i - \mu_i(\boldsymbol{\beta})] + b_i(\boldsymbol{\beta})\left\{[Y_i - \mu_i(\boldsymbol{\beta})]^2 - \sigma_i(\boldsymbol{\beta})^2\right\},$$

where $a_i(\boldsymbol{\beta}), b_i(\boldsymbol{\beta})$ are specified nonrandom functions. With this estimating function, the information in the variance concerning the parameters $\boldsymbol{\beta}$ is being used to improve efficiency. Among quadratic estimating functions, it can be shown that (2.51) is optimal in the sense of producing estimators that are asymptotic efficient (Crowder 1987). In general, to choose the optimal estimating function, the first four moments of the data must be known, which may seem unlikely, but this approach may be contrasted with the use of the score as estimating function which effectively requires all of the moments to be known. There are two problems with using quadratic estimating functions. First, consistency requires the first two moments to be correctly specified. Second, to estimate the covariance matrix of the estimator, the skewness and kurtosis must be estimated, and these may be highly unstable. We return to this topic in Sect. 9.10.

2.9 Hypothesis Testing

Throughout the book, we emphasize estimation over hypothesis testing, for reasons discussed in Chap. 4, but in this section describe the rationale and machinery of frequentist hypothesis testing.

2.9.1 Motivation

A common aim of statistical analysis is to judge the evidence from the data in support of a particular hypothesis, defined through specific parameter values. Hypothesis tests have historically been used for various purposes, including:

- Determining whether a set of data is *consistent* with a particular hypothesis
- Making a *decision* as to which of two hypotheses is best supported by the data

We assume there exists a test statistic $T = T(Y)$ with large values of T suggesting departures from H_0. In Sects. 2.9.3–2.9.5, three specific recipes are described, namely, score, Wald, and likelihood ratio test statistics. We define the p-value, or *significance level*, as

$$p = p(Y) = \Pr[\, T(Y) > T(y) \mid H_0 \,],$$

so that, intuitively, if this probability is "small," the data are inconsistent with H_0. If $T(Y)$ is continuous, then under H_0, the p-value $p(Y)$ follows the distribution $U(0, 1)$. Consequently, the significance level is the observed $p(y)$. The distribution of $T(Y)$ under H_0 may be known analytically or may be simulated to produce a Monte Carlo or bootstrap test.

The nomenclature associated with the broad topic of hypothesis testing is confusing, but we distinguish three procedures:

1. A *pure significance test* calculates p but does not reject H_0 and is often viewed as an exploratory tool.
2. A *test of significance* sets a cutoff value α (e.g., $\alpha = 0.05$) and rejects H_0 if $p < \alpha$ corresponding to $T > T_\alpha$. The latter is known as the *critical region*.
3. A *hypothesis test* goes one step further and specifies an *alternative hypothesis*, H_1. One then reports whether H_0 is rejected or not. The null hypothesis has

special position as the "status quo," and conventionally the phrase "accept H_0" is not used because not rejecting may be due to low power (perhaps because of a small sample size) as opposed to H_0 being true.

Rejecting H_0 when it is true is known as a type I error, and a type II error occurs when H_0 is not rejected when it is in fact false. To evaluate the probability of a type II error, specific alternative values of the parameters need to be considered. The *power* is defined as the probability of rejecting H_0 when it is false. We emphasize that a test of significance may reject H_0 for general departures, while a hypothesis test rejects in the specific direction of H_1.

A key point is that the consistency of the data with H_0 is being evaluated, and there is no reference to the probability of the null hypothesis being true. As usual in frequentist inference, H_0 is a fixed unknown and probability statements cannot be assigned to it.[4]

2.9.2 Preliminaries

We consider a p-dimensional vector of parameters $\boldsymbol{\theta}$ and consider two testing situations. In the first, we consider the *simple* null hypothesis $H_0 : \boldsymbol{\theta} = \boldsymbol{\theta}_0$ versus the alternative $H_1 : \boldsymbol{\theta} \neq \boldsymbol{\theta}_0$. In the second, we consider a partition of the parameter vector $\boldsymbol{\theta} = [\boldsymbol{\theta}_1, \boldsymbol{\theta}_2]$, where the dimensions of $\boldsymbol{\theta}_1$ and $\boldsymbol{\theta}_2$ are $p - r$ and r, respectively, and a *composite* null. Specifically, in the composite case, we compare the hypotheses:

$$H_0 : \boldsymbol{\theta}_1 \text{ unrestricted}, \boldsymbol{\theta}_2 = \boldsymbol{\theta}_{20},$$
$$H_1 : \boldsymbol{\theta} = [\boldsymbol{\theta}_1, \boldsymbol{\theta}_2] \neq [\boldsymbol{\theta}_1, \boldsymbol{\theta}_{20}].$$

As a simple example, in a regression context, let $\boldsymbol{\theta} = [\theta_1, \theta_2]$ with θ_1 the intercept and θ_2 the slope. We may then be interested in $H_0 : \theta_2 = 0$ with θ_1 unspecified. In both the simple and composite situations, the unrestricted MLE under the alternative is denoted $\widehat{\boldsymbol{\theta}}_n = [\widehat{\boldsymbol{\theta}}_{n1}, \widehat{\boldsymbol{\theta}}_{n2}]$.

For simplicity of exposition, unless stated otherwise, we suppose that the responses $Y_i, i = 1, \ldots, n$, are independent and identically distributed. Consequently we have $p(\boldsymbol{y} \mid \boldsymbol{\theta}) = \prod_{i=1}^{n} p(y_i \mid \boldsymbol{\theta})$. The extension to the nonidentically distributed situation, as required for regression, is straightforward. The $p \times 1$ score vector is

$$\boldsymbol{S}_n(\boldsymbol{\theta}) = \sum_{i=1}^{n} \frac{\partial l_i(\boldsymbol{\theta})}{\partial \boldsymbol{\theta}}$$

---

[4] As described in Chap. 3, in the Bayesian approach to hypothesis testing, a prior distribution is placed on the alternatives (and on the null), allowing the calculation of the probability of H_0 given the data, relative to other hypotheses under consideration.

where $l_i(\boldsymbol{\theta})$ is the log-likelihood contribution from observation i, $i = 1, \ldots, n$. Let $\boldsymbol{S}_n(\boldsymbol{\theta}) = [\boldsymbol{S}_{n1}(\boldsymbol{\theta}), \boldsymbol{S}_{n2}(\boldsymbol{\theta})]^{\mathrm{T}}$ be a partition of the score vector with $\boldsymbol{S}_{n1}(\boldsymbol{\theta})$ of dimension $(p-r) \times 1$ and $\boldsymbol{S}_{n2}(\boldsymbol{\theta})$ of dimension $r \times 1$. Under the composite null, let $\widehat{\boldsymbol{\theta}}_n^0 = [\widehat{\boldsymbol{\theta}}_{n10}, \boldsymbol{\theta}_{20}]$ denote the MLE, where $\widehat{\boldsymbol{\theta}}_{n10}$ is found from the estimating equation

$$\boldsymbol{S}_{n1}(\widehat{\boldsymbol{\theta}}_{n10}, \boldsymbol{\theta}_{20}) = \boldsymbol{0}.$$

In general, $\widehat{\boldsymbol{\theta}}_{n10} \neq \widehat{\boldsymbol{\theta}}_{n1}$.

In the independent and identically distributed case, $\boldsymbol{I}_n(\boldsymbol{\theta}) = n\boldsymbol{I}_1(\boldsymbol{\theta})$ is the information in a sample of size n. Suppressing the dependence on $\boldsymbol{\theta}$, let

$$\boldsymbol{I}_1 = \begin{bmatrix} \boldsymbol{I}_{11} & \boldsymbol{I}_{12} \\ \boldsymbol{I}_{21} & \boldsymbol{I}_{22} \end{bmatrix}$$

denote a partition of the expected information matrix, where $\boldsymbol{I}_{11}$, $\boldsymbol{I}_{12}$, $\boldsymbol{I}_{21}$, and $\boldsymbol{I}_{22}$ are of dimensions $(p-r) \times (p-r)$, $(p-r) \times r$, $r \times (p-r)$, and $r \times r$, respectively. The inverse of $\boldsymbol{I}_1$ is

$$\boldsymbol{I}_1^{-1} = \begin{bmatrix} \boldsymbol{I}_{11 \cdot 2}^{-1} & -\boldsymbol{I}_{11 \cdot 2}^{-1} \boldsymbol{I}_{12} \boldsymbol{I}_{22}^{-1} \\ -\boldsymbol{I}_{22 \cdot 1}^{-1} \boldsymbol{I}_{21} \boldsymbol{I}_{11}^{-1} & \boldsymbol{I}_{22 \cdot 1}^{-1} \end{bmatrix}$$

where

$$\boldsymbol{I}_{11 \cdot 2} = \boldsymbol{I}_{11} - \boldsymbol{I}_{12} \boldsymbol{I}_{22}^{-1} \boldsymbol{I}_{21}$$

$$\boldsymbol{I}_{22 \cdot 1} = \boldsymbol{I}_{22} - \boldsymbol{I}_{21} \boldsymbol{I}_{11}^{-1} \boldsymbol{I}_{12}$$

using results from Appendix B.

2.9.3 Score Tests

We begin with the simple null $H_0 : \boldsymbol{\theta} = \boldsymbol{\theta}_0$. Recall the asymptotic distribution of the score, given in (2.17):

$$n^{-1/2} \boldsymbol{S}_n(\boldsymbol{\theta}) \to_d \mathrm{N}_p[\boldsymbol{0}, \boldsymbol{I}_1(\boldsymbol{\theta})].$$

Therefore, under the null hypothesis

$$\boldsymbol{S}_n(\boldsymbol{\theta}_0)^{\mathrm{T}} \boldsymbol{I}_1^{-1}(\boldsymbol{\theta}_0) \boldsymbol{S}_n(\boldsymbol{\theta}_0)/n \to_d \chi_p^2. \qquad (2.52)$$

Intuitively, if the elements of $\boldsymbol{S}_n(\boldsymbol{\theta}_0)$ are large, this means that the components of the gradient at $\boldsymbol{\theta}_0$ are large. The latter occurs when $\boldsymbol{\theta}_0$ is "far" from the estimator $\widehat{\boldsymbol{\theta}}_n$ for which $\boldsymbol{S}_n(\widehat{\boldsymbol{\theta}}_n) = \boldsymbol{0}$. In (2.52), the matrix $\boldsymbol{I}_1^{-1}(\boldsymbol{\theta}_0)$ is scaling the gradient distance. The information may be evaluated at the MLE, $\widehat{\boldsymbol{\theta}}_n$, rather than at $\boldsymbol{\theta}_0$, since $\boldsymbol{I}_1(\widehat{\boldsymbol{\theta}}_n) \to_p \boldsymbol{I}_1(\boldsymbol{\theta}_0)$, by the weak law of large numbers.

Under the composite null hypothesis, $H_0 : \boldsymbol{\theta}_1$ unrestricted, $\boldsymbol{\theta}_2 = \boldsymbol{\theta}_{20}$:

$$\boldsymbol{S}_n(\widehat{\boldsymbol{\theta}}_n^0)^\mathsf{T} \boldsymbol{I}_1^{-1}(\widehat{\boldsymbol{\theta}}_n^0) \boldsymbol{S}_n(\widehat{\boldsymbol{\theta}}_n^0)/n \to_d \chi_r^2.$$

As a simplification, we can express this statistic in terms of partitioned information matrices. Since r elements of the score vector are zero, that is, $\boldsymbol{S}_{n2}(\widehat{\boldsymbol{\theta}}_n^0) = \boldsymbol{0}$, we have

$$\boldsymbol{S}_{n1}(\widehat{\boldsymbol{\theta}}_n^0)^\mathsf{T} \boldsymbol{I}_{11\cdot 2}^{-1}(\widehat{\boldsymbol{\theta}}_n^0) \boldsymbol{S}_{n1}(\widehat{\boldsymbol{\theta}}_n^0)/n \to_d \chi_r^2.$$

Hence, the model only needs to be fitted under the null. Each of the score statistics remains asymptotically valid on replacement of the expected information by the observed information.

2.9.4 Wald Tests

Under the simple null hypothesis $H_0 : \boldsymbol{\theta} = \boldsymbol{\theta}_0$, the Wald statistic is based upon the asymptotic distribution

$$\sqrt{n}(\widehat{\boldsymbol{\theta}}_n - \boldsymbol{\theta}_0) \to_d N_p \left[\boldsymbol{0}, \boldsymbol{I}_1(\boldsymbol{\theta}_0)^{-1} \right], \qquad (2.53)$$

and the Wald statistic is the quadratic form based on (2.53):

$$\sqrt{n}(\widehat{\boldsymbol{\theta}}_n - \boldsymbol{\theta}_0)^\mathsf{T} \boldsymbol{I}_1(\boldsymbol{\theta}_0) \sqrt{n}(\widehat{\boldsymbol{\theta}}_n - \boldsymbol{\theta}_0) \to_d \chi_p^2. \qquad (2.54)$$

An alternative form that is often used in practice is

$$\sqrt{n}(\widehat{\boldsymbol{\theta}}_n - \boldsymbol{\theta}_0)^\mathsf{T} \boldsymbol{I}_1(\widehat{\boldsymbol{\theta}}_n) \sqrt{n}(\widehat{\boldsymbol{\theta}}_n - \boldsymbol{\theta}_0) \to_d \chi_p^2,$$

which again follows because $\boldsymbol{I}_1(\widehat{\boldsymbol{\theta}}_n) \to_p \boldsymbol{I}_1(\boldsymbol{\theta}_0)$, by the weak law of large numbers.

Under a composite null hypothesis, the Wald statistic is based on the marginal distribution of $\widehat{\boldsymbol{\theta}}_{n2}$:

$$\sqrt{n}(\widehat{\boldsymbol{\theta}}_{n2} - \boldsymbol{\theta}_{20})^\mathsf{T} \boldsymbol{I}_{11\cdot 2}(\widehat{\boldsymbol{\theta}}_n^0) \sqrt{n}(\widehat{\boldsymbol{\theta}}_{n2} - \boldsymbol{\theta}_{20}) \to_d \chi_r^2.$$

The observed information may replace the expected information in either form of the Wald statistic.

2.9.5 Likelihood Ratio Tests

Finally, we consider the likelihood ratio statistic which, under a simple null, is

$$2 \left[l_n(\widehat{\boldsymbol{\theta}}_n) - l_n(\boldsymbol{\theta}_0) \right].$$

Unlike the score and Wald statistics, the asymptotic distribution is not an obvious quadratic form, and so we provide a sketch proof of the asymptotic distribution under H_0. A second-order Taylor expansion of $l_n(\boldsymbol{\theta}_0)$ about $\widehat{\boldsymbol{\theta}}_n$ gives

$$l_n(\boldsymbol{\theta}_0) = l_n(\widehat{\boldsymbol{\theta}}_n) + (\boldsymbol{\theta}_0 - \widehat{\boldsymbol{\theta}}_n)^{\mathrm{T}} \frac{\partial l_n(\boldsymbol{\theta})}{\partial \boldsymbol{\theta}}\bigg|_{\widehat{\boldsymbol{\theta}}_n} + \frac{1}{2}(\boldsymbol{\theta}_0 - \widehat{\boldsymbol{\theta}}_n)^{\mathrm{T}} \frac{\partial^2 l_n(\boldsymbol{\theta})}{\partial \boldsymbol{\theta} \partial \boldsymbol{\theta}^{\mathrm{T}}}\bigg|_{\widetilde{\boldsymbol{\theta}}} (\boldsymbol{\theta}_0 - \widehat{\boldsymbol{\theta}}_n),$$

where $\widetilde{\boldsymbol{\theta}}$ is between $\boldsymbol{\theta}_0$ and $\widehat{\boldsymbol{\theta}}_n$. The middle term on the right-hand side is zero, and

$$\frac{1}{n} \frac{\partial^2 l_n(\boldsymbol{\theta})}{\partial \boldsymbol{\theta} \partial \boldsymbol{\theta}^{\mathrm{T}}}\bigg|_{\widetilde{\boldsymbol{\theta}}} \to_p -\boldsymbol{I}_1(\boldsymbol{\theta}_0).$$

Hence,

$$-2\left[l_n(\boldsymbol{\theta}_0) - l_n(\widehat{\boldsymbol{\theta}}_n)\right] = 2\left[l_n(\widehat{\boldsymbol{\theta}}_n) - l_n(\boldsymbol{\theta}_0)\right]$$
$$\approx n(\widehat{\boldsymbol{\theta}}_n - \boldsymbol{\theta}_0)^{\mathrm{T}} \boldsymbol{I}_1(\boldsymbol{\theta}_0)(\widehat{\boldsymbol{\theta}}_n - \boldsymbol{\theta}_0),$$

and so

$$2\left[l_n(\widehat{\boldsymbol{\theta}}_n) - l_n(\boldsymbol{\theta}_0)\right] \to_d \chi_p^2. \tag{2.55}$$

Similarly, under a composite null hypothesis:

$$2\left[l_n(\widehat{\boldsymbol{\theta}}_n) - l_n(\widehat{\boldsymbol{\theta}}_n^0)\right] \to_d \chi_r^2.$$

2.9.6 Quasi-likelihood

We briefly consider the quasi-likelihood model described in Sect. 2.5. The score test can be based on the quasi-score statistic $\boldsymbol{U}_n(\boldsymbol{\beta}) = \boldsymbol{D}^{\mathrm{T}} \boldsymbol{V}^{-1}(\boldsymbol{Y} - \boldsymbol{\mu})/\alpha$, with the information in a sample of size n being $\boldsymbol{D}^{\mathrm{T}} \boldsymbol{V}^{-1} \boldsymbol{D}/\alpha$. The latter is also used in the calculation of a Wald statistic since it supplies the required standard errors. Similarly, a quasi-likelihood ratio test can be performed using $l_n(\widehat{\boldsymbol{\theta}}_n, \alpha)$, the form of which is given in (2.32). Unknown α can be accommodated by substitution of a consistent estimator $\widehat{\alpha}$. For example, we might estimate α via the Pearson statistic estimator (2.31).

If one wished to account for estimation of α, then one possibility is to assume that $(n-p) \times \widehat{\alpha}$ follows a χ_{n-p}^2 distribution and then evaluate significance based on the ratio of scaled χ^2-squared random variables, to give an F distribution under the null (see Appendix B). Outside of the normal linear model, this seems a dubious exercise, however, since the numerator and denominator will not be independent, and either of the χ^2 approximations could be poor. The use of an F statistic is conservative, however (so that significance will be reduced over the use of the plug-in χ^2 approximation).

2.9.7 Comparison of Test Statistics

The score test statistic is invariant under reparameterization, provided that the expected, rather than the observed, information is used. The score statistic may also be evaluated without second derivatives if $S_n(\boldsymbol{\theta}_0)S_n(\boldsymbol{\theta}_0)^{\mathrm{T}}$ is used, which may be useful if these derivatives are complex, or unavailable. The score statistic requires the value of the score at the null, but the MLE under the alternative is not required.

Confidence intervals can be derived directly from the Wald statistic so that there is a direct link between estimation and testing. Interpretation is also straightforward; in particular, statistical versus practical significance can be immediately considered. A major drawback of the Wald statistic is that it is not invariant to the parameterization chosen, which ties in with our earlier observation (Sect. 2.3) that asymptotic confidence intervals are more accurate on some scales than on others. The Wald statistic uses the MLE but not the value of the maximized likelihood.

The likelihood ratio statistic is invariant under reparameterization. Confidence intervals derived from likelihood ratio tests always preserve the support of the parameter, unlike score- and Wald-based intervals (unless a suitable parameterization is adopted). Similar to the attainment of the Cramér–Rao lower bound (Appendix F), there is an elegant theory under which the likelihood ratio test statistic emerges as the *uniformly most powerful* (UMP) test, via the famous Neyman–Pearson lemma; see, for example, Schervish (1995). The likelihood ratio test requires the fitting of two models.

The score, Wald, and likelihood ratio test statistics are asymptotically equivalent but are not equally well behaved in finite samples. In general, and by analogy with the asymptotic optimality of the MLE, the likelihood ratio statistic is often recommended for use in regular models. If $\widehat{\boldsymbol{\theta}}_n$ and $\boldsymbol{\theta}_0$ are close, then the three statistics will tend to agree.

Chapter 4 provides an extended discussion and critique of hypothesis testing.

Example: Poisson Mean

We illustrate the use of the three statistics in a simple context. Suppose we have data $Y_i \mid \lambda \sim_{iid} \text{Poisson}(\lambda)$, $i = 1, \ldots, n$, and we are interested in $H_0 : \lambda = \lambda_0$. The log-likelihood, score, and information are

$$l_n(\lambda) = -n\lambda + n\overline{Y} \log \lambda,$$

$$S_n(\lambda) = -n + \frac{n\overline{Y}}{\lambda} = \frac{n(\overline{Y} - \lambda)}{\lambda},$$

$$I_n(\lambda) = \frac{n}{\lambda}.$$

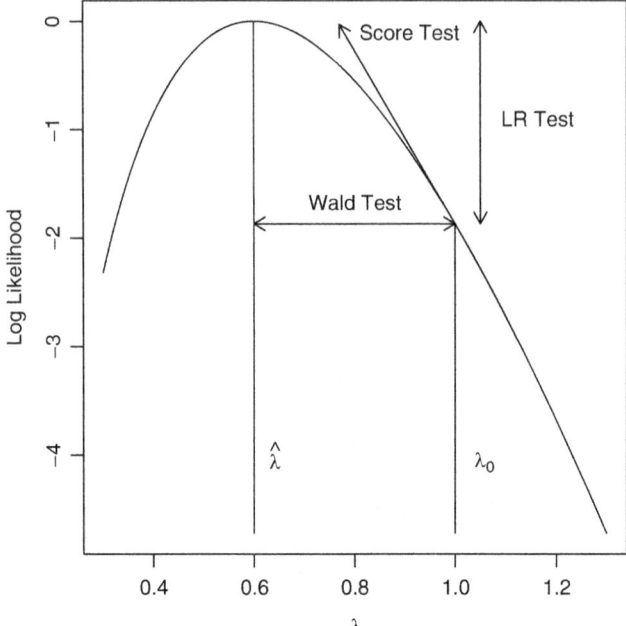

Fig. 2.4 Geometric interpretation of score, Wald, and likelihood ratio (LR) statistics, for Poisson data and a test of $H_0 : \lambda_0 = 1$, with data resulting in $\widehat{\lambda} = \overline{y} = 0.6$

The score and Wald statistics follow from (2.52) and (2.54) and both lead to

$$\frac{n(\overline{Y} - \lambda_0)^2}{\lambda_0} \to_d \chi_1^2$$

under the null. From (2.55), the likelihood ratio statistic is

$$2n \left[\overline{Y}(\log \overline{Y} - \log \lambda_0) - (\overline{Y} - \lambda_0) \right] \to_d \chi_1^2.$$

Suppose we observe $\sum_{i=1}^{20} y_i = 12$ events in $n = 20$ trials so that $\widehat{\lambda} = \overline{y} = 0.6$. Assume we are interested in testing the null hypothesis $H_0 : \lambda_0 = 1.0$. The score and Wald statistics are 3.20 and the likelihood ratio statistic is 3.74, with associated observed significance levels of 7.3% and 5.4%, respectively. Figure 2.4 plots the log-likelihood against λ for these data. The (unscaled) statistics are indicated on the figure. The score test is based on the gradient at λ_0, the Wald statistic is the squared horizontal distance between $\widehat{\lambda}$ and λ_0, and the likelihood ratio test statistic is two times the vertical distance between $l(\widehat{\lambda})$ and $l(\lambda_0)$.

We now reparameterize to $\theta = \log \lambda$, so that the null becomes $H_0 : \theta = \theta_0 = 0$. The likelihood ratio statistic is invariant to parameterization, and the score statistic turns out to be the same as previously in this example, since the observed and expected information are equal. The forms of the Wald, score, and likelihood ratio statistics, for general θ_0, are

$$n(\log \overline{Y} - \theta_0)^2 \exp(\theta_0)$$
$$n\left[\overline{Y} - \exp(\theta_0)\right]^2 \exp(-\theta_0)$$
$$2n\left\{\overline{Y}(\widehat{\theta} - \theta_0) - [\exp(\widehat{\theta}) - \exp(\theta_0)]\right\}$$

with numeric values of 5.22, 3.20 and 3.74, respectively, in the example.

2.10 Concluding Remarks

In Sect. 1.2, we emphasized that model formulation should begin with the model that is felt most appropriate for the context, before proceeding to determine the behavior of inferential procedures under this model. In this chapter we have seen that likelihood-based inference is asymptotically efficient *if* the model is correct. Hence, if one has strong belief in the assumed model, then a likelihood approach is appealing, particularly if the score equations are of linear exponential family form, since in this case consistent estimators of the parameters in the assumed regression model are obtained. If the likelihood is not of linear exponential form, then there are no guarantees of consistency under model misspecification. So far as estimation of the standard error is concerned, in situations in which n is sufficiently large for asymptotic inference to be accurate, sandwich estimation or the bootstrap may be used to provide consistent model-free standard errors, so long as the observations are uncorrelated. The relevance of asymptotic calculations for particular sample sizes may be investigated via simulation. In general, sandwich estimation is a very simple, broadly applicable and appealing technique.

In many instances the context and/or questions of interest may determine the mean function and perhaps give clues to the mean–variance relationship. The form of the data may suggest viable candidates for the full probability model. A caveat to this is that models such as the Poisson or exponential for which there is no dispersion parameter should be used with extreme caution since there is no mechanism to "soak up" excess variability. In practice, if the data exhibit overdispersion, as is often the case, then this will lead to confidence intervals that are too short. Information on the mean and variance may be used within a quasi-likelihood approach to define an estimator, and if n is sufficiently large, sandwich estimation can provide reliable standard errors. Experience of particular models may help to determine whether the assumption of a particular likelihood with the desired mean and variance functions is likely to be much less reliable than a quasi-likelihood approach. The choice of how parametric one wishes to be will often come down to personal taste.

We finally note that the efficiency-robustness trade-off will be weighted in different directions depending on the nature of the analysis. In an exploratory setting, one may be happy to proceed with a likelihood analysis, while in a confirmatory setting, one may want to be more conservative.

2.11 Bibliographic Notes

Numerous accounts of the theory behind frequentist inference are available, Cox and Hinkley (1974) remains a classic text. Casella and Berger (1990) also provides an in-depth discussion of frequentist estimation and hypothesis testing. A mathematically rigorous treatment of the estimating functions approach is provided by van der Vaart (1998). A gentler and very readable presentation of a reduced amount of material is Ferguson (1996). Further discussion of estimating functions, particularly for quasi-likelihood, may be found in Heyde (1997) and Crowder (1986).

Likelihood was introduced by Fisher (1922, 1925b), and quasi-likelihood by Wedderburn (1974). Asymptotic details for quasi-likelihood are described in McCullagh (1983), while Gauss–Markov theorems detailing optimality are described in Godambe and Heyde (1987) and Heyde (1997). Firth (1993) provides an excellent review of quasi-likelihood.

Crowder (1987) gives counterexamples that reveal situations in which quasi-likelihood is unreliable. Linear and quadratic estimating functions are described by Firth (1987) and Crowder (1987). Firth (1987) also investigates the efficiency of quasi-likelihood estimators and concludes that such estimators are robust to "moderate departures" from the likelihood corresponding to the score.

The form of the sandwich estimator was given in Huber (1967). White (1980) implemented the technique for the linear model, and Royall (1986) provides a clear and simple account with many examples. Carroll et al. (1995, Appendix A.3) gives a very readable review of sandwich estimation.

Efron (1979) introduced the bootstrap, and subsequently there has been a huge literature on its theoretical properties and practical use. Bickel and Freedman (1981) and Singh (1981) provide early theoretical discussions; see also van der Vaart (1998). Book-length treatments include Efron and Tibshirani (1993) and Davison and Hinkley (1997).

The score test was introduced in Rao (1948) as an alternative to the likelihood ratio and Wald tests introduced in Neyman and Pearson (1928) and Wald (1943), respectively. Consequently, the score test is sometimes known as the Rao score test. Cox and Hinkley (1974) provide a general discussion of hypothesis testing. Peers (1971) compares the power of score, Wald, and likelihood ratio tests. An excellent expository article on the three statistics, emphasizing a geometric perspective, may be found in Buse (1982).

2.12 Exercises

2.1 Suppose $Y_1, Y_2 \mid \theta \sim_{iid} U(\theta - 0.5, \theta + 0.5)$. Show that $\Pr(\min\{Y_1, Y_2\} < \theta < \max\{Y_1, Y_2\} \mid \theta) = 0.5$, so that $[\min\{Y_1, Y_2\}, \max\{Y_1, Y_2\}]$ is a 50% confidence interval for θ. Suppose we observe a particular interval with

$\max\{Y_1, Y_2\} - \min\{Y_1, Y_2\} \geq 0.5$. Show that in this case we know with probability 1 that this interval contains θ.[5]

2.2 Consider a single observation from a Poisson distribution: $Y \mid \theta \sim \text{Poisson}(\theta)$.

   (a) Suppose we wish to estimate $\exp(-3\theta)$. Show that the UMVUE is $(-2)^y$ for $y = 0, 1, 2, \ldots$ Is this a reasonable estimator?
   (b) Suppose we wish to estimate θ^2. Show that $T(T-1)/n^2$ is the UMVUE for θ^2. By examining the case $T = 1$ comment on whether this is a sensible estimator.

2.3 Let $Y_i \mid \sigma^2 \sim_{iid} N(\mu, \sigma^2)$ with μ known.

   (a) Show that the distribution $p(y \mid \sigma^2)$ is a one-parameter exponential family member.
   (b) Show that $\widehat{\sigma}^2 = \frac{1}{n}\sum_{i=1}^{n}(Y_i - \mu)^2$ is an unbiased estimator of σ^2 and evaluate its variance.
   (c) Consider estimators of the form $\widetilde{\sigma}_a^2 = a\sum_{i=1}^{n}(Y_i - \mu)^2$. Determine the value of a that minimizes the mean squared error.
   (d) The use of mean squared error to judge an estimator is appropriate for a quadratic loss function, in this case $L(\widetilde{\sigma}_a^2, \sigma^2) = (\widetilde{\sigma}_a^2 - \sigma^2)^2$. Since $\sigma^2 > 0$, there is an asymmetry in this loss function. Hence, explain why downward bias in an estimator of σ^2 can be advantageous.
   (e) Show that $\widehat{\sigma}^2$ is optimal amongst estimators $\widetilde{\sigma}_a^2$ with respect to the *Stein* loss function
   $$L_s(\widetilde{\sigma}_a^2, \sigma^2) = \left(\frac{\widetilde{\sigma}_a^2}{\sigma^2}\right) - \log\left(\frac{\widetilde{\sigma}_a^2}{\sigma^2}\right) - 1.$$

2.4 Suppose $Y_i \mid \theta_i \sim_{ind} \text{Poisson}(\theta_i)$ with $\theta_i \sim_{ind} \text{Ga}(\mu_i b, b)$ for $i = 1, \ldots, n$.

   (a) Show that $E[Y_i] = \mu_i$ and $\text{var}(Y_i) = \mu_i(1 + b^{-1})$.
   (b) Show that the marginal distribution of $Y_i \mid \mu_i, b$ is negative binomial.
   (c) Suppose $\log \mu_i = \beta_0 + \beta_1 x_i$. Write down the likelihood function $L(\boldsymbol{\beta}, b)$, log-likelihood function $l(\boldsymbol{\beta}, b)$, score function $\boldsymbol{S}(\boldsymbol{\beta}, b)$, and expected information matrix $\boldsymbol{I}(\boldsymbol{\beta}, b)$.

2.5 Consider the exponential regression problem with independent responses
$$p(y_i \mid \lambda_i) = \lambda_i e^{-\lambda_i y_i}, \quad y_i > 0$$
and $\log \lambda_i = -\beta_0 - \beta_1 x_i$ for given covariates x_i, $i = 1, \ldots, n$. We wish to estimate the 2×1 regression parameter $\boldsymbol{\beta} = [\beta_0, \beta_1]^T$ using MLE.

---

[5]This exercise shows that although the confidence interval has the correct frequentist coverage when averaging over all possible realizations of data, for some data we know with probability 1 that the *specific* interval created contains the parameter. The probability distribution of the data in this example is not regular (since the support of the data depends on the unknown parameter), and so we might anticipate difficulties. Conditioning on an ancillary statistic resolves the problems; see Davison (2003, Example 12.3).

Table 2.5 Survival times y_i and concentrations of a contaminant x_i for $i = 1, \ldots, 15$

i	1	2	3	4	5	6	7	8	9	10	11	12	13	14	15
x_i	6.1	4.2	0.5	8.8	1.5	9.2	8.5	8.7	6.7	6.5	6.3	6.7	0.2	8.7	7.5
y_i	0.8	3.5	12.4	1.1	8.9	2.4	0.1	0.4	3.5	8.3	2.6	1.5	16.6	0.1	1.3

(a) Find expressions for the likelihood function $L(\boldsymbol{\beta})$, log-likelihood function $l(\boldsymbol{\beta})$, score function $\boldsymbol{S}(\boldsymbol{\beta})$, and Fisher's information matrix $\boldsymbol{I}(\boldsymbol{\beta})$.

(b) Find expressions for the maximum likelihood estimate $\widehat{\boldsymbol{\beta}}$. If no closed-form solution exists, then instead provide a functional form that could be simply implemented.

(c) For the data in Table 2.5, numerically maximize the likelihood function to obtain estimates of $\boldsymbol{\beta}$. These data consist of the survival times (y) of rats as a function of concentrations of a contaminant (x). Find the asymptotic covariance matrix for your estimate using the information $\boldsymbol{I}(\boldsymbol{\beta})$. Provide a 95% confidence interval for each of β_0 and β_1.

(d) Plot the log-likelihood function $l(\beta_0, \beta_1)$ and compare with the log of the asymptotic normal approximation to the sampling distribution of the MLE.

(e) Find the maximum likelihood estimate $\widehat{\beta_0}$ under the null hypothesis $H_0 : \beta_1 = 0$.

(f) Perform score, likelihood ratio, and Wald tests of the null hypothesis $H_0 : \beta_1 = 0$ with $\alpha = 0.05$. In each case, explicitly state the formula you use to compute the test statistic.

(g) Summarize the results of the estimation and hypothesis testing carried out above. In particular, address the question of whether increasing concentrations of the contaminant are associated with a rat's life expectancy.

2.6 Consider the so-called Neyman–Scott problem (Neyman and Scott 1948) in which $Y_{ij} \mid \mu_i, \sigma^2 \sim_{ind} N(\mu_i, \sigma^2)$, $i = 1, \ldots, n$, $j = 1, 2$. Obtain the MLE of σ^2 and show that it is inconsistent. Why does the inconsistency arise in this example?

2.7 Consider the example discussed at the end of Sect. 2.4.3 in which the true distribution is gamma, but the assumed likelihood is exponential.

(a) Evaluate the form of the sandwich estimator of the variance, and compare with the form of the model-based estimator.

(b) Simulate data from Ga(4,2) and Ga(10,2) distributions, for $n = 10$ and $n = 30$, and obtain the MLEs and sandwich and model-based variance estimates. Compare these variances with the empirical variances observed in the simulations.

(c) Provide figures showing the log of the gamma densities of the previous part, plotted against y, along with the "closest" exponential densities.

2.8 Consider the Poisson-gamma random effects model given by (2.33) and (2.34), which leads to a negative binomial marginal model with the variance a quadratic function of the mean. Design a simulation study, along the lines of that which

2.12 Exercises

produced Table 2.3, to investigate the efficiency and robustness under the Poisson model, quasi-likelihood (with variance proportional to the mean), the negative binomial model, and sandwich estimation. Use a loglinear model

$$\log \mu_i = \beta_0 + \beta_1 x_i,$$

with $x_i \sim_{iid} N(0,1)$, for $i = 1, \ldots, n$, and $\beta_0 = 2$, $\beta_1 = \log 2$. You should repeat the simulation for different values of both n and the negative binomial overdispersion parameter b. Report the 95% confidence interval coverages for β_0 and β_1, for each model.

2.9 A *pivotal* bootstrap interval is evaluated as follows. Let $R_n = \widehat{\theta}_n - \theta$ be a *pivot*, and $H(r) = \Pr_F(R_n \leq r)$ be the distribution function of the pivot. Now define an interval $C_n = [a_n, b_n]$ where

$$a_n = \widehat{\theta}_n - H^{-1}\left(1 - \frac{\alpha}{2}\right)$$

$$b_n = \widehat{\theta}_n - H^{-1}\left(\frac{\alpha}{2}\right).$$

(a) Show that

$$\Pr(a_n \leq \theta_n \leq b_n) = 1 - \alpha$$

so that C_n is an exact $100(1-\alpha)\%$ confidence interval for θ.

(b) Hence, show that the confidence interval is $C_n = [\widehat{a}_n, \widehat{b}_n]$ where

$$\widehat{a}_n = \widehat{\theta}_n - \widehat{H}^{-1}\left(1 - \frac{\alpha}{2}\right) = \widehat{\theta}_n - r^\star_{1-\alpha/2}$$
$$= 2\widehat{\theta}_n - \theta^\star_{1-\alpha/2}$$
$$\widehat{b}_n = \widehat{\theta}_n - \widehat{H}^{-1}\left(\frac{\alpha}{2}\right) = \widehat{\theta}_n - r^\star_{\alpha/2}$$
$$= 2\widehat{\theta}_n - \theta^\star_{\alpha/2}$$

where $r^\star_\gamma$ denotes the γ sample quantile of the B bootstrap samples $[R^\star_{n1}, \ldots, R^\star_{nB}]$ and $\theta^\star_\gamma$ the γ sample quantile of $[\widehat{\theta}^\star_{n1}, \ldots, \widehat{\theta}^\star_{nB}]$. [Hint: To evaluate a_n and b_n, we need to know H, which is unknown, but may be estimated based on the bootstrap estimates

$$\widehat{H}(r) = \frac{1}{B}\sum_{b=1}^{B} I(R^\star_{nb} \leq r)$$

where $R^\star_{nb} = \widehat{\theta}^\star_{nb} - \widehat{\theta}_n$, $b = 1, \ldots, B$.]

Chapter 3
Bayesian Inference

3.1 Introduction

In the Bayesian approach to inference, all *unknown* quantities contained in a probability model for the observed data are treated as random variables. This is in contrast to the frequentist view described in Chap. 2 in which parameters are treated as fixed *constants*. Specifically, with respect to the inferential targets of Sect. 2.1, the fixed but unknown parameters and hypotheses are viewed as random variables under the Bayesian approach. Additionally, the unknowns may include missing data, or the true covariate value in an errors-in-variables setting.

The structure of this chapter is as follows. In Sect. 3.2 we describe the constituents of the posterior distribution and its summarization and in Sect. 3.3 consider the asymptotic properties of Bayesian estimators. Section 3.4 examines prior specification, and in Sect. 3.5 issues relating to model misspecification are discussed. Section 3.6 describes one approach to accounting for model uncertainty via Bayesian model averaging. As we see in Sect. 3.2, to implement the Bayesian approach, integration over the parameter space is required, and historically this has proved a significant hurdle to the routine use of Bayesian methods. Consequently, we discuss implementation issues in some detail. In Sect. 3.7, we provide a description of so-called conjugate situations in which the required integrals are analytically tractable, before providing an overview of analytical and numerical integration techniques, importance sampling, and direct sampling from the posterior. One particular technique, Markov chain Monte Carlo (MCMC), has greatly extended the range of models that may be analyzed with Bayesian methods, and Sect. 3.8 is devoted to a description of MCMC. Section 3.9 considers the important topic of *exchangeability*, and in Sect. 3.10 hypothesis testing via so-called Bayes factors is discussed. Section 3.11 considers a hybrid approach to inference in which the likelihood is taken as the sampling distribution of an estimator and is combined with a prior via Bayes theorem. Concluding remarks appears in Sect. 3.12, including a comparison of frequentist and Bayesian approaches, and the chapter ends with bibliographic notes in Sect. 3.13.

3.2 The Posterior Distribution and Its Summarization

Let $\boldsymbol{\theta} = [\theta_1, \ldots, \theta_p]^\mathrm{T}$ denote all of the unknowns of the model, which we continue to refer to as parameters, and $\boldsymbol{y} = [y_1, \ldots, y_n]^\mathrm{T}$ the vector of observed data. Also let $\mathcal{I}$ represent all relevant information that is currently available to the individual who is carrying out the analysis, in addition to $\boldsymbol{y}$. In the following description, we assume for simplicity that each element of $\boldsymbol{\theta}$ is continuous.

Bayesian inference is based on the *posterior* probability distribution of $\boldsymbol{\theta}$ after observing $\boldsymbol{y}$, which is given by Bayes theorem:

$$p(\boldsymbol{\theta} \mid \boldsymbol{y}, \mathcal{I}) = \frac{p(\boldsymbol{y} \mid \boldsymbol{\theta}, \mathcal{I}) \pi(\boldsymbol{\theta} \mid \mathcal{I})}{p(\boldsymbol{y} \mid \mathcal{I})}. \tag{3.1}$$

There are two key ingredients: the *likelihood* function $p(\boldsymbol{y} \mid \boldsymbol{\theta}, \mathcal{I})$ and the *prior* distribution $\pi(\boldsymbol{\theta} \mid \mathcal{I})$. The latter represents the probability beliefs for $\boldsymbol{\theta}$ held *before* observing the data $\boldsymbol{y}$. Both are dependent upon the current information $\mathcal{I}$. Different individuals will have different information $\mathcal{I}$, and so in general their prior distributions (and possibly their likelihood functions) may differ. The denominator in (3.1), $p(\boldsymbol{y} \mid \mathcal{I})$, is a normalizing constant which ensures that the right-hand side integrates to one over the parameter space. Though of crucial importance, for notational convenience, from this point onwards we suppress the dependence on $\mathcal{I}$, to give

$$p(\boldsymbol{\theta} \mid \boldsymbol{y}) = \frac{p(\boldsymbol{y} \mid \boldsymbol{\theta}) \pi(\boldsymbol{\theta})}{p(\boldsymbol{y})},$$

where the normalizing constant is

$$p(\boldsymbol{y}) = \int_\theta p(\boldsymbol{y} \mid \boldsymbol{\theta}) \pi(\boldsymbol{\theta}) \, d\boldsymbol{\theta}, \tag{3.2}$$

and is the marginal probability of the observed data given the model, that is, the likelihood and the prior. Ignoring this constant gives

$$p(\boldsymbol{\theta} \mid \boldsymbol{y}) \propto p(\boldsymbol{y} \mid \boldsymbol{\theta}) \times \pi(\boldsymbol{\theta})$$

or, more colloquially,

$$\text{Posterior} \propto \text{Likelihood} \times \text{Prior}.$$

The use of the posterior distribution for inference is very intuitively appealing since it probabilistically combines the information on the parameters contained in the data and in the prior.

The manner by which inference is updated from prior to posterior extends naturally to the sequential arrival of data. Suppose first that $\boldsymbol{y}_1$ and $\boldsymbol{y}_2$ represent the current totality of data. Then the posterior is

3.2 The Posterior Distribution and Its Summarization

$$p(\boldsymbol{\theta} \mid \boldsymbol{y}_1, \boldsymbol{y}_2) = \frac{p(\boldsymbol{y}_1, \boldsymbol{y}_2 \mid \boldsymbol{\theta})\pi(\boldsymbol{\theta})}{p(\boldsymbol{y}_1, \boldsymbol{y}_2)}. \tag{3.3}$$

Now consider a previous occasion at which only $\boldsymbol{y}_1$ was available. The posterior based on these data only is

$$p(\boldsymbol{\theta} \mid \boldsymbol{y}_1) = \frac{p(\boldsymbol{y}_1 \mid \boldsymbol{\theta})\pi(\boldsymbol{\theta})}{p(\boldsymbol{y}_1)}.$$

After observing $\boldsymbol{y}_1$ and before observing $\boldsymbol{y}_2$, the "prior" for $\boldsymbol{\theta}$ corresponds to the posterior $p(\boldsymbol{\theta} \mid \boldsymbol{y}_1)$, since this distribution represents the current beliefs concerning $\boldsymbol{\theta}$. We then update via

$$p(\boldsymbol{\theta} \mid \boldsymbol{y}_1, \boldsymbol{y}_2) = \frac{p(\boldsymbol{y}_2 \mid \boldsymbol{y}_1, \boldsymbol{\theta})\pi(\boldsymbol{\theta} \mid \boldsymbol{y}_1)}{p(\boldsymbol{y}_2 \mid \boldsymbol{y}_1)}. \tag{3.4}$$

Factorizing the right-hand side of (3.3) gives

$$p(\boldsymbol{\theta} \mid \boldsymbol{y}_1, \boldsymbol{y}_2) = \frac{p(\boldsymbol{y}_2 \mid \boldsymbol{y}_1, \boldsymbol{\theta})}{p(\boldsymbol{y}_2 \mid \boldsymbol{y}_1)} \times \frac{p(\boldsymbol{y}_1 \mid \boldsymbol{\theta})\pi(\boldsymbol{\theta})}{p(\boldsymbol{y}_1)},$$

which equals the right-hand side of (3.4). Hence, consistent inference based on $\boldsymbol{y}_1$ and $\boldsymbol{y}_2$ is reached regardless of whether we produce the posterior in one or two stages. In the case of conditionally independent observations,

$$p(\boldsymbol{y}_1, \boldsymbol{y}_2 \mid \boldsymbol{\theta}) = p(\boldsymbol{y}_1 \mid \boldsymbol{\theta})p(\boldsymbol{y}_2 \mid \boldsymbol{\theta})$$

in (3.3) and

$$p(\boldsymbol{y}_2 \mid \boldsymbol{y}_1, \boldsymbol{\theta}) = p(\boldsymbol{y}_2 \mid \boldsymbol{\theta})$$

in (3.4).

At first sight, the Bayesian approach to inference is deceptively straightforward, but there are a number of important issues that must be considered in practice. The first, clearly vital, issue is prior specification. Second, once prior and likelihood ingredients have been decided upon, we need to summarize the (usually) multivariate posterior distribution, and as we will see, this summarization requires integration over the parameter space, which may be of high dimension. Finally, a Bayesian analysis must address the effect that possible model misspecification has on inference. Prior specification is taken up in Sect. 3.4 and model misspecification in Sect. 3.5. Next, posterior summarization is described.

Typically the posterior distribution $p(\boldsymbol{\theta} \mid \boldsymbol{y})$ is multivariate, and marginal distributions for parameters of interest will be needed. The univariate marginal distribution for θ_i is

$$p(\theta_i \mid \boldsymbol{y}) = \int_{\boldsymbol{\theta}_{-i}} p(\boldsymbol{\theta} \mid \boldsymbol{y}) \, d\boldsymbol{\theta}_{-i}, \tag{3.5}$$

where $\boldsymbol{\theta}_{-i}$ is the vector $\boldsymbol{\theta}$ excluding θ_i, that is, $\boldsymbol{\theta}_{-i} = [\theta_1, \ldots, \theta_{i-1}, \theta_{i+1}, \ldots, \theta_p]$. While examining the complete distribution will often be informative, reporting summaries of this distribution is also useful. To this end moments and quantiles may be calculated. For example, the posterior mean is

$$\mathrm{E}[\theta_i \mid \boldsymbol{y}] = \int_{\theta_i} \theta_i p(\theta_i \mid \boldsymbol{y})\, d\theta_i. \tag{3.6}$$

The $100 \times q\%$ quantile, $\theta_i(q)$, with $0 < q < 1$ is found by solving

$$q = \Pr[\theta_i \leq \theta_i(q)] = \int_{-\infty}^{\theta_i(q)} p(\theta_i \mid \boldsymbol{y})\, d\theta_i. \tag{3.7}$$

The posterior median $\theta_i(0.5)$ is often an adequate summary of the location of the posterior marginal distribution.

Formally, the choice between posterior means and medians can be made by viewing point estimation as a decision problem. For simplicity suppose that θ is univariate and the action, a, is to choose a point estimate for θ. Let $L(\theta, a)$ denote the loss associated with choosing action a when θ is the true state of nature. The (posterior) expected loss of an action a is

$$\overline{L}(a) = \int_{\theta} L(\theta, a) p(\theta \mid \boldsymbol{y})\, d\theta \tag{3.8}$$

and the optimal choice is the action that minimizes the expected loss. Different loss functions lead to different estimates (Exercise 3.1). For example, minimizing (3.8) with the quadratic loss $L(\theta, a) = (\theta - a)^2$ leads to reporting the posterior mean, $\widehat{a} = \mathrm{E}[\theta \mid \boldsymbol{y}]$. The linear loss,

$$L(\theta, a) = \begin{cases} c_1(a - \theta) & \theta \leq a \\ c_2(\theta - a) & \theta > a \end{cases},$$

corresponds to a loss which is proportional to c_1 if we overestimate and to c_2 if we underestimate. This function leads to $\widehat{a}$ such that

$$\Pr(\theta \leq \widehat{a} \mid \boldsymbol{y}) = \frac{c_2}{c_1 + c_2} = \frac{c_2/c_1}{1 + c_2/c_1},$$

that is, $\widehat{a} = \theta\left(\frac{c_2}{c_1+c_2}\right)$, so that presenting a quantile is the optimal action. Notice that only the ratio of losses is required. When $c_1 = c_2$, under- and overestimation are deemed equally hazardous, and the median of the posterior should be reported.

A $100 \times p\%$ equi-tailed *credible interval* ($0 < p < 1$) is provided by

$$[\,\theta_i(\{1-p\}/2),\ \theta_i(\{1+p\}/2)\,].$$

This interval is the one that is usually reported in the majority of Bayesian analyses carried out, since it is the easiest to calculate. However, in cases where the posterior is skewed, one may wish to instead calculate a *highest posterior density* (HPD) interval in which points inside the interval have higher posterior density than those outside the interval. Such an interval is also the shortest credible interval.

Another useful inferential quantity is the *predictive* distribution for unobserved (e.g., future) observations z. Under conditional independence, so that $p(z \mid \theta, y) = p(z \mid \theta)$, this distribution is

$$p(z \mid y) = \int_\theta p(z \mid \theta) p(\theta \mid y) \, d\theta. \tag{3.9}$$

This derivation clearly assumes that the likelihood for the original data y is also appropriate for the unobserved observations z.

The Bayesian approach therefore provides very natural inferential summaries. However, these summaries require the evaluation of integrals, and for most models, these integrals are analytically intractable. Methods for implementation are considered in Sects. 3.7 and 3.8.

3.3 Asymptotic Properties of Bayesian Estimators

Although Bayesian purists would not be concerned with the frequentist properties of Bayesian procedures, personally I find it reassuring if, for a particular model, a Bayesian estimator can be shown to be, as a minimum, consistent. Efficiency is also an interesting concept to examine.

We informally give a number of results, before referencing more rigorous treatments. We only consider parameter vectors of finite dimension. An important condition that we assume in the following is that the prior distribution is positive in a neighborhood of the true value of the parameter.

The famous Bernstein–von Mises theorem states that, with increasing sample size, the posterior distribution tends to a normal distribution whose mean is the MLE and whose variance–covariance matrix is the inverse of Fisher's information. Let θ be the true value of a p-dimensional parameter, and suppose we are in the situation in which the data are independent and identically distributed. Denote the posterior mean by $\widetilde{\theta}_n = \widetilde{\theta}_n(Y_n) = \mathrm{E}[\theta \mid Y_n]$ and the MLE by $\widehat{\theta}_n$. Then,

$$\sqrt{n}(\widetilde{\theta}_n - \theta) = \sqrt{n}(\widetilde{\theta}_n - \widehat{\theta}_n) + \sqrt{n}(\widehat{\theta}_n - \theta)$$

and we know that $\sqrt{n}(\widehat{\theta}_n - \theta) \to_d \mathrm{N}_p[0, I(\theta)^{-1}]$, where $I(\theta)$ is the information in a sample of size 1 (Sect. 2.4.1). It can be shown that $\sqrt{n}(\widetilde{\theta}_n - \widehat{\theta}_n) \to_p 0$ and so

$$\sqrt{n}(\widetilde{\theta}_n - \theta) \to_d \mathrm{N}_p[0, I(\theta)^{-1}].$$

Hence, $\widetilde{\theta}_n$ is $\sqrt{n}$-consistent and asymptotically efficient. It is important to emphasize that the effect of the prior diminishes as $n \to \infty$. As van der Vaart (1998, p. 140) dryly notes, "Apparently, for an increasing number of observations one's prior beliefs are erased (or corrected) by the observations."

The Bernstein–von Mises theorem is so-called because of the papers by Bernstein (1917) and von Mises (1931), though the theorem has been refined by a number of authors. For references and a recent treatment, see van der Vaart (1998, Sect. 10.2). An early paper on consistency of Bayesian estimators is Doob (1948) and again there have been many refinements; see van der Vaart (1998, Sect. 10.4). An important assumption is that the parameter space is finite. Diaconis and Freedman (1986) describe the problems that can arise in the infinite-dimensional case.

3.4 Prior Choice

The specification of the prior distribution is clearly a necessary and crucial aspect of the Bayesian approach. With respect to prior choice, an important first observation is that for all $\boldsymbol{\theta}$ for which $\pi(\boldsymbol{\theta}) = 0$, we necessarily have $p(\boldsymbol{\theta} \mid \boldsymbol{y}) = 0$, regardless of any realization of the observed data, which clearly illustrates that great care should be taken in excluding parts of the parameter space a priori.

We distinguish between two types of prior specification. In the first, which we label as *baseline prior* specification, we presume an analysis is required in which the prior distribution has "minimal impact," so that the information in the likelihood dominates the posterior. An alternative label for such an analysis is *objective Bayes*. For an interesting discussion of the merits of this approach, see Berger (2006). Other labels that have been put forward for such prior specification include reference, noninformative and nonsubjective. Such priors may be used in situations (for example, in a regulatory setting) in which one must be as "objective" as possible. There is a vast literature on the construction of objective Bayesian procedures, with an aim often being to define procedures which have good frequentist properties.

An analysis with a baseline prior may be the only analysis performed or, alternatively, may provide an analysis with which other analyses in which *substantive priors* are specified may be compared. Such substantive priors constitute the second type of specification in which the incorporation of contextual information is required. Once we have a candidate substantive prior, it is often beneficial to simulate hypothetical data sets from the prior and examine these realizations to see if they conform to what is desirable. A popular label for analyses for which the priors are, at least in part, based on subject matter information is *subjective Bayes*.

3.4.1 Baseline Priors

On first consideration it would seem that the specification of a baseline prior is straightforward since one can take

$$\pi(\boldsymbol{\theta}) \propto 1, \tag{3.10}$$

so that the posterior distribution is simply proportional to the likelihood $p(\boldsymbol{y} \mid \boldsymbol{\theta})$. There are two major difficulties with the use of (3.10), however.

3.4 Prior Choice

The first difficulty is that (3.10) provides an improper specification (i.e. it does not integrate to a positive constant $< \infty$) unless the range of each element of $\boldsymbol{\theta}$ is finite. In some instances this may not be a practical problem if the posterior corresponding to the prior is proper and does not exhibit any aberrant behavior (examples of such behavior are presented shortly). A posterior arising from an improper prior may be justified as a limiting case of proper priors, though some statisticians are philosophically troubled by this argument. Another justification for an improper prior is that such a choice may be thought of as approximating a prior that is "locally uniform" close to regions where the likelihood is non-negligible (so that the likelihood dominates) and decreasing to zero outside of this region. Great care must be taken to ensure that the posterior corresponding to an improper prior choice is proper. For nonlinear models, for example, improper priors should never be used (as an example shortly demonstrates). It is difficult to give general guidelines as to when a proper posterior will result from an improper prior. For example, improper priors for the regression parameters in a generalized linear model (which are considered in detail in Chap. 6) will often, but not always, lead to a proper posterior.

Example: Binomial Model

Suppose $Y \mid p \sim \text{Binomial}(n, p)$, with an improper uniform prior on the logit of p, which we denote $\theta = \log[p/(1-p)]$. Then, $\pi(\theta) \propto 1$ implies a prior on p of

$$\pi(p) \propto [p(1-p)]^{-1},$$

which is, of course, also improper.[1] With this prior an improper posterior results if $y = 0$ (or $y = n$) since the non-integrable spike at $p = 0$ (or $p = 1$) remains in the posterior. Note that this prior results in the MLE being recovered as the posterior mean.

Example: Nonlinear Regression Model

To illustrate the non-propriety in a nonlinear situation, consider the simple model

$$Y_i \mid \theta \sim_{ind} \text{N}\left[\exp(-\theta x_i), \sigma^2\right], \tag{3.11}$$

for $i = 1, \ldots, n$, with $\theta > 0$ and σ^2 assumed known. With an improper uniform prior on θ, $\pi(\theta) = 1$, we label the resulting (unnormalized) "posterior" as

$$q(\theta \mid \boldsymbol{y}) = p(\boldsymbol{y} \mid \theta) \times \pi(\theta) = \exp\left[-\frac{1}{2\sigma^2} \sum_{i=1}^{n} \left(y_i - e^{-\theta x_i}\right)^2\right].$$

---

[1] This prior is sometimes known as *Haldane's prior* (Haldane 1948).

As $\theta \to \infty$,

$$q(\theta \mid \boldsymbol{y}) \to \exp\left[-\frac{1}{2\sigma^2}\sum_{i=1}^{n} y_i^2\right], \quad (3.12)$$

a constant, so that the posterior is improper, because the tail is non-integrable, that is,

$$\int_{\theta_c}^{\infty} q(\theta \mid \boldsymbol{y}) = \infty$$

for all $\theta_c > 0$. Intuitively, the problem is that as $\theta \to \infty$ the corresponding nonlinear curve does not move increasingly away from the data, but rather to the asymptote $E[Y \mid \theta] = 0$. The result is that a finite sum of squares results in (3.12), even in the limit. By contrast, there are no asymptotes in a linear model, and so as the parameters increase or decrease to $\pm\infty$, the fitted line moves increasingly far from the data which results in an infinite sum of squares in the limit, in which case the likelihood, and therefore the posterior, is zero. □

To summarize, it is ill-advised to think of improper priors as a default choice. Rather, improper priors should be used with care, and it is better to assume that they will lead to problems until the contrary can be shown. The safest strategy is clearly to specify proper priors, and this is the approach generally taken in this book.

The second difficulty with (3.10) is that if we reparameterize the model in terms of $\phi = g(\theta)$, where $g(\cdot)$ is a one-one mapping, then the prior for ϕ corresponding to (3.10) is

$$\pi(\phi) = \left|\frac{d\boldsymbol{\theta}}{d\phi}\right|,$$

so that, unless g is a linear transformation, the prior is no longer constant. We have just seen an example of this with the binomial model. As another example, consider a variance σ^2, with prior $\pi(\sigma^2) \propto 1$. This choice implies a prior for the standard deviation, $\pi(\sigma) \propto \sigma$, which is nonconstant. The problem is that we cannot be "flat" on different nonlinear scales. This issue indicates that a desirable property in constructing baseline priors is their invariance to parameterization in order to obtain the same prior regardless of the starting parameterization.

A number of methods have been proposed for the specification of baseline or non-informative priors (we avoid the latter term since it is arguable that priors are ever non-informative). Jeffreys (1961, Sect. 3.10) suggested the use of

$$\pi(\boldsymbol{\theta}) \propto |\boldsymbol{I}(\boldsymbol{\theta})|^{1/2}, \quad (3.13)$$

where $\boldsymbol{I}(\boldsymbol{\theta})$ is Fisher's expected information. This prior has the desirable property of invariance to reparameterization. The invariance holds in general but is obvious in the case of univariate θ. If $\phi = g(\theta)$,

$$I_\phi(\phi) = I_\theta(\theta) \times \left(\frac{d\theta}{d\phi}\right)^2, \quad (3.14)$$

3.4 Prior Choice

where the subscripts now emphasize the parameterization. Consequently, if we start with

$$\pi_\phi(\phi) \propto I_\phi(\phi)^{1/2}$$

this implies

$$\pi_\theta(\theta) \propto I_\phi \left[g^{-1}(\phi)\right]^{1/2} \left|\frac{d\phi}{d\theta}\right| = I_\theta(\theta)^{1/2}$$

from (3.14). Hence, prior (3.13) results if we use the prescription of Jeffreys, but begin with ϕ. In the case of $Y \mid p \sim \text{Binomial}(n,p)$ the information is $I(p) = n/[p(1-p)]$ (Sect. 2.4.1). Therefore, Jeffreys prior is $\pi(p) \propto [p(1-p)]^{-1/2}$. This prior has the advantage of producing a proper posterior when $y = 0$ or $y = n$, a property not shared by Haldane's prior.

Unfortunately, the application of the above procedure to multivariate θ can lead to posterior distributions that have undesirable characteristics. For example, in the Neyman–Scott problem, the use of Jeffreys prior gives, as $n \to \infty$, a limiting posterior mean that is inconsistent, in a frequentist sense (see Exercise 3.3).

A refinement of Jeffreys approach for selecting priors on a more objective basis is provided by *reference priors*. We briefly describe this approach heuristically; more detail can be found in Bernardo (1979) and Berger and Bernardo (1992). For any prior/likelihood distribution, suppose we can calculate the expected information concerning a parameter of interest that will be provided by the data. The more informative the prior, the less information the data will provide. An infinitely large sample would provide all of the missing information about the quantity of interest, and the reference prior is chosen to maximize this missing information.

3.4.2 Substantive Priors

The specification of substantive priors is obviously context specific, but we give a number of general considerations. Specific models will be considered in subsequent chapters. In this section we will discuss some general techniques but will not describe prior elicitation in any great detail; see Kadane and Wolfson (1998), O'Hagan (1998), and Craig et al. (1998) and the ensuing discussion for more on this topic which can be thought of as the measurement of probabilities.

When specifying a substantive prior, it is obvious that we need a clear understanding of the meaning of the parameters of the model for which we are specifying priors, and this can often be achieved by reparameterization.

Example: Linear Regression

Consider the simple linear regression $\text{E}[Y \mid x] = \gamma_0 + \gamma_1 x$. Interpretation is often easier if we reparameterize as

$$\mathrm{E}[Y \mid z] = \beta_0 + \beta_1(z - \bar{z})$$

where $z = c \times x$ and c is chosen so that the units of z are convenient. Under this parameterization, β_0 is the expected response at $z = \bar{z}$. It will often be easier to specify a prior for β_0 than for γ_0, the average response at $x = 0$, which may be meaningless. The slope parameter, β_1, is the change in expected response corresponding to a c-unit increase in x (1-unit increase in z).

Example: Exponential Regression

It may be easier to specify priors on observable quantities, before transforming back to the parameters. For the nonlinear model (3.11), we might specify a prior for the expected response at $x = \widehat{x}$, $\phi = \exp(-\theta \widehat{x})$ to give a prior $\pi_\phi(\phi)$. The prior for θ is

$$\pi_\theta(\theta) = \pi_\phi\left[\exp(-\theta \widehat{x})\right] \times \widehat{x}\exp(-\theta \widehat{x}),$$

the last term corresponding to the Jacobian of the transformation $\phi \to \theta$. As an example, one might assume a $\mathrm{Be}(a,b)$ prior for ϕ, with a and b chosen to give a 90% interval for ϕ. □

While the axioms of probability are uncontroversial, the interpretation of probability has been contested for centuries. In the frequentist approach of Chap. 2, probability was defined in an objective frequentist sense. If the event A is of interest and an experiment is repeated n times resulting in n_A occasions on which A occurs, then

$$P(A) = \lim_{n \to \infty} \frac{n_A}{n}.$$

In contrast, in the subjective Bayesian worldview, probabilities are viewed as subjective and conditional upon an individual's experiences and knowledge, although one may of course base subjective probabilities upon frequencies. Cox and Hinkley (1974, p. 53) state, with reference to the use of Bayes theorem, "If the prior distribution arises from a physical random mechanism with known properties, this argument is entirely uncontroversial," but continue, "A frequency prior is, however, rarely available. To apply the Bayesian approach more generally a wider concept of probability is required ... the prior distribution is taken as measuring the investigator's subjective opinion about the parameter from evidence other than the data under analysis."

As alluded to by this last quote, an obvious procedure is to base the prior distribution upon previously collected data. Ideally, preliminary modeling of such data should be carried out to acknowledge sampling error. If one believed that the data-generation mechanism for both sets of data was comparable, then it would be logical to base the posterior on the combined data (and then once again one has to decide on how to pick a prior distribution). Often such comparability is not reasonable, and a conservative approach is to take the prior as the posterior based

3.4 Prior Choice

on the additional data, but with an inflated variance, to accommodate the additional uncertainty. This approach acknowledges nonsystematic differences, but systematic differences (in particular, biases in one or both studies) may also be present, and this is more difficult to deal with.

Roughly speaking, so long as the prior does not assign zero mass to any region, the likelihood will dominate with increasing sample size (as we saw in Sect. 3.3), so that prior choice becomes decreasingly important. A very difficult problem in prior choice is the specification of the *joint distribution* over multiple parameters. In some contexts one may be able to parameterize the model so that one believes a priori that the components are independent, but in general this will not be possible.

Due to the difficulties of prior specification, a common approach is to carry out a *sensitivity analysis* in which a range of priors are considered and the *robustness* of inference to these choices is examined. An alternative is to *model average* across the different prior models; see Sect. 3.10.

3.4.3 Priors on Meaningful Scales

As we will see in Chaps. 6 and 7, loglinear and linear logistic forms are extremely useful regression models, taking the forms

$$\log \mu = \beta_0 + \beta_1 x_1 + \ldots + \beta_k x_k$$

$$\log \left(\frac{\mu}{1-\mu} \right) = \beta_0 + \beta_1 x_1 + \ldots + \beta_k x_k$$

retrospectively, where $\mu = E[Y]$. Both forms are examples of generalized linear models (GLMs) which are discussed in some detail in Chap. 6.

Often there will be sufficient information in the data for $\beta = [\beta_0, \beta_1, \ldots, \beta_k]^T$ to be analyzed using independent normal priors with large variances (unless, for example, there are many correlated covariates). The use of an improper prior for β will often lead to a proper posterior though care should be taken. Chapter 5 discusses prior choice for the linear model and Chap. 6 for GLMs, and Sect. 6.8 provides an example of an improper posterior that arises in the context of a Poisson model with a linear link.

If we wish to use informative priors for β, we may specify independent normal priors, with the parameters for each component being obtained via specification of two quantiles with associated probabilities. For loglinear and logistic models, these quantiles may be given on the exponentiated scale since these are more interpretable (as the rate ratio and odds ratio, respectively). If θ_1, θ_2 are the quantiles and p_1, p_2 are the associated probabilities, then the parameters of the normal prior are

$$\mu = \frac{z_1 \theta_2 - z_2 \theta_1}{z_1 - z_2} \tag{3.15}$$

$$\sigma = \frac{\theta_1 - \theta_2}{z_1 - z_2} \tag{3.16}$$

Fig. 3.1 The beta prior, Be(2.73, 5.67), which gives Pr($p < 0.1$) = 0.05 and Pr($p < 0.6$) = 0.95

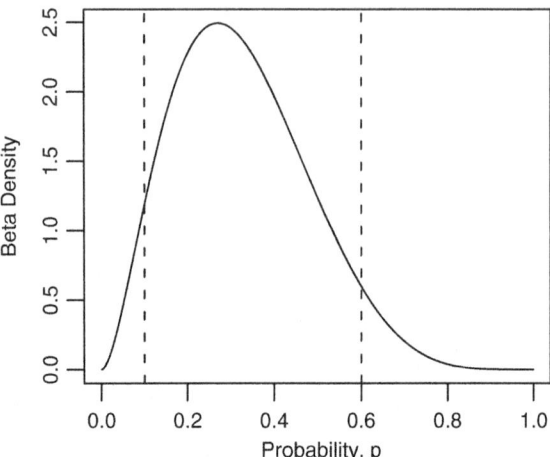

where z_1 and z_2 are the quantiles of a standard normal random variable. For example, in an epidemiological context with a Poisson regression model, we may wish to specify a prior on a relative risk parameter, $\exp(\beta_1)$ which has a median of 1 (corresponding to no association) and a 95% point of 3 (if we think it is unlikely that the relative risk associated with a unit increase in exposure exceeds 3). If we take $\theta_1 = \log(1)$ and $\theta_2 = \log(3)$, along with $p_1 = 0.5$ and $p_2 = 0.95$, then we obtain $\beta_1 \sim N(0, 0.668^2)$. In general, less care is required in prior choice for intercepts in GLMs since they are very accurately estimated with even small amounts of data.

Many candidate prior distributions contain two parameters. For example, a beta prior may be used for a probability and lognormal or gamma distributions may be used for positive parameters such as measures of scale. A convenient way to choose these parameters is to, as above, specify two quantiles with associated probabilities and then solve for the two parameters. For example, suppose we wish to specify a beta prior, Be(a_1, a_2), for a probability p, such that the p_1 and p_2 quantiles are q_1 and q_2. Then we may solve

$$[p_1 - \Pr(p < q_1 \mid a_1, a_2)]^2 + [p_2 - \Pr(p < q_2 \mid a_1, a_2)]^2 = 0$$

for a_1, a_2. For example, taking $p_1 = 0.05, p_2 = 0.95, q_1 = 0.1, q_2 = 0.6$ yields $a_1 = 2.73, a_2 = 5.67$, and Fig. 3.1 shows the resulting density.

3.4.4 Frequentist Considerations

We briefly give a simple example to illustrate the frequentist bias-variance trade-off of prior specification, by examining the mean squared error (MSE) of a Bayesian estimator. Consider data Y_i, $i = 1, \ldots, n$, with Y_i independently and identically

3.4 Prior Choice

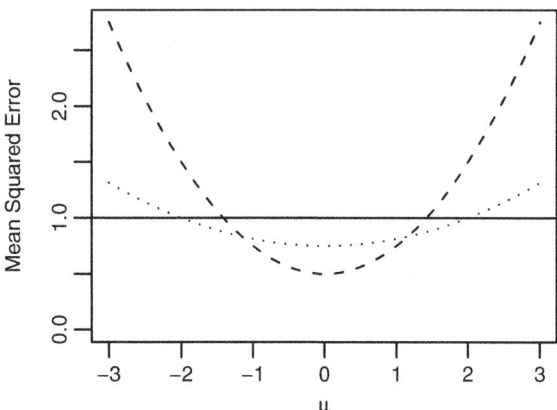

Fig. 3.2 Mean squared error of the posterior mean estimator when $\sqrt{n}(\overline{Y}_n - \mu) \to_d N(0, \sigma^2)$ with σ^2 known and prior $\mu \sim N(m, v)$. The *dashed line* represents the case with $v = 1$ and the *dotted line* when $v = 3$, as a function of the parameter μ. The mean squared error of the sample mean is the *solid horizontal line*

distributed with $E[Y_i \mid \mu] = \mu$ and $\text{var}(Y_i \mid \mu) = \sigma^2$ with σ^2 known. The asymptotic distribution of the sample mean is

$$\sqrt{n}(\overline{Y}_n - \mu) \to_d N(0, \sigma^2).$$

We treat this distribution as the likelihood and examine a Bayesian analysis with prior

$$\mu \sim N(m, v).$$

The posterior is

$$\mu \mid \boldsymbol{Y}_n \to_d N\left(w_n \overline{Y}_n + (1 - w_n)m, w_n \frac{\sigma^2}{n}\right)$$

where

$$w_n = \frac{nv}{nv + \sigma^2}.$$

We first observe that the posterior mean estimator is consistent since $w_n \to 1$ as $n \to \infty$), so long as $v > 0$, but the estimator has finite sample bias if $v^{-1} \neq 0$. The mean squared error of the posterior mean estimator is

$$\text{MSE} = \text{Variance} + \text{Bias}^2$$
$$= w_n \frac{\sigma^2}{n} + [w_n \mu + (1 - w_n)m - \mu]^2$$
$$= w_n \frac{\sigma^2}{n} + (1 - w_n)^2 (m - \mu)^2.$$

Figure 3.2 illustrates the MSE as a function of μ for two different prior distributions that are both centered at zero but have different variances of $v = 1, 3$. For simplicity we have chosen $\sigma^2/n = 1$ with $n = 9$ (so that the MSE of the sample mean is 1, and is indicated as the solid horizontal line). The trade-off when specifying the variance of the prior is clear; if the true μ is close to m, then reductions in MSE are achieved with a small v, though the range of μ over which an improved MSE is achieved is narrower than with the wider prior. At values of μ of $m \pm \sqrt{v + \sigma^2/n}$, the MSE of the sample mean and Bayesian estimator are equal. The variance of the estimator is given by the lowest point of the MSE curves, and the bias dominates for large $|\mu|$.

Example: Lung Cancer and Radon

As an example of prior specification, we return to the simple model considered repeatedly in Chap. 2 with likelihood

$$Y_i \mid \boldsymbol{\beta} \sim_{ind} \text{Poisson}\left[E_i \exp(\beta_0 + \beta_1 x_i) \right],$$

where recall that Y_i are counts of lung cancer incidence in Minnesota in 1998–2002, and x_i is a measure of residential radon in county i, $i = 1, \ldots, n$. The obvious improper prior here is $\pi(\boldsymbol{\beta}) \propto 1$ (and results in a proper posterior for this likelihood).

To specify a substantive prior, we need to have a clear interpretation of the parameters, and β_0 and β_1 are not the most straightforward quantities to contemplate. Hence, we reparameterize the model as

$$Y_i \mid \boldsymbol{\theta} \sim_{ind} \text{Poisson}\left(E_i \theta_0 \theta_1^{x_i - \overline{x}} \right),$$

where $\boldsymbol{\theta} = [\theta_0, \theta_1]^{\text{T}}$ so that

$$\theta_0 = \text{E}[Y/E \mid x = \overline{x}] = \exp(\beta_0 + \beta_1 \overline{x})$$

is the expected standardized mortality ratio in an area with average radon. The standardization that leads to expected numbers E implies we would expect θ_0 to be centered around 1. The parameter $\theta_1 = \exp(\beta_1)$ is the relative risk associated with a one-unit increase in radon. Due to ecological bias, studies often show a negative association between lung cancer incidence and radon (and it is this ecological association we are estimating for this illustration and not the individual-level association). We take lognormal priors for θ_0 and θ_1 and use (3.15) and (3.16) to deduce the lognormal parameters. For θ_0 we take a lognormal prior with 2.5% and 97.5% quantiles of 0.67 and 1.5 to give $\mu = 0, \sigma = 0.21$. For θ_1 we assume the relative risk associated with a one-unit increase in radon is between 0.8 and 1.2 with probability 0.95, to give $\mu = -0.02, \sigma = 0.10$. We return to this example later in the chapter.

3.5 Model Misspecification

The behavior of Bayesian estimators under misspecification of the likelihood has received less attention than frequentist estimators. Recall the result concerning the behavior of the MLE $\widehat{\boldsymbol{\theta}}_n$ under model misspecification summarized in (2.27), which we reproduce here:

$$\sqrt{n}\,(\widehat{\boldsymbol{\theta}}_n - \boldsymbol{\theta}_{\scriptscriptstyle\mathrm{T}}) \to_d \mathrm{N}_p\left[\mathbf{0}, \boldsymbol{J}^{-1}\boldsymbol{K}(\boldsymbol{J}^{\scriptscriptstyle\mathrm{T}})^{-1}\right]$$

where

$$\boldsymbol{J} = \mathrm{E}_{\scriptscriptstyle\mathrm{T}}\left[\frac{\partial^2}{\partial\boldsymbol{\theta}\partial\boldsymbol{\theta}^{\scriptscriptstyle\mathrm{T}}}\log p(Y \mid \boldsymbol{\theta}_{\scriptscriptstyle\mathrm{T}})\right]$$

$$\boldsymbol{K} = \mathrm{E}_{\scriptscriptstyle\mathrm{T}}\left[\left(\frac{\partial}{\partial\boldsymbol{\theta}}\log p(Y \mid \boldsymbol{\theta}_{\scriptscriptstyle\mathrm{T}})\right)\left(\frac{\partial}{\partial\boldsymbol{\theta}}\log p(Y \mid \boldsymbol{\theta}_{\scriptscriptstyle\mathrm{T}})\right)^{\scriptscriptstyle\mathrm{T}}\right]$$

with $\boldsymbol{\theta}_{\scriptscriptstyle\mathrm{T}}$ the true $\boldsymbol{\theta}$ and $p(Y \mid \boldsymbol{\theta})$ the assumed model. Let $\widetilde{\boldsymbol{\theta}}_n = \mathrm{E}[\boldsymbol{\theta} \mid \boldsymbol{Y}_n]$ be the posterior mean which we here view as a function of $\boldsymbol{Y}_n = [Y_1, \ldots, Y_n]^{\scriptscriptstyle\mathrm{T}}$. From Sect. 3.3, $\sqrt{n}(\widetilde{\boldsymbol{\theta}}_n - \widehat{\boldsymbol{\theta}}_n) \to_p \mathbf{0}$, and hence

$$\sqrt{n}(\widetilde{\boldsymbol{\theta}}_n - \boldsymbol{\theta}_{\scriptscriptstyle\mathrm{T}}) \to_d \mathrm{N}_p\left[\mathbf{0}, \boldsymbol{J}^{-1}\boldsymbol{K}(\boldsymbol{J}^{\scriptscriptstyle\mathrm{T}})^{-1}\right].$$

This has important implications since it shows that, asymptotically, the effect of model misspecification on the posterior mean is the same as its effect on the MLE. If the likelihood is of linear exponential family form, correct specification of the mean function leads to consistent estimation of the parameters in the mean model (see Sect. 6.5.1 for details). As with the reported variance of the MLE, the spread of the posterior distribution could be completely inappropriate, however. While sandwich estimation can be used to "correct" the variance estimator for the MLE, there is no such simple solution for the posterior mean, or other Bayesian summaries.

With respect to model misspecification, the emphasis in the Bayesian literature has been on sensitivity analyses, or on embedding a particular likelihood or prior choice within a larger class. Embedding an initial model within a continuous class is a conceptually simple approach. For example, a Poisson model may be easily extended to a negative binomial model.

A difficulty with considering model classes with large numbers of unknown parameters is that uncertainty on parameters of interest will be increased if a simple model is closer to the truth. In particular, model expansion may lead to a decrease in precision, as we now illustrate. As we have seen, as n increases, the prior effect is negligible and the posterior variance is given by the inverse of Fisher's information, (Sect. 3.3). Suppose that we have k parameters in an original model, and we are considering an expanded model with p parameters, and let

$$\begin{bmatrix} \boldsymbol{I}_{11} & \boldsymbol{I}_{12} \\ \boldsymbol{I}_{21} & \boldsymbol{I}_{22} \end{bmatrix},$$

where I_{11} is a $k \times k$ matrix corresponding to the information on the parameters of the simpler model (which includes the parameters of interest), and I_{22} is the $(p-k) \times (p-k)$ information matrix concerning the additional parameters in the enlarged model. In the simpler model, the information on the parameters of interest is I_{11}, while for the enlarged model, it is

$$I_{11} - I_{12}I_{22}^{-1}I_{21},$$

which is never greater than I_{11}. This is an oversimplified discussion (as we shall see in Sect. 5.9), but it highlights that there can be a penalty to pay for specifying an overly complex model.

3.6 Bayesian Model Averaging

If a discrete number of models are considered, then model averaging provides an alternative means of assessing model uncertainty. The Bayesian machinery handles multiple models in a very straightforward fashion since essentially the unknown model is treated as an additional discrete parameter. Let $M_1, \ldots, M_J$ denote the J models under consideration and $\boldsymbol{\theta}_j$ the parameters of the jth model. Suppose, for illustration, there is a parameter of interest ϕ (which we assume is univariate) that is well defined for each of the J models under consideration. The posterior for ϕ is a mixture over the J individual model posteriors:

$$p(\phi \mid \boldsymbol{y}) = \sum_{j=1}^{J} p(\phi \mid M_j, \boldsymbol{y}) \Pr(M_j \mid \boldsymbol{y})$$

where

$$p(\phi \mid M_j, \boldsymbol{y}) = \int p(\phi \mid \boldsymbol{\theta}_j, M_j, \boldsymbol{y}) p(\boldsymbol{\theta}_j \mid M_j, \boldsymbol{y}) \, d\boldsymbol{\theta}_j$$

$$= \frac{1}{p(\boldsymbol{y} \mid M_j)} \int p(\phi \mid \boldsymbol{\theta}_j, M_j, \boldsymbol{y}) p(\boldsymbol{y} \mid \boldsymbol{\theta}_j, M_j) p(\boldsymbol{\theta}_j \mid M_j) \, d\boldsymbol{\theta}_j,$$

$$\Pr(M_j \mid \boldsymbol{y}) = \frac{p(\boldsymbol{y} \mid M_j) \Pr(M_j)}{p(\boldsymbol{y})}$$

$$= \frac{\int p(\boldsymbol{y} \mid \boldsymbol{\theta}_j, M_j) p(\boldsymbol{\theta}_j \mid M_j) \, d\boldsymbol{\theta}_j \Pr(M_j)}{p(\boldsymbol{y})}$$

and with $\Pr(M_j)$ the prior belief in model j and $p(\boldsymbol{\theta}_j \mid M_j)$ the prior on the parameters of model M_j. The marginal probabilities of the data under the different models are calculated as

$$p(\boldsymbol{y} \mid M_j) = \int p(\boldsymbol{y} \mid \boldsymbol{\theta}_j, M_j) p(\boldsymbol{\theta}_j \mid M_j) \, d\boldsymbol{\theta}_j,$$

3.6 Bayesian Model Averaging

with

$$p(\boldsymbol{y}) = \sum_{j=1}^{J} p(\boldsymbol{y} \mid M_j) \Pr(M_j).$$

To summarize the posterior for ϕ, we might report the posterior mean

$$\mathrm{E}[\phi \mid \boldsymbol{y}] = \sum_{j=1}^{J} \mathrm{E}[\phi \mid \boldsymbol{y}, M_j] \times \Pr(M_j \mid \boldsymbol{y}),$$

which is simply the average of the posterior means across models, weighted by the posterior weight received by each model. The posterior variance is

$$\mathrm{var}(\phi \mid \boldsymbol{y}) = \sum_{j=1}^{J} \mathrm{var}(\phi \mid \boldsymbol{y}, M_j) \times \Pr(M_j \mid \boldsymbol{y})$$

$$+ \sum_{j=1}^{J} \{\mathrm{E}[\phi \mid \boldsymbol{y}, M_j] - \mathrm{E}[\phi \mid \boldsymbol{y}]\}^2 \times \Pr(M_j \mid \boldsymbol{y})$$

which averages the posterior variances concerning ϕ in each model, with the addition of a term that accounts for between-model uncertainty in the mean.

Although model averaging is very appealing in principle, in practice there are many difficult choices, including the choice of the class of models to consider and the priors over both the models and the parameters of the models. Summarization can also be difficult because the parameter of interest may have different interpretations in different models. For example, in a regression setting, suppose we fit the single model

$$\mathrm{E}[Y \mid x_1, x_2] = \beta_0 + \beta_1 x_1 + \beta_2 x_2$$

with β_1 the parameter of interest. The interpretation of β_1 is as the average change in response corresponding to a unit increase in x_1, with x_2 held constant. If we average over this model and the model with x_1 only, then the usual "x_2 held constant" qualifier is not accurate, so a phrase such as "allowing for the possibility of x_2 in the model" may be instead used. Performing model averaging over models which represent different scientific theories is also not appealing if the search for a causal explanation is sought. If prediction is the aim, then model averaging is much more appealing since parameter interpretation is often irrelevant (see Chap. 12). Another disadvantage of model averaging is that it may encourage the user to believe they have accounted for "all" uncertainty in which covariates to include in the model which is a dangerous conclusion to draw.

3.7 Implementation

In this section we provide an overview of methods for evaluating the integrals required for performing Bayesian inference. We begin, in Sect. 3.7.1, by describing so-called conjugate situations in which the prior and likelihood combination is constructed in order for the posterior to be of the same form as the prior. Unfortunately, in a regression setting, conjugate analyses are rarely available beyond the linear model. In Sect. 3.7.2 the analytical Laplace approximation is described. Quadrature methods are considered in Sect. 3.7.3 before we turn to a method that combines Laplace and numerical integration in a very clever way, in Sect. 3.7.4, to give a method known as the *integrated nested Laplace approximation* (INLA). More recently developed sampling-based (Monte Carlo) approaches have transformed the practical application of Bayesian methods, and we therefore describe these approaches in some detail. In Sect. 3.7.5, importance sampling Monte Carlo is considered, and in Sects. 3.7.6 and 3.7.7, direct sampling from the posterior is described. MCMC algorithms are particularly important, and to these we devote Sect. 3.8.

Beyond the crucial importance of integration in Bayesian inference, this material is also relevant in a frequentist context. Specifically, in Part III of this book, we will consider nonlinear and generalized linear mixed effects models for which integration over the random effects is required in order to obtain the likelihood for the fixed effects.

3.7.1 Conjugacy

So-called *conjugate prior* distributions allow analytical evaluation of many of the integrals required for Bayesian inference, at least for certain convenient parameters. A conjugate prior is such that $p(\boldsymbol{\theta} \mid \boldsymbol{y})$ and $p(\boldsymbol{\theta})$ belong to the same family. We assume $\dim(\boldsymbol{\theta}) = p$. This definition is not adequate since it will always be true given a suitable definition of the family of distributions. To obtain a more useful class, we first note that if $\boldsymbol{T}(\boldsymbol{Y})$ denotes a *sufficient statistic* for a particular likelihood $p(\cdot \mid \boldsymbol{\theta})$, then

$$p(\boldsymbol{\theta} \mid \boldsymbol{y}) = p(\boldsymbol{\theta} \mid \boldsymbol{t}) \propto p(\boldsymbol{t} \mid \boldsymbol{\theta})p(\boldsymbol{\theta}).$$

This allows a definition of a conjugate family in terms of likelihoods that admit a sufficient statistic of fixed dimension.

The p-parameter exponential family of distributions has the form:

$$p(y_i \mid \boldsymbol{\theta}) = f(y_i)g(\boldsymbol{\theta}) \exp\left[\boldsymbol{\lambda}(\boldsymbol{\theta})^{\mathrm{T}}\boldsymbol{u}(y_i)\right],$$

where, in general, $\boldsymbol{\lambda}(\boldsymbol{\theta})$ and $\boldsymbol{u}(y_i)$ have the same dimension as $\boldsymbol{\theta}$ and $\boldsymbol{\lambda}(\boldsymbol{\theta})$ is called the *natural parameter* (and in a linear exponential family, we have $\boldsymbol{u}(y_i) = y_i$). For n independent and identically distributed observations from $p(\cdot \mid \boldsymbol{\theta})$,

3.7 Implementation

Table 3.1 Conjugate priors and associated posterior distributions, for various likelihood choices

Prior	Likelihood	Posterior
$\theta \sim \text{N}(m, v)$	$\overline{Y} \mid \theta \sim \text{N}(\theta, \sigma^2/n)$ σ^2 known	$\theta \mid y \sim \text{N}[w\overline{y} + (1-w)m, w\sigma^2/n]$ with $w = v/(v + \sigma^2/n)$
$\theta \sim \text{Be}(a, b)$	$Y \mid \theta \sim \text{Bin}(n, \theta)$	$\theta \mid y \sim \text{Be}(a + y, b + n - y)$
$\theta \sim \text{Ga}(a, b)$	$Y \mid \theta \sim \text{Poisson}(\theta)$	$\theta \mid y \sim \text{Ga}(a + y, b + 1)$
$\theta \sim \text{Ga}(a, b)$	$Y \mid \theta \sim \text{Exp}(\theta)$	$\theta \mid y \sim \text{Ga}(a + y, b + 1)$

$$p(\boldsymbol{y} \mid \boldsymbol{\theta}) = \left[\prod_{i=1}^{n} f(y_i)\right] g(\boldsymbol{\theta})^n \exp\left[\boldsymbol{\lambda}(\boldsymbol{\theta})^{\text{T}} \boldsymbol{t}(\boldsymbol{y})\right],$$

where

$$\boldsymbol{t}(\boldsymbol{y}) = \sum_{i=1}^{n} \boldsymbol{u}(y_i).$$

The conjugate prior density is defined as

$$p(\boldsymbol{\theta}) = c(\eta, \boldsymbol{v}) \times g(\boldsymbol{\theta})^\eta \exp\left[\boldsymbol{\lambda}(\boldsymbol{\theta})^{\text{T}} \boldsymbol{v}\right],$$

where η and $\boldsymbol{v}$ are specified, a priori. The resulting posterior distribution is

$$p(\boldsymbol{\theta} \mid \boldsymbol{y}) = c(\eta + n, \boldsymbol{v} + \boldsymbol{t}) \times g(\boldsymbol{\theta})^{\eta+n} \exp\left\{\boldsymbol{\lambda}(\boldsymbol{\theta})^{\text{T}}[\boldsymbol{v} + \boldsymbol{t}(\boldsymbol{y})]\right\},$$

demonstrating conjugacy. Comparison with $p(y_i \mid \boldsymbol{\theta})$ indicates that η may be viewed as a *prior sample size* giving rise to a *sufficient statistic* $\boldsymbol{v}$.

The above derivations are often not required if one wishes to simply obtain the conjugate distribution for a given likelihood, since it can be determined quickly via inspection of the kernel of the likelihood. The predictive distribution is often more complex to derive, however, but is straightforward under the above formulation. In the case of a conjugate prior, for new observations $\boldsymbol{Z} = [Z_1, ..., Z_m]$ arising as an independent and identically distributed sample from $p(Z \mid \boldsymbol{\theta})$, the *predictive distribution* is

$$p(\boldsymbol{z} \mid \boldsymbol{y}) = \left[\prod_{i=1}^{m} f(z_i)\right] \frac{c[\eta + n, \boldsymbol{v} + \boldsymbol{t}(\boldsymbol{y})]}{c[\eta + n + m, \boldsymbol{v} + \boldsymbol{t}(\boldsymbol{y}, \boldsymbol{z})]}.$$

Table 3.1 gives the conjugate choices for a variety of likelihoods.

Beyond the normal linear model, the direct practical use of conjugacy in a regression setting is limited, but as we will see subsequently, the material of this section is very useful when implementing direct sampling or MCMC approaches.

Example: Binomial Likelihood

Suppose we have a single observation from a binomial distribution, $Y \mid \theta \sim$ Binomial(n, θ):

$$p(y \mid \theta) = \binom{n}{y} \theta^y (1-\theta)^{n-y}.$$

By direct inspection we recognize that the conjugate prior is a beta distribution, but for illustration we follow the more long-winded route. In exponential family form,

$$p(y \mid \theta) = \binom{n}{y} (1-\theta)^n \exp\left[y \log\left(\frac{\theta}{1-\theta}\right)\right],$$

or, in terms of the natural parameter $\lambda = \lambda(\theta) = \log[\theta/(1-\theta)]$,

$$p(y \mid \lambda) = \binom{n}{y} [1 + \exp(\lambda)]^{-n} \exp(y\lambda).$$

The conjugate prior for λ is therefore identified as

$$\pi(\lambda) = c(\eta, v)[1 + \exp(\lambda)]^{-\eta} \exp[v\lambda] \tag{3.17}$$

so that the prior for θ is

$$\pi(\theta) = c(\eta, v)(1-\theta)^\eta \exp\left[v \log \frac{\theta}{1-\theta}\right] \frac{1}{\theta(1-\theta)}$$

$$= \frac{\Gamma(\eta+2)}{\Gamma(v+1)\Gamma(\eta-v+1)} \theta^{v-1}(1-\theta)^{\eta-v-1},$$

the Be(a, b) distribution with parameters $a = v, b = \eta - v$. An interpretation of these parameters is that a prior sample size $\eta = a + b$ yields the prior sufficient statistic $v = a$. It follows immediately that the posterior is Be$(a + y, b + n - y)$.

We write

$$E[\theta \mid y] = \frac{a+y}{a+b+n}$$

$$= \frac{y}{n} w + \frac{a}{a+b}(1-w)$$

where $w = n/(a+b+n)$, so that the posterior mean is a weighted combination of the MLE, $\widehat{\theta} = y/n$, and the prior mean. Similarly,

$$\text{mode}[\theta \mid y] = \frac{a+y-1}{a+b+n-2}$$

$$= \frac{y}{n} w^\star + \frac{a-1}{a+b-2}(1-w^\star),$$

3.7 Implementation

where $w^* = n/(a+b+n-2)$, so that the posterior mode is a weighted combination of the prior mode (if it exists) and the MLE. The choice of a uniform distribution, $a = b = 1$, results in the posterior mode equaling the MLE, as expected in this one-dimensional example.

The marginal distribution of the data, given likelihood and prior, is the beta-binomial distribution

$$\Pr(y) = \binom{n}{y} \frac{\Gamma(a+b)}{\Gamma(a)\Gamma(b)} \times \frac{\Gamma(a+y)\Gamma(b+n-y)}{\Gamma(a+b+n)},$$

for $y = 0, \ldots, n$. If $a = b = 1$, the prior predictive is uniform over the space of outcomes: $p(y) = (n+1)^{-1}$ for $y = 0, 1, \ldots, n$, in line with intuition.

The mean of the prior predictive is

$$\mathrm{E}[Y] = \mathrm{E}_\theta[\mathrm{E}(Y \mid \theta)] = n \times \frac{a}{a+b},$$

with variance

$$\mathrm{var}(Y) = \mathrm{var}_\theta[\mathrm{E}(Y \mid \theta)] + \mathrm{E}_\theta[\mathrm{var}(Y \mid \theta)] = n\mathrm{E}(\theta)[1 - \mathrm{E}(\theta)] \times \frac{a+b+n}{a+b+1},$$

illustrating the overdispersion relative to $\mathrm{var}(Y \mid \theta) = n\theta(1-\theta)$, if $n > 1$. If $n = 1$, there is no overdispersion since we have a single Bernoulli random variable for which the variance is always determined by the mean.

The predictive distribution for a new trial, in which $Z = 0, 1, \ldots, m$ denotes the number of successes and m the number of trials, is

$$p(z \mid y) = \binom{m}{z} \frac{\Gamma(a+b+n)}{\Gamma(a+y)\Gamma(b+n-y)} \times \frac{\Gamma(a+b+z)\Gamma(b+n-y+m-z)}{\Gamma(a+b+n+m)},$$

which is another version of the beta-binomial distribution and is an overdispersed binomial for which

$$\mathrm{E}[Z \mid y] = m \times \mathrm{E}[\theta \mid y] = m \times \frac{a+y}{a+b+n},$$

and

$$\mathrm{var}(Z \mid y) = m \times \mathrm{E}(\theta \mid y) \times [1 - \mathrm{E}(\theta \mid y)] \times \frac{a+b+n+m}{a+b+n+1}.$$

As $n \to \infty$, with y/n fixed, the predictive $p(z \mid y)$ approaches the binomial distribution $\mathrm{Bin}(m, y/n)$. This makes sense since, under correct model specification, for large n we effectively know θ, and so binomial variability is the only uncertainty that remains.

3.7.2 Laplace Approximation

In this section let

$$I = \int_{-\infty}^{\infty} \exp[\,nh(\theta)\,]\,d\theta, \tag{3.18}$$

denote a generic integral of interest, and we suppose initially that θ is a scalar. Depending on the form of $h(\cdot)$, (3.18) can correspond to the evaluation of a variety of quantities of interest including $p(\boldsymbol{y})$ and posterior moments. The n appearing in (3.18) is included solely to make the asymptotic arguments more transparent.

Let $\widetilde{\theta}$ denote the mode of $h(\cdot)$. We carry out a Taylor series expansion about $\widetilde{\theta}$, assuming that $h(\cdot)$ is sufficiently well behaved for this operation; in particular we assume that at least two derivatives exist. The expansion is

$$nh(\theta) = n \sum_{k=0}^{\infty} \frac{(\theta - \widetilde{\theta})^k}{k!} h^{(k)}(\widetilde{\theta}),$$

where $h^{(k)}(\widetilde{\theta})$ represents the kth derivative of $h(\cdot)$ evaluated at $\widetilde{\theta}$. Hence,

$$\begin{aligned} I &= \int_{-\infty}^{\infty} \exp\left[n \sum_{k=0}^{\infty} \frac{(\theta - \widetilde{\theta})^k}{k!} h^{(k)}(\widetilde{\theta}) \right] d\theta \\ &\approx \exp\left[nh(\widetilde{\theta}) \right] \int_{-\infty}^{\infty} \exp\left[\frac{nh^{(2)}}{2}(\widetilde{\theta})(\theta - \widetilde{\theta})^2 \right] d\theta, \end{aligned}$$

where we have ignored quadratic terms and above in the Taylor series and exploited $h^{(1)}(\widetilde{\theta}) = 0$. Writing $\widetilde{v} = -1/h^{(2)}(\widetilde{\theta})$ gives the estimate

$$\widehat{I} = \exp\left[nh(\widetilde{\theta}) \right] \left(\frac{2\pi\widetilde{v}}{n} \right)^{1/2}, \tag{3.19}$$

which is known as the *Laplace approximation*. The error is such that

$$\frac{I}{\widehat{I}} = 1 + O(n^{-1}).$$

Suppose we wish to evaluate the posterior expectation of a positive function of interest $\phi(\theta)$, that is,

$$\begin{aligned} \mathrm{E}[\phi(\theta) \mid \boldsymbol{y}] &= \frac{\int \exp[\log \phi(\theta) + \log p(\boldsymbol{y} \mid \theta) + \log \pi(\theta) + \log(d\theta/d\phi)]\,d\theta}{\int \exp[\log p(\boldsymbol{y} \mid \theta) + \log \pi(\theta)]\,d\theta} \\ &= \frac{\int \exp[nh_1(\theta)]\,d\theta}{\int \exp[nh_2(\theta)]\,d\theta}. \end{aligned}$$

where the Jacobian has been included in the numerator of the first line. Application of (3.19) to numerator and denominator gives

$$\widehat{\mathrm{E}}[\phi(\theta) \mid y] = \frac{\widetilde{v}_1}{\widetilde{v}_0} \frac{\exp[nh_1(\widetilde{\theta}_1)]}{\exp[nh_0(\widetilde{\theta}_0)]}$$

where $\widetilde{\theta}_j$ is the mode of $h_j(\cdot)$ and $\widetilde{v}_j = -1/h_j^{(2)}(\widetilde{\theta}_j)$, $j = 0, 1$. Further,

$$\widehat{\mathrm{E}}[\phi(\theta) \mid y] = \mathrm{E}[\phi(\theta) \mid y][1 + O(n^{-2})],$$

since errors in the numerator and denominator cancel (Tierney and Kadane 1986). If ϕ is not positive then a simple solution is to add a large constant to ϕ, apply Laplace's method, and subtract the constant.

Now consider multivariate $\boldsymbol{\theta}$ with $\dim(\boldsymbol{\theta}) = p$ and with required integral

$$I = \int_{-\infty}^{\infty} \cdots \int_{-\infty}^{\infty} \exp[nh(\boldsymbol{\theta})] \, d\theta_1 \cdots d\theta_p.$$

The above argument may be generalized to give the Laplace approximation

$$\widehat{I} = \exp\left[nh(\widetilde{\boldsymbol{\theta}})\right] \left(\frac{2\pi}{n}\right)^{p/2} |\widetilde{\boldsymbol{v}}|^{1/2}, \qquad (3.20)$$

where $\widetilde{\boldsymbol{\theta}}$ is the maximum of $h(\cdot)$ and $\widetilde{\boldsymbol{v}}$ is the $p \times p$ matrix whose (i,j)th element is

$$-\left.\frac{\partial^2 h}{\partial \theta_i \partial \theta_j}\right|_{\widetilde{\boldsymbol{\theta}}}.$$

An important drawback of analytic approximations is the difficulty in performing error assessment, so that in practice one does not know the accuracy of approximation. The evaluation of derivatives can also be analytically and numerically troublesome. These shortcomings apart, however, we will see that these approximations are useful as components of other approaches, such as the scheme described in Sect. 3.7.4, and for suggesting proposals for importance sampling and MCMC algorithms.

3.7.3 Quadrature

We consider numerical integration rules for approximating integrals of the form

$$I = \int f(t) \, dt,$$

via the weighted sum

$$\hat{I} = \sum_{i=1}^{m} w_i f(t_i),$$

where the points t_i and weights w_i define the integration rule. So-called *Gauss* rules are optimal rules (in a sense we will define shortly) that are constructed to integrate weighted functions of polynomials accurately. Specifically, if $p(t)$ is a polynomial of degree $2m - 1$, then the Gauss rule (t_i, w_i) is such that

$$\sum_{i=1}^{m} w_i p(t_i) = \int w(t) p(t)\, dt.$$

It can be shown that no rule has this property for polynomials of degree $2m$, showing the optimality of Gauss rules. Different classes of rule emerge for different choices of weight function. We describe Gauss–Hermite rules that correspond to the weight function

$$w(t) = \exp(-t^2) \tag{3.21}$$

which is of obvious interest in a statistics context. If the integral is of the form

$$I = \int g(t) \exp(-t^2)\, dt$$

and $f(t)$ can be well approximated by a polynomial of degree $2m - 1$, we would expect an m-point Gauss–Hermite rule to be accurate.

The points of the Gauss–Hermite rule are the zeroes of the Hermite polynomials $H_m(t)$ with weights

$$w_i = \frac{2^{m-1} m! \sqrt{\pi}}{m^2 [H_{m-1}(t_i)]^2}.$$

In general, the points of the rule need to be located and scaled appropriately. Suppose that μ and σ are the approximate mean and standard deviation of θ, and let $t = (\theta - \mu)/\sqrt{2}\sigma$. The integral of interest is

$$I = \int f(\theta)\, d\theta = \int g(\mu + \sqrt{2}\sigma t) \sqrt{2}\sigma e^{-t^2}\, dt$$

and applying the transformation yields

$$\hat{I} = \sum_{i=1}^{m} w_i^\star g(t_i^\star),$$

where $w_i^\star = w_i \sqrt{2}\sigma$ and $t_i^\star = \mu + \sqrt{2}\sigma t_i$.

In practice μ and σ are unknown but may be estimated at the same time as I is evaluated to give an *adaptive Gauss–Hermite* rule (Naylor and Smith 1982).

Suppose $\boldsymbol{\theta}$ is two-dimensional, and we wish to evaluate

$$I = \int f(\boldsymbol{\theta})\,d\boldsymbol{\theta} = \int\int f(\theta_1,\theta_2)\,d\theta_2\,d\theta_1 = \int f^\star(\theta_1)\,d\theta_1$$

where

$$f^\star(\theta_1) = \int f(\theta_1,\theta_2)\,d\theta_2.$$

We form

$$\widehat{I} = \sum_{i=1}^{m_1} w_i \widehat{f^\star}(\theta_{1i}),$$

with

$$\widehat{f^\star}(\theta_{1i}) = \sum_{j=1}^{m_2} u_j f(\theta_{1i},\theta_{2j})$$

to give

$$\widehat{I} = \sum_{i=1}^{m_1}\sum_{j=1}^{m_2} w_i u_j f(\theta_{1i},\theta_{2j}),$$

which is known as a *Cartesian Product* rule. Such rules can provide very accurate integration with relatively few points, but the number of points required is prohibitive in high dimensions since for p parameters and m points, a total of m^p points are required. Consequently, these rules tend to be employed when $p \leq 10$.

In common with the Laplace method, quadrature methods do not provide an estimate of the error of the approximation. In practice, consistency of the estimates across increasing grid sizes may be examined.

3.7.4 Integrated Nested Laplace Approximations

We briefly review the INLA computational approach which combines Laplace approximations and numerical integration in a very efficient manner; see Rue et al. (2009) for a more extensive treatment. Consider a model with parameters $\boldsymbol{\theta}_1$ that are assigned normal priors, with the remaining parameters being denoted $\boldsymbol{\theta}_2$ with $G = \dim(\boldsymbol{\theta}_1)$ and $V = \dim(\boldsymbol{\theta}_2)$. Assume for ease of explanation that the normal prior is centered at zero with variance–covariance matrix $\boldsymbol{\Sigma}$, $\mathrm{N}_G(\boldsymbol{0},\boldsymbol{\Sigma})$, where $\boldsymbol{\Sigma}$ depends on elements in $\boldsymbol{\theta}_2$. Many models fall into this class including generalized linear models (Chap. 6) and generalized linear mixed models (Chap. 9). The posterior is

$$\pi(\boldsymbol{\theta}_1, \boldsymbol{\theta}_2 \mid \boldsymbol{y}) \propto \pi(\boldsymbol{\theta}_1 \mid \boldsymbol{\theta}_2)\pi(\boldsymbol{\theta}_2) \prod_{i=1}^{n} p(\boldsymbol{y}_i \mid \boldsymbol{\theta}_1, \boldsymbol{\theta}_2)$$

$$\propto \pi(\boldsymbol{\theta}_2) \mid \boldsymbol{\Sigma}(\boldsymbol{\theta}_2) \mid^{-1/2} \exp\left[-\frac{1}{2}\boldsymbol{\theta}_1^{\mathrm{T}}\boldsymbol{\Sigma}(\boldsymbol{\theta}_2)^{-1}\boldsymbol{\theta}_1 + \sum_{i=1}^{n} \log p(\boldsymbol{y}_i \mid \boldsymbol{\theta}_1, \boldsymbol{\theta}_2)\right].$$

(3.22)

Of particular interest are the posterior univariate marginal distributions $\pi(\theta_{1g} \mid \boldsymbol{y})$, $g = 1, \ldots, G$, and $\pi(\theta_{2v} \mid \boldsymbol{y})$, $v = 1, \ldots, V$. The "normal" parameters $\boldsymbol{\theta}_1$ are dealt with by analytical approximations (as applied to the term in the exponent of (3.22), conditional on specific values of $\boldsymbol{\theta}_2$). Numerical integration techniques are applied to $\boldsymbol{\theta}_2$, so that V should not be too large for accurate inference (Sect. 3.7.3). For elements of $\boldsymbol{\theta}_1$ we write

$$\pi(\theta_{1g} \mid \boldsymbol{y}) = \int \pi(\boldsymbol{\theta}_1 \mid \boldsymbol{\theta}_2, \boldsymbol{y}) \times \pi(\boldsymbol{\theta}_2 \mid \boldsymbol{y})\, d\boldsymbol{\theta}_2$$

which may be evaluated via the approximation

$$\widetilde{\pi}(\theta_{1g} \mid \boldsymbol{y}) = \int \widetilde{\pi}(\theta_{1g} \mid \boldsymbol{\theta}_2, \boldsymbol{y}) \times \widetilde{\pi}(\boldsymbol{\theta}_2 \mid \boldsymbol{y})\, d\boldsymbol{\theta}_2$$

$$\approx \sum_{k=1}^{K} \widetilde{\pi}(\theta_{1g} \mid \boldsymbol{\theta}_2^{(k)}, \boldsymbol{y}) \times \widetilde{\pi}(\boldsymbol{\theta}_2^{(k)} \mid \boldsymbol{y}) \times \Delta_k \quad (3.23)$$

for a set of weights Δ_k, $k = 1, \ldots, K$. Laplace or related analytical approximations are applied to carry out the integration (over $\boldsymbol{\theta}_{1g'}$, $g' \neq g$) required for evaluation of $\widetilde{\pi}(\theta_{1g} \mid \boldsymbol{\theta}_2, \boldsymbol{y})$. To produce the grid of points $\{\boldsymbol{\theta}_2^{(k)}, k = 1, \ldots, K\}$ over which numerical integration is performed, the mode of $\widetilde{\pi}(\boldsymbol{\theta}_2 \mid \boldsymbol{y})$ is located and the Hessian is approximated, from which the grid of points $\{\boldsymbol{\theta}_2^{(k)}, k = 1, \ldots, K\}$, with associated weights Δ_k, is created and used in (3.23), as was described in Sect. 3.7.3. The output of INLA consists of posterior marginal distributions, which can be summarized via means, variances, and quantiles.

3.7.5 Importance Sampling Monte Carlo

The first sampling-based technique we describe directly estimates the required integrals. To motivate importance sampling Monte Carlo, consider the one-dimensional integral

$$I = \int_0^1 f(\theta)\, d\theta = \mathrm{E}[f(\theta)],$$

3.7 Implementation

where the expectation is with respect to the uniform distribution, $U(0,1)$. This formulation suggests the obvious estimator

$$\widehat{I}_m = \frac{1}{m} \sum_{t=1}^{m} f(\theta^{(t)}),$$

with $\theta^{(t)} \sim_{iid} U(0,1)$, $t = 1, \ldots, m$. By the central limit theorem (Appendix G),

$$\sqrt{m}(\widehat{I}_m - I) \to_d N[0, \text{var}(f)],$$

where $\text{var}(f) = E[f(\theta)^2] - I^2$ and we have assumed the latter exists. The form of the variance reveals that the efficiency of the method is determined by how variable the function f is, with respect to the uniform distribution over $[0, 1]$. If f were constant, we would have zero variance!

To achieve an approximately constant function, we can trivially rewrite the integral as

$$I = \int f(\theta)\, d\theta = \int \frac{f(\theta)}{g(\theta)} g(\theta)\, d\theta = E_g\left[\frac{f(\theta)}{g(\theta)}\right], \quad (3.24)$$

where we no longer restrict θ to lie in $(0, 1)$. Define the estimator

$$\widehat{I}_m = \frac{1}{m} \sum_{t=1}^{m} \frac{f(\theta^{(t)})}{g(\theta^{(t)})},$$

where $\theta^{(t)} \sim_{iid} g(\cdot)$, with

$$\sqrt{m}(\widehat{I}_m - I) \to_d N[0, \text{var}(f/g)],$$

and

$$\text{var}(f/g) = E_g\left[\left(\frac{f}{g}\right)^2\right] - I^2.$$

The latter may be estimated by

$$\widehat{\text{var}}(f/g) = \frac{1}{m} \sum_{t=1}^{m} \left(\frac{f(\theta^{(t)})}{g(\theta^{(t)})}\right)^2 - \widehat{I}_m^2.$$

Consequently, the aim is to find a density that closely mimics f (up to proportionality), so that the Monte Carlo estimator will have low variance because samples from *important* regions of the parameter space (where the function is large) are being drawn, hence the label *importance sampling Monte Carlo*. A great strength of importance sampling is that it produces not only an estimate of I but a measure of uncertainty also. Specifically, we may construct the 95% confidence interval

$$\left[\widehat{I}_m - 1.96\frac{\sqrt{\widehat{\text{var}}(f/g)}}{\sqrt{m}}, \widehat{I}_m + 1.96\frac{\sqrt{\widehat{\text{var}}(f/g)}}{\sqrt{m}}\right]. \qquad (3.25)$$

It may seem strange to be utilizing an asymptotic frequentist interval estimate when evaluating an integral for Bayesian inference, but in this context the "sample size" m is controlled by the user and is large so that an asymptotic interval is uncontroversial (since a flat prior on I would give the same Bayesian interval).

Efficient use of importance sampling critically depends on finding a suitable $g(\cdot)$. From the form of $\text{var}(f/g)$, it is clear that if the support of θ is infinite, $g(\cdot)$ must dominate in the tails; otherwise, the variance will be infinite and the estimate will not be useful in practice (even though the estimator is unbiased). It is also desirable to have a $g(\cdot)$ which is computationally inexpensive to sample from. Student's t, or mixtures of Student's t distributions (West 1993), perhaps with iteration to tune the proposal, are popular.

3.7.6 Direct Sampling Using Conjugacy

The emergence of methods to sample from the posterior distribution have revolutionized the practical applicability of the Bayesian inferential approach. Such methods utilize the duality between samples and densities: Given a sample, we can reconstruct the density and functions of interest, and given an arbitrary density, we can almost always generate a sample, given the range of generic random variate generators available. With respect to the latter, the ability to obtain *direct* samples from a distribution decreases as the dimensionality of the parameter space increases, and MCMC methods provide an attractive alternative. However, as discussed in Sect. 3.8, a major practical disadvantage to the use of MCMC is that the generated samples are dependent which complicates the calculation of Monte Carlo standard errors. Automation of MCMC algorithms is also not straightforward since an assessment of the convergence of the Markov chain is required. Further, it is not straightforward to calculate marginal densities such as (3.5) with MCMC. For problems with small numbers of parameters, direct sampling methods provide a strong competitor to MCMC, primarily because independent samples from the posterior are provided and no assessment of convergence is required.

Suppose we have generated independent samples $\{\boldsymbol{\theta}^{(t)}, t = 1, \ldots, m\}$ from $p(\boldsymbol{\theta} \mid \boldsymbol{y})$, with $\boldsymbol{\theta}^{(t)} = [\theta_1^{(t)}, \ldots, \theta_p^{(t)}]$; we describe how such samples may be used for inference. The univariate marginal posterior for $p(\theta_j \mid \boldsymbol{y})$ may be approximated by the histogram constructed from the points $\theta_j^{(t)}$, $t = 1, \ldots, m$. Posterior means $\text{E}[\theta_j \mid \boldsymbol{y}]$ may be approximated by

$$\widehat{\text{E}}[\theta_j \mid \boldsymbol{y}] = \frac{1}{m}\sum_{t=1}^{m} \theta_j^{(t)},$$

3.7 Implementation

with other moments following in an obvious fashion. Coverage probabilities of the form $\Pr(a < \theta_j < b \mid y)$ are estimated by

$$\widehat{\Pr}(a < \theta_j < b \mid y) = \frac{1}{m} \sum_{t=1}^{m} I\left(a < \theta_j^{(t)} < b\right),$$

with $I(\cdot)$ representing the indicator function which is 1 if its argument is true and 0 otherwise. The central limit theorem (Appendix G) allows the accuracy of these approximations to be simply determined since the samples are independent.

We discuss how to estimate the standard error associated with the estimate

$$\widehat{\mu}_m = \frac{1}{m} \sum_{t=1}^{m} \theta^{(t)} \qquad (3.26)$$

of $\mu = \mathrm{E}[\theta \mid y]$. By the strong law of large numbers, $\widehat{\mu}_m \to_{a.s.} \mu$ as $m \to \infty$, and the central limit theorem (Appendix G) gives

$$\sqrt{m}(\widehat{\mu}_m - \mu) \to_d \mathrm{N}(0, \sigma^2)$$

where $\sigma^2 = \mathrm{var}(\theta \mid y)$ (assuming this variance exists). The Monte Carlo standard error is $\sigma/\sqrt{m}$, with consistent estimate of σ:

$$\widehat{\sigma}_m = \sqrt{\frac{1}{m} \sum_{t=1}^{m} (g(\boldsymbol{\theta}^{(t)}) - \widehat{\mu}_m)^2}.$$

By Slutsky's theorem (Appendix G)

$$\frac{\widehat{\mu}_m - \mu}{\widehat{\sigma}_m/\sqrt{m}} \to_d \mathrm{N}(0, 1)$$

as $m \to \infty$. An asymptotic confidence interval for μ is therefore

$$\widehat{\mu}_m \pm 1.96 \times \frac{\widehat{\sigma}_m}{\sqrt{m}}.$$

We may also wish to obtain standard errors for functions that are not simple expectations. For example, consider the posterior variance of a univariate parameter θ:

$$\sigma^2 = \mathrm{var}(\theta \mid y) = \mathrm{E}[(\theta - \mu)^2 \mid y].$$

where $\mu = \mathrm{E}[\theta \mid y]$. An obvious estimator is

$$\widehat{\sigma}_m^2 = \frac{1}{m} \sum_{t=1}^{m} (\theta^{(t)} - \widehat{\mu}_m)^2$$

where $\widehat{\mu}_m$ is given by (3.26). Now,

$$\sqrt{m}\left(\begin{bmatrix}\widehat{\mu}_m\\\widehat{\sigma}_m^2\end{bmatrix}-\begin{bmatrix}\mu\\\sigma^2\end{bmatrix}\right)\to_d \mathrm{N}_2\left(\begin{bmatrix}0\\0\end{bmatrix},\begin{bmatrix}\sigma^2 & \mu_3^*\\\mu_3^* & \mu_4^*-\sigma^4\end{bmatrix}\right)$$

where $\mu_j^* = \mathrm{E}[(\theta-\mu)^j \mid \boldsymbol{y}]$ is the jth central moment, $j = 3, 4$ (where we assume that these quantities exist). The standard error of $\widehat{\sigma}^2$ is estimated by

$$\sqrt{\frac{\widehat{\mu}_{4,m}^* - \widehat{\sigma}_m^4}{m}} \qquad (3.27)$$

where $\widehat{\mu}_{4,m}^* = \frac{1}{m}\sum_{t=1}^{m}(\theta^{(t)} - \widehat{\mu}_m)^4$ which can, unfortunately, be highly unstable. Therefore, accurate interval estimates for σ^2 require larger sample sizes than are needed for accurate estimates for μ.

Once samples from $p(\boldsymbol{\theta} \mid \boldsymbol{y})$ are obtained, it is straightforward to convert to samples for a parameter of interest $g(\boldsymbol{\theta})$ via $g(\boldsymbol{\theta}^{(t)})$. This property is important in a conjugate setting since although we have analytical tractability for one set of parameters, we may be interested in functions of interest that are not so convenient. For example, with likelihood $Y \mid \theta \sim \mathrm{Binomial}(n, \theta)$ and prior $\theta \sim \mathrm{Be}(a, b)$, we know that $\theta \mid y \sim \mathrm{Be}(a + y, b + n - y)$. However, suppose we are interested in the odds $g(\theta) = \theta/(1-\theta)$. Given samples $\theta^{(t)}$ from the beta posterior, we can simply form $g(\theta^{(t)}) = \theta^{(t)}/(1-\theta^{(t)})$, $t = 1, \ldots, m$. As an aside, in this setting, for a Bayesian analysis with a proper prior, the realizations $Y = 0$ or $Y = n$ do not cause problems, in contrast to the frequentist case in which the MLE for $g(\theta)$ is undefined.

3.7.7 Direct Sampling Using the Rejection Algorithm

The *rejection algorithm* is a generic and widely applicable method for generating samples from arbitrary probability distributions.

Theorem (Rejection Sampling).
Suppose we wish to sample from the distribution

$$f(x) = \frac{f^*(x)}{\int f^*(x)\, dx},$$

and we have a proposal distribution $g(\cdot)$ for which

$$M = \sup_x \frac{f^*(x)}{g(x)} < \infty.$$

Then the algorithm:

1. Generate $U \sim U(0, 1)$ and, independently, $X \sim g(\cdot)$.

3.7 Implementation

2. Accept X if

$$U < \frac{f^\star(X)}{Mg(X)},$$

otherwise return to 1,

produces accepted points with distribution $f(x)$, and the acceptance probability is

$$p_a = \frac{\int f^\star(x)\, dx}{M}.$$

Proof. The following is based on Ripley (1987). We have

$$\Pr(X \leq x \cap \text{ acceptance }) = \Pr(X \leq x) \Pr(\text{ acceptance } | X \leq x)$$

$$= \int_{-\infty}^{x} g(y) \Pr(\text{ acceptance } | y)\, dy$$

$$= \int_{-\infty}^{x} g(y) \frac{f^\star(y)}{Mg(y)} dy = \int_{-\infty}^{x} \frac{f^\star(y)}{M} dy.$$

The probability of acceptance is

$$\Pr(\text{acceptance}) = \int_{-\infty}^{\infty} \frac{f^\star(y)}{M} dy = p_a.$$

The number of iterations until accepting a point is a geometric random variable with probability p_a. The expected number of iterations until acceptance is p_a^{-1}. It follows that

$$\Pr(X \leq x \mid \text{ acceptance}) = \sum_{i=1}^{\infty} \Pr(\text{ acceptance on the } i\text{th trial })$$

$$= \sum_{i=1}^{\infty} (1 - p_a)^{i-1} \int_{-\infty}^{x} \frac{f^\star(y)}{M} dy = \frac{1}{p_a} \int_{-\infty}^{x} \frac{f^\star(y)}{M} dy$$

$$= \frac{M}{\int_{-\infty}^{\infty} f^\star(y)} \int_{-\infty}^{x} \frac{f^\star(y)}{M} dy = \int_{-\infty}^{x} f(y) dy,$$

as required. □

We describe a rejection algorithm that is convenient for generating samples from the posterior (Smith and Gelfand 1992). Let $\boldsymbol{\theta}$ denote the unknown parameters, and assume that we can evaluate the maximized likelihood

$$M = \sup_{\boldsymbol{\theta}} p(\mathbf{y} \mid \boldsymbol{\theta}) = p(\mathbf{y} \mid \widehat{\boldsymbol{\theta}})$$

where $\widehat{\boldsymbol{\theta}}$ is the MLE. The algorithm then proceeds as follows:

1. Generate $U \sim U(0, 1)$ and, independently, sample from the prior, $\boldsymbol{\theta} \sim \pi(\boldsymbol{\theta})$.
2. Accept $\boldsymbol{\theta}$ if
$$U < \frac{p(\boldsymbol{y} \mid \boldsymbol{\theta})}{M},$$
otherwise return to 1.

The probability that a point is accepted is
$$p_a = \frac{\int p(\boldsymbol{y} \mid \boldsymbol{\theta}) \pi(\boldsymbol{\theta}) \, d\boldsymbol{\theta}}{M} = \frac{p(\boldsymbol{y})}{M}.$$

This algorithm can be very easy to implement since finding the MLE can often be carried out routinely. We need then only generate points from the prior and evaluate the likelihood at these points. Rejection sampling from the prior is very intuitive; the prior supplies the points which are then "filtered out" via the likelihood.

The empirical rejection rate can be used to derive the normalizing constant as
$$\widetilde{p}(\boldsymbol{y}) = M \times \widehat{p}_a \qquad (3.28)$$
which may be useful for model assessment/selection (Sect. 3.10). If we desire m samples from the posterior, the number of generations required from the prior $\pi(\cdot)$ is $m + m^\star$ (where $m^\star$ is the number of rejected points), and $m^\star$ is a negative binomial random variable (Appendix D). The MLE of p_a is $m/(m + m^\star)$.

An alternative importance sampling estimator of the normalizing constant that is more efficient than (3.28) is
$$\widehat{p}(\boldsymbol{y}) = \frac{1}{m + m^\star} \sum_{t=1}^{m+m^\star} p(\boldsymbol{y} \mid \boldsymbol{\theta}^{(t)}), \qquad (3.29)$$
where $\boldsymbol{\theta}^{(t)} \sim_{iid} \pi(\cdot)$, $t = 1, \ldots, m + m^\star$. Notice that there is no rejection of points associated with this calculation so that all $m + m^\star$ prior points are used. Although (3.29) is the more efficient estimator, (3.28) provides an alternative estimator as a by-product that is useful for code checking. The estimator (3.28) assumes that all normalizing constants are included in M. If the maximization has been carried out with respect to $M^\star = p^\star(\boldsymbol{y} \mid \boldsymbol{\theta})$ where $p^\star(\boldsymbol{y} \mid \boldsymbol{\theta}) = p(\boldsymbol{y} \mid \boldsymbol{\theta})/c$, then we must instead use the estimate
$$\widetilde{p}(\boldsymbol{y}) = c \times M^\star \times \widehat{p}_a. \qquad (3.30)$$

Posterior moments can be estimated directly as averages of the accepted points, or we may implement importance sampling estimators that use all points generated from the prior. For example, the posterior mean
$$\mathrm{E}[\boldsymbol{\theta} \mid \boldsymbol{y}] = \frac{\int \boldsymbol{\theta} p(\boldsymbol{y} \mid \boldsymbol{\theta}) \pi(\boldsymbol{\theta}) \, d\boldsymbol{\theta}}{\int p(\boldsymbol{y} \mid \boldsymbol{\theta}) \pi(\boldsymbol{\theta}) \, d\boldsymbol{\theta}} = \frac{\mathrm{E}\left[\boldsymbol{\theta} p(\boldsymbol{y} \mid \boldsymbol{\theta})\right]}{\mathrm{E}\left[p(\boldsymbol{y} \mid \boldsymbol{\theta})\right]}$$

3.7 Implementation

may be estimated by

$$\widehat{E}[\theta \mid y] = \frac{\frac{1}{m+m^*} \sum_{t=1}^{m+m^*} \theta^{(t)} p(y \mid \theta^{(t)})}{\frac{1}{m+m^*} \sum_{t=1}^{m+m^*} p(y \mid \theta^{(t)})},$$

where $\theta^{(t)} \sim_{iid} \pi(\cdot), t = 1, \ldots, m + m^*$.

Clearly we need a proper prior distribution to implement the above algorithm. The efficiency of the algorithm will depend on the correspondence between the likelihood and the prior, as measured through $p(y)$. For large n, the algorithm will become less efficient since the likelihood becomes increasingly concentrated, and so prior points are less likely to be accepted (which is another manifestation of the prior becoming less important with increasing sample size, Sect. 3.3).

The rejection algorithm that samples from the prior does not need the functional form of the prior to be available. As an example, Wakefield (1996) used a predictive distribution from a Bayesian analysis as the prior for the analysis of a separate dataset; samples from the predictive distribution could be simply generated, even though no closed form was available for this distribution.

Example: Poisson Likelihood, Lognormal Prior

We illustrate some of the technique described in the previous sections using a Poisson likelihood with data from a geographical cluster investigation carried out in the United Kingdom (Black 1984). The Sellafield nuclear site is located in the northwest of England on the coast of West Cumbria. Initially, the site produced plutonium for defense purposes and subsequently carried out the reprocessing of spent fuel from nuclear power stations in Britain and abroad and stored and discharged to sea low-level radioactive waste. Seascale is a village 3 km to the south of Sellafield and had $y = 4$ cases of lymphoid malignancy among 0–14 year olds during 1968–1982, compared with $E = 0.25$ expected cases (based on the number of children in the region and registration rates for the overall northern region of England). A question here is whether such a large number of cases could have reasonably occurred by chance. There is substantial information available on the incidence of childhood leukemia across the United Kingdom as a whole.

We assume the model $Y \mid \theta \sim \text{Poisson}[E \exp(\theta)]$, where θ is the log relative risk (the ratio of the risk in the study region, to that in the northern region), the MLE of which is $\widehat{\theta} = \log(16) = 2.77$ with asymptotic standard error 0.25. We assume an $N(\mu, \sigma^2)$ normal prior for θ, which is equivalent to a lognormal prior $\text{LogNorm}(\mu, \sigma^2)$ for $\exp(\theta)$. To choose the prior parameters, we assume, for illustration, that the median relative risk is 1 and the 90% point of the prior is 10, which leads, from (3.15) and (3.16), to $\mu = 0$ and $\sigma^2 = 1.38^2$.

We will estimate

$$I_r = \int_{-\infty}^{\infty} \theta^r \Pr(y \mid \theta)\pi(\theta) \, d\theta$$

$$= \frac{E^y(2\pi b^2)^{-1/2}}{y!} \int \exp\left[r\log\theta - E\exp(\theta) + \theta y - \frac{(\theta - a)^2}{2b^2}\right] d\theta$$

$$= \frac{E^y(2\pi b^2)^{-1/2}}{y!} \int \exp[h_r(\theta)] \, d\theta$$

for $r = 0, 1, 2$, to give the normalizing constant and posterior mean and variance as

$$p(y) = I_0$$

$$E[\theta \mid y] = \frac{I_1}{I_0}$$

$$\mathrm{var}(\theta \mid y) = \frac{I_2}{I_0} - \frac{I_1^2}{I_0^2}.$$

We choose to calculate the posterior variance not because it is a quantity of particular interest but because it provides a summary that is not particularly easy to estimate and so reveals some of the complications of the various methods.

To apply the Laplace method, we first give the first and second derivatives of $h_r(\theta)$:

$$h_r^{(1)}(\theta) = \frac{r}{\theta} - E\exp(\theta) + y - \frac{\theta - a}{b^2}$$

$$h_r^{(2)}(\theta) = -\frac{r}{\theta^2} - E\exp(\theta) - \frac{1}{b^2},$$

for $r = 0, 1, 2$. The estimates based on the Laplace approximation are shown in Table 3.2. The mean and variance are accurately estimated, but the variance is underestimated for these data. We implemented Gauss–Hermite rules using $m = 5, 10, 15, 20$ points, with the grid centered and scaled by the Laplace approximations of the mean and variance of the posterior. Table 3.2 shows that $\Pr(y)$ and $E[\theta \mid y]$ are well estimated across all grid sizes, while there is more variability in the estimate of $\mathrm{var}(\theta \mid y)$, though it is more accurately estimated then with the Laplace approximation.

We now turn to importance sampling. We have

$$I_r = \int_{-\infty}^{\infty} f_r(\theta) \, d\theta = E\left[\frac{f_r(\theta)}{g(\theta)}\right],$$

with $f_r(\theta) = \theta^r \Pr(y \mid \theta)\pi(\theta)$.

We take as proposal, $g(\cdot)$, a normal distribution scaled via the Laplace estimates of location and scale. Table 3.2 shows estimates resulting from the use of $m = 5,000$ points and the estimator

3.7 Implementation

Table 3.2 Laplace, Gauss–Hermite, and Monte Carlo approximations for Poisson lognormal model with an observed count of $y =$ and an expected count of $E = 0.25$

	Pr(y) ($\times 10^2$)	E[$\theta \mid y$]	var($\theta \mid y$)
Truth	1.37	2.27	0.329
Laplace	1.35	2.29	0.304
Gauss–Hermite $m = 5$	1.36	2.27	0.328
Gauss–Hermite $m = 10$	1.37	2.27	0.331
Gauss–Hermite $m = 15$	1.37	2.27	0.331
Gauss–Hermite $m = 20$	1.37	2.27	0.331
Importance sampling	1.37 [1.35,1.38]	2.27 [2.24,2.29]	0.336 [0.310,0.362]
Rejection algorithm	1.37	2.27 [2.25,2.28]	0.332 [0.319,0.346]
Metropolis–Hastings	–	2.27 [2.22,2.32]	0.328 [0.294,0.361]

The importance sampling and rejection algorithms are based on samples of size $m = 5{,}000$. The Metropolis–Hastings algorithm was run for 51,000 iterations, with the first 1,000 discarded as burn-in. 95% confidence intervals for the relevant estimates are displayed (where available) in square brackets in the last three lines of the table

$$\widehat{I}_r = \frac{1}{m} \sum_{t=1}^{m} \frac{f_r(\theta^{(t)})}{g(\theta^{(t)})}$$

where $\theta^{(t)}$ are independent samples from the normal proposal. The variance of the estimator is

$$\text{var}\left(\widehat{I}_r\right) = \frac{\text{var}(f_r/g)}{m}$$

The delta method can be used to produce measures of accuracy for the posterior mean and variance, though these measures are a little cumbersome. The variance of the normalizing constant is

$$\text{var}\left[\widehat{\text{Pr}}(y)\right] = \text{var}(\widehat{I}_0).$$

To evaluate the variances of the posterior mean and posterior variance estimates we need the multivariate delta method. We must also include covariance terms if the same samples are used to evaluate all three integrals. The formulas are:

$$\text{var}\left[\widehat{\text{E}}(\theta \mid y)\right] = \text{var}\left(\frac{\widehat{I}_1}{\widehat{I}_0}\right)$$

$$\approx \frac{\text{var}(\widehat{I}_1)}{\widehat{I}_0^2} + \frac{\widehat{I}_1^2 \text{var}(\widehat{I}_0)}{\widehat{I}_0^4} - \frac{2\widehat{I}_1}{\widehat{I}_0^3} \text{cov}(\widehat{I}_0, \widehat{I}_1)$$

$$\text{var}\left[\widehat{\text{var}}(\theta \mid y)\right] = \text{var}\left(\frac{\widehat{I}_2}{\widehat{I}_0} - \frac{\widehat{I}_1^2}{\widehat{I}_0^2}\right)$$

$$\approx \left(-\frac{\widehat{I}_2}{\widehat{I}_0^2} + \frac{2\widehat{I}_1^2}{\widehat{I}_0^3}\right)^2 \text{var}(\widehat{I}_0) + \left(\frac{-2\widehat{I}_1}{\widehat{I}_0^2}\right)^2 \text{var}(\widehat{I}_1) + \left(\frac{1}{\widehat{I}_0}\right)^2 \text{var}(\widehat{I}_2)$$

Fig. 3.3 Histogram representations of posterior distributions in the Sellafield example for (**a**) the log relative risk θ and (**b**) the relative risk $\exp(\theta)$, with priors superimposed as *solid lines*. The prior on θ is normal, so that the prior on $\exp(\theta)$ is lognormal

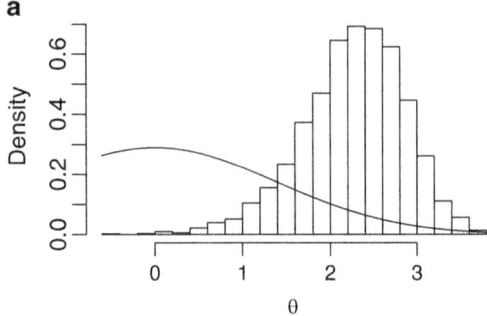

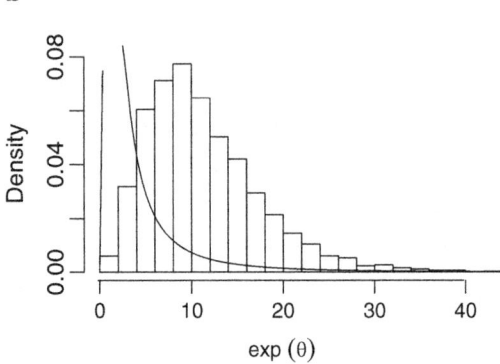

$$+ 2\left(\frac{-\widehat{I}_2}{\widehat{I}_0^2} + \frac{2\widehat{I}_1^2}{\widehat{I}_0^3}\right)\left(\frac{-2\widehat{I}_1}{\widehat{I}_0^2}\right)\operatorname{cov}\left(\widehat{I}_0, \widehat{I}_1\right)$$

$$+ 2\left(\frac{-\widehat{I}_2}{\widehat{I}_0^2} + \frac{2\widehat{I}_1^2}{\widehat{I}_0^3}\right)\left(\frac{1}{\widehat{I}_0}\right)\operatorname{cov}\left(\widehat{I}_0, \widehat{I}_2\right)$$

$$+ 2\left(\frac{-2\widehat{I}_1}{\widehat{I}_0^2}\right)\left(\frac{1}{\widehat{I}_0}\right)\operatorname{cov}\left(\widehat{I}_1, \widehat{I}_2\right).$$

Using these forms we obtain the interval estimates displayed in Table 3.2. The estimates of each of the three summaries are accurate though the interval estimate for the posterior variance is quite wide, because of the inherent instability associated with estimating the standard error.

Finally we implement a rejection algorithm, sampling from the prior distribution and estimating $\Pr(y)$ using the importance sampling estimator, (3.29). The mean and variance of the samples was used to evaluate $E[\theta \mid y]$ and $\mathrm{var}(\theta \mid y)$, with the standard error of the latter based on (3.27). The acceptance probability was 0.07, the small value being explained by the discrepancy between the prior and the likelihood, which is illustrated in Fig. 3.3(a) which gives a histogram representation, based on 5000 points, of $p(\theta \mid y)$, along with the prior drawn as a solid curve.

Panel (b) displays the marginal posterior distribution of the relative risk, $p(e^\theta \mid y)$, which is of more substantive interest, and is simply produced via exponentiation of the θ samples. The rejection estimates in Table 3.2 have relatively narrow interval estimates.

3.8 Markov Chain Monte Carlo

3.8.1 *Markov Chains for Exploring Posterior Distributions*

The fundamental idea behind MCMC is to construct a Markov chain over the parameter space, with invariant distribution the posterior distribution of interest. Specifically, consider a random variable x with support $\mathbb{R}^p$ and density $\pi(\cdot)$. We give a short summary of the essence of discrete time Markov chain theory.

A sequence of random variables $\boldsymbol{X}^{(0)}, \boldsymbol{X}^{(1)}, \ldots$ is called a Markov chain on a state space $\mathbb{R}^p$ if for all t and for all measurable sets A:

$$\Pr\left(\boldsymbol{X}^{(t+1)} \in A \mid \boldsymbol{X}^{(t)}, \boldsymbol{X}^{(t-1)}, \ldots, \boldsymbol{X}^{(0)}\right) = \Pr\left(\boldsymbol{X}^{(t+1)} \in A \mid \boldsymbol{X}^{(t)}\right)$$

so that the probability of moving to any set A at time $t+1$ only depends on where we are at time t. Furthermore, for a *homogeneous* Markov chain,

$$\Pr\left(\boldsymbol{X}^{(t+1)} \in A \mid \boldsymbol{X}^{(t)}\right) = \Pr\left(\boldsymbol{X}^{(1)} \in A \mid \boldsymbol{X}^{(0)}\right).$$

If there exists $p(\boldsymbol{x}, \boldsymbol{y})$ such that

$$\Pr(\boldsymbol{X}_1 \in A \mid \boldsymbol{x}) = \int_A p(\boldsymbol{x}, \boldsymbol{y}) \, d\boldsymbol{y},$$

then $p(\boldsymbol{x}, \boldsymbol{y})$ is called the *transition kernel density*. A probability distribution $\pi(\cdot)$ on $\mathbb{R}^p$ is called an *invariant distribution* of a Markov chain with transition kernel density $p(\boldsymbol{x}, \boldsymbol{y})$ if so-called *global balance* holds:

$$\pi(\boldsymbol{y}) = \int_{\mathbb{R}^p} \pi(\boldsymbol{x}) p(\boldsymbol{x}, \boldsymbol{y}) \, d\boldsymbol{x}.$$

A Markov chain is called *reversible* if

$$\pi(\boldsymbol{x}) p(\boldsymbol{x}, \boldsymbol{y}) = \pi(\boldsymbol{y}) p(\boldsymbol{y}, \boldsymbol{x}) \qquad (3.31)$$

for $\boldsymbol{x}, \boldsymbol{y} \in \mathbb{R}^p$, $\boldsymbol{x} \neq \boldsymbol{y}$. It can shown (Exercise 3.5) that if (3.31) holds, then $\pi(\cdot)$ is the invariant distribution which is useful since (3.31) can be easy to check.

A key idea is that if we have an invariant distribution, then we can evaluate long-term, or ergodic, averages from realizations of the chain. This is crucial for making inference in a Bayesian setting since it means we can estimate quantities of interest such as posterior means and medians. In Markov chain theory, conditions on the transition kernel under which invariant distributions exist is an important topic. Within an MCMC context, this is not important since the posterior distribution is the invariant distribution and we are concerned with constructing Markov chains (transition kernels) with $\pi(\cdot)$ as invariant distribution. Only very mild conditions are typically required to ensure that $\pi(\cdot)$ is the invariant distribution, typically *aperiodocity* and *irreducibility*. A chain is *periodic* if there are places in the parameter space that can only be reached at certain regularly spaced times; otherwise, it is *aperiodic*. A Markov chain with invariant distribution $\pi(\cdot)$ is *irreducible* if for any starting point, there is positive probability of entering any set to which $\pi(\cdot)$ assigns positive probability.

Suppose that $x^{(1)}, \ldots, x^{(m)}$ represents the sample path of the Markov chain. Then expectations with respect to the invariant distribution

$$\mu = \mathrm{E}[g(x)] = \int g(x)\pi(x)\,dx$$

may be approximated by $\widehat{\mu}_m = \frac{1}{m}\sum_{t=1}^{m} g(x^{(t)})$. Monte Carlo standard errors are more difficult to obtain than in the independent sampling case. The Markov chain law of large numbers (the ergodic theorem) tells us that

$$\widehat{\mu}_m \to_{a.s.} \mu$$

as $m \to \infty$, and the Markov chain central limit theorem states that

$$\sqrt{m}(\widehat{\mu}_m - \mu) \to_d \mathrm{N}(0, \tau^2)$$

where

$$\tau^2 = \mathrm{var}\left[g(x^{(t)})\right] + 2\sum_{k=1}^{\infty} \mathrm{cov}\left[g(x^{(t)}), g(x^{(t+k)})\right] \quad (3.32)$$

and the summation term accounts for the dependence in the chain. Chan and Geyer (1994) provide assumptions for validity of this form. Section 3.8.6 describes how τ^2 may be estimated in practice. We now describe algorithms that define Markov chains that are well suited to Bayesian computation.

3.8.2 The Metropolis–Hastings Algorithm

The Metropolis–Hastings algorithm (Metropolis et al. 1953; Hastings 1970) provides a very flexible method for defining a Markov chain. At iteration t of the Markov chain's evolution, suppose the *current* point is $x^{(t)}$. The following steps provide the new point $x^{(t+1)}$:

1. Sample a point y from a *proposal* distribution $q(\cdot \mid x^{(t)})$.
2. Calculate the acceptance probability:

$$\alpha(x^{(t)}, y) = \min\left[\frac{\pi(y)}{\pi(x^{(t)})} \times \frac{q(x^{(t)} \mid y)}{q(y \mid x^{(t)})}, 1\right]. \qquad (3.33)$$

3. Set

$$x^{(t+1)} = \begin{cases} y & \text{with probability } \alpha(x^{(t)}, y) \\ x^{(t)} & \text{otherwise.} \end{cases}$$

In a Bayesian context, the term $\pi(y)/\pi(x^{(t)})$ in (3.33) is the ratio of the posterior densities at the proposed to the current point; since we are taking the ratio, the normalizing constant in the posterior cancels, which is crucial since this is typically unavailable. The second term in (3.33) is the ratio of the density of moving from $y \to x^{(t)}$ to the density of moving from $x^{(t)} \to y$, and it is this term that guarantees global balance and hence that the Markov chain has the correct invariant distribution; see Exercise 3.6. In an *independence* chain, the proposal distribution does not depend on the current point, that is, $q(y \mid x^{(t)})$ is independent of $x^{(t)}$. We now consider a special case of the algorithm that is particularly easy to implement and widely used.

3.8.3 The Metropolis Algorithm

Suppose the proposal distribution is *symmetric* in the sense that

$$g(y \mid x^{(t)}) = g(x^{(t)} \mid y).$$

In this case the product of ratios in (3.33) simplifies to

$$\alpha(x^{(t)}, y) = \min\left[\frac{\pi(y)}{\pi(x^{(t)})}, 1\right]$$

so that only the ratio of target posterior densities is required. In the *random walk*, Metropolis algorithm $q(y \mid x^{(t)}) = q(|y - x^{(t)}|)$, with common choices for $q(\cdot)$ being normal or uniform distributions. In a range of circumstances, an acceptance probability of around 30% is optimal (Roberts et al. 1997), which may be obtained by tuning the proposal density, the variance in a normal proposal, for example. The balancing act is between having high acceptance rates with small movement and having low acceptance rates with large movement.

3.8.4 The Gibbs Sampler

We describe a particularly popular algorithm for simulating from a Markov chain, the Gibbs sampler. We describe two flavors: the sequential Gibbs sampler and the random scan Gibbs sampler. In the following, let x_{-i} represent the vector x with the ith variable removed, that is, $x_{-i} = [x_1, \ldots, x_{i-1}, x_{i+1}, \ldots, x_p]$.

The *sequential scan Gibbs sampling* algorithm starts with some initial value $\boldsymbol{x}^{(0)}$ and then, with current point $\boldsymbol{x}^{(t)} = [x_1^{(t)}, \ldots, x_p^{(t)}]$, undertakes the following p steps to produce a new point $\boldsymbol{x}^{(t+1)} = [x_1^{(t+1)}, \ldots, x_p^{(t+1)}]$:

- Sample $x_1^{(t+1)} \sim \pi_1\left(x_1 \mid \boldsymbol{x}_{-1}^{(t)}\right)$
- Sample $x_2^{(t+1)} \sim \pi_2\left(x_2 \mid x_1^{(t+1)}, x_3^{(t)}, \ldots, x_p^{(t)}\right)$

$$\vdots$$

- Sample $x_p^{(t+1)} \sim \pi_p\left(x_p \mid \boldsymbol{x}_{-p}^{(t+1)}\right)$.

The beauty of the Gibbs sampler is that the often hard problem of sampling for the full p-dimensional variable $\boldsymbol{x}$ has been broken into sampling for each of the p variables in turn via the *conditional distributions*.

We now illustrate that the Gibbs sampling algorithm produces a transition kernel density that gives the required stationary distribution. We do this by showing that each component is a Metropolis–Hastings step. Consider a single component move in the Gibbs sampler from the current point $\boldsymbol{x}^{(t)}$ to the new point $\boldsymbol{x}^{(t+1)}$, with $\boldsymbol{x}^{(t+1)}$ obtained by replacing the ith component in $\boldsymbol{x}^{(t)}$ with a draw from the full conditional $\pi\left(x_i \mid \boldsymbol{x}_{-i}^{(t)}\right)$. We view this move in light of the Metropolis–Hastings algorithm in which the proposal density is the full conditional itself. Then the Metropolis–Hastings acceptance ratio becomes

$$\alpha(\boldsymbol{x}^{(t)}, \boldsymbol{x}^{(t+1)}) = \min\left[\frac{\pi\left(x_i^{(t+1)}, \boldsymbol{x}_{-i}^{(t)}\right) \pi\left(x_i^{(t)} \mid \boldsymbol{x}_{-i}^{(t+1)}\right)}{\pi\left(x_i^{(t)}, \boldsymbol{x}_{-i}^{(t)}\right) \pi\left(x_i^{(t+1)} \mid \boldsymbol{x}_{-i}^{(t)}\right)}, 1\right]$$

$$= \min\left[\frac{\pi\left(\boldsymbol{x}_{-i}^{(t)}\right)}{\pi\left(\boldsymbol{x}_{-i}^{(t)}\right)}, 1\right] = 1$$

because $\pi\left(\boldsymbol{x}_{-i}^{(t)}\right) = \pi\left(x_i^{\star}, \boldsymbol{x}_{-i}^{(t)}\right) / \pi\left(x_i^{\star} \mid \boldsymbol{x}_{-i}^{(t)}\right)$.

Consequently, when we use full conditionals as our proposals in the Metropolis–Hastings step, we always accept. This means that drawing from a full conditional distribution produces a Markov chain with stationary distribution $\pi(\boldsymbol{x})$. Clearly, we cannot keep updating only the ith component, because we will not be able to explore the whole state space this way, that is, we do not have an irreducible Markov chain. Therefore, we can update each component in turn, though this is not the only way to execute Gibbs sampling (though it is the easiest to implement and the most common approach). We can also randomly select an component to update. This is called *random scan* Gibbs sampling:

- Sample a component i by drawing a random variable with probability mass function $[\alpha_1, \ldots, \alpha_p]$ where $\alpha_i > 0$ and $\sum_{i=1}^{p} \alpha_i = 1$.
- Sample $x_i^{(t+1)} \sim \pi_i\left(x_i \mid \boldsymbol{x}_{-i}^{(t)}\right)$.

Roberts and Sahu (1997) examine the convergence rate of the sequential and random scan Gibbs sampling schemes and show that the sequential scan version has a better rate of convergence in the Gaussian models they examine.

In many cases, conjugacy (Sect. 3.7.1) can be exploited to derive the conditional distributions. Many examples of this are given in Chaps. 5 and 8. It is also common for sampling from a full conditional distribution to not require knowledge of the normalizing constant of the target distribution. For example, we saw in Sect. 3.7.7 that rejection sampling does not require the normalizing constant.

3.8.5 Combining Markov Kernels: Hybrid Schemes

Suppose we can construct m transition kernels, each with invariant distribution $\pi(\cdot)$. There are two simple ways to combine these transition kernels. First, we can construct a Markov chain, where at each step we sequentially generate new states from all kernels in a predetermined order. As long as the new Markov chain is irreducible, then it will have the required invariant distribution, and we can, for example, use the ergodic theorem on the samples from the new Markov chain. Hence, we can combine Gibbs and Metropolis–Hastings steps. One popular form is *Metropolis within Gibbs* in which all components with recognizable conditionals are sampled with Gibbs steps with Metropolis–Hastings for the remainder. In the second method of combining Markov kernels, we first create a probability vector $[\alpha_1, \ldots, \alpha_m]$, then randomly select kernel i with probability α_i, and then use this kernel to move the Markov chain.

In general, one can be creative in the construction of a Markov chain, but care must be taken to ensure the proposed chain is "legal," in the sense of having the required stationary distribution. As an example, a chain with a Metropolis step that keeps proposing points until the kth point, with $k \geq 1$, is accepted does not have the correct invariant distribution.

A final warning is that care is required to ensure that the posterior of interest is proper since there is no built in check when an MCMC scheme is implemented. For example, one may be able to construct a set of proper conditional distributions for Gibbs sampling, even when the joint posterior distribution is not proper; see, for example, Hobert and Casella (1996).

3.8.6 Implementation Details

Although theoretically not required, many users remove an initial number of iterations, the rationale being that inferential summaries should not be influenced by initial points that might be far from the main mass of the posterior distribution. Inference is then based on samples collected subsequent to this "burn-in" period.

In order to obtain valid Monte Carlo standard errors for empirical averages, some estimate for τ^2 in (3.32) is required. Time series methods exist to estimate τ^2, but we describe a simple approach based on batch means (Glynn and Iglehart 1990). The basic idea is to split the output of length m into K batches each of length B, with B chosen to be large enough so that the batch means have low serial correlation; B should not be too large, however, because we want K to be large enough to provide a reliable estimate of τ^2. The mean of the function of interest is then estimated within each of the batches:

$$\widehat{\mu}_k = \frac{1}{B} \sum_{t=(k-1)B+1}^{KB} g(\boldsymbol{x}^{(t)})$$

for $k = 1, \ldots, K$. The combined estimate of the mean is the average of the batch means

$$\widehat{\mu} = \frac{1}{K} \sum_{k=1}^{K} \widehat{\mu}_k.$$

Then $\sqrt{B}(\widehat{\mu}_k - \mu)$, $k = 1, \ldots, K$ are approximately independently distributed as $N(0, \tau^2)$, and so τ^2 can be estimated by

$$\widehat{\tau}^2 = \frac{B}{K-1} \sum_{k=1}^{K} (\widehat{\mu}_k - \widehat{\mu})^2$$

and

$$\widehat{\text{var}}(\widehat{\mu}) = \frac{\widehat{\tau}^2}{K} = \frac{B}{K(K-1)} \sum_{k=1}^{K} (\widehat{\mu}_k - \widehat{\mu})^2.$$

Normal or Students t confidence intervals can be calculated based on the square root of this quantity. The construction of these intervals has the advantage of being simple, but the output should be viewed with caution as the above derivation contains a number of approximations.

MCMC approaches provide no obvious estimator of the normalizing constant $p(\boldsymbol{y})$, but a number of indirect methods have been proposed (Meng and Wong 1996; DiCiccio et al. 1997)

Aside from directly calculating integrals, we may also form graphical summaries of parameters of interest, essentially using the dependent samples in the same way that we would independent samples. For example, a histogram of $x_i^{(t)}$ provides an estimate of the posterior marginal distribution, $\pi_i(x_i)$, $i = 1, \ldots, p$.

In practice, there are a number of important issues that require thought when implementing MCMC. A crucial question is how large m should be in order to obtain a reliable Monte Carlo estimate. The Markov chain will display better mixing properties if the parameters are approximately independent in the posterior. In an extreme case, if we have independence, then

$$\pi(x_1,\ldots,x_p) = \prod_{i=1}^{p} \pi(x_i)$$

and Gibbs sampling via the conditional distributions $\pi(x_i)$, $i = 1, \ldots, p$, equates to direct sampling from the posterior.

Dependence in the Markov chain may be greatly reduced by sampling simultaneously for variables that are highly depend, a strategy known as *blocking*. Reparameterization may also be helpful in this regard. As the blocks become larger, the acceptance rate (if a Metropolis-Hastings algorithm is used) may be reduced to an unacceptably low level in which case there is a trade-off with respect to the size of blocks to use. Some chains may be very *slow mixing*, and an examination of autocorrelation aids in deciding on the number of iterations required. If storage of samples is an issue, then one may decide to "thin" the chain by only collecting samples at equally spaced intervals.

A number of methods have been proposed for "diagnosing convergence." Trace plots provide a useful method for detecting problems with MCMC convergence and mixing. Ideally, trace plots of unnormalized log posterior and model parameters should look like stationary time series. Slowly mixing Markov chains produce trace plots with high autocorrelation, which can be further visualized by plotting the autocorrelation at different lags. Slow mixing does not imply lack of convergence, however, but that more samples will be required for accurate inference (as can be seen from (3.32)). When examining trace plots and autocorrelations, it is clearer to work with parameters transformed to $\mathbb{R}$. Running multiple chains from different starting points is also very useful since one may compare inference between the different chains. Gelman and Rubin (1992) provide one popular convergence diagnostic based on multiple chains. As with the use of diagnostics in regression modeling, convergence diagnostics may detect evidence of poor behavior, but there is no guarantee of good behavior of the chain, even if all convergence diagnostics appear reasonable.

Example: Poisson Likelihood, Lognormal Prior

Recall the Poisson lognormal example in which $y = 4$ and $E = 0.25$ with a single parameter, the log relative risk θ. Gibbs sampling corresponds to direct sampling from the univariate posterior for θ, which we have already illustrated using the rejection algorithm.

We implement a random walk Metropolis algorithm using a normal kernel and the asymptotic variance of the MLE for θ multiplied by 3 as the variance of the proposal, to achieve a reasonable acceptance probability of 0.32. This multiplier was found by trial and error, based on preliminary runs of the Markov chain. It is important to restart the chain when the proposal is changed based on past realizations to ensure the chain is still Markovian. Table 3.2 gives estimates of the posterior mean and variance based on a run length of 51,000, with the first 1,000 discarded as a burn-in. The confidence interval for the estimates of the posterior mean and posterior variance is based on the batch means method, with $K = 50$ batches of size $B = 1,000$.

Example: Lung Cancer and Radon

We return to the lung cancer and radon example, first introduced in Sect. 1.3.3, to demonstrate the use of the Metropolis random walk algorithm in a situation with more than one parameter. For direct comparison with methods applied in Chap. 2, we assume an improper flat prior on $\boldsymbol{\beta} = [\beta_0, \beta_1]$ so that the posterior $p(\boldsymbol{\beta} \mid \boldsymbol{y})$ is proportional to the likelihood.

We begin by implementing a Metropolis random walk algorithm based on a pair of univariate normal distributions. In this example, the Gibbs sampler is less appealing since the required conditional distributions do not assume known forms. The first step is to initialize $\beta_0^{(0)} = \widehat{\beta}_j$, where $\widehat{\beta}_j$, $j = 0, 1$, are the MLEs. We then iterate, at iteration t, between:

1. Generate $\beta_0^\star \sim \mathrm{N}(\beta_0^{(t)}, c_0 \widehat{V}_0)$, where $\widehat{V}_0$ is the asymptotic variance of $\widehat{\beta}_0$. Calculate the acceptance probability:

$$\alpha_0(\beta_0^\star, \beta_0^{(t)}) = \min\left[\frac{p(\beta_0^\star, \beta_1^{(t)} \mid \boldsymbol{y})}{p(\beta_0^{(t)}, \beta_1^{(t)} \mid \boldsymbol{y})}, 1\right]$$

and set

$$\beta_0^{(t+1)} = \begin{cases} \beta_0^\star & \text{with probability } \alpha_0(\beta_0^\star, \beta_0^{(t)}), \\ \beta_0^{(t)} & \text{otherwise.} \end{cases}$$

2. Generate $\beta_1^\star \sim \mathrm{N}(\beta_1^{(t)}, c_1 \widehat{V}_1)$, where $\widehat{V}_1$ is the asymptotic variance of $\widehat{\beta}_1$. Calculate the acceptance probability:

$$\alpha_1(\beta_1^\star, \beta_1^{(t)}) = \min\left[\frac{p(\beta_0^{(t+1)}, \beta_1^\star \mid \boldsymbol{y})}{p(\beta_0^{(t+1)}, \beta_1^{(t)} \mid \boldsymbol{y})}, 1\right]$$

and set

$$\beta_1^{(t+1)} = \begin{cases} \beta_1^\star & \text{with probability } \alpha_1(\beta_1^\star, \beta_1^{(t)}), \\ \beta_1^{(t)} & \text{otherwise.} \end{cases}$$

The constants c_0 and c_1 are chosen to provide a trade-off between gaining a high proportion of acceptances and moving around the support of the parameter space; this is illustrated in Fig. 3.4 where the realized parameters from the first 1,000 iterations of two Markov chains are plotted. In panels (a) and (d), we chose

3.8 Markov Chain Monte Carlo

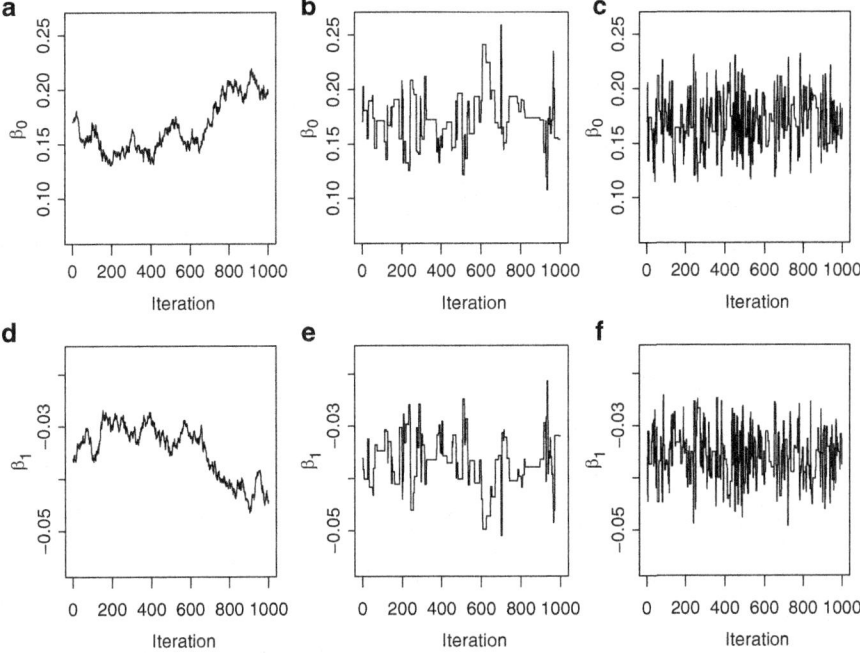

Fig. 3.4 Sample paths from Metropolis–Hastings algorithms for β_0 (*top row*) and β_1 (*bottom row*) for the lung cancer and radon data. In the *left column* the proposal random walk has small variance; in the *center column* large variance and in the *right column*, we use a bivariate proposal

$c_0 = c_1 = c = 0.1$ and in panels (b) and (e) $c_0 = c_1 = c = 2$. For $c = 0.1$ the acceptance rate is 0.90, but movement around the space is slow, as indicated by the meandering nature of the chain, while for $c = 2$ the moves tend to be larger, but the chain sticks at certain values, as seen by the horizontal runs of points (the acceptance rate is 0.14).

Figure 3.6a shows a scatterplot representation of the joint distribution $p(\beta_0, \beta_1 | y)$ and clearly shows the strong negative dependence; the asymptotic correlation between the MLEs $\widehat{\beta}_0$ and $\widehat{\beta}_1$ is -0.90, and the posterior correlation between β_0 and β_1 is -0.90 also (the correspondence between these correlations is not surprising since the sample size is large and the prior is flat). The strong negative dependence is evident in each of the first two columns of Fig. 3.4. Figure 3.5 shows the autocorrelations between sampled parameters at lags of between 1 and 40. The top row is for β_0, and the bottom is for β_1. In panels (a) and (d), the autocorrelations are high because of the small movements of the chain.

The dependence in the chain may be reduced via reparameterization or by generation from a bivariate proposal. We implement the latter with variance–covariance matrix equal to $c \times \text{var}(\widehat{\boldsymbol{\beta}})$. The acceptance rate for the bivariate proposal with $c = 2$ is 0.29, which is reasonable. We then iterate the following:

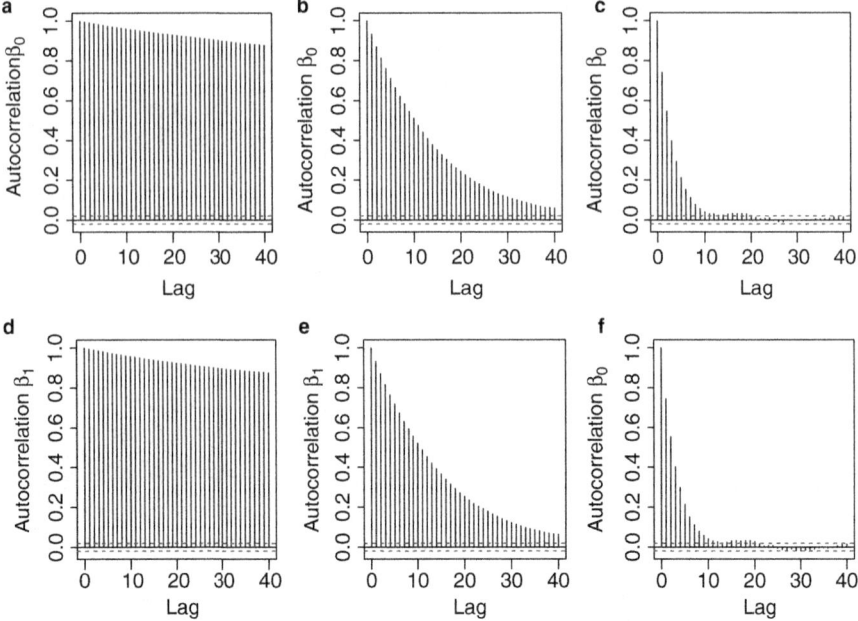

Fig. 3.5 Autocorrelation functions for β_0 (*top row*) and β_1 (*bottom row*) for the lung cancer and radon data. *First column*: univariate random walk, $c = 0.1$, *second column*: univariate random walk, $c = 2$, *third column*: bivariate random walk, $c = 2$

1. Generate $\boldsymbol{\beta}^\star \sim \mathrm{N}_2(\boldsymbol{\beta}^{(t)}, c\widehat{\boldsymbol{V}})$, where $\widehat{\boldsymbol{V}}$ is the asymptotic variance of the MLE $\widehat{\boldsymbol{\beta}}$.
2. Calculate the acceptance probability

$$\alpha(\boldsymbol{\beta}^\star, \boldsymbol{\beta}^{(t)}) = \min\left[\frac{p(\boldsymbol{\beta}^\star \mid \boldsymbol{y})}{p(\boldsymbol{\beta}^{(t)} \mid \boldsymbol{y})}, 1\right]$$

and set

$$\boldsymbol{\beta}^{(t+1)} = \begin{cases} \boldsymbol{\beta}^\star & \text{with probability } \alpha(\boldsymbol{\beta}^\star, \boldsymbol{\beta}^{(t)}), \\ \boldsymbol{\beta}^{(t)} & \text{otherwise.} \end{cases}$$

Note that the choice of c and the dependence in the chain do not jeopardize the invariant distribution, but rather the length of chain until practical convergence is reached and the number of points required for summarization. More points are required when there is high positive dependence in successive iterates, which is clear from (3.32). The final column of Fig. 3.4 shows the sample path from the bivariate proposal, with good movement and little dependence between the parameters. Panels (c) and (f) show that the autocorrelation is also greatly reduced.

3.8 Markov Chain Monte Carlo

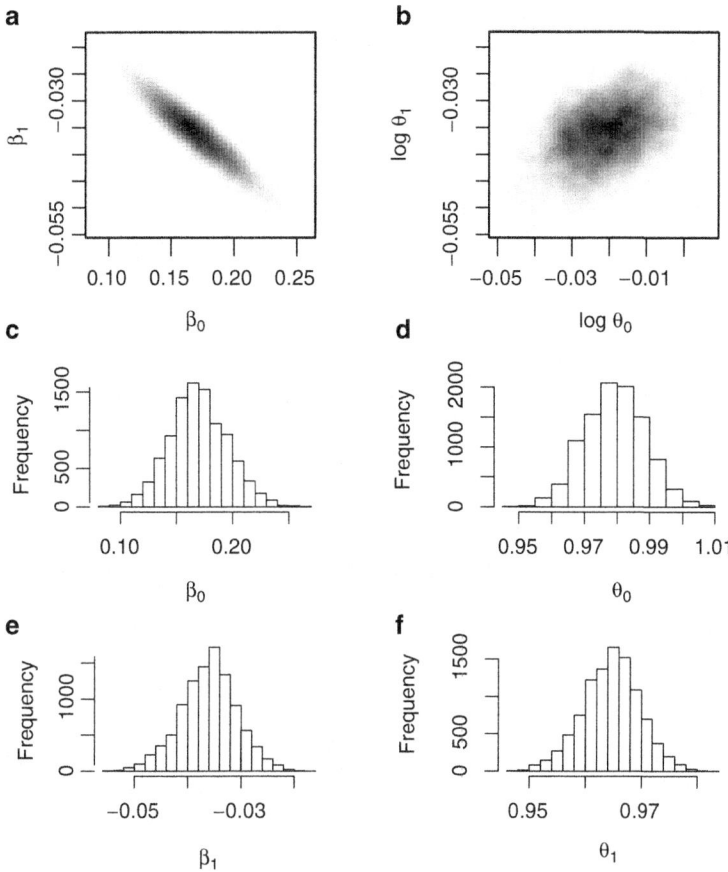

Fig. 3.6 Posterior summaries for the lung cancer and radon data: (**a**) $p(\beta_0, \beta_1 \mid y)$, (**b**) $p(\log \theta_0, \log \theta_1 \mid y)$, (**c**) $p(\beta_0 \mid y)$, (**d**) $p(\theta_0 \mid y)$, (**e**) $p(\beta_1 \mid y)$, (**f**) $p(\theta_1 \mid y)$

Figure 3.6 shows inference for the reparameterized model

$$Y_i \mid \boldsymbol{\theta} \sim_{ind} \text{Poisson}(E_i \theta_0 \theta_1^{x_i - \bar{x}})$$

where $\theta_0 = \exp(\beta_0 + \beta_1 \bar{x}) > 0$ and $\theta_1 = \exp(\beta_1) > 0$ along with summaries for the β_0, β_1 parameterization. Figure 3.6(b) shows the bivariate posterior for $\log \theta_0, \log \theta_1$ and demonstrates that the parameters are virtually independent (the correlation is -0.03). By comparison there is strong negative dependence between β_0 and β_1 (panel (a)). Panels (d) and (f) show histogram representations of the posteriors of interest $p(\theta_0 \mid y)$ and $p(\theta_1 \mid y)$.

The posterior median (95% credible interval) for $\exp(\beta_1)$ is 0.965 [0.954, 0.975] which is almost identical to the asymptotic inference under a Poisson model (Table 2.4), which is again not surprising given the large sample size.

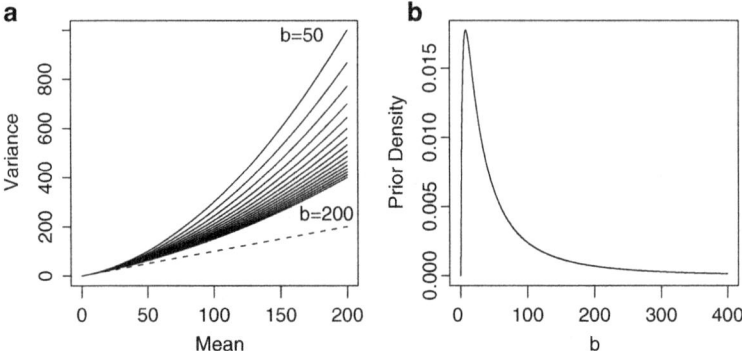

Fig. 3.7 (a) Mean–variance relationships, in the negative binomial model, for values of b between 50 and 200, in increments of 10 units. The *dashed line* is the line of equality corresponding to the Poisson model, which is recovered as $b \to \infty$. (b) Lognormal prior for b

The Poisson model should be used with caution since the variance is determined by the mean, with no additional parameter to soak up excess-Poisson variability, which is often present in practice. To overcome this shortcoming we provide a Bayesian analysis with a negative binomial likelihood, parameterized so that

$$E[Y_i \mid \beta, b] = \mu_i(\beta), \quad \text{var}(Y_i \mid \beta, b) = \mu_i(\beta)[1 + \mu(\beta)/b]. \tag{3.34}$$

We will continue with an improper flat prior for β, but a prior for b requires more thought. To determine a prior, we plot the mean–variance relationship in Fig. 3.7a, for different values of b. In this example the regression model does not include information on confounders such as smoking. The absence of these variables will certainly lead to bias in the estimate of $\exp(\beta_1)$ due to confounding, but with respect to b, we might expect considerable excess-Poisson variability due to missing variables. The sample average of the observed counts is 158, and we specify a lognormal prior for b by giving two quantiles of the overdispersion, $\mu(1 + \mu/b)$, at $\mu = 158$, and then solve for b. Specifically, we suppose that there is a 50% chance that the overdispersion is less than $1.5 \times \mu$ and a 95% chance that it is less than $5 \times \mu$. Formulas (3.15) and (3.16) give a lognormal prior with parameters 3.68 and 1.26^2 and 5%, 50%, and 95% quantiles of 4.9, 40, and 316, respectively. Figure 3.7(b) gives the resulting lognormal prior density.

A random walk Metropolis algorithm with a normal proposal was constructed for β_0, β_1, b with the variance–covariance matrix taken as 3 times the asymptotic variance–covariance matrix ($\widehat{b}$ is asymptotically independent of $\widehat{\beta}_0$ and $\widehat{\beta}_1$), based on the expected information. The posterior median and 95% credible interval for $\exp(\beta_1)$ are 0.970 [0.955,0.987], and for b the summaries are 57.8 [34.9,105]. The MLE is $\widehat{b} = 61.3$, with asymptotic 95% confidence interval (calculated on the $\log b$ scale and then exponentiated) of [35.4,106]. Therefore, likelihood and Bayesian inference for b are in close agreement for these data. Histograms of samples from

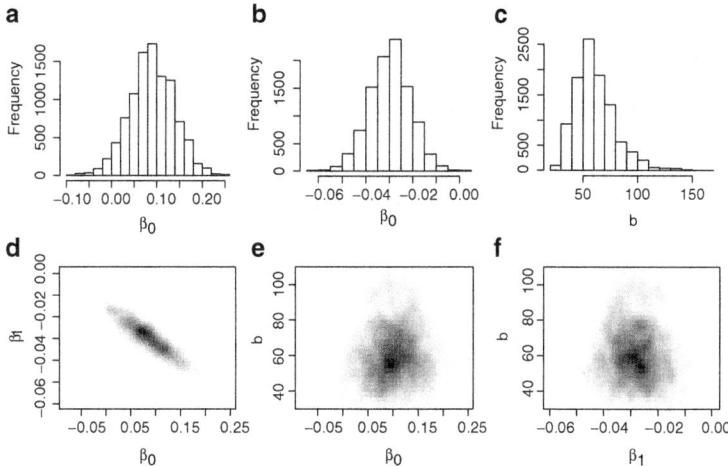

Fig. 3.8 Univariate and bivariate summaries of the posterior $p(\beta_0, \beta_1, b \mid y)$, arising from the negative binomial model

the univariate posteriors for β_0, β_1, and b are shown in the first row of Fig. 3.8, while bivariate scatterplots are shown in the second row. The posterior marginals for β_0 and β_1 are very symmetric, while that for b is slightly skewed.

3.8.7 Implementation Summary

While MCMC has revolutionized Bayesian inference in terms of the breadth of applications and complexity of models that can now be considered, other methods may still be preferable in some situations, in particular when the number of parameters is small. Direct sampling from the posterior is particularly appealing since one retains all of the advantages of sample-based inference (e.g., the ability to simply examine generic functions of interest), without the need to worry about the convergence issues associated with MCMC. Quadrature methods are also appealing for low-dimensional problems, since they are highly efficient. The latter is particularly important if the calculation of the likelihood is expensive. Importance sampling Monte Carlo methods are appealing in that error assessment may be carried out; analytical approximations are, in general, poor in this respect.

INLA is very attractive due to its speed of computation, though a reliable measure of accuracy is desirable and there are practical situations in which the method is not accurate. For example, the method is less accurate for binomial data with small denominators (Fong et al. 2010). In exploratory situations, one may always use quick methods such as INLA for initial modeling, with more computationally demanding approaches being used when a set of finals models are honed in upon.

INLA is also useful for performing simulation studies to examine the properties of model summaries. In general, comparing results across different methods is a good idea. When deciding upon a method of implementation, there is often a clear trade-off between efficiency and the time taken to code prospective methods. MCMC methods are often easy to implement, but are not always the most efficient (at least not for basic schemes) and are difficult to automate. For many high-dimensional problems, MCMC may be the only method that is feasible, although INLA may be available if the model is of the required form (a small number of "non-Gaussian" parameters).

An important paper in the history of MCMC is that of Green (1995) in which *reversible jump MCMC* was introduced. This method can be used in situations in which the parameter space is of varying dimension across different models.

3.9 Exchangeability

We now provide a brief discussion of de Finetti's celebrated representation theorem which describes the form of the marginal distribution of a collection of random variables, under certain assumptions. As we will see, this provides one way in which important modeling questions can be framed. We first require the introduction of a very important concept in Bayesian inference, *exchangeability*.

Definition. Let $p(y_1, \ldots, y_n)$ be the joint density of $Y_1, \ldots, Y_n$. If

$$p(y_1, \ldots, y_n) = p(y_{\pi(1)}, \ldots, y_{\pi(n)})$$

for all permutations, π, of $\{1, 2, \ldots, n\}$, then $Y_1, \ldots, Y_n$ are (finitely) *exchangeable*.

This definition essentially says that the labels identifying the individual components are uninformative. Obviously if a collection of n random variables is exchangeable, this implies that the marginal distribution of all single random variables are the same, as are the marginal distributions for all pairs, all triples, etc. A collection of random variables is infinitely exchangeable if every finite subcollection is exchangeable.

As a simple example, consider Bernoulli random variables, Y_i, for $i = 1, 2, 3 = n$. Under exchangeability,

$$\begin{aligned} \Pr(Y_1 = a, Y_2 = b, Y_3 = c) &= \Pr(Y_1 = a, Y_2 = c, Y_3 = b) \\ &= \Pr(Y_1 = b, Y_2 = a, Y_3 = c) \\ &= \Pr(Y_1 = b, Y_2 = c, Y_3 = a) \\ &= \Pr(Y_1 = c, Y_2 = a, Y_3 = b) \\ &= \Pr(Y_1 = c, Y_2 = b, Y_3 = a) \end{aligned}$$

for all $a, b, c = 0, 1$.

3.9 Exchangeability

Result. If $\theta \sim p(\theta)$ and $Y_1, \ldots, Y_n$ are conditionally independent and identically distributed given θ, then $Y_1, \ldots, Y_n$ are exchangeable.

Proof. By definition:

$$p(y_1, \ldots, y_n) = \int p(y_1, \ldots, y_n \mid \boldsymbol{\theta}) \pi(\boldsymbol{\theta}) \, d\boldsymbol{\theta}$$

$$= \int \left[\prod_{i=1}^n p(y_i \mid \boldsymbol{\theta}) \right] \pi(\boldsymbol{\theta}) \, d\boldsymbol{\theta}$$

$$= \int \left[\prod_{i=1}^n p(y_{\pi(i)} \mid \boldsymbol{\theta}) \right] \pi(\boldsymbol{\theta}) \, d\boldsymbol{\theta}$$

$$= p(y_{\pi(1)}, \ldots, y_{\pi(n)})$$

We now present the converse of this result.

Theorem. *de Finetti's representation theorem for 0/1 random variables.*

If $Y_1, Y_2, \ldots$ is an infinitely exchangeable sequence of 0/1 random variables, there exists a distribution $\pi(\cdot)$ such that the joint mass function $\Pr(y_1, \ldots, y_n)$ has the form

$$\Pr(y_1, \ldots, y_n) = \int_0^1 \prod_{i=1}^n \theta^{y_i}(1-\theta)^{1-y_i} \pi(\theta) \, d\theta,$$

where

$$\int_0^\theta \pi(u) \, du = \lim_{n \to \infty} \Pr\left(\frac{Z_n}{n} \leq \theta\right),$$

with $Z_n = Y_1 + \ldots + Y_n$, and $\theta = \lim_{n \to \infty} Z_n/n$.

Proof. The following is based on Bernardo and Smith (1994). Let $z_n = y_1 + \ldots + y_n$ be the number of 1's (which we label "successes") in the first n observations. Then, due to exchangeability,

$$\Pr(y_1 + \ldots + y_n = z_n) = \binom{n}{z_n} \Pr(Y_{\pi(1)}, \ldots, Y_{\pi(n)}),$$

for all permutations π of $\{1, 2, \ldots, n\}$ such that $y_{\pi(1)} + \ldots + y_{\pi(n)} = z_n$. We can embed the event $y_1 + \ldots + y_n = z_n$ within a longer sequence and

$$\Pr(Y_1 + \ldots + Y_n = z_n) = \sum_{Z_N = z_n}^{N-(n-z_n)} \Pr(z_n, z_N) = \sum_{Z_N = z_n}^{N-(n-z_n)} \Pr(z_n \mid z_N) \Pr(z_N),$$

where $\Pr(z_N)$ is the "prior" belief in the number of successes out of N. To obtain the conditional probability, we observe that it is "as if" we have a population of N

items of which z_N are successes and $N - z_N$ failures, from which we draw n items. The distribution of $z_n \mid z_N$ successes is therefore hypergeometric so that

$$\Pr(y_1 + \ldots + y_n = z_n) = \sum_{z_N = z_n}^{N-(n-z_n)} \frac{\binom{z_N}{z_n}\binom{N-z_N}{n-z_n}}{\binom{N}{n}} \Pr(z_N).$$

We now let $\Pi(\theta)$ be the step function which is 0 for $\theta < 0$ and has jumps of $\Pr(z_N)$ at $\theta = z_N/N$, $z_N = 0, \ldots, N$. We now let $N \to \infty$. Then the hypergeometric distribution tends to a binomial distribution with parameters n and θ and the prior $\Pr(z_N)$ is translated into a prior for θ, which we write as $\pi(\theta)$. Consequently,

$$\Pr(y_1 + \ldots + y_n = z_n) \to \binom{n}{z_n} \int \theta^{z_n}(1-\theta)^{n-z_n} \pi(\theta)\, d\theta,$$

as $N \to \infty$. □

The implications of this theorem are of great significance. By the strong law of large numbers, $\theta = \lim_{n \to \infty} Z_n/n$, so that $\pi(\cdot)$ represents our beliefs about the limiting relative frequency of 1's. Hence, we have an interpretation of θ. Further, we may view the Y_i as conditional independent, Bernoulli random variables, conditional on the random variable θ.

In conventional language, we have a *likelihood function*

$$\Pr(y_1, \ldots, y_n \mid \theta) = \prod_{i=1}^{n} p(y_i \mid \theta) = \prod_{i=1}^{n} \theta^{y_i}(1-\theta)^{1-y_i},$$

where the *parameter* θ is assigned a *prior distribution* $\pi(\theta)$.

In general, if $Y_1, Y_2, \ldots$ is an infinitely exchangeable sequence of random variables, there exists a probability density function $\pi(\cdot)$ such that

$$p(y_1, \ldots, y_n) = \int \prod_{i=1}^{n} p(y_i \mid \boldsymbol{\theta})\pi(\boldsymbol{\theta})\, d\boldsymbol{\theta}, \qquad (3.35)$$

with $p(Y \mid \boldsymbol{\theta})$ denoting the density function corresponding to the "unknown parameter" $\boldsymbol{\theta}$. A sketch proof of (3.35) may be found in Bernardo and Smith (1994). This result tells us that a conditional independence model can be justified via an exchangeability argument. In this general case, further assumptions on $Y_1, Y_2, \ldots$ are required to identify $p(Y \mid \boldsymbol{\theta})$. Bernardo and Smith (1994) present the assumptions that lead to a number of common modeling choices. For example, suppose that $Y_1, Y_2, \ldots$ is an infinitely exchangeable sequence of random variables such that $Y_i > 0$, $i = 1, 2, \ldots$. Further, suppose that for any event A in $\mathbb{R} \times \ldots \times \mathbb{R}$, and for all n,

$$\Pr[(y_1, \ldots, y_n) \in A] = \Pr[(y_1, \ldots, y_n) \in A + \boldsymbol{a}]$$

for all $a \in \mathbb{R} \times \ldots \times \mathbb{R}$ such that $a^\mathsf{T} 1 = 0$ and $A + a$ is an event in $\mathbb{R} \times \ldots \times \mathbb{R}$. Then the joint density for $y_1, \ldots, y_n$ is

$$p(y_1, \ldots, y_n) = \int_0^\infty \prod_{i=1}^n \theta^{-1} \exp(-\theta^{-1} y_i) \times \pi(\theta) \, d\theta$$

where $\int_0^\infty \pi(u) \, du = \lim_{n \to \infty} \Pr(\bar{y}_n \leq \theta)$ and $\bar{y}_n = (y_1 + \ldots + y_n)/n$. For a proof, see Diaconis and Ylvisaker (1980). Hence, a belief in exchangeability and a "lack of memory" property leads to the integral of the predictive distribution being the marginal distribution that is constructed from the product of a conditionally independent set of *exponential* random variables and a prior. The parameter is identified as the sample mean from a large number of observations.

This kind of approach is of theoretical interest, but in practice the choice of likelihood will often be based more directly on the context and previous experience with similar data types. Exchangeability is very useful in practice for prior specification, however. Before one uses a particular conditional independence model, one can think about whether all units are deemed exchangeable. If some collection of units are distinguishable, then one should not assume conditional independence for all units, and one may instead separate the units into groups within which exchangeability holds. For further discussion, see Sect. 8.6.

In terms of modeling, if we believe that a sequence of random variables is exchangeable, this allows us to write down a conditional independence model. We emphasize that independence is a very different assumption since it implies that we learn nothing from past observations:

$$p(y_{m+1}, \ldots, y_n \mid y_1, \ldots, y_m) = p(y_{m+1}, \ldots, y_n)$$

In a regression context, the situation is slightly more complicated. Informally, exchangeability within covariate-defined groups gives the usual conditional independence model, where we now condition on parameters and covariates; Bernardo and Smith (1994, Sect. 4.64) contains details.

3.10 Hypothesis Testing with Bayes Factors

We now turn to a description of Bayes factors, which are the conventional Bayesian method for comparison of hypotheses/models. Let the observed data be denoted $y = [y_1, \ldots, y_n]$, and assume two hypotheses of interest, H_0 and H_1. The application of Bayes theorem gives the probability of the hypothesis H_0, given data y, as

$$\Pr(H_0 \mid y, H_0 \cup H_1) = \frac{p(y \mid H_0) \Pr(H_0 \mid H_0 \cup H_1)}{p(y \mid H_0 \cup H_1)}$$

Table 3.3 Losses corresponding to the decision δ, when the truth is H and L_I and L_{II} are the losses associated with type I and II errors, respectively

		Decision	
$L(\delta, H)$		$\delta = 0$	$\delta = 1$
Truth H	H_0	0	L_I
	H_1	L_{II}	0

where

$$p(\boldsymbol{y} \mid H_0 \cup H_1) = p(\boldsymbol{y} \mid H_0)\Pr(H_0 \mid H_0 \cup H_1) + p(\boldsymbol{y} \mid H_1)\Pr(H_1 \mid H_0 \cup H_1)$$

is the probability of the data averaged over H_0 and H_1. The prior probability that H_0 is true, given one of H_0 and H_1 is true, is $\Pr(H_0 \mid H_0 \cup H_1)$, and $\Pr(H_1 \mid H_0 \cup H_1) = 1 - \Pr(H_0 \mid H_0 \cup H_1)$ is the prior on the alternative hypothesis. This simple calculation makes it clear that to evaluate the probability that the null is true, one is actually calculating the probability of the null *given* that H_0 or H_1 is true. Therefore, we are calculating the "relative truth"; H_0 may provide a poor fit to the data, but H_1 may be even worse. Although conditioning on $H_0 \cup H_1$ is crucial to interpretation, we will drop it for compactness of notation.

If we wish to compare models H_0 and H_1, then a natural measure is given by the *posterior odds*

$$\frac{\Pr(H_0 \mid \boldsymbol{y})}{\Pr(H_1 \mid \boldsymbol{y})} = \frac{p(\boldsymbol{y} \mid H_0)}{p(\boldsymbol{y} \mid H_1)} \times \frac{\Pr(H_0)}{\Pr(H_1)}, \quad (3.36)$$

where the *Bayes factor*

$$\mathrm{BF} = \frac{p(\boldsymbol{y} \mid H_0)}{p(\boldsymbol{y} \mid H_1)}$$

is the ratio of the marginal distributions of the data under the two models, and $\Pr(H_0)/\Pr(H_1)$ is the *prior odds*. Care is required in the choice of priors when Bayes factors are calculated; see Sect. 4.3.2 for further discussion.

Depending on the nature of the analysis, we may: simply report the Bayes factor; or we may place priors on the hypotheses and calculate the posterior odds of H_0; or we may go a step further and derive a decision rule. Suppose we pursue the latter and let $\delta = 0/1$ represent the decision to pick H_0/H_1. With respect to Table 3.3, the posterior expected loss associated with decision δ is

$$\mathrm{E}[L(\delta, H)] = L(\delta, H_0)\Pr(H_0 \mid \boldsymbol{y}) + L(\delta, H_1)\Pr(H_1 \mid \boldsymbol{y})$$

3.10 Hypothesis Testing with Bayes Factors

so that for the two possible decisions (accept/reject H_0) the expected losses are

$$E[L(\delta = 0, H)] = 0 \times \Pr(H_0 \mid y) + L_{\text{II}} \times \Pr(H_1 \mid y)$$
$$E[L(\delta = 1, H)] = L_{\text{I}} \times \Pr(H_0 \mid y) + 0 \times \Pr(H_1 \mid y).$$

To find the decision that minimizes posterior expected cost, let $v = \Pr(H_1 \mid y)$ so that

$$E[L(\delta = 0, H)] = L_{\text{II}} \times v \qquad (3.37)$$
$$E[L(\delta = 1, H)] = L_{\text{I}} \times (1 - v). \qquad (3.38)$$

We should choose $\delta = 1$ if $L_{\text{II}} \times v \geq L_{\text{I}}(1 - v)$, that is, if $v/(1-v) \geq L_{\text{I}}/L_{\text{II}}$, or $v \geq L_{\text{I}}/(L_{\text{I}} + L_{\text{II}})$. Hence, we report H_1 if

$$\Pr(H_1 \mid y) \geq \frac{L_{\text{I}}}{L_{\text{I}} + L_{\text{II}}} = \frac{1}{1 + L_{\text{II}}/L_{\text{I}}},$$

illustrating that we only need to specify the ratio of losses. If incorrect decisions are equally costly, we should therefore report the hypothesis that has the greatest posterior probability, in line with intuition. These calculations can clearly be extended to three or more hypotheses. The models that represent each hypothesis need not be nested as with likelihood ratio tests, though careful prior choice is required so as to not inadvertently favor one model over another. One remedy to this difficulty is described in Sect. 6.16.3.

To evaluate the Bayes factor, we need to calculate the normalizing constants under H_0 and H_1. A generic normalizing constant is

$$I = p(y) = \int p(y \mid \boldsymbol{\theta})\pi(\boldsymbol{\theta}) \, d\boldsymbol{\theta}. \qquad (3.39)$$

We next derive a popular approximation to the Bayes factor. The integral (3.39) is an integral of the form (3.18) with

$$nh(\boldsymbol{\theta}) = \log p(y \mid \boldsymbol{\theta}) + \log \pi(\boldsymbol{\theta}).$$

Letting $\widetilde{\boldsymbol{\theta}}$ denote the posterior mode, we may apply (3.20) with $nh(\widetilde{\boldsymbol{\theta}}) = \log p(y \mid \widetilde{\boldsymbol{\theta}}) + \log \pi(\widetilde{\boldsymbol{\theta}})$ to give the Laplace approximation

$$\log p(y) = \log p(y \mid \widetilde{\boldsymbol{\theta}}) + \log \pi(\widetilde{\boldsymbol{\theta}}) + \frac{p}{2}\log 2\pi - \frac{p}{2}\log n + \frac{1}{2}\log |\widetilde{v}|.$$

As n increases, the prior contribution will become negligible, and the posterior mode will be close to the MLE $\widehat{\boldsymbol{\theta}}$. Dropping terms of $O(1)$, we obtain the crude approximation

$$-2\log p(y) \approx -2\log p(y \mid \widehat{\boldsymbol{\theta}}) + p\log n.$$

Let hypothesis H_j be indexed by parameters $\boldsymbol{\theta}_j$ of length p_j and $\widehat{\boldsymbol{\theta}}_j$ denote the MLEs for $j = 0, 1$. Without loss of generality, assume $p_0 \leq p_1$. We may approximate twice the log Bayes factor by

$$2\left[\log p(\boldsymbol{y} \mid H_0) - \log p(\boldsymbol{y} \mid H_1)\right]$$
$$= 2\left[\log p(\boldsymbol{y} \mid \widehat{\boldsymbol{\theta}}_0) - \log p(\boldsymbol{y} \mid \widehat{\boldsymbol{\theta}}_1)\right] + (p_1 - p_0)\log n$$
$$= 2\left[l(\widehat{\boldsymbol{\theta}}_0) - l(\widehat{\boldsymbol{\theta}}_1)\right] + (p_1 - p_0)\log n \quad (3.40)$$

which is the log-likelihood ratio statistic (see Sect. 2.9.5) with the addition of a term that penalizes complexity; (3.40) is known as the Bayesian information criteria (BIC). The Schwarz criterion (Schwarz 1978) is the BIC divided by 2. If the maximized likelihoods are approximately equal, then model H_0 is preferred if $p_0 < p_1$, as it contains fewer parameters. As n increases, the penalty term increases in size showing the difference in behavior with frequentist tests in which significance levels are often kept constant with respect to sample size. A more detailed comparison of Bayesian and frequentist approaches to hypothesis testing will be carried out in Chap. 4.

3.11 Bayesian Inference Based on a Sampling Distribution

We now describe an approach to Bayesian inference which is pragmatic and computationally simple and allows frequentist summaries to be embedded within a Bayesian framework. This is useful in situations in which one would like to examine the impact of prior specification. It is also appealing to examine frequentist procedures with no formal Bayesian justification from a Bayesian slant. Suppose we are in a situation in which the sample size n is sufficiently large for accurate asymptotic inference and suppose we have a parameter $\boldsymbol{\theta}$ of length p. The sampling distribution of the estimator is

$$\widehat{\boldsymbol{\theta}}_n \mid \boldsymbol{\theta} \sim N_p(\boldsymbol{\theta}, \boldsymbol{V}_n),$$

where $\boldsymbol{V}_n$ is assumed known. The notation here is sloppy; it would be more accurate to state the distribution as

$$\boldsymbol{V}_n^{-1/2}(\widehat{\boldsymbol{\theta}}_n - \boldsymbol{\theta}) \sim N_p(\boldsymbol{0}, \boldsymbol{I}).$$

Appealing to conjugacy, it is then convenient to combine this "likelihood" with the prior $\boldsymbol{\theta} \sim N_p(\boldsymbol{m}, \boldsymbol{W})$ to give the posterior

$$\boldsymbol{\theta} \mid \widehat{\boldsymbol{\theta}}_n \sim N_p(\boldsymbol{m}_n^\star, \boldsymbol{W}_n^\star) \quad (3.41)$$

where

$$W_n^\star = (W^{-1} + V_n^{-1})^{-1}$$
$$m_n^\star = W_n^\star(W^{-1}m + V_n^{-1}\widehat{\theta}_n)$$

The posterior distribution is therefore easy to determine since we only require a point estimate $\widehat{\theta}_n$, with an associated variance–covariance matrix, and specification of the prior mean and variance–covariance matrix.

An even more straightforward approach, when a single parameter is of interest, is to ignore the remaining nuisance parameters and focus only on this single estimate and standard error. There are a number of advantages to this approach, not least of which is the removal of the need for prior specification over the nuisance parameters. Let θ denote the parameter of interest and $\boldsymbol{\alpha}$ the $(p \times 1)$ vector of nuisance parameters. Following Wakefield (2009a), we give a derivation beginning with the asymptotic distribution (we drop the explicit dependence on n for notational convenience):

$$\begin{bmatrix} \widehat{\boldsymbol{\alpha}} \\ \widehat{\theta} \end{bmatrix} \sim N_{p+1}\left(\begin{bmatrix} \boldsymbol{\alpha} \\ \theta \end{bmatrix}, \begin{bmatrix} I_{00} & I_{01} \\ I_{01}^{\mathrm{T}} & I_{11} \end{bmatrix}^{-1}\right) \quad (3.42)$$

where I_{00} is the $p \times p$ expected information matrix for $\boldsymbol{\alpha}$, I_{11} is the information concerning θ, and I_{01} is the $p \times 1$ vector of cross terms. We now reparameterize the model and consider $(\boldsymbol{\alpha}, \theta) \to (\boldsymbol{\gamma}, \theta)$ where

$$\boldsymbol{\gamma} = \boldsymbol{\alpha} + \frac{I_{01}}{I_{00}}\theta$$

which yields

$$\begin{bmatrix} \widehat{\boldsymbol{\gamma}} \\ \widehat{\theta} \end{bmatrix} \sim N_{p+1}\left(\begin{bmatrix} \boldsymbol{\gamma} \\ \theta \end{bmatrix}, \begin{bmatrix} I_{00}^\star & \mathbf{0} \\ \mathbf{0}^{\mathrm{T}} & I_{11} \end{bmatrix}^{-1}\right) \quad (3.43)$$

where $\widehat{\boldsymbol{\gamma}} = \widehat{\boldsymbol{\alpha}} + (I_{01}/I_{00})\widehat{\theta}$ and $\mathbf{0}$ is a $p \times 1$ vector of zeros. Hence, asymptotically, the "likelihood" factors into independent pieces

$$p(\widehat{\boldsymbol{\gamma}}, \widehat{\theta} \mid \boldsymbol{\gamma}, \theta) = p(\widehat{\boldsymbol{\gamma}} \mid \boldsymbol{\gamma}) \times p(\widehat{\theta} \mid \theta).$$

We now assume independent priors on $\boldsymbol{\gamma}$ and θ, $\pi(\boldsymbol{\gamma}, \theta) = \pi(\boldsymbol{\gamma})\pi(\theta)$, to give

$$p(\boldsymbol{\gamma}, \theta \mid \widehat{\boldsymbol{\gamma}}, \widehat{\theta}) = p(\widehat{\boldsymbol{\gamma}} \mid \boldsymbol{\gamma})\pi(\boldsymbol{\gamma})p(\widehat{\theta} \mid \theta)\pi(\theta)$$
$$= p(\boldsymbol{\gamma} \mid \widehat{\boldsymbol{\gamma}})p(\theta \mid \widehat{\theta})$$

so that the posterior factors also and we can concentrate on $p(\theta \mid \widehat{\theta})$ alone. The simple model

$$\widehat{\theta} \mid \theta \sim N(\theta, V)$$
$$\theta \sim N(m, W)$$

therefore results in the posterior

$$\theta \mid \widehat{\theta} \sim N\left[(W^{-1} + V^{-1})^{-1}(W^{-1}m + V^{-1}\widehat{\theta}), (W^{-1} + V^{-1})^{-1}\right]. \quad (3.44)$$

The above approach is similar to the "null orthogonality" reparameterization of Kass and Vaidyanathan (1992). The reparameterization is also that which is used when the linear model

$$Y_i = \alpha + x_i\theta + \epsilon_i$$

is written as

$$Y_i = \gamma + (x_i - \overline{x})\theta + \epsilon_i$$

which, of course, yields uncorrelated least squares estimators $\widehat{\gamma}, \widehat{\theta}$. The reparameterization trick works because of the assumption of independent priors on γ and θ which, of course, does not imply independent priors on α and θ. However, we emphasize that we do not need to explicitly specify priors on γ, because the terms involving γ cancel in the calculation.

Bayes factors can also be simply evaluated under either of the approximations, (3.41) or (3.44). To illustrate for the latter, suppose θ is univariate, and we wish to compare the hypotheses

$$H_0 : \theta = 0, \quad H_1 : \theta \neq 0,$$

with the prior under the alternative, $\theta \sim N(0, W)$. The Bayes factor is

$$\begin{aligned} BF &= \frac{p(\widehat{\theta} \mid \theta_0)}{\int p(\widehat{\theta} \mid \theta)\pi(\theta)\,d\theta} \\ &= \sqrt{\frac{V+W}{V}} \exp\left[-\frac{1}{2}\frac{\widehat{\theta}^2}{V}\frac{W}{V+W}\right]. \end{aligned} \quad (3.45)$$

This approach allows a Bayesian interpretation of published results, since all that is required for calculation of (3.45) is $\widehat{\theta}$ and V, which may be derived from a confidence interval or the estimate with its associated standard error.

More controversially, an advantage of the use of the asymptotic distribution of the MLE only is that the Bayes factor calculation may be based on nonstandard likelihoods or estimating functions which do not have formal Bayesian justifications. For example, the estimate and standard error may arise from conditional or marginal likelihoods (as described in Sect. 2.4.2), or using sandwich estimates of the variance. As discussed in Chap. 2, a strength of modern frequentist methods based on estimating functions is that estimators are produced that are consistent under much milder assumptions than were used to derive the estimators (e.g., the estimator may be based on a score equation, but the variance estimate may not require the likelihood to be correctly specified). The use of a consistent variance estimate with (3.45) allows the benefits of frequentist sandwich estimation and

Bayesian prior specification to be combined. Bayesian hypothesis testing may also be based on frequentist summaries. Exercises 3.10 and 3.11 give further details on the approach described in this section, including the extension to having estimators and standard errors from multiple studies.

3.12 Concluding Remarks

Bayesian analyses should not be restricted to convenient likelihoods and likelihood/prior combinations; this is especially true with the advent of modern computational approaches. However, one still needs to be careful that the sampling scheme (i.e., the design) is acknowledged by the likelihood specification and that the likelihood/prior combination leads to a proper posterior.

We now follow up on Sect. 1.6 and describe situations in which frequentist and Bayesian methods are likely to agree and when one is preferable over the other. We concentrate on estimation since point and interval estimation are directly comparable under the two paradigms. For model comparison, the objectives of Bayes factors and hypothesis tests are fundamentally different (see, e.g., Berger (2003)), and so comparison is more difficult. Chapter 4 compares and critiques frequentist and Bayesian approaches to hypothesis testing.

On a philosophical level, the Bayesian approach is satisfying since one simply follows the rules of probability as applied to the unknowns whether they be parameters or hypotheses. This is in stark contrast to the frequentist approach in which the parameters are fixed. Consequently, credible intervals are probabilistic and easily interpretable, and posterior distributions on parameters of interest are obtained through marginalization. Another appealing characteristic is that the Bayesian approach to inference may be formally derived via decision theory; see, for example, Bernardo and Smith (1994). A concept that has received a lot of discussion is the *likelihood principle* (Berger and Wolpert 1988; Royall 1997) which states that the likelihood function contains all relevant information. So two sets of data with proportional likelihoods should lead to the same conclusion. The likelihood principle leads one toward a Bayesian approach since all frequentist criteria invalidate this principle, and a true likelihood approach as followed by, for example, Royall (1997) is difficult to calibrate. The likelihood principle is a cornerstone of many Bayesian developments, but in this book we follow a far more pragmatic approach and so do not provide further details on this topic.

In contrast, the frequentist approach is more difficult to justify on philosophical grounds. Instead, much theory has been developed in terms of optimality within a frequentist set of guidelines. For example, as discussed in Sect. 2.8, there is a Gauss–Markov theorem for linear estimating functions (Godambe and Heyde 1987; McCullagh 1983), while Crowder (1987) considers the optimality of quadratic estimating functions.

We have seen that, so long as the prior does not exclude regions of the parameter space, Bayesian estimators have similar frequentist properties to MLEs. The greatest drawback of the Bayesian approach is the need to specify both a likelihood and a prior distribution. Sensitivity to each of these components can be examined, but carrying out such an endeavor in practice is difficult and one is then faced with the difficulty of how results should be reported. The frequentist approach to model misspecification is quite different, and the use of sandwich estimation to give a consistent standard error is very appealing. There is no Bayesian approach analogous to sandwich estimation, but see Szpiro et al. (2010) for some progress on a Bayesian justification of sandwich estimation.

For small n, Bayesian methods are desirable; in an extreme case if the number of parameters exceeds n, then a Bayesian approach (or some form of penalization, see Chaps. 10–12) must be followed. In this situation there is no way that the likelihood can be checked and inference will be sensitive to both likelihood and prior choices. When the model is very complex, then Bayesian methods are again advantageous since they allow a rigorous treatment of nuisance parameters; MCMC has allowed the consideration of more and more complicated hierarchical models, for example. Spatial models, particularly those that exploit Markov random field second stages, provide a good example of models that are very naturally analyzed using MCMC or INLA, where the conditional independencies may be exploited; see Sect. 9.7 for an illustrative example. Unfortunately, assessments of the effects of model misspecification are difficult for such complex models; instead sensitivity studies are again typically carried out. Consistency results under model misspecification are difficult to come by for complex models (such as those discussed in Chap. 9). Bayesian methods are also appealing in situations in which the maximum likelihood estimator provides a poor summary of the likelihood, for example, in variance components problems.

If n is sufficiently large for asymptotic normality of the sampling distribution to be accurate, then frequentist methods have advantages over Bayesian alternatives. In particular, as just mentioned, sandwich estimation can be used to provide a consistent estimator of the variance–covariance matrix of the estimator. Hence, if the estimator is consistent, reliable confidence coverage will be guaranteed. We stress that n needs to be sufficiently large for the sandwich estimator to be stable. A typical Bayesian approach would be to increase model complexity, often through the introduction of random effects. The difficulty with this is that although more flexibility is achieved, a specific form needs to be assumed for the mean–variance relationship, in contrast to sandwich estimation.

We briefly mention two topics which have not been discussed in this chapter. The *linear Bayesian* method (Goldstein and Wooff 2007) is an appealing approach in which Bayesian inference is carried out on the basis of expectation rather than probability. The appeal comes from the removal of the need to specify complete prior distributions, rather the means and variances of the parameters only require specification. The deviance information criterion (DIC) is a popular approach for comparison of models that was introduced by Spiegelhalter et al. (1998). The method is controversial, however, as the discussion of the aforementioned paper makes clear; see also Plummer (2008).

3.13 Bibliographic Notes

Bayes' original paper was published posthumously as Bayes (1763). The book by Jeffreys was highly influential: the original edition was published in 1939 and the third edition as Jeffreys (1961). Other influential works include Savage (1972) and translations of de Finetti's books, De Finetti (1974, 1975).

Bernardo and Smith (1994) provide a thorough description of the decision-theoretic justification of the Bayesian approach. O'Hagan and Forster (2004) give a good overview of Bayesian methodology and Gelman et al. (2004) and Carlin and Louis (2009) descriptions with a more practical flavor. Robert (2001) provides a decision-theoretic approach. Hoff (2009) is an excellent introductory text.

Approaches to addressing the sensitivity of inference to different prior choices, are described in O'Hagan (1994, Chap. 7). A good overview of methods for integration is provided by Evans and Swartz (2000). Lindley (1980), Tierney and Kadane (1986), and Kass et al. (1990) provide details of the Laplace method in a Bayesian context. Devroye (1986) provides an excellent and detailed overview of random variate generation. Smith and Gelfand (1992) emphasize the duality between samples and densities and illustrate the use of simple rejection algorithms in a Bayesian context. Gamerman and Lopes (2006) provides an introduction to MCMC; an up-to- date summary may be found in Brooks et al. (2011). Computational techniques that have not been discussed include reversible jump Markov chain Monte Carlo (Green 1995) which may be used when the parameter space changes dimension across models, variational approximations (Jordan et al. 1999; Ormerod and Wand 2010), and approximate Bayesian computation (ABC) (Beaumont et al. 2002; Fearnhead and Prangle 2012). Kass and Raftery (1995) give a review of Bayes factors, including a discussion of computation and prior choice. Johnson (2008) discusses the use of Bayes factors based on summary statistics.

3.14 Exercises

3.1 Derive the posterior mean and posterior quantiles as the solution to quadratic and linear loss, respectively, as described in Sect. 3.2.

3.2 Consider a random sample $Y_i \mid \theta \sim_{iid} N(\theta, \sigma^2), i = 1, \ldots, n$, with θ unknown and σ^2 known.

   (a) By writing the likelihood in exponential family form, obtain the conjugate prior and hence the posterior distribution.
   (b) Using the conjugate formulation, derive the predictive distribution for a new univariate observation Z from $N(\theta, \sigma^2)$, assumed conditionally independent of $Y_1, \ldots, Y_n$.

3.3 Consider the Neyman–Scott problem in which $Y_{ij} \mid \mu_i, \sigma^2 \sim_{ind} N(\mu_i, \sigma^2)$, $i = 1, \ldots, n, j = 1, 2$.

Table 3.4 Case-control data: $Y = 1$ corresponds to the event of esophageal cancer, and $X = 1$ exposure to greater than 80 g of alcohol per day

	$X = 0$	$X = 1$	
$Y = 1$	104	96	200
$Y = 0$	666	109	775

(a) Show that Jeffreys prior in this case is

$$\pi(\mu_1, \ldots, \mu_n, \sigma^2) \propto \sigma^{-n-2}.$$

(b) Derive the posterior distribution corresponding to this prior and show that

$$\mathrm{E}[\sigma^2 \mid \boldsymbol{y}] = \frac{1}{2(n-1)} \sum_{i=1}^{n} \frac{(Y_{i1} - Y_{i2})^2}{2}.$$

(c) Hence, using Exercise 2.6, show that $\mathrm{E}[\sigma^2 \mid \boldsymbol{y}] \to \sigma^2/2$ as $n \to \infty$, so that the posterior mean is inconsistent.

(d) Examine the posterior distribution corresponding to the prior

$$\pi(\mu_1, \ldots, \mu_n, \sigma^2) \propto \sigma^{-2}.$$

(e) Is the posterior mean for σ^2 consistent in this case?

3.4 Consider the data given in Table 3.4, which are a simplified version of those reported in Breslow and Day (1980). These data arose from a case-control study (Sect. 7.10) that was carried out to investigate the relationship between esophageal cancer and various risk factors. There are 200 cases and 775 controls. Disease status is denoted Y with $Y = 0/1$ corresponding to without/with disease, and alcohol consumption is represented by X with $X = 0/1$ denoting $< 80\,\mathrm{g}/ \geq 80\,\mathrm{g}$ on average per day. Let the probabilities of high alcohol consumption in the cases and controls be denoted

$$p_1 = \Pr(X = 1 \mid Y = 1) \quad \text{and} \quad p_2 = \Pr(X = 1 \mid Y = 0),$$

respectively. Further, let X_1 be the number exposed from n_1 cases and X_2 be the number exposed from n_2 controls. Suppose $X_i \mid p_i \sim \mathrm{Binomial}(n_i, p_i)$ in the case ($i = 1$) and control ($i = 2$) groups.

(a) Of particular interest in studies such as this is the *odds ratio* defined by

$$\theta = \frac{\Pr(Y = 1 \mid X = 1)/\Pr(Y = 0 \mid X = 1)}{\Pr(Y = 1 \mid X = 0)/\Pr(Y = 0 \mid X = 0)}.$$

3.14 Exercises

Show that the odds ratio is equal to

$$\theta = \frac{\Pr(X=1 \mid Y=1)/\Pr(X=0 \mid Y=1)}{\Pr(X=1 \mid Y=0)/\Pr(X=0 \mid Y=0)} = \frac{p_1/(1-p_1)}{p_2/(1-p_2)}.$$

(b) Obtain the MLE and a 90% confidence interval for θ, for the data of Table 3.4.

(c) We now consider a Bayesian analysis. Assume that the prior distribution for p_i is the beta distribution $\text{Be}(a, b)$ for $i = 1, 2$. Show that the posterior distribution $p_i \mid x_i$ is given by the beta distribution $\text{Be}(a + x_i, b + n_i - x_i)$, $i = 1, 2$.

(d) Consider the case $a = b = 1$. Obtain expressions for the posterior mean, mode, and standard deviation. Evaluate these posterior summaries for the data of Table 3.4. Report 90% posterior credible intervals for p_1 and p_2.

(e) Obtain the asymptotic form of the posterior distribution and obtain 90% credible intervals for p_1 and p_2. Compare this interval with the exact calculation of the previous part.

(f) Simulate samples $p_1^{(t)}, p_2^{(t)}$, $t = 1, \ldots, T = 1,000$ from the posterior distributions $p_1 \mid x_1$ and $p_2 \mid x_2$. Form histogram representations of the posterior distributions using these samples, and obtain sample-based 90% credible intervals.

(g) Obtain samples from the posterior distribution of $\theta \mid x_1, x_2$ and provide a histogram representation of the posterior. Obtain the posterior median and 90% credible interval for $\theta \mid x_1, x_2$ and compare with the likelihood analysis.

(h) Suppose the rate of esophageal cancer is 17 in 100,000. Describe how this information may be used to evaluate

$$q_1 = \Pr(Y=1 \mid X=1) \quad \text{and} \quad q_0 = \Pr(Y=1 \mid X=0).$$

3.5 Prove that if global balance, as given by (3.31), holds then $\pi(\cdot)$ is the invariant distribution, that is,

$$\pi(A) = \int_{\mathbb{R}^p} \pi(\boldsymbol{x}) P(\boldsymbol{x}, A) \, d\boldsymbol{x},$$

for all measurable sets A.

3.6 Prove that the Metropolis–Hastings algorithm, defined through (3.33), has invariant distribution $\pi(\cdot)$, by showing that detailed balance (3.31) holds.

3.7 We consider the data described in the example at the end of Sect. 3.7.7 concerning the leukemia count, Y, assumed to follow a Poisson distribution with mean $E \times \delta$. Consider the $y = 4$ observed leukemia cases in Seascale, with expected number of cases $E = 0.25$. Previously in this chapter, a lognormal prior was assumed for δ. In this exercise, a conjugate gamma prior will be used.

(a) Show that with a Ga(a,b) prior, the posterior distribution for δ is a gamma distribution also. Hence, determine the posterior mean, mode, and variance. Show that the posterior mean can be written as a weighted combination of the MLE and the prior mean. Similarly write the posterior mode as a weighted combination of the MLE and the prior mode.

(b) Determine the form of the prior predictive $\Pr(y)$ and show that it corresponds to a negative binomial distribution.

(c) Obtain the predictive distribution $\Pr(z \mid y)$ for the number of cases z in a future period of time with expected number of cases $E^\star$.

(d) Obtain the posterior distribution under gamma prior distributions with parameters $a = b = 0.1$, $a = b = 1.0$, and $a = b = 10$. Determine the 5%, 50%, and 95% posterior quantiles in each case and comment on the sensitivity to the prior.

3.8 Consider a situation in which the likelihood may be summarized as

$$\sqrt{n}(\overline{Y}_n - \mu) \to_d N(0, \sigma^2),$$

where $\overline{Y}_n = \frac{1}{n}\sum_{i=1}^n Y_i$, with σ^2 known, and the prior for μ is the Cauchy distribution with parameters 0 and 1, that is,

$$p(\mu) = \frac{1}{\pi(1+\mu^2)}, \quad -\infty < \mu < \infty.$$

We label this likelihood-prior combination as model M_c.

(a) Describe a rejection algorithm for obtaining samples from the posterior distribution, with the proposal density taken as the prior.

(b) Implement the rejection algorithm for the case in which $\bar{y} = 0.2$, $\sigma^2 = 2$ and $n = 10$. Provide a histogram representation of the posterior, and evaluate the posterior mean and variance. Also obtain an estimate of the normalizing constant, $p(\boldsymbol{y} \mid M_c)$.

(c) Describe an importance sampling algorithm for evaluating $p(\boldsymbol{y} \mid M_c)$, $E[\mu \mid \boldsymbol{y}, M_c]$, and $\text{var}(\mu \mid \boldsymbol{y}, M_c)$.

(d) For the data of part (b), implement the importance sampling algorithm, and calculate $p(\boldsymbol{y} \mid M_c)$ and $E[\mu \mid \boldsymbol{y}, M_c]$ and $\text{var}(\mu \mid \boldsymbol{y}, M_c)$.

(e) Now assume that the prior for μ is the normal distribution $N(0, 0.4)$. Denote this model M_n. Obtain the form of the posterior distribution in this case.

(f) For the data of part (b), obtain the normalizing constant $p(\boldsymbol{y} \mid M_n)$ and the posterior mean and variance. Compare these summaries with those obtained under the Cauchy prior. Interpret the ratio

$$\frac{p(\boldsymbol{y} \mid M_n)}{p(\boldsymbol{y} \mid M_c)},$$

that is, the Bayes factor, for these data.

3.14 Exercises

Table 3.5 Genetic data from an experiment carried out by Mendel that concerned the numbers of peas that were classified by their shape and color

Round yellow	Wrinkled yellow	Round green	Wrinkled green	Total
n_1	n_2	n_3	n_4	n_+
315	101	108	32	556

3.9 The data in Table 3.5 result from one of the famous experiments carried out by Mendel in which pure bred peas with wrinkled green seeds were crossed with pure bred peas with wrinkled green seeds. These data are given on page 15 of the English translation (Mendel 1901) of Mendel (1866). All of the first-generation hybrids had round yellow seeds (since this characteristic is dominant), but when these plants were self-pollinated, four different phenotypes (characteristics) were observed and are displayed in Table 3.5.

A model for these data is provided by the multinomial $M_4(n_+, \boldsymbol{p})$ where $\boldsymbol{p} = [p_1, p_2, p_3, p_4]^\mathrm{T}$, and p_j denotes the probability of falling in cell j, $j = 1, \ldots, 4$, that is,

$$\Pr(\boldsymbol{N} = \boldsymbol{n} \mid \boldsymbol{p}) = \frac{n_+!}{\prod_{j=1}^4 n_j!} \prod_{j=1}^4 p_j^{n_j},$$

where $\boldsymbol{N} = [N_1, \ldots, N_4]^\mathrm{T}$ and $\boldsymbol{n} = [n_1, \ldots, n_4]^\mathrm{T}$. In this exercise a Bayesian analysis of these data will be carried out using the conjugate Dirichlet prior distribution, $\mathrm{Dir}(a_1, a_2, a_3, a_4)$:

$$p(\boldsymbol{p}) = \frac{\Gamma\left(\sum_{j=1}^4 a_j\right)}{\prod_{j=1}^4 \Gamma(a_j)} \prod_{j=1}^4 p_j^{a_j - 1},$$

where $a_j > 0$, $j = 1, \ldots, 4$, are specified a priori.

(a) Show that the marginal prior distributions for p_j are the beta distributions $\mathrm{Be}(a_j, a - a_j)$, where $a = \sum_{j=1}^4 a_j$.
(b) Obtain the distributional form, and the associated parameters, of the posterior distribution $p(\boldsymbol{p} \mid \boldsymbol{n})$.
(c) For the genetic data and under a prior for $\boldsymbol{p}$ that is uniform over the simplex (i.e., $a_1 = a_2 = a_3 = a_4 = 1$), evaluate $\mathrm{E}[p_j \mid \boldsymbol{n}]$ and $\mathrm{s.d.}(p_j \mid \boldsymbol{n})$, $j = 1, \ldots, 4$.
(d) Obtain histogram representations and 90% credible intervals for $p_j \mid \boldsymbol{n}$, $j = 1, \ldots, 4$.
(e) Determine the form of the predictive distribution for $[N_1, N_2, N_3, N_4]$ given $n_+ = \sum_j n_j$. Describe how a sample from this predictive distribution could be obtained.

A particular model of interest is that which states that genes are inherited independently of each other, so that the ratio of counts is 9:3:3:1, or

$$H_0 : p_{10} = \frac{9}{16}, p_{20} = \frac{3}{16}, p_{30} = \frac{3}{16}, p_{40} = \frac{1}{16}.$$

The evidence in favor of this model, versus the alternative of $H_1 : p$ unspecified, will now be determined.

(f) For the data in Table 3.5, carry out a likelihood ratio test comparing H_0 and H_1.
(g) Obtain analytical expressions for $\Pr(n \mid H_0)$ and $\Pr(n \mid H_1)$.
(h) Evaluate the Bayes factor $\Pr(n \mid H_0)/\Pr(n \mid H_1)$ for the genetic data. Comment on the evidence for/against H_0 and compare with the conclusion from the likelihood ratio test statistic.

3.10 With respect to Sect. 3.11, consider the "likelihood," $\widehat{\theta} \mid \theta \sim N(\theta, V)$ and the prior $\theta \sim N(0, W)$. Show that $\theta \mid \widehat{\theta} \sim N(r\widehat{\theta}, rV)$ where $r = W/(V + W)$.

3.11 Again consider the situation discussed in Sect. 3.11 in which a Bayesian analysis is carried out based not on the full data but rather on summary statistics.

(a) Suppose data are to be combined from two studies with a common underlying parameter θ. The estimates from the two studies are $\widehat{\theta}_1, \widehat{\theta}_2$ with standard errors $\sqrt{V_1}$ and $\sqrt{V_2}$ (with the two estimators being conditionally independent given θ). Show that the Bayes factor that summarizes the evidence from the two studies, that is,

$$\frac{p(\widehat{\theta}_1, \widehat{\theta}_2 \mid H_0)}{p(\widehat{\theta}_1, \widehat{\theta}_2 \mid H_1)},$$

takes the form

$$\text{BF}(\widehat{\theta}_1, \widehat{\theta}_2) = \sqrt{\frac{W}{RV_1V_2}} \exp\left[-\frac{1}{2}\left(Z_1^2 RV_2 + 2Z_1 Z_2 R\sqrt{V_1 V_2} + Z_2^2 RV_1\right)\right]$$

where $R = W/(V_1 W + V_2 W + V_1 V_2)$ and $Z_1 = \widehat{\theta}_1/\sqrt{V_1}$ and $Z_2 = \widehat{\theta}_2/\sqrt{V_2}$ are the usual Z-statistics.

(b) Suppose now there are K studies with estimates $\widehat{\theta}_k$ and asymptotic variances V_k, $k = 1, \ldots, K$, and again assume a common underlying parameter θ. Show that the Bayes factor

$$\frac{p(\widehat{\theta}_1, \ldots, \widehat{\theta}_K \mid H_0)}{p(\widehat{\theta}_1, \ldots, \widehat{\theta}_K \mid H_1)},$$

3.14 Exercises

takes the form

$$\mathrm{BF}(\widehat{\theta}_1,\ldots,\widehat{\theta}_K)$$

$$= \frac{\prod_{k=1}^{K}(2\pi V_k)^{-1/2}\exp\left(-\frac{\widehat{\theta}_k^2}{2V_k}\right)}{\int \prod_{k=1}^{K}(2\pi V_k)^{-1/2}\exp\left(-\frac{\left(\widehat{\theta}_k-\theta\right)^2}{2V_k}\right)(2\pi W)^{-1/2}\exp\left(-\frac{\theta^2}{2W}\right)d\theta}$$

$$= \sqrt{W\left(W^{-1}+\sum_{k=1}^{K}V_k^{-1}\right)}\exp\left[-\frac{1}{2}\left(\sum_{k=1}^{K}\frac{\widehat{\theta}_k}{V_k}\right)^2\left(W^{-1}+\sum_{k=1}^{K}V_k^{-1}\right)^{-1}\right].$$

Further, show that the posterior summarizing beliefs about θ given the K estimates is

$$\theta\mid\widehat{\theta}_1,\ldots,\widehat{\theta}_K\sim\mathrm{N}(\mu,\sigma^2)$$

where

$$\mu = \left(\sum_{k=1}^{K}\frac{\widehat{\theta}_k}{V_k}\right)\left(W^{-1}+\sum_{k=1}^{K}V_k^{-1}\right)^{-1}$$

and

$$\sigma^2 = \left(W^{-1}+\sum_{k=1}^{K}V_k^{-1}\right)^{-1}.$$

Chapter 4
Hypothesis Testing and Variable Selection

4.1 Introduction

In Sects. 2.9 and 3.10, we briefly described the frequentist and Bayesian machinery for carrying out hypothesis testing. In this chapter we extend this discussion, with an emphasis on critiquing the various approaches and on hypothesis testing in a regression setting. We examine both single and multiple hypothesis testing situations; Sects. 4.2 and 4.3 consider the frequentist and Bayesian approaches, respectively. Section 4.4 describes the well-known Jeffreys–Lindley paradox that highlights the starkly different conclusions that can occur when frequentist and Bayesian hypothesis testing is carried out. This is in contrast to estimation, in which conclusions are often in agreement. In Sects. 4.5–4.7, various aspects of multiple testing are considered. The discussion includes situations in which the number of tests is known a priori and variable selection procedures in which the number of tests is driven by the data. Section 4.9 provides a discussion of the impact on inference that the careless use of variable selection can have. Section 4.10 describes a pragmatic approach to variable selection. Concluding remarks appear in Section 4.11.

4.2 Frequentist Hypothesis Testing

Early in this chapter we will consider a univariate parameter $\theta \in \mathbb{R}$. Suppose we are interested in evaluating the evidence in the data with respect to the null hypothesis:

$$H_0 : \theta = \theta_0$$

using a statistic T. By convention, large values are less likely under the null. The observed value of the test statistic is t_{obs}. As discussed in Sect. 2.9, there are various possibilities for T including squared Wald, likelihood ratio, and score statistics. Under regularity conditions, $T \to_d \chi_1^2$ under the null, as $n \to \infty$. If n is not large, or regularity conditions are violated, permutation or Monte Carlo tests (perhaps based on bootstrap samples, as described in Sect. 2.7) can often be performed to derive the empirical distribution of the test statistic under the null. A type I error is said to occur when we reject H_0 when it is in fact true, while a type II error is to not reject H_0 when it is false.

4.2.1 Fisherian Approach

Under the null, for continuous sample spaces, the tail-area probability $\Pr(T > t \mid H_0)$ is uniform. This is not true for discrete sample spaces, but in the following, unless stated otherwise, we will assume we are in situations in which uniformity holds. Let

$$p = \Pr(T > t_{\text{obs}} \mid H_0)$$

denote the observed p-value, the probability of observing t_{obs}, *or a more extreme value*, with repeated sampling under the null.

Fisher advocated the pure test of significance, in which the observed p-value is reported as the measure of evidence against the null (Fisher 1925a), with H_0 being rejected if p is small. Alternative hypotheses are not explicitly considered and so there is no concept of rejecting the null in favor of a specific alternative; ideally, the test statistic will be chosen to have high power under plausible alternatives, however.

4.2.2 Neyman–Pearson Approach

In contrast to the procedure of Fisher, the Neyman–Pearson approach is to specify an alternative hypothesis, H_1, with H_0 nested in H_1. The celebrated Neyman–Pearson lemma of Neyman and Pearson (1933) proved that, for fixed type I error

$$\alpha = \Pr(T > t_{\text{fix}} \mid H_0),$$

the most powerful procedure is provided by the likelihood ratio test (Sect. 2.9.5). The decision rule is to reject the null if $p < \alpha$. Due to the *fixed* threshold, this procedure controls the type I error at α.

4.2.3 Critique of the Fisherian Approach

A common explanation for seeing a "small" p-value is that *either H_0 is not true or H_0 is true and we have been "unlucky."* A major practical difficulty is on defining "small." Put another way, how do we decide on a *threshold* for significance?

4.2 Frequentist Hypothesis Testing

The p-value is uniform under the null, but with a large sample size, we will be able to detect very subtle departures from the null and so will often obtain small p-values because the null is rarely "true." To rectify this a confidence interval for θ is often reported, along with the p-value, so that the scientific significance of the departure of θ from θ_0 can be determined. The ability to detect smaller and smaller differences from the null with increasing sample size suggests that the p-value threshold rule used in practice should decrease with increasing n, but there are no universally recognized rules. In a hypothesis testing context a natural definition of consistency is that the rule for rejection is such that the probability of the correct decision being made tends to 1 as the sample size increases. So the current use of p-values, in which typically 0.05 or 0.01 is used as a threshold for rejection, regardless of sample size, is *inconsistent*; by construction, the probability of rejecting the null when it is true does not decrease to zero with increasing sample size. By contrast, the type II error will typically decrease to zero with increasing sample size. A more balanced approach than placing special emphasis on the type I error would be to have both type I and type II errors decrease to zero as n increases.

There are two common misinterpretations of p-values. The most basic is to interpret a p-value as the probability of the null given the data, which is a serious misconception. Probabilities of the truth of hypotheses are only possible under a Bayesian approach. More subtly, using the *observed* value of the test statistic t_{obs} does not allow one to say that following the general procedure will result in control of the type I error at p, because the threshold is data-dependent and not fixed. The key observation is that the p-value is associated with, "observing t_{obs}, or a more extreme value," so that the tail area begins at the *observed* value of the statistic. For example, if $p = 0.013$, we cannot say that the procedure controls the type I error at 1.30%. Such control of the type I error is provided by a fixed α level procedure which is based on a fixed threshold, t_{fix} with $\alpha = \Pr(T > t_{fix} \mid H_0)$.

There is some merit in the consideration of a tail area when one wishes to control the type I error rate, but when no such control is sought, the use of a tail area seems simply of mathematically convenience. As an alternative the ordinate $p(T = t_{obs} \mid H_0)$ may be considered, which brings one closer to a Bayesian formulation (see Sect. 4.3.1), but from a frequentist perspective, it is not clear how to scale the observed statistic without an alternative hypothesis.

4.2.4 Critique of the Neyman–Pearson Approach

As with the use of p-values we need to decide on a size α for the test. The historical emphasis has been on fixing α and then evaluating power, but as with a threshold for p-values, practical guidance on how α should depend on sample size is important but lacking. With an α level that does not change with sample size, one is implicitly accepting that type II errors become more important with increasing sample size, and in a manner which is implied rather than chosen by the investigator. Pearson (1953, p. 68) expressed the desirability of a decreasing α as sample size increases:

"... the quite legitimate device of reducing α as n increases." As we have already noted, a fixed significance level with respect to n gives an inconsistent procedure.

By merely stating that $p < \alpha$, information is lost, but if we state an observed p-value, then we lose control of the type I error because control requires a fixed binary decision rule. The procedure must also be viewed in the light of both H_0 and H_1 being "wrong" since no model is a correct specification of the data-generating process.

For discrete data, the discreteness of the statistic causes difficulties, particularly for small sample sizes. To achieve exact level α tests, so-called randomization rules have been suggested. Under such rules, the same set of data may give different conclusions depending on the result of the randomization, which is clearly undesirable.

4.3 Bayesian Hypothesis Testing with Bayes Factors

4.3.1 Overview of Approaches

In the Bayesian approach, all unknowns in a model are treated as random variables, even though they relate to quantities that are in reality fixed. Therefore, the "true" hypothesis is viewed as an unknown parameter for which the posterior is derived, once the alternatives have been specified. The latter step is essential since we require a sample space of hypotheses. In the case of two hypotheses, we have the following candidate data-generating mechanisms:

$$H_0 \;\Rightarrow\; \beta_0 \mid H_0 \;\Rightarrow\; \boldsymbol{y} \mid \beta_0$$
$$H_1 \;\Rightarrow\; \beta_1 \mid H_1 \;\Rightarrow\; \boldsymbol{y} \mid \beta_1.$$

The posterior probability of H_j is, via Bayes theorem,

$$\Pr(H_j \mid \boldsymbol{y}) = \frac{p(\boldsymbol{y} \mid H_j) \times \pi_j}{p(\boldsymbol{y})}$$

with π_j the prior probability of hypothesis H_j, $j = 0, 1$. The likelihood of the data is

$$p(\boldsymbol{y} \mid H_j) = \int p(\boldsymbol{y} \mid \beta_j) p(\beta_j \mid H_j) \, d\beta_j \qquad (4.1)$$

with $p(\beta_j \mid H_j)$ the prior distribution over the parameters associated with hypothesis H_j, $j = 0, 1$, and

$$p(\boldsymbol{y}) = p(\boldsymbol{y} \mid H_0) \times \pi_0 + p(\boldsymbol{y} \mid H_1) \times \pi_1.$$

4.3 Bayesian Hypothesis Testing with Bayes Factors

The posterior odds in favor of H_0 is therefore

$$\text{Posterior Odds} = \frac{\Pr(H_0 \mid \boldsymbol{y})}{\Pr(H_1 \mid \boldsymbol{y})} = \text{Bayes factor} \times \text{Prior Odds} \quad (4.2)$$

where the

$$\text{Bayes factor} = \frac{p(\boldsymbol{y} \mid H_0)}{p(\boldsymbol{y} \mid H_1)}, \quad (4.3)$$

and the prior odds are π_0/π_1 with $\pi_1 = 1 - \pi_0$. The Bayes factor is the ratio of the density of the data under the null to the density under the alternative and is an intuitively appealing summary of the information the data provide concerning the hypotheses. The Bayes factor was discussed previously in Sect. 3.10. From (4.2), we also see that

$$\text{Bayes Factor} = \frac{\text{Posterior Odds}}{\text{Prior Odds}},$$

which emphasizes that the Bayes factor summarizes the information in the data and does not involve the prior beliefs about the hypotheses. As can be seen in (4.1), priors on the parameters are involved in each of the numerator and denominator of the Bayes factor, since these provide the distributions over which the likelihoods are averaged.

When it comes to reporting/making decisions, various approaches based on Bayes factors are available for different contexts. Most simply, one may just report the Bayes factor. Kass and Raftery (1995), following Jeffreys (1961), present a guideline for the interpretation of Bayes factors. For example, if the negative log base 10 Bayes factor lies between 1 and 2 (so that the data are 10–100 times more likely under the alternative, as compared to the null), then there is said to be *strong* evidence against the null hypothesis. Such thresholds may be useful in some situations, but in general one would like the guidelines to be context driven. Going beyond the consideration of the Bayes factor only, one may include prior probabilities on the null and alternative, to give the posterior odds (4.2). Stating the posterior probabilities may be sufficient, but one may wish to derive a formal rule for deciding upon which of H_0 or H_1 to report.

Recall from Sect. 3.10 that, under a Bayesian *decision theory* approach to hypothesis testing, the "decision" δ is taken that minimizes the posterior expected loss. Following the notation of Table 3.3, the losses associated with type I and type II errors are L_I and L_II, respectively. Minimization of the posterior expected loss then results in the rule to choose $\delta = 1$ if

$$\frac{\Pr(H_1 \mid \boldsymbol{y})}{\Pr(H_0 \mid \boldsymbol{y})} \geq \frac{L_\text{I}}{L_\text{II}},$$

or equivalently if

$$\Pr(H_1 \mid \boldsymbol{y}) \geq \frac{1}{1 + L_\text{II}/L_\text{I}}. \quad (4.4)$$

For example, if a type I error is four times as bad as a type II error, we should report H_1 only if $\Pr(H_1 \mid y) \geq 0.8$. In contrast, if the balance of losses is reversed, and a type II error is four times as costly as a type I error, we report H_1 if $\Pr(H_1 \mid y) \geq 0.2$.

Discreteness of the sample space does not pose any problems for a Bayesian analysis, since one need only consider the data actually observed and not other hypothetical realizations.

4.3.2 Critique of the Bayes Factor Approach

As always with the Bayesian approach, we need to specify priors for all of the unknowns, which here correspond to each of the hypotheses and all parameters (including nuisance parameters) that are contained within the models defined under the two hypotheses. It turns out that placing improper priors upon the parameters that are the focus of the hypothesis test leads to anomalous behavior of the Bayes factor. We give an informal discussion of the fundamental difference between estimation and hypothesis testing with respect to the choice of improper priors. Suppose we have a model that depends on a univariate unknown parameter, θ with improper prior $p(\theta) = c$, for arbitrary $c > 0$. The posterior, upon which estimation is based, is

$$\frac{p(y \mid \theta)p(\theta)}{\int p(y \mid \theta)p(\theta)\, d\theta} \tag{4.5}$$

and so the arbitrary constant in the prior cancels in both numerator and denominator. Now suppose we are interested in comparison of the hypotheses $H_0 : \theta = \theta_0$, $H_1 : \theta \neq \theta_0$ with $\theta \in \mathbb{R}$. The Bayes factor is

$$\frac{p(y \mid H_0)}{p(y \mid H_1)} = \frac{p(y \mid \theta_0)}{\int p(y \mid \theta)p(\theta)\, d\theta},$$

so that the denominator of the Bayes factor depends, crucially, upon c. Hence, in this setting the Bayes factors with an improper prior on θ is not well defined.

Specifying prior distributions for all of the parameters under each hypothesis can be difficult, but Sect. 3.11 describes a strategy based on test statistics which requires a prior distribution for the parameter of interest only.

In principle, one can compare non-nested models using a Bayesian approach, but in practice great care must be taken in specifying the priors under the two hypotheses, in order to not inadvertently favor one hypothesis over another. One possibility is to specify priors on functions of the parameters that are meaningful under both hypotheses; for an example of this approach, see Sect. 6.16.

As with the Neyman–Pearson approach, all of the calculations have to be conditioned upon $H_0 \cup H_1$. In a Bayesian context, we need to emphasize that we are obtaining the posterior probability of the null given one of the null or alternative is true and under the assumed likelihood and priors. Consequently, posterior probabilities on hypotheses must be viewed in a relative, rather than an absolute, sense since the truth will rarely correspond to H_0 or H_1. Hence, the precise interpretation is that the posterior probability of H_0 is the posterior probability of H_0, *given* that one of H_0 or H_1 is true.

If one follows the decision theory route, one must also specify the ratio of losses which is usually difficult. In general, Bayes factor calculation requires analytically intractable integrals over the null and alternative parameter spaces, to give the two normalizing constants $p(\boldsymbol{y} \mid H_0)$ and $p(\boldsymbol{y} \mid H_1)$. Further, Markov chain Monte Carlo approaches do not simply supply these normalizing constants. Analytical approximations exist under certain conditions, see Sect. 3.10.

4.3.3 A Bayesian View of Frequentist Hypothesis Testing

We consider an artificial situation in which the only available data in a Bayesian analysis corresponds to knowing that the event $T > t_{\text{fix}}$ has occurred. This means that the likelihood of the data, $\Pr(\text{data} \mid H_0)$ coincides with the α level. To obtain $\Pr(H_0 \mid \text{data})$ we must specify the alternative hypothesis. We consider the simple case in which the model contains a single parameter θ with null $H_0 : \theta = \theta_0$ and alternative $H_1 : \theta = \theta_1$. Then

$$\Pr(H_0 \mid \text{data}) = \frac{\Pr(\text{data} \mid H_0) \times \pi_0}{\Pr(\text{data} \mid H_0) \times \pi_0 + \Pr(\text{data} \mid H_1) \times \pi_1} \quad (4.6)$$

where $\pi_j = \Pr(H_j), j = 0, 1$. Dividing by $\Pr(H_1 \mid \text{data})$ gives

$$\text{Posterior Odds} = \frac{\Pr(\text{data} \mid H_0)}{\Pr(\text{data} \mid H_1)} \times \text{Prior Odds}$$

$$= \frac{\alpha}{\text{power at } \theta_1} \times \text{Prior Odds} \quad (4.7)$$

which depends, in addition to the α level, on the *prior* on H_0, π_0, and on the *power*, $\Pr(\text{data} \mid H_1)$. Equation (4.7) implies that, for two studies that report a result as significant at the same α level, the one with the greater power will, in a Bayesian formulation, provide greater evidence against the null. The power is never explicitly considered when reporting under the Fisherian or Neyman–Pearson approaches. An important conclusion is that to make statements about the "evidence" that the data contain with respect to a hypothesis, as summarized in an α level, one would want to know the power or, as a minimum, the sample size (since this is an important component of the power).

The prior is also important which seems, as already noted, reasonable when one considers the usual interpretation of a tail area in terms of "either H_0 is true and we were unlucky or H_0 is not true." A prior on H_0 is very useful in weighing these two possibilities. A key observation is that although a particular dataset may be unlikely

Fig. 4.1 Lower bound for $\Pr(H_0 \mid \text{data})$, under three prior specifications, as a function of the p-value

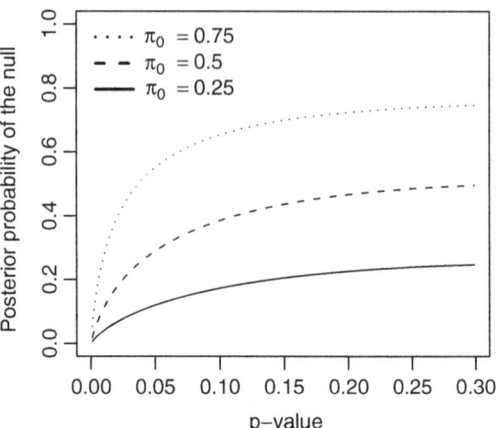

under the null, it may also be unlikely under chosen alternatives, so that there may be insufficient evidence to reject the null, at least in comparison to these alternatives.

Sellke et al. (2001) summarize a number of different arguments that lead to the following, quite remarkable, result. For a p-value $p < e^{-1} = 0.368$:

$$\Pr(H_0 \mid \text{data}) \geq \left[1 - \left(\frac{1}{ep \log p} \times \frac{\pi_1}{\pi_0}\right)^{-1}\right]^{-1}. \tag{4.8}$$

Hence, given a p-value, one may calculate a lower bound on the posterior probability of the null. Figure 4.1 illustrates this lower bound, as a function of the p-value, for three different prior probabilities, π_0. We see, for example, that with a p-value of 0.05 and a prior probability on the null of $\pi_0 = 0.75$, we obtain $\Pr(H_0 \mid \text{data}) \geq 0.55$.

The discussion of Sect. 4.2.3, combined with the implications of (4.7) and (4.8), might prompt one to ask why p-values are still in use today, in particular with the almost ubiquitous application of a 0.05 or 0.01 decision threshold. With these thresholds, which are often required for the publication of results, the relationship (4.8), with $\pi_0 = 0.5$, gives $\Pr(H_0 \mid \text{data}) \geq 0.29$ and 0.11 with $p = 0.05$ and 0.01, respectively. Rejection of H_0 with such probabilities may not be unreasonable in some circumstances but the difference between the p-value and $\Pr(H_0 \mid \text{data})$ is apparent.

Small prior probabilities, π_0, were not historically the norm since, particularly in experimental situations, data would not be collected if there were little chance the alternative were true.

In some disciplines scientists may calibrate p-values to the sample sizes with which they are familiar, as no doubt Fisher did when the 0.05 rule emerged. For example, in Tables 29 and 30 of Statistical Methods for Research Workers (Fisher 1990), the sample sizes were 30 and 17, and Fisher discusses the 0.05 limit in each case, though in both cases he concentrates more on the context than on the absolute value of 0.05.

4.4 The Jeffreys–Lindley Paradox

Poor calibration of p-values could be one of the reasons why so many "findings" are not reproducible, along with the other usual suspects of confounding, data dredging, multiple testing, and poorly measured covariates.

4.4 The Jeffreys–Lindley Paradox

We now discuss a famous example in which Bayesian and frequentist approaches to hypothesis testing give starkly different conclusions. The example has been considered by many authors, but Lindley (1957) and Jeffreys (1961) provide early discussions; see also Bartlett (1957). To illustrate the so-called Jeffreys–Lindley "paradox," we assume that $\overline{Y}_n \mid \theta \sim \mathrm{N}(\theta, \sigma^2/n)$ with σ^2 known and θ unknown. Suppose the null is $H_0 : \theta = 0$, with alternative $H_1 : \theta \neq 0$. Let

$$\overline{y}_n = z_{1-\alpha/2} \times \sigma/\sqrt{n}$$

where α is the level of the test and $\Pr(Z < z_{1-\alpha/2}) = 1 - \alpha/2$, with $Z \sim \mathrm{N}(0,1)$. We define $\overline{y}_n$ in this manner, so that for different values of n the α level remains constant. For a Bayesian analysis, assume $\pi_0 = \Pr(H_0)$, and under the alternative $\theta \sim \mathrm{N}(0, \tau^2)$. In the early discussions of the paradox, a uniform prior over a finite range was assumed, but the message of the paradox is unchanged with the use of a normal prior. Then

$$\Pr(H_0 \mid \overline{y}_n) = \frac{\text{Bayes Factor} \times \text{Prior Odds}}{1 + \text{Bayes Factor} \times \text{Prior Odds}}$$

where the Bayes factor is

$$\text{Bayes Factor} = \frac{p(\overline{y}_n \mid H_0)}{p(\overline{y}_n \mid H_1)} \qquad (4.9)$$

and the Prior Odds $= \pi_0/(1 - \pi_0)$. The prior predictive distributions, the ratios of whose densities give the Bayes factor (4.9), are

$$\overline{y}_n \mid H_0 \sim \mathrm{N}(0, \sigma^2/n) \qquad (4.10)$$

$$\overline{y}_n \mid H_1 \sim \mathrm{N}(0, \sigma^2/n + \tau^2). \qquad (4.11)$$

Figure 4.2 shows these two densities, as a function of $\overline{y}_n$, for $\sigma^2 = 1, \tau^2 = 0.2^2$, and $n = 100$. An α level of 0.05 gives $\overline{y}_n = 1.96 \times \sigma/\sqrt{n} = 0.20$, the value indicated in the figure with a dashed-dotted vertical line. For this value, the Bayes factor equals 0.48, so that the data are roughly twice as likely under the alternative as compared to the null. The Sellke et al. (2001) bound on the Bayes factor is $\text{BF} \geq -ep \log p$ which for $p = 0.05$ gives $\text{BF} \geq 0.41$.

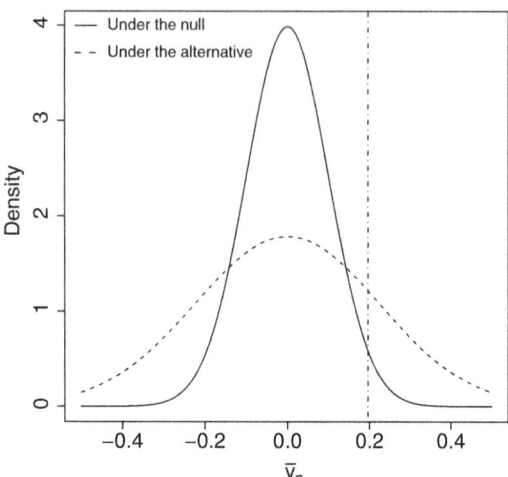

Fig. 4.2 Numerator (*solid line*) and denominator (*dashed line*) of the Bayes factor for $n = 100$. The model is $\bar{Y}_n \mid \theta \sim N(\theta, \sigma^2/n)$ with $\sigma^2 = 1$. The null and alternative are $H_0 : \theta = 0$ and $H_1 : \theta \neq 0$, and the prior under the alternative is $\theta \sim N(0, \tau^2)$ with $\tau^2 = 0.2^2$. The *dashed-dotted vertical line* corresponds to $\bar{y}_n = 0.20$ which for this n gives $\alpha = 0.05$

The Bayes factor is the ratio of (4.10) and (4.11):

$$\text{Bayes Factor} = \frac{(2\pi\sigma^2/n)^{-1/2} \exp\left[-\frac{\bar{y}_n^2}{2\sigma^2/n}\right]}{(2\pi[\sigma^2/n + \tau^2])^{-1/2} \exp\left[-\frac{\bar{y}_n^2}{2(\sigma^2/n + \tau^2)}\right]}$$

$$= \sqrt{\frac{\sigma^2/n + \tau^2}{\sigma^2/n}} \exp\left[-\frac{z_{1-\alpha/2}^2}{2} \frac{\tau^2}{\tau^2 + \sigma^2/n}\right]. \quad (4.12)$$

This last expression reveals that, as $n \to \infty$, the Bayes factor $\to \infty$, so that $\Pr(H_0 \mid \bar{y}_n) \to 1$. Therefore, the "paradox" is that for a level of significance α, chosen to be arbitrarily small, we can find datasets which make the posterior probability of the null arbitrarily close to 1, for some n. Hence, frequentist and Bayes procedures can, for sufficiently large sample size, come to opposite conclusions with respect to a hypothesis test.

Figure 4.3 plots the posterior probability of the null as a function of n for $\sigma^2 = 1, \tau^2 = 0.2^2, \pi_0 = 0.5, \alpha = 0.05$. From the starting position of 0.5 (the prior probability, indicated as a dashed line), the curve $\Pr(H_0 \mid \bar{y}_n)$ initially falls, reaching a minimum at around $n = 100$, and then increases towards 1, illustrating the "paradox." For large values of n, $\bar{y}_n$ is very close to the null value of 0, but there is high power to detect any difference from 0, and so an α of 0.05 is not difficult to achieve. The Bayes factor also incorporates the density under the alternative and values close to 0 are more likely under the null, as illustrated in Fig. 4.2.

We now consider a Bayesian analysis of the above problem but assume that the data appear only in the form of knowing that $|\bar{Y}_n| \geq \bar{y}_n$, a censored observation. This is clearly not the usual situation since a Bayesian would condition on the *actual* value observed, but it does help to understand the paradox. The Bayes factor is

4.4 The Jeffreys–Lindley Paradox

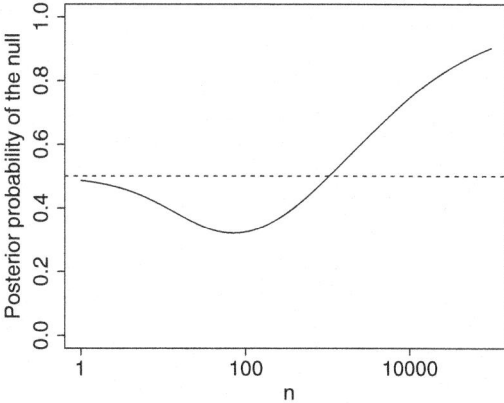

Fig. 4.3 Posterior probability of the null versus sample size, for a fixed α level of 0.05. The model is $\overline{Y}_n \mid \theta \sim \mathrm{N}(\theta, \sigma^2/n)$ with $\sigma^2 = 1$. The null and alternative are $H_0 : \theta = 0$ and $H_1 : \theta \neq 0$, and the prior under the alternative is $\theta \sim \mathrm{N}(0, \tau^2)$ with $\tau^2 = 0.2^2$

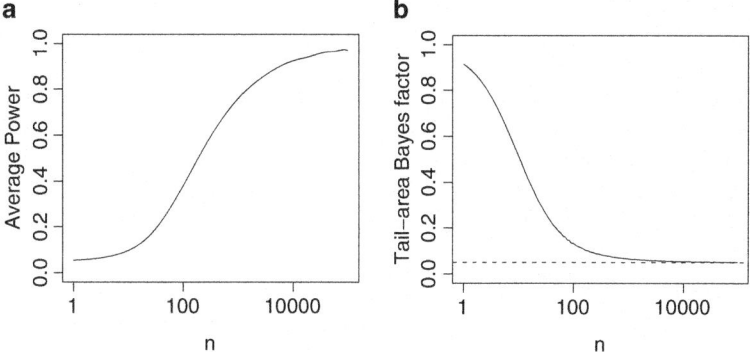

Fig. 4.4 Bayes factor based on a tail area with null and alternative of $H_0 : \theta = 0$ and $H_1 : \theta \neq 0$: (a) Average power, which corresponds to the denominator of the Bayes factor, under a $\mathrm{N}(0, 0.2^2)$ prior and for a fixed α level of 0.05 and (b) Bayes factor based on the tail area, with $\alpha = 0.05$; the *horizontal dashed line* indicates a tail-area Bayes factor value of 0.05

$$\frac{\Pr(|\overline{Y}_n| \geq \overline{y}_n | H_0)}{\Pr(|\overline{Y}_n| \geq \overline{y}_n | H_1)} = \frac{\alpha}{\int \Pr(|\overline{Y}_n| \geq \overline{y}_n | \theta) p(\theta) \, d\theta},$$

that is, the type I error rate divided by the power averaged over the prior $p(\theta)$. Figure 4.4a gives the average power as a function of n. We see a monotonic increase with sample size towards the value 1, as we would expect with fixed α.

Since the Bayes factor is the ratio of α to the average power, we see in Fig. 4.4b that the Bayes factor based on the tail-area information is monotonic decreasing towards α as n increases (and with $\pi_0 = 0.5$, this gives the posterior probability of the null also). For our present purposes, the calculation with the tail area illustrates that when a Bayesian analysis conditions on a tail area, the conclusions are in line with a frequentist analysis.

The difference in behavior between a genuine Bayesian analysis that conditions on the actual statistic and that based on conditioning on a tail area is apparent. As noted by Lindley (1957, p. 189–190), "... the paradox arises because the significance level argument is based on the area under a curve and the Bayesian argument is based on the ordinate of the curve."

Ignoring now the comparison with tests of significance, it is informative to examine the Bayes factor for fixed $\overline{y}_n$. Upon rearrangement of (4.12),

$$\text{Bayes Factor} = \sqrt{\frac{\sigma^2 + n\tau^2}{\sigma^2}} \exp\left[-\frac{\overline{y}_n^2}{2} \frac{n/\sigma^2}{1 + \sigma^2/n\tau^2}\right].$$

As $\tau^2 \to \infty$, the Bayes Factor $\to \infty$ so that $\Pr(H_0 \mid \overline{y}_n) \to 1$, which is at first sight counter intuitive since increasing τ^2 places *less* prior mass close to $\theta = 0$. However, this behavior occurs because averaging with respect to the prior on θ with large τ^2 produces a small $\Pr(\overline{y}_n \mid H_1)$, because the prior under the alternative is spreading mass very thinly across a large range; $\tau^2 \gg 0$ suggests very little prior belief in any $\theta \neq 0$. Hence, even if the data point strongly to a particular $\theta \neq 0$, we still prefer H_0. More generally, $\tau^2 \gg 0$ should not be interpreted as "ignorance" since it supports very *big* effects. Said another way, as $\tau^2 \to 0$, the Bayes factor favors the alternative, even though as τ^2 gets smaller and smaller the prior under the alternative becomes more and more concentrated about the null.

4.5 Testing Multiple Hypotheses: General Considerations

In the following sections we examine how inference proceeds when more than a single hypothesis test is performed. There are many situations in which multiple hypothesis testing arises, but we concentrate on just two. In the first, which we refer to as a *fixed number of tests* scenario, we suppose that the number of hypotheses to be tested is known a priori, and is not data driven, which makes the task of evaluating the properties of proposed solutions (both frequentist and Bayesian) more straightforward. This case is discussed in Sect. 4.6. As an example, we will shortly introduce a running example that concerns comparing, between two populations, expression levels for $m = 1,000$ gene transcripts (during transcription, a gene is transcribed into (mutiple) RNA transcripts). In the second situation, which we refer to as *variable selection*, and which is discussed in Sect. 4.7, the number of hypotheses to be tested is random, which makes the evaluation of properties more difficult.

One of the biggest abuses of statistical techniques is the unprincipled use of model selection. Two examples of this are separately testing the significance of a large number of variables and then reporting only those that are nominally "significant" (the problem considered in Sect. 4.6), and testing multiple confounders to see which ones to control for (the problem considered in Sect. 4.7). In each of these cases, even if the exact procedure is described, unless care is exercised, interpretation is extremely difficult.

4.6 Testing Multiple Hypotheses: Fixed Number of Tests

Suppose we wish to examine the association between a response and m different covariates. In a typical epidemiological study, many potential risk factors are measured, and an exploratory, hypothesis-generating procedure may systematically examine the association between the outcome and each of the risk factors. In general, the covariates may not be independent, which complicates the analysis. Another fixed number of tests scenario is when m responses are examined with respect to a single covariate. Recently, there has been intense interest in so-called high throughput techniques in which thousands, or tens of thousands, of variables are measured, often as a screening exercise in which the aim is to see which of the variables are associated with some biological endpoint. For example, one may examine whether the expression levels of many thousands of genes are elevated or reduced in samples from cancer patients, as compared to cancer-free individuals.

When m tests are preformed, the aim is to decide which of the nulls should be rejected. Table 4.1 shows the possibilities when m tests are performed and K are flagged as requiring further attention. Here m_0 is the number of *true nulls*, B is the number of *type I errors*, and C is the number of *type II errors*, and each of these quantities is unknown. The aim is to select a rule on the basis of some criterion and this in turn will determine K. The internal cells of Table 4.1 are random variables, whose distribution depends on the rule by which K is derived.

Example: Microarray Data

To illustrate the multiple testing problem in a two-group setting, we examine a subset of microarray data presented by Kerr (2009). The data we analyze consist of expression levels on $m = 1,000$ transcripts measured in Epstein-Barr virus-transformed lymphoblastic cell line tissue, in each of two populations. Each transcript was measured on 60 individuals of European ancestry (CEU) and 45 ethnic Chinese living in Beijing (CHB). The data have been normalized, and $\log_2$ transformed, so that a one-unit difference between recorded values corresponds to a doubling of expression level.

Let $\overline{Y}_{ki}$ be the measured expression level for transcript i in population k, with $i = 1, \ldots, m$, and $k = 0/1$ representing the CEU/CHB populations. Then define

Table 4.1 Possibilities when m tests are performed and K are flagged as worthy of further attention

	Not flagged	Flagged	
H_0	A	B	m_0
H_1	C	D	m_1
	$m-K$	K	m

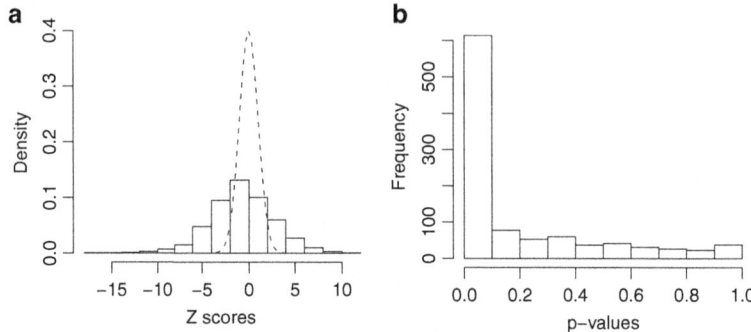

Fig. 4.5 (**a**) Z scores and (**b**) p-values, for 1,000 transcripts in the microarray data

$Y_i = \overline{Y}_{1i} - \overline{Y}_{0i}$ and let s_{ki}^2 be the sample variance in population k, for transcript i, $i = 1, \ldots, m$. We now assume

$$Y_i \mid \mu_i \sim_{iid} N(\mu_i, \sigma_i^2)$$

where $\sigma_i^2 = s_{1i}^2/60 + s_{0i}^2/45$ is the sample variance, which is reliably estimated for the large sample sizes in the two populations and therefore assumed known. The null hypotheses of interest are that the difference in the average expression level between the two populations is zero. We let $H_i = 0$ correspond to the null for transcript i, that is, $\mu_i = 0$ for $i = 1, \ldots, m$. Figure 4.5a gives a histogram of the Z scores Y_i/σ_i, along with the reference $N(0, 1)$ distribution. Clearly, unless there are problems with the model formulation, there are a large number of transcripts that are differentially expressed between the two populations, as confirmed by the histogram of p-values displayed in Fig. 4.5b.

4.6.1 Frequentist Analysis

In a single test situation we have seen that the historical emphasis has been on control of the type I error rate. We let $H_i = 0/1$ represent the hypotheses for the $i = 1, \ldots, m$ tests. In a multiple testing situation there are a variety of criteria that may be considered. With respect to Table 4.1, the *family-wise error rate* (FWER) is the probability of making *at least* one type I error, that is, $\Pr(B \geq 1 \mid H_1 = 0, \ldots, H_m = 0)$. Intuitively, this is a sensible criteria if one has a strong prior belief that all (or nearly all) of the null hypotheses are true, since in such a situation making at least one type I error should be penalized (this is made more concrete in Sect. 4.6.2). In contrast, if one believes that a number of the nulls are likely to be false, then one would be prepared to accept a greater number of type I errors, in exchange for discovering more true associations. As in all hypothesis testing situations, we want a method for trading off type I and type II errors.

4.6 Testing Multiple Hypotheses: Fixed Number of Tests

Table 4.2 True FWER as a function of the correlation ρ between two bivariate normal test statistics

ρ	True FWER
0	0.0497
0.3	0.0484
0.5	0.0465
0.7	0.0430
0.9	0.0362

Let B_i be the event that the ith null is incorrectly rejected, so that, with respect to Table 4.1, B, the random variable representing the number of incorrectly rejected nulls, corresponds to $\cup_{i=1}^{m} B_i$. With a common level for each test $\alpha^\star$, the FWER is

$$\alpha_{\text{F}} = \Pr(B \geq 1 \mid H_1 = 0, \ldots, H_m = 0) = \Pr\left(\cup_{i=1}^{m} B_i \mid H_1 = 0, \ldots, H_m = 0\right)$$

$$\leq \sum_{i=1}^{m} \Pr(B_i \mid H_1 = 0, \ldots, H_m = 0)$$

$$= m\alpha^\star. \tag{4.13}$$

The *Bonferroni* method takes $\alpha^\star = \alpha_{\text{F}}/m$ to give FWER $\leq \alpha_{\text{F}}$. For example, to control the FWER at a level of $\alpha = 0.05$ with $m = 10$ tests, we would take $\alpha^\star = 0.05/10 = 0.005$. Since it controls the FWER, the Bonferroni method is stringent (i.e., conservative in the sense that the bar is set high for rejection) and so can result in a loss of power in the usual situation in which the FWER is set at a low value, for example 0.05. A little more conservatism is also introduced via the inequality, (4.13). The Sidák correction, which we describe shortly, overcomes this aspect.

If the test statistics are independent,

$$\Pr(B \geq 1) = 1 - \Pr(B = 0)$$
$$= 1 - \Pr\left(\cap_{i=1}^{m} B_i'\right)$$
$$= 1 - \prod_{i=1}^{m} \Pr(B_i')$$
$$= 1 - (1 - \alpha^\star)^m.$$

Consequently, to achieve FWER $= \alpha_{\text{F}}$ we may take $\alpha^\star = 1 - (1 - \alpha_{\text{F}})^{1/m}$, the so-called Sidák correction (Sidák 1967).

With dependent tests, the Bonferroni approach is even more conservative; we demonstrate with $m = 2$ and bivariate normal test statistics with correlation ρ. Suppose we wish to achieve a FWER of 0.05. Table 4.2 gives the FWER achieved using Bonferroni and illustrates how the test becomes more conservative as the correlation increases. The situation becomes worse as m increases in size. The k-FWER criteria (Lehmann and Romano 2005) extends FWER to the incorrect rejection of k or more nulls (Exercise 4.2).

A simple remedy to the conservative nature of the control of FWER is to increase α_F. An intuitive measure to calibrate a procedure is via the expected number of false discoveries:

$$\text{EFD} = m_0 \times \alpha^\star$$
$$\leq m \times \alpha^\star$$

where $\alpha^\star$ is the level for each test. If m_0 is close to m, this inequality will be practically useful. As an example, one could specify $\alpha^\star$ such that the EFD ≤ 1 (say), by choosing $\alpha^\star = 1/m$.

Recently there has been interest in a criterion that is particularly useful in multiple testing situations. We first define the false discovery proportion (FDP) as the proportion of incorrect rejections:

$$\text{FDP} = \begin{cases} \frac{B}{K} & \text{if } B > 0 \\ 0 & \text{if } B = 0. \end{cases}$$

Then the *false discovery rate* (FDR), the expected proportion of rejected nulls that are actually true, is

$$\text{FDR} = \text{E}[\,\text{FDP}\,] = \text{E}[\,B/K \mid B > 0\,]\Pr(B > 0).$$

Consider the following procedure for independent p-values, each of which is uniform under the null:

1. Let $P_{(1)} < \ldots < P_{(m)}$ denote the ordered p-values.
2. Define $l_i = i\alpha/m$ and $R = \max\{i : P_{(i)} < l_i\}$ where α is the value at which we would like FDR control.
3. Define the p-value threshold as $p_T = P_{(R)}$.
4. Reject all hypotheses for which $P_i \leq P_T$, that is, set $H_i = 1$ in such cases, $i = 1, \ldots, m$.

Benjamini and Hochberg (1995) show that if this procedure is applied, then regardless of how many nulls are true (m_0) and regardless of the distribution of the p-values when the null is false,

$$\text{FDR} \leq \frac{m_0}{m}\alpha < \alpha.$$

We say that the FDR is controlled at α.

Example: Hypothetical Data

We simulate data from $m = 100$ hypothetical tests in which $m_0 = 95$ tests are null, to give $m_1 = 5$ tests for which the alternative is true. Figure 4.6 displays the sorted observed $-\log_{10}(p\text{-values})$ versus the expected $-\log_{10}(p\text{-values})$, along

4.6 Testing Multiple Hypotheses: Fixed Number of Tests

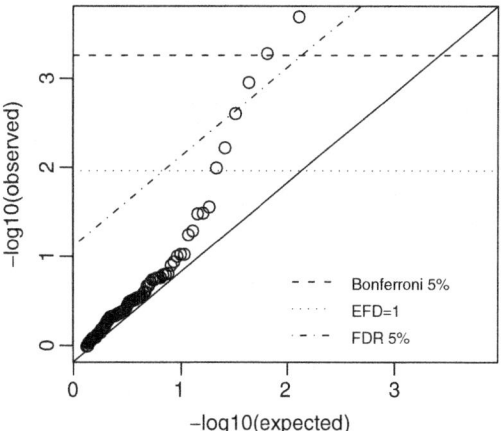

Fig. 4.6 Observed versus expected $-\log_{10}(p\text{-values})$ for a simulated set of data with 95 nulls and 5 alternatives. Three criteria for rejection, based on Bonferroni, the expected number of false discoveries (EFD), and the false discovery rate (FDR), are included on the plot

with a line of equality (solid line). Also displayed are three approaches to calling significance. The top dashed line corresponds to a Bonferroni correction at the 5% level (so that the line is at $-\log_{10}(0.05/100) = 3.30$). This criterion calls a single test as significant illustrating the conservative nature of the control of FWER at a low value. If we choose instead to control the expected number of false discoveries at 1, then the dotted line at $-\log_{10}(1/100) = 2$ results. We see that all 5 true alternatives are selected, along with a single false positive. Finally, we examine those hypotheses that would be rejected if we control the FDR at $\alpha = 0.05$, via the Benjamini–Hochberg procedure. On the log to the base 10 scale, the potential thresholds $l_i = i\alpha/m$, $i = 1, \ldots, m$ correspond to a line with slope 1 and intercept $-\log_{10}(\alpha)$. The dotted-dashed line gives the FDR threshold (recall the FDR is an expectation) corresponding to $\alpha = 0.05$. The use of this threshold gives three p-values as significant, for an empirical FDR of zero.

□

The algorithm of Benjamini and Hochberg (1995) begins with a desired FDR and then provides the p-value threshold. Storey (2002) proposed an alternative method by which, for any fixed rejection region, a criteria closely related to FDR, the *positive false discovery rate* pFDR $= E[B/K \mid K > 0]$, may be estimated. We assume rejection regions of the form $T > t_{\text{fix}}$ and consider the pFDR associated with regions of this form, which we write as pFDR(t_{fix}). We define, for $i = 1, \ldots, m$ tests, the random variables $H_i = 0/1$ corresponding to null/alternative hypotheses and test statistics T_i. Then, with $\pi_0 = \Pr(H = 0)$ and $\pi_1 = 1 - \pi_0$ independently for all tests,

$$\text{pFDR}(t_{\text{fix}}) = \frac{\Pr(T > t_{\text{fix}} \mid H = 0) \times \pi_0}{\Pr(T > t_{\text{fix}} \mid H = 0) \times \pi_0 + \Pr(T > t_{\text{fix}} \mid H = 1) \times \pi_1}.$$

Note the similarity with (4.6). Consideration of the false discovery odds:

$$\frac{\text{pFDR}(t_{\text{fix}})}{1 - \text{pFDR}(t_{\text{fix}})} = \frac{\Pr(T > t_{\text{fix}} \mid H = 0)}{\Pr(T > t_{\text{fix}} \mid H = 1)} \times \frac{\pi_0}{\pi_1}$$

explicitly shows the weighted trade-off of type I and type II errors, with weights determined by the prior on the null/alternative; this expression mimics (4.7). Storey (2003) rigorously shows that

$$\text{pFDR}(t_{\text{fix}}) = \Pr(H = 0 \mid T > t_{\text{fix}}).$$

giving a Bayesian interpretation. In terms of p-values, the rejection region corresponding to $T > t_{\text{fix}}$ is of the form $[0, \gamma]$. Let P be the random p-value resulting from a test. Under the null, $P \sim U(0, 1)$, and so

$$\begin{aligned}\text{pFDR}(t_{\text{fix}}) &= \frac{\Pr(P \leq \gamma \mid H = 0) \times \pi_0}{\Pr(P \leq \gamma)} \\ &= \frac{\gamma \times \pi_0}{\Pr(P \leq \gamma)}.\end{aligned} \quad (4.14)$$

From this expression, the crucial role of π_0 is evident. Storey (2002) estimates (4.14), using uniformity of p-values under the null, to produce the estimates

$$\begin{aligned}\widehat{\pi}_0 &= \frac{\#\{p_i > \lambda\}}{m(1 - \lambda)} \\ \widehat{\Pr}(P \leq \gamma) &= \frac{\#\{p_i \leq \gamma\}}{m}\end{aligned} \quad (4.15)$$

with λ chosen via the bootstrap to minimize the mean-squared error for prediction of the pFDR. The expression (4.15) calculates the empirical proportion of p-values to the right of λ and then inflates this to account for the proportion of null p-values in $[0, \lambda]$.

This method highlights the benefits of using the totality of p-values to estimate fundamental quantities of interest such as π_0. In general, information in all of the data may also be exploited, and in Sect. 4.6.2, we describe a Bayesian mixture model that uses the totality of data.

The *q-value* is the minimum FDR that can be attained when a particular test is called significant. We give a derivation of the q-value and, following Storey (2002), first define a set of nested rejection regions $\{t_\alpha\}_{\alpha=0}^1$ where α is such that $\Pr(T > t_\alpha \mid H = 0) = \alpha$. Then

$$p\text{-value}(t) = \inf_{t_\alpha : t \in t_\alpha} \Pr(T > t_\alpha \mid H = 0)$$

is the p-value corresponding to an observed statistic t. The q-value is defined as

$$q\text{-value}(t) = \inf_{t_\alpha : t \in t_\alpha} \Pr(H = 0 \mid T > t_\alpha). \quad (4.16)$$

4.6 Testing Multiple Hypotheses: Fixed Number of Tests

Therefore, for each observed statistic t_i, there is an associated q-value. It can be shown that (Exercise 4.3)

$$\Pr(H_0 \mid T > t_{\text{obs}}) < \Pr(H_0 \mid T = t_{\text{obs}}) \qquad (4.17)$$

so that the evidence for H_0 given the exact ordinate is always greater than that corresponding to the tail area.

When one decides upon a value of FDR (or pFDR) to use in practice, the sample size should again be taken into account, since for large sample size, one would not want to tolerate as large an FDR as with a small sample size. Again, we would prefer a procedure that was consistent. However, as in the single test situation, there is no prescription for deciding how the FDR should decrease with increasing sample size.

Example: Microarray Data

Returning to the microarray example, application of the Bonferroni correction to control the FWER at 0.05 produces a list of 220 significant transcripts. In this context, it is likely that there are a large proportion of non-null transcripts (Storey et al. 2007) and there are relatively large sample sizes for each test (so the power is good), and so this choice is likely to be very conservative. The procedure of Benjamini and Hochberg with FDR control at 0.05 gives 480 significant transcripts. Applying the method of Storey gives an estimate of the proportion of nulls as $\widehat{\pi}_0 = 0.33$. At a pFDR threshold of 0.05, 603 transcripts are highlighted.

4.6.2 Bayesian Analysis

In some situations, a Bayesian analysis of m tests may proceed in exactly the same fashion as with a single test, that is, one can apply the same procedure m times; see Wakefield (2007a) for an example. In this case the priors on each of the m null hypotheses will be independent. In other situations, however, one may often wish to jointly model the data so that the totality of information can be used to estimate parameters that are common to all tests.

In terms of reporting, as with a single test (as considered in Sect. 4.3), the Bayes factors

$$\text{Bayes Factor}_i = \frac{p(\boldsymbol{y}_i \mid H_i = 0)}{p(\boldsymbol{y}_i \mid H_i = 1)}, \qquad (4.18)$$

$i = 1, \ldots, m$ are a starting point. These Bayes factors may then be combined with prior probabilities $\pi_{0i} = \Pr(H_i = 0)$, to give

$$\text{Posterior Odds}_i = \text{Bayes Factor}_i \times \text{Prior Odds}_i, \qquad (4.19)$$

where Prior Odds$_i = \pi_{0i}/(1 - \pi_{0i})$.

Proceeding to a decision theory approach. Suppose for simplicity common losses, L_I and L_II, associated with type 1 and type 2 errors, for each test. The aim is to define a rule for deciding which of the m null hypotheses to reject. The operating characteristics, in terms of "false discovery" and "non-discovery," corresponding to this rule may then be determined. The loss associated with a particular set of decisions $\boldsymbol{\delta} = [\delta_1, \ldots, \delta_m]$ and hypotheses $\boldsymbol{H} = [H_1, \ldots, H_m]$ is the expectation over the posterior

$$\mathrm{E}[L(\boldsymbol{\delta}, \boldsymbol{H})] = L_\text{I} \sum_{i=1}^{m} \left[\delta_i \Pr(H_i = 0 \mid \boldsymbol{y}_i) + \frac{L_\text{II}}{L_\text{I}}(1 - \delta_i) \Pr(H_i = 1 \mid \boldsymbol{y}_i) \right]$$

$$= L_\text{I} \left[\mathrm{EFP} + \frac{L_\text{II}}{L_\text{I}} \times \mathrm{EFN} \right]$$

where EFP is the expected number of false positives and EFN is the expected number of false negatives. These characteristics of the procedure are given, respectively, by

$$\mathrm{EFD} = \sum_{i=1}^{m} \delta_i \Pr(H_i = 0 \mid \boldsymbol{y}_i)$$

$$\mathrm{EFN} = \sum_{i=1}^{m} (1 - \delta_i) \Pr(H_i = 1 \mid \boldsymbol{y}_i),$$

where $\Pr(H_i = 0 \mid \boldsymbol{y}_i)$ and $\Pr(H_i = 1 \mid \boldsymbol{y}_i)$ are the posterior probabilities on the null and alternative. We should report test i as significant if

$$\Pr(H_i = 1 \mid \boldsymbol{y}_i) \geq \frac{1}{1 + L_\text{II}/L_\text{I}},$$

which is identical to the expression derived for a single test, (4.4).

Define $K = \sum_{i=1}^{m} \delta_i$ as the number of rejected tests. Then dividing EFD by K gives an estimate, based on the posterior, of the proportion of false discoveries, and dividing EFN by $m - K$ gives a posterior estimate of the proportion of false non-discoveries. Hence, for a given ratio of losses, we can determine the expected number of false discoveries and false non-discoveries, and the FDR and FNR. As n_i, the sample size associated with test i, increases, under correct specification of the model, the power for each test increases, and so EFD/K and $\mathrm{EFN}/(m-K)$ will tend to zero (assuming the model is correct). This is in contrast to the frequentist approach in which a fixed (independent of sample size) FDR rule is used so that the false non-discovery rate does not decrease to zero (even when the model is true).

Notice that the use of Bayes factors does not depend on the number of tests, m, so that, for example, we could analyze the data in the same way regardless of whether m is 1 or 1,000,000. Similarly, for the assumed independent priors, the posterior probabilities do not depend on m, and for the loss structure considered, the decision

4.6 Testing Multiple Hypotheses: Fixed Number of Tests

does not depend on m. Hence, the Bayes procedure gives thresholds that depend on n (since the Bayes factor will depend on sample size, see Exercise 4.1 for an example) but not on m, while the contrary is true for many frequentist procedures such as Bonferroni.

There is a prior that results in a Bayesian Bonferroni-type correction. If the prior probabilities of each of the nulls are independent with $\pi_{0i} = \pi_0$ for $i = 1, \ldots, m$. Then the prior probability that all nulls are true is

$$\Pi_0 = \Pr(H_1 = 0, \ldots, H_m = 0) = \pi_0^m$$

which we refer to as prior P_1. For example, if $\pi_0 = 0.5$ and $m = 10$, $\Pi_0 = 0.00098$, which may not reflect the required prior belief. Suppose instead that we wish to fix the prior probability that all of the nulls are true at Π_0. A simple way of achieving this is to take $\pi_{0i} = \Pi_0^{1/m}$, a prior specification we call P_2. Westfall et al. (1995) show that for independent tests

$$\alpha_{\text{B}} = \Pr(H_i = 0 \mid \boldsymbol{y}_i, P_2) \approx m \times \Pr(H_i = 0 \mid \boldsymbol{y}_i, P_1) = m \times \alpha_{\text{B}}^\star$$

so that a Bayesian version of Bonferroni is recovered.

An alternative approach is to specify a full model for the totality of data. These data can then be exploited to estimate common parameters. In particular, the proportion of null tests π_0 can be estimated, which is crucial for inference since posterior odds and decisions are (unsurprisingly) highly sensitive to the value of π_0. The decision is still based on the posterior, and there continues to be a trade-off between false positive and false negatives depending on the decision threshold used. We illustrate using the microarray data.

Example: Microarray Data

Recall that we assume $Y_i \mid \mu_i \sim_{ind} \text{N}(\mu_i, \sigma_i^2)$, $i = 1, \ldots, m$ where $m = 1{,}000$. We first describe a Bayesian analysis in which the m transcripts are analyzed separately. We assume under the null that $\mu_i = 0$, while under the alternative $\mu_i \sim_{iid} \text{N}(0, \tau^2)$ with τ^2 fixed. For illustration, we assume that for non-null genes, a fold change in the mean greater than 10%, that is, $\log_2 \mu_i > 0.138$, only occurs with probability 0.025. Given

$$\Pr\left(-\infty < \frac{\mu_i}{\tau} < \frac{\log_2(1.1)}{\tau}\right) = 0.975$$

we can solve for τ to give

$$\tau = \frac{\log_2(1.1)}{\Phi^{-1}(0.975)} = 0.070,$$

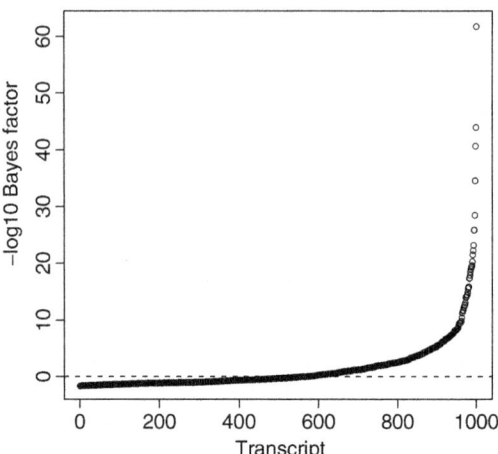

Fig. 4.7 Ordered $-\log_{10}(\text{Bayes factors})$ for the microarray data. The *dashed line* at 0 is for reference

where $\Phi(\cdot)$ is the distribution function of a standard normal random variable. The prior on μ_i is therefore

$$\mu_i = \begin{cases} 0 & \text{with probability } \pi_0 \\ N(0, 0.138^2) & \text{with probability } \pi_1 = 1 - \pi_0 \end{cases}.$$

The Bayes factor for the ith transcript is

$$\text{Bayes Factor}_i = \sqrt{\frac{\sigma_i^2 + \tau^2}{\tau^2}} \exp\left[-\frac{Z_i^2}{2} \frac{\tau^2}{\sigma_i^2 + \tau^2}\right] \quad (4.20)$$

where $Z_i = Y_i/\sigma_i$ is the Z score for the ith transcript. Therefore, we see that the Bayes factor depends on the power through σ_i^2 (which itself depends on the sample sizes), as well as on the Z-score, while the p-value depends on the latter only. In Fig. 4.7, we plot the ordered $-\log_{10}(\text{Bayes factors})$ (so that high values correspond to evidence against the null). A reference line of 0 is indicated and, using this reference, for 487 transcripts the data are more likely under the alternative than under the null.

To obtain the posterior odds, we need to specify a prior for the null. We assume $\pi_0 = \Pr(H_i = 0)$ so that the prior is the same for all transcripts. The posterior odds are the product of the Bayes factor and the prior odds and are highly sensitive to the choice of π_0. For illustration, suppose the decision rule is to call a transcript significant if the posterior odds of $H = 0$ are less than 1 (which corresponds to a ratio of losses, $L_{II}/L_I = 1$). Figure 4.8 plots the number of such significant transcripts under this rule, as a function of the prior, π_0. The sensitivity to the choice of π_0 is evident. To overcome this problem, we now describe a joint model for the data on all $m = 1{,}000$ transcripts that allows estimation of parameters that are common across transcripts, including π_0. Notice that for virtually the complete range of π_0 more transcripts would be called as significant under the Bayes rule than under the FWER.

4.6 Testing Multiple Hypotheses: Fixed Number of Tests

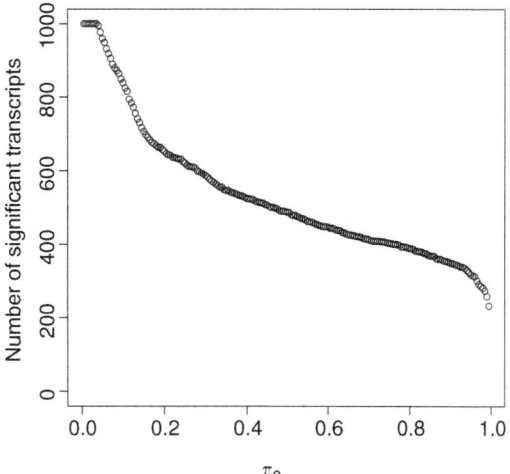

Fig. 4.8 Number of significant transcripts in the microarray data, as measured by the posterior probability on the null $\Pr(H_i = 0 \mid \boldsymbol{y}) < 0.5$, $i = 1, \ldots, 1,000$, as a function of the common prior probability on the null, $\pi_0 = \Pr(H_i = 0)$

We specify a mixture model for the collection $[\mu_1, \ldots, \mu_m]$, with

$$\mu_i = \begin{cases} 0 & \text{with probability } \pi_0 \\ \mathrm{N}(\delta, \tau^2) & \text{with probability } \pi_1 = 1 - \pi_0 \end{cases}.$$

We use mixture component indicators $H_i = 0/1$ to denote the zero/normal membership model for transcript i. Collapsing over μ_i gives the three-stage model:

Stage One:

$$Y_i \mid H_i, \delta, \tau, \pi_0 \sim_{ind} \begin{cases} \mathrm{N}(0, \sigma_i^2) & \text{if } H_i = 0 \\ \mathrm{N}(\delta, \sigma_i^2 + \tau^2) & \text{if } H_i = 1. \end{cases}$$

Stage Two: $\quad H_i \mid \pi_1 \sim_{iid} \mathrm{Bernoulli}(\pi_1), i = 1, \ldots, m.$

Stage Three: Independent priors on the common parameters:

$$p(\delta, \tau, \pi_0) = p(\delta) p(\tau) p(\pi_0).$$

We illustrate the use of this model with

$$p(\delta) \propto 1,$$
$$p(\tau) \propto 1/\tau$$
$$p(\pi_0) = 1,$$

so that we have improper priors for δ and τ^2. The latter choice still produces a proper posterior because we have fixed variances at the first stage of the model (see Sect. 8.6.2 for further discussion). Implementation is via a Markov chain Monte Carlo algorithm (see Sect. 3.8). Exercise 4.4 derives details of the algorithm.

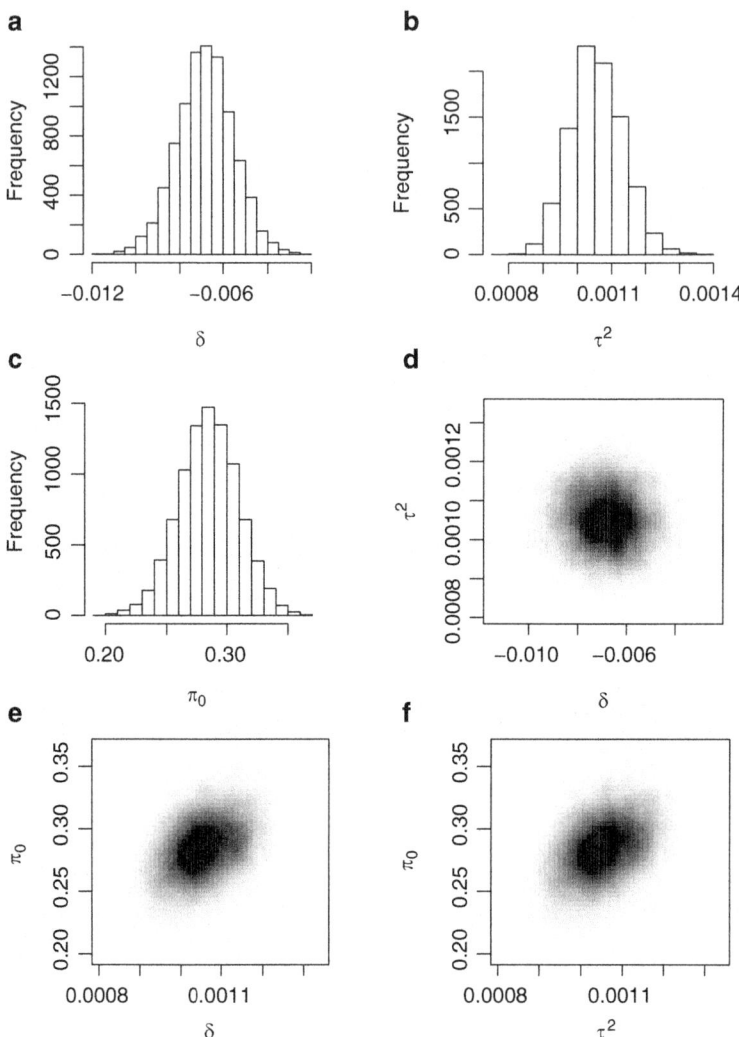

Fig. 4.9 Posterior distributions for selected parameters of the mixture model, for the microarray data: (**a**) $p(\delta \mid y)$, (**b**) $p(\tau^2 \mid y)$, (**c**) $p(\pi_0 \mid y)$, (**d**) $p(\delta, \tau^2 \mid y)$, (**e**) $p(\delta, \pi_0 \mid y)$, (**f**) $p(\tau^2, \pi_0 \mid y)$

The posterior median and 95% interval for δ ($\times 10^{-3}$) is $-6.8\,[-9.4, -0.40]$, while for τ^2 ($\times 10^{-3}$), we have 1.1 [0.92,1.2]. Of more interest are the posterior summaries for π_0: 0.29 [0.24,0.33], giving a range that is consistent with the pFDR estimate of 0.33. Figure 4.9 displays univariate and bivariate posterior distributions. The distributions resemble normal distributions, reflecting the large samples within populations and the number of transcripts.

4.6 Testing Multiple Hypotheses: Fixed Number of Tests

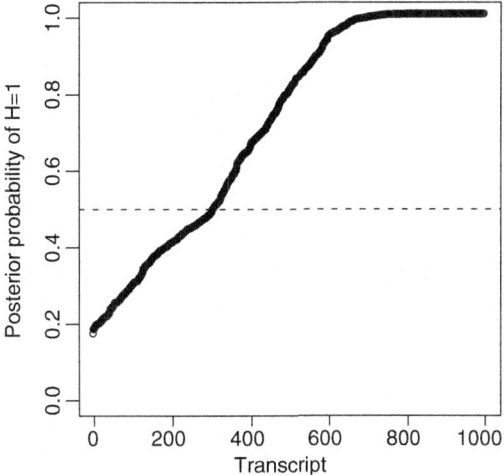

Fig. 4.10 Posterior probabilities $\Pr(H_i = 1 \mid \boldsymbol{y})$, from the mixture model for the microarray data, for each of the $i = 1, \ldots, 1{,}000$ transcripts, ordered in terms of increasing posterior probability on the alternative

For transcript i, we may evaluate the posterior probabilities of the alternative

$$\begin{aligned}
\Pr(H_i = 1 \mid y_i) &= \mathrm{E}[H_i \mid \boldsymbol{y}] \\
&= \mathrm{E}_{\delta, \tau^2, \pi_0 \mid y}\left[\Pr(H_i \mid \delta, \tau^2, \pi_0)\right] \\
&= \mathrm{E}_{\delta, \tau^2, \pi_0 \mid y}\left[\Pr(H_i = 1 \mid \boldsymbol{y}, \delta, \tau^2, \pi_0)\right] \\
&= \mathrm{E}_{\delta, \tau^2, \pi_0 \mid y}\left[\frac{p(\boldsymbol{y} \mid H_i = 1, \delta, \tau^2) \times \pi_1}{p(\boldsymbol{y} \mid H_i = 1, \delta, \tau^2) \times \pi_1 + p(\boldsymbol{y} \mid H_i = 0) \times \pi_0}\right]
\end{aligned} \quad (4.21)$$

where

$$p(\boldsymbol{y} \mid H_i = 1, \delta, \tau^2, \pi_0) = [2\pi(\sigma_i^2 + \tau^2)]^{-1/2} \exp\left[-\frac{(y_i - \delta)^2}{2(\sigma_i^2 + \tau^2)}\right]$$

$$p(\boldsymbol{y} \mid H_i = 0, \delta, \tau^2, \pi_0) = [2\pi\sigma_i^2]^{-1/2} \exp\left[-\frac{y_i^2}{2\sigma_i^2}\right].$$

Expression (4.21) averages $\Pr(H_i = 1 \mid \boldsymbol{y}, \delta, \tau^2, \pi_0)$ with respect to the posterior $p(\delta, \tau^2, \pi_0 \mid \boldsymbol{y})$ and may be simply evaluated via

$$\frac{1}{T} \sum_{t=1}^{T} \frac{p(\boldsymbol{y} \mid H_i = 1, \delta^{(t)}, \tau^{2(t)}) \pi_1^{(t)}}{p(\boldsymbol{y} \mid H_i = 1, \delta^{(t)}, \tau^{2(t)}, \pi_0^{(t)}) \pi_1^{(t)} + p(\boldsymbol{y} \mid H_i = 0) \pi_0^{(t)}}$$

given samples $\delta^{(t)}, \tau^{2(t)}, \pi_0^{(t)}, t = 1, \ldots, T$, from the Markov chain.

Figure 4.10 displays the ordered posterior probabilities, $\Pr(H_i = 1 \mid \boldsymbol{y})$, $i = 1, \ldots, m$, along with a reference line of 0.5. Using this line as a threshold, 689 transcripts are flagged as "significant," and the posterior estimate of the proportion of false discoveries is 0.12. Interestingly, the posterior estimate of the proportion of false negatives (i.e., non-discoveries) is 0.35. The latter figure is rarely reported but is a useful summary. Previously, using a pFDR threshold of 0.05, there were 603 significant transcripts. Interestingly, using a rule that picked the 603 transcripts whose posterior probability on the alternative was highest yielded an estimate of the posterior probability of the proportion of false discoveries as 0.07, which is not very different from the pFDR estimate. This is reassuring for both the Bayesian and the pFDR approaches.

For this example, sensitivity analyses might relax the independence between transcripts and, more importantly, the normality assumption for the random effects μ_i.

The Bayes factor, (4.20), was derived under the assumption of a normal sampling likelihood. In general, if we have large sample sizes, we may take as likelihood the sampling distribution of an estimator and combine this with a normal prior, to give a closed-form estimator. The latter is an approximation to a Bayesian analysis with weakly informative priors on the nuisance parameters and was described in Sect. 3.11, with Bayes factor (3.45).

4.7 Testing Multiple Hypotheses: Variable Selection

A ubiquitous issue in regression modeling is deciding upon which covariates to include in the model. It is useful to distinguish three scenarios:

1. *Confirmatory:* In which a summary of the strength of association between a response and covariates is required. We include in this category the situation in which an a priori hypothesis concerning a particular response/covariate relationship is of interest; additional variables have been measured and we wish, for example, to know which to adjust for in order to reduce confounding.
2. *Exploration:* In which the aim is to gain clues about structure in the data. A particular example is when one wishes to gain leads as to which covariates are associated with a response, perhaps to guide future study design.
3. *Prediction:* In which we are not explicitly concerned with association but merely with predicting a response based on a set of covariates. In this case, we are not interested in the numerical values of parameters but rather in the ability to predict new outcomes. Chapters 10–12 examines prediction in detail, including the assessment of predictive accuracy.

For exploration, formal inference is not required and so we will concentrate on the confirmatory scenario. As we will expand upon in Sect. 5.9, a trade-off must be made when deciding on variables for inclusion and it is often not desirable to fit the full model. To summarize the discussion, as we include more covariates in the model, bias in estimates is reduced, but variability may be increased, depending on how strong a predictor the covariate is and on its association with other covariates.

Example: Prostate Cancer

To illustrate a number of the methods available for variable selection, we consider a dataset originally presented by Stamey et al. (1989) and introduced in Sect. 1.3.1. The data were collected on $n = 97$ men before radical prostatectomony. We take as response the log of prostate-specific antigen (PSA) which was being forwarded in the paper as a preoperative marker, that is, a predictor of the clinical stage of cancer. The authors examined log PSA as a function of eight covariates: log(can vol); log(weight) (where weight is prostate weight); age; log(BPH); SVI; log(cap pen); the Gleason score, referred to as gleason; and percentage Gleason score 4 or 5, referred to as PGS45.

Figure 1.1 shows the relationships between the response and each of the covariates and indicates what look like a number of strong associations, while Fig. 1.2 gives some idea of the dependencies among the more strongly associated covariates. After Sect. 4.9, we will return to this example, after describing a number of methods for selecting variables in Sect. 4.8 and discussing model uncertainty in Sect. 4.9.

4.8 Approaches to Variable Selection and Modeling

We now review a number of approaches to variable selection. Let k be the number of covariates, and for ease of exposition, assume each covariate is either binary or continuous, so that the association is summarized by a univariate parameter. We also exclude interactions so that the largest model contains $k + 1$ regression coefficients. Allowing for the inclusion/exclusion of each covariate only, there are 2^k possible models, a number which increases rapidly with k. For example, with $k = 20$ there are $1,048,576$ possible models. The number of models increases even more rapidly with the number of covariates, if we allow variables with more than two levels and/or interactions.

The hierarchy principle states that if an interaction term is included in the model, then the constituent main effects should be included also. If we do not apply the hierarchy principle, there are 2^{2^k-1} *interaction* models (i.e., models that include main effects and/or interactions), where k is the number of variables. For example, $k = 2$ leads to 8 models. Denoting the variables by A and B, these models are

$$1, \quad A, \quad B, \quad A+B, \quad A+B+A.B, \quad A+A.B, \quad B+A.B, \quad A.B.$$

The class of hierarchical models includes all models that obey the hierarchy principle. Applying the hierarchy principle in the $k = 2$ case reduces the number from 8 to 5, as we lose the last three models in the above list. With $k = 5$ variables, there are 2,147,483,648 interaction models, illustrating the sharp increase in the number of models with k. There is no general rule for counting the number of models that satisfy the hierarchy principle for a given dimension. For some discussion, see Darroch et al. (1980, Sect. 6). The latter include a list of the number of hierarchical models for $k = 1, \ldots, 5$; for $k = 5$, the number of hierarchical models is 7,580.

We begin by illustrating the problems of variable selection with a simple example.

Example: Confounder Adjustment

Suppose the true model is

$$y_i = \beta_0 + \beta_1 x_{1i} + \beta_2 x_{2i} + \epsilon_i, \qquad (4.22)$$

with $\epsilon_i \mid \sigma^2 \sim_{iid} N(0, \sigma^2)$, $i = 1, \ldots, n$. We take x_1 as the covariate of interest, so that estimation of β_1 is the focus. However, we decide to "control" for the possibility of $\beta_2 \neq 0$ via a test. For simplicity, we assume that σ^2 is known and assess significance by examining whether a 95% confidence interval for β_2 contains zero (which is equivalent to a two-sided hypothesis test with $\alpha = 0.05$). If the interval contains zero, then the model,

$$E[Y_i \mid x_{1i}, x_{2i}] = \beta_0^* + \beta_1^* x_{1i},$$

is fitted; otherwise, we fit (4.22). We illustrate the effects of this procedure through a simulation in which we take $\beta_0 = \beta_1 = \beta_2 = 1$, $\sigma^2 = 3^2$, and $n = 10$. The covariates x_1 and x_2 are simulated from a bivariate normal with means zero, variances one and correlation 0.7.

In Fig. 4.11a, we display the sampling distribution of $\widehat{\beta}_1$ given the fitting of model (4.22). The mean and standard deviation of the distribution of $\widehat{\beta}_1$ are 1.00 and 1.23, respectively. Unbiasedness follows directly from least squares/likelihood theory (Sect. 5.6).

Figure 4.11b displays the sampling distribution of the *reported* estimator when we allow for the possibility of adjustment according to a test of $\beta_2 \neq 0$. The mean and standard deviation of the distribution of the reported estimator of β_1 are 1.23 and 1.01, respectively, showing positive bias and a reduced variance. This distribution is a mixture of the sampling distribution of $\widehat{\beta}_1$ (the estimator obtained from the full model), and the sampling distribution of $\widehat{\beta}_1^*$, with the mixing weight on the latter corresponding to one minus the power of the test of $\beta_2 = 0$. The sampling distribution of $\widehat{\beta}_1^*$ is shifted because the effects of both x_1 and x_2, are being included in the estimate and the distribution is shifted to the right because x_1 and x_2 are positively correlated. Using the conditional mean of a bivariate normal (given as (D.1) in Appendix D) we have

4.8 Approaches to Variable Selection and Modeling

$$E[Y \mid x_1] = \beta_0 + \beta_1 x_1 + \beta_2 E[X_2 \mid x_1]$$
$$= \beta_0 + (\beta_1 + 0.7) \times x_1$$
$$= \beta_0^\star + \beta_1^\star x_1$$

illustrating the bias,

$$E[\widehat{\beta_1^\star}] - \beta_1 = 0.7 \qquad (4.23)$$

when the reduced model is fitted. Allowing for the possibility of adjustment gives an estimator with a less extreme bias, since sometimes the full model is fitted (if the null is rejected). The reason for the lower *reported* variance in the potentially adjusted analysis is the bias-variance trade-off intrinsic to variable selection. In model (4.22), the information concerning β_1 and β_2 is entangled because of the correlation between x_1 and x_2, which results in a higher variance. Section 5.9 provides further discussion. The reported variance is not appropriate, however, since it does not acknowledge the model building process, an issue we examine in Sect. 4.9. As $n \to \infty$, the power of the test to reject $\beta_2 = 0$ tends to 1, and we recover an unbiased estimator with an appropriate variance.

□

4.8.1 Stepwise Methods

A number of methods have been proposed that proceed in a stepwise fashion, adding or removing variables from a current model. We describe three of the most historically popular approaches.

Forward selection begins with the null model, $E[Y \mid \boldsymbol{x}] = \beta_0$, and then fits each of the models

$$E[Y \mid \boldsymbol{x}] = \beta_0 + \beta_j x_j,$$

$j = 1, \ldots, k$. Subject to a minimal requirement (i.e., a particular p-value threshold), the model that contains the covariate that provides the greatest "improvement" in fit is then carried forward. This procedure is then iterated until no covariates meet the minimal requirement (i.e., all the p-values are greater than the threshold), or all the variables are in the model.

Backward elimination has the same flavor but begins with the full model, and then removes, at each stage, the covariate that is contributing least to the fit. For example, the variable with the largest p-value, so long as it is bigger than some prespecified value, is removed from the model.

Each of these approaches can miss important models. For example, in forward selection, x_1 may be the "best" single variable, but x_1 and any other variable may be "worse" than x_2 and x_3 together (say), but the latter combination will never be considered. Related problems can occur with backward elimination. Such considerations lead to *Efroymson's algorithm* (Efroymson 1960) in which forward selection is followed by backward elimination. The initial steps are identical to

Fig. 4.11 (a) Sampling distribution of $\hat{\beta}_1$, controlling for x_2, and (b) sampling distribution of $\hat{\beta}_1$, given the possibility of controlling for x_2

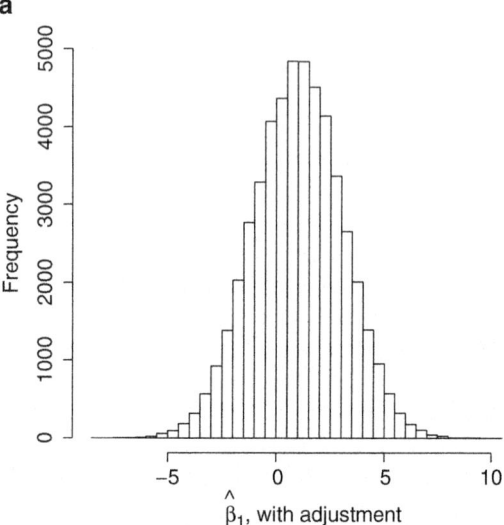

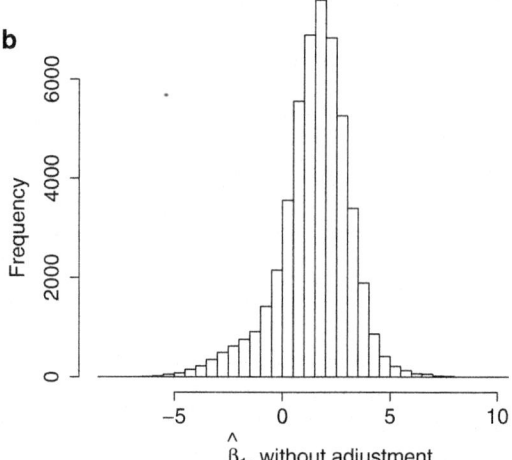

forward selection, but with three or more variables in the model, the loss of fit of each of the variables (excluding the last one added) is examined, in order to avoid the scenario just described, since in this case if the order of variables being added was x_1, x_2, x_3, it would then be possible for x_1 to be removed. The "p-value to enter" value (i.e., the threshold for forward selection) is chosen to be smaller than the "p-value to remove" value (i.e., the threshold for backward elimination), to prevent cycling in which a variable is continually added and then removed. The choice of inclusion/exclusion values is contentious for forward selection, backward elimination and Efroymson's algorithm.

4.8 Approaches to Variable Selection and Modeling

The Efroymson procedure, although overcoming some of the deficiencies of forward selection and backwards elimination, can still miss important models. The overall frequentist properties of any subset selection approach are difficult to determine, as we discuss in Sect. 4.9.

Each of the stepwise approaches may miss important models. A popular alternative is to examine all possible models and to then select the "best" model. We next provide a short summary of some of the criteria that have been suggested for this selection.

4.8.2 All Possible Subsets

We first consider linear models and again suppose there are k potential regressors, with the full model of the form

$$y = x\beta + \epsilon \tag{4.24}$$

with $E[\epsilon] = 0$, $\text{var}(\epsilon) = \sigma^2 I_n$, and where y is $n \times 1$, x is $n \times (k+1)$, and β is $(k+1) \times 1$.

The R^2 measure of variance explained is

$$R^2 = 1 - \frac{\text{RSS}}{\text{CTSS}}$$

where the residual and corrected total sum of squares are given, respectively, by

$$\text{RSS} = (y - x\widehat{\beta})^{\text{T}}(y - x\widehat{\beta})$$
$$\text{CTSS} = (y - 1\overline{y})^{\text{T}}(y - 1\overline{y}).$$

Consequently, R^2 can be interpreted as measuring the closeness of the fit to the data, with $R^2 = 1$ for a perfect fit (RSS = 0) and $R^2 = 0$ if the model does not improve upon the intercept only model. In terms of a comparison of nested models, the R^2 measure is nondecreasing in the number of variables, and so picking the model with the smallest R^2 will always produce the full model.

Let P represent a model constructed from covariates whose indices are a subset of $\{1, 2, \ldots, k\}$, with $p = |P| + 1$ regression coefficients in this model. The number of parameters p accounts for the inclusion of an intercept so that in the full model $p = k + 1$. Suppose the fit of model P yields estimator $\widehat{\beta}_P$ and residual sum of squares RSS_P. For model comparison, a more useful measure than R^2 is the adjusted R^2 which is defined as

$$R_a^2 = 1 - \frac{\text{RSS}_P/(n-p)}{\text{CTSS}/(n-1)}$$
$$= 1 - (1 - R^2)\left(\frac{n-1}{n-p}\right).$$

Maximization of R_a^2 leads to the model that produces the smallest estimate of σ^2 across models.

A widely used statistic, known as Mallows C_P, was introduced by Mallows (Mallows 1973).[1] For the model associated with the subset P

$$C_P = \frac{\text{RSS}_P}{\widehat{\sigma}^2} - (n - 2p) \tag{4.25}$$

where $\text{RSS}_P = (\boldsymbol{y} - \boldsymbol{x}\widehat{\boldsymbol{\beta}}_P)^\text{T}(\boldsymbol{y} - \boldsymbol{x}\widehat{\boldsymbol{\beta}}_P)$ is the residual sum of squares and $\widehat{\sigma}^2 = \text{RSS}_k/(n-k-1)$ is the error variance estimate from the full model that contains all k covariates. This criteria may be derived via consideration of the prediction error that results from choosing the model under consideration (as we show in Sect. 10.6.1). It is usual to plot C_P versus p and for a good model C_P will be close to, or below, p, since $\text{E}[\text{RSS}_P] = (n - p)\sigma^2$ and so $\text{E}[C_P] = p$ for a good model.

Lindley (1968) showed that Mallows C_P can also be derived from a Bayesian decision approach to multiple regression in which, among other assumptions, the aim is prediction and the X's are random and multivariate normal.

We now turn to more general models than (4.24). Consideration of the likelihood alone is not useful since the likelihood increases as parameters are added to the model, as we saw with the residual sum of squares in linear models. A number of *penalized likelihood* statistics have been proposed that penalize models for their complexity. A large number of statistics have been proposed, but we concentrate on just two, AIC and BIC. *An Information Criteria* (AIC, Akaike 1973) is a generalization of Mallows C_P and is defined as

$$\text{AIC} = -2l(\widehat{\boldsymbol{\beta}}_P) + 2p \tag{4.26}$$

where $l(\widehat{\boldsymbol{\beta}}_P)$ denotes the maximized log-likelihood of, and p the number of parameters in, model P. A derivation of AIC is presented in Sect. 10.6.5. We have already encountered the Bayesian information criterion (BIC) in Sect. 3.10 as an approximation to a Bayes factor. The BIC is given by

$$\text{BIC} = -2l(\widehat{\boldsymbol{\beta}}_P) + p \log n.$$

For the purposes of model selection, one approach is to choose between models by selecting the one with the minimum AIC or BIC. In general, BIC penalizes larger models more heavily than AIC, so that in practice AIC tends to pick models that are more complicated. As an indication, for a single parameter ($p = 1$ in (4.26)), the significance level is $\alpha = 0.157$ corresponding to $\Pr(\chi_1^2 < 2)$, which is a very liberal threshold. Given regularity conditions, BIC is *consistent* (Haughton 1988, 1989; Rao and Wu 1989), meaning if the correct model is in the set being considered, it will be picked with a probability that approaches 1 with increasing sample size,

---

[1] Named in honor of Cuthbert Daniel with whom Mallows initially discussed the use of the C_P statistic.

while AIC is not. The appearance of n in the penalty term of BIC is not surprising, since this is required for consistency.

4.8.3 Bayesian Model Averaging

Rather than select a single model, Bayesian model averaging (BMA) places priors over the candidate models, and then inference for a function of interest is carried out by averaging over the posterior model probabilities. Section 3.6 described this approach in detail, and we will shortly demonstrate its use with the prostate cancer data.

4.8.4 Shrinkage Methods

An alternative approach to selecting a model is to consider the full model but to allow shrinkage of the least squares estimates. Ridge regression and the lasso fit within this class of approaches and are considered in detail in Sects. 10.5.1 and 10.5.2, respectively. Such methods are often used in situations in which the data are sparse (in the sense of k being large relative to n).

4.9 Model Building Uncertainty

If a single model is selected on the basis of a stepwise method or via a search over all models, then bias will typically result. Interval estimates, whether they be based on Bayesian or frequentist approaches, will tend to be too narrow since they are produced by conditioning on the final model and hence do not reflect the mechanism by which the model was selected; see Chatfield (1995) and the accompanying discussion.

To be more explicit, let P denote the procedure by which a final model M is selected, and suppose it is of interest to examine the properties of an estimator $\widehat{\phi}$ of a univariate parameter ϕ, for example, a regression coefficient associated with a covariate of interest. The usual frequentist unbiasedness results concern the expectation of an estimator within a fixed model. We saw an example of bias following variable selection, with the bias given by (4.23). In general, the estimator obtained from a selection procedure will not be unbiased with respect to the final model chosen, that is,

$$\mathrm{E}[\widehat{\phi} \mid P] = \mathrm{E}_{M \mid P}[\,\mathrm{E}(\widehat{\phi} \mid M)\,] \quad (4.27)$$

$$\neq \mathrm{E}(\widehat{\phi} \mid \widehat{M}), \quad (4.28)$$

where $\widehat{M}$ is the final model chosen. In addition,

$$\text{var}(\widehat{\phi} \mid P) = \text{E}_{M\mid P}[\text{var}(\widehat{\phi} \mid M)] + \text{var}_{M\mid P}(\text{E}[\widehat{\phi} \mid M]) \qquad (4.29)$$

$$\neq \text{var}(\widehat{\phi} \mid \widehat{M}) \qquad (4.30)$$

where the latter approximates the first term of (4.29) only. Hence, in general, the reported variance conditional on a chosen model will be an underestimate. The bias and variance problems arise because the procedure by which $\widehat{M}$ was chosen is not being acknowledged.

From a Bayesian standpoint, the same problem exists because the posterior distribution should reflect all sources of uncertainty and a priori all possible models that may be entertained should be explicitly stated, with prior distributions being placed upon different models and the parameters of these models. Model averaging should then be carried out across the different possibilities, a process which is fraught with difficulties not least in placing "comparable" priors over what may be fundamentally different objects (see Sect. 6.16.3 for an approach to rectifying this problem). Suppose there are m potential models and that $p_j = \text{Pr}(M_j \mid \boldsymbol{y})$ is the posterior probability of model j, $j = 1, \ldots, m$. Then

$$\text{E}[\phi \mid \boldsymbol{y}] = \sum_{j=1}^{k} \text{E}[\phi \mid M_j, \boldsymbol{y}] \times p_j$$

$$\neq \text{E}[\phi \mid \widehat{M}, \boldsymbol{y}], \qquad (4.31)$$

where the latter is that which would be reported, based on a single model $\widehat{M}$. The "bias" is $\text{E}[\phi \mid \widehat{M}, \boldsymbol{y}] - \text{E}[\phi \mid \boldsymbol{y}]$. In addition,

$$\text{var}(\phi \mid \boldsymbol{y}) = \sum_{j=1}^{m} \text{var}(\phi \mid M_j, \boldsymbol{y}) \times p_j + \sum_{j=1}^{m} (\text{E}[\phi \mid M_j, \boldsymbol{y}] - \text{E}[\phi \mid \boldsymbol{y}])^2 \times p_j$$

$$(4.32)$$

$$\neq \text{var}(\phi \mid \widehat{M}, \boldsymbol{y}), \qquad (4.33)$$

so that the variance in the posterior acknowledges both the weighted average of the within-model variances, via the first term in (4.32), and the weighted contributions to the between-model variability, via the second term. Note the analogies between the frequentist and Bayesian biases, (4.28) and (4.31), and the reported variances, (4.30) and (4.33).

The fundamental message here is that carrying out model selection leads to estimators whose frequency properties are not those of the estimators without any tests being performed (Miller 1990; Breiman and Spector 1992) and Bayesian single model summaries are similarly misleading. This problem is not unique to

Table 4.3 Parameter estimates, standard errors, and T statistics for the prostate cancer data. The full model and models chosen by stepwise/BIC and C_P/AIC are reported

	Variable	Full model			Stepwise/BIC model			C_P/AIC model		
		Est.	(Std. err.)	T stat.	Est.	Std. err.	T stat.	Est.	Std. err.	T stat.
1	log(can vol)	0.59	(0.088)	6.7	0.55	(0.075)	7.4	0.57	(0.075)	7.6
2	log(weight)	0.46	(0.17)	2.7	0.51	(0.15)	3.9	0.42	(0.17)	2.5
3	age	−0.020	(0.011)	−1.8	–	–	–	−0.015	(0.011)	−1.4
4	log(BPH)	0.11	(0.058)	1.8	–	–	–	0.11	(0.058)	1.9
5	SVI	0.77	(0.24)	3.1	0.67	(0.21)	3.2	0.72	(0.21)	3.5
6	log(cap pen)	−0.11	(0.091)	−1.2	–	–	–	–	–	–
7	gleason	0.045	(0.16)	0.29	–	–	–	–	–	–
8	PGS45	0.0045	(0.0044)	1.0	–	–	–	–	–	–
σ		0.78	–	–	0.72	–	–	0.71	–	–

variable selection. Similar problems occur when other forms of model refinement are entertained, such as transformations of y and/or x, or experimenting with a variety of variance models and error distributions.

Example: Prostate Cancer

We begin by fitting the full model containing all eight variables. Table 4.3 gives the coefficients, standard errors, and T statistics. For this example, the forward selection and backward elimination stepwise procedures all lead to the same model containing the three variables log(can vol), log(weight), and SVI. The p-value thresholds were chosen to be 0.05. The standard errors associated with the significant variables all decrease for the reduced model when compared to the full model. This behavior reflects the bias-variance trade-off whereby a reduced model may have increased precision because of the fewer competing explanations for the data (for more discussion, see Sect. 5.9). We emphasize, however, that uncertainty in the model search is not acknowledged in the estimates of standard error. We see that the estimated standard deviation is also smaller in the reduced model.

Turning now to methods that evaluate all subsets, Figure 4.12 plots the C_P statistic versus the number of parameters in the model. For clarity, we do not include models with less than four parameters in the plot, since these were not competitive. Recall that we would like models with a small number of parameters whose C_P value is close to or less than the line of equality. The variable plotting labels are given in Table 4.3. For these data, we pick out the model with variables labeled 1, 2, 3, 4, and 5 since this corresponds to a model that is close to the line in Fig. 4.12 and has relatively few parameters. The five variables are log(can vol), log(weight), age, log(BPH), and SVI, so that age and log(BPH) are added to the stepwise model.

Carrying out an exhaustive search over all main effects models, using the adjusted R^2 to pick the best model (which recall is equivalent to picking that model with

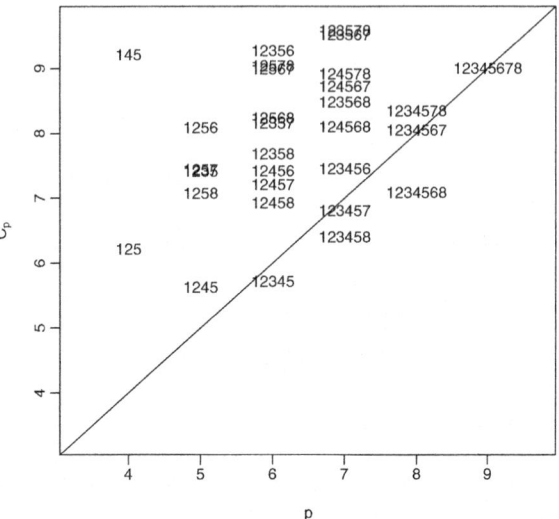

Fig. 4.12 Mallows' C_P statistic plotted versus p, where $p-1$ is the number of covariates in the model, for the prostate cancer data. The line of equality is indicated, for a good model $E[C_P] \approx p$, where the expectation is over repeated sampling. The variable labels are given in Table 4.3

the smallest $\widehat{\sigma}^2$), gives a model with seven variables (gleason is the variable not included). The estimate of the error variance is $\widehat{\sigma} = 0.70$. The minimum BIC model was the same model as picked by the stepwise procedures.

We used Bayesian model averaging with, for illustration, equal weights on each of the 2^8 models and weakly informative priors. The most probable model has posterior probability 0.20 and contains log(can vol), log(weight), and SVI, while the second replaces log(weight) with log(BPH) and has posterior probability 0.09. The third most probable model adds log(BPH) to the most probable model and has probability 0.037. Cumulatively across models, the posterior probability that log(can vol) is in the model is close to 1, with the equivalent posterior probabilities for SVI, log(weight), and log(BPH) being 0.69, 0.66, and 0.27, respectively. A more detailed practical examination of BMA is presented at the end of Chap. 10.

4.10 A Pragmatic Compromise to Variable Selection

One solution to deciding upon which variables for inclusion in a regression model is to never refine the model for a given dataset. This approach is philosophically pure but pragmatically dubious (unless one is in the context of, say, a randomized experiment) since we may obtain appropriate inference for a model that is a very poor description of the phenomenon under study. It is hard to state general strategies, but on some occasions, it may be safest, and the most informative, to report multiple models.

We consider situations that are not completely confirmatory and not completely exploratory. Rather we would like to obtain a good description of the phenomena under study and also have some faith in reported interval estimates. The philosophy suggested here is to think as carefully as possible about the model before the analysis proceeds. In particular, context-specific models should be initially posited. Hopefully the initial model provides a good description, but after fitting the model, model checking should be carried out and the model may be refined in the face of *clear* model inadequacy, with refinement ideally being carried out within distinct a priori known classes. A key requirement is to describe the procedure followed when the results are reported.

If a model is chosen because it is clearly superior to the alternatives then, roughly speaking, inference may proceed as if the final model were the one that was chosen initially. This is clearly a subjective procedure but can be informally justified via either frequentist or Bayesian approaches. From a frequentist viewpoint, it may be practically reasonable to assume, with respect to (4.28), that $\mathrm{E}[\phi \mid P] \approx \mathrm{E}[\phi \mid \widehat{M}]$ because $\widehat{M}$ would be almost always chosen in repeated sampling under these circumstances. In a similar vein, under a Bayesian approach, the above procedure is consistent in which model averaging in which the posterior model weight on the chosen model is close to 1 (since alternative models are only rejected on the basis of clear inadequacy), that is, with reference to (4.31), $\mathrm{E}[\phi \mid \boldsymbol{y}] \approx \mathrm{E}[\phi \mid \widehat{M}, \boldsymbol{y}]$, because $\Pr(\widehat{M} \mid \boldsymbol{y}) \approx 1$. The aim is to provide probability statements, from either philosophical standpoints that are "honest" representations of uncertainty.

The same heuristic applies more broadly to examination of model choice, beyond which variables to put in the mean model. As an example of when the above procedure should not be applied, examining quantile–quantile plots of residuals for different Student's t distributions and picking the one that produces the straightest line would not be a good idea.

4.11 Concluding Comments

In this chapter, we have discussed frequentist and Bayesian approaches to hypothesis testing. With respect to variable selection, we make the following tentative conclusions. For pure confirmatory studies, one should not carry out model selection and use instead background context to specify the model. Prediction is a totally different enterprise and is the subject of Chaps. 10–12. In exploratory studies, stepwise and all subsets may point to important models, but attaching (frequentist or Bayesian) probabilistic statements to interval estimates is difficult. For studies somewhere between pure confirmation and exploratory, one should attempt to minimize model selection, as described in Sect. 4.10.

From a Bayesian or a frequentist perspective, regardless of the criteria used in a multiple hypothesis testing situation, it is essential to report the exact procedure followed, to allow critical interpretation of the results.

We have seen that when a point null, such as $H_0 : \theta = 0$, is tested, then frequentist and Bayesian procedures may well differ considerably in their conclusions. This is in contrast to the testing of a one-sided null such as $H_0 : \theta \leq 0$; see Casella and Berger (1987) for discussion. We conclude that hypothesis testing is difficult regardless of the frequentist or Bayesian persuasion of the analysis. A particular difficulty is how to calibrate the decision rule; many would agree that the Bayesian approach is the most natural since it directly estimates $\Pr(H = 0 \mid y)$, but this estimate depends on the choices for the alternative hypotheses (so is a relative rather than an absolute measure) and on all of the prior specifications. The practical interpretation of the p-value depends crucially on the power (sample size and observed covariate distribution in a regression setting) and reporting point and interval estimates alongside a p-value or an α level is strongly recommended.

Model choice is a fundamentally more difficult endeavor than estimation since we rarely, if ever, specify an exactly true model. In contrast, estimation is concerned with parameters (such as averages or linear associations with respect to a population) and these quantities are well defined (even if the models within which they are embedded are mere approximations).

4.12 Bibliographic Notes

There is a vast literature contrasting Bayesian and frequentist approaches to hypothesis testing, and we mention just a few references. Berger (2003) summarizes and contrasts the Fisherian (p-values), Neyman (α levels), and Jeffreys (Bayes factors) approaches to hypothesis testing, and Goodman (1993) provides a very readable, nontechnical commentary. Loss functions more complex than those considered in Sect. 4.3 are discussed in, for example, Inoue and Parmigiani (2009).

The running multiple hypothesis testing example concerned the analysis of multiple transcripts from a microarray experiment. The analysis of such data has received a huge amount of attention; see, for example, Kerr (2009) and Efron (2008).

4.13 Exercises

4.1 Consider the simple situation in which $Y_i \mid \theta \sim_{iid} N(\theta, \sigma^2)$ with σ^2 known. The MLE $\widehat{\theta} = \overline{Y} \sim N(\theta, V)$ with $V = \sigma^2/n$. The null and alternative hypotheses are $H_0 : \theta = 0$ and $H_1 : \theta \neq 0$, and under the alternative, assume $\theta \sim N(0, W)$. Consider the case $W = \sigma^2$:

  (a) Derive the Bayes factor for this situation.
  (b) Suppose that the prior odds are PO $= \pi_0/(1 - \pi_0)$, with π_0 the prior on the null, and let $R = L_{\text{II}}/L_{\text{I}}$ be the ratio of losses of type II to type I errors. Show that this setup leads to a decision rule to reject H_0 of the form

$$\sqrt{1+n} \times \exp\left(-\frac{Z^2}{2}\frac{n}{1+n}\right) \times \text{PO} < R \qquad (4.34)$$

where $Z = \widehat{\theta}/\sqrt{V}$ is the usual Z-statistic.

4.13 Exercises

(c) Rearrangement of (4.34) gives a Wald statistic threshold of

$$Z^2 > \frac{2(1+n)}{n} \log\left(\frac{\text{PO}}{\text{R}}\sqrt{1+n}\right).$$

Form a table of the p-values corresponding to this threshold, as a function of π_0 and n and with $R = 1$. Hence, comment on the use of 0.05 as a threshold.

4.2 The k-FWER criteria controls the probability of rejecting k or more true null hypotheses, with $k = 1$ giving the usual FWER criteria. Show that the procedure that rejects only the null hypotheses H_i, $i = 1, \ldots, m$ for those p-values with $p_i \leq k\alpha/m$, controls the k-FWER at level α.

4.3 Prove expression (4.17).

4.4 In this question, an MCMC algorithm for the Bayesian mixture model described in Sect. 4.6.2 will be derived and applied to "pseudo" gene expression data that is available on the book website.

The three-stage model is:

Stage One:

$$Y_i \mid H_i, \delta, \tau, \pi_0 \sim_{ind} \begin{cases} \text{N}(0, \sigma_i^2) & \text{if } H_i = 0 \\ \text{N}(\delta, \sigma_i^2 + \tau^2) & \text{if } H_i = 1 \end{cases}.$$

Stage Two: $H_i \mid \pi_1 \sim_{iid} \text{Bernoulli}(\pi_1)$.

Stage Three: Independent priors on the common parameters:

$$p(\delta, \tau, \pi_0) \propto 1/\tau.$$

Derive the form of the conditional distributions

$$\delta \mid \tau^2, \pi_0, \boldsymbol{H}$$
$$\tau^2 \mid \delta, \pi_0, \boldsymbol{H}$$
$$\pi_0 \mid \tau^2, \delta, \boldsymbol{H}$$
$$H_i \mid \delta, \tau^2, \pi_0, H_i, \quad i = 1, \ldots, m,$$

where $\boldsymbol{H} = [H_1, \ldots, H_m]$. The form for τ^2 requires a Metropolis–Hastings step (as described in Sect. 3.8.2).

Implement this algorithm for the gene expression data on the book website.

Part II
Independent Data

Chapter 5
Linear Models

5.1 Introduction

In this chapter we consider linear regression models. These models have received considerable attention because of their mathematical and computational convenience and the relative ease of parameter interpretation. We discuss a number of issues that require consideration in order to perform a successful linear regression analysis. These issues are relevant irrespective of the inferential paradigm adopted and so apply to both frequentist and Bayesian analyses.

The structure of this chapter is as follows. We begin in Sect. 5.2 by describing a motivating example, before laying out the linear model specification in Sect. 5.3. A justification for linear modeling is provided in Sect. 5.4. In Sect. 5.5, we discuss parameter interpretation, and in Sects. 5.6 and 5.7, we describe, respectively, frequentist and Bayesian approaches to inference. In Sect. 5.8, the analysis of variance is briefly discussed. Section 5.9 provides a discussion of the bias-variance trade-off that is encountered when one considers which covariates to include in the mean model. In Sect. 5.10, we examine the robustness of the least squares estimator to model assumptions; this estimator can be motivated from estimating function, likelihood, and Bayesian perspectives. The assessment of assumptions is considered in Sect. 5.11. Section 5.12 returns to the motivating example. Concluding remarks are provided in Sect. 5.13 with references to additional material in Sect. 5.14.

5.2 Motivating Example: Prostate Cancer

Throughout this chapter we use the prostate cancer data of Sect. 1.3.1 to illustrate the main points. These data consist of nine measurements taken on 97 men. Along with the response, the log of prostate-specific antigen (PSA), there are eight covariates. As an illustrative inferential question, we consider estimation of the linear association between log(PSA) and the log of cancer volume, with possible

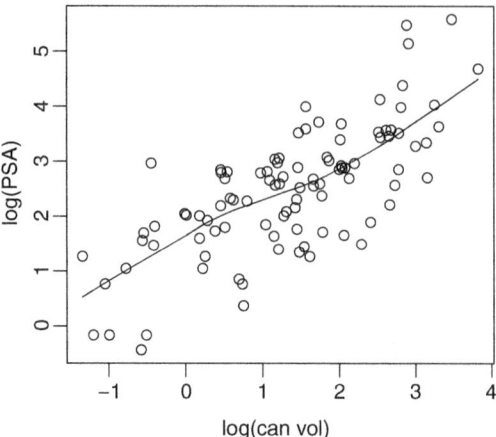

Fig. 5.1 Log of prostate-specific antigen versus log cancer volume, with smoother superimposed

adjustment for other "important" variables. Figure 5.1 plots log(PSA) versus log cancer volume, along with a smoother. The relationship looks linear, but Figs. 1.1 and 1.2 showed that log(PSA) was also associated with a number of the additional seven covariates and that there are strong associations between the eight covariates themselves. Consequently, we might question whether some or all of the other seven variables should be added to the model.

5.3 Model Specification

A multiple linear regression model takes the form

$$Y_i = \beta_0 + \beta_1 x_{i1} + \ldots + \beta_k x_{ik} + \epsilon_i, \tag{5.1}$$

where we begin by assuming that the error terms are uncorrelated with $E[\epsilon_i] = 0$ and $\text{var}(\epsilon_i) = \sigma^2$. In a *simple* linear regression model, $k = 1$ so that we have a single covariate. Linearity here is with respect to the parameters, and so variables may undergo nonlinear transforms from their original scale, before inclusion in (5.1).

In matrix form we write

$$Y = x\beta + \epsilon, \tag{5.2}$$

where

$$Y = \begin{bmatrix} Y_1 \\ Y_2 \\ \vdots \\ Y_n \end{bmatrix}, \quad x = \begin{bmatrix} 1 & x_{11} & \ldots & x_{1k} \\ 1 & x_{21} & \ldots & x_{2k} \\ \vdots & \vdots & \ddots & \vdots \\ 1 & x_{n1} & \ldots & x_{nk} \end{bmatrix}, \quad \beta = \begin{bmatrix} \beta_0 \\ \beta_1 \\ \vdots \\ \beta_k \end{bmatrix}, \quad \epsilon = \begin{bmatrix} \epsilon_1 \\ \epsilon_2 \\ \vdots \\ \epsilon_n \end{bmatrix}.$$

5.3 Model Specification

with $E[\epsilon] = \mathbf{0}$ and $\text{var}(\epsilon) = \sigma^2 \mathbf{I}_n$. We will also sometimes write

$$Y_i = \boldsymbol{x}_i \boldsymbol{\beta} + \epsilon_i,$$

where $\boldsymbol{x}_i = [1 \ x_{i1} \ \ldots \ x_{ik}]$ for $i = 1, \ldots, n$.

The covariates may be continuous or discrete. Discrete variables with a finite set of values are known as *factors*, with the values being referred to as *levels*. The levels may be ordered, and the ordering may or not be based upon numerical values. For example, dose levels of a drug are associated with numerical values but may be viewed as factor levels. Suppose x represents dose, with levels 0, 1, and 5. There are two alternative models that are immediately suggested for such an x variable. First, we may use a simple linear model in x:

$$E[Y \mid x] = \beta_0 + \beta_1 x. \tag{5.3}$$

Second, we may adopt the model

$$E[Y \mid x] = \alpha_0 \times I(x = 0) + \alpha_1 \times I(x = 1) + \alpha_2 \times I(x = 5), \tag{5.4}$$

where the indicator function

$$I(x = \widetilde{x}) = \begin{cases} 0 & \text{if } x \neq \widetilde{x} \\ 1 & \text{if } x = \widetilde{x} \end{cases}$$

and ensures that the appropriate level of x is picked. The mean function (5.4) allows for nonlinearity in the modeled association between Y and the observed x values, but does not allow interpolation to unobserved values of x. In contrast, (5.3) allows interpolation but imposes linearity. For an *ordinal* variable, the order of categories matters, but there are not specific values associated with each level (though values will be assigned as labels for computation). An example of an ordinal value is a pain score with categories none/mild/medium/severe. Alternatively, the levels may be nominal (such as female/male). The coding of factors is discussed in Sect. 5.5.2. Covariates may be of inherent were specific interest or may be included in the model in order to control for sources of variability or, more specifically, confounding; Sect. 5.9 provides more discussion.

The lower-/uppercase notation adopted here explicitly emphasizes that the covariates $\boldsymbol{x}$ are viewed as fixed while the responses $\boldsymbol{Y}$ are random variables. This is true regardless of whether the covariates were fixed by design or were random with respect to the sampling scheme. In the latter case it is assumed that the distribution of $\boldsymbol{x}$ does not carry information concerning $\boldsymbol{\beta}$ or σ^2, so that it is *ancillary* (Appendix F). Specifically, letting $\boldsymbol{\gamma}$ denote parameters associated with a model for $\boldsymbol{x}$, we assume that

$$p(\boldsymbol{y}, \boldsymbol{x} \mid \boldsymbol{\beta}, \sigma^2, \boldsymbol{\gamma}) = p(\boldsymbol{y} \mid \boldsymbol{x}, \boldsymbol{\beta}, \sigma^2) \times p(\boldsymbol{x} \mid \boldsymbol{\gamma}), \tag{5.5}$$

so that conditioning on $\boldsymbol{x}$ does not incur a loss in information with respect to $\boldsymbol{\beta}$. Hence, we can ignore the second term on the right-hand side of (5.5).

Random covariates, as just discussed, should be distinguished from inaccurately measured covariates. We will assume throughout that the x values are measured *without error*, an assumption that must always be critically assessed. In an observational setting in particular, it is common for elements of x to be measured with at least some error, but, informally speaking, we hope that these errors are small relative to the ranges; if this is not the case, then we must consider so-called *errors-in-variables* models; methods for addressing this problem are extensively discussed in Carroll et al. (2006).

5.4 A Justification for Linear Modeling

In this section we discuss the assumption of linearity. In general, there is no reason to expect the effects of continuous covariates to be causally linear,[1] but if we have a "true" model, $E[Y \mid x] = f(x)$, then a first-order Taylor series expansion about a point x_0 gives

$$f(x) \approx f(x_0) + \left.\frac{df}{dx}\right|_{x_0} (x - x_0)$$
$$= \beta_0 + \beta_1(x - x_0)$$

so that, at least for x values close to x_0, we have an approximately linear relationship.

As an example, Fig. 5.2 shows the height of 50 children plotted against their age. The true nonlinear form from which these data were generated is the so-called Jenss curve:

$$E[Y \mid x] = \beta_0 + \beta_1 x - \exp(\beta_2 + \beta_3 x),$$

where Y is the height of the child at year x. This model was studied by Jenss and Bayley (1937), and the parameter values for the simulation were taken from Dwyer et al. (1983). The solid line on Fig. 5.2 is the curve from which these data were simulated, and the dotted and dashed lines are the least squares fits using data from ages less than 1.5 years only and greater than 4.5 years only, respectively. At younger ages, the association is approximately linear, and similarly for older ages, but a single linear curve does not provide a good description over the complete age range.

---

[1] In fact, as illustrated in Example 1.3.4, many physical phenomena are driven by differential equations with nonlinear models arising as solutions to these equations.

Fig. 5.2 Illustration of linear approximations to a nonlinear growth curve model

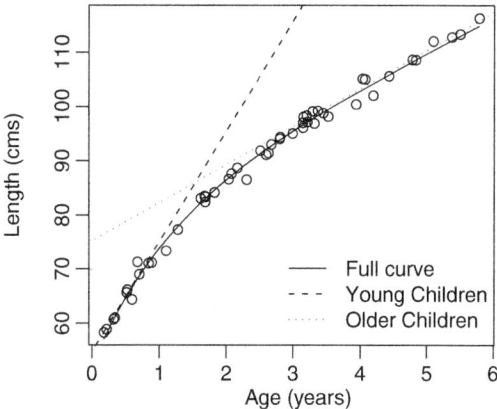

5.5 Parameter Interpretation

Before considering inference, we discuss parameter interpretation for the linear model. This topic is of vital importance in many settings, in order to report analyses in a meaningful manner. Interpretation is of far less concern in situations in which we simply wish to carry out prediction; methods for this endeavor are described in Chaps. 10–12. In a Bayesian analysis the specification of informative prior distributions requires a clear understanding of the meaning of parameters.

5.5.1 Causation Versus Association

We begin with the simple linear regression model

$$\mathrm{E}[Y \mid x] = \beta_0 + \beta_1 x. \tag{5.6}$$

Here we have explicitly conditioned upon x which is an important distinction since, for example, the models

$$\mathrm{E}[Y] = \mathrm{E}[\mathrm{E}(Y \mid x)] = \beta_0 \tag{5.7}$$

and

$$\mathrm{E}[Y \mid x] = \beta_0, \tag{5.8}$$

are very different. In (5.7) no assumptions are made, and we are simply saying that there is an average response in the population. However, (5.8) states that the expected response does not vary with x, which is a very strong assumption. Consequently, care should be taken to understand which situation is being considered.

We first consider the intercept parameter β_0 in (5.6), which is the expected response at $x = 0$. The latter expectation may make little sense (e.g., suppose the response is blood pressure and the covariate is weight), and there are a number of reasons to instead use the model

$$E[Y \mid x] = \beta_0^\star + \beta_1(x - x^\star), \tag{5.9}$$

within which $\beta_0^\star$ is the expected response at $x = x^\star$. By choosing $x^\star$ to be a meaningful value, we will, for example, be able to specify a prior for $\beta_0^\star$ more easily in a Bayesian analysis (see Sect. 3.4.2 for further discussion). Choosing $x^\star = \bar{x}$ is dataset specific (which does not allow simple comparison of estimates across studies) but provides a number of statistical advantages. Of course, models (5.6) and (5.9) provide identical inference since they are simply two parameterizations of the same model.

In both (5.6) and (5.9), the *mathematical* interpretation of the parameter β_1 is that it represents the additive change in the expected response for a unit increase in x. Notice that the interpretation of β_1 depends on the scales of measurement of both x and Y. More generally, $c\beta_1$ represents the additive change in the expected response for a c unit change in x. A difficulty with such interpretations is that it is inviting to think that if we were to provide an intervention and, for example, increase x by one unit for every individual in a population, then the expected response would change by β_1. The latter is a *causal* interpretation and is not appropriate in most situations, and never in observational studies, because unmeasured variables that are associated with both Y and x will be contributing to the observed association, $\widehat{\beta}_1$, between Y and x. In a designed experiment in which everything proceeds as planned, x is randomly assigned to each individual, and we may interpret β_1 as the expected change in the response for an individual following an intervention in which x were increased by one unit. Even in this ideal situation we need to know that the randomization was successfully implemented. It is also preferable to have large sample sizes so that any chance imbalance in variables between groups (as defined by different x values) is small.

We illustrate the problems with a simple idealized example. Suppose the "true" model is

$$E[Y \mid x, z] = \beta_0 + \beta_1 x + \beta_2 z, \tag{5.10}$$

and let

$$E[Z \mid x] = a + bx, \tag{5.11}$$

describe the linear association between x and z. Then, if Y is regressed on x, only

$$\begin{aligned} E[Y \mid x] &= E_{Z \mid x}[E(Y \mid x, Z)] \\ &= \beta_0 + \beta_1 x + \beta_2 E[Z \mid x] \\ &= \beta_0^\star + \beta_1^\star x \end{aligned} \tag{5.12}$$

where

$$\begin{aligned} \beta_0^\star &= \beta_0 + a\beta_2 \\ \beta_1^\star &= \beta_1 + b\beta_2 \end{aligned} \tag{5.13}$$

5.5 Parameter Interpretation

showing that, when we observe x only and fit model (5.12), our estimate of β_1^* reflects not just the effect of x but, in addition, the effect of z mediated through its association with x. If $b = 0$, so that X and Z are uncorrelated, or if $\beta_2 = 0$, so that Z does not affect Y, then there will be no bias. Here "bias" refers to estimation of β_1, and not to β_1^*. So for bias to occur in a linear model, Z must be related to both Y and X which, roughly speaking, is the definition of a *confounder*. The simulation at the end of Sect. 4.8 illustrated this phenomenon. A major problem in observational studies is that unmeasured confounders can always distort the true association. This argument reveals the beauty of randomization in which, by construction, there cannot be systematic differences between groups of units randomized to different x levels.

To rehearse this argument in a particular context, suppose Y represents the proportion of individuals with lung cancer in a population of individuals with smoking level (e.g., pack years) x. We know that alcohol consumption, z, is also associated with lung cancer, but it is unmeasured. In addition, X and Z are positively correlated. If we fit model (5.12), that is, regress Y on x only, then the resultant $\widehat{\beta}_1^*$ is reflecting not only the effect of smoking but that of alcohol also through its association with smoking. Specifically, since $b > 0$ (individuals who smoke are more likely to have increased alcohol consumption), then (5.13) indicates that $\widehat{\beta}_1^*$ will overestimate the true smoking effect β_1. If we were to intervene in our study population and (somehow) decrease smoking levels by one unit, then we would not expect the lung cancer incidence to decrease by β_1^* because alcohol consumption in the population has remained constant (assuming the imposed reduction does not change alcohol patterns). Rather, from (5.10), the expected decrease in the fraction with lung cancer will be β_1 if there were no other confounders (which of course is not the case). The interpretation of β_1 is the following. If we were to examine two groups of individuals within the study population with levels of smoking of $x+1$ and x, then we would expect lung cancer incidence to be $\widehat{\beta}_1^*$ higher in the group with the higher level of smoking.

To summarize, great care must be taken with parameter interpretation in observational studies because we are estimating associations and not causal relationships. The parameter estimate associated with x reflects not only the "true" effect of x but also the effects of all other unmeasured variables that are related to both x and Y.

5.5.2 Multiple Parameters

In the model

$$\mathrm{E}[Y \mid x_1, \ldots, x_k] = \beta_0 + \beta_1 x_1 + \ldots + \beta_k x_k,$$

the parameter β_j is the additive change in the average response associated with a unit change in x_j, with all other variables held constant.

In some situations the parameters of a model may be very difficult to interpret. Consider the quadratic model:

$$E[Y \mid x] = \beta_0 + \beta_1 x + \beta_2 x^2.$$

In this model, interpretation of β_1 (and β_2) is difficult because we cannot change x by one unit and simultaneously hold x^2 constant. An alternative parameterization that is easier to interpret is $\boldsymbol{\gamma} = [\gamma_0, \gamma_1, \gamma_2]$, where $\gamma_0 = \beta_0$, $\gamma_1 = -\beta_1/2\beta_2$, and $\gamma_2 = \beta_0 - \beta_1^2/4\beta_2$. Here γ_1 is the x value representing the turning point of the quadratic, and γ_2 is the expected value of the curve at this point.

We now discuss parameterizations that may be adopted when coding factors. We begin with a simple example in which we examine the association between a response Y and a two-level factor x_1, which we refer to as gender, and code as $x_1 = 0/1$, for female/male. The obvious formulation of the model is

$$E[Y \mid x_1] = \begin{cases} \beta_0' + \beta_1' & \text{if } x_1 = 0 \text{ (female)}, \\ \beta_0' + \beta_2' & \text{if } x_1 = 1 \text{ (male)}. \end{cases}$$

The parameters in this model are clearly not *identifiable*; the data may be summarized as two means, but the model contains three parameters. This nonidentifiability is sometimes referred to as (intrinsic) *aliasing*, and the solution is to place a constraint on the parameters.

In the *sum-to-zero* parameterization, we impose the constraint $\beta_1' + \beta_2' = 0$, to give the model

$$E[Y \mid x_1] = \begin{cases} \beta_0'' - \beta_1'' & \text{if } x_1 = 0 \text{ (female)}, \\ \beta_0'' + \beta_1'' & \text{if } x_1 = 1 \text{ (male)}. \end{cases}$$

In this case $E[Y \mid \boldsymbol{x}] = \boldsymbol{x}\boldsymbol{\beta}''$, where the rows of the design matrix are $\boldsymbol{x} = [1, -1]$ if female and $\boldsymbol{x} = [1, 1]$ if male. We write

$$E[Y] = E[Y \mid x_1 = 0] \times p_0 + E[Y \mid x_1 = 1] \times (1 - p_0)$$
$$= \beta_0'' + \beta_1''(1 - 2p_0),$$

where p_0 is the proportion of females in the population. We therefore see that β_0'' is the expected response if $p_0 = 1/2$, and

$$E[Y \mid x_1 = 1] - E[Y \mid x_1 = 0] = 2\beta_1'',$$

is the expected difference in responses between males and females.

An alternative parameterization imposes the *corner-point* constraint and assigns $\beta_1' = 0$ so that

$$E[Y \mid x_1] = \begin{cases} \beta_0 & \text{if } x_1 = 0 \text{ (female)}, \\ \beta_0 + \beta_1 & \text{if } x_1 = 1 \text{ (male)}. \end{cases}$$

5.5 Parameter Interpretation

For this parameterization, $E[Y \mid x] = x\beta$, where $x = [1, 0]$ if female and $x = [1, 1]$ if male. In this model, β_0 is the expected response for females, and β_1 is the additive change in the expected response for males, as compared to females.

A final model is

$$E[Y \mid x_1] = \begin{cases} \beta_0^\dagger & \text{if } x_1 = 0 \text{ (female)} \\ \beta_1^\dagger & \text{if } x_1 = 1 \text{ (male)}. \end{cases}$$

In this case $E[Y \mid x] = x\beta^\dagger$ where $x = [1, 0]$ if female and $x = [0, 1]$ if male so that $\beta_0^\star$ is the expected response for a female and $\beta_1^\star$ is the expected response for a male. We stress that inference for each of the formulations is identical; all that changes is parameter interpretation.

The benefits or otherwise of alternative parameterizations should be considered in the light of their extension to the case of more than two levels and to multiple factors. For example, the $[\beta_0^\dagger, \beta_1^\dagger]$ parameterization does not generalize well to a situation in which there are multiple factors and we do not wish to assume a unique mean for each combination of factors (i.e., a non-saturated model). It is obviously important to determine the default parameterization adopted in any particular statistical package so that parameter interpretation can be accurately carried out.

In this book we adopt the corner-point parameterization. Unlike the sum-to-zero constraint, this parameterization is not symmetric, since the first level of each factor is afforded special status, but parameter interpretation is relatively straightforward. If possible, one should define the factors so that the first level is the most natural "baseline." We illustrate the use of this parameterization with an example concerning two factors, x_1 and x_2, with x_1 having 3 levels, coded as 0, 1, 2, and x_2 having 4 levels coded as 0, 1, 2, 3. The coding for the no interaction (main effects[2] only) model is

$$E[Y \mid x_1, x_2] = \begin{cases} \mu & \text{if } x_1 = 0, x_2 = 0, \\ \mu + \alpha_j & \text{if } x_1 = j, j = 1, 2, x_2 = 0, \\ \mu + \beta_k & \text{if } x_1 = 0, x_2 = k, k = 1, 2, 3, \\ \mu + \alpha_j + \beta_k & \text{if } x_1 = j, j = 1, 2, x_2 = k, k = 1, 2, 3. \end{cases}$$

As shorthand, we write this model as

$$E[Y \mid x_1 = j, x_2 = k] = \mu + \alpha_j \times I(x_1 = j) + \beta_k \times I(x_2 = k),$$

for $j = 0, 1, 2$, $k = 0, 1, 2, 3$, with $\alpha_0 = \beta_0 = 0$.

---

[2] This terminology is potentially deceptive since "effects" invite a causal interpretation.

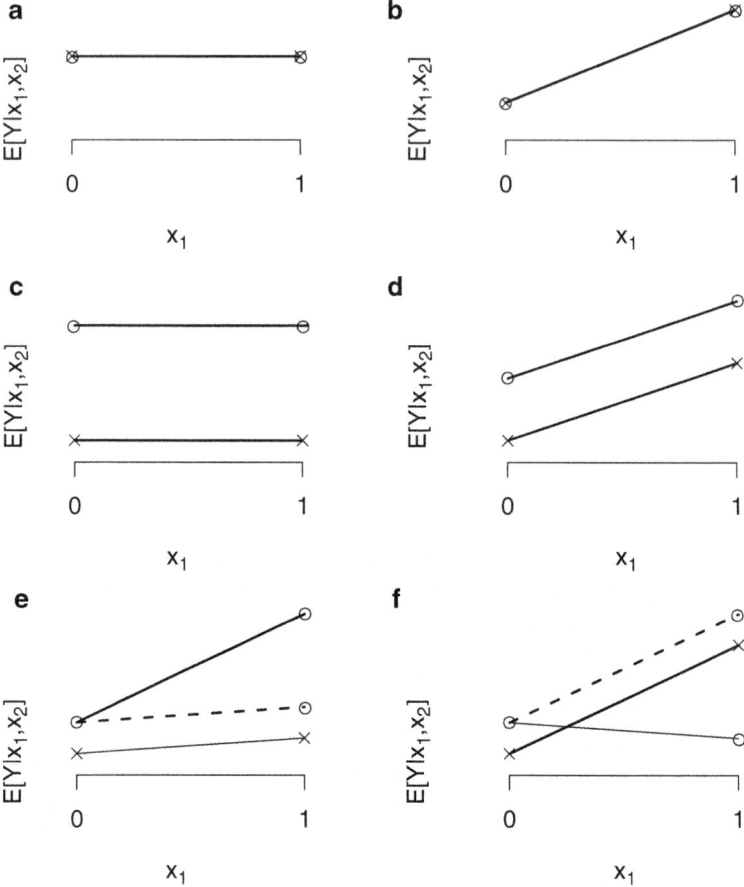

Fig. 5.3 Expected values for various models with two binary factors x_1 and x_2, "×" represents $x_2 = 0$ and "∘" $x_2 = 1$: (**a**) Null model, (**b**) x_1 main effect only, (**c**) x_2 main effect only, (**d**) x_1 and x_2 main effects, (**e**) interaction model 1, (**f**) interaction model 2. The *dashed lines* in panels (**e**) and (**f**) denote the expected response under the main effects only model

When one or more of the covariates are factors, interest may focus on interactions. To illustrate, suppose first we have two binary factors, x_1 and x_2 each coded as $0, 1$. The most general form for the mean is the saturated model

$$E[Y \mid x_1, x_2] = \mu + \alpha_1 \times I(x_1 = 1) + \beta_1 \times I(x_2 = 1) + \gamma_{11} \times I(x_1 = 1, x_2 = 1)$$

(5.14)

where we have four unknown parameters and the responses may be summarized as four mean values. Figure 5.3 shows a variety of scenarios that may occur with this model. Panel (a) shows the null model in which the response does not depend

5.5 Parameter Interpretation

Table 5.1 Corner-point notation for two-factor model with interaction

		x_2			
		0	1	2	3
x_1	0	μ	$\mu + \beta_1$	$\mu + \beta_2$	$\mu + \beta_3$
	1	$\mu + \alpha_1$	$\mu + \alpha_1 + \beta_1 + \gamma_{11}$	$\mu + \alpha_1 + \beta_2 + \gamma_{12}$	$\mu + \alpha_1 + \beta_3 + \gamma_{13}$
	2	$\mu + \alpha_2$	$\mu + \alpha_2 + \beta_1 + \gamma_{21}$	$\mu + \alpha_2 + \beta_2 + \gamma_{22}$	$\mu + \alpha_2 + \beta_3 + \gamma_{23}$

on either variable, and panels (b) and (c) main effects due to x_1 only and to x_2 only, respectively. In panel (d) the response depends on both factors in a simple additive main effects only fashion (which is characterized by the parallel lines on the plot). The association with x_2 is the same for both levels of x_1 and $\gamma_{11} = 0$ in (5.14). Panels (e) and (f) show two different interaction scenarios. In panel (e), when $x_1 = 1$ and $x_2 = 1$ simultaneously, the expected response is greater than that predicted by the main effects only model (which is shown as a dashed line). In panel (f), the effect of the interaction is to reduce the association due to x_2. For the $x_1 = 0$ population, individuals with $x_2 = 1$ have an increased expected response over individuals with $x_2 = 0$. In the $x_1 = 1$ population, this association is reversed. In the saturated model (5.14), γ_{11} is measuring the difference between the average in the $x_1 = 1, x_2 = 1$ population and that predicted by the main effects only model. In the saturated model, α_1 is the expected change in the response between the $x_1 = 1$ and the $x_1 = 0$ populations when $x_2 = 0$, $\alpha_1 + \gamma_{11}$ is this same comparison when $x_2 = 1$.

In this example we have a *two-way* (also known as *first-order*) interaction (a terminology that extends in an obvious fashion to three or more factor). If an interaction exists in a model, then all main effects that are involved in the interaction will often be included in the model, which is known as the *hierarchy principle* (see Sect. 4.8 for further discussion). Following this principle aids in interpretation, but there are situations in which one would not restrict oneself to this subset of models. For example, in a prediction setting (Chaps. 12–10), we may ignore the hierarchy principle.

Table 5.1 illustrates the corner-point parameterization for the case in which there are two factors with three and four levels and all two-way interactions are present. The main effects model is obtained by setting $\gamma_{jk} = 0$ for $j = 1, 2, k = 1, 2, 3$. This notation extends to generalized linear models, as we see in Chap. 6.

5.5.3 Data Transformations

Model (5.1) assumes uncorrelated errors with constant variance. If there is evidence of nonconstant variance, the response may be transformed to achieve constant variance, though this changes other characteristics of the model. Historically, this was a popular approach due to the lack of easily implemented alternatives to the linear model with constant variance, and it is still useful in some instances.

For example, for positive data taking the log transform and fitting linear models is a common strategy. An alternative approach that is often preferable is to retain the mean–variance relationship and model on the original scale of the response (using a generalized linear model, for example see Chap. 6).

Suppose we have

$$E[Y] = \mu_y$$

and

$$\text{var}(Y) = \sigma_y^2 g(\mu_y),$$

so that the mean–variance relationship is determined by $g(\cdot)$, which is assumed known, at least approximately. Consider the transformed random variable, $Z = h(Y)$. Taking the approximation

$$Z \approx h(\mu_y) + (Y - \mu_y) h'(\mu_y),$$

where $h'(\mu_y) = \frac{dh}{dy}\big|_{\mu_y}$, produces

$$E[Z] \approx h(\mu_y),$$

and

$$\text{var}(Z) \approx \sigma_y^2 g(\mu_y) h'(\mu_y)^2.$$

To obtain independence between the variance and the mean, we therefore require

$$h(\cdot) = \int g(y)^{-1/2} \, dy. \tag{5.15}$$

For example, a commonly encountered relationship for positive responses is $\text{var}(Y) = \sigma_y^2 \mu_y^2$, so that the coefficient of variation (which is the standard deviation divided by the mean) is constant. In this case, the suggested transformation, from (5.15), is $Z = \log Y$. As a second example, if $\text{var}(Y) = \sigma_y^2 \mu_y$, the recommended transformation is $Z = \sqrt{Y}$.

Transformations of Y, and/or covariates, may also be taken in order to obtain an approximately linear association, though it is advisable to do this before seeing the scatterplot of y versus x, since data dredging is a bad idea, as discussed in Sect. 4.10.

Parameter interpretation is usually less straightforward if we have transformed the response and/or the covariates, as we illustrate with a series of examples. In this section, for clarity, we explicitly state the base of the logarithm. Suppose we fit the model

$$\log_e Y = \beta_0 + \beta_1 x + \epsilon, \tag{5.16}$$

or equivalently

$$Y = \exp(\beta_0 + \beta_1 x + \epsilon) = \exp(\beta_0 + \beta_1 x) \delta, \tag{5.17}$$

5.5 Parameter Interpretation

where $\delta = \exp(\epsilon)$. The expectation of Y depends on the distribution of ϵ, but the median of $Y \mid x$ is $\exp(\beta_0 + \beta_1 x)$, so long as the median of ϵ is zero. It will often be more appropriate to report associations in terms of the median for a positive response; $\exp(\beta_0)$ is the median response when $x = 0$, and $\exp(\beta_1)$ is the ratio of median responses corresponding to a unit increase in x. We may interpret the intercept in terms of the expected value for specific distributional choices for ϵ. For example, if $\epsilon \mid x \sim N(0, \sigma^2)$, then since Y is lognormal (Appendix D),

$$E[Y \mid x] = \exp(\beta_0 + \beta_1 x + \sigma^2/2),$$

giving $E[Y \mid x = 0] = \exp(\beta_0 + \sigma^2/2)$ and

$$\frac{E[Y \mid x+1]}{E[Y \mid x]} = \exp(\beta_1), \qquad (5.18)$$

so that $\exp(\beta_1)$ can be interpreted as the *ratio* of expected responses between subpopulations whose x values differ by one unit. The interpretation (5.18) is true for other distributions, so long as $E[\exp(\epsilon) \mid x]$ does not depend on x. In general, if (5.18) holds, $\exp(c\beta_1)$ is the ratio of expected responses between subpopulations with covariate values $x + c$ and x. An alternative interpretation follows from observing that

$$\frac{d}{dx} E[Y \mid x] = \beta_1 E[Y \mid x],$$

so that the rate of change of the mean function with respect to x is proportional to the mean, with proportionality constant β_1.

Model (5.16), with the assumption of normal errors, is useful if the standard deviation on the original scale is proportional to the mean (to give a constant coefficient of variation) since, evaluating the variance of a lognormal distribution (Appendix D),

$$\mathrm{var}(Y \mid x) = E[Y \mid x]^2 \left[\exp(\sigma^2) - 1\right],$$

and if σ^2 is small, $\exp(\sigma^2) \approx 1 + \sigma^2$, and so

$$\mathrm{var}(Y \mid x) \approx E[Y \mid x]^2 \sigma^2,$$

showing that for this model we have, approximately, a constant coefficient of variation. Hence, log transformation of the response is often useful for strictly positive responses, which ties in with the example following (5.15).

A model that looks similar to (5.17) is

$$Y = E[Y \mid x] + \epsilon = \exp(\beta_0 + \beta_1 x) + \epsilon. \qquad (5.19)$$

In this model we have additive errors, whereas in the previous case, the errors were multiplicative. For the additive model, $\exp(\beta_0)$ is the expected value at $x = 0$, and $\exp(\beta_1)$ is the ratio of expected responses between subpopulations whose x values differ by one unit, regardless of the error distribution (so long as it has zero mean). In model (5.19), we may question whether additive errors are reasonable given that the mean function is always positive, though if the responses are well away from zero, there may not be a problem. Model (5.19) is nonlinear in the parameters, whereas (5.16) is linear, which has implications for inference and computation, as discussed in Chap. 6.

We now consider the model

$$Y = \beta_0 + \beta_1 \log_{10} x + \epsilon \qquad (5.20)$$

which can be useful if linearity of the mean is reasonable on a log scale. For example, if we have dose levels of a drug x of 1, 10, 100, and 1,000, then we would be very surprised if changing x from 1 to 2 produces the same change in the expected response as increasing x from 1,000 to 1,001. Modeling on the original scale might also result in extreme x values that are overly influential, though the appropriateness of the description of the relationship between Y and x should drive the decision as to which scale to model on. For model (5.20), the obvious mathematical interpretation is that β_1 represents the difference in the expected response for individuals whose $\log_{10} x$ values differ by one unit. A more substantive interpretation follows from observing that

$$\mathrm{E}[Y \mid cx] - \mathrm{E}[Y \mid x] = \beta_1 \log_{10} c$$

so that for a $c = 10$-fold increase in x, the expected responses differ by β_1. Therefore, taking $\log_{10} x$ gives an associated coefficient that gives the same change in the average when going from 1 to 10, as when going from 100 to 1,000.

Similarly, if we consider a linear model in $\log_2 x$, then $k\beta_1$ is the additive difference between the expected response for two subpopulations with covariates $2^k x$ and x. For example, if one subpopulation has twice the covariate of another, the difference in the expected response is β_1. In general, if we reparameterize via $\log_a x$ (to give β_1 as the change corresponding to an a-fold change), then the effect of a b-fold change is $\beta_1 \log_a b$. As an example, if we initially assume the model

$$\mathrm{E}[Y \mid x] = \beta_0 + \beta_1 \log_e x,$$

then $\beta_1 \log_e 10 = 2.30 \times \beta_1$ is the expected change for a 10-fold change in x.

We now consider the model with both Y and x transformed

$$\log_e Y = \beta_0 + \beta_1 \log_{10} x + \epsilon.$$

Under this specification, $\exp(\beta_1)$ represents the multiplicative change in the median response corresponding to a 10-fold increase in x.

Example: Prostate Cancer

For the prostate data, a simple linear regression model that does not adjust for additional variables is

$$\log(\text{PSA}) = \beta_0 + \beta_1 \times \log_e(\text{can vol}) + \epsilon$$

where the errors ϵ are uncorrelated with $\text{E}[\epsilon] = 0$ and $\text{var}(\epsilon) = \sigma^2$. In this model, $\exp(\beta_1)$ is the multiplicative change in median PSA associated with an e-fold change in cancer volume. Perhaps more usefully, $2.30 \times \beta_1$ is the multiplicative change in median PSA associated with a 10-fold increase in cancer volume.

5.6 Frequentist Inference

5.6.1 Likelihood

Consider the model

$$Y = x\beta + \epsilon,$$

with $\epsilon \sim \text{N}_n(0, \sigma^2 \mathbf{I}_n)$, $x = [1, x_1, \ldots, x_k]$, and $\beta = [\beta_0, \beta_1, \ldots, \beta_k]^\text{T}$. The complete parameter vector is $\theta = [\beta, \sigma]$ and is of dimension $p \times 1$ where $p = k+2$. The likelihood function is

$$L(\theta) = (2\pi\sigma^2)^{-n/2} \exp\left[-\frac{1}{2\sigma^2}(y - x\beta)^\text{T}(y - x\beta)\right],$$

with log likelihood

$$l(\theta) = -\frac{n}{2}\log(2\pi\sigma^2) - \frac{1}{2\sigma^2}(y - x\beta)^\text{T}(y - x\beta),$$

which yields the score equations (estimating functions)

$$S_1(\theta) = \frac{\partial l}{\partial \beta} = -\frac{1}{\sigma^2} x^\text{T}(Y - x\beta) \tag{5.21}$$

$$S_2(\theta) = \frac{\partial l}{\partial \sigma} = -\frac{n}{\sigma} + \frac{1}{\sigma^3}(Y - x\beta)^\text{T}(Y - x\beta). \tag{5.22}$$

Setting (5.21) and (5.22) to zero (and assuming $x^\text{T} x$ is of full rank) gives the MLEs

$$\widehat{\beta} = (x^\text{T} x)^{-1} x^\text{T} Y$$

$$\widehat{\sigma} = \left[\frac{1}{n}(Y - x\widehat{\beta})^\text{T}(Y - x\widehat{\beta})\right]^{1/2}.$$

We now examine the properties of these estimators, beginning with $\widehat{\beta}$:

$$E[\widehat{\beta}] = (x^\mathrm{T}x)^{-1}x^\mathrm{T}E[Y]$$
$$= (x^\mathrm{T}x)^{-1}x^\mathrm{T}x\beta$$
$$= \beta$$

so that $\widehat{\beta}$ is an unbiased estimator for all n. Though S_2 is an unbiased estimating function, $\widehat{\sigma}$ is a nonlinear function of S_2 and so has finite sample bias (but is asymptotically unbiased).

Asymptotic variance estimators are obtained from the information matrix:

$$\boldsymbol{I}(\boldsymbol{\theta}) = -\mathrm{E}\left[\frac{\partial \boldsymbol{S}}{\partial \boldsymbol{\theta}}\right] = \begin{bmatrix} \boldsymbol{I}_{11} & \boldsymbol{I}_{12} \\ \boldsymbol{I}_{21} & \boldsymbol{I}_{22} \end{bmatrix}$$

where $\boldsymbol{S} = [\boldsymbol{S}_1, S_2]^\mathrm{T}$, and

$$\boldsymbol{I}_{11} = \frac{\partial \boldsymbol{S}_1}{\partial \boldsymbol{\beta}} = \frac{x^\mathrm{T}x}{\sigma^2}$$

$$\boldsymbol{I}_{12} = \boldsymbol{I}_{21}^\mathrm{T} = \frac{\partial \boldsymbol{S}_1}{\partial \sigma} = \boldsymbol{0}$$

$$I_{22} = \frac{\partial S_2}{\partial \sigma} = \frac{2n}{\sigma^2}.$$

Taking $\mathrm{var}(\widehat{\boldsymbol{\theta}}) = \boldsymbol{I}(\boldsymbol{\theta})^{-1}$ gives

$$\mathrm{var}(\widehat{\beta}) = \sigma^2(x^\mathrm{T}x)^{-1}$$
$$\mathrm{var}(\widehat{\sigma}) = \frac{\sigma^2}{2n}.$$

In practice, σ^2 is replaced by its estimator to give

$$\widehat{\mathrm{var}}(\widehat{\beta}) = \widehat{\sigma}^2(x^\mathrm{T}x)^{-1}$$
$$\widehat{\mathrm{var}}(\widehat{\sigma}) = \frac{\widehat{\sigma}^2}{2n}.$$

For $\widehat{\beta}$ to be unbiased, we need only assume $\mathrm{E}[Y \mid x] = x\beta$, while for $\mathrm{var}(\widehat{\beta}) = \sigma^2(x^\mathrm{T}x)^{-1}$, we require $\mathrm{var}(Y \mid x) = \sigma^2 \mathbf{I}_n$, but *not* normality of errors. The expression for the variance is also exact for finite n.

The asymptotic distribution of the MLE based on n observations, $\widehat{\beta}_n$, is

$$(x^\mathrm{T}x)^{1/2}(\widehat{\beta}_n - \beta) \to_d \mathrm{N}_{k+1}(\mathbf{0}, \sigma^2 \mathbf{I}_{k+1}), \tag{5.23}$$

5.6 Frequentist Inference

and (by Slutsky's theorem, Appendix G) is still valid if σ is replaced by a consistent estimator. It should be stressed that normality of Y is not required, just n sufficiently large for the central limit theorem to apply. Since $\widehat{\beta}_n$ is a linear combination of independent observations, the central limit theorem may be directly applied. Another way of viewing this asymptotic derivation is of replacing the likelihood $p(\boldsymbol{y} \mid \boldsymbol{\beta})$ by $p(\widehat{\boldsymbol{\beta}}_n \mid \boldsymbol{\beta})$.

For $\widehat{\sigma}$ to be asymptotically unbiased, we require $\text{var}(\boldsymbol{Y} \mid \boldsymbol{x}) = \sigma^2 \mathbf{I}_n$, so that the estimating function for σ, (5.22), is unbiased. For $\widehat{\text{var}}(\widehat{\sigma}) = \widehat{\sigma}^2/2n$ to hold, we need the third and fourth moments to be correct and equal to zero and σ^2, respectively, as with the normal distribution. The dependence on higher-order moments results in inference for σ being intrinsically more hazardous than inference for $\boldsymbol{\beta}$.

Intervals for β_j, the jth components of $\boldsymbol{\beta}$, are based upon the statistic

$$\frac{\widehat{\beta}_j - \beta_j}{\widehat{\text{s.e.}}(\widehat{\beta}_j)},$$

where the standard error in the denominator is $\widehat{\sigma}$ times the square root of the (j,j)th element of $(\boldsymbol{x}^\mathsf{T}\boldsymbol{x})^{-1}$. The robustness to non-normality of the data is in part due to the standardization via the estimated standard error. In particular, we only require $\widehat{\sigma} \to_p \sigma$. An asymptotic $100 \times (1-\alpha)\%$ confidence interval for β_j is

$$\widehat{\beta}_j \pm z_{\alpha/2} \times \widehat{\text{s.e.}}(\widehat{\beta}_j)$$

where $z_{\alpha/2} = \Phi(\alpha/2)$.

If we wish to make inference about σ^2, then we might be tempted to construct a confidence interval for σ^2 by leaning on $\epsilon_i \mid \sigma^2 \sim_{iid} \text{N}(0, \sigma^2)$. This leads to

$$\frac{\text{RSS}}{\sigma^2} \sim \chi^2_{n-k-1}, \tag{5.24}$$

where $\text{RSS} = \sum_{i=1}^{n}(Y_i - \boldsymbol{x}_i\widehat{\boldsymbol{\beta}})^2$ is the residual sum of squares. Intervals obtained in this manner are extremely non-robust to departures from normality; however, see van der Vaart (1998, p. 27). The chi-square statistic does not standardize in any way, and any attempt to do so would require an estimate of the fourth moment of the error distribution, an endeavor that will be difficult due to the inherent variability in an estimate of the kurtosis (for a normal distribution, the kurtosis is zero, and so we do not require an estimate). Consequently, an interval (or test) based on (5.24) should not be used in practice unless we have strong evidence to suggest normality (or close to normality) of errors.

If the errors are such that $\boldsymbol{\epsilon} \mid \sigma^2 \sim \text{N}_n(\mathbf{0}, \sigma^2\mathbf{I}_n)$, then combining (5.23) with (5.24) gives, using (E.2) of Appendix E, the distribution

$$\widehat{\boldsymbol{\beta}} \sim \text{T}_{k+1}\left[\boldsymbol{\beta}, s^2(\boldsymbol{x}^\mathsf{T}\boldsymbol{x})^{-1}, n-k-1\right], \tag{5.25}$$

a $(k+1)$-dimensional Student's t distribution with location $\boldsymbol{\beta}$, scale matrix $s^2(\boldsymbol{x}^T\boldsymbol{x})^{-1}$, and $n-k-1$ degrees of freedom (Sect. D). A $100 \times (1-\alpha)\%$ confidence interval for β_j follows as

$$\widehat{\beta}_j \pm t^{n-k-1}_{\alpha/2} \times \widehat{\text{s.e.}}(\widehat{\beta}_j)$$

where $t^{n-k-1}_{\alpha/2}$ is the $\alpha/2$ percentage point of a standard t random variable with $n - k - 1$ degrees of freedom. A more reliable approach to the construction of confidence intervals for elements of $\boldsymbol{\beta}$ is to use the bootstrap or sandwich estimation, though if n is small, the latter are likely to be unstable. For small n, a Bayesian approach may be taken, though there is no way that the distributional assumption made for the data (i.e., the likelihood) can be reliably assessed.

We have just discussed the non-robustness of (5.24) to normality. It is perhaps surprising then that confidence intervals constructed from (5.25) are used, since they are derived directly from (5.24). However, the resultant intervals are conservative in the sense that they are wider than those constructed from (5.23), explaining their widespread use.

For a test of $H_0 : \beta_j = c$, $j = 1, \ldots, k$, we may derive a t-test. Under H_0,

$$T = \frac{\widehat{\beta}_j - c}{S_j^{1/2}\widehat{\sigma}} \sim T_{n-k-1}, \tag{5.26}$$

where S_j is the (j,j)th element of $(\boldsymbol{x}^T\boldsymbol{x})^{-1}$ and T_{n-k-1} denotes the univariate t distribution with $n - k - 1$ degrees of freedom, location $\widehat{\beta}_j$, and scale $S_j\widehat{\sigma}^2$. Although $\widehat{\sigma}$ can be very unstable, (5.26) it is an example of a self-normalized sum and so is asymptotically normal (Giné et al. 1997). The test with $c = 0$ is equivalent to the partial F statistic

$$F = \frac{\text{FSS}(\beta_j \mid \beta_0, \ldots, \beta_{j-1}, \beta_{j+1}, \ldots, \beta_k)/1}{\text{RSS}(\boldsymbol{\beta})/(n-k-1)},$$

where $\text{RSS}(\boldsymbol{\beta})$ is the residual sum of squares given the regression model $E[Y \mid \boldsymbol{x}] = \boldsymbol{x}\boldsymbol{\beta}$ and the fitted sum of squares

$$\text{FSS}(\beta_j \mid \beta_0, \ldots, \beta_{j-1}, \beta_{j+1}, \ldots, \beta_k) = \text{RSS}(\beta_0, \ldots, \beta_{j-1}, \beta_{j+1}, \ldots, \beta_k) - \text{RSS}(\boldsymbol{\beta}),$$

is equal to the change in residual sum of squares when β_j is dropped from the model. The "partial" here refers to the occurrence of β_l, $l \neq j$ in the model. Under H_0, $F \sim F_{1,n-k-1}$. The link with (5.26) is that $F = T^2$ with T evaluated at $c = 0$.

Let $\boldsymbol{\beta} = [\boldsymbol{\beta}_1, \boldsymbol{\beta}_2]$ be a partition with $\boldsymbol{\beta}_1 = [\beta_0, \ldots, \beta_q]$ and $\boldsymbol{\beta}_2 = [\beta_{q+1}, \ldots, \beta_k]$, with $0 \leq q < k$. Interest may focus on simultaneously testing whether a set of parameters is equal to zero, via a test of the null

$$H_0 : \boldsymbol{\beta}_1 \text{ unrestricted}, \; \boldsymbol{\beta}_2 = \boldsymbol{0} \quad \text{versus} \quad H_1 : \boldsymbol{\beta} = [\boldsymbol{\beta}_1, \boldsymbol{\beta}_2] \neq [\boldsymbol{\beta}_1, \boldsymbol{0}].$$

5.6 Frequentist Inference

Under H_0, the partial F statistic

$$F = \frac{\text{FSS}(\beta_{q+1},\ldots,\beta_k \mid \beta_0,\beta_1,\ldots,\beta_q)/(k-q)}{\text{RSS}/(n-k-1)} = \frac{\text{FSS}(\boldsymbol{\beta}_2 \mid \boldsymbol{\beta}_1)/(k-q)}{\text{RSS}/(n-k-1)} \quad (5.27)$$

is distributed as $F_{k-q,n-k-1}$ (Appendix D). Note that

$$\text{FSS}(\boldsymbol{\beta}_2 \mid \boldsymbol{\beta}_1) \neq \text{FSS}(\boldsymbol{\beta}_2),$$

unless $[x_1,\ldots,x_q]$ is orthogonal to $[x_{q+1},\ldots,x_k]$. Such derivations are crucial to the mechanics of analysis of variance models, which we describe in Sect. 5.8.

Extending the above with $q = -1$ so that all $k+1$ parameters are being considered, the $100 \times (1-\alpha)\%$ confidence interval for $\boldsymbol{\beta}$ is the ellipsoid

$$(\boldsymbol{\beta} - \widehat{\boldsymbol{\beta}})^{\mathrm{T}} \boldsymbol{x}^{\mathrm{T}} \boldsymbol{x} (\boldsymbol{\beta} - \widehat{\boldsymbol{\beta}}) \leq (k+1) s^2 F_{k+1,n-k-1}(1-\alpha) \quad (5.28)$$

where $s^2 = \text{RSS}/(n-k-1)$ and $F_{k+1,n-k-1}(1-\alpha)$ is the $1-\alpha$ point of the F distribution with $k+1, n-k-1$ degrees of freedom. The total sum of squares (TSS) may be partitioned as

$$\begin{aligned}
\text{TSS} &= (\boldsymbol{y} - \boldsymbol{x}\widehat{\boldsymbol{\beta}})^{\mathrm{T}}(\boldsymbol{y} - \boldsymbol{x}\widehat{\boldsymbol{\beta}}) \\
&= (\boldsymbol{y} - \boldsymbol{x}\boldsymbol{\beta} + \boldsymbol{x}\boldsymbol{\beta} - \boldsymbol{x}\widehat{\boldsymbol{\beta}})^{\mathrm{T}}(\boldsymbol{y} - \boldsymbol{x}\boldsymbol{\beta} + \boldsymbol{x}\boldsymbol{\beta} - \boldsymbol{x}\widehat{\boldsymbol{\beta}}) \\
&= (\boldsymbol{y} - \boldsymbol{x}\boldsymbol{\beta})^{\mathrm{T}}(\boldsymbol{y} - \boldsymbol{x}\boldsymbol{\beta}) + (\boldsymbol{\beta} - \widehat{\boldsymbol{\beta}})^{\mathrm{T}} \boldsymbol{x}^{\mathrm{T}} \boldsymbol{x} (\boldsymbol{\beta} - \widehat{\boldsymbol{\beta}}) \\
&= \text{RSS} + \text{FSS}.
\end{aligned}$$

Such expressions are specific to the linear model.

We now consider prediction of both an expected and an observed response. The latter require consideration of what we term *measurement error*, though we recognize that the errors in the model in general represent not only discrepancies arising from the measurement instrument but all manner of additional errors and sources of model misspecification. For inference concerning the *expected* response at covariate vector $\boldsymbol{x}_0$, we define $\theta = \boldsymbol{x}_0 \boldsymbol{\beta}$. Then $\widehat{\theta} = \boldsymbol{x}_0 \widehat{\boldsymbol{\beta}}$ and under correct first and second moment specification and via the central limit theorem:

$$[\boldsymbol{x}_0 (\boldsymbol{x}^{\mathrm{T}} \boldsymbol{x})^{-1} \boldsymbol{x}_0^{\mathrm{T}}]^{-1/2} (\widehat{\theta}_n - \theta) \to_d \text{N}(0, \sigma^2) \quad (5.29)$$

from which confidence intervals may be constructed. For prediction of an *observed* response at $\boldsymbol{x}_0$, we define $\phi = \boldsymbol{x}_0 \boldsymbol{\beta} + \epsilon$ with estimator $\widehat{\phi} = \boldsymbol{x}_0 \widehat{\boldsymbol{\beta}} + \widehat{\epsilon}$. It is now crucial to make a distributional assumption for the errors. Under $\epsilon \sim \text{N}(0, \sigma^2)$,

$$[1 + \boldsymbol{x}_0 (\boldsymbol{x}^{\mathrm{T}} \boldsymbol{x})^{-1} \boldsymbol{x}_0^{\mathrm{T}}]^{-1/2} (\widehat{\phi} - \phi) \sim \text{N}(0, \sigma^2). \quad (5.30)$$

The accuracy of intervals based on this form will be extremely sensitive to the normality assumption.

5.6.2 Least Squares Estimation

We describe an intuitive method of estimation with a long history and attractive properties. In *ordinary* least squares, the estimator is chosen to minimize the residual sum of squares

$$\mathrm{RSS}(\beta) = \sum_{i=1}^{n}(Y_i - \boldsymbol{x}_i\beta)^2 = (\boldsymbol{Y} - \boldsymbol{x}\beta)^{\mathrm{T}}(\boldsymbol{Y} - \boldsymbol{x}\beta).$$

Differentiation (and scaling for convenience) gives

$$-\frac{1}{2}\frac{\partial}{\partial \beta}\mathrm{RSS} = \boldsymbol{G}(\beta) = \boldsymbol{x}^{\mathrm{T}}(\boldsymbol{Y} - \boldsymbol{x}\beta) \tag{5.31}$$

with solution

$$\widehat{\beta} = (\boldsymbol{x}^{\mathrm{T}}\boldsymbol{x})^{-1}\boldsymbol{x}^{\mathrm{T}}\boldsymbol{Y},$$

so long as $\boldsymbol{x}^{\mathrm{T}}\boldsymbol{x}$ is of full rank. If we assume $\mathrm{E}[Y \mid \boldsymbol{x}] = \boldsymbol{x}\beta$, then $\mathrm{E}[\boldsymbol{G}(\beta)]=0$ and so (5.31) corresponds to an estimating equation, and we may apply the nonidentically distributed version of Result 2.1, summarized in (2.13), with

$$\boldsymbol{A}_n = \mathrm{E}\left[\frac{\partial \boldsymbol{G}}{\partial \beta}\right] = -\boldsymbol{x}^{\mathrm{T}}\boldsymbol{x}$$

$$\boldsymbol{B}_n = \mathrm{var}(\boldsymbol{G}) = \boldsymbol{x}^{\mathrm{T}}\mathrm{var}(\boldsymbol{Y})\boldsymbol{x}.$$

Consequently, to obtain the variance–covariance matrix of $\widehat{\beta}$, we need to specify $\mathrm{var}(\boldsymbol{Y})$. Assuming $\mathrm{var}(\boldsymbol{Y}) = \sigma^2 \boldsymbol{I}_n$ gives $\boldsymbol{B} = \sigma^2 \boldsymbol{x}^{\mathrm{T}}\boldsymbol{x}$ and

$$(\boldsymbol{x}^{\mathrm{T}}\boldsymbol{x})^{1/2}(\widehat{\beta} - \beta) \to_d \mathrm{N}_{k+1}(\boldsymbol{0}, \sigma^2 \boldsymbol{I}_{k+1}).$$

More generally, sandwich estimation may be applied, as we discuss in Sect. 5.6.4.

In the method of *generalized least squares*, we assume $\mathrm{E}[Y \mid \boldsymbol{x}] = \boldsymbol{x}\beta$ and $\mathrm{var}(\boldsymbol{Y} \mid \boldsymbol{x}) = \sigma^2 \boldsymbol{V}$ where $\boldsymbol{V}$ is a known matrix (*weighted least squares* corresponds to diagonal $\boldsymbol{V}$) and consider the function

$$\mathrm{RSS}_G(\beta) = (\boldsymbol{Y} - \boldsymbol{x}\beta)^{\mathrm{T}}\boldsymbol{V}^{-1}(\boldsymbol{Y} - \boldsymbol{x}\beta).$$

Minimization of $\mathrm{RSS}_G(\beta)$ yields the estimating function

$$\boldsymbol{G}_G(\beta) = \boldsymbol{x}^{\mathrm{T}}\boldsymbol{V}^{-1}(\boldsymbol{Y} - \boldsymbol{x}\beta),$$

and corresponding estimator

$$\widehat{\beta}_G = (\boldsymbol{x}^{\mathrm{T}}\boldsymbol{V}^{-1}\boldsymbol{x})^{-1}\boldsymbol{x}^{\mathrm{T}}\boldsymbol{V}^{-1}\boldsymbol{Y},$$

5.6 Frequentist Inference

with asymptotic distribution

$$(\boldsymbol{x}^\mathsf{T}\boldsymbol{V}^{-1}\boldsymbol{x})^{1/2}(\widehat{\boldsymbol{\beta}}_G - \boldsymbol{\beta}) \to_d \mathrm{N}_{k+1}(\boldsymbol{0}, \sigma^2 \mathbf{I}_{k+1}). \quad (5.32)$$

This estimator also arises from a likelihood with $\epsilon \sim \mathrm{N}_n(\boldsymbol{0}, \sigma^2 \boldsymbol{V})$ with $\boldsymbol{V} = \mathbf{I}_n$ giving the ordinary least squares estimator, as expected. An unbiased estimator of σ^2 is

$$\widehat{\sigma}_G^2 = \frac{1}{n-k-1}(\boldsymbol{Y} - \boldsymbol{x}\widehat{\boldsymbol{\beta}})^\mathsf{T} \boldsymbol{V}^{-1}(\boldsymbol{Y} - \boldsymbol{x}\widehat{\boldsymbol{\beta}}), \quad (5.33)$$

(see Exercise 5.1) and may be substituted for σ^2 in (5.32).

Given a particular dataset with n cases, a natural question is as follows: What is the practical significance of a central limit theorem and the associated regularity conditions? In the simple linear regression context, we require

$$\max_{1 \le i \le n}(x_i - \overline{x})^2 / \sum_{j=1}^n (x_j - \overline{x})^2 \to 0, \quad (5.34)$$

as $n \to \infty$. Intuitively, the imaginary way in which the number of data points is going to infinity is such that no single x value can dominate. In Sect. 5.10 we will present a number of simulations showing the behavior of the least squares estimator as a function of n, the distribution of the errors, and the distribution of the x values. Such simulations give one an indication of when asymptotic normality "kicks in." The required conditions indicate the sorts of x distributions that are more or less desirable for valid asymptotic inference. A crucial observation is that reliable asymptotic inference via (5.32) requires the mean–variance relationship to be correctly specified. We now present a theorem that provides one justification for the use of the least squares estimator.

5.6.3 The Gauss–Markov Theorem

Definition. The best linear unbiased estimator (BLUE) of $\boldsymbol{\beta}$:

- Is a linear function of $\boldsymbol{Y}$, so that the estimator can be written $\boldsymbol{B}^\mathsf{T}\boldsymbol{Y}$, for an $n \times (k+1)$ matrix $\boldsymbol{B}$
- Is unbiased so that $\mathrm{E}[\boldsymbol{B}^\mathsf{T}\boldsymbol{Y}] = \boldsymbol{\beta}$
- Has the smallest variance among all linear estimators

We now state and prove a celebrated theorem.

The Gauss–Markov Theorem: Consider the linear model $\mathrm{E}[\boldsymbol{Y}] = \boldsymbol{x}\boldsymbol{\beta}$, where $\boldsymbol{Y}$ is $n \times 1$, $\boldsymbol{x}$ is $n \times (k+1)$, and $\boldsymbol{\beta}$ is $(k+1) \times 1$. Suppose further that $\mathrm{cov}(\boldsymbol{Y}) = \sigma^2 \mathbf{I}_n$. Then $\widehat{\boldsymbol{\beta}} = (\boldsymbol{x}^\mathsf{T}\boldsymbol{x})^{-1}\boldsymbol{x}^\mathsf{T}\boldsymbol{Y}$ is the best linear unbiased estimator (BLUE) of $\boldsymbol{c}^\mathsf{T}\boldsymbol{\beta}$.

Proof. The estimator $\widehat{\beta} = (x^\mathsf{T} x)^{-1} x^\mathsf{T} Y$ is clearly linear, and we have already shown it is unbiased. We therefore only need to show the variance is smallest among linear unbiased estimators.

Let $\widetilde{\beta} = AY$ be another linear unbiased estimator with A a $(k+1) \times n$ matrix. Since the estimator is unbiased, $\mathrm{E}[\widetilde{\beta}] = A\mathrm{E}[Y] = Ax\beta$ for any β, which implies $Ax = \mathbf{I}_{k+1}$. Now

$$\mathrm{var}(\widetilde{\beta}) - \mathrm{var}(\widehat{\beta}) = A\sigma^2 \mathbf{I}_{k+1} A^\mathsf{T} - \sigma^2 (x^\mathsf{T} x)^{-1}$$
$$= \sigma^2 \left[AA^\mathsf{T} - Ax(x^\mathsf{T} x)^{-1} x^\mathsf{T} A^\mathsf{T} \right].$$

At this point we define $h = x(x^\mathsf{T} x)^{-1} x^\mathsf{T}$, which is known as the *hat* matrix (see Sect. 5.11.2). The hat matrix is symmetric and idempotent so that $h^\mathsf{T} = h$ and $hh^\mathsf{T} = h$. Further, $\mathbf{I}_n - h$ inherits these properties. Using these facts, we can write

$$\mathrm{var}(\widetilde{\beta}) - \mathrm{var}(\widehat{\beta}) = \sigma^2 A(\mathbf{I}_n - h) A^\mathsf{T}$$
$$= \sigma^2 A(\mathbf{I}_n - h)(\mathbf{I}_n - h)^\mathsf{T} A^\mathsf{T}$$

and this $(k+1) \times (k+1)$ matrix is positive definite, establishing that $\widehat{\beta}$ has the smallest variance among linear unbiased estimators. □

This result shows that $\widehat{\beta}$, which is the least squares estimate, the maximum likelihood estimate with a normal model, and the Bayesian posterior mean with normal model and improper prior $\pi(\beta, \sigma^2) \propto \sigma^{-2}$ (as we show in Sect. 5.7), is optimal among linear estimators. We emphasize that, in the above theorem, only first and second moment assumptions were used with no distributional assumptions being required.

5.6.4 Sandwich Estimation

We have already examined the properties of the ordinary least squares/maximum likelihood estimator $\widehat{\beta} = (x^\mathsf{T} x)^{-1} xY$ and have seen that $\mathrm{var}(\widehat{\beta}) = (x^\mathsf{T} x)^{-1} \sigma^2$, if $\mathrm{var}(Y \mid x) = \sigma^2 \mathbf{I}_n$. Suppose that the correct variance model is $\mathrm{var}(Y \mid x) = \sigma^2 V$ so that the model from which the estimator was derived was incorrect. Then the estimator is still unbiased, but the appropriate variance estimator is

$$\mathrm{var}(\widehat{\beta}) = (x^\mathsf{T} x)^{-1} x^\mathsf{T} \mathrm{var}(Y \mid x) x (x^\mathsf{T} x)^{-1}$$
$$= (x^\mathsf{T} x)^{-1} x^\mathsf{T} V x (x^\mathsf{T} x)^{-1} \sigma^2, \tag{5.35}$$

Expression (5.35) can also be derived directly from the estimating function

$$G(\beta) = x^\mathsf{T}(Y - x\beta),$$

5.6 Frequentist Inference

since we know

$$(A_n^{-1} B_n A_n^{\mathsf{T}\,-1})^{1/2}(\widehat{\boldsymbol{\beta}} - \boldsymbol{\beta}) \to_d N_{k+1}(\mathbf{0}_n, \mathbf{I}_n),$$

where

$$B_n = \mathrm{var}(\boldsymbol{G}) = \boldsymbol{x}^{\mathsf{T}} \boldsymbol{V} \boldsymbol{x} \sigma^2$$

$$A_n = \mathrm{E}\left[\frac{\partial \boldsymbol{G}}{\partial \boldsymbol{\beta}}\right] = -\boldsymbol{x}^{\mathsf{T}} \boldsymbol{x},$$

to give

$$\mathrm{var}(\widehat{\boldsymbol{\beta}}) = (\boldsymbol{x}^{\mathsf{T}} \boldsymbol{x})^{-1} \boldsymbol{x}^{\mathsf{T}} \boldsymbol{V} \boldsymbol{x} (\boldsymbol{x}^{\mathsf{T}} \boldsymbol{x})^{-1} \sigma^2.$$

We now describe a sandwich estimator of the variance that relaxes the constant variance assumption but assumes uncorrelated responses. When the variance is not constant, the ordinary least squares estimator is consistent (since the mean specification is correct), but the usual standard errors will be inappropriate.

Consider the estimating function $\boldsymbol{G}(\boldsymbol{\beta}) = \boldsymbol{x}^{\mathsf{T}}(\boldsymbol{Y} - \boldsymbol{x}\boldsymbol{\beta})$. The "bread" of the sandwich $\boldsymbol{A}^{-1}$ remains unchanged since $\boldsymbol{A}$ does not depend on Y. The "filling" becomes

$$B = \mathrm{var}(\boldsymbol{G}) = \boldsymbol{x}^{\mathsf{T}} \mathrm{var}(\boldsymbol{Y}) \boldsymbol{x} = \sum_{i=1}^{n} \sigma_i^2 \boldsymbol{x}_i^{\mathsf{T}} \boldsymbol{x}_i, \tag{5.36}$$

where $\sigma_i^2 = \mathrm{var}(Y_i)$ and we have assumed that the data are uncorrelated. Unfortunately, σ_i^2 is unknown, but various simple estimation techniques are available. An obvious estimator stems from setting $\widehat{\sigma}_i^2 = (Y_i - \boldsymbol{x}_i \boldsymbol{\beta})^2$ to give

$$\widehat{B}_n = \sum_{i=1}^{n} \boldsymbol{x}_i^{\mathsf{T}} \boldsymbol{x}_i (Y_i - \boldsymbol{x}_i \widehat{\boldsymbol{\beta}})^2, \tag{5.37}$$

and its use provides a consistent estimator of (5.36). However, this variance estimator has finite sample downward bias.

For linear regression, the MLE

$$\widehat{\sigma}^2 = \frac{1}{n} \sum_{i=1}^{n} (Y_i - \boldsymbol{x}_i \widehat{\boldsymbol{\beta}})^2 = \frac{1}{n} \sum_{i=1}^{n} \widehat{\sigma}_i^2,$$

is downwardly biased (as we saw in Sect. 5.6.1), with bias $-(k+1)\sigma^2/n$, which suggests using

$$\widehat{B}_n = \frac{n}{n-k-1} \sum_{i=1}^{n} \boldsymbol{x}_i^{\mathsf{T}} \boldsymbol{x}_i (Y_i - \boldsymbol{x}_i \widehat{\boldsymbol{\beta}})^2. \tag{5.38}$$

This simple correction provides an estimator of the variance that has finite bias, since the bias in $\widehat{\sigma}^2$ changes as a function of the design points $\boldsymbol{x}_i$, but will often improve on (5.37). In linear regression, if $\text{var}(Y_i) = \sigma^2$, then $E[(Y_i - \boldsymbol{x}_i\widehat{\boldsymbol{\beta}})^2] = \sigma^2(1 - h_{ii})$ where h_{ii} is the ith diagonal element of the hat matrix $\boldsymbol{x}(\boldsymbol{x}^\mathsf{T}\boldsymbol{x})^{-1}\boldsymbol{x}^\mathsf{T}$ (we derive this result in Sect. 5.11.2). Therefore, another suggested correction is

$$\widehat{\boldsymbol{B}}_n = \sum_{i=1}^{n} \boldsymbol{x}_i^\mathsf{T}\boldsymbol{x}_i \frac{(Y_i - \boldsymbol{x}_i\widehat{\boldsymbol{\beta}})^2}{(1 - h_{ii})}. \quad (5.39)$$

For each of (5.37), (5.38), and (5.39), the variance of the estimator $\widehat{\boldsymbol{\beta}}$ is consistently estimated by $\widehat{\boldsymbol{A}}_n^{-1}\widehat{\boldsymbol{B}}_n\widehat{\boldsymbol{A}}_n^{-1}$.

We report the results of a small simulation study, in which we examine the performance of the sandwich estimator as a function of n, the distribution of x, and the variance estimator. We carry out six sets of simulations with the x distribution either uniform on (0,1) or exponential with rate parameter 1, and $\text{var}(Y \mid x) = E[Y \mid x]^q \times \sigma^2$ with $q = 0, 1, 2$, so that the variance of the errors is constant, increases in proportion to the mean, or increases in proportion to the square of the mean. The errors are normally distributed and uncorrelated in all cases (Sect. 5.10 considers the impact of other forms of model misspecification).

In Table 5.2, we see that, as expected, confidence intervals obtained directly from the usual variance of the ordinary least squares estimator, that is, $(\boldsymbol{x}^\mathsf{T}\boldsymbol{x})^{-1}\widehat{\sigma}^2$, give accurate coverage when the error variance is constant. When the x distribution is uniform, the coverage is accurate even under variance model misspecification. There is poor coverage for the exponential distribution, however, which worsens with increasing n. The coverage of the sandwich estimator confidence intervals requires large samples to obtain accurate coverage for the exponential x model. There is a clear efficiency loss when using sandwich estimation, if the variance of the errors is constant. The downward bias of the sandwich estimator based on the unadjusted residuals is apparent, though this bias decreases with increasing n. Working with residuals standardized by $n/(n - k - 1)$, (5.38), improves the coverage, while the use of the hat matrix version, (5.39), improves performance further.

If the errors are correlated, the sandwich estimators of the variance considered here will not be consistent. Chapter 8 provides a description of sandwich estimators for the correlated data situation that may be used when there is replication across "clusters."

Example: Prostate Cancer

We fit the model

$$\log y_i = \beta_0 + \beta_1 \log_{10}(x_i) + \epsilon_i \quad (5.40)$$

where y_i is PSA and x_i is the cancer volume for individual i and ϵ_i are assumed uncorrelated with constant variance σ^2.

5.6 Frequentist Inference

Table 5.2 Confidence interval coverage of nominal 95% intervals under a model-based variance estimator in which the variance is assumed independent of the mean and under three sandwich estimators given by (5.37)–(5.39)

n	Model-based	Sandwich 1	Sandwich 2	Sandwich 3
5	95	84	90	93
10	95	88	91	92
25	94	92	93	94
50	95	94	94	94
100	95	95	95	95
250	95	95	95	95
$\text{var}(Y \mid x) = \sigma^2$, x uniform				
5	95	82	88	92
10	95	85	88	91
25	95	89	91	92
50	95	91	92	93
100	95	93	93	94
250	95	94	94	94
$\text{var}(Y \mid x) = \sigma^2$, x exponential				
5	95	83	89	92
10	95	89	92	93
25	95	92	94	94
50	95	94	95	95
100	95	95	95	95
250	95	95	95	95
$\text{var}(Y \mid x) = \text{E}[Y \mid x] \times \sigma^2$, x uniform				
5	92	76	83	89
10	90	77	82	87
25	87	83	85	88
50	85	87	88	90
100	85	90	91	92
250	83	93	93	93
$\text{var}(Y \mid x) = \text{E}[Y \mid x] \times \sigma^2$, x exponential				
5	95	83	89	92
10	95	89	92	93
25	95	92	93	94
50	94	94	94	94
100	95	94	95	95
250	95	95	95	95
$\text{var}(Y \mid x) = \text{E}[Y \mid x]^2 \times \sigma^2$, x uniform				
5	89	70	78	86
10	81	71	75	82
25	75	78	80	85
50	73	85	86	88
100	71	89	90	91
250	68	92	92	93
$\text{var}(Y \mid x) = \text{E}[Y \mid x]^2 \times \sigma^2$, x exponential				

The true values are $\beta_0 = 1$, $\beta_1 = 1$, and all results are based on 10,000 simulations. In all cases, the errors are normally distributed and uncorrelated. The true variance model and distribution of x are given in the last line of each block

Table 5.3 Least squares/maximum likelihood parameter estimates and model-based and sandwich estimates of the standard errors, for the prostate cancer data

Parameter	Estimate	Model-based standard error	Sandwich standard error
β_0	1.51	0.122	0.123
β_1	0.719	0.0682	0.0728

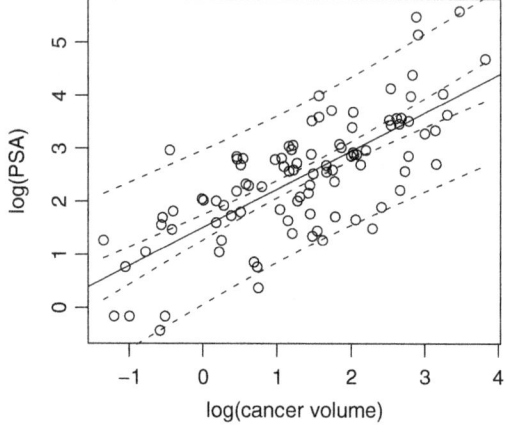

Fig. 5.4 Log of prostate-specific antigen versus log of cancer volume, along with the least squares/maximum likelihood fit, and 95% pointwise confidence intervals for the expected linear association (*narrow bands*) and for a new observation (*wide bands*)

where y_i is PSA and x_i is cancer volume for individual i and ϵ_i are assumed uncorrelated with constant variance σ^2. Table 5.3 gives summaries of the linear association under model-based and sandwich variance estimates. The point estimates and model-based standard error estimates arise from either ML estimation (assuming normality of errors) or ordinary least squares estimation of β. The sandwich estimates of the standard errors relax the constancy of variance assumption but assume uncorrelated errors. The standard error of the intercept is essentially unchanged under sandwich estimation, when compared to the model-based version, while that for the slope is slightly increased. The sample size of $n = 97$ is large enough to guarantee asymptotic normality of the estimator. For a 10-fold increase in cancer volume (in cc), there is a $\exp(\widehat{\beta}_1) = 2.1$ increase in PSA concentration.

Figure 5.4 plots the log of PSA versus the log of cancer volume and superimposes the estimated linear association, along with pointwise 95% confidence intervals for the expected linear association and for a new observation (assuming normally distributed data). There does not appear to be any deviation in random scatter of the data around the line (a residual plot would give a clearer way of assessing the nonconstant variance assumption, as we will see in Sect. 5.10). In Fig. 5.5(a), we plot PSA versus log cancer volume and clearly see the variance of PSA increasing with increasing cancer volume on this scale. Figure 5.5(b) plots PSA versus cancer volume. It is very difficult to assess the goodness of fit of the fitted relationship

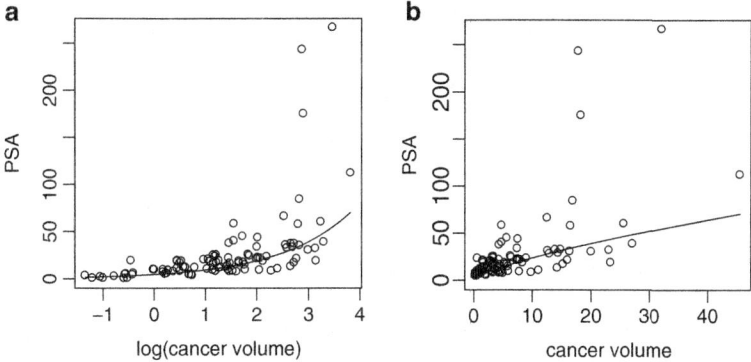

Fig. 5.5 (**a**) Prostate-specific antigen versus log cancer volume, (**b**) Prostate-specific antigen versus cancer volume. In each case, the least squares/maximum likelihood fit is included

or assumptions concerning the mean–variance relationship when the response and covariate are on their original scales. In both plots, the fitted line is from the fitting of model (5.40).

5.7 Bayesian Inference

We now consider Bayesian inference for the linear model. As with likelihood inference, we are required to specify the probability of the data and we assume $Y \mid \beta, \sigma^2 \sim N_n(x\beta, \sigma^2 I_n)$. The posterior distribution is

$$p(\beta, \sigma^2 \mid y) \propto L(\beta, \sigma^2) \times \pi(\beta, \sigma^2). \tag{5.41}$$

Closed-form posterior distributions for β and σ^2 are only available under restricted prior distributions. In particular, consider the improper prior distribution

$$\pi(\beta, \sigma^2) = p(\beta) \times p(\sigma^2) \propto \sigma^{-2} \tag{5.42}$$

Under this prior and likelihood combination, the posterior is, up to proportionality,

$$p(\beta, \sigma^2 \mid y) \propto (\sigma^2)^{-(n+2)/2} \exp\left[-\frac{1}{2\sigma^2}(y - x\beta)^\mathrm{T}(y - x\beta)\right]. \tag{5.43}$$

To derive $p(\beta \mid y)$, we need to integrate σ^2 from the joint distribution. To achieve this, it is useful to use an equality derived earlier, (2.23):

$$(y - x\beta)^\mathrm{T}(y - x\beta) = s^2(n - k - 1) + (\widehat{\beta} - \beta)^\mathrm{T} x^\mathrm{T} x (\widehat{\beta} - \beta),$$

where $\widehat{\boldsymbol{\beta}}$ is the ML/LS estimate. Substitution into (5.43) gives

$$p(\boldsymbol{\beta} \mid \boldsymbol{y}) \propto \int (\sigma^2)^{-(n+2)/2} \exp\left\{-\frac{1}{2\sigma^2}\left[s^2(n-k-1)\right.\right.$$
$$\left.\left. + (\widehat{\boldsymbol{\beta}}-\boldsymbol{\beta})^{\mathsf{T}}\boldsymbol{x}^{\mathsf{T}}\boldsymbol{x}(\widehat{\boldsymbol{\beta}}-\boldsymbol{\beta})\right]\right\} d\sigma^2.$$

The integrand here is the kernel of an inverse gamma distribution (Appendix D) for σ^2 and so has a known normalizing constant, the substitution of which gives

$$p(\boldsymbol{\beta} \mid \boldsymbol{y}) \propto \Gamma\left(\frac{n}{2}\right)\left[\frac{s^2(n-k-1)+(\widehat{\boldsymbol{\beta}}-\boldsymbol{\beta})^{\mathsf{T}}\boldsymbol{x}^{\mathsf{T}}\boldsymbol{x}(\widehat{\boldsymbol{\beta}}-\boldsymbol{\beta})}{2}\right]^{-n/2}$$
$$\propto \left[1 + \frac{(\widehat{\boldsymbol{\beta}}-\boldsymbol{\beta})^{\mathsf{T}}s^{-2}\boldsymbol{x}^{\mathsf{T}}\boldsymbol{x}(\widehat{\boldsymbol{\beta}}-\boldsymbol{\beta})}{n-k-1}\right]^{-(n-k-1+k+1)/2}$$

after some simplification. By inspection we recognize that this expression is the kernel of a $(k+1)$-dimensional t distribution (Appendix D) with location $\widehat{\boldsymbol{\beta}}$, scale matrix $s^2(\boldsymbol{x}^{\mathsf{T}}\boldsymbol{x})^{-1}$, and $n-k-1$ degrees of freedom, that is,

$$\boldsymbol{\beta} \mid \boldsymbol{y} \sim \mathrm{T}_{k+1}\left[(\boldsymbol{x}^{\mathsf{T}}\boldsymbol{x})^{-1}\boldsymbol{x}^{\mathsf{T}}\boldsymbol{y}, (\boldsymbol{x}^{\mathsf{T}}\boldsymbol{x})^{-1}s^2, n-k-1\right]. \quad (5.44)$$

Consequently, under the prior (5.42), the Bayesian posterior mean $\mathrm{E}[\boldsymbol{\beta} \mid \boldsymbol{y}]$ corresponds to the MLE, and $100(1-\alpha)\%$ credible intervals are identical to $100(1-\alpha)\%$ confidence intervals, though of course the two intervals have very different interpretations.

Asymptotically, as with likelihood estimation, it is the covariance model $\mathrm{var}(\boldsymbol{Y} \mid \boldsymbol{x})$ that is most important for valid inference, and normality of the error terms is unimportant. One way of thinking about this is as replacing $\boldsymbol{y} \mid \boldsymbol{\beta}, \sigma^2$ by

$$\widehat{\boldsymbol{\beta}} \mid \boldsymbol{\beta}, \widehat{\sigma}^2 \sim \mathrm{N}_{k+1}\left[\boldsymbol{\beta}, \widehat{\sigma}^2(\boldsymbol{x}^{\mathsf{T}}\boldsymbol{x})^{-1/2}\right].$$

We may derive the marginal posterior distribution of σ^2 as

$$\sigma^2 \mid \boldsymbol{y} \sim (n-p-1)s^2 \times \chi^{-2}_{n-k-1}, \quad (5.45)$$

a scaled inverse chi-squared distribution. As in the frequentist development, inference for σ^2 is likely to be highly sensitive to the normality assumption.

Although we can obtain analytic forms for $p(\boldsymbol{\beta} \mid \boldsymbol{y})$ and $p(\sigma^2 \mid \boldsymbol{y})$ under the prior (5.42), closed forms will not be available for general functions of interest. Direct sampling from the posterior may be utilized for inference in this case though. A sample from the *joint* distribution $p(\boldsymbol{\beta}, \sigma^2 \mid \boldsymbol{y})$ can be generated using the composition method (Sect. 3.8.4) via the factorization

5.7 Bayesian Inference

$$p(\boldsymbol{\beta}, \sigma^2 \mid \boldsymbol{y}) = p(\sigma^2 \mid \boldsymbol{y}) \times p(\boldsymbol{\beta} \mid \sigma^2, \boldsymbol{y}),$$

where $\boldsymbol{\beta} \mid \sigma^2, \boldsymbol{y} \sim \mathrm{N}_{k+1}\left[\widehat{\boldsymbol{\beta}}, \sigma^2 (\boldsymbol{x}^\mathsf{T}\boldsymbol{x})^{-1}\right]$, and $\sigma^2 \mid \boldsymbol{y}$ is given by (5.45). Independent samples are generated via the pair of distributions

$$\sigma^{2(t)} \sim p(\sigma^2 \mid \boldsymbol{y})$$
$$\boldsymbol{\beta}^{(t)} \sim p(\boldsymbol{\beta} \mid \sigma^{2(t)}, \boldsymbol{y}),$$

for $t = 1, \ldots, T$. Samples for functions of interest $\phi = g(\boldsymbol{\beta}, \sigma^2)$ are then available as $\phi^{(t)} = g(\boldsymbol{\beta}^{(t)}, \sigma^{2(t)})$.

The conjugate prior (Sect. 3.7.1) here takes the form $\pi(\boldsymbol{\beta}, \sigma^2) = \pi(\boldsymbol{\beta} \mid \sigma^2) \pi(\sigma^2)$ with $\boldsymbol{\beta} \mid \sigma^2 \sim \mathrm{N}_{k+1}(\boldsymbol{m}, \sigma^2 \boldsymbol{V})$ and $\sigma^{-2} \sim \mathrm{Ga}(a, b)$. However, this specification is not that useful in practice since the prior for $\boldsymbol{\beta}$ depends on σ^2. In particular, for smaller and smaller σ^2, the prior for $\boldsymbol{\beta}$ becomes increasingly concentrated about $\boldsymbol{m}$ which would not seem realistic in many contexts.

Under other prior distributions, analytic/numerical approximations or sampling-based techniques are required. An obvious prior choice is

$$\boldsymbol{\beta} \sim \mathrm{N}(\boldsymbol{m}, \boldsymbol{V}), \quad \sigma^{-2} \sim \mathrm{Ga}(a, b)$$

which gives the posterior

$$p(\boldsymbol{\beta}, \sigma^2 \mid \boldsymbol{y}) \propto l(\boldsymbol{\beta}, \sigma^2) \pi(\boldsymbol{\beta}) \pi(\sigma^2)$$

which is intractable, unless $\boldsymbol{V}^{-1}$ is the $(k+1) \times (k+1)$ matrix of zeroes, which is the improper prior case, (5.42), already considered. Although the posterior is not available in closed form under this prior, it is straightforward to construct a blocked Gibbs sampling algorithm (Sect. 3.8.4). Specifically, letting $L(\boldsymbol{\beta}, \sigma^2)$ denote the likelihood, one iterates between the pair of conditional distributions:

$$p(\boldsymbol{\beta} \mid \boldsymbol{y}, \sigma^2) \propto L(\boldsymbol{\beta}, \sigma^2) \pi(\boldsymbol{\beta})$$
$$\sim \mathrm{N}(\boldsymbol{m}^\star, \boldsymbol{V}^\star) \qquad (5.46)$$
$$p(\sigma^{-2} \mid \boldsymbol{y}, \boldsymbol{\beta}) \propto L(\boldsymbol{\beta}, \sigma^2) \pi(\sigma^{-2})$$
$$\sim \mathrm{Ga}\left(a + \frac{n}{2}, b + \frac{1}{2}(\boldsymbol{y} - \boldsymbol{x}\boldsymbol{\beta})^\mathsf{T}(\boldsymbol{y} - \boldsymbol{x}\boldsymbol{\beta})\right) \qquad (5.47)$$

where

$$\boldsymbol{m}^\star = \boldsymbol{W} \times \widehat{\boldsymbol{\beta}} + (\mathbf{I}_{k+1} - \boldsymbol{W}) \times \boldsymbol{m}$$
$$\boldsymbol{V}^\star = \boldsymbol{W} \times \mathrm{var}(\widehat{\boldsymbol{\beta}})$$

and

$$\boldsymbol{W} = (\boldsymbol{x}^\mathsf{T}\boldsymbol{x} + \boldsymbol{V}^{-1}\sigma^2)^{-1}(\boldsymbol{x}^\mathsf{T}\boldsymbol{x}).$$

Conditional conjugacy is exploited in this derivation; for details, see Exercise 5.4. For general prior distributions, the Gibbs sampler is less convenient because the conditional distributions will be of unrecognizable form, but Metropolis–Hastings steps (Sect. 3.8.2) for $\beta \mid y, \sigma^2$ and $\sigma^{-2} \mid y, \beta$ are straightforward to construct.

5.8 Analysis of Variance

The analysis of variance, or ANOVA, is a method by which the variability in the response is partitioned into components due to the various classifying variables and due to error. At one level, the ANOVA model is just a special case of a multiple linear regression model, but ANOVA does not simply have a role as an "outgrowth" of linear models. Rather Cox and Reid (2000, p. 245) state that ANOVA has a role "in clarifying the structure of sets of data, especially relatively complicated mixtures of crossed and nested data. This indicates what contrasts can be estimated and the relevant basis for estimating error. From this viewpoint the analysis of variance table comes first, then the linear model, not *vice-versa*." A study of the analysis of variance is intrinsically linked to the study of the design of experiments. Numerous books exist on ANOVA and the design of experiments; here we only give a brief discussion and introduce the main concepts. Specifically, we distinguish between *crossed* and *nested* (or hierarchical) designs and *fixed* and *random* effects modeling.

5.8.1 One-Way ANOVA

Consider the data in Table 5.4, taken from Davies (1967), which consist of the yield (in grams) from six randomly chosen batches of raw material, with five replicates each. The aim of this experiment was to find out to what extent batch-to-batch variation is responsible for variation in the final product yield.

Data such as these correspond to the simplest situation in which we have a single factor and a one-way classification. We may model the yield Y_{ij} in the jth sample from batch i as

$$Y_{ij} = \mu + \alpha_i + \epsilon_{ij}, \tag{5.48}$$

Table 5.4 Yield of dyestuff in grams of standard color, in each of six batches

Replicate observation	Batch					
	1	2	3	4	5	6
1	1,545	1,540	1,595	1,445	1,595	1,520
2	1,440	1,555	1,550	1,440	1,630	1,455
3	1,440	1,490	1,605	1,595	1,515	1,450
4	1,520	1,560	1,510	1,465	1,635	1,480
5	1,580	1,495	1,560	1,545	1,625	1,445

5.8 Analysis of Variance

with $\epsilon_{ij} \mid \sigma^2 \sim_{iid} N(0, \sigma^2)$, $i = 1, \ldots, a$, $j = 1, \ldots, n$. We need a constraint to prevent aliasing (Sect. 5.5.2), with two possibilities being the sum-to-zero constraint, $\sum_{i=1}^{a} \alpha_i = 0$, and corner-point constraint: $\alpha_1 = 0$. Model (5.48) is an example of a multiple linear regression with mean

$$E[Y \mid x] = x\beta$$

in which

$$Y = \begin{bmatrix} Y_{11} \\ \vdots \\ Y_{1n} \\ Y_{21} \\ \vdots \\ Y_{2n} \\ \vdots \\ Y_{a1} \\ \vdots \\ Y_{an} \end{bmatrix}, \quad x = \begin{bmatrix} 1 & 1 & 0 & \ldots & 0 \\ \cdots & \cdots & \cdots & \cdots & \cdots \\ 1 & 1 & 0 & \ldots & 0 \\ 1 & 0 & 1 & \ldots & 0 \\ \cdots & \cdots & \cdots & \cdots & \cdots \\ 1 & 0 & 1 & \ldots & 0 \\ \cdots & \cdots & \cdots & \cdots & \cdots \\ 1 & 0 & 0 & \ldots & 1 \\ \cdots & \cdots & \cdots & \cdots & \cdots \\ 1 & 0 & 0 & \ldots & 1 \end{bmatrix}, \quad \beta = \begin{bmatrix} \mu \\ \alpha_1 \\ \vdots \\ \alpha_a \end{bmatrix},$$

and where we adopt the corner-point constraint. Suppose we are interested in whether there are differences between the strengths from different looms. No differences correspond to the null hypothesis:

$$H_0 : \alpha_1 = \ldots = \alpha_a = 0. \tag{5.49}$$

Carrying out $a(a-1)/2$ t-tests leads to multiple testing problems (Sect. 4.5). Viewing this problem from a frequentist perspective and with $a = 6$ batches, we have 15 tests of pairs of batches, and with an individual type I error of 0.05, this gives an overall type I error of $1 - 0.95^{10} = 0.54$. As an alternative, we may test (5.49) using an F test (Sect. 5.6.1). Specifically, the F statistic is given by

$$F = \frac{\text{FSS}(\alpha \mid \mu)/(a-1)}{\text{RSS}(\alpha)/a(n-1)} \tag{5.50}$$

where

$$\text{FSS}(\alpha \mid \mu) = \text{RSS}(\mu) - \text{RSS}(\mu, \alpha)$$

is the fitted sum of squares that results when $\alpha = [\alpha_1, \ldots, \alpha_a]$ is added to the model containing μ only. In (5.50), the F statistic is the ratio of two so-called mean squares, which are average sum of squares, and under H_0, since the contributions in numerator and denominator are independent, $F \sim F_{a-1, a(n-1)}$. The ANOVA table associated with the test is given in Table 5.5. This table lays out the quantities that require calculation and shows the decomposition of the total sum of squares into

Table 5.5 ANOVA table for the one-way classification. The F statistic is for a test of $H_0 : \alpha_1 = \alpha_2 = \ldots = \alpha_a = 0$; DF is short for degrees of freedom and EMS for the expected mean square, which is E[SS/DF]

Source	Sum of squares	DF	EMS	F statistic
Between batches	$SS_1 = n \sum_{i=1}^{a} (\overline{Y}_{i\cdot} - \overline{Y}_{\cdot\cdot})^2$	$a - 1$	$\sigma^2 + n \frac{\sum_{i=1}^{a} \alpha_i^2}{a-1}$	$\frac{SS_1/(a-1)}{SS_2/a(n-1)}$
Error	$SS_2 = \sum_{i=1}^{a} \sum_{j=1}^{n} (Y_{ij} - \overline{Y}_{i\cdot})^2$	$a(n-1)$	σ^2	
Total	$SS_T = \sum_{i=1}^{a} \sum_{j=1}^{n} (Y_{ij} - \bar{Y}_{\cdot\cdot})^2$	$an - 1$		

Table 5.6 One-way ANOVA table for the dyestuff data; DF is shorthand for degrees of freedom

Source	Sum of squares	DF	Mean square	F statistic
Between batches	56,358	5	11,272	4.60 (0.0044)
Error	58,830	24	2,451	
Total	115,188	29		

The quantity in brackets in the final column is the p-value

that due to groups (batches in this example) and that due to error. The intuition behind the F test is that if there are no group effects, then the average sum of squares corresponding to the groups will, in expectation, equal the error variance. Consequently, we see in Table 5.5 that the expected mean square is simply σ^2 when $\alpha_1 = \ldots = \alpha_a = 0$. The success of the F test depends on the fact that we may decompose the overall sum of squares into the sum of the constituent parts corresponding to different components, and these follow independent χ^2 random variables.

Table 5.6 gives the numerical values for the dyestuff data of Table 5.4 and results in a very small p-value. As discussed in Sect. 4.2, the calibration of p-values is difficult, but for this relatively small sample size, a p-value of 0.0044 strongly suggests that the null is very unlikely to be true, and we would conclude that there are significant differences between batch means for these data. A Bayesian approach to testing may be based on Bayes factors. In this linear modeling context, there are close links between the Bayes factor and the F statistic (O'Hagan 1994, Sect. 9.34), though as usual the interpretations of the two quantities differ considerably. It is straightforward to extend the F test to the case of different sample sizes within looms, that is, to the case of general n_i, $i = 1, \ldots, a$.

If we are interested in the overall average yield, we would not want to ignore batch effects if present (even if they are not of explicit interest), because a model with no batch effects would not allow for the positive correlations that are induced between yields within the same batch. This issue is discussed in far greater detail in Chap. 8.

5.8 Analysis of Variance

Table 5.7 Data on clotting times (in minutes) for eight subjects, each of whom receives four treatments

Subject	Treatment 1	2	3	4	Mean
1	8.4	9.4	9.8	12.2	9.95
2	12.8	15.2	12.9	14.4	13.82
3	9.6	9.1	11.2	9.8	9.92
4	9.8	8.8	9.9	12.0	10.12
5	8.4	8.2	8.5	8.5	8.40
6	8.6	9.9	9.8	10.9	9.80
7	8.9	9.0	9.2	10.4	9.38
8	7.9	8.1	8.2	10.0	8.55
Mean	9.30	9.71	9.94	11.02	9.99

5.8.2 Crossed Designs

We now consider two factors, which we label A and B, with a and b levels, respectively. If each level of A is *crossed* with each level of B, we have a *factorial* design. Suppose that there are n replicates within each of the ab cells. The interaction model is

$$Y_{ijk} = \mu + \alpha_i + \beta_j + \gamma_{ij} + \epsilon_{ijk},$$

for $i = 1, \ldots, a$, $j = 1, \ldots, b$, and $k = 1, \ldots, n$. This model contains $1 + a + b + ab$ parameters, while the data supply only ab sample means. Therefore, it is clear that constraints on the parameters are required. In the corner-point parameterization (Sect. 5.5.2), the $1 + a + b$ constraints are

$$\alpha_1 = \beta_1 = \gamma_{11} = \ldots = \gamma_{1b} = \gamma_{21} = \ldots \gamma_{a1} = 0.$$

Alternatively, we may adopt the sum-to-zero constraints:

$$\sum_{i=1}^{a} \alpha_i = \sum_{j=1}^{b} \beta_j = \sum_{i=1}^{a} \gamma_{ij} = \sum_{j=1}^{b} \gamma_{ij} = 0.$$

Table 5.7 reproduces data from Armitage and Berry (1994) in which clotting times of plasma are analyzed. These data are from a crossed design in which each of $a = 8$ subjects received $b = 4$ treatments. The design is crossed since each patient receives each of the treatments. These data also provide an example of a *randomized block design* in which the aim is to provide a more homogeneous experimental setting within which to compare the treatments. Ignoring the blocking factor increases the unexplained variability and reduces efficiency. Section 8.3 provides further discussion.

Table 5.8 ANOVA table for the two-way crossed classification with one observation per cell; DF is short for degrees of freedom and EMS for the expected mean square

Source	Sum of squares	DF	EMS	F statistic
Factor A	$SS_A = b\sum_{i=1}^{a}(\overline{Y}_{i.} - \overline{Y}_{..})^2$	$a-1$	$\frac{SS_A}{a-1}$	$\frac{\sigma^2 + b\sum_{i=1}^{a}\alpha_i^2}{a-1}$
Factor B	$SS_B = a\sum_{j=1}^{b}(\overline{Y}_{.j} - \overline{Y}_{..})^2$	$b-1$	$\frac{SS_B}{b-1}$	$\frac{\sigma^2 + a\sum_{j=1}^{b}\beta_j^2}{b-1}$
Error	$SS_E =$ $\sum_{i=1}^{a}\sum_{j=1}^{b}(Y_{ij} - \overline{Y}_{i.} - \overline{Y}_{.j} + \overline{Y}_{..})^2$	$(a-1)(b-1)$	$\frac{SS_E}{(a-1)}$	σ^2
Total	$SS_T = \sum_{i=1}^{a}\sum_{j=1}^{b}(Y_{ij} - \overline{Y}_{..})^2$	$ab-1$		

Table 5.9 ANOVA table for the plasma clotting time data in Table 5.7; DF is short for degrees of freedom. The quantity in brackets in the final column is the p-value

Source of variation	Sum of squares	DF	Mean square	F statistic
Treatment	13.0	3	4.34	6.62 (0.0026)
Subjects	79.0	7	11.3	17.2 (2.2×10^{-7})
Error	13.8	21	0.656	
Total	105.8	31		

There are no replicates within each of the 8×4 cells in Table 5.7, and so it is not possible to examine interactions between subjects and treatments. Consequently, we concentrate on the main effects only model:

$$Y_{ij} = \mu + \alpha_i + \beta_j + \epsilon_{ij}, \tag{5.51}$$

for $i = 1, \ldots, 4; j = 1, \ldots, 8$ and with $\epsilon_{ij} \mid \sigma^2 \sim_{iid} N(0, \sigma^2)$. Here we adopt the corner-point parameterization with $\alpha_1 = 0$ and $\beta_1 = 0$. Table 5.8 gives the generic ANOVA table for a two-way classification with no replicates, and Table 5.9 gives the numerical values for the plasma data. For these data, primary interest is in treatment effects (the α_i's), and Table 5.9 shows the steps to obtaining a p-value of 0.0026 for the null of $H_0 : \alpha_2 = \alpha_3 = \alpha_4 = 0$ which, for this small sample size, points strongly towards the null being unlikely. In passing, we note that there are large between-subject differences for these data, so that the crossed design is very efficient.

We now examine treatment differences using estimation. Under the improper prior

$$p(\mu, \boldsymbol{\alpha}, \boldsymbol{\beta}, \sigma^2) \propto \frac{1}{\sigma^2}$$

interval estimates obtained from Bayesian, likelihood, and least squares analyses are identical. We take a Bayesian stance and report the posterior distribution for each of the treatment effects. We let $\boldsymbol{\theta} = [\mu, \boldsymbol{\alpha}, \boldsymbol{\beta}]$ where $\boldsymbol{\alpha} = [\alpha_2, \alpha_3, \alpha_4]$ and $\boldsymbol{\beta} = [\beta_2, \ldots, \beta_8]$. The joint posterior for $\boldsymbol{\theta}$ is multivariate Student's t, with $n - k - 1 = 32 - 11 = 21$ degrees of freedom, posterior mean $\widehat{\boldsymbol{\theta}}$ (the least squares estimate) and posterior scale matrix, $(\boldsymbol{x}^T\boldsymbol{x})^{-1}\widehat{\sigma}^2$, where $\widehat{\sigma}^2$ is the usual

5.8 Analysis of Variance

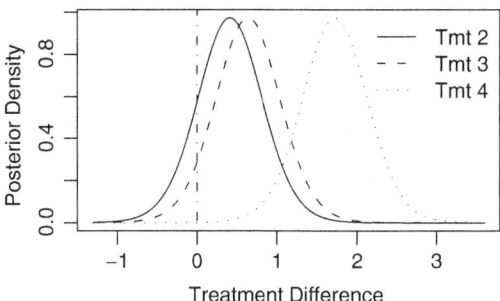

Fig. 5.6 Marginal posterior distributions for the treatment contrasts, with treatment 1 as the baseline, for the plasma clotting time data in Table 5.7

unbiased estimator of the residual error variance. Since treatment 1 is the reference we examine treatment differences with respect to this baseline group. Figure 5.6 gives the posterior distributions for $\alpha_2, \alpha_3, \alpha_4$. The posterior probabilities that the average responses under treatments 2, 3, and 4 are greater than zero are 0.16, 0.065, and 0.00017, respectively. Consequently, we conclude that there is strong evidence that treatment 4 differs from treatment 1, with decreasingly lesser evidence of differences between treatment 1 and treatments 3 and 2.

5.8.3 Nested Designs

For a design with two factors, suppose that Y_{ijk} denotes a response at level i of factor A and level j of factor B, with replication indexed by k. In a nested design, in contrast to a crossed design, $j = 1$ in level 1 of factor A has no meaningful connection with $j = 1$ in level 2 of factor A. In the context of the previous example, suppose each of eight patients received a single treatment each, but with k replicate measurements. In this case, we again have two factors, treatments and patients, but the patient effects are *nested* within treatments. A nested model for two factors is

$$Y_{ijk} = \mu + \alpha_i + \beta_{j(i)} + \epsilon_{ijk},$$

with $i = 1, \ldots, a$ indexing factor A and $j = 1, \ldots, b$ factor B. In the nested patient/treatment example, A represents treatment and B patient, and so $\beta_{j(i)}$ represents the change in expected response for patient j within level i of treatment. Notice that there is no interaction in the model, because factor B is nested within factor A, and not crossed, and so there is no way of estimating the usual interactions. In a sense, $\beta_{j(i)}$ is an interaction parameter since it is the patient effect specific to a particular treatment. Table 5.10 gives the ANOVA table for this design.

Table 5.10 ANOVA table for the two-way nested classification; DF is short for degrees of freedom and EMS for the expected mean square

Source	Sum of squares	DF	EMS	F statistic
Factor A	$SS_A = bn \sum_{i=1}^{a}(\overline{Y}_{i..} - \overline{Y}_{...})^2$	$a-1$	$\frac{SS_A}{a-1}$	$\frac{\sigma^2 + bn\sum_{i=1}^{a}\alpha_i^2}{a-1}$
Factor B (within A)	$SS_B = n\sum_{i=1}^{a}\sum_{j=1}^{b}(\overline{Y}_{ij.} - \overline{Y}_{i..})^2$	$a(b-1)$	$\frac{SS_B}{(a(b-1)}$	$\frac{\sigma^2 + a\sum_{j=1}^{b}\beta_j^2}{b-1}$
Error	$SS_E = \sum_{i=1}^{a}\sum_{j=1}^{b}\sum_{i=1}^{n}(Y_{ijk} - \overline{Y}_{ij.})^2$	$ab(n-1)$	$\frac{SS_E}{(a-1)(b-1)}$	σ^2
Total	$SS_T = \sum_{i=1}^{a}\sum_{j=1}^{b}\sum_{i=1}^{n}(Y_{ijk} - \overline{Y}_{...})^2$	$abn-1$		

5.8.4 Random and Mixed Effects Models

The examples we have presented so far are known, in the frequentist literature, as *fixed effects* ANOVA models since the parameters, for example, the α_i's in the one-way classification, are viewed as nonrandom. An alternative *random effects* approach is to view these parameters as a sample from a probability distribution, with the usual choice being $\alpha_i \mid \sigma_\alpha^2 \sim_{iid} N(0, \sigma_\alpha^2)$. From a frequentist perspective, the choice is based on whether the units that are selected can be viewed as being a random sample from some larger distribution of effects. Often, patients in a trial may be regarded as a random sample from some population, while treatment effects may be regarded as fixed effects. In this case, we have a *mixed effects* model. Model (5.51) was used for the data in Table 5.7 with the α_i and β_j being treated as fixed effects. Alternatively, we could use a mixed effects model with the individual effects α_i being treated as random effects and the β_j, representing treatment effects, being seen as fixed effects.

From a Bayesian perspective, the distinction being fixed and random effects is less distinct since all unknowns are viewed as random variables. However, the prior choice reflects the distinction. For example, in model (5.51), the "fixed effects" corresponding to treatments may be assigned independent prior distributions $\beta_j \sim N(0, V)$ where V is fixed, while the "random effects" corresponding to patients may be assigned the prior $\alpha_i \mid \sigma_\alpha^2 \sim_{iid} N(0, \sigma_\alpha^2)$ with σ_α^2 assigned a prior and estimated from the data.

A full description of estimation for random and mixed effects models will be postponed until Chap. 8, though here we briefly describe likelihood-based inference for the one-way model (5.48). Readers who have not previously encountered random effects models may wish to skip the remainder of this section and return after consulting Chap. 8. The one-way model is

$$Y_{ij} = \mu + \alpha_i + \epsilon_{ij},$$

5.9 Bias-Variance Trade-Off

Table 5.11 ANOVA table for test of $H_0 : \sigma_\alpha^2 = 0$; DF is short for degrees of freedom and EMS for the expected mean square

Source	Sum of squares	DF	EMS
Between batches	$n\sum_{i=1}^{a}(\bar{Y}_{i\cdot} - \bar{Y}_{\cdot\cdot})^2$	$a-1$	$\sigma^2 + n\sigma_\alpha^2$
Error	$\sum_{i=1}^{a}\sum_{j=1}^{n}(Y_{ij} - \bar{Y}_{i\cdot})^2$	$a(n-1)$	σ^2
Total	$\sum_{i=1}^{a}\sum_{j=1}^{n}(Y_{ij} - \bar{Y}_{\cdot\cdot})^2$	$an-1$	

where we have the usual assumption $\epsilon_{ij} \mid \sigma^2 \sim_{iid} N(0, \sigma^2)$, $j = 1, \ldots, n$, and add $\alpha_i \mid \sigma_\alpha^2 \sim_{iid} N(0, \sigma_\alpha^2)$, $i = 1, \ldots, a$ as the random effects distribution. We no longer need a constraint on the α_i's in the random effects model since these parameters are "tied together" via the normality assumption. A primary question of interest is often whether there are between-unit differences, and this can be examined via the hypothesis $H_0 : \sigma_\alpha^2 = 0$. In the one-way classification, this test turns out to be equivalent to the F test given previously in Sect. 5.8.1, though this equivalence is not true for more complex models. The ANOVA table given in Table 5.11 is very similar to that for the fixed effects model form in Table 5.5, though we highlight the difference in the final column.

Estimation via a likelihood approach proceeds by integrating the α_i from the model to give the marginal distribution

$$p(y_i \mid \mu, \sigma^2, \sigma_\alpha^2) = \int p(y_i \mid \mu, \alpha_i, \sigma^2) \times p(\alpha_i \mid \sigma_\alpha^2) \, d\alpha_i,$$

and results in

$$y_i \mid \mu, \sigma^2, \sigma_\alpha^2 \sim_{iid} N(\mu \mathbf{1}_r, \sigma^2 \mathbf{I}_r + \sigma_\alpha^2 \mathbf{J}_r),$$

where $\mathbf{1}_r$ is the $r \times 1$ vector of 1's, $\mathbf{I}_r$ is the $r \times r$ identity matrix, and $\mathbf{J}_r$ is the $r \times r$ matrix of 1's. This likelihood can be maximized with respect to $\mu, \sigma_\alpha^2, \sigma^2$, and asymptotic standard errors may be calculated from the information matrix. A Bayesian approach combines the marginal likelihood with a prior $\pi(\mu, \sigma_\alpha^2, \sigma^2)$.

5.9 Bias-Variance Trade-Off

Chapter 4 gave an extended discussion of model formulation and model selection, and the example at the end of Sect. 4.8 acted as a prelude to this section in which we describe the bias-variance trade-off that is encountered when we consider which variables to include in a model.

Suppose the true model is

$$\mathbf{Y} = \mathbf{x}\boldsymbol{\beta} + \boldsymbol{\epsilon},$$

where Y is $n \times 1$, x is $n \times (k+1)$, β is $(k+1) \times 1$, and the errors are such that $E[\epsilon] = 0$ and $\text{var}(\epsilon) = \sigma^2 I_n$. We have seen that the estimator

$$\widehat{\beta} = (x^T x)^{-1} x^T Y,$$

arises from ordinary least squares, likelihood (with normal errors, or large n), and Bayesian (with normal errors and prior (5.42), or large n) considerations. Asymptotically,

$$(x^T x)^{1/2}(\widehat{\beta}_n - \beta) \to_d N_{k+1}(0, \sigma^2 I_n)$$

where we assume $x^T x$ is of full rank. Since $x^T x$ is positive definite (all proper variance–covariance matrices are positive definite), we can find a unique Cholesky decomposition that is an upper-triangular matrix U such that $(x^T x)^{-1} = UU^T$. Proofs of the matrix results in this section may be found in Schott (1997, p.139–140). This decomposition leads to

$$\text{var}(\widehat{\beta}_j) = \sigma^2 \sum_{l=1}^{k+1} U_{jl}^2,$$

with $U_{jl} = 0$ if $j > l$.

We now split the collection of predictors into two groups, $x = [x_A, x_B]$, and examine the implications of regressing on a subset of predictors. Let $\beta = [\beta_A, \beta_B]^T$ where x_A is $n \times (q+1)$ with $q < k$ and β_A is $(q+1) \times 1$. Now suppose we fit the model

$$E[Y \mid x_A, x_B] = x_A \beta_A^\star$$

where we distinguish between $\beta_A^\star$ and β_A since the interpretation of the two sets of parameters differs. In particular, each coefficient in β_A has an interpretation as the linear association of the corresponding variable, controlling for all of the other variables in x. For coefficients in $\beta_A^\star$, control is only for variables in x_A. The estimator in the reduced model is

$$\widehat{\beta}_A^\star = (x_A^T x_A)^{-1} x_A^T Y,$$

and

$$\begin{aligned} E[\widehat{\beta}_A^\star] &= (x_A^T x_A)^{-1} x_A^T E[Y] \\ &= (x_A^T x_A)^{-1} x_A^T (x_A \beta_A + x_B \beta_B) \\ &= \beta_A + (x_A^T x_A)^{-1} x_A^T x_B \beta_B, \end{aligned} \quad (5.52)$$

5.9 Bias-Variance Trade-Off

so that the second term is the bias arising from omission of the last $k-q$ covariates. This defines the quantity that is being consistently estimated by $\widehat{\beta}_{\text{A}}^{\star}$. An alternative, less direct, derivation follows from the results of Sect. 2.4.3 in which we showed that the Kullback–Leibler distance between the true model and the reduced (assumed) model is that which is being minimized.

From (5.52), we see that the bias is zero if x_{A} and x_{B} are orthogonal, or if $\beta_{\text{B}} = 0$. Consequently, for bias to result, we need x_{B} to be associated with both the response Y and at least one of the variables in x_{A}. These requirements, roughly speaking, are the conditions for x_{B} to be considered a confounder. More precisely, Rothman and Greenland (1998) give the following criteria for a confounder:

1. A confounding variable must be associated with the response.
2. A confounding variable must be associated with the variable of interest in the population from which the data are sampled.
3. A confounding variable must not be affected by the variable of interest or the response. In particular it cannot be an intermediate step in the causal path between the variable of interest and the response.

At first sight, this result suggests that we should include as many variables as possible in the mean model, since this will reduce bias. But the splitting of the mean squared error of an estimator into the sum of the squared bias and the variance shows that this is only half of the story. Unfortunately, including variables that are not associated (or have a weak association only) with Y can increase the variance of the estimator (or equivalently, the posterior variance), as we now demonstrate.

We write

$$(x_{\text{A}}^{\text{T}} x_{\text{A}})^{-1} = U_{\text{A}} U_{\text{A}}^{\text{T}}$$

where U_{A} is upper-triangular and consists of the first $q+1$ rows and columns of U. Denoting the jth element of the estimators from the reduced and full models as $\widehat{\beta}_{\text{A}j}^{\star}$ and $\widehat{\beta}_{\text{A}j}$, retrospectively, we have

$$\text{var}(\widehat{\beta}_{\text{A}j}^{\star}) = \sigma^2 \sum_{l=1}^{q+1} U_{jl}^2$$

$$\leq \text{var}(\widehat{\beta}_{\text{A}j}),$$

for $j = 0, 1, \ldots, q$, with equality if and only if x_{A} and x_{B} are orthogonal.

Hence, if σ^2 is fixed across analyses, we conclude that adding covariates decreases precision. Intuitively this is because there is only so much information within a dataset, and if we add in variables that are related to Y and are not orthogonal to existing variables, the associations are not so accurately estimated since there are now competing explanations for the data.

Another layer of complexity is added when we take into account estimation of σ^2 since the *estimated* standard errors of the estimator now depend on $\widehat{\sigma}^2$. The usual unbiased estimator is given by the residual sum of squares divided by the degrees of freedom. The former is nonincreasing as covariates are added to the model, and the latter is decreasing. Consequently, as variables are entered into the model in terms of their "significance," a typical pattern is for $\widehat{\sigma}^2$ to decrease with the addition of important covariates, with an increase then occurring as variables that are almost unrelated are added (due to the decrease in the denominator of the estimator).

To expand on this further, consider the "true" model in which we assume for simplicity that β_B is univariate so that x_B is $n \times 1$:

$$y = x_A \beta_A + x_B \beta_B + \epsilon$$

where $E[\epsilon] = 0$ and $var(\epsilon) = \sigma^2 I_n$. We now fit the model

$$Y = x_A \beta_A^\star + \epsilon^\star,$$

so that x_B is omitted. Then, viewing X_B as random (since it is unobserved), we obtain

$$var(Y \mid x_A) = \sigma^2 I_n + \beta_B^2 var(X_B \mid x_A),$$

showing the form of the increase in residual variance (unless $\beta_B = 0$) when variables related to the response are added to the model. If x_A and x_B are collinear, the variance of X_B does not depend on x_A.

We expand on the development of this section, with a slight change of notation, via the "true" model

$$Y_i = \beta_0 + \beta_A(x_i - \overline{x}) + \beta_B(z_i - \overline{z}) + \epsilon_i,$$

and fitted model

$$Y_i = \beta_0^\star + \beta_A^\star(x_i - \overline{x}) + \epsilon_i.$$

Then, $\widehat{\beta}_0 = \widehat{\beta}_0^\star = \overline{Y}$ (since the covariates are centered in each model), and so each is an unbiased estimator of the intercept:

$$E[\widehat{\beta}_0] = E[\widehat{\beta}_0^\star] = \beta_0.$$

From (5.52),

$$E[\widehat{\beta}_A^\star] = \beta_A + \beta_B \times \frac{S_{xz}}{S_{xx}}$$

$$= \beta_A + \beta_B \times \rho_{xz} \left(\frac{S_{zz}}{S_{xx}}\right)^{1/2} \qquad (5.53)$$

5.9 Bias-Variance Trade-Off

where

$$S_{xx} = \sum_{i=1}^{n}(x_i - \overline{x})^2, \quad S_{xz} = \sum_{i=1}^{n}(x_i - \overline{x})(z_i - \overline{z}) \quad S_{zz} = \sum_{i=1}^{n}(z_i - \overline{z})^2$$

and

$$\rho_{xz} = \frac{S_{xz}}{(S_{xx}S_{zz})^{1/2}}.$$

We have seen (5.53) before in a slightly different form, namely (5.11) in the context of confounding. In the full model we have

$$(\boldsymbol{x}^{\mathrm{T}}\boldsymbol{x})^{-1} = \begin{bmatrix} 1/n & 0 & 0 \\ 0 & S_{zz}/D & -S_{xz}/D \\ 0 & -S_{xz}/D & S_{xx}/D \end{bmatrix},$$

where $D = S_{xx}S_{zz} - S_{xz}^2$, giving

$$\mathrm{var}(\widehat{\beta}_{\mathrm{A}}) = \frac{\sigma^2}{S_{xx} - S_{xz}^2/S_{zz}}$$

$$\geq \frac{\sigma^2}{S_{xx}} = \mathrm{var}(\widehat{\beta}_{\mathrm{A}}^\star),$$

with equality if and only if $S_{xz} = 0$ (so that X and Z are orthogonal), assuming that σ^2 is known.

When deciding upon the number of covariates for inclusion in the mean model, there are therefore competing factors to consider. The bias in the estimator cannot increase as more variables are added, but the precision of the estimator may increase or decrease, depending on the strength of the associations of the variables that are candidates for inclusion. The unexplained variation in the data (measured through $\widehat{\sigma}^2$) may be reduced, but the uncertainty in which of the covariates to assign the variation in the response to is increased. If the number of potential additional variables is large, the loss of precision may be considerable.

Section 4.8 described and critiqued various approaches to variable selection, emphasizing that the strategy taken is highly dependent on the context and in particular whether the aim is exploratory, confirmatory, or predictive. Chapter 12 considers the latter case in detail.

Example: Prostate Cancer

In this section we briefly illustrate the ideas of the previous section using two covariates from the PSA dataset, log(can vol) which we denote x_2 and log(cap pen) which we denote x_1. Let $\boldsymbol{x} = [x_1, x_2]$ and recall Y is log(PSA). Figure 5.7(a)

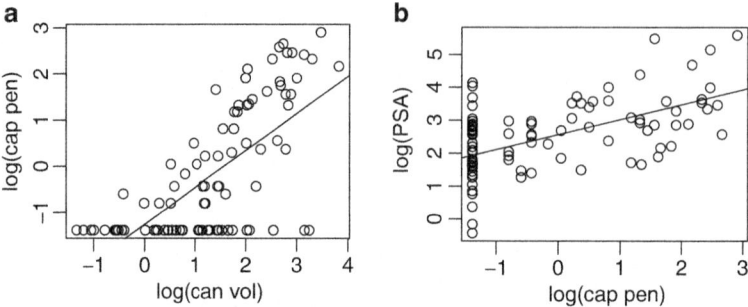

Fig. 5.7 (a) Association between log capsular penetration and log cancer volume, with fitted line, (b) association between log prostate-specific antigen and log capsular penetration, with fitted line

plots x_2 versus x_1 and illustrates the strong association between these variables. Figure 5.7(b) plots Y versus x_1, and we see an association here too. We obtain the following estimates:

$$E[Y \mid \boldsymbol{x}] = \beta_0^\star + \beta_1^\star x_1 \tag{5.54}$$
$$= 1.51 + 0.72 \times x_1 \tag{5.55}$$
$$E[Y \mid \boldsymbol{x}] = \beta_0 + \beta_1 x_1 + \beta_2 x_2 \tag{5.56}$$
$$= 1.61 + 0.66 \times x_1 + 0.080 \times x_2 \tag{5.57}$$
$$E[x_2 \mid x_1] = a + bx_1$$
$$= -12.6 + 0.80 \times x_1 \tag{5.58}$$

We first confirm, using (5.12) and (5.11), that the estimate associated with log(can vol) in model (5.54) combines the effect of this variable and log(cap pen):

$$\widehat{\beta_1^\star} = \widehat{\beta}_1 + \widehat{b} \times \widehat{\beta}_2$$
$$= 0.66 + 0.80 \times 0.08 = 0.72,$$

with $\widehat{b}$ from (5.58), to give the estimate appearing in (5.55). The standard error associated with x_1 in model (5.54) is 0.068, while in the full model (5.56), it increases to 0.092 due to the association observed in Fig. 5.7a between x_1 and x_2.

5.10 Robustness to Assumptions

In this section we investigate the behavior of the estimator

$$\widehat{\boldsymbol{\beta}} = (\boldsymbol{x}^\mathsf{T}\boldsymbol{x})^{-1}\boldsymbol{x}^\mathsf{T}\boldsymbol{Y},$$

5.10 Robustness to Assumptions

under departures from the assumptions that lead to

$$(\boldsymbol{x}^\mathrm{T}\boldsymbol{x})^{1/2}(\widehat{\boldsymbol{\beta}}_n - \boldsymbol{\beta}) \to_d \mathrm{N}_{k+1}(\mathbf{0}_{k+1}, \sigma^2 \mathbf{I}_{k+1}).$$

Correct inference arises from normality of the estimator, and the error terms should have constant variance and absence of correlation. Normality of the estimator occurs with a sufficiently large sample size or if the error terms are normal. Judging when the sample size is large enough can be assessed through simulation, and there is an interplay between sample size and the closeness of the error distribution to normality. We present results examining the effect of departures on confidence interval coverage, but these are identical to Bayesian credible intervals under the improper prior (5.42). Regardless of the distribution of the errors and the mean–variance relationship, we always obtain an unbiased estimator, hence the emphasis on confidence interval coverage.

5.10.1 Distribution of Errors

We begin by examining the effect of non-normality of the errors and simulate data from a linear model with errors that are uncorrelated and with constant variance. The distribution of the errors is taken as either normal, Laplacian, Student's t with 3 degrees of freedom, or lognormal. We examine the behavior of the least squares estimator for β_1, with $n = 5$ and $n = 20$, and two distributions for the covariate, either $x_i \sim_{iid} \mathrm{U}(0,1)$ or $x_i \sim_{iid} \mathrm{Ga}(1,1)$ (an exponential distribution), for $i = 1, \ldots, n$. The latter was chosen to examine the effects of a skewed covariate distribution.

Table 5.12 presents the 95% confidence interval coverage for β_1; based on 10,000 simulations, the true value is $\beta_1 = 0$. For the normal error distributions, the coverage should be exactly 95%, but we include simulation-based results to give an indication of the Monte Carlo error. In all cases the coverage probabilities are good, showing the robustness of inference in this simple scenario. When the number of covariates, k is large relative to n, more care is required, especially if the distributions of the covariate are very skewed. Lumley et al. (2002) discuss the validity of the least squares estimator when the data are not normal.

5.10.2 Nonconstant Variance

We have already considered the robustness of inference to nonconstant error variance in Sect. 5.6.4, in the context of sandwich estimation. Table 5.2 showed that confidence interval coverage will be poor when an incorrect mean–variance relationship is assumed. Sandwich estimation provides a good frequentist alternative estimation strategy, so long as the sample size is large enough for the variance of

Table 5.12 Coverage of 95% confidence intervals for β_1 for various error distributions, distributions of the covariate, and sample sizes n. The entries are based on 10,000 simulations

Error distribution	Distribution of x	n	Coverage
Normal N(0, 1)	Uniform	5	95
Normal N(0, 1)	Uniform	20	94
Normal N(0, 1)	Exponential	5	95
Normal N(0, 1)	Exponential	20	95
Laplacian Lap(0, 1)	Uniform	5	95
Laplacian Lap(0, 1)	Uniform	20	95
Laplacian Lap(0, 1)	Exponential	5	94
Laplacian Lap(0, 1)	Exponential	20	95
Student $T(0, 1, 3)$	Uniform	5	95
Student $T(0, 1, 3)$	Uniform	20	95
Student $T(0, 1, 3)$	Exponential	5	95
Student $T(0, 1, 3)$	Exponential	20	95
Lognormal LN(0, 1)	Uniform	5	95
Lognormal LN(0, 1)	Uniform	20	96
Lognormal LN(0, 1)	Exponential	5	94
Lognormal LN(0, 1)	Exponential	20	95

the estimator to be reliably estimated. The bootstrap (Sect. 2.7) provides another method for reliable variance estimation, again when the sample size is not small.

5.10.3 Correlated Errors

Finally we investigate the effect on coverage of correlated error terms. A simple scenario to imagine is (x, y) pairs collected on consecutive days. We assume an AR(1) autoregression model of order 1 (Sect. 8.4.2) which results in $\epsilon \mid \sigma^2 \sim \mathrm{N}(\mathbf{0}_n, \sigma^2 V)$, where V is the $n \times n$ matrix

$$V = \begin{bmatrix} 1 & \rho & \rho^2 & \cdots & \rho^{n-1} \\ \rho & 1 & \rho & \cdots & \rho \\ \vdots & \vdots & \vdots & \ddots & \vdots \\ \rho^{n-1} & \rho^{n-2} & \rho & \cdots & 1 \end{bmatrix}$$

and with ρ the correlation between errors on successive days. Table 5.13 gives the 95% confidence interval coverage (arising from a model in which the errors are assumed uncorrelated) as a function of sample size, the distribution of x (uniform or exponential), and strength of correlation. The table clearly shows that correlated errors can drastically impact confidence interval coverage, with the coverage becoming increasingly bad as the sample size increases.

5.11 Assessment of Assumptions

Table 5.13 95% confidence interval for the slope parameter β_1 as a function of the autocorrelation parameter ρ and the sample size n. The entries are based upon 10,000 simulations and are calculated under a model in which the errors are assumed uncorrelated

Distribution of x	Correlation ρ	n	Coverage
Uniform	0.1	5	94
Uniform	0.1	20	93
Uniform	0.1	50	92
Uniform	0.5	5	89
Uniform	0.5	20	76
Uniform	0.5	50	75
Uniform	0.95	5	79
Uniform	0.95	20	36
Uniform	0.95	50	26
Exponential	0.1	5	94
Exponential	0.1	20	93
Exponential	0.1	50	93
Exponential	0.5	5	89
Exponential	0.5	20	79
Exponential	0.5	50	77
Exponential	0.95	5	81
Exponential	0.95	20	41
Exponential	0.95	50	32

Intuitively, one might expect that in this situation the standard errors based on $(x^\mathsf{T} x)^{-1}\sigma^2$ would always underestimate the true standard error of the estimator. In the scenario described above, the effect of correlated errors depends critically upon the correlation among the x variables across time, however. If the x-variable is slowly varying over time, then the standard errors will be underestimated, but if the variable is changing rapidly, then the true standard errors may be smaller than those reported. This is because if there is high positive correlation, then the difference in the error terms on consecutive days is small, and so if Y changes, it must be due to changes in x. For further discussion, see Sect. 8.3.

5.11 Assessment of Assumptions

In this section we will describe a number of approaches for assessing the assumptions required for valid inference.

5.11.1 Review of Assumptions

We consider the linear regression model:

$$Y = x\beta + \epsilon$$

where Y is $n \times 1$, x is $n \times (k+1)$, β is $(k+1) \times 1$, and ϵ is $n \times 1$, with $\epsilon \mid \sigma^2 \sim N_n(0, \sigma^2 I_n)$. Under these assumptions, we have seen that the estimator $\widehat{\beta} = (x^T x)^{-1} x^T Y$, with $\text{var}(\widehat{\beta}) = (x^T x)^{-1} \sigma^2$, emerges from likelihood, least squares, and Bayesian approaches. The standard errors and confidence intervals we report are valid if:

- The error terms have constant variance. If sandwich estimation is used, then this assumption may be relaxed, so long as we have a large sample size.
- The error terms are uncorrelated.
- The estimator is normally distributed, so that we can effectively replace the likelihood $Y \mid \beta, \sigma^2$ by $\widehat{\beta} \mid \beta \sim N_p \left[\beta, (x^T x)^{-1} \widehat{\sigma}^2 \right]$. This occurs if the error terms are normally distributed and/or the sample size n is sufficiently large for the central limit theorem to ensure that the estimator is normally distributed.

As we saw in Sect. 5.10, confidence interval coverage can be very poor if the error variance is nonconstant and/or the errors are correlated. Normality of errors is not a big issue with the linear model with respect to estimation (which explains the popularity of least squares), unless the sample size is very small (relative to the number of predictors) or the distribution of the x values is very skewed. For validity of a predictive interval for an observable, we need to make a further assumption concerning the distribution of the error terms, however. This interval is given by (5.30) under the assumption of normal errors.

From a frequentist perspective and given the assumed mean model, $E[Y \mid x] = x\beta$, the estimator $\widehat{\beta}$ is an unbiased estimator of β. For example, in simple linear regression, $\widehat{\beta}_1$ is an unbiased estimator of the linear association in a population, regardless of the true relationship between response and covariate. The assumed mean model may be a poor description, however, and we will usually wish to examine the appropriateness of the model to decide on whether linearity holds.

Another aspect of model checking is scrutinizing the data for *outlying* or *influential* points. It is difficult to define exactly what is meant by an outlier, and we content ourselves with a fuzzy description of an outlier as "a data point that is unusual relative to the others." Single outlying observations may stand out in the plots described below. The presence of multiple outliers is more troublesome due to *masking*, in which the presence of an outlier is hidden by other outliers.

5.11.2 Residuals and Influence

In general, model checking may be carried out *locally*, using informal techniques such as residual plots, or *globally* using formal testing procedures; we concentrate on the former. The *observed error* is given by

$$e_i = Y_i - \widehat{Y}_i, \tag{5.59}$$

5.11 Assessment of Assumptions

where $\widehat{Y}_i = x_i\widehat{\beta}$, while the *true error* is

$$\epsilon_i = Y_i - \mathrm{E}[Y_i \mid x_i].$$

In *residual analysis* we examine the observed residuals for discrepancies from the assumed model. We define *residuals* as

$$e = [e_1, \ldots, e_n]^{\mathrm{T}} = Y - \widehat{Y} = (\mathbf{I}_n - h)Y, \qquad (5.60)$$

where $h = x(x^{\mathrm{T}}x)^{-1}x^{\mathrm{T}}$ is the *hat* (or projection) matrix encountered in Sect. 5.6.3. The hat matrix is symmetric, $h^{\mathrm{T}} = h$, and idempotent, $hh^{\mathrm{T}} = h$. We want to examine the relationship between e and ϵ so we can use the former to assess whether assumptions concerning the latter hold.

Substitution of

$$Y = x\beta + \epsilon$$

into (5.60) gives

$$e = (\mathbf{I}_n - h)\epsilon, \qquad (5.61)$$

or

$$e_i = \epsilon_i - \sum_{j=1}^{n} h_{ij}\epsilon_j, \qquad (5.62)$$

showing that the estimated residuals differ from the true residuals, complicating residual analysis.

We examine the moments of the error terms. The residuals e are random variables since they are a function of the random variables ϵ. We have

$$\mathrm{E}[e] = (\mathbf{I}_n - h)\mathrm{E}[\epsilon] = \mathbf{0}_n$$

and the variance–covariance matrix is

$$\mathrm{var}(e) = (\mathbf{I}_n - h)(\mathbf{I}_n - h)^{\mathrm{T}}\sigma^2 = (\mathbf{I}_n - h)\sigma^2,$$

so that fitting the model has induced dependence in the residuals. In particular,

$$\mathrm{var}(e_i) = (1 - h_{ii})\sigma^2,$$

since for a symmetric and idempotent matrix $h_{ii} = \sum_{j=1}^{n} h_{ij}^2$ (see Schott 1997, p. 374), and

$$\mathrm{cov}(e_i, e_j) = -h_{ij},$$

showing that the observed errors have correlation given by

$$\mathrm{corr}(e_i, e_j) = -\frac{h_{ij}}{[(1-h_{ii})(1-h_{jj})]^{1/2}}.$$

Consequently, even if the model is correctly specified, the residuals have nonconstant variance and are correlated. We may write

$$\widehat{Y}_i = h_{ii}Y_i + \sum_{j=1, j\neq i}^{n} h_{ij}Y_j, \tag{5.63}$$

so that if h_{ii} is large relative to the other elements in the ith row of $\boldsymbol{h}$, then the ith fitted value will be largely influenced by Y_i; h_{ii} is known as the *leverage*. Note that the leverage depends on the design matrix (i.e., the $\boldsymbol{x}$'s) only. Exercise 5.8 shows that $\mathrm{tr}(\boldsymbol{h}) = k+1$ so the average leverage is at least $(k+1)/n$. If $h_{ii} = 1$, $\widehat{y}_i = \boldsymbol{x}_i \widehat{\boldsymbol{\beta}}$ and the ith observation is fitted exactly, using a single degree of freedom for this point alone, which is not desirable.

Based on these results we may define *standardized residuals*:

$$e_i^* = \frac{Y_i - \widehat{Y}_i}{\widehat{\sigma}(1 - h_{ii})^{1/2}}, \tag{5.64}$$

for $i = 1, \ldots, n$, and where $\widehat{\sigma}$ is an unbiased estimator of σ. These residuals have mean $\mathrm{E}[\widehat{\sigma} e_i^*] = 0$ and variance $\mathrm{var}(\widehat{\sigma} e_i^*) = \sigma^2$, but they are not independent since they are based on $n - k - 1$ independent quantities. Often the $(1 - h_{ii})^{1/2}$ terms in the denominator of (5.64) are ignored.

For the simple linear regression model,

$$h_{ii} = \frac{1}{n} + \frac{(x_i - \overline{x})^2}{\sum_{k=1}^{n}(x_k - \overline{x})^2}$$

and

$$h_{ij} = \frac{1}{n} + \frac{(x_i - \overline{x})(x_j - \overline{x})}{\sum_{k=1}^{n}(x_k - \overline{x})^2}.$$

Therefore, with respect to (5.63), we see that an extreme x_i value produces a fitted value $\widehat{Y}_i$ that is more heavily influenced by the observed value of Y_i. Such x_i values also influence other fitted values, particularly those with x values not close to $\overline{x}$. The two constraints on the model are

$$\sum_{i=1}^{n} e_i = \sum_{i=1}^{n} Y_i - \widehat{Y}_i = 0$$

$$\sum_{i=1}^{n} e_i x_i = \sum_{i=1}^{n} (Y_i - \widehat{Y}_i) x_i = 0$$

which induces correlation in the e_i's.

5.11.3 Using the Residuals

The constancy of variance assumption may be assessed by plotting the residuals, e_i versus the fitted values $\widehat{Y}_i$ with a random scatter suggesting no cause for concern. Examination may be simpler if squared residuals e_i^2 or absolute values of the residuals $|e_i|$ are plotted versus the fitted values $\widehat{Y}_i$. These plots are useful since departures from constant variance often correspond to a mean–variance relationship which, given sufficient data and range of the mean function, will hopefully reveal itself in these plots. If the variance increases with the mean, plotting e_i versus $\widehat{Y}_i$ will reveal a funnel shape with the wider end of the funnel to the right of the plot. For the plots using the squared or absolute residuals, interpretation may be improved with the addition of a smoother.

When one of the columns of x represents time, we may plot the residuals versus time and assess dependence between error terms. Dependence may also be detected using scatterplots of lagged residuals, for example, by plotting e_i versus e_{i-1} for $i = 2, \ldots, n$. Independent residuals should produce a plot with a random scatter of points. The autocorrelation at different lags may also be estimated for equally spaced data in time, while for unequally spaced data, a semi-variogram may be constructed. The latter is described in the context of longitudinal data in Sect. 8.8.

To assess normality of the residuals, we may construct a normal QQ plot. We first order the residuals and call these $e_{(i)}$, $i = 1, \ldots, n$. The expected order statistic of size n from a normal distribution is given (approximately) by

$$f_{(i)} = \Phi^{-1}\left(\frac{i - 0.5}{n}\right), i = 1, \ldots, n,$$

where $\Phi(\cdot)$ is the cumulative distribution function of the standard normal distribution, that is, if $Z \sim N(0,1)$ then $\Phi(z) = \Pr(Z < z)$. We then plot $e_{(i)}$ versus $f_{(i)}$. If the normality assumption is reasonable, the points should lie approximately on a straight line. If we plot the ordered standardized residuals $e^*_{(i)}$ versus $f_{(i)}$, then, in addition, the line should have slope one. Deciding on whether the points are suitably close to linear is difficult and may be aided by simulating multiple datasets from which intervals may be derived for each i. Care must be taken in interpretation as (5.62) shows that the observed residuals are a linear combination of the error terms and hence may exhibit *supernormality*, that is, even if ϵ_i is not normal, $\sum_{j=1}^{n} h_{ij}\epsilon_j$ may tend toward normality (and dominate the first term, ϵ_i).

Figure 5.8 shows what we might expect to see under various distributional assumptions. QQ normal plots for normal. Laplacian, Student's t_3, and lognormal error distributions are displayed in the four rows, with sample sizes of $n = 10, 25, 50, 200$ across columns. The characteristic skewed shape of the lognormal distribution is revealed for all sample sizes, but it is difficult to distinguish between the Laplacian and the normal, even for a large sample size. For small n, interpretation is very difficult.

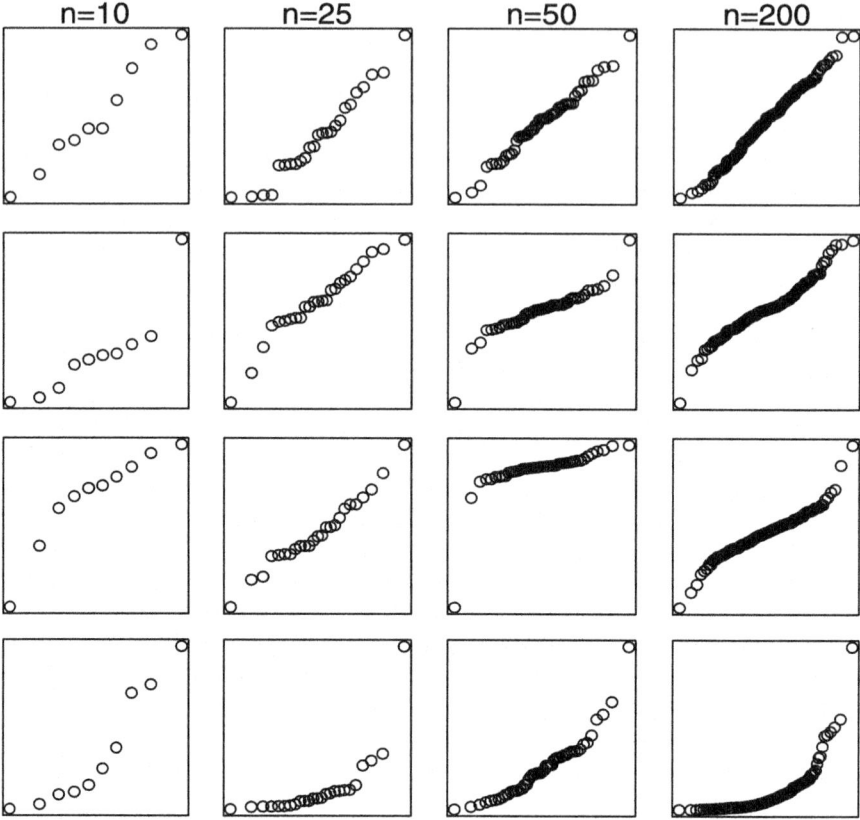

Fig. 5.8 Normal scores plot for various distributions and sample sizes. Columns 1–4 represent sample sizes of 10, 25, 50, and 200. Rows 1–4 correspond to errors generated from normal, Laplacian, Student's t_3, and lognormal distributions, respectively. In each plot, the expected residuals are plotted on the x-axis, and the observed ordered residuals on the y-axis

In general, simulation may be used to examine the behavior of plots when the model is true. QQ plots may be constructed to assess any distributional assumption, by an appropriate choice of $f_{(i)}$. The Bayesian approach to inference allow alternative likelihoods to the normal to be fitted relatively easily under an MCMC implementation. We have concentrated on frequentist residuals, but all of the above plots may be based on Bayesian residuals. For example, we can obtain samples from the posterior distribution of β and σ and then substitute these samples into

$$e_i^\star = \frac{y_i - \boldsymbol{x}_i \boldsymbol{\beta}}{\sigma(1 - h_{ii})^{1/2}}, \tag{5.65}$$

to produce samples from the posterior distribution of the residuals. The posterior mean or median of the $e_i^\star$ can then be calculated and examined. More simply, one

Table 5.14 Parameter estimates and standard errors (model-based and sandwich) for the prostate cancer data

Variable	Estimate	Standard error Model-based	Sandwich
log(can vol)	0.59	0.088	0.077
log(weight)	0.45	0.17	0.19
age	−0.020	0.011	0.0094
log(BPH)	0.11	0.058	0.057
SVI	0.77	0.24	0.21
log(cap pen)	−0.11	0.091	0.079
gleason	0.045	0.16	0.13
PGS45	0.0045	0.0044	0.0042
$\widehat{\sigma}$	0.78	–	–

could substitute the posterior means or medians of β and σ into (5.65). An early use of Bayesian residuals analysis was provided by Chaloner and Brant (1988).

A major problem with residual analysis, unless one is in purely exploratory mode, is that if the assumptions are found wanting and we change the model, what are the frequentist properties in terms of bias, the coverage of intervals, and the α level of tests? Recall the discussion of Chap. 4. To avoid changing the model, including transforming x and/or y, one should try and think as much as possible about a suitable model, *before* the data are analyzed. As always the exact procedure followed should be reported, so that inferential summaries can be more easily interpreted. The same problems exist for a Bayesian analysis, since one should specify a priori all models that one envisages fitting (which may not be feasible in advance), with subsequent averaging across models (Sect. 3.6).

5.12 Example: Prostate Cancer

We return to the PSA data and provide a more comprehensive analysis. We fit the full (main effects only) model

$$\log \text{PSA} = \beta_0 + \beta_1 \times \log(\text{can vol}) + \beta_2 \times \log(\text{weight}) + \beta_3 \times \text{age} + \beta_4 \times \log(\text{bph})$$
$$+ \beta_5 \times \text{svi} + \beta_6 \times \log(\text{cap pen}) + \beta_7 \times \text{gleason} + \beta_8 \times \text{PGS45} + \epsilon,$$

with $\epsilon \,|\, \sigma^2 \sim_{iid} N(0, \sigma^2)$. The resultant least squares parameter estimates and standard errors are given in Table 5.14. This table includes the sandwich standard errors, to address the possibility of nonconstant variance error terms. These are virtually identical to the model-based standard errors. This is not surprising given Fig. 5.9(a), which plots the absolute value of the residuals against the fitted values, and indicates that the constant variance assumption appears reasonable.

With $n - k - 1 = 88$, we do not require normality of errors, but for illustration we include a QQ normal plot in Fig. 5.9(b) and see that the errors are close to normal. Figures 5.9(c) and (d) plot the residuals versus two of the more important covariates,

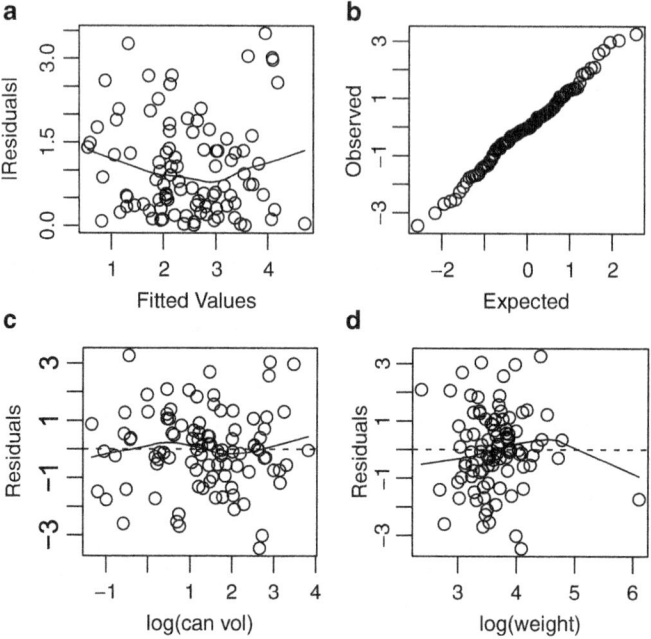

Fig. 5.9 Diagnostic plots in the prostate cancer study: (**a**) absolute values of residuals versus fitted values, with smoother, (**b**) normal QQ plot of residuals; (**c**) residuals versus log cancer volume, with smoother, (**d**) residuals versus log weight, with smoother

log cancer volume and log weight, with smoothers added. In each case, we see no strong evidence of nonlinearity.

We now discuss a Bayesian analysis of these data. With the improper prior (5.42), we saw in Sect. 5.7 that inference was identical with the frequentist approach so that the estimates and (model-based) standard errors in Table 5.14 are also posterior means and posterior standard deviations. Figure 5.10 displays the marginal posterior densities (which are located and scaled Student's t distributions with 88 degrees of freedom) for the eight coefficients. In this plot, for comparability, we scale each of the x variables to lie on the range (0,1).

Turning now to an informative prior distribution, without more specific knowledge, we let $\boldsymbol{\beta}^\star = [\beta_0^\star, \ldots, \beta_8^\star]^T$ represent the vector of coefficients associated with the standardized covariates on (0,1). The prior is taken as $\pi(\boldsymbol{\beta}^\star)\pi(\sigma^2)$ with

$$\pi(\boldsymbol{\beta}^\star) = \prod_{j=0}^{8} \pi(\beta_j^\star) \qquad (5.66)$$

and $\pi(\beta_0^\star) \propto 1$ (an improper prior). For the regression coefficients $\beta_j^\star \sim_{iid} N(0, V)$ with the standard deviations, $\sqrt{V}$, chosen in the following way. For the prostate

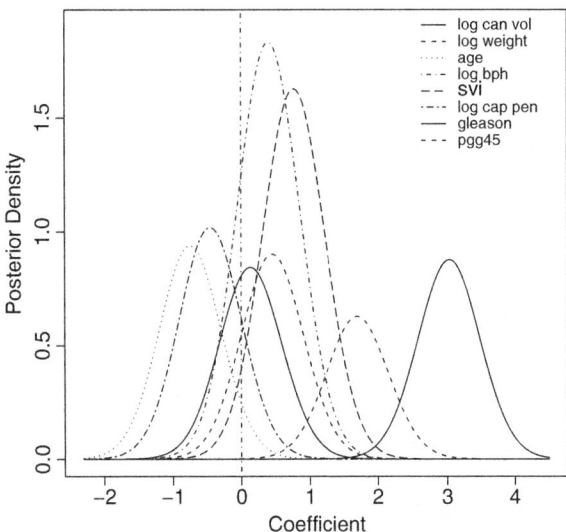

Fig. 5.10 Marginal posterior distributions of regression coefficients associated with the eight (standardized) covariates, for the prostate cancer data

data, we believe that it is unlikely that any of the standardized covariates, over the range (0,1), will change the median PSA by more than 10 units. The way we include this information in the prior is by assuming that the $1.96 \times \sqrt{V}$ point of the prior corresponds to the maximum value we believe is a priori plausible, that is, we set $\beta_j^\star = \log(10)$ equal to this point. For σ^2, we assume the improper choice $\pi(\sigma^2) \propto \sigma^{-2}$.

Figure 5.11 shows the 95% credible intervals under the flat and informative priors, and we see the general shrinkage towards zero (the prior mean). On average there is around a 10% reduction in the posterior standard deviations (and hence the credible intervals) under the informative prior, which shows how the use of informative priors can aid in the bias-variance trade-off. The above analysis is closely related to *ridge regression*, as will be discussed in Sect. 10.5.1.

5.13 Concluding Remarks

In this chapter we have concentrated on the linear model

$$Y = x\beta + \epsilon$$

where β is $n \times (k+1)$ and $\epsilon \sim N_n(0_n, \sigma^2 I_n)$. Although the range of models that are routinely available for fitting has expanded greatly (see Chaps. 6 and 7), the linear model continues to be popular. There are good reasons for this, since parameter interpretation is straightforward and the estimators commonly used are linear in the data and therefore possess desirable robustness properties.

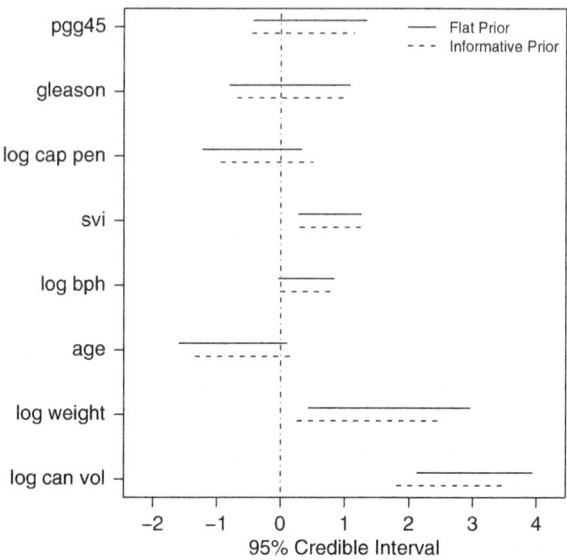

Fig. 5.11 95% credible intervals for regression coefficients corresponding to standardized covariates, under flat and informative priors, for the prostate cancer data

Unless n is not large, or there is substantial prior information, the point estimate

$$\widehat{\boldsymbol{\beta}} = (\boldsymbol{x}^\mathrm{T}\boldsymbol{x})^{-1}\boldsymbol{x}^\mathrm{T}\boldsymbol{y}$$

and $100(1-\alpha)\%$ interval estimate

$$\widehat{\beta}_j \pm t_{1-\alpha/2}^{n-k-1} \times \widehat{\mathrm{s.e.}}(\widehat{\beta}_j),$$

where $t_{1-\alpha/2}^{n-k-1}$ is the $100(1-\alpha/2)\%$ point of a Student's t distribution with $n-k-1$ degrees of freedom, emerges from likelihood, ordinary least squares, and Bayesian analyses. These summaries are robust to a range of distributions for the error terms, so long as n is large. Nonconstant error variance and correlated errors can both seriously damage the appropriateness of the interval estimate, however. With larger sample sizes, sandwich estimation provides a good approach for guarding against nonconstant error variance.

5.14 Bibliographic Notes

McCullagh and Nelder (1989, Chap. 3) provide an extended discussion on parameterization issues, including aliasing, and the interpretation of parameters. For more discussion of conditions for asymptotic normality for simple linear regression, see (van der Vaart 1998, p.21). Firth (1987) discusses the loss of precision when the data are not normally distributed and shows that the skewness of the true distribution of the errors is an important factor. The theory presented in Lehmann (1986,

p. 209–211) indicates that dependence in the residuals can cause real problems for estimation of appropriate standard errors. Further details of residual analysis may be found in Cook and Weisberg (1982).

The classic frequentist text on the analysis of variance is Scheffé (1959), while Searle et al. (1992) provide a more recent treatment. An interesting discussion, from a Bayesian slant, is provided by Gelman and Hill (2007, Chap. 22).

Numerous texts have been written on the linear model; see, for example, Ravishanker and Dey (2002) and Seber and Lee (2003) for the theory and Faraway (2004) for a more practical slant.

5.15 Exercises

5.1 Consider the model

$$Y = x\beta + \epsilon,$$

where Y is the $n \times 1$ vector of responses, x is the $n \times (k+1)$ design matrix, $\beta = [\beta_0, \ldots, \beta_k]$, and $E[\epsilon] = 0$, $\text{var}(\epsilon) = \sigma^2 V$ where V is a *known* correlation matrix V.

(a) By considering the sum of squares,

$$\text{RSS}_v = (Y - x\beta)^T V^{-1} (Y - x\beta).$$

show that the *generalized* least squares estimator is

$$\widehat{\beta}_v = (x^T V^{-1} x)^{-1} x^T V^{-1} Y,$$

provided the necessary inverse exists.

(b) Derive the distribution of $\widehat{\beta}_v$.

(c) Show that $\widehat{\sigma}_v^2$, as defined in (5.33), is an unbiased estimator of σ^2.

5.2 Suppose $\widehat{\beta}_1 \neq \widehat{\beta}_2$ are two different least squares estimates of β. Show there are infinitely many least squares estimates of β.

5.3 Let $Y_i = \beta_0 + \beta_1 x_i + \epsilon_i$, $i = 1, \ldots, n$, where $E[\epsilon_i] = 0$, $\text{var}(\epsilon_i) = \sigma^2$ and $\text{cov}(\epsilon_i, \epsilon_j) = 0$ for $i \neq j$. Prove that the least squares estimates of β_0 and β_1 are uncorrelated if and only if $\bar{x} = 0$.

5.4 Consider the simple linear regression model

$$Y_i = \beta_0 + \beta_1 x_i + \epsilon_i,$$

with $\epsilon_i \mid \sigma^2 \sim_{iid} N(0, \sigma^2)$, $i = 1, \ldots, n$. Suppose the prior distribution is of the form

$$\pi(\beta_0, \beta_1, \sigma^2) = \pi(\beta_0, \beta_1) \times \sigma^{-2}, \qquad (5.67)$$

where the prior for $[\beta_0, \beta_1]$ is

$$\begin{bmatrix} \beta_0 \\ \beta_1 \end{bmatrix} \sim N_2 \left(\begin{bmatrix} m_0 \\ m_1 \end{bmatrix}, \begin{bmatrix} v_{00} & v_{01} \\ v_{01} & v_{11} \end{bmatrix} \right).$$

In this exercise the conditional distributions required for Gibbs sampling (Sect. 3.8.4) will be derived.

(a) Write down the form of the posterior distribution (up to proportionality) and derive the conditional distributions $p(\beta_0 \mid \beta_1, \sigma^2, \boldsymbol{y})$, $p(\beta_1 \mid \beta_0, \sigma^2, \boldsymbol{y})$, and $p(\sigma^2 \mid \beta_0, \beta_1, \boldsymbol{y})$. Hence, give details of the Gibbs sampling algorithm.

(b) Another blocked Gibbs sampling algorithm (Sect. 3.8.6) would simulate from the distributions $p(\boldsymbol{\beta} \mid \sigma^2, \boldsymbol{y})$ and $p(\sigma^2 \mid \boldsymbol{\beta}, \boldsymbol{y})$. Derive these distributions, given in (5.46) and (5.47), and hence describe the form of the Gibbs sampling algorithm.

5.5 The algorithm derived in Exercise 5.4(b) will now be implemented for the prostate cancer data of Sect. 1.3.1. These data are available in the R package lasso2 and are named Prostate. Take Y as log prostate specific antigen and x as log cancer volume. Implement the blocked Gibbs sampling algorithm using the prior (5.67), with $m_0 = m_1 = 0$, $v_{00} = v_{11} = 2$, and $v_{01} = 0$. Run two chains, one with starting values corresponding to the unbiased estimates of the parameters and one starting from a point randomly generated from the prior $\pi(\beta_0, \beta_1)$. Report:

(a) Histogram representations of the univariate marginal distributions $p(\beta_0 \mid \boldsymbol{y})$, $p(\beta_1 \mid \boldsymbol{y})$, and $p(\sigma \mid \boldsymbol{y})$ and scatterplots of the bivariate marginal distributions $p(\beta_0, \beta_1 \mid \boldsymbol{y})$, $p(\beta_0, \sigma \mid \boldsymbol{y})$, and $p(\beta_1, \sigma \mid \boldsymbol{y})$.
(b) The posterior means, standard deviations, and 10%, 50%, 90% quantiles for β_0, β_1, and σ.
(c) $\Pr(\beta_1 > 0.5 \mid \boldsymbol{y})$.
(d) Justify your choice of "burn-in" period (Sect. 3.8.6). For example, you may present the trace plots $\beta_0^{(t)}$, $\beta_0^{(t)}$, $\log \sigma^{2(t)}$ versus t.
(e) Confirm the results you have obtained using INLA or WinBUGS.

5.6 In this question, parameter interpretation will be considered. Consider a continuous univariate response y, with two potential covariates, a continuous variable x_1, and a binary factor x_2. The x variables will be referred to as age and gender, respectively. Consider the four models:

Model A

$$y = \begin{cases} \theta_0 + \epsilon, & \text{for men } (x_2 = 0) \\ \theta_1 + \epsilon, & \text{for women } (x_2 = 1). \end{cases}$$

Model B

$$y = \theta_0 + \theta_1 x_1 + \epsilon.$$

5.15 Exercises

Model C

$$y = \begin{cases} \theta_0 + \theta_1 x_1 + \epsilon, & \text{for men } (x_2 = 0) \\ \theta_2 + \theta_1 x_1 + \epsilon, & \text{for women } (x_2 = 1). \end{cases}$$

Model D

$$y = \begin{cases} \theta_0 + \theta_1 x_1 + \epsilon, & \text{for men } (x_2 = 0) \\ (\theta_0 + \phi_0) + \theta_1 x_1 + \epsilon, & \text{for women } (x_2 = 1). \end{cases}$$

Model E

$$y = \begin{cases} \theta_0 + \theta_1 x_1 + \epsilon, & \text{for men } (x_2 = 0), \text{ and} \\ \theta_0 + \theta_2 x_1 + \epsilon, & \text{for women } (x_2 = 1). \end{cases}$$

Model F

$$y = \begin{cases} \theta_0 + \theta_1 x_1 + \epsilon, & \text{for men } (x_2 = 0), \\ \theta_2 + \theta_3 x_1 + \epsilon, & \text{for women } (x_2 = 1). \end{cases}$$

For each model, the error terms ϵ are assumed to have zero mean.

(a) For each model, provide a careful interpretation of the parameters and give a description of the assumed form of the relationship.
(b) Which of the above models are equivalent?

5.7 Let $Y_1, \ldots, Y_n$ be distributed as $Y_i \mid \theta, \sigma^2 \sim_{ind} N(i\theta, i^2\sigma^2)$ for $i = 1, \ldots, n$. Find the generalized least squares estimate of θ and prove that the variance of this estimate is σ^2/n.

5.8 Suppose that the design matrix x of dimension $n \times (k+1)$ has rank $k+1$ and let $h = x(x^T x)^{-1} x^T$ represent the hat matrix. Show that $\text{tr}(h) = (k+1)$.

5.9 Consider the model

$$Y_i = \beta_0 + \beta_1 X_i + \epsilon_i$$

for $i = 1, \ldots, n$, where

$$\begin{bmatrix} X_i \\ \epsilon_i \end{bmatrix} \sim N_2 \left(\begin{bmatrix} \mu_x \\ 0 \end{bmatrix}, \begin{bmatrix} \sigma_x^2 & 0 \\ 0 & \sigma_\epsilon^2 \end{bmatrix} \right),$$

to give

$$\begin{bmatrix} Y_i \\ X_i \end{bmatrix} \sim N_2 \left(\begin{bmatrix} \mu_y \\ \mu_x \end{bmatrix}, \begin{bmatrix} \sigma_y^2 & \sigma_{xy} \\ \sigma_{xy} & \sigma_x^2 \end{bmatrix} \right)$$

where $\sigma_y^2 = \beta_1^2 \sigma_x^2 + \sigma_\epsilon^2$, $\mu_y = \beta_0 + \beta_1 \mu_x$ and $\sigma_{xy} = \beta_1 \sigma_x^2$.

(a) Derive $E[Y_i \mid x_i]$ and $\text{var}(Y_i \mid x_i)$.

Now suppose one does not observe x_i, $i = 1, \ldots, n$ but instead $w_i = x_i + u_i$ where

$$\begin{bmatrix} X_i \\ \epsilon_i \\ U_i \end{bmatrix} \sim N_3 \left(\begin{bmatrix} \mu_x \\ 0 \\ 0 \end{bmatrix}, \begin{bmatrix} \sigma_x^2 & 0 & 0 \\ 0 & \sigma_\epsilon^2 & 0 \\ 0 & 0 & \sigma_\epsilon^2 \end{bmatrix} \right).$$

Assume that Y_i is conditionally independent of W_i, that is, $E[Y_i \mid x_i, u_i] = E[Y_i \mid x_i]$. Suppose the true model is $E[Y_i \mid x_i] = \beta_0 + \beta_1 x_i$ but the observed data are $[w_i, y_i]$, $i = 1, \ldots, n$.

(b) Relate $E[Y_i \mid w_i]$ to $E[x_i \mid w_i]$.
(c) What is the joint distribution of X_i and W_i and what is $E[X_i \mid w_i]$?
(d) Using your answers to (b) and (c), show that $E[Y_i \mid w_i] = \beta_0^\star + \beta_1^\star x_i$.
(e) What is the relationship between $\beta_0^\star, \beta_1^\star$ and β_0, β_1?

Chapter 6
General Regression Models

6.1 Introduction

In this chapter we consider the analysis of data that are not well-modeled by the linear models described in Chap. 5. We continue to assume that the responses are (conditionally) independent. We describe two model classes, *generalized linear models* (GLMs) and what we refer to as *nonlinear models*. In the latter, a response Y is assumed to be of the form $Y = \mu(\bm{x}, \bm{\beta}) + \epsilon$ with $\mu(\bm{x}, \bm{\beta})$ nonlinear in $\bm{x}$ and the errors ϵ independent with zero mean.

In Sect. 6.2 we introduce a motivating pharmacokinetic dataset that we will subsequently analyze using both GLMs and nonlinear models. Section 6.3 considers GLMs, which were introduced as an extension to linear models and have received considerable attention due to their computational and mathematical convenience. While computational advances have unshackled the statistician from the need to restrict attention to GLMs, they still provide an extremely useful class. Parameter interpretation for GLMs is discussed in Sect. 6.4. Sections 6.5, 6.6, 6.7, and 6.8 describe, respectively, likelihood inference, quasi-likelihood inference, sandwich estimation, and Bayesian inference for the GLM. Section 6.9 considers the assessment of the assumptions required for reliable inference in GLMs. In Sect. 6.10, we introduce nonlinear regression models, with identifiability discussed in Sect. 6.11. We then describe likelihood and least squares approaches to inference in Sects. 6.12 and 6.13 and sandwich estimation in Sect. 6.14. A geometrical comparison of linear and nonlinear least squares is provided in Sect. 6.15. Bayesian inference is described in Sect. 6.16 and Sect. 6.17 concentrates on the examination of assumptions. Concluding comments appear in Sect. 6.18 with bibliographic notes in Sect. 6.19.

In Chap. 7 we discuss models for binary data; models for such data could have been included in this chapter but are considered separately since there are a number of wrinkles that deserve specific attention.

6.2 Motivating Example: Pharmacokinetics of Theophylline

In Table 1.2 we displayed pharmacokinetic data on the sampling times and measured concentrations of the drug theophylline, collected from a subject who received an oral dose of 4.53 mg/kg. These data are plotted in Fig. 6.1, along with fitted curves from various approaches to modeling that we describe subsequently. We will fit both a nonlinear (so-called, compartmental) model to these data and a GLM. Let x_i and y_i represent the sampling time and concentration in sample i, respectively, for $i = 1, \ldots, n = 10$.

In Sect. 1.3.4, we detailed the aims of a pharmacokinetic study and described in some detail compartmental models that have been successfully used for modeling concentration–time data. Let $\mu(x)$ represent the deterministic model relating the response to time, x; $\mu(x)$ will usually be the mean response, though may correspond to the median response, depending on the assumed error structure. Notationally we have suppressed the dependence of $\mu(x)$ on unknown parameters. For the data considered here, a starting point for $\mu(x)$ is

$$\mu(x) = \frac{Dk_a}{V(k_a - k_e)} \left[\exp(-k_e x) - \exp(-k_a x) \right] \tag{6.1}$$

where $k_a > 0$ is the absorption rate constant, $k_e > 0$ is the elimination rate constant, and $V > 0$ is the (apparent) volume of distribution (that converts total amount of drug into concentration). This model was motivated in Sect. 1.3.4. A stochastic component may be added to (6.1) in a variety of ways, but one simple approach is via

$$y(x) = \mu(x) + \delta(x), \tag{6.2}$$

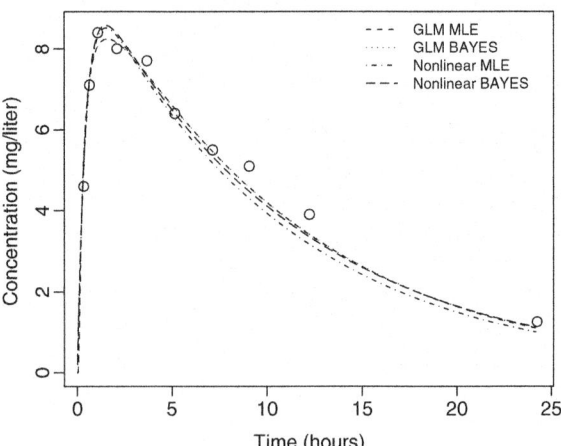

Fig. 6.1 Theophylline data, along with fitted curves under various models and inferential approaches. Four curves are included, corresponding to MLE and Bayes analyses of GLM and nonlinear models. The two nonlinear curves are indistinguishable

6.2 Motivating Example: Pharmacokinetics of Theophylline

where $E[\delta(x)] = 0$ and $\text{var}[\delta(x)] = \sigma^2 \mu(x)^2$ with $\delta(x)$ at different times x being independent. The variance model produces a constant coefficient of variation (defined as the ratio of the standard deviation to the mean), which is often observed in practice for pharmacokinetic data. Combining (6.1) and (6.2) gives an example of a three parameter nonlinear model. An approximately constant coefficient of variation can also be achieved by taking

$$\log y(x) = \log \mu(x) + \epsilon(x),$$

with $E[\epsilon(x)] = 0$ and $\text{var}[\epsilon(x)] = \sigma^2$. In this case, $\mu(x)$ represents the median concentration at time x (Sect. 5.5.3).

Model (6.1) is sometimes known as the *flip-flop* model, because there is an identifiability problem in that the same curve is achieved with each of the parameter sets $[V, k_a, k_e]$ and $[Vk_e/k_a, k_e, k_a]$. Recall from Sect. 2.4.1 that identifiability is required for consistency and asymptotic normality of the MLE. Often, identifiability is achieved by enforcing $k_a > k_e > 0$, since the absorption rate is greater than the elimination rate for most drugs. Such identifiability issues are not a rare phenomenon for nonlinear models, and will receive further attention in Sect. 6.11.

Model (6.1) may be written in the alternative form

$$\mu(x) = \frac{Dk_a}{V(k_a - k_e)} \left[\exp(-k_e x) - \exp(-k_a x) \right]$$
$$= \exp(\beta_0 + \beta_1 x) \left\{ 1 - \exp[-(k_a - k_e)x] \right\}, \quad (6.3)$$

where $\beta_0 = \log[Dk_a/V(k_a - k_e)]$ and $\beta_1 = -k_e$. As an alternative to the compartmental model, (6.1), we will also consider the fractional polynomial model (as introduced by Nelder 1966) given by

$$\mu(x) = \exp\left(\beta_0 + \beta_1 x + \beta_2 / x\right). \quad (6.4)$$

Comparison with (6.3) shows that β_2 is the parameter that is determining the absorption phase. This model only makes sense if it produces both an increasing absorption phase and a decreasing elimination phase, which correspond, retrospectively, to $\beta_2 < 0$ and $\beta_1 < 0$. When combined with an appropriate choice for the stochastic component, model (6.4) falls within the GLM class, as we see shortly.

In a pharmacokinetic study, as discussed in Sect. 1.3.4, interest often focuses on certain derived parameters. Of specific interest are $x_{1/2}$, the elimination half-life, which is the time it takes for the drug concentration to drop by 50% (for times sufficiently large for elimination to be the dominant process); $x_{\max}$, the time to maximum concentration; $\mu(x_{\max})$, the maximum concentration; and Cl, the clearance, which is the amount of blood cleared of drug in unit time.

With respect to model (6.1), the derived parameters of interest, in terms of $[V, k_a, k_e]$, are

$$x_{1/2} = \frac{\log 2}{k_e}$$

$$x_{\max} = \frac{1}{k_a - k_e} \log\left(\frac{k_a}{k_e}\right)$$

$$\mu(x_{\max}) = \frac{Dk_a}{V(k_a - k_e)} [\exp(-k_e x_{\max}) - \exp(-k_a x_{\max})]$$

$$= \frac{D}{V}\left(\frac{k_a}{k_e}\right)^{k_a/(k_a-k_e)}$$

$$Cl = \frac{D}{\text{AUC}}$$

$$= V \times k_e$$

where AUC is the area under the concentration–time curve between 0 and ∞. With respect to model (6.4), as functions of $\boldsymbol{\beta} = [\beta_0, \beta_1, \beta_2]$,

$$x_{1/2} = -\frac{\log 2}{\beta_1}$$

$$x_{\max} = \left(\frac{\beta_2}{\beta_1}\right)^{1/2}$$

$$\mu(x_{\max}) = D \exp\left[\beta_0 - 2(\beta_1\beta_2)^{1/2}\right]$$

$$Cl = \frac{\sqrt{\beta_1/\beta_2}}{2\exp(\beta_0)K_1[2(\beta_1\beta_2)^{1/2}]}, \qquad (6.5)$$

where $K_s(x)$ denotes a modified Bessel function of the second kind of order s. Consequently, for both models, the quantities of interest are nonlinear functions of the original parameters, which has implications for inference.

6.3 Generalized Linear Models

Generalized linear models (GLMs) were introduced by Nelder and Wedderburn (1972) and provide a class with relatively broad applicability and desirable statistical properties. For a GLM:

- The responses y_i follow an exponential family, so that the distribution is of the form

$$p(y_i \mid \theta_i, \alpha) = \exp\left(\frac{y_i\theta_i - b(\theta_i)}{\alpha} + c(y_i, \alpha)\right), \qquad (6.6)$$

6.3 Generalized Linear Models

Table 6.1 Characteristics of some common GLMs. The notation is as in (6.6). The canonical parameter is θ, the mean is $E[Y] = \mu$, and the variance is $\text{var}(Y) = \alpha V(\mu)$

Distribution	$N(\mu, \sigma^2)$	Poisson(μ)	Bernoulli(μ)	$Ga(1/\alpha, 1/[\mu\alpha])$
Mean $E[Y \mid \theta]$	θ	$\exp(\theta)$	$\frac{\exp(\theta)}{1+\exp(\theta)}$	$-\frac{1}{\theta}$
Variance $V(\mu)$	1	μ	$\mu(1-\mu)$	μ^2
$b(\theta)$	$\theta^2/2$	$\exp(\theta)$	$\log(1+e^\theta)$	$-\log(-\theta)$
$c(y, \alpha)$	$-\frac{1}{2}\left[\frac{y^2}{2} + \log(2\pi\alpha)\right]$	$-\log y!$	1	$\frac{\log(y/\alpha)}{\alpha} - \log y + \log \Gamma(\alpha)$

for functions $b(\cdot)$, $c(\cdot, \cdot)$ and where θ_i and α are scalars. It is straightforward to show (using the results of Sect. 2.4) that

$$E[Y_i \mid \theta_i, \alpha] = \mu_i$$
$$= b'(\theta_i)$$

and

$$\text{var}(Y_i \mid \theta_i, \alpha) = \alpha b''(\theta_i)$$
$$= \alpha V(\mu_i),$$

for $i = 1, \ldots, n$. We assume $\text{cov}(Y_i, Y_j \mid \theta_i, \theta_j, \alpha) = 0$, for $i \neq j$ (Chap. 9 provides the extension to dependent data).

- A *link function* $g(\cdot)$ provides the connection between the mean function $\mu_i = E[Y_i \mid \theta_i, \alpha]$ and the *linear predictor* $\boldsymbol{x}_i\boldsymbol{\beta}$ via

$$g(\mu_i) = \boldsymbol{x}_i\boldsymbol{\beta},$$

where $\boldsymbol{x}_i$ is a $(k+1) \times 1$ vector of explanatory variables (including a 1 for the intercept) and $\boldsymbol{\beta} = [\beta_0, \beta_1, \ldots, \beta_k]^\mathrm{T}$ is a $(k+1) \times 1$ vector of regression parameters.

To summarize, a GLM assumes a linear relationship on a transformed mean scale (which, as we shall see, offers certain computational and statistical advantages) and an exponential family form for the distribution of the response.

If α is known, then (6.6) is a one-parameter exponential family model. If α is unknown, then the distribution may or may not be a two-parameter exponential family model. So-called *canonical links* have $\theta_i = \boldsymbol{x}_i\boldsymbol{\beta}$ and provide simplifications in terms of computation.

GLMs are very useful pedagogically since they separate the deterministic and stochastic components of the model, and this aspect was emphasized in the abstract of Nelder and Wedderburn (1972): "The implications of the approach in designing statistics courses are discussed."

Table 6.1, adapted from Table 2.1 of McCullagh and Nelder (1989), characterizes a number of common GLMs. Another example which is not listed in the table, is the inverse Gaussian distribution; Exercise 6.1 derives the detail for this case.

Example: Pharmacokinetics of Theophylline

Model (6.3) is an example of a GLM with a log link:

$$\log \mu(\boldsymbol{x}) = \beta_0 + \beta_1 x_1 + \beta_2 x_2 \tag{6.7}$$

where $\boldsymbol{x} = [1, x_1, x_2]$ and $x_2 = 1/x_1$.

Turning to the stochastic component, as noted in Sect. 6.2, the error terms often display a constant coefficient of variation. With this in mind, we may combine (6.7) with a gamma distribution via

$$Y(\boldsymbol{x}) \mid \boldsymbol{\beta}, \alpha \sim_{ind} \text{Ga}\{\alpha^{-1}, [\mu(\boldsymbol{x})\alpha]^{-1}\}, \tag{6.8}$$

to give $E[Y(\boldsymbol{x})] = \mu(\boldsymbol{x})$ and $\text{var}[Y(\boldsymbol{x})] = \alpha\mu(\boldsymbol{x})^2$ so that $\alpha^{1/2}$ is the coefficient of variation. Lindsey et al. (2000) examine various distributional choices for pharmacokinetic data and found the gamma assumption to be reasonable in their examples. It is interesting to note that for the gamma distribution, the reciprocal transform is the canonical link, but this option is not statistically appealing since it does not constrain the mean function to be positive. In the pharmacokinetic context the reciprocal link also results in a concentration–time curve that is not integrable between 0 and ∞ so that the fundamental clearance parameter is undefined. One disadvantage of the loglinear GLM defined above, compared to the nonlinear compartmental model we discuss later, is that if multiple doses are considered, the mean function does not correspond to a GLM.

Example: Lung Cancer and Radon

In Sect. 1.3.3 we described data on lung cancer incidence in counties in Minnesota, with Y_i the number of cases, x_i the average radon, and E_i the expected number of cases, in area i, $i = 1, \ldots, n$. These data were examined repeatedly in Chaps. 2 and 3.

A starting model is $Y_i \mid \boldsymbol{\beta} \sim_{ind} \text{Poisson}[E_i \exp(\beta_0 + \beta_1 x_i)]$, which we write as

$$\log \Pr(Y = y_i \mid \boldsymbol{\beta}) = y_i \log \mu_i - \mu_i - \log y_i!$$

with $\log \mu_i = \log E_i + \beta_0 + \beta_1 x_i$, to give a GLM with a (canonical) log link. As discussed in Chaps. 2 and 3, this model is fundamentally inadequate because $\alpha = 1$, and so there is no parameter to allow for excess-Poisson variation. The latter can be modeled using the negative binomial model of Sect. 6.3 or the quasi-likelihood approach described in Sect. 6.6.

With unknown scale parameter, the negative binomial is not a GLM. We consider the case of known b (which will rarely be of interest in a practical setting). For consistency with its use in Chap. 2, we label the scale parameter of the negative binomial model as b. In the following, care should therefore be taken to discriminate between $b(\cdot)$, as in (6.6), and the scale parameter, b. From (2.40),

$$\log \Pr(Y = y_i \mid \mu_i) = b^{-1}\left[y_i b \log\left(\frac{\mu_i}{\mu_i + b}\right) - b^2 \log(\mu_i + b)\right]$$
$$+ \log \Gamma(y_i + b) - \log \Gamma(b) - \log y_i! - b(b+1)\log b$$

which is of the form (6.6) with

$$\theta_i = b\log\left(\frac{\mu_i}{\mu_i + b}\right),$$
$$b(\theta_i) = b^2 \log(\mu_i + b),$$
$$c(y_i, b) = \log \Gamma(y_i + b) - \log \Gamma(b) - \log y_i! - b(b+1)\log b,$$

so that

$$\mathrm{E}[Y_i \mid \mu_i] = \mu_i = b'(\theta_i)$$
$$= \frac{be^{\theta_i/b}}{1 - e^{\theta_i/b}},$$
$$\mathrm{var}(Y_i \mid \mu_i) = b \times b''(\theta_i)$$
$$= \mu_i + \mu_i^2/b.$$

The canonical link is

$$\theta_i = b\log\left(\frac{\mu_i}{\mu_i + b}\right) = \boldsymbol{x}\boldsymbol{\beta},$$

which depends on b. The negative binomial distribution is described in detail by Cameron and Trivedi (1998).

6.4 Parameter Interpretation

Interpretation of the regression parameters in a GLM is link function specific. The linear link was discussed in Chap. 5, and the log link was considered repeatedly (in the context of the lung cancer and radon data) in Chaps. 2 and 3. We provide an interpretation of binary data link functions, such as the logistic, in Chap. 7. Linearity on some scale offers advantages, as illustrated by the following example.

Consider the log linear model:

$$\log \mu(x) = \beta_0 + \beta_1 x_1 + \beta_2 x_2.$$

The parameter $\exp(\beta_1)$ has a relatively straightforward interpretation, being the multiplicative change in the average response associated with a one-unit increase in x_1, with x_2 held constant.

In contrast, for general nonlinear models, the parameters often define particular functions of the response covariate curve or fundamental quantities that define the system under study. We saw an example of this in Sect. 6.2, in which the nonlinear concentration–time curve (6.1) was defined in terms of the volume of distribution V and the absorption and elimination rate constants k_a and k_e. Alternatively, we could define the model in terms of characteristics of the curve, for example, the half-life, $x_{1/2}$, the time to maximum concentration, $x_{\max}$, and the maximum concentration, $\mu(x_{\max})$. We now discuss inference for the GLM.

6.5 Likelihood Inference for GLMs

6.5.1 Estimation

We first derive the score vector and information matrix. For an independent sample from the exponential family (6.6)

$$l(\boldsymbol{\theta}) = \sum_{i=1}^{n} l_i(\boldsymbol{\theta}) = \sum_{i=1}^{n} \frac{y_i \theta_i - b(\theta_i)}{\alpha} + c(y_i, \alpha),$$

where $\boldsymbol{\theta} = \boldsymbol{\theta}(\boldsymbol{\beta}) = [\theta_1(\boldsymbol{\beta}), \ldots, \theta_n(\boldsymbol{\beta})]$ is the vector of canonical parameters. Using the chain rule, the score function is

$$\boldsymbol{S}(\boldsymbol{\beta}) = \frac{\partial l}{\partial \boldsymbol{\beta}} = \sum_{i=1}^{n} \frac{dl_i}{d\theta_i} \frac{d\theta_i}{d\mu_i} \frac{\partial \mu_i}{\partial \boldsymbol{\beta}}$$

$$= \sum_{i=1}^{n} \frac{Y_i - b'(\theta_i)}{\alpha} \frac{1}{V_i} \frac{\partial \mu_i}{\partial \boldsymbol{\beta}}, \quad (6.9)$$

where $\text{var}(Y_i \mid \boldsymbol{\beta}) = \alpha V_i$ and

$$\frac{d^2 b}{d\theta_i^2} = \frac{d\mu_i}{d\theta_i} = V_i,$$

for $i = 1, \ldots, n$. Hence,

$$\boldsymbol{S}(\boldsymbol{\beta}) = \sum_{i=1}^{n} \left(\frac{\partial \mu_i}{\partial \boldsymbol{\beta}}\right)^{\text{T}} \frac{[Y_i - \text{E}(Y_i \mid \mu_i)]}{\text{var}(Y_i \mid \mu_i)}$$

$$= \boldsymbol{D}^{\text{T}} \boldsymbol{V}^{-1} [\boldsymbol{Y} - \boldsymbol{\mu}(\boldsymbol{\beta})] / \alpha, \quad (6.10)$$

6.5 Likelihood Inference for GLMs

where D is the $n \times (k+1)$ matrix with elements $\partial \mu_i / \partial \beta_j$, $i = 1, \ldots, n$, $j = 0, \ldots, k$, and V is the $n \times n$ diagonal matrix with ith diagonal element V_i. Consequently, an estimator $\widehat{\beta}_n$ defined through $S(\widehat{\beta}_n) = 0$ will be consistent so long as the mean function is correctly specified, since the estimating function is unbiased in this case. For canonical links, for which $\theta_i = x_i \beta$,

$$\sum_{i=1}^n \frac{\partial l_i}{\partial \beta} = \sum_{i=1}^n \frac{dl_i}{d\theta_i} \frac{\partial \theta_i}{\partial \beta} = \frac{1}{\alpha} \sum_{i=1}^n x_i^\mathrm{T} [Y_i - \mu_i(\beta)]$$

so that the sufficient statistics

$$\sum_{i=1}^n x_i^\mathrm{T} Y_i = \sum_{i=1}^n x_i^\mathrm{T} \mu_i(\widehat{\beta})$$

are recovered at the MLE, $\widehat{\beta}$.

From Result 2.1, the MLE has asymptotic distribution

$$I_n(\beta)^{1/2} (\widehat{\beta}_n - \beta) \to_d N_{k+1}(0, I_{k+1}),$$

where the expected information is

$$I_n(\beta) = \mathrm{E}[S(\beta) S(\beta)^\mathrm{T}] = D^\mathrm{T} V^{-1} D / \alpha.$$

In practice we use

$$I_n(\widehat{\beta}_n) = \widehat{D}^\mathrm{T} \widehat{V}^{-1} \widehat{D} / \alpha,$$

where $\widehat{V}$ and $\widehat{D}$ are evaluated at $\widehat{\beta}_n$. The variance of the estimator is

$$\widehat{\mathrm{var}}(\widehat{\beta}) = \alpha \left(\widehat{D}^\mathrm{T} \widehat{V}^{-1} \widehat{D} \right)^{-1} \tag{6.11}$$

and is consistently estimated if the second moment is correctly specified.

The information matrix may be written in a particularly simple and useful form, as we now show. We first let $\eta_i = g(\mu_i)$ denote the linear predictor. The score, (6.9), may be written, for parameter j, $j = 0, 1, \ldots, k$, as

$$S_j(\beta) = \frac{\partial l}{\partial \beta_j} = \sum_{i=1}^n \frac{(Y_i - \mu_i)}{\alpha V_i} \frac{d\mu_i}{d\eta_i} \frac{\partial \eta_i}{\partial \beta_j}$$

$$= \sum_{i=1}^n \frac{(Y_i - \mu_i)}{\alpha V_i} \frac{d\mu_i}{d\eta_i} x_{ij}. \tag{6.12}$$

Hence, element (j, j') of the expected information is

$$-\sum_{i=1}^{n} \text{E}\left[\frac{\partial^2 l_i}{\partial \beta_j \partial \beta_{j'}}\right] = \sum_{i=1}^{n} \text{E}\left[\left(\frac{\partial l_i}{\partial \beta_j}\right)\left(\frac{\partial l_i}{\partial \beta_{j'}}\right)\right]$$

$$= \sum_{i=1}^{n} \text{E}\left[\frac{(Y_i - \mu_i)x_{ij}}{\alpha V_i} \frac{d\mu_i}{d\eta_i} \frac{(Y_i - \mu_i)x_{ij'}}{\alpha V_i} \frac{d\mu_i}{d\eta_i}\right]$$

$$= \sum_{i=1}^{n} \frac{x_{ij}x_{ij'}}{\alpha V_i}\left(\frac{d\mu_i}{d\eta_i}\right)^2.$$

The information matrix therefore takes the form

$$I(\boldsymbol{\beta}) = \boldsymbol{x}^{\text{T}} \boldsymbol{W}(\boldsymbol{\beta}) \boldsymbol{x} \qquad (6.13)$$

where $\boldsymbol{W}$ is the diagonal matrix with elements

$$w_i = \frac{(d\mu_i/d\eta_i)^2}{\alpha V_i},$$

$i = 1, \ldots, n$.

When α is unknown, it may be estimated using maximum likelihood or the method of moments estimator

$$\widehat{\alpha} = \frac{1}{n - k - 1} \sum_{i=1}^{n} \frac{(Y_i - \widehat{\mu}_i)^2}{V(\widehat{\mu}_i)}, \qquad (6.14)$$

where $\widehat{\mu}_i = \widehat{\mu}_i(\widehat{\boldsymbol{\beta}})$. Section 2.5 contained the justification for this estimator, which has the advantage of being, in general, a consistent estimator in a broader range of circumstances than the MLE. The method of moments approach is routinely used for normal and gamma data. As usual, there will be an efficiency loss when compared to the use of the MLE if the distribution underlying the derivation of the latter is "true."

The use of (6.10) is appealing since it depends on only the first two moments so that consistency of $\widehat{\boldsymbol{\beta}}_n$ does not depend on the distribution of the data. Accurate asymptotic confidence interval coverage depends only on correct specification of the mean–variance relationship. Section 6.7 describes how the latter requirement may be relaxed.

If the score is of the form (6.6), that is, if the score arises from an exponential family, it is not necessary to have a mean function of GLM form (i.e., a linear predictor on some scale). So, for example, the nonlinear models considered later in the chapter, when embedded within an exponential family, also share consistency of estimation (so long as regularity conditions are satisfied).

6.5.2 Computation

Computation is relatively straightforward for GLMs, since the form of a GLM yields a log-likelihood surface that is well behaved, for all but pathological datasets. In particular, a variant of the Newton–Raphson method (a generic method for root-finding), known as *Fisher scoring*, may be used to find the MLEs. We briefly digress to describe the Newton–Raphson method. Let $S(\beta)$ represent a $p \times 1$ vector of functions that are themselves functions of a $p \times 1$ vector β. We wish to find β such that $S(\beta) = 0$. A first-order Taylor series expansion about $\beta^{(0)}$ gives

$$S(\beta) \approx S(\beta^{(0)}) + (\beta - \beta^{(0)})^\mathrm{T} S'(\beta^{(0)}).$$

Setting the left-hand side to zero yields

$$\beta = \beta^{(0)} - S'(\beta^{(0)})^{-1} S(\beta^{(0)}).$$

The Newton–Raphson method iterates the step:

$$\beta^{(t+1)} = \beta^{(t)} - S'(\beta^{(t)})^{-1} S(\beta^{(t)}),$$

for $t = 0, 1, 2, \ldots$ The Fisher scoring method is the Newton–Raphson method applied to the score equation, but with the observed information, $S'(\beta)$, replaced by the expected information $\mathrm{E}[S'(\beta)] = -I(\beta)$ to give

$$\beta^{(t+1)} = \beta^{(t)} + I(\beta^{(t)})^{-1} S(\beta^{(t)}),$$

so that a new estimate is calculated based on the score and information evaluated at the previous estimate. Recall that for a GLM, $I(\beta) = x^\mathrm{T} W(\beta) x$. Using this form, and (6.12), we write

$$\beta^{(t+1)} = (x^\mathrm{T} W^{(t)} x)^{-1} x^\mathrm{T} W^{(t)} \left[x \beta^{(t)} + (W^{(t)})^{-1} u^{(t)} \right]$$
$$= (x^\mathrm{T} W^{(t)} x)^{-1} x^\mathrm{T} W^{(t)} z^{(t)} \quad (6.15)$$

where $u^{(t)}$ and $z^{(t)}$ are $n \times 1$ vectors with ith elements

$$u_i^{(t)} = \frac{(Y_i - \mu_i^{(t)})}{\alpha V_i^{(t)}} \left. \frac{d\mu_i}{d\eta_i} \right|_{\beta^{(t)}},$$

and

$$z_i^{(t)} = x_i \beta^{(t)} + (Y_i - \mu_i^{(t)}) \left. \frac{d\eta_i}{d\mu_i} \right|_{\beta^{(t)}},$$

Table 6.2 Point and 90% interval estimates for the theophylline data of Table 1.2, under various models and estimation techniques. CV is the coefficient of variation and is expressed as a percentage. The Bayesian point estimates correspond to the posterior medians

Model	$x_{1/2}$	$x_{\max}$	$\mu(x_{\max})$	CV ($\times 100$)
GLM MLE	7.23 [6.89,7.59]	1.60 [1.52,1.69]	8.25 [7.95,8.56]	4.38 [3.04,6.33]
GLM sandwich	7.23 [6.97,7.50]	1.60 [1.57,1.64]	8.25 [8.02,8.48]	4.38 [3.04,6.33]
Nonlinear MLE	7.54 [7.09,8.01]	1.51 [1.36,1.66]	8.59 [7.99,9.24]	6.32 [4.38,9.13]
Nonlinear sandwich	7.54 [7.11,7.98]	1.51 [1.43,1.58]	8.59 [8.11,9.10]	6.32 [4.38,9.13]
Prior	8.00 [5.30,12.0]	1.50 [0.75,3.00]	9.00 [6.80,12.0]	5.00 [2.50,10.0]
GLM Bayes	7.26 [6.93,7.74]	1.60 [1.51,1.68]	8.24 [7.89,8.54]	5.21 [3.72,7.86]
Nonlinear Bayes	7.57 [7.15,8.04]	1.50 [1.36,1.66]	8.59 [8.22,8.94]	6.01 [4.34,8.93]

respectively. The Fisher scoring updates (6.15) therefore have the form of a weighted least squares solution to

$$(z^{(t)} - x\beta)^{\mathrm{T}} W^{(t)} (z^{(t)} - x\beta) \qquad (6.16)$$

with "working" or "adjusted" response $z^{(t)}$. This method is therefore known as *iteratively reweighted least squares* (IRLS). For canonical links, the observed and expected information coincide so that the Fisher scoring and Newton–Raphson methods are identical.

The existence and uniqueness of estimates have been considered by a number of authors; early references are Wedderburn (1976) and Haberman (1977).

Example: Pharmacokinetics of Theophylline

Fitting the gamma model (6.8) with mean function (6.7) gives MLEs for $[\beta_0, \beta_1, \beta_2]$ of $[2.42, -0.0959, -0.246]$. The fitted curve is shown in Fig. 6.1. The method of moments estimate of the coefficient of variation, $100\sqrt{\alpha}$, is 5.3%, while the MLE is 4.4%. Asymptotic standard errors for $[\beta_0, \beta_1, \beta_2]$, based on the method of moments estimator for α, are $[0.033, 0.0028, 0.018]$. The point estimates of β are identical, regardless of the estimate used for α, because the root of the score is independent of α in a GLM, as is clear from (6.10).

The top row of Table 6.2 gives MLEs for the derived parameters, along with asymptotic 90% confidence intervals, derived using the delta method. All are based upon the method of moments estimator for α. The parameters of interest are all positive, and so the intervals were obtained on the log scale and then exponentiated. Deriving an interval estimate for the clearance parameter using the delta method is more complex. Working with $\theta = \log Cl$, we have

$$\mathrm{var}(\widehat{\theta}) = [D_0 \; D_1 \; D_2] V^\star \begin{bmatrix} D_0 \\ D_1 \\ D_2 \end{bmatrix}$$

6.5 Likelihood Inference for GLMs

where, from (6.5),

$$D_0 = \frac{\partial \theta}{\partial \beta_0} = 1$$

$$D_1 = \frac{\partial \theta}{\partial \beta_1} = \frac{1}{\beta_1} + \sqrt{\frac{\beta_2}{\beta_1}} \frac{K_0\left(2\sqrt{\beta_1 \beta_2}\right)}{K_1\left(2\sqrt{\beta_1 \beta_2}\right)}$$

$$D_2 = \frac{\partial \theta}{\partial \beta_2} = \sqrt{\frac{\beta_1}{\beta_2}} \frac{K_0\left(2\sqrt{\beta_1 \beta_2}\right)}{K_1\left(2\sqrt{\beta_1 \beta_2}\right)},$$

and $V^\star$ is the variance–covariance matrix of $\widehat{\boldsymbol{\beta}}$ as given by (6.11). For the theophylline data, the MLE is $\widehat{Cl} = 0.042$ with asymptotic 90% confidence interval [0.041,0.044]. Inference for the clearance parameter using the sampling-based Bayesian approach that we describe shortly is straightforward, once samples are generated from the posterior.

Example: Poisson Data with a Linear Link

We now describe a GLM that is a little more atypical and reveals some of the subtleties of modeling that can occur. In the context of a spatial study, suppose that, in a given time period, Y_{i0} represents the number of counts of a (statistically) rare disease in an unexposed group of size N_{i0}, while Y_{i1} represents the number of counts of a rare disease in an exposed group of size N_{i1}, all in area i, $i = 1, \ldots, n$. Suppose also that we only observe the sum of the disease counts, $Y_i = Y_{i0} + Y_{i1}$, along with N_{i0} and N_{i1}. If we had observed Y_{i0}, Y_{i1}, we would fit the model $Y_{ij} \mid \boldsymbol{\beta}^\star \sim_{ind} \text{Poisson}(N_{ij}\beta_j^\star)$ so that $0 < \beta_j^\star < 1$ is the probability of disease in exposure group j, with $j = 0/1$ representing unexposed/exposed and $\boldsymbol{\beta}^\star = [\beta_0^\star, \beta_1^\star]$. Then, writing $x_i = N_{1i}/N_i$ as the proportion of exposed individuals, the distribution of the total disease counts is

$$Y_i \mid \boldsymbol{\beta}^\star \sim_{ind} \text{Poisson}\left\{N_i[(1-x_i)\beta_0^\star + x_i\beta_1^\star]\right\}, \qquad (6.17)$$

so that we have a Poisson GLM with a linear link function. Since the parameters $\beta_0^\star$ and $\beta_1^\star$ are the probabilities (or risks) of disease for unexposed and exposed individuals, respectively, a parameter of interest is the relative risk, $\beta_1^\star/\beta_0^\star$.

We illustrate the fitting of this model using data on the incidence of lip cancer in men in $n = 56$ counties of Scotland over the years 1975–1980. These data were originally reported by Kemp et al. (1985) and have been subsequently reanalyzed by numerous others, see, for example, Clayton and Kaldor (1987). The covariate x_i is the proportion of individuals employed in agriculture, fishing, and farming in county i. We let Y_i represent the number of cases in county i. Model (6.17) requires some adjustment, since the only available data here, in addition to x_i, are the expected numbers E_i that account for the age breakdown in county i (see Sect. 1.3.3). We briefly describe the model development in this case, since it requires care and reveals assumptions that may otherwise be unapparent.

Let Y_{ijk} be the number of cases, from a population of N_{ijk} in county i, exposure group j, and age stratum k, $i = 1, \ldots, n$, $j = 0, 1$, $k = 1, \ldots, K$. An obvious starting model for a rare disease is

$$Y_{ijk} \mid p_{ijk} \sim_{ind} \text{Poisson}(N_{ijk} p_{ijk}).$$

This model contains far too many parameters, p_{ijk}, to estimate, and so we simplify by assuming

$$p_{ijk} = \beta_j \times p_k, \tag{6.18}$$

across all areas i. Consequently, p_k is the probability of disease in age stratum k and $\beta_j > 0$ is the relative risk adjustment in exposure group j, and we are assuming that the exposure effect is the same across areas and across age stratum. The age-specific probabilities p_k are assumed known (e.g., being based on rates from a larger geographic region). The numbers of exposed individuals in each age stratum are unknown, and we therefore make the important assumption that the proportion of exposed and unexposed individuals is constant across age stratum, that is, $N_{i0k} = N_{ik}(1 - x_i)$ and $N_{i1k} = N_{ik} x_i$. This assumption is made since N_{i0k} and N_{i1k} are unavailable and is distinct from assumption (6.18) which concerns the underlying disease model. Summing across stratum and exposure groups gives

$$Y_i \mid \boldsymbol{\beta} \sim_{ind} \text{Poisson}\left(\beta_0 (1 - x_i) \sum_{k=1}^{K} N_{ik} p_k + \beta_1 x_i \sum_{k=1}^{K} N_{ik} p_k\right).$$

Letting $E_i = \sum_{k=1}^{K} N_{ik} p_k$ represent the expected number of cases, and simplifying the resultant expression gives

$$Y_i \mid \boldsymbol{\beta} \sim_{ind} \text{Poisson}\left\{E_i [(1 - x_i) \beta_0 + x_i \beta_1]\right\}. \tag{6.19}$$

Under this model,

$$\mathrm{E}\left[\frac{Y_i}{E_i}\right] = \beta_0 + (\beta_1 - \beta_0) x_i, \tag{6.20}$$

illustrating that the mean model for the standardized morbidity ratio (SMR), Y_i/E_i, is linear in x. Figure 6.2 plots the SMRs Y_i/E_i versus x_i, with a linear fit added, and we see evidence of increasing SMR with increasing x.

Fitting the Poisson linear link model gives estimates (asymptotic standard errors) for β_0 and β_1 of 0.45 (0.043) and 10.1 (0.77). The fitted line (6.20) is superimposed on Fig. 6.2. The estimate of the relative risk β_1/β_0 is 22.7 with asymptotic standard

Fig. 6.2 Plot of standardized morbidity ratio versus proportion exposed for lip cancer incidence in 56 counties of Scotland. The linear model fit is indicated

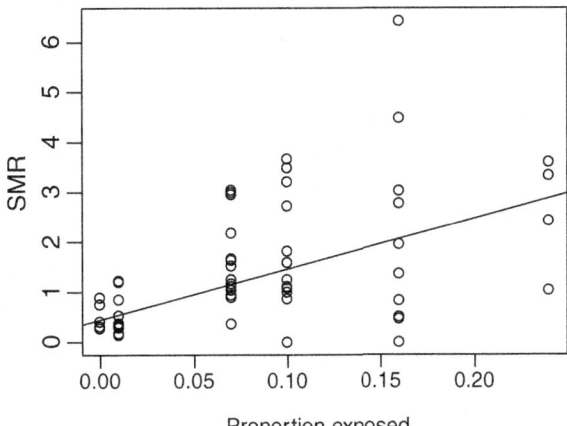

error 3.39. The latter is a model-based estimate and in particular depends on there being no excess-Poisson variation, which is highly dubious for applications such as this, because of all of the missing auxiliary information, including data on smoking.

6.5.3 Hypothesis Testing

Suppose that $\dim(\boldsymbol{\beta}) = k+1$ and let $\boldsymbol{\beta} = [\boldsymbol{\beta}_1, \boldsymbol{\beta}_2]$ be a partition with $\boldsymbol{\beta}_1 = [\beta_0, \ldots, \beta_q]$ and $\boldsymbol{\beta}_2 = [\beta_{q+1}, \ldots, \beta_k]$, with $0 \leq q < k$. Interest focuses on testing whether the subset of $k - q$ parameters are equal to zero via a test of the null

$$H_0 : \boldsymbol{\beta}_1 \text{ unrestricted, } \boldsymbol{\beta}_2 = \boldsymbol{\beta}_{20}$$
$$H_1 : \boldsymbol{\beta} = [\boldsymbol{\beta}_1, \boldsymbol{\beta}_2] \neq [\boldsymbol{\beta}_1, \boldsymbol{\beta}_{20}]. \tag{6.21}$$

As outlined in Sect. 2.9, there are three main frequentist approaches to hypothesis testing, based on Wald, score, and likelihood ratio tests. We concentrate on the latter. For the linear model, the equivalent approach is based on an F test (Sect. 5.6.1), which formally accounts for estimation of the scale parameter.

The log-likelihood is

$$l(\boldsymbol{\beta}) = \sum_{i=1}^{n} \frac{y_i \theta_i - b(\theta_i)}{\alpha} + c(y_i, \alpha),$$

with α the scale parameter. We let $\boldsymbol{\theta} = \boldsymbol{\theta}(\boldsymbol{\beta}) = [\theta_1(\boldsymbol{\beta}), \ldots, \theta_n(\boldsymbol{\beta})]$ denote the vector of canonical parameters. Under the null, from Sect. 2.9.5,

$$2\left[l(\widehat{\boldsymbol{\beta}}) - l(\widehat{\boldsymbol{\beta}}^{(0)})\right] \to_d \chi^2_{k-q},$$

where $\widehat{\boldsymbol{\beta}}$ is the unrestricted MLE and $\widehat{\boldsymbol{\beta}}^{(0)} = [\widehat{\boldsymbol{\beta}}_{10}, \boldsymbol{\beta}_{20}]$ is the MLE under the null.

In some circumstances, one may assess the *overall* fit of a particular model via comparison of the likelihood of this model with the maximum attainable log-likelihood which occurs under the *saturated model*. We write $\widetilde{\boldsymbol{\theta}} = [\widetilde{\theta}_1, \ldots, \widetilde{\theta}_n]$ to represent the MLEs under the saturated model. Similarly, let $\widehat{\boldsymbol{\theta}} = [\widehat{\theta}_1, \ldots, \widehat{\theta}_n]$ denote the MLEs under a reduced model containing $q+1$ parameters. The log-likelihood ratio statistic of H_0 : reduced model, H_1 : saturated model is

$$2\left[l(\widetilde{\boldsymbol{\theta}}) - l(\widehat{\boldsymbol{\theta}})\right] = \frac{2}{\alpha}\sum_{i=1}^{n}\left[Y_i(\widetilde{\theta}_i - \widehat{\theta}_i) - b(\widetilde{\theta}_i) + b(\widehat{\theta}_i)\right] = \frac{D}{\alpha}, \quad (6.22)$$

where D is known as the *deviance* (associated with the saturated model) and D/α is the *scaled deviance*. If the saturated model has a fixed number of parameters, p, then, under the reduced model,

$$\frac{D}{\alpha} \rightarrow_d \chi^2_{p-q-1}.$$

In general, this result is rarely used, though cross-classified discrete data provide one instance in which the overall fit of a model can be assessed in this way. An alternative measure of the overall fit is the Pearson statistic

$$X^2 = \sum_{i=1}^{n}\frac{(Y_i - \widehat{\mu}_i)^2}{V(\widehat{\mu}_i)}, \quad (6.23)$$

with $X^2 \rightarrow_d \chi^2_{p-q-1}$ under the null. Again, the saturated model should contain a fixed number of parameters (as $n \rightarrow \infty$).

Consider again the nested testing situation with hypotheses, (6.21). We describe an attractive additivity property of the likelihood ratio test statistic for nested models. Let $\widehat{\boldsymbol{\beta}}^{(0)}, \widehat{\boldsymbol{\beta}}^{(1)}$ and $\widehat{\boldsymbol{\beta}}^{(s)}$ represent the MLEs of $\boldsymbol{\beta}$ under the null, alternative, and saturated models, respectively. Suppose that the dimensionality of $\widehat{\boldsymbol{\beta}}^{(j)}$ is q_j with $0 < q_0 < q_1 < p$. Under H_0,

$$2\left[l(\widehat{\boldsymbol{\beta}}^{(1)}) - l(\widehat{\boldsymbol{\beta}}^{(0)})\right] = 2\left\{l(\widehat{\boldsymbol{\beta}}^{(s)}) - l(\widehat{\boldsymbol{\beta}}^{(0)}) - [l(\widehat{\boldsymbol{\beta}}^{(s)}) - l(\widehat{\boldsymbol{\beta}}^{(1)})]\right\}$$

$$= \frac{1}{\alpha}(D_0 - D_1) \rightarrow_d \chi^2_{q_1-q_0},$$

where D_j is the deviance representing the fit under hypothesis j, relative to the saturated model, $j = 0, 1$. The Pearson statistic does not share this additivity property.

6.5 Likelihood Inference for GLMs

For a GLM, in contrast to the linear model (see Sect. 5.8), even if a covariate is orthogonal to all other covariates, its significance will still depend on which covariates are currently in the model.

Example: Normal Linear Model

We consider the model $Y \mid \boldsymbol{\beta} \sim N_n(\boldsymbol{x}\boldsymbol{\beta}, \sigma^2 \mathbf{I}_n)$. The log-likelihood is

$$l(\boldsymbol{\beta}, \sigma) = -n \log \sigma - \frac{1}{2\sigma^2}(\boldsymbol{y} - \boldsymbol{x}\boldsymbol{\beta})^\mathsf{T}(\boldsymbol{y} - \boldsymbol{x}\boldsymbol{\beta}),$$

with α in the GLM formulation being replaced by σ^2. Again, let $\boldsymbol{\beta} = [\boldsymbol{\beta}_1, \boldsymbol{\beta}_2]$ where $\boldsymbol{\beta}_1 = [\beta_0, \ldots, \beta_q]$ and $\boldsymbol{\beta}_2 = [\beta_{q+1}, \ldots, \beta_k]$, and consider the null H_0 : $\boldsymbol{\beta}_1$ unrestricted, $\boldsymbol{\beta}_2 = \boldsymbol{\beta}_{20}$. Under this null, from (6.22),

$$D = \sum_{i=1}^n \left(Y_i - \boldsymbol{x}_i \widehat{\boldsymbol{\beta}}^{(0)}\right)^2$$

where $\boldsymbol{x}_i \widehat{\boldsymbol{\beta}}^{(0)}$ are the fitted values for the ith case, based on the MLEs under the reduced model, H_0. In this case, the asymptotic distribution is exact since

$$\frac{\sum_{i=1}^n (Y_i - \boldsymbol{x}_i \widehat{\boldsymbol{\beta}}^{(0)})^2}{\sigma^2} \sim \chi^2_{n-q+1}. \tag{6.24}$$

This result is almost never directly useful, however, since σ^2 is rarely known.

In terms of comparing the nested hypotheses $H_0 : \boldsymbol{\beta}_1$ unrestricted, $\boldsymbol{\beta}_2 = \boldsymbol{\beta}_{20}$, and $H_1 : \boldsymbol{\beta} = [\boldsymbol{\beta}_1, \boldsymbol{\beta}_2] \neq [\boldsymbol{\beta}_1, \boldsymbol{\beta}_{20}]$, the likelihood ratio statistic is

$$\frac{1}{\sigma^2}(D_0 - D_1) = \frac{1}{\sigma^2}\left[\sum_{i=1}^n (Y_i - \boldsymbol{x}_i \widehat{\boldsymbol{\beta}}^{(0)})^2 - \sum_{i=1}^n (Y_i - \boldsymbol{x}_i \widehat{\boldsymbol{\beta}}^{(1)})^2\right]$$

$$= \frac{\mathrm{RSS}_0 - \mathrm{RSS}_1}{\sigma^2} = \frac{\mathrm{FSS}_{01}}{\sigma^2} \tag{6.25}$$

where $\boldsymbol{x}\widehat{\boldsymbol{\beta}}^{(j)}$ are the fitted values corresponding to the MLEs under model j, RSS_j is the residual sum of squares for model j, $j = 0, 1$, and FSS_{01} is the fitted sum of squares due to the additional parameters present in H_1.

In practice if n is large, we may use (6.25) with σ^2 replaced by a consistent estimator $\widehat{\sigma}^2$. Alternatively, the ratios of scaled versions of (6.25) and (6.24) may be taken to produce an F-statistic by which statistical significance may be assessed, as described in Sect. 5.6.1.

Example: Lung Cancer and Radon

Under a Poisson model, the deviance and scaled deviance are identical since $\alpha = 1$. For a Poisson model with MLE $\widehat{\beta}$, the deviance is

$$2 \sum_{i=1}^{n} \left[(\mu_i(\widehat{\beta}) - y_i) + y_i \log \left(\frac{y_i}{\mu_i(\widehat{\beta})} \right) \right]$$

and if the sum of the observed and fitted counts agree, then we obtain the intuitive distance measure

$$2 \sum_{i=1}^{n} y_i \log \left(\frac{y_i}{\mu_i(\widehat{\beta})} \right).$$

For the Minnesota data, suppose we wish to test $H_0 : \beta_0$ unrestricted, $\beta_1 = 0$ versus $H_1 : [\beta_0, \beta_1] \neq [\beta_0, 0]$, in the model $\mu_i = E_i \exp(\beta_0 + \beta_1 x_i)$. The likelihood ratio statistic is

$$T = 2 \sum_{i=1}^{n} y_i \log \left(\frac{\mu_i(\widehat{\beta})}{\mu_i(\widehat{\beta}^{(0)})} \right),$$

since $\sum_{i=1}^{n} \mu_i(\widehat{\beta}) = \sum_{i=1}^{n} \mu_i(\widehat{\beta}^{(0)})$, and where $\widehat{\beta}$ and $\widehat{\beta}^{(0)}$ are the MLEs under the null and alternative hypotheses. Under H_0, $T \to_d \chi_1^2$.

For the Minnesota data $T = 46.2$ to give an extremely small p-value. The estimate (standard error) of β_1 is -0.036 (0.0054) so that for a one-unit increase in average radon, there is an associated drop in relative risk of lung cancer of 3.6%.

6.6 Quasi-likelihood Inference for GLMs

Section 2.5 provided an extended discussion of quasi-likelihood, and here we recap the key points. GLMs that do not contain a scale parameter are particularly vulnerable to variance model misspecification, specifically the presence of overdispersion in the data. The Poisson and binomial models are especially susceptible in this respect.

Rather than specify a complete probability model for the data, quasi-likelihood proceeds by specifying the mean and variance as

$$\mathrm{E}[Y_i \mid \beta] = \mu_i(\beta)$$
$$\mathrm{var}(Y_i \mid \beta) = \alpha V(\mu_i),$$

with $\mathrm{cov}(Y_i, Y_j \mid \beta) = 0$. From these specifications, the quasi-score is defined as in (2.30) and coincides with the score function (6.10). Hence, the maximum quasi-likelihood estimator $\widehat{\beta}$ is identical to the MLE due to the multiplicative form of the variance model. Estimation of α may be carried out using the form (6.14) or via

6.6 Quasi-likelihood Inference for GLMs

$$\widehat{\alpha} = \frac{D}{n-k-1},$$

where D is the deviance and $\dim(\beta) = k+1$. Asymptotic inference is based on

$$(D^{\mathsf{T}}V^{-1}D/\alpha)^{1/2}(\widehat{\beta}_n - \beta) \to_d N_{k+1}(0, I_{k+1}).$$

In practice, D and V are evaluated at $\widehat{\beta}_n$, and $\widehat{\alpha}$ replaces α.

Hypothesis tests follow in an obvious fashion, with adjustment for $\widehat{\alpha}$. Specifically, if as before

$$l(\beta, \alpha) = \int_y^\mu \frac{y-t}{\alpha V(t)} dt,$$

then if $l(\beta) = l(\beta, \alpha = 1)$ represents the likelihood upon which the quasi-likelihood is based (e.g., a Poisson or binomial likelihood),

$$l(\beta) = l(\beta, \alpha) \times \alpha \qquad (6.26)$$

and to test $H_0 : \beta_1$ unrestricted, $\beta_2 = \beta_{20}$, we may use the quasi-likelihood ratio test statistic

$$2\left[l(\widehat{\beta}, \widehat{\alpha}) - l(\widehat{\beta}^{(0)}, \widehat{\alpha})\right] \to_d \chi^2_{k-q-1},$$

or equivalently

$$2\left[l(\widehat{\beta}) - l(\widehat{\beta}^{(0)})\right] \to_d \widehat{\alpha} \times \chi^2_{k-q}. \qquad (6.27)$$

If, as is usually the case, $\widehat{\alpha} > 1$, then larger differences in the log-likelihood are required to attain the same level of significance, as compared to the $\alpha = 1$ case.

Example: Lung Cancer and Radon

Fitting the quasi-likelihood model

$$E[Y_i \mid \beta] = E_i \exp(\beta_0 + \beta_1 x_i) \qquad (6.28)$$

$$\mathrm{var}(Y_i \mid \beta) = \alpha E[Y_i \mid \beta], \qquad (6.29)$$

yields identical point estimates for β to the Poisson model, with scale parameter estimate $\widehat{\alpha} = 2.81$, obtained via (6.14). Therefore, with respect to $H_0 : \beta_0$ unrestricted, $\beta_1 = 0$, the quasi log-likelihood ratio statistic is $46.2/2.81 = 16.5$ so that the significance level is vastly reduced, though still strongly suggestive of a nonzero slope.

6.7 Sandwich Estimation for GLMs

The asymptotic variance–covariance for $\widehat{\boldsymbol{\beta}}$, which is given by (6.11), is appropriate only if the first two moments are correctly specified. In general, as detailed in Sect. 2.6, $\text{var}(\widehat{\boldsymbol{\beta}}) = \boldsymbol{A}^{-1}\boldsymbol{B}(\boldsymbol{A}^{\mathrm{T}})^{-1}$ where

$$\boldsymbol{A} = \mathrm{E}\left[\frac{\partial \boldsymbol{G}}{\partial \boldsymbol{\beta}}\right] = \boldsymbol{D}^{\mathrm{T}}\boldsymbol{V}^{-1}\boldsymbol{D}, \tag{6.30}$$

regardless of the distribution of the data (so long as the mean is correctly specified), and

$$\boldsymbol{B} = \text{var}\left[\boldsymbol{G}(\boldsymbol{\beta})\right] = \boldsymbol{D}^{\mathrm{T}}\boldsymbol{V}^{-1}\text{var}(\boldsymbol{Y})\boldsymbol{V}^{-1}\boldsymbol{D},$$

where $\boldsymbol{G}(\boldsymbol{\beta}) = \boldsymbol{S}(\boldsymbol{\beta})/n$. Under the assumption of uncorrelated errors,

$$\widehat{\boldsymbol{B}} = \sum_{i=1}^{n}\left(\frac{\partial \mu_i}{\partial \boldsymbol{\beta}}\right)^{\mathrm{T}} \frac{\text{var}(Y_i)}{V_{ii}^2}\left(\frac{\partial \mu_i}{\partial \boldsymbol{\beta}}\right) \tag{6.31}$$

where a naive estimator of $\text{var}(Y_i)$ is

$$\widehat{\sigma}_i^2 = (Y_i - \widehat{\mu}_i)^2, \tag{6.32}$$

which has finite sample bias. Combination of (6.31) and (6.32) provides a consistent estimator of the variance and therefore asymptotically corrects confidence interval coverage (so long as independence of responses holds).

Bootstrap methods (Sect. 2.7.2) may also be used to provide inference that is robust to certain aspects of model misspecification, provided n is sufficiently large. The resampling residuals method may be applied, but the meaning of residuals is ambiguous in GLMs (Sect. 6.9), and this method does not correct for mean–variance misspecification, which is a major drawback. The resampling cases approach corrects for this aspect. Davison and Hinkley (1997, Sect. 7.2) discuss both resampling residuals and resampling cases in the context of GLMs.

Example: Pharmacokinetics of Theophylline

Table 6.2 gives confidence intervals for $x_{1/2}$, $x_{\max}$ and $\mu(x_{\max})$, based on sandwich estimation. In each case, the interval estimates are a little shorter than the model-based estimates. This could be due to either instability in the sandwich estimates with a small sample size ($n = 10$) or to the gamma mean–variance assumption being inappropriate.

6.8 Bayesian Inference for GLMs

We now consider Bayesian inference for the GLM. The posterior is

$$p(\boldsymbol{\beta}, \alpha \mid \boldsymbol{y}) \propto l(\boldsymbol{\beta}, \alpha)\pi(\boldsymbol{\beta}, \alpha)$$

where it is usual to assume prior independence between the regression coefficients $\boldsymbol{\beta}$ and the scale parameter α, that is, $\pi(\boldsymbol{\beta}, \alpha) = \pi(\boldsymbol{\beta})\pi(\alpha)$.

6.8.1 Prior Specification

Recall that $\boldsymbol{\beta} = [\beta_0, \beta_1, \ldots, \beta_k]$. Often, β_j, $j = 0, 1, \ldots, k$, is defined on $\mathbb{R}$, and so a multivariate normal prior for $\boldsymbol{\beta}$ is the obvious choice. Furthermore, independent priors are frequently defined for each component. As a limiting case, the improper prior $\pi(\boldsymbol{\beta}) \propto 1$ results. However, care should be taken with this choice since it may lead to an improper posterior. With canonical links, impropriety only occurs for pathological datasets (see the binomial model example of Sect. 3.4), but for noncanonical links, innocuous datasets may lead to impropriety, as the Poisson data with a linear link example described below illustrates. If the scale parameter $\alpha > 0$ is unknown, gamma or lognormal distributions provide obvious choices.

Poisson Data with a Linear Link

Recall the Poisson model with a linear link function

$$Y_i \mid \boldsymbol{\beta} \sim_{ind} \text{Poisson}\{E_i[(1 - x_i)\beta_0 + x_i\beta_1]\}$$

and suppose we assume an improper uniform prior for $\beta_0 > 0$, that is,

$$\pi(\beta_0) \propto 1.$$

We define $e^\gamma = \beta_1/\beta_0 > 0$ as the parameter of interest and write

$$\mu_i = \beta_0 E_i[(1 - x_i) + x_i \exp(\gamma)] = \beta_0 \mu_i^\star.$$

The marginal posterior for γ is

$$p(\gamma \mid \boldsymbol{y}) = \int p(\beta_0, \gamma \mid \boldsymbol{y})\, d\beta_0$$

$$\propto \int l(\beta_0, \gamma)\, d\beta_0 \times \pi(\gamma)$$

$$\propto \int \exp\left(-\beta_0 \sum_{i=1}^{n} \mu_i^{\star y_i}\right) \beta_0^{\sum_{i=1}^{n} y_i} \prod_{i=1}^{n} \mu_i^{\star y_i}\, d\beta_0 \times \pi(\gamma)$$

$$\propto \prod_{i=1}^{n} \left(\frac{E_i[(1-x_i) + x_i e^{\gamma}]}{\sum_{i=1}^{n} E_i[(1-x_i) + x_i e^{\gamma}]}\right)^{y_i} \times \pi(\gamma) \qquad (6.33)$$

$$= l(\gamma) \times \pi(\gamma), \qquad (6.34)$$

where the last line follows from the previous on recognizing that the integrand is the kernel of a Ga $(\sum_{i=1}^{n} y_i, \sum_{i=1}^{n} \mu_i^{\star})$ distribution. The "likelihood," $l(\gamma)$ in (6.34), is of multinomial form with the total number of cases y_+ distributed among the n areas with probabilities proportional to $E_i[(1-x_i) + x_i \exp(\gamma)]$ so that, for example, larger E_i and larger x_i (if $\gamma > 0$) lead to a larger allocation of cases to area i. The likelihood contribution to the posterior tends to the constant

$$\prod_{i=1}^{n} \left(\frac{E_i(1-x_i)}{\sum_{i=1}^{n} E_i(1-x_i)}\right)^{y_i} \qquad (6.35)$$

as $\gamma \to -\infty$, showing that, in general, a proper prior is required (since the tail will be non-integrable). The constant (6.35) is nonzero unless $x_i = 1$ in any area with $y_i \neq 0$. The reason for the impropriety is that in the limit as $\gamma \to -\infty$, the relative risk $\exp(\gamma) \to 0$ so that exposed individuals cannot get the disease, which is not inconsistent with the observed data, unless all individuals in area i are exposed, $x_i = 1$, and $y_i \neq 0$ in that area since then clearly (under the assumed model) the cases are due to exposure. A similar argument holds as $\gamma \to \infty$, with replacement of $1 - x_i$ by x_i in (6.35) providing the limiting constant.

Figure 6.3 illustrates this behavior for the Scottish lip cancer example, for which $x_i = 0$ in five areas. The log-likelihood has been scaled to have maximum 0, and the constant (6.35) is indicated with a dashed horizontal line. The MLE $\widehat{\gamma} = \log(22.7)$ is indicated as a vertical dotted line.

6.8.2 Computation

Unfortunately, when continuous covariates are present in the model, conjugate analysis is unavailable. However, sampling-based approaches are relatively easy to implement. In particular, if informative priors are available, then the rejection algorithm of Sect. 3.7.6 is straightforward to implement with sampling from the prior.

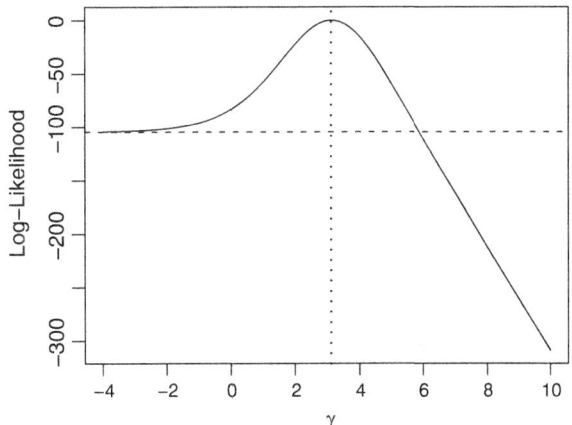

Fig. 6.3 Log-likelihood for the log relative risk parameter γ, for the Scottish lip cancer data. The *dashed horizontal line* is the constant to which the log-likelihood tends to as $\gamma \to -\infty$

MCMC (Sect. 3.8) is obviously a candidate for computation and was illustrated for Poisson and negative binomial models in Chap. 3. The INLA method described in Sect. 3.7.4 may also be used.

As described in Sect. 3.3, there is asymptotic equivalence between the sampling distribution of the MLE and the posterior distribution. Hence, Bayes estimators for β are consistent due to the form of the likelihood, so long as the priors are nonzero in a neighborhood of the true values of β.

6.8.3 Hypothesis Testing

A simple method for examining hypotheses involving a single parameter, $H_0 : \beta_j = 0$ versus $H_1 : \beta_j \neq 0$, with any remaining parameters unrestricted, is to evaluate the posterior tail probability $\Pr(\beta_j > 0 \mid y)$, with values close to 0 or 1 indicating that the null is unlikely to be true. Bayes factors (which were discussed in Sects. 3.10 and 4.3) provide a more general tool for comparing hypotheses (by analogy with the likelihood ratio statistic, though of course, as usual, interpretation is very different):

$$\text{BF} = \frac{p(y \mid H_0)}{p(y \mid H_1)}.$$

The use of Bayes factors will be illustrated in Sect. 6.16.3. As discussed in Sect. 4.3.2, great care is required in the specification of priors when model comparison is carried out using Bayes factors.

6.8.4 Overdispersed GLMs

Quasi-likelihood provides a simple procedure by which frequentist inference may accommodate overdispersion in GLMs. No such simple remedy exists within the Bayesian framework. An alternative method of increasing the flexibility of GLMs is through the introduction of random effects. We have already seen an example of this in Sect. 2.5 when the negative binomial model was derived via the introduction of gamma random effects into a Poisson model.

Example: Lung Cancer and Radon

The Bayesian Poisson model was fitted in Chap. 3 using a Metropolis–Hastings implementation. Here the use of the INLA method of Sect. 3.7.4, with improper flat priors on β_0, β_1, gives a 95% interval estimate for the relative risk $\exp(\beta_1)$ of [0.954,0.975] which is identical to that based on asymptotic likelihood inference (the posterior mean and MLE both equal -0.036, and the posterior standard deviation and standard error both equal 0.0054).

Example: Pharmacokinetics of Theophylline

With respect to the gamma GLM with $\mu(x) = \exp(\beta_0 + \beta_1 x + \beta_2/x)$, the interpretation of β_0 and β_2 in particular is not straightforward, which makes prior specification difficult. As an alternative, we specify prior distributions on the half-life $x_{1/2}$, time to maximum $x_{\max}$, maximum concentration $\mu(x_{\max})$, and coefficient of variation, $\sqrt{\alpha}$. We choose independent lognormal priors for these four parameters. For a generic parameter θ, denote the prior by $\theta \sim \text{LogNorm}(\mu, \sigma)$. To obtain the moments of these distributions, we specify the prior median θ_m and the 95% point of the prior θ_u. We then solve for the moments via

$$\mu = \log(\theta_m), \quad \sigma = \frac{\log(\theta_u) - \mu}{1.645}, \tag{6.36}$$

as described in Sect. 3.4.2. Based on a literature search, we assume prior 50% (95%) points of 8 (12), 1.5 (3), and 9 (12) for $x_{1/2}$, $x_{\max}$, and $\mu(x_{\max})$, respectively. For the coefficient of variation, the corresponding values are 0.05 (0.10). The third line of Table 6.2 summarizes these priors. To examine the posterior, we use a rejection algorithm, as described in Sect. 3.7.6. We sample from the prior on the parameters of interest and then back-solve for the parameters that describe the likelihood. For the loglinear model, the transformation to β is via

6.8 Bayesian Inference for GLMs

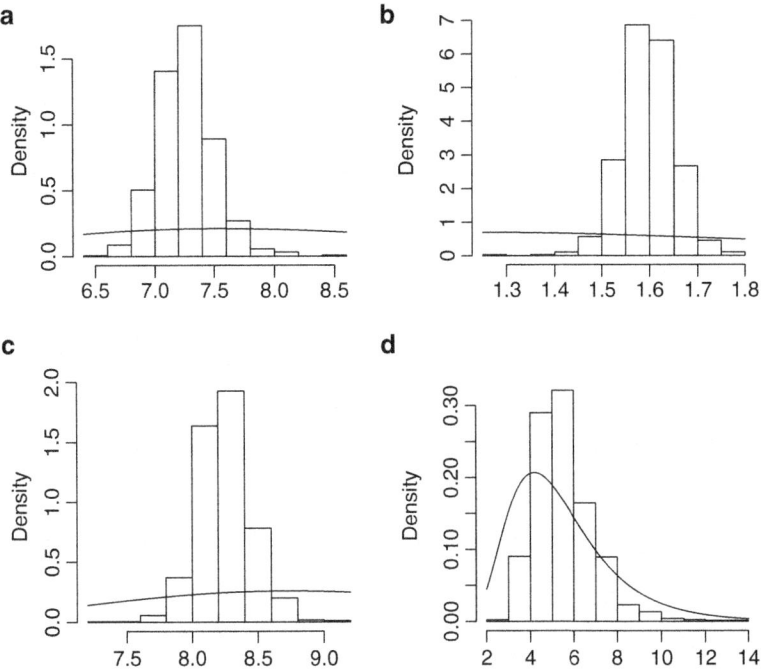

Fig. 6.4 Histogram representations of posterior distributions from the GLM for the theophylline data for (**a**) half-life, (**b**) time to maximum, (**c**) maximum concentration, and (**d**) coefficient of variation, with priors superimposed as *solid lines*

$$\beta_1 = -\frac{\log 2}{x_{1/2}}$$
$$\beta_2 = \beta_1 x_{\max}^2$$
$$\beta_0 = \log \mu(x_{\max}) + 2(\beta_1 \beta_2)^{1/2}.$$

Table 6.2 summarizes inference for the parameters of interest, via medians and 90% interval estimates. Point and interval estimates show close correspondence with the frequentist summaries. Figure 6.4 gives the posterior distributions for the half-life, the time to maximum concentration, the maximum concentration, and the coefficient of variation (expressed as a percentage). The prior distributions are also indicated as solid curves. We see some skewness in each of the posteriors, which is common for nonlinear parameters unless the data are abundant.

Inference for the clearance parameter is relatively straightforward, since one simply substitutes samples for β into (6.5). Figure 6.5 gives a histogram representation of the posterior distribution. The posterior median of the clearance is 0.042 with 90% interval [0.041,0.044]; these summaries are identical to the likelihood-based counterparts. We see that the posterior shows little skewness; the clearance

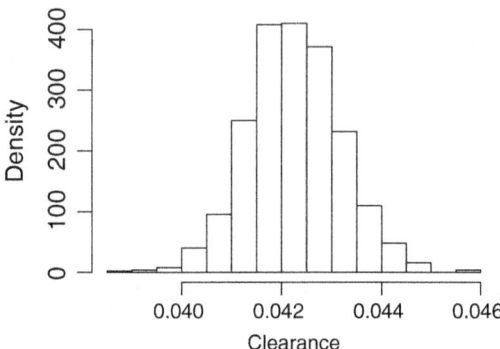

Fig. 6.5 Posterior distribution of the clearance parameter from the GLM fitted to the theophylline data

parameter is often found to be well behaved, since it is a function of the area under the curve, which is reliably estimated so long as the tail of the curve is captured.

6.9 Assessment of Assumptions for GLMs

The assessment of assumptions for GLMs is more difficult than with linear models. The definition of a residual is more ambiguous, and for discrete data in particular, the interpretation of residuals is far more difficult, even when the model is correct. Various attempts have been made to provide a general definition of residuals that possess zero mean, constant variance, and a symmetric distribution. In general, the latter two desiderata are in conflict.

When first examining the data, one may plot the response, transformed to the linear predictor scale, against covariates. For example, with Poisson data and canonical log link, one may plot $\log y$ versus covariates x.

The obvious definition of a residual is

$$e_i = Y_i - \widehat{\mu}_i$$

but clearly in a GLM, such residuals will generally have unequal variances so that some form of standardization is required. Pearson residuals, upon which we concentrate, are defined as

$$e_i^\star = \frac{Y_i - \widehat{\mu}_i}{\sqrt{\widehat{\mathrm{var}}(Y_i)}} = \frac{Y_i - \widehat{\mu}_i}{\widehat{\sigma}_i},$$

where $\widehat{\mathrm{var}}(Y_i) = \widehat{\alpha} V(\widehat{\mu}_i)$ and $\widehat{\mu}_i$ are the fitted values from the model. Squaring and summing these residuals reproduce Pearson's χ^2 statistic:

$$X^2 = \sum_{i=1}^{n} e_i^{\star 2},$$

6.9 Assessment of Assumptions for GLMs

as previously introduced, (6.23). For Pearson residuals, $\mathrm{E}[\widehat{\sigma}_i e_i^\star] = 0$ and $\mathrm{E}[e_i^{\star 2}] = 1$, but the third moment is not equal to zero in general so that the residuals are skewed. As an example, for Poisson data, $\mathrm{E}[e^{\star 3}] = \mu^{-1/2}$. Clearly for normal data, Pearson residuals have zero skewness.

Deviance residuals are given by

$$e_i^\star = \mathrm{sign}(Y_i - \widehat{\mu}_i)\sqrt{D_i}$$

so that $D = \sum_{i=1}^n e_i^{\star 2}$, as defined in Sect. 6.5.3. As an example, for a Poisson likelihood, the deviance residuals are

$$e_i^\star = \mathrm{sign}(y_i - \widehat{\mu}_i)\{2[y_i \log(y_i/\widehat{\mu}_i) - y_i + \widehat{\mu}_i]\}^{1/2}.$$

For discrete data with small means, residuals are extremely difficult to interpret since the response can only take on a small number of discrete values. One strategy to aid in interpretation is to simulate data with the same design (i.e., x values) and under the parameter estimates from the fitted model. One may then examine residual plots to see their form when the model is known.

As with linear model residuals (Sect. 5.11), Pearson or deviance residuals can be plotted against covariates to suggest possible model forms. They may also be plotted against fitted values or some function of the fitted values to access mean–variance relationships. If the spread is not constant, then this suggests that the assumed mean–variance relationship is not correct. McCullagh and Nelder (1989, p. 398–399) recommend plotting against the fitted values transformed to the "constant-information" scale. For example, for Poisson data, the suggestion is to plot the residuals against $2\sqrt{\widehat{\mu}}$. Residuals can also be examined for outliers/points of high influence.

For the linear model, the diagonal elements of the hat matrix, $\boldsymbol{h} = \boldsymbol{x}(\boldsymbol{x}^\mathrm{T}\boldsymbol{x})^{-1}\boldsymbol{x}^\mathrm{T}$, correspond to the *leverage* of response i, with $h_{ii} = 1$ if $\widehat{y}_i = \boldsymbol{x}_i\widehat{\boldsymbol{\beta}}$ (Sect. 5.11.2). Consideration of (6.15) reveals that for a GLM we may define a hat matrix as $\boldsymbol{h} = \boldsymbol{w}^{1/2}\boldsymbol{x}(\boldsymbol{x}^\mathrm{T}\boldsymbol{w}\boldsymbol{x})^{-1}\boldsymbol{x}^\mathrm{T}\boldsymbol{w}^{1/2}$, from which the diagonal elements may be extracted and, once again, large values of h_{ii} indicate that the fit is sensitive to y_i in some way. As with the linear model, responses with h_{ii} close to 1 have high influence. Unlike the linear case, $\boldsymbol{h}$ depends on the response through $\boldsymbol{w}$. Another useful standardized version of residuals is

$$e_i^\star = \frac{Y_i - \widehat{\mu}_i}{\sqrt{(1 - h_{ii})\widehat{\mathrm{var}}(Y_i)}},$$

for $i = 1, \ldots, n$.

It is approximately true that

$$\boldsymbol{V}^{-1/2}(\widehat{\boldsymbol{\mu}} - \boldsymbol{\mu}) \approx \boldsymbol{h}\boldsymbol{V}^{-1/2}(\boldsymbol{Y} - \boldsymbol{\mu})$$

(McCullagh and Nelder 1989, p. 397), and so

$$V^{-1/2}(Y - \mu) \approx (I - h)V^{-1/2}(Y - \widehat{\mu})$$

which shows the effect of estimation of μ on properties of the residuals.

Example: Pharmacokinetics of Theophylline

We fit the gamma GLM $Y_i \mid \beta, \alpha \sim_{ind} \text{Ga}\left[\alpha^{-1}, (\alpha\mu_i)^{-1}\right]$ using MLE and calculate Pearson residuals

$$e_i^\star = \frac{Y_i - \widehat{\mu}_i}{\sqrt{\widehat{\alpha}}\,\widehat{\mu}_i}.$$

In Fig. 6.6(a), these residuals are plotted versus time x_i and show no obvious systematic pattern, though interpretation is difficult, given the small number of data points and the spacing of these points over time. Figure 6.6(b) plots $|e_i^\star|$ against fitted values to attempt to discover any unmodeled mean–variance relationship, and again no strong signal is apparent.

Example: Lung Cancer and Radon

As we have seen, fitting the quasi-likelihood model given by the mean and variance specifications (6.28) and (6.29) yields $\widehat{\alpha} = 2.76$, illustrating a large amount of overdispersion. The quasi-MLE for β_1 is -0.035, with standard error 0.0088. We compare with a negative binomial model having the same loglinear mean model and

$$\text{var}(Y_i) = \mu_i(1 + \mu_i/b). \tag{6.37}$$

Previously, a negative binomial model was fitted to these data using a frequentist approach in Sect. 2.5 and a Bayesian approach in Sect. 3.8 The negative binomial MLE is -0.029, with standard error 0.0082, illustrating that there is some sensitivity to the model fitted.

For these data, the MLE is $\widehat{b} = 61.3$ with standard error 17.3. Figure 6.7 shows the fitted quadratic relationship (6.37) for these data. We also plot the quasi-likelihood fitted variance function. At first sight, it is surprising that the latter is not steeper, but the jittered fitted values included at the top of the plot are mostly concentrated on smaller values. The few larger values are very influential in producing a small estimated value of b (which corresponds to a large departure from the linear mean–variance model).

6.9 Assessment of Assumptions for GLMs

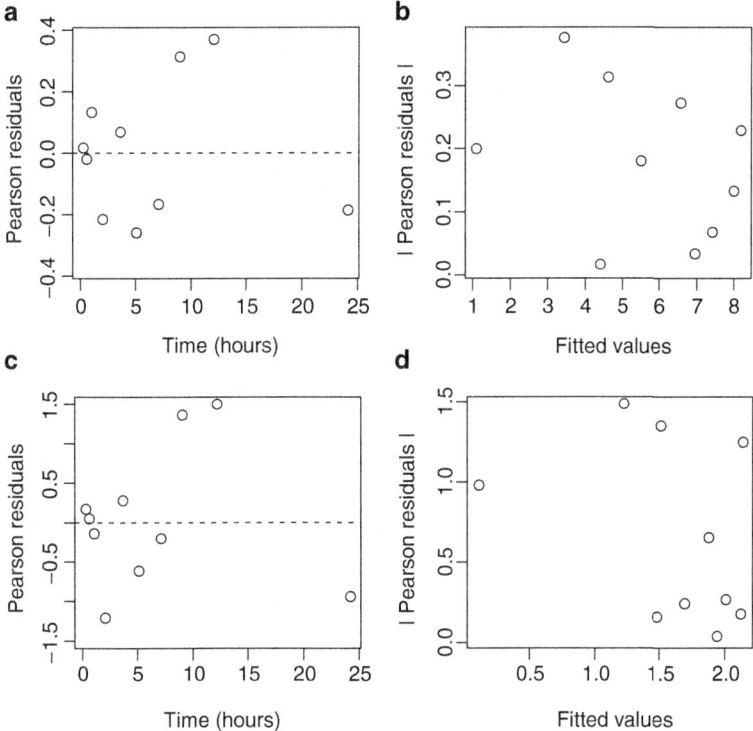

Fig. 6.6 Pearson residual plots for the theophylline data: (**a**) residuals versus time for the GLM, (**b**) absolute values of residuals versus fitted values for the GLM, (**c**) residuals versus time for the nonlinear compartmental model, and (**d**) absolute values of residuals versus fitted values for the nonlinear compartmental model

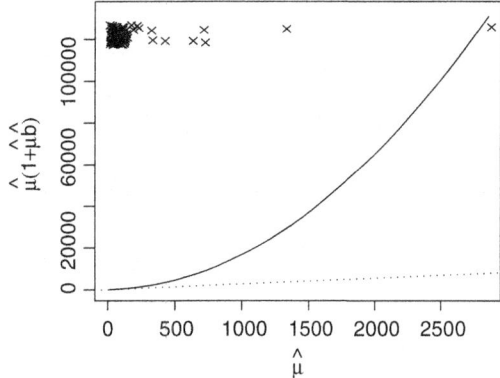

Fig. 6.7 The *solid line* shows the fitted negative binomial variance function, $\widehat{\text{var}}(Y) = \hat{\mu}(1 + \hat{\mu}/\hat{b})$ plotted versus $\hat{\mu}$ for the lung cancer and radon data. The *dotted line* corresponds to the fitted quasi-likelihood model, $\widehat{\text{var}}(Y) = \hat{b} \times \hat{\mu}$

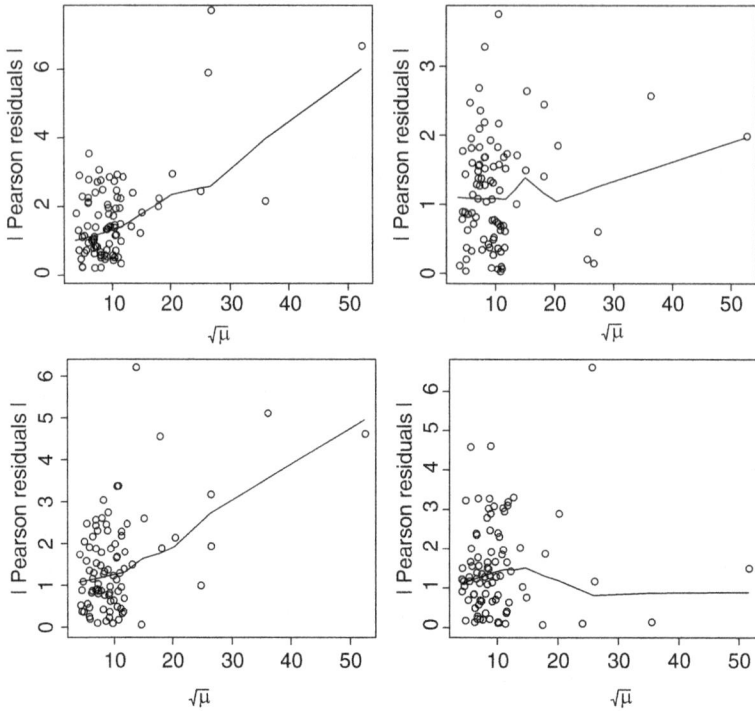

Fig. 6.8 Absolute values of Poisson Pearson residuals versus $\sqrt{\mu}$ when the true mean–variance relationship is quadratic, but we analyze as if linear, for four simulated datasets with the same expected numbers and covariate values as in the lung cancer and radon data

To attempt to determine which variance function is more appropriate, we simulate data under the negative binomial model using $\{E_i, x_i, i = 1, \ldots, n\}$ and $[\widehat{\beta}, \widehat{b}\,]$.

We then fit a Poisson model (which provides identical fitted values as from a quasi-likelihood model), form residuals $(y - \widehat{\mu})/\sqrt{\widehat{\mu}}$, that is, residuals from a Poisson model, and then plot the absolute value versus $\sqrt{\widehat{\mu}}$ to see if we can detect a trend. In the majority of simulations, the inadequacy of assuming the variance is proportional to the mean is apparent; this endeavor is greatly helped by having just a few points with very large fitted values. Specifically, the upward trend indicates that the Poisson linear mean–variance assumption is not strong enough. Figure 6.8 shows four representative plots. Figure 6.9 gives the equivalent plot from the real data. This plot shows a similar behavior to the simulated data, and so we tentatively conclude that the quadratic mean–variance relationship is more appropriate for these data. Cox (1983) provides further discussion of the effects on estimation of different forms of overdispersion, including an extended discussion of excess-Poisson variation.

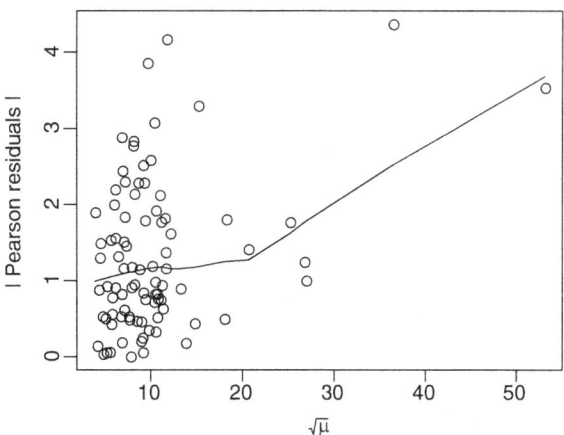

Fig. 6.9 Absolute values of Poisson Pearson residuals versus $\sqrt{\mu}$ for the lung cancer and radon data

6.10 Nonlinear Regression Models

We now consider models of the form

$$Y_i = \mu_i(\boldsymbol{\beta}) + \epsilon_i, \tag{6.38}$$

for $i = 1, \ldots, n$, where $\mu_i(\boldsymbol{\beta}) = \mu(\boldsymbol{x}_i, \boldsymbol{\beta})$ is nonlinear in $\boldsymbol{x}_i$, $\boldsymbol{\beta}$ is assumed to be of dimension $k + 1$, $\mathrm{E}[\epsilon_i \mid \mu_i] = 0$, $\mathrm{var}(\epsilon_i \mid \mu_i) = \sigma^2 f(\mu_i)$, and $\mathrm{cov}(\epsilon_i, \epsilon_j \mid \mu_i, \mu_j) = 0$. Such models are often used for positive responses, and if such data are modeled on the original scale, it is common to find that the variance is of the form $f(\mu) = \mu$ or $f(\mu) = \mu^2$. An alternative approach that is appropriate for the latter case is to assume constant errors on the log-transformed response scale (see Sect. 5.5.3). More generally, we might assume that $\mathrm{var}(\epsilon_i \mid \boldsymbol{\beta}, \boldsymbol{x}_i) = \sigma^2 g_1(\boldsymbol{\beta}, \boldsymbol{x}_i)$, with $\mathrm{cov}(\epsilon_i, \epsilon_j \mid \boldsymbol{\beta}, \boldsymbol{x}_i, \boldsymbol{x}_j) = g_2(\boldsymbol{\beta}, \boldsymbol{x}_i, \boldsymbol{x}_j)$. When data are measured over time, serial correlation can be a particular problem. We concentrate on the simpler second moment structure here.

Example: Michaelis–Menten Model

A nonlinear form that is used to model the kinetics of many enzymes has mean

$$\mu(z) = \frac{\alpha_0 z}{\alpha_1 + z},$$

a nonlinear model. Parameter interpretation is obtained by recognizing that as $z \to \infty$, $\mu(z) \to \alpha_0$ and at α_1, $\mu(\alpha_1) = \alpha_0/2$. A possible model for such data is

$$Y(z) = \mu(z) + \epsilon(z),$$

with $\mathrm{E}[\epsilon(z)] = 0$, $\mathrm{var}[\epsilon(z)] = \sigma^2 \mu(z)^r$, with $r = 0, 1,$ or 2. An alternative approach is to write

$$\frac{1}{\mu(x)} = \beta_0 + \beta_1 x$$

where

$$x = 1/z$$
$$\beta_0 = 1/\alpha_0$$
$$\beta_1 = \alpha_1/\alpha_0,$$

which is a GLM with reciprocal link.

6.11 Identifiability

For many nonlinear models, *identifiability* is an issue, by which we mean that the same curve may be obtained with different sets of parameter values. We have already seen one example of this for the nonlinear model fitted to the theophylline data (Sect. 6.2). As a second example, consider the sum-of-exponentials model

$$\mu(x, \boldsymbol{\beta}) = \beta_0 \exp(-x\beta_1) + \beta_2 \exp(-x\beta_3), \qquad (6.39)$$

where $\boldsymbol{\beta} = [\beta_0, \beta_1, \beta_2, \beta_3]$ and $\beta_j > 0$, $j = 0, 1, 2, 3$. The same curve results under the parameter sets $[\beta_0, \beta_1, \beta_2, \beta_3]$ and $[\beta_2, \beta_3, \beta_0, \beta_1]$, and so we have non-identifiability. In the previous "flip-flop" model (Sect. 6.2), identifiability could be imposed through a substantive assumption such as $k_a > k_e > 0$, and for model (6.39), we may enforce (say) $\beta_3 > \beta_1 > 0$ and work with the set

$$\boldsymbol{\gamma} = [\log \beta_0, \log(\beta_3 - \beta_1), \log \beta_2, \log \beta_1]$$

which constrains $\beta_0 > 0$, $\beta_2 > 0$, and $\beta_1 > \beta_3 > 0$. If a Bayesian approach is followed, a second possibility is to retain the original parameter set, but assign one set of curves zero mass in the prior. The latter option is less appealing since it can lead to a discontinuity in the prior.

6.12 Likelihood Inference for Nonlinear Models

6.12.1 *Estimation*

To obtain the likelihood function, a probability model for the data must be fully specified. A common choice is

$$Y_i \mid \boldsymbol{\beta}, \sigma \sim_{ind} \mathrm{N}[\mu_i(\boldsymbol{\beta}), \sigma^2 \mu_i(\boldsymbol{\beta})^r],$$

for $i = 1, \ldots, n$, and with $r = 0, 1$, or 2 being common choices. The corresponding likelihood function is

$$l(\boldsymbol{\beta}, \sigma) = -n \log \sigma - \frac{r}{2} \sum_{i=1}^{n} \log \mu_i(\boldsymbol{\beta}) - \frac{1}{2\sigma^2} \sum_{i=1}^{n} \frac{[Y_i - \mu_i(\boldsymbol{\beta})]^2}{\mu_i^r(\boldsymbol{\beta})}. \tag{6.40}$$

Differentiation with respect to $\boldsymbol{\beta}$ and σ yields, with a little rearrangement, the score equations

$$\boldsymbol{S}_1(\boldsymbol{\beta}, \sigma) = \frac{\partial l}{\partial \boldsymbol{\beta}}$$

$$= \frac{r}{2\sigma^2} \sum_{i=1}^{n} \frac{\partial \mu_i}{\partial \boldsymbol{\beta}} \frac{1}{\mu_i(\boldsymbol{\beta})} \left\{ \frac{[Y_i - \mu_i(\boldsymbol{\beta})]^2}{\mu_i^r(\boldsymbol{\beta})} - \sigma^2 \right\} + \frac{1}{\sigma^2} \sum_{i=1}^{n} \frac{[Y_i - \mu_i(\boldsymbol{\beta})]}{\mu_i(\boldsymbol{\beta})^r} \frac{\partial \mu_i}{\partial \boldsymbol{\beta}} \tag{6.41}$$

$$S_2(\boldsymbol{\beta}, \sigma) = \frac{\partial l}{\partial \sigma}$$

$$= -\frac{n}{\sigma} + \frac{1}{\sigma^3} \sum_{i=1}^{n} \frac{[Y_i - \mu_i(\boldsymbol{\beta})]^2}{\mu_i^r(\boldsymbol{\beta})}.$$

Notice that this pair of quadratic estimating functions (Sect. 2.8) are such that $\mathrm{E}[\boldsymbol{S}_1] = 0$ and $\mathrm{E}[S_2] = 0$ if the first two moments are correctly specified, in which case consistency of $\boldsymbol{\beta}$ results. It is important to emphasize that if $r > 0$, we require the second moment to be correctly specified in order to produce a consistent estimator of $\boldsymbol{\beta}$. If $r = 0$, the first term of (6.41) disappears, and we require the first moment only for consistency. In general, the MLEs $\widehat{\boldsymbol{\beta}}$ are not available in closed form, but numerical solutions are usually straightforward (e.g., via Gauss–Newton methods or variants thereof) and are available in most statistical software. The MLE for σ^2 is

$$\widehat{\sigma}^2 = \frac{1}{n} \sum_{i=1}^{n} \frac{[Y_i - \mu_i(\widehat{\boldsymbol{\beta}})]^2}{\mu_i^r(\widehat{\boldsymbol{\beta}})}, \tag{6.42}$$

but, by analogy with the linear model case, it is more usual to use the degrees of freedom adjusted estimator

$$\tilde{\sigma}^2 = \frac{1}{n-k-1} \sum_{i=1}^n \frac{[Y_i - \mu_i(\widehat{\beta})]^2}{\mu_i^r(\widehat{\beta})}. \tag{6.43}$$

For a nonlinear model, $\tilde{\sigma}^2$ has finite sample bias but is often preferred to (6.42) because of better small sample performance.

Under the usual regularity conditions,

$$I(\theta)^{1/2}(\widehat{\theta}_n - \theta) \to_d N_{k+1}(0, I_{k+1}).$$

where $\theta = [\beta, \sigma]$ and $I(\theta)$ is Fisher's expected information. In the case of $r = 0$, we obtain

$$l(\beta, \sigma) = -n \log \sigma - \frac{1}{2\sigma^2} \sum_{i=1}^n [Y_i - \mu_i(\beta)]^2$$

$$S_1(\beta, \sigma) = \frac{1}{\sigma^2} \sum_{i=1}^n [Y_i - \mu_i(\beta)] \frac{\partial \mu_i}{\partial \beta} \tag{6.44}$$

$$S_2(\beta, \sigma) = -\frac{n}{\sigma} + \frac{1}{\sigma^3} \sum_{i=1}^n [Y_i - \mu_i(\beta)]^2$$

$$I_{11} = -E\left[\frac{\partial S_1}{\partial \beta}\right] = \frac{1}{\sigma^2} \sum_{i=1}^n \left(\frac{\partial \mu_i}{\partial \beta}\right)^T \left(\frac{\partial \mu_i}{\partial \beta}\right)$$

$$I_{12} = -E\left[\frac{\partial S_1}{\partial \sigma}\right] = 0$$

$$I_{21} = -E\left[\frac{\partial S_2}{\partial \beta}\right] = 0^T$$

$$I_{22} = -E\left[\frac{\partial S_2}{\partial \sigma}\right] = \frac{2n}{\sigma^2}.$$

Asymptotically,

$$\frac{\sum_{i=1}^n [Y_i - \mu(\widehat{\beta})]^2}{\sigma^2} \to_d \chi^2_{n-k-1} \tag{6.45}$$

which may be used to construct approximate F tests, as described in Sect. 6.12.2. If r is unknown, then it may also be estimated by deriving the score from the likelihood (6.40), though an abundance of data will be required. Estimation of the power in a related variance model is carried out in the example at the end of Sect. 9.20.

Example: Pharmacokinetics of Theophylline

We let y_i represent the log concentration and assume the model $y_i \mid \boldsymbol{\beta}, \sigma^2 \sim_{ind}$ $N[\mu_i(\boldsymbol{\beta}), \sigma^2]$, $i = 1, \ldots, n$, where

$$\mu_i(\boldsymbol{\beta}) = \log\left\{\frac{Dk_a}{V(k_a - k_e)}\left[\exp(-k_e x) - \exp(-k_a x)\right]\right\} \quad (6.46)$$

with $\boldsymbol{\beta} = [\beta_0, \beta_1, \beta_2]$ and $\beta_0 = V$, $\beta_1 = k_a$, $\beta_2 = k_e$. We fit this model using maximum likelihood estimation for $\boldsymbol{\beta}$ and the moment estimator (6.43) for σ^2. The results are displayed in Table 6.2, with the fitted curve displayed on Fig. 6.1. Confidence intervals, based on the asymptotic distribution of the MLE, were calculated for the parameters of interest using the delta method. These parameters are all positive, and so the intervals were obtained on the log-transformed scale and then exponentiated.

In Fig. 6.10, slices through the three-dimensional likelihood surface are displayed. The two-dimensional surfaces are evaluated at the MLE of the third variable. A computationally expensive alternative would be to profile with respect to the third parameter, as described in Sect. 2.4.2. In the left column the range of each variable is taken as three times the asymptotic standard errors, and the surfaces are very well behaved. By contrast, in the right column of the figure, the range is ± 30 standard errors, and here we see very irregular shapes, with some of the contours remaining open. Such shapes are typical when nonlinear models are fitted and are not in general only apparent at points far from the maximum of the likelihood.

6.12.2 Hypothesis Testing

As usual, hypothesis tests may be carried out using Wald, score, or likelihood ratio statistics, and again we concentrate on the latter. Suppose that $\dim(\boldsymbol{\beta}) = k + 1$ and let $\boldsymbol{\beta} = [\boldsymbol{\beta}_1, \boldsymbol{\beta}_2]$ be a partition with $\boldsymbol{\beta}_1 = [\beta_0, \ldots, \beta_q]$ and $\boldsymbol{\beta}_2 = [\beta_{q+1}, \ldots, \beta_k]$, with $0 \leq q < k$. Interest focuses on testing whether a subset of $k - q$ parameters are equal to zero via a test of the null

$$H_0 : \boldsymbol{\beta}_1 \text{ unrestricted}, \; \boldsymbol{\beta}_2 = \boldsymbol{\beta}_{20} \quad \text{versus} \quad H_1 : \boldsymbol{\beta} = [\boldsymbol{\beta}_1, \boldsymbol{\beta}_2] \neq [\boldsymbol{\beta}_1, \boldsymbol{\beta}_{20}].$$

Asymptotically, and with known σ,

$$2\left[l(\widehat{\boldsymbol{\beta}}^{(1)}, \sigma^2) - l(\widehat{\boldsymbol{\beta}}^{(0)}, \sigma^2)\right] \to_d \chi^2_{k-q-1}$$

where $\widehat{\boldsymbol{\beta}}^{(0)}$ and $\widehat{\boldsymbol{\beta}}^{(1)}$ are the MLEs under null and alternative, respectively, and $l(\boldsymbol{\beta}, \sigma^2)$ is given by (6.40). Unlike the normal linear model, this result is only

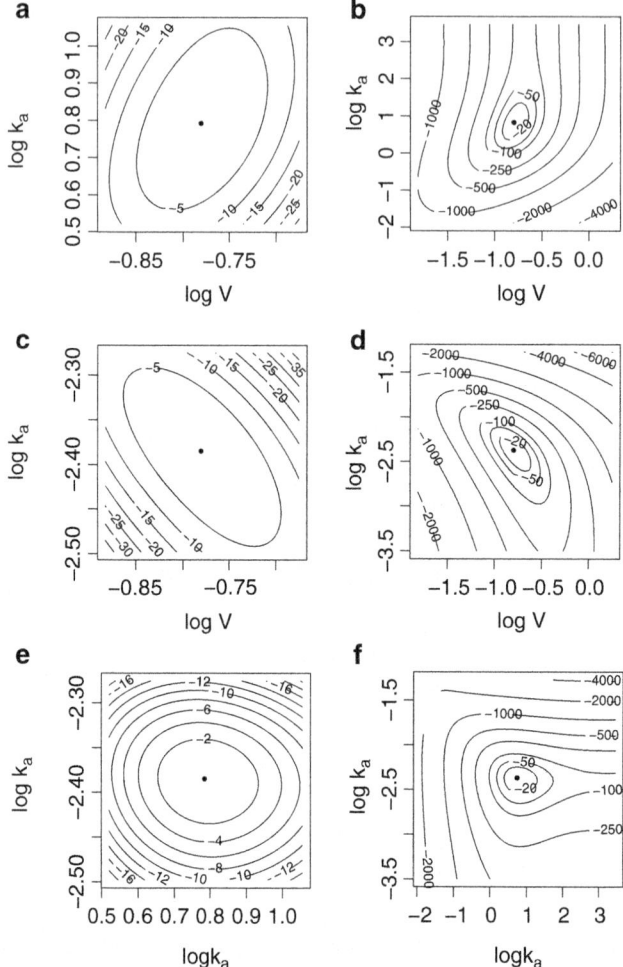

Fig. 6.10 Likelihood contours for the theophylline data with the range of each parameter being the MLE ± 3 standard errors in the *left column* and ± 30 standard errors in the *right column*; (**a**) and (**b**) $\log k_a$ versus $\log V$, (**c**) and (**d**) $\log k_e$ versus $\log V$, and (**e**) and (**f**) $\log k_e$ versus $\log k_a$. On each plot, the *filled circle* represents the MLE. In each panel, the third variable is held at its maximum value

asymptotically valid for a normal nonlinear model. For the usual case of unknown σ^2, one may substitute an estimate or use an F test with degrees of freedom $k-q-1$ and $n-k-1$, though the numerator and denominator sums of squares are only asymptotically independent. The denominator sum of squares is given in (6.45). More cautiously, one may assess the significance using Monte Carlo simulation under the null.

6.13 Least Squares Inference

We first consider model (6.38) with $E[\epsilon_i \mid \mu_i] = 0$, $\text{var}(\epsilon_i \mid \mu_i) = \sigma^2$, and $\text{cov}(\epsilon_i, \epsilon_j \mid \mu_i, \mu_j) = 0$. In this case we may obtain ordinary least squares estimates, $\widehat{\boldsymbol{\beta}}$, that minimize the sum of squares

$$\sum_{i=1}^{n}[Y_i - \mu_i(\boldsymbol{\beta})]^2 = [\boldsymbol{Y} - \boldsymbol{\mu}(\boldsymbol{\beta})]^{\text{T}}[\boldsymbol{Y} - \boldsymbol{\mu}(\boldsymbol{\beta})].$$

Differentiation with respect to $\boldsymbol{\beta}$, and letting $\boldsymbol{D}$ be the $n \times (k+1)$ dimensional matrix with element (i, j), $\partial \mu_i / \partial \beta_j$, yields the estimating function

$$\sum_{i=1}^{n}[Y_i - \mu_i(\boldsymbol{\beta})]\frac{\partial \mu_i}{\partial \boldsymbol{\beta}} = \boldsymbol{D}^{\text{T}}(\boldsymbol{Y} - \boldsymbol{\mu})$$

which is identical to (6.44) and is optimal within the class of linear estimating functions, under correct specification of the first two moments.

If we now assume uncorrelated errors with $\text{var}(\epsilon_i \mid \mu_i) = \sigma^2 \mu_i^r(\boldsymbol{\beta})$, then the method of *generalized least squares* estimates $\widehat{\boldsymbol{\beta}}$ by temporarily forgetting that the variance depends on $\boldsymbol{\beta}$. This is entirely analogous to the motivation for quasi-likelihood; see the discussion centered around (2.28) in Sect. 2.5.1. We therefore minimize

$$\sum_{i=1}^{n}\frac{[Y_i - \mu_i(\boldsymbol{\beta})]^2}{\mu_i^r(\boldsymbol{\beta})} = [\boldsymbol{Y} - \boldsymbol{\mu}(\boldsymbol{\beta})]^{\text{T}}\boldsymbol{V}(\boldsymbol{\beta})^{-1}[\boldsymbol{Y} - \boldsymbol{\mu}(\boldsymbol{\beta})],$$

where $\boldsymbol{V}$ is the $n \times n$ diagonal matrix with diagonal elements $\mu_i^r(\boldsymbol{\beta})$, $i = 1, \ldots, n$. The estimating function is

$$\sum_{i=1}^{n}\frac{[Y_i - \mu_i(\boldsymbol{\beta})]^2}{\mu_i^r(\boldsymbol{\beta})}\frac{\partial \mu_i}{\partial \boldsymbol{\beta}} = \boldsymbol{D}^{\text{T}}\boldsymbol{V}^{-1}(\boldsymbol{Y} - \boldsymbol{\mu}),$$

which is identical to that under quasi-likelihood (6.10). Inference may be based on the asymptotic result

$$(\boldsymbol{D}^{\text{T}}\boldsymbol{V}^{-1}\boldsymbol{D}/\sigma^2)^{1/2}(\widehat{\boldsymbol{\beta}}_n - \boldsymbol{\beta}) \to_d N_{k+1}(\boldsymbol{0}, \boldsymbol{I}_{k+1}). \qquad (6.47)$$

If the normal model is true, then the GLS estimator is not as efficient as that obtained from a likelihood approach but is more reliable under model misspecification. Therefore, the approach that is followed should depend on how much faith we have in the assumed model.

In Sect. 9.10, we will discuss further the trade-offs encountered when one wishes to exploit the additional information concerning β contained within the variance function.

6.14 Sandwich Estimation for Nonlinear Models

The sandwich estimator of the variance is again available and takes exactly the same form as with the GLM. In particular, consider the estimating function

$$G(\beta) = D^{\mathrm{T}} V^{-1}(Y - \mu),$$

with D an $n \times (k+1)$ matrix with elements $\partial \mu_i / \partial \beta_j$, $i = 1, \ldots, n$, $j = 0, \ldots, k+1$ and V the diagonal matrix with elements $V_{ii} = \mu_i(\beta)^r$ with $r \geq 0$ known. This estimating equation arises from likelihood considerations if $r = 0$ or, more generally, from GLS. With this form for $G(\cdot)$, (6.30), (6.31), and (6.32) all hold.

Example: Pharmacokinetics of Theophylline

We now let y_i be the concentration and consider the model with first two moments

$$\mathrm{E}[Y_i \mid \beta, \sigma^2] = \mu_i(\beta) = \frac{Dk_a}{V(k_a - k_e)} \left[\exp(-k_e x) - \exp(-k_a x) \right],$$

$$\mathrm{var}(Y_i \mid \beta, \sigma^2) = \sigma^2 \mu_i(\beta)^2,$$

for $i = 1, \ldots, n$. One possibility for fitting is generalized least squares. As an alternative, we may assume $Y_i \mid \beta, \sigma^2 \sim_{ind} \mathrm{N}[\mu_i(\beta), \sigma^2 \mu_i(\beta)^2]$, $i = 1, \ldots, n$ and proceed with maximum likelihood estimation. Table 6.3 gives estimates of the above model under GLS and MLE, along with likelihood estimation for the model,

$$\log y_i \mid \beta, \tau^2 \sim_{ind} \mathrm{N} \left\{ \log[\mu_i(\beta)], \tau^2 \right\}.$$

There are some differences in the table, but overall the estimates and standard errors are in reasonable agreement. Table 6.2 gives confidence intervals for $x_{1/2}$, $x_{\max}$, and $\mu(x_{\max})$ based on sandwich estimation. As with the GLM analysis, the interval estimates are a little shorter.

Table 6.3 Point estimates and asymptotic standard errors for the theophylline data, under various models and estimation techniques. In all cases the coefficient of variation is approximately constant

Model	log V	log k_a	log k_e
MLE log scale	−0.78 (0.035)	0.79 (0.089)	−2.39 (0.037)
GLS original scale	−0.77 (0.030)	0.81 (0.055)	−2.39 (0.032)
MLE original scale	−0.74 (0.025)	0.85 (0.069)	−2.45 (0.044)

6.15 The Geometry of Least Squares

In this section we briefly discuss the geometry of least squares to gain insight into the fundamental differences between linear and nonlinear fitting.

We consider minimization of

$$(y - \mu)^T (y - \mu) \tag{6.48}$$

where y and μ are $n \times 1$ vectors. We first examine the linear model, $\mu = x\beta$, where x is $n \times (k+1)$ and β is $(k+1) \times 1$. For fixed x, the so-called *solution locus* maps out the fitted values $x\widetilde{\beta}$ for all values of $\widetilde{\beta}$ and is a $(k+1)$-dimensional hyperplane of infinite extent. Differentiation of (6.48) gives

$$x^T(y - x\widehat{\beta}) = x^T e = 0$$

where $\widehat{\beta} = (x^T x)^{-1} x^T y$ and e is the $n \times 1$ vector of residuals. So the sum of squares is minimized when the vector $(y - x\beta)$ is orthogonal to the hyperplane that constitutes the solution locus. The fitted values are

$$\widehat{y} = x\widehat{\beta} = x(x^T x)^{-1} x^T y = hy,$$

and are the orthogonal projection of y onto the plane spanned by the columns of x, with h the matrix that represents this projection.

For a nonlinear model, the solution locus is a curved $(k+1)$-dimensional surface, possibly with finite extent. In contrast to the linear model, equally spaced points on lines in the parameter space do not map to equally spaced points on the solution locus but rather to unequally spaced points on curves.

These observations have several implications. In terms of inference, recall from Sect. 5.6.1, in particular equation (5.27) with $q = -1$, that for a linear model, a $100(1 - \alpha)\%$ confidence interval for β is the ellipsoid

$$(\beta - \widehat{\beta})^T x^T x (\beta - \widehat{\beta}) \leq (k+1) s^2 F_{k+1, n-k-1}(1 - \alpha).$$

Geometrically, the region has this form because the solution locus is a plane and the residual vector is orthogonal to the plane so that values of β map onto a disk. For nonlinear models, asymptotic inference for β results from

$$(\boldsymbol{\beta} - \widehat{\boldsymbol{\beta}})^{\mathrm{T}} \widehat{\boldsymbol{V}}^{-1} (\boldsymbol{\beta} - \widehat{\boldsymbol{\beta}}) \leq (k+1) s^2 F_{k+1, n-k-1}(1-\alpha),$$

where $\widehat{\mathrm{var}}(\widehat{\boldsymbol{\beta}}) = \widehat{\sigma}^2 \widehat{\boldsymbol{V}}$, with $\widehat{\sigma}^2 = s^2$. The approximation occurs because the solution locus is curved, and equi-spaced points in the parameter space map to unequally spaced points on curved lines on the solution locus. Intuitively, inference will be more accurate if the relevant part of the solution locus is flat and if parallel equi-spaced lines in the parameter space map to parallel equi-spaced lines on the solution locus. The curvature and lack of equally spaced points manifest itself in contours of equal likelihood being banana-shaped and perhaps "open" (so that they do not join). The right column of Fig. 6.10 gives examples of this behavior. Another important aspect is that reparameterization of the model can alter the behavior of points mapped onto the solution locus, but cannot affect the curvature of the locus. Hence, the curvature of the solution locus has been referred to as the *intrinsic curvature* (Beale 1960; Bates and Watts 1980), while the aspect that is parameterization dependent is the *parameter-effects curvature* (Bates and Watts 1980). We note that the solution locus does not depend on the observed data but only on the model and design. As $n \to \infty$, the surface becomes increasingly locally linear and inference correspondingly more accurate.

We illustrate with a simple fictitious example with $n = 2$, $\boldsymbol{x} = [1, 2]$, and $\boldsymbol{y} = [0.2, 0.7]$. We compare two models, each with a single parameter, the linear zero intercept model

$$\mu = x\beta, \qquad -\infty < \beta < \infty,$$

and the (simplified) nonlinear Michaelis–Menten model

$$\mu = x/(x + \theta), \qquad \theta > 0.$$

Figure 6.11(a) plots the data versus the two fitted curves (obtained via least squares), while panel (b) plots the solution locus for the linear model, which in this case is a line (since $k = 0$). The point $[x_1 \widehat{\beta}, x_2 \widehat{\beta}]$ with least squares estimate

$$\widehat{\beta} = \sum_{i=1}^{2} x_i y_i / \sum_{i=1}^{2} x_i^2 = 0.32,$$

is the fitted point and is indicated as a solid circle. The dashed line is the vector joining $[y_1, y_2]$ to the fitted point and is perpendicular to the curved solution locus. The circles indicated on the solution locus correspond to changes in β of 0.1 and are equi-spaced on the locus. The final aspect to note is that the locus is of infinite extent.

Panel (c) of Fig. 6.11 plots the solution locus for the Michaelis–Menten model, for which $\widehat{\theta} = 1.70$. The vector joining $[y_1, y_2]$ to the fitted values $[x_1/(x_1 + \widehat{\theta}), x_2/(x_2 + \widehat{\theta})]$ is perpendicular to the curved solution locus, but we see that points on the

6.15 The Geometry of Least Squares

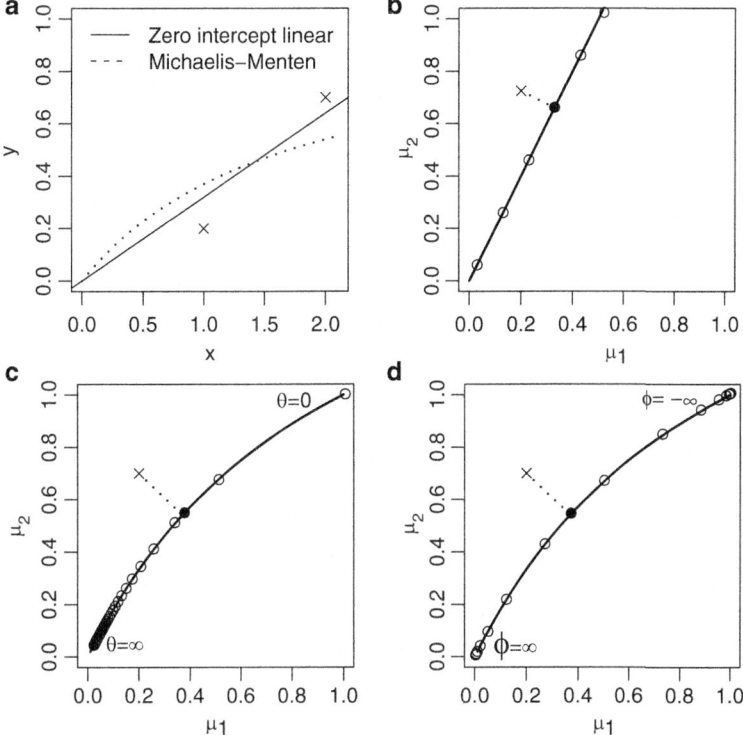

Fig. 6.11 (a) Fictitious data with $x = [1, 2]$ and $y = [0.2, 0.7]$, and fitted lines (b) solution locus for the zero intercept linear model with the observed data indicated as a cross and the fitted value as a *filled circle*, (c) solution locus for the Michaelis–Menten model with the observed data indicated as a *cross* and the fitted value as a *filled circle*, and (d) solution locus for the Michaelis–Menten model under a second parametrization with the observed data indicated as a *cross* and the fitted value as a *filled circle*

latter are not equally spaced. Also, the solution locus is of finite extent moving from the point $[0, 0]$ for $\theta = \infty$ to the point $(1,1)$ for $\theta = 0$ (these are the asymptotes of the model). Finally, panel (d) reproduces panel (c) with the Michaelis–Menten model reparameterized as $\left[x_1/[x_1 + \exp(\widehat{\phi})], x_2/[x_2 + \exp(\widehat{\phi})]\right]$, with $\phi = \log \theta$. The spacing of points on the solution locus is quite different under the new parameterization. The points are more equally spaced close to the fitted value, indicating that asymptotic standard errors are more likely to be accurate under this parametrization.

6.16 Bayesian Inference for Nonlinear Models

Bayesian inference for nonlinear models is based on the posterior distribution

$$p(\beta, \sigma^2 \mid y) \propto l(\beta)\pi(\beta, \sigma^2).$$

We discuss in turn prior specification, computation, and hypothesis testing.

6.16.1 Prior Specification

We begin by assuming independent priors on β and σ^2:

$$\pi(\beta, \sigma^2) = \pi(\beta)\pi(\sigma^2).$$

The prior on σ^2 is a less critical choice, and $\sigma^{-2} \sim \text{Ga}(a, b)$ is an obvious candidate. The choice $a = b = 0$, which gives the improper prior $\pi(\sigma^2) \propto 1/\sigma^2$, will often be a reasonable option. If the variance model is of the form $\text{var}(Y_i) = \sigma^2 \mu_i(\beta)^r$, then clearly substantive prior beliefs will depend on r so that we must specify the conditional form $\pi(\sigma^2 \mid r)$, since the scale of σ^2 depends on the choice for r.

So far as a prior for β is concerned, great care must be taken to ensure that the resultant posterior is proper; Sect. 3.4 provided an example of the problems that can arise with a nonlinear model. In general, models must be considered on a case-by-case basis. However, a parameter, θ (say), corresponding to an asymptote (so that $\mu \to a$ as $\theta \to \infty$), will generally require proper priors because the likelihood tends to the constant

$$\exp\left[-\frac{1}{2\sigma^2}\sum_{i=1}^{n}(y_i - a)^2\right]$$

as $\theta \to \infty$ and not zero as is necessary to ensure propriety.

6.16.2 Computation

Unfortunately, closed-form posterior distributions do not exist with a nonlinear mean function, but sampling-based methods are again relatively straightforward to implement. A pure Gibbs sampling strategy (Sect. 3.8.4) is not so appealing since the conditional distribution, $\beta \mid y, \sigma$, will not have a familiar form. However, Metropolis–Hastings algorithms (Sect. 3.8.2) will be easy to construct. If an informative prior is present, direct sampling via a rejection algorithm, with the prior as a proposal, may present a viable option.

6.16.3 Hypothesis Testing

As with GLMs (Sect. 6.8.3), posterior tail areas and Bayes factors are available to test hypotheses/compare models.

Example: Pharmacokinetics of Theophylline

We report a Bayesian analysis of the theophylline data and specify lognormal priors for $x_{1/2}$, $x_{\max}$, and $\mu(x_{\max})$ using the same specification as with the GLM analysis. Samples from the posterior for $[V, k_a, k_e]$ are obtained from the rejection algorithm. Specifically, we sample from the prior on the parameters of interest and then backsolve for the parameters that describe the likelihood. For the compartmental model, we transform back to the original parameters via

$$k_e = (\log 2)/x_{1/2}$$

$$0 = x_{\max}(k_a - k_e) - \log\left(\frac{k_a}{k_e}\right) \qquad (6.49)$$

$$V = \frac{D}{\mu(x_{\max})}\left(\frac{k_a}{k_e}\right)^{k_a/(k_a - k_e)}$$

so that k_a is not directly available but must be obtained as the root of (6.49).

Table 6.2 summarizes inference for the parameters of interest with the interval estimates and medians being obtained as the sample quantiles. Figure 6.12 shows the posteriors for functions of interest under the nonlinear model. The posteriors are skewed for all functions of interest. These figures and Table 6.2 show that Bayesian inference for the GLM and nonlinear model are very similar. Frequentist and Bayesian methods are also in close agreement for these data, which is reassuring.

Recall that the parameter sets $[V, k_a, k_e]$ and $[Vk_e/k_a, k_e, k_a]$ produce identical curves for the compartmental model (6.1). One solution to this identifiability problem is to enforce $k_a > k_e > 0$, for example, by parameterizing in terms of $\log k_e$ and $\log(k_a - k_e)$. Pragmatically, not resorting to this parameterization is reasonable, so long as k_a and k_e are not close. Figure 6.13 shows the bivariate posterior distribution $p(k_a, k_e \mid y)$, and we see that $k_a \gg k_e$ for these data, and so there is no need to address the identifiability issue.

Another benefit of specifying the prior in terms of model-free parameters is that models may be compared using Bayes factors on an "even playing field," in the sense that the prior input for each model is identical. For more discussion of this issue, see Pérez and Berger (2002). To illustrate, we compare the GLM and nonlinear compartmental models. The normalizing constants for these models are 0.00077 and 0.00032, respectively, as estimated via importance sampling with the prior as

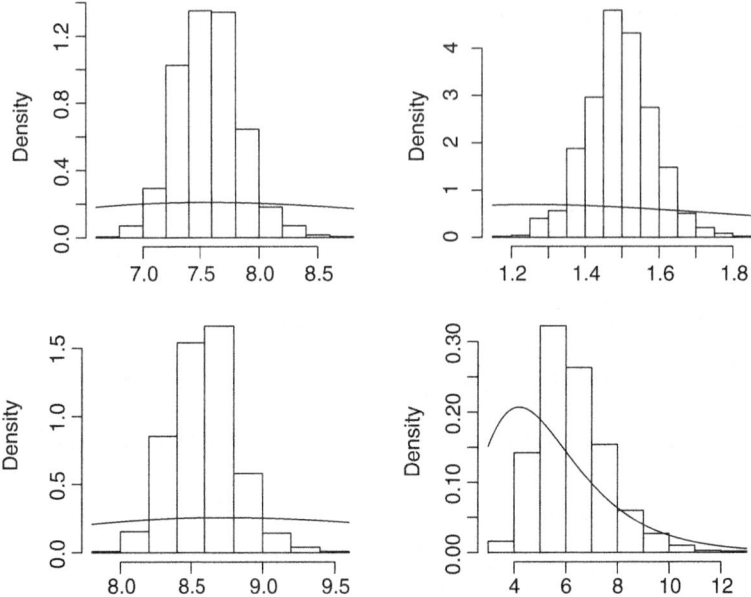

Fig. 6.12 Histogram representations of posterior distributions from the nonlinear compartmental model for the theophylline data for the (**a**) half-life, (**b**) time to maximum, (**c**) maximum concentration, and (**d**) coefficient of variation, with priors superimposed as *solid lines*

Fig. 6.13 Image plot of samples from the joint posterior distribution of the absorption and elimination rate constants, k_a and k_e, for the theophylline data

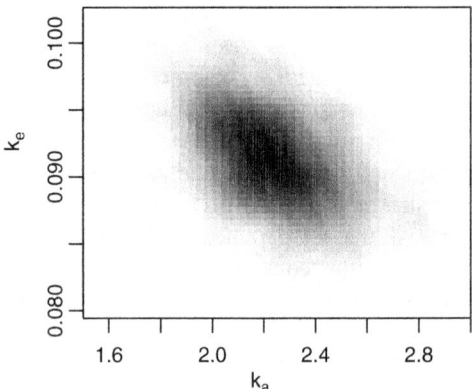

proposal and using (3.28). Consequently, the Bayes factor comparing the GLM to the nonlinear model is 2.4 so that the data are just over twice as likely under the GLM, but this is not strong evidence. For these data, based on the above analyses, we

6.16 Bayesian Inference for Nonlinear Models

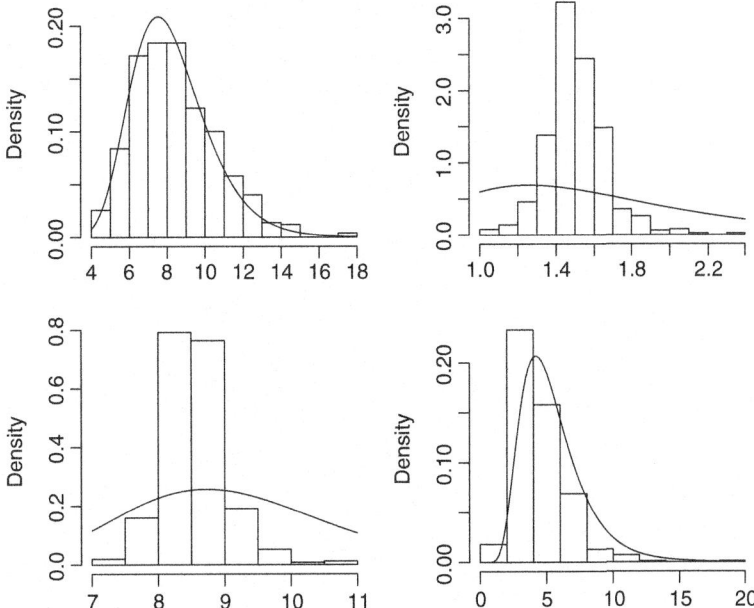

Fig. 6.14 Histogram representations of posterior distributions from the nonlinear compartmental models for the reduced theophylline dataset of $n = 3$ points for the (**a**) half-life, (**b**) time to maximum, (**c**) maximum concentration, and (**d**) coefficient of variation, with priors superimposed as *solid lines*

conclude that both the GLM and the nonlinear models provide adequate fits to the data, and there is little difference between the frequentist and Bayesian approaches to inference.

We now demonstrate the benefits of a Bayesian approach with substantive prior information, when the data are sparse. To this end, we consider a reduced dataset consisting of the first $n = 3$ concentrations only. Clearly, a likelihood or least squares approach is not possible in this case, since the number of parameters (three regression parameters plus a variance) is greater than the number of data points. We fit the nonlinear model with the same priors as used previously and with computation carried out with the rejection algorithm. Figure 6.14 shows the posterior distributions, with the priors also indicated. As we might expect, there is no/little information in the data concerning the terminal half-life $\log k_e/2$ or the standard deviation σ. In contrast, the data are somewhat informative with respect to the time to maximum concentration, and the maximum concentration.

6.17 Assessment of Assumptions for Nonlinear Models

In contrast to GLMs, residuals are unambiguously defined for nonlinear models as

$$e_i^\star = \frac{y_i - \widehat{\mu}_i}{\sqrt{\widehat{\mathrm{var}}(Y_i)}}, \qquad (6.50)$$

which we refer to as Pearson residuals. These residuals may be used in the usual ways; see Sects. 5.11.3 and 6.9. In particular, the residuals may be plotted versus covariates to assess the mean model, and the absolute values of the residuals may be plotted versus the fitted values $\widehat{\mu}_i$ to assess the appropriateness of the mean–variance model. For a small sample size, normality of the errors will aid in accurate asymptotic inference and may be assessed via a normal QQ plot, as described in Sect. 5.11.3.

Example: Pharmacokinetics of Theophylline

Letting y_i represent the log concentration at time x_i, we examine the Pearson residuals, as given by (6.50), obtained following likelihood estimation with the model $y_i \mid \boldsymbol{\beta}, \sigma^2 \sim_{ind} N(\mu_i, \sigma^2)$, with μ_i given by (6.46), for $i = 1, \ldots, n$. Figure 6.6(c) plots $e_i^\star$ versus x_i and shows no gross inadequacy of the mean model. Panel (d), which plots $|e_i^\star|$ versus x_i, similarly shows no great problem with the mean–variance relationship. Figure 6.15 gives a normal QQ plot of the residuals and indicates no strong violation of normality. In all cases, interpretation is hampered by the small sample size.

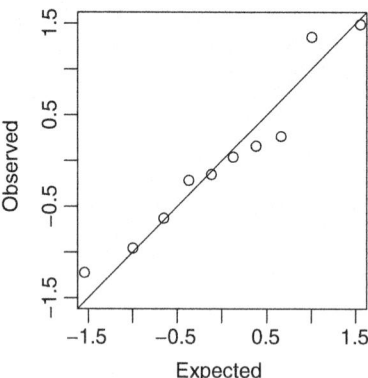

Fig. 6.15 Normal QQ plot for the theophylline data and model (6.46)

6.18 Concluding Remarks

Within the broad class of general regression models, the use of GLMs offers certain advantages in terms of computation and interpretation, though one should not restrict attention to this class. Many results and approaches used for linear models hold approximately for GLMs. For example, the influence of points was defined through the weight matrix used in the "working response" approach implicit in the IRLS algorithm (Sect. 6.5.2). The form of GLMs, in particular the linearity of the score with respect to the responses, is such that asymptotic inference is accurate for relatively small n.

Care is required in the fitting of, and inference for, nonlinear models. For example, models must be examined to see if the parameters are uniquely identified. For both GLMs and nonlinear models, the examination of residual plots is essential to determine whether the assumed model is appropriate, but such plots are difficult to interpret because the behavior of residuals is not always obvious, even if the fitted model is correct. The use of a distribution from the exponential family is advantageous in that results on consistency of estimators follow easily, as discussed in Sect. 6.5.1. The identifiability of nonlinear models should always be examined, and one should be wary of the accuracy of asymptotic inference for small sample sizes. The parameterization adopted is also important, as discussed in Sect. 6.15.

6.19 Bibliographic Notes

The most comprehensive and interesting description of GLMs remains McCullagh and Nelder (1989). An excellent review is also given by Firth (1993). Sandwich estimation for GLMs is discussed by Kauermann and Carroll (2001).

Nonlinear models are discussed by Bates and Watts (1988) and Chap. 2 of Davidian and Giltinan (1995), with an emphasis on generalized least squares. Book-length treatments on nonlinear models are provided by Gallant (1987); Seber and Wild (1989); see also Carroll and Ruppert (1988).

Gibaldi and Perrier (1982) provide a comprehensive account of pharmacokinetic models and principles and Godfrey (1983) an account of compartmental modeling in general. Wakefield et al. (1999) provide a review of pharmacokinetic and pharmacodynamic modeling including details on both the biological and statistical aspects of such modeling. The model given by (6.7) and (6.8) was suggested by Wakefield (2004) and was developed more extensively in Salway and Wakefield (2008).

6.20 Exercises

6.1 A random variable Y is inverse Gaussian if its density is of the form

$$p(y \mid \lambda, \delta) = \left(\frac{\delta}{2\pi y^3}\right)^{1/2} \exp\left[\frac{-\delta(y-\lambda)^2}{2\lambda^2 y}\right],$$

for $y > 0$.

(a) Show that the inverse Gaussian distribution is a member of the exponential family and identify θ, α, $b(\theta)$, $a(\alpha)$, and $c(y,\alpha)$.
(b) Give forms for $E[Y \mid \theta,\alpha]$ and $\text{var}(Y \mid \theta,\alpha)$ and determine the canonical link function.

6.2 Table 6.4 reproduces data, from Altham (1991), of counts of T_4 cells/mm^3 in blood samples from 20 patients in remission from Hodgkin's disease and from 20 additional patients in remission from disseminated malignancies. A question of interest here is whether there is a difference in the distribution of cell counts between the two diseases. A quantitative assessment of any difference is also desirable.

(a) Carry out an exploratory examination of these data and provide an informative graphical summary of the two distributions of responses.
(b) These data may be examined: (1) on their original scale, (2) $\log_e$ transformed, and (3) square root transformed. Carefully define a difference in location parameter in each of the designated scales. What are the considerations when choosing a scale? Obtain 90% confidence interval for each of the difference parameters.
(c) Fit Poisson, gamma, and inverse Gaussian models to the cell count data, assuming canonical links in each case.
(d) Using the asymptotic distribution of the MLE, give 90% confidence intervals for the difference parameters in each of the three models. Under each of the models, would you conclude that the means of the two groups are equal?

6.3 The data in Table 6.5, taken from Wakefield et al. (1994), were collected following the administration of a single 30 mg dose of the drug cadralazine

Table 6.4 Counts of T_4 cells/mm^3 in blood samples from 20 patients in remission from Hodgkin's disease and 20 other patients in remission from disseminated malignancies

Hodgkin's disease	396	568	1,212	171	554	1,104	257	435	295	397
Non-Hodgkin's disease	375	375	752	208	151	116	736	192	315	1,252
Hodgkin's disease	288	1,004	431	795	1,621	1,378	902	958	1,283	2,415
Non-Hodgkin's disease	675	700	440	771	688	426	410	979	377	503

6.20 Exercises

Table 6.5 Concentrations y_i of the drug cadralazine as a function of time x_i, obtained from a subject who was administered a dose of 30 mg. These data are from Wakefield et al. (1994)

Observation number i	Time (hours) x_i	Concentration (mg/liter) y_i
1	2	1.63
2	4	1.01
3	6	0.73
4	8	0.55
5	10	0.41
6	24	0.01
7	28	0.06
8	32	0.02

to a cardiac failure patient. The response y_i represents the drug concentration at time x_i, $i = 1, \ldots, 8$. The most straightforward model for these data is to assume

$$\log y_i = \mu(\boldsymbol{\beta}) + \epsilon_i = \log\left[\frac{D}{V}\exp(-k_e x_i)\right] + \epsilon_i,$$

where $\epsilon_i \mid \sigma^2 \sim_{iid} N(0, \sigma^2)$, $\boldsymbol{\beta} = [V, k_e]$ and the dose is $D = 30$. The parameters are the volume of distribution $V > 0$ and the elimination rate k_e.

(a) For this model, obtain expressions for:

   (i) The log-likelihood function $L(\boldsymbol{\beta}, \sigma^2)$
   (ii) The score function $\boldsymbol{S}(\boldsymbol{\beta}, \sigma^2)$
   (iii) The expected information matrix $\boldsymbol{I}(\boldsymbol{\beta}, \sigma^2)$

(b) Obtain the MLE and provide an asymptotic 95% confidence interval for each element of $\boldsymbol{\beta}$.
(c) Plot the data, along with the fitted curve.
(d) Using residuals, examine the appropriateness of the assumptions of the above model. Does the model seem reasonable for these data?
(e) The clearance $Cl = V \times k_e$ and elimination half-life $x_{1/2} = \log 2 / k_e$ are parameters of interest in this experiment. Find the MLEs of these parameters along with asymptotic 95% confidence intervals.

A Bayesian analysis will now be carried out, assuming independent lognormal priors for V, k_e and an independent inverse gamma prior for σ^2. For the latter, assume the improper prior $\pi(\sigma^2) \propto \sigma^{-2}$.

(f) Assume that the 50% and 90% points for V are 20 and 40 and that for k_e, these points are 0.12 and 0.25. Solve for the lognormal parameters using the method of moments equations (6.36).
(g) Implement an MCMC Metropolis–Hastings algorithm (Sect. 3.8.2). Report the median and 90% interval estimates for each of V, k_e, Cl, and $x_{1/2}$. Pro-

vide graphical summaries of each of the univariate and bivariate posterior distributions.

6.4 Let Y_i represent a count and $\boldsymbol{x}_i = [x_{i1}, \ldots, x_{ik}]$ a covariate vector for individual i, $i = 1, \ldots, n$. Assume that $Y_i \mid \mu_i \sim_{iid} \text{Poisson}(\mu_i)$, with

$$\mu_i = \text{E}[Y_i \mid \gamma_{0i}, \gamma_1, \ldots, \gamma_k] = \exp(\gamma_{0i} + \gamma_1 x_{i1} + \ldots + \gamma_k x_{ik}), \quad (6.51)$$

where the intercept is a *random effect* (see Chap. 9) that varies according to

$$\gamma_{0i} \mid \gamma_0, \tau^2 \sim_{iid} \text{N}(\gamma_0, \tau^2).$$

(a) Give an interpretation of each of the parameters γ_0 and γ_1.
(b) Suppose we fit an alternative Poisson model with mean

$$\mu_i^\star = \text{E}[Y_i \mid \beta_0, \beta_1, \ldots, \beta_k] = \exp(\beta_0 + \beta_1 x_{i1} + \ldots + \beta_k x_{ik}). \quad (6.52)$$

Evaluate
$$\text{E}[Y_i \mid \tau^2, \gamma_0, \gamma_1, \ldots, \gamma_k],$$

and hence, by comparison with $\text{E}[Y_i \mid \beta_0, \beta_1, \ldots, \beta_k]$, equate γ_j to β_j, $j = 0, 1, \ldots, k$.
(c) Evaluate $\text{var}(Y_i \mid \tau^2, \gamma_0, \gamma_1, \ldots, \gamma_k)$ and compare this expression with $\text{var}(Y_i \mid \beta_0, \beta_1, \ldots, \beta_k)$.
(d) Suppose one is interested in the parameters $\gamma_1, \ldots, \gamma_k$. Use your answers to the previous two parts to discuss the implications of fitting model (6.52) when the true model is (6.51).
(e) Now consider an alternative random effects structure in which

$$\delta_i \mid a, b \sim_{iid} \text{Ga}(a, b),$$

where $\delta_i = \exp(\gamma_{0i})$. Evaluate the marginal mean $\text{E}[Y_i \mid a, b, \gamma_1, \ldots, \gamma_k]$ and marginal variance $\text{var}(Y_i \mid a, b, \gamma_1, \ldots, \gamma_k)$.
(f) Compare the expressions for the mean and variance under the normal and gamma formulations.
(g) For the Poisson-Gamma model, calculate the form of the likelihood

$$L(\gamma_1, \ldots, \gamma_k, a, b) = \prod_{i=1}^{n} \int \text{Pr}(y_i \mid \gamma_{0i}, \gamma_1, \ldots, \gamma_k) \pi(\gamma_{0i} \mid a, b) \, d\gamma_{0i}.$$

Derive expressions for the score and information matrix and hence describe how inference may be performed from a likelihood standpoint.

6.20 Exercises

Table 6.6 Concentrations y_i of the drug theophylline as a function of time x_i obtained from a subject who was administered an oral dose of size 4.40 mg/kg

Observation number i	Time (hours) x_i	Concentration (mg/liter) y_i
1	0.27	1.72
2	0.52	7.91
3	1.00	8.31
4	1.92	8.33
5	3.50	6.85
6	5.02	6.08
7	7.03	5.40
8	9.00	4.55
9	12.00	3.01
10	24.30	0.90

6.5 Table 6.6 gives concentration–time data for an individual who was given a dose of 4.40 mg/kg of the drug theophylline. In this chapter we have analyzed the data from another of the individuals in the same trial.

(a) For the data in Table 6.6,[1] fit the gamma GLM given by (6.7) and (6.8) using maximum likelihood and report the MLEs and standard errors.

(b) Obtain MLEs and standard errors for the parameters of interest $x_{1/2}$, $x_{\max}$, $\mu(x_{\max})$, and Cl.

(c) Let z_i represent the log concentration and consider the model $z_i \mid \beta, \sigma^2 \sim_{ind} \text{N}[\mu_i(\beta), \sigma^2]$, $i = 1, \ldots, n$, where $\mu_i(\beta)$ is given by the compartmental model (6.46). Fit this model using maximum likelihood and report the MLEs and standard errors.

(d) Obtain the MLEs and standard errors for the parameters of interest $x_{1/2}$, $x_{\max}$, $\mu(x_{\max})$, and Cl.

(e) Compare these summaries with those obtained under the GLM.

(f) Examine the fit of the two models and discuss which provides the better fit.

---

[1] These data correspond to individual 2 in the Theoph data, which are available in R.

Chapter 7
Binary Data Models

7.1 Introduction

In this chapter we consider the modeling of binary data. Such data are ubiquitous in many fields. Binary data present a number of distinct challenges, and so we devote a separate chapter to their modeling, though we lean heavily on the methods introduced in Chap. 6 on general regression modeling. It is perhaps surprising that the simplest form of outcome can pose difficulties in analysis, but a major problem is the lack of information contained within a variable that can take one of only two values. This can lead to a number of problems, for example, in assessing model fit. Another major complication arises because models for probabilities are generally nonlinear, which can lead to curious behavior of estimators in the presence of confounders. Difficulties in interpretation also arise, even when independent regressors are added to the model.

The outline of this chapter is as follows. We give some motivating examples in Sect. 7.2, and in Sect. 7.3, describe the genesis of the binomial model, which is a natural candidate for the analysis of binary data. Generalized linear models for binary data are examined in Sect. 7.4. The binomial model has a variance determined by the mean, with no additional parameter to accommodate excess-binomial variation, and so Sect. 7.5 describes methods for dealing with such variation. For reasons that will become apparent, we will focus on logistic regression models, beginning with a detailed description in Sect. 7.6. This section includes discussions of estimation from likelihood, quasi-likelihood, and Bayesian perspectives. Conditional likelihood and "exact" inference are the subject of Sect. 7.7. Assessing the adequacy of binary models is discussed in Sect. 7.8. Summary measures that exhibit nonobvious behavior are the subject of Sect. 7.9. Case-control studies are a common design, which offer interesting inferential challenges with respect to inference, and are described in Sect. 7.10. Concluding comments appear in Sect. 7.11. Section 7.12 gives references to more in-depth treatments of binary modeling and to source materials.

7.2 Motivating Examples

7.2.1 Outcome After Head Injury

We will illustrate methods for binary data using the data first encountered in Sect. 1.3.2. The binary response is outcome after head injury (dead/alive), with four discrete covariates: pupils (good/poor), coma score (depth of coma, low/high), hematoma present (no/yes), and age (categorized as 1–25, 26–54, ≥55). These data were presented in Table 1.1, but it is difficult to discern patterns from this table. In general, cross-classified data such as these may be explored by looking at marginal and conditional tables of counts or frequencies. Figure 7.1 displays conditional frequencies, with panel (a) corresponding to low coma score and panel (b) to high coma score. These plots suggest that the probability of death increases with age, that a low coma score is preferable to a high coma score, and that good pupils are beneficial. The association with the hematoma variable is less clear. The sample sizes are lost in these plots, which makes interpretation more difficult.

7.2.2 Aircraft Fasteners

Montgomery and Peck (1982) describe a study in which the compressive strength of fasteners used in the construction of aircraft was examined. Table 7.1 gives the total number of fasteners tested and the number of failures at a range of pressure loads. We see that the proportion failing increases with load. For these data we will aim to find a curve to adequately model the relationship between the probability of fastener failure and load pressure.

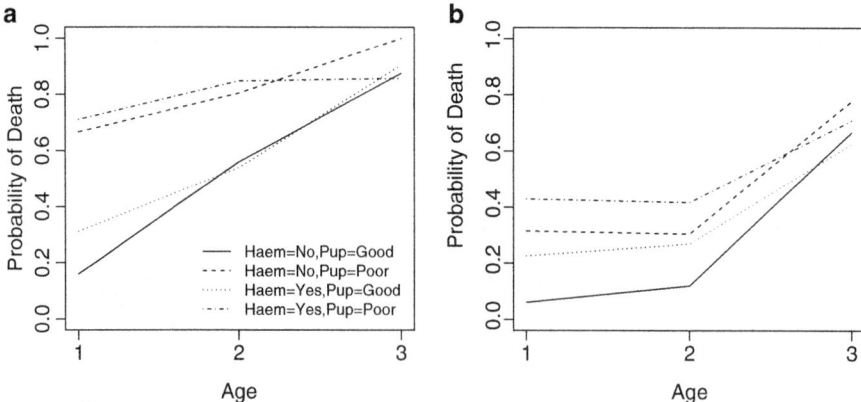

Fig. 7.1 Probability of death after head injury as a function of age, hematoma score, and pupils: Panels (**a**) and (**b**) are for low and high coma scores, respectively

Table 7.1 Number of aircraft fastener failures at specified pressure loads

Load (psi)	Failures	Sample size	Proportion failing
2,500	10	50	0.20
2,700	17	70	0.24
2,900	30	100	0.30
3,100	21	60	0.35
3,300	18	40	0.45
3,500	43	85	0.51
3,700	54	90	0.60
3,900	33	50	0.66
4,100	60	80	0.75
4,300	51	65	0.78

7.2.3 Bronchopulmonary Dysplasia

We describe data from van Marter et al. (1990) and subsequently analyzed by Pagano and Gauvreau (1993) on the absence/presence of bronchopulmonary dysplasia (BPD) as a function of birth weight (in grams) for $n = 223$ babies. BPD is a chronic lung disease that affects premature babies. In this study, BPD was defined as a function of both oxygen requirement and compatible chest radiograph, with 147 of the babies having neither characteristic by day 28 of life. We take as illustrative aim the prediction of BDP using birth weight, the rationale being that if a good predictive model can be found, then measures could be taken to decrease the probability of BPD. There are a number of caveats that should be attached to this analysis. First, these data are far from a random sample of births, as they are sampled from intubated infants with weights less than 1,751 g (so that all of the babies are of low birth weight). In general, an estimate of the incidence of BPD is difficult to tie down, in part, because of changes in the definition of the condition. Allen et al. (2003) provide a discussion of this issue and report that, of preterm infants with birth weights less than 1,000 g, 30% develop BPD. Second, a number of additional covariates would be available in a serious attempt at prediction, including gender and the medication used by the mothers.

Figure 7.2 displays the BPD indicator, plotted as short vertical lines at 0 and 1, as a function of birth weight. Visual assessment suggests that children with lower birth weight tend to have an increased chance of BPD. It is hard to discern the shape of the association from the raw binary data alone, however, since one is trying to compare the distributions of zeros and ones, which is difficult. This example is distinct from the aircraft fasteners because the latter contained multiple responses at each x value. Binning on the basis of birthweight and plotting the proportions with BPD in each bin would provide a more informative plot.

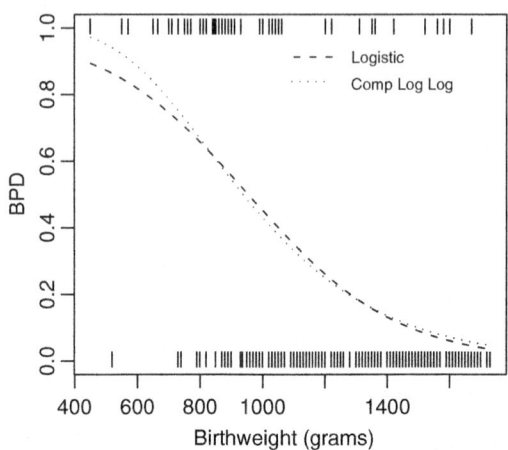

Fig. 7.2 Indicator of bronchopulmonary dysplasia (BPD), as a function of birth weight. The *short vertical lines* at 0 and 1 indicate the observed birth weights for non-BPD and BPD infants, respectively. The *dashed curve* corresponds to a logistic regression fit and the *dotted curve* to a complementary log–log regression fit

7.3 The Binomial Distribution

7.3.1 Genesis

In the following we will refer to the basic sampling unit as an individual. Let Z denote the *Bernoulli* random variable with

$$\Pr(Z = z \mid p) = p^z (1-p)^{1-z},$$

$z = 0, 1$, and

$$p = \Pr(Z = 1 \mid p),$$

for $0 < p < 1$. For concreteness, we will call the $Z = 1$ outcome a positive response. A random variable taking two values *must* have a Bernoulli distribution, and all moments are determined as functions of p. In particular, $\text{var}(Z \mid p) = p(1-p)$ so that there is no concept of underdispersion or overdispersion for a Bernoulli random variable.

Suppose there are N individuals, and let Z_j denote the outcome for the jth individual, $j = 1, \ldots, N$. Also let $Y = \sum_{j=1}^{N} Z_j$ be the total number of individuals with a positive outcome, and suppose that each has equal probabilities, that is, $p = p_1 = \ldots = p_N$. Under the assumption that the Bernoulli random variables are *independent*,

$$Y \mid p \sim \text{Binomial}(N, p)$$

so that

$$\Pr(Y = y \mid p) = \binom{N}{y} p^y (1-p)^{1-y}, \tag{7.1}$$

for $y = 0, 1, \ldots, N$.

7.3 The Binomial Distribution

Constant $p = p_j$, $j = 1, \ldots, N$, over the N individuals is not necessary for Y to follow a binomial distribution. Suppose that individual j has probability p_j drawn at random from a distribution with mean $\bar{p}$. In this case,

$$E[Z_j] = E[E(Z_j \mid p_j)] = \bar{p}$$

and

$$Y \mid \bar{p} \sim \text{Binomial}(N, \bar{p}). \tag{7.2}$$

Crucial to this derivation is the assumption that p_j are *independent* draws from the distribution with mean $\bar{p}$, which means that the Z_j are also independent for $j = 1, \ldots, N$. Alternative scenarios are described in the context of overdispersion in Sect. 7.5.

We give a second derivation of the binomial distribution. Suppose $Y_j \mid \lambda_j \sim_{ind}$ Poisson(λ_j), $j = 1, 2$ are independent Poisson random variables with rates λ_j. Then,

$$Y_1 \mid Y_1 + Y_2, p \sim \text{Binomial}(Y_1 + Y_2, p),$$

with $p = \lambda_1/(\lambda_1 + \lambda_2)$ (Exercise 7.3).

7.3.2 Rare Events

Suppose that $Y \mid p \sim \text{Binomial}(N, p)$ and that $p \to 0$ and $N \to \infty$, with $\lambda = Np$ fixed (or tending to a constant). Then Exercise 7.1 shows that, in the limit, $Y \mid \lambda \sim \text{Poisson}(\lambda)$. Approximating the binomial distribution with a Poisson has a number of advantages. Computationally, the Poisson model can be more stable than the binomial model. Also, $\lambda > 0$ can be modeled via a loglinear form which provides a more straightforward interpretation than the logistic form, $\log[p/(1-p)]$. The following example illustrates one use of this result for obtaining a closed-form distribution when counts are summed.

Example: Lung Cancer and Radon

In Sect. 1.3.3 we introduced the lung cancer dataset, with Y_i being the number of cases in area i. A possible model for these data is

$$Y_i \mid \theta_i \sim \text{Poisson}(E_i \theta_i), \tag{7.3}$$

where E_i is the expected number of cases based on the age and gender breakdown of area i and θ_i is the relative risk associated with the area, for $i = 1, \ldots, n$.

A formal derivation of this model is as follows (see Sect. 6.5 for a related discussion). Let Y_{ij} be the disease counts in area i and age-gender stratum j and

N_{ij} the associated population, $i = 1, \ldots, n, j = 1, \ldots, J$. In the Minnesota study, we have $J = 36$, corresponding to male/female and 18 age bands: 0–4, 5–9,..., 80–84, 85+. We only have access to the total counts in the area, Y_i, and so we require a model for this sum. One potential model is $Y_{ij} \mid p_{ij} \sim \text{Binomial}(N_{ij}, p_{ij})$, with p_{ij} the probability of lung cancer diagnosis in area i, stratum j. With binomial Y_{ij}, the distribution of $Y_i = \sum_{j=1}^{J} Y_{ij}$ is a convolution, which is unfortunately awkward to work with. For example, for $J = 2$,

$$\Pr(y_i \mid p_{i1}, p_{i2})$$
$$= \sum_{y_{i1}=l_i}^{u_i} \binom{N_{i1}}{y_{i1}} \binom{N_{i2}}{y_i - y_{i1}} p_{i1}^{y_{i1}} (1-p_{i1})^{N_{i1}-y_{i1}} p_{i2}^{y_i - y_{i1}} (1-p_{i2})^{N_{i2}-y_i+y_{i1}}$$

where $l_i = \max(0, y_i - N_{i2})$, $u_i = \min(N_{i1}, y_i)$, gives the range of admissible values that y_{i1} can take, given the margins $Y_i, N_i - Y_{i1} - Y_{i2}, N_{i1}, N_{i2}$. Lung cancer is statistically rare, and so we can use the Poisson approximation to give $Y_{ij} \mid p_{ij} \sim \text{Poisson}(N_{ij} p_{ij})$. The distribution of the sum Y_i is then straightforward:

$$Y_i \mid p_{i1}, \ldots, p_{iJ} \sim \text{Poisson}\left(\sum_{j=1}^{J} N_{ij} p_{ij}\right). \tag{7.4}$$

There are insufficient data to estimate the $n \times J$ probabilities p_{ij}, and so it is common to assume $p_{ij} = \theta_i \times q_j$, where q_j are a set of known reference stratum-specific rates and θ_i is an area-specific term that summarizes the deviation of the risks in area i from the reference rates. Therefore, this model assumes that the effect on risk of being in area i is the same across stratum. Usually, the q_j are assumed known. Consequently, (7.4) simplifies to $Y_i \mid \theta_i \sim \text{Poisson}\left(\theta_i \sum_{j=1}^{J} N_{ij} q_j\right)$, and substituting the expected numbers $E_i = \sum_{j=1}^{J} N_{ij} q_j$ produces model (7.3).

7.4 Generalized Linear Models for Binary Data

7.4.1 Formulation

Let $Z_{ij} = 0/1$ denote the absence/presence of the binary characteristic of interest in each of the $j = 1, \ldots, N_i$ trials, with $i = 1, \ldots, n$ different "conditions." Let $Y_i = \sum_{j=1}^{N_i} Z_{ij}$ denote the number of positive responses and $N = \sum_{i=1}^{n} N_i$ the total number of trials. Further, suppose there are k explanatory variables recorded for each condition, and let $\boldsymbol{x}_i = [1, x_{i1}, \ldots, x_{ik}]$ denote the row vector of dimension $1 \times (k+1)$ for $i = 1, \ldots, n$. We now wish to model the probability of a positive response $p(\boldsymbol{x}_i)$, as a function of $\boldsymbol{x}_i$, in order to identify structure within the data.

7.4 Generalized Linear Models for Binary Data

We might naively model the observed proportion via the linear model

$$\frac{Y_i}{N_i} = \boldsymbol{x}_i \boldsymbol{\beta} + \epsilon_i,$$

for $i = 1, \ldots, n$. There are a number of difficulties with such an approach. First, the observed proportions must lie in the range $[0, 1]$, while the modeled probability $\boldsymbol{x}_i \boldsymbol{\beta}$ is unrestricted. We could attempt to put constraints on the parameters in order to alleviate this drawback, but this is inelegant and soon becomes cumbersome with multiple explanatory variables. The resultant inference is also difficult due to the restricted ranges. The second difficulty is that we saw in Sect. 5.6.4 that in the usual linear model framework, an appropriate mean–variance model is crucial for well-calibrated inference (unless sandwich estimation is turned to). A linear model is usually associated with error terms with constant variance, but this is not appropriate here since

$$\mathrm{var}\left(\frac{Y_i}{N_i}\right) = \frac{p(\boldsymbol{x}_i)[1 - p(\boldsymbol{x}_i)]}{N_i}$$

so that the variance changes with the mean. The generalized linear model, introduced and discussed in Sect. 6.3, can rectify these deficiencies. For sums of binary variables, the binomial model is a good starting point.

The binomial model is a member of the exponential family, specifically $Y \mid p \sim \mathrm{Binomial}(N, p)$, that is, (7.1), translates to

$$p(y \mid p) = \exp\left[y \log\left(\frac{p}{1-p}\right) + N \log(1-p)\right], \tag{7.5}$$

which provides the stochastic element of the model. For the deterministic part, we specify a monotonic, differentiable link function:

$$g[p(\boldsymbol{x})] = \boldsymbol{x}\boldsymbol{\beta}. \tag{7.6}$$

The exponential family is appealing from a statistical standpoint since correct specification of the mean function leads to consistent inference, since the score function is linear in the data (this function is given for the logistic model in (7.12)). With a GLM, the computation is also usually straightforward (Sect. 6.5.2). Non-linear models can also be considered, however, if warranted by the application. For example, Diggle and Rowlingson (1994) considered modeling disease risk as a function of distance x from a point source of pollution. These authors desired a model for which disease risk returned to baseline as $x \to \infty$ and suggested a model for the odds of the form

$$\frac{\Pr(Z=1 \mid x)}{\Pr(Z=0 \mid x)} = \beta_0 \left[1 + \beta_1 \exp(-\beta_2 x^2)\right],$$

with β_0 corresponding to baseline odds, β_1 corresponding to the excess odds at $x = 0$ (i.e., at the point source), and β_2 determining the speed at which the odds decline to baseline. Such nonlinear models are computationally more difficult to fit but produce consistent parameter estimates, if combined with an exponential family.

7.4.2 Link Functions

From (7.5) we see that the so-called *canonical* link is the logit $\theta = \log[p/(1-p)]$. We will see that *logistic regression models* of the form

$$\log\left[\frac{p(\boldsymbol{x})}{1 - p(\boldsymbol{x})}\right] = \boldsymbol{x}\boldsymbol{\beta} \tag{7.7}$$

offer a number of advantages in terms of computation and inference. This link function is by far the most popular in practice, and so Sect. 7.6 is dedicated to logistic regression modeling.

Other link functions that may be used for binomial data include the *probit*, *complimentary log–log*, and *log–log* links. The probit link is

$$\Phi^{-1}[p(\boldsymbol{x})] = \boldsymbol{x}\boldsymbol{\beta},$$

where $\Phi[\cdot]$ is the distribution function of a standard normal random variable. This link function generally produces similar inference to the logistic link function. The logistic and probit link functions are symmetric in the sense that $g(p) = -g(1-p)$.

The complementary log–log link function is

$$\log\{-\log[1 - p(\boldsymbol{x})]\} = \boldsymbol{x}\boldsymbol{\beta}, \tag{7.8}$$

to give

$$p(\boldsymbol{x}) = 1 - \exp[-\exp(\boldsymbol{x}\boldsymbol{\beta})],$$

which is not symmetric. Hence, the log–log link model

$$-\log\{-\log[p(\boldsymbol{x})]\} = \boldsymbol{x}\boldsymbol{\beta}$$

with

$$p(\boldsymbol{x}) = \exp[-\exp(-\boldsymbol{x}\boldsymbol{\beta})]$$

may also be used and will not produce the same inference as (7.8). If $g_{\text{CLL}}(\cdot)$ and $g_{\text{LL}}(\cdot)$ represent the complementary log–log and log–log links, respectively, then the two are related via $g_{\text{CLL}}(p) = -g_{\text{LL}}(1-p)$.

7.5 Overdispersion

Overdispersion is a phenomena that occurs frequently in applications and, in the binomial data context, describes a situation in which the variance $\text{var}(Y_i \mid p_i)$ exceeds the binomial variance $N_i p_i (1 - p_i)$.

Often overdispersion occurs due to clustering in the population from which the individuals were drawn. To motivate a variance model, suppose for simplicity that the N_i individuals for whom we measure outcomes in trial i are actually broken into C_i clusters of size k_i so that $N_i = C_i \times k_i$. These clusters may correspond to families, geographical areas, genetic subgroups, etc. Within the cth cluster, the number of positive responders Y_{ic} has distribution $Y_{ic} \mid p_{ic} \sim_{ind} \text{binomial}(k_i, p_{ic})$, where each p_{ic} is drawn independently from some distribution, for $c = 1, \ldots, C_i$. Let P_{ic} represent a random variable with

$$E[P_{ic}] = p_i$$

$$\text{var}(P_{ic}) = \tau_i^2 p_i (1 - p_i),$$

where the variance is written in this form for convenience (as we see shortly). In the following we will use expressions for iterated expectation, variance, and covariance, as described in Appendix B. Then, letting $Y_i = \sum_{c=1}^{C_i} Y_{ic}$,

$$E[Y_i] = E\left[\sum_{c=1}^{C_i} Y_{ic}\right] = \sum_{c=1}^{C_i} E_{P_{ic}}\left[E(Y_{ic} \mid P_{ic})\right] = \sum_{c=1}^{C_i} E_{P_{ic}}[k_i P_{ic}] = N_i p_i.$$

Turning to the variance,

$$\text{var}(Y_i) = \text{var}\left(\sum_{c=1}^{C_i} Y_{ic}\right) = \sum_{c=1}^{C_i} \text{var}(Y_{ic}),$$

since the counts are independent, as each p_{ic} is drawn independently. Continuing with this calculation and exploiting the iterated variance formula,

$$\text{var}(Y_i) = \sum_{c=1}^{C_i} \{\text{E}\left[\text{var}(Y_{ic} \mid p_{ic})\right] + \text{var}\left(\text{E}[Y_{ic} \mid p_{ic}]\right)\}$$

$$= \sum_{c=1}^{C_i} \{\text{E}_{P_{ic}}[k_i P_{ic}(1 - P_{ic})] + \text{var}_{P_{ic}}(k_i P_{ic})\}$$

$$= \sum_{c=1}^{C_i} \{k_i p_i - k_i \left[\text{var}(P_{ic}) + \text{E}[P_{ic}]^2\right] + k_i^2 \tau_i^2 p_i(1 - p_i)\}$$

$$= N_i p_i (1 - p_i) \times \left[1 + (k_i - 1)\tau_i^2\right]$$

$$= N_i p_i (1 - p_i) \sigma_i^2.$$

Hence, the within-trial clustering has induced excess-binomial variation. Suppose each cluster is of size $k_i = 1$ (i.e., $C_i = N_i$); then we recover the binomial case (7.2). The above derivation requires $1 \leq \sigma_i^2 \leq k_i \leq N_i$, since $0 \leq \sigma_i^2 \leq 1$ (McCullagh and Nelder 1989, Sect. 4.5.1). If we were to assume a second moment model with a common $\sigma_i^2 = \sigma^2$ to give

$$\text{var}(Y_i) = N_i p_i (1 - p_i) \sigma^2 \tag{7.9}$$

then the constraint becomes $\sigma^2 \leq N_i$, which is unfavorable, but will rarely be a problem in practice.

If we have a single cluster, that is, $C_i = 1$, then $k_i = N_i$ and

$$\text{var}(Y_i) = N_i p_i (1 - p_i) \times \left[1 + (N_i - 1)\tau_i^2\right]. \tag{7.10}$$

Suppose $Z_{ij}, j = 1, \ldots, N_i$ are the binary outcomes within-trial i so that $Y_i = \sum_{j=1}^{N_i} Z_{ij}$. Then, for the case of a single cluster ($C_i = 1$),

$$\text{cov}(Z_{ij}, Z_{ik}) = \text{E}[\text{cov}(Z_{ij}, Z_{ik} \mid p_{i1})] + \text{cov}(\text{E}[Z_{ij} \mid p_{ij}], \text{E}[Z_{ik} \mid p_{ik}])$$

$$= \text{cov}_{P_{i1}}(P_{i1}, P_{i1})$$

$$= \text{var}(P_{i1}) = \tau_i^2 p_i (1 - p_i),$$

so that τ_i^2 is the correlation between any two outcomes in trial i.

We now discuss a closely related scenario in which we start by assuming that outcomes within a trial have correlation τ_i^2. Then (Exercise 7.4),

$$\text{var}(Y_i) = N_i p_i (1 - p_i) \times \left[1 + (N_i - 1)\tau_i^2\right]. \tag{7.11}$$

Notice that, unlike the derivation leading to (7.10), underdispersion can occur if $\tau_i^2 < 0$. The equality of (7.10) and (7.11) shows that the effect of either a random response probability or positively correlated outcomes within a trial is

7.5 Overdispersion

indistinguishable marginally (unless one is willing to make assumptions about the within-trial distribution, but such assumptions are uncheckable).

Inferentially, two approaches are suggested. We could specify the first two moments only and use quasi-likelihood. This route is taken in Sect. 7.6.3. Alternatively, one can assume a specific distributional form and then proceed with parametric inference, as we now illustrate.

The most straightforward way to model overdispersion parametrically is to assume the binomial probability arises from a conjugate beta model. This model is

$$Y_i \mid q_i \sim \text{Binomial}(N_i, q_i)$$

$$q_i \sim \text{Beta}(a_i, b_i),$$

where we can parameterize as $a_i = dp_i$, $b_i = d(1 - p_i)$ so that

$$p_i = \frac{a_i}{d}$$

$$\text{var}(p_i) = \frac{p_i(1 - p_i)}{d + 1}.$$

An obvious choice of mean model is the linear logistic model

$$p_i = \frac{\exp(\boldsymbol{x}_i \boldsymbol{\beta})}{1 + \exp(\boldsymbol{x}_i \boldsymbol{\beta})}.$$

Notice that $d = 0$ corresponds to the binomial model. Integration over the random effects results in the beta-binomial marginal model:

$$\Pr(Y_i=y_i) = \binom{N_i}{y_i} \frac{\Gamma(a_i + b_i)}{\Gamma(a_i)\Gamma(b_i)} \frac{\Gamma(a_i + y_i)\Gamma(b_i + N_i - y_i)}{\Gamma(a_i + b_i + N_i)}, \quad y_i = 0, 1, \ldots, N_i.$$

The marginal moments are

$$\text{E}[Y_i] = N_i p_i = N_i \left(\frac{a_i}{a_i + b_i}\right)$$

$$\text{var}(Y_i) = N_i p_i (1 - p_i) \left(\frac{a_i + b_i + N_i}{a_i + b_i + 1}\right),$$

confirming that there is no overdispersion when $N_i = 1$. This variance is also equal to (7.10), with the assumption of constant τ_i^2 on recognizing that $\tau^2 = (a_i + b_i + 1)^{-1} = 1/(d+1)$. Unfortunately, the log-likelihood $l(\boldsymbol{\beta}, d)$ is not easy to deal with due to the gamma functions. More seriously, the beta-binomial distribution is not of exponential family form and does not possess the consistency properties of distributions within this family.

Liang and McCullagh (1993) discuss the modeling of overdispersed binary data. In particular, they suggest plotting residuals

$$\frac{y_i - N_i \widehat{p}_i}{\sqrt{N_i \widehat{p}_i (1 - \widehat{p}_i)}}$$

against N_i in order to see whether there is any association, which may help to choose between models (7.9) and (7.10).

7.6 Logistic Regression Models

7.6.1 Parameter Interpretation

We write the probability of $Y = 1$ as $p(\boldsymbol{x})$ to emphasize the dependence on covariates $\boldsymbol{x}$. Model (7.7) is equivalent to saying that the *odds* of a positive outcome may be modeled in a multiplicative fashion, that is,

$$\frac{p(\boldsymbol{x})}{1 - p(\boldsymbol{x})} = \exp(\boldsymbol{x}\boldsymbol{\beta}) = \exp(\beta_0) \prod_{j=1}^{k} \exp(x_j \beta_j).$$

Less intuition is evident on the probability scale for which

$$p(\boldsymbol{x}) = \frac{\exp(\boldsymbol{x}\boldsymbol{\beta})}{1 + \exp(\boldsymbol{x}\boldsymbol{\beta})}.$$

The transformation used here is known as the expit transform (and is the inverse of the logit transform). The expression for the probability makes it clear that we have enforced $0 < p(\boldsymbol{x}) < 1$.

For clarity, we discuss interpretation in the situation in which $p(\boldsymbol{x})$ is the probability of a disease, given exposure $\boldsymbol{x}$. Consider first the logistic regression model in the case where the exposures have no effect on the probability of disease:

$$\log\left[\frac{p(\boldsymbol{x})}{1 - p(\boldsymbol{x})}\right] = \beta_0.$$

In this case, β_0 is the log odds of disease for all levels of the exposures $\boldsymbol{x}$. Equivalent statements are that $\exp(\beta_0)$ is the odds of disease and $\exp(\beta_0)/[1 + \exp(\beta_0)]$ is the probability of disease, regardless of the levels of $\boldsymbol{x}$.

Now consider the situation of a single exposure x for an individual with probability $p(x)$ and

$$\log\left[\frac{p(x)}{1 - p(x)}\right] = \beta_0 + \beta_1 x.$$

7.6 Logistic Regression Models

The parameter $\exp(\beta_0)$ is the odds of disease at exposure $x = 0$, that is, the odds for an unexposed individual. The parameter $\exp(\beta_1)$ is the odds ratio for a unit increase in x. For example, if $\exp(\beta_1) = 2$, the odds of disease double for a unit increase in exposure. If x is a binary exposure, coded as 0/1, then $\exp(\beta_1)$ is the ratio of odds when going from unexposed to exposed:

$$\frac{p(1)/[1-p(1)]}{p(0)/[1-p(0)]} = \frac{\exp(\beta_0 + \beta_1)}{\exp(\beta_0)} = \exp(\beta_1).$$

For a rare disease, the odds ratio and relative risk, which is given by $p(x)/p(x-1)$ for a univariate exposure, are approximately equal, with the relative risks being easier to interpret (see Sect. 7.10.2 for a more detailed discussion).

Logistic regression models may be defined for multiple factors and continuous variables in an exactly analogous fashion to the multiple linear models considered in Chap. 5. We simply include on the right-hand side of (7.6) the relevant design matrix and associated parameters. This is a benefit of the GLM framework in which we have linearity on some scale, though, with noncanonical link functions, parameter interpretation is usually more difficult.

The logistic model may be derived in terms of the so-called *tolerance distributions*. Let $U(x)$ denote an underlying continuous measure of the disease state at exposure x. We observe a binary version, $Y(x)$, of this variable which is related to $U(x)$ via

$$Y(x) = \begin{cases} 0 & \text{if } U(x) \le c \\ 1 & \text{if } U(x) > c, \end{cases}$$

for some threshold c. Suppose that the continuous measure follows a logistic distribution: $U(x) \sim \text{logistic}\,[\mu(x), 1]$. This distribution is given by

$$p(u \mid \mu, \sigma) = \frac{\exp\{(u-\mu)/\sigma\}}{\sigma\{1 + \exp[(u-\mu)/\sigma]\}^2}, \quad -\infty < u < \infty.$$

The logistic distribution function, for the case $\sigma = 1$, is

$$\Pr\,[U(x) < u] = \frac{\exp(u-\mu)}{1 + \exp(u-\mu)}, \quad -\infty < u < \infty.$$

From this model for $U(x)$, we can obtain the probability of the discrete outcome as

$$p(x) = \Pr\,[Y(x) = 1] = \Pr\,[U(x) > c] = \frac{\exp(\mu(x) - c)}{1 + \exp(\mu(x) - c)},$$

which is equivalent to

$$\log\left[\frac{p(x)}{1-p(x)}\right] = \mu(x) - c.$$

So far we have not specified how the exposure x changes the distribution of the continuous latent variable $U(x)$. We assume that the effect of exposure to x is to move the location of the underlying variable $U(x)$ in a linear fashion via $\mu(x) = a + bx$, but while keeping the variance constant. We then obtain

$$\log\left[\frac{p(x)}{1-p(x)}\right] = \beta_0 + \beta_1 x,$$

where $\beta_0 = a - c$ and $\beta_1 = b$, that is, a linear logistic regression model.

The probit and complementary log–log links may similarly be derived from normal and extreme-value[1] tolerance distributions, respectively.

7.6.2 Likelihood Inference for Logistic Regression Models

We consider the logistic regression model

$$\log\left[\frac{p_i(\boldsymbol{\beta})}{1-p_i(\boldsymbol{\beta})}\right] = \boldsymbol{x}_i\boldsymbol{\beta},$$

where $\boldsymbol{x}_i$ is a $1 \times (k+1)$ vector of covariates measured on the ith individual and $\boldsymbol{\beta}$ is the $(k+1) \times 1$ vector of associated parameters. We write $p_i(\boldsymbol{\beta})$ to emphasize that the probability of a positive response is a function of $\boldsymbol{\beta}$. For the general binomial model the log-likelihood is

$$l(\boldsymbol{\beta}) = \sum_{i=1}^n Y_i \log p_i(\boldsymbol{\beta}) + \sum_{i=1}^n (N_i - Y_i)\log\left[1 - p_i(\boldsymbol{\beta})\right],$$

with score function

$$\boldsymbol{S}(\boldsymbol{\beta}) = \sum_{i=1}^n \frac{\partial p_i(\boldsymbol{\beta})}{\partial \boldsymbol{\beta}} \frac{[Y_i - N_i p(\widehat{\boldsymbol{\beta}})]}{p(\boldsymbol{\beta})[1 - p(\widehat{\boldsymbol{\beta}})]}. \quad (7.12)$$

Letting $\boldsymbol{\mu}$ represent the $n \times 1$ vector with ith element $\mu_i = N_i p_i(\boldsymbol{\beta})$ allows (7.12) to be rewritten as

$$\boldsymbol{S}(\boldsymbol{\beta}) = \boldsymbol{D}^\mathrm{T}\boldsymbol{V}^{-1}\left[\boldsymbol{Y} - \boldsymbol{\mu}(\boldsymbol{\beta})\right], \quad (7.13)$$

where $\boldsymbol{D}$ is the $n \times (k+1)$ matrix with (i,j)th element $\partial \mu_i/\partial \beta_j$, $i = 1,\ldots,n$, $j = 0,\ldots,k$ and $\boldsymbol{V}$ is the $n \times n$ diagonal matrix with ith diagonal element $N_i p(\boldsymbol{x}_i)\left[1 - p(\boldsymbol{x}_i)\right]$. From Sect. 6.5.1,

---

[1] u has an extreme-value distribution if its distribution function is of the form $F(u) = 1 - \exp\{-\exp[(u-\mu)/\sigma]\}$.

7.6 Logistic Regression Models

$$I_n(\boldsymbol{\beta})^{1/2}(\widehat{\boldsymbol{\beta}}_n - \boldsymbol{\beta}) \to_d N_{k+1}(\mathbf{0}, \mathbf{I}_{k+1}),$$

where $I_n(\boldsymbol{\beta}) = \boldsymbol{D}^\mathsf{T} \boldsymbol{V}^{-1} \boldsymbol{D}$. For the logistic model,

$$\frac{\partial \mu_i}{\partial \beta_j} = x_{ij} N_i p_i (1 - p_i)$$

$$V_{ii} = N_i p_i (1 - p_i).$$

Consequently, the score takes a particularly simple form:

$$\boldsymbol{S}(\boldsymbol{\beta}) = \boldsymbol{x}^\mathsf{T} [\boldsymbol{Y} - \boldsymbol{\mu}(\boldsymbol{\beta})].$$

Hence, at the maximum, $\boldsymbol{x}^\mathsf{T}\boldsymbol{Y} = \boldsymbol{x}^\mathsf{T}\boldsymbol{\mu}(\widehat{\boldsymbol{\beta}})$ so that selected sums of the outcomes (as defined by the design matrix) are preserved. In addition, element (j, j') of $I_n(\boldsymbol{\beta})$ takes the form

$$\sum_{i=1}^n x_{ij} x_{ij'} N_i p_i (1 - p_i).$$

We now turn to hypothesis testing and consider a model with $0 < q \le k$ parameters and fitted probabilities $\widehat{\boldsymbol{p}}$. The log-likelihood is

$$l(\widehat{\boldsymbol{p}}) = \sum_{i=1}^n \left[y_i \log \widehat{p}_i + (N_i - y_i) \log(1 - \widehat{p}_i) \right],$$

with the maximum attainable value occurring at $\widetilde{p}_i = y_i / N_i$. The deviance is

$$D = 2\left[l(\widetilde{\boldsymbol{p}}) - l(\widehat{\boldsymbol{p}}) \right]$$

$$= 2 \sum_{i=1}^n \left[y_i \log\left(\frac{y_i}{\widehat{y}_i} \right) + (N_i - y_i) \log\left(\frac{N_i - y_i}{N_i - \widehat{y}_i} \right) \right], \qquad (7.14)$$

where $\widetilde{\boldsymbol{p}}$ is the vector of probabilities, $\widetilde{p}_i$, $i = 1, \ldots, n$. Notice that the deviance will be small when $\widehat{y}_i$ is close to y_i. The above form may also be derived directly from (6.22) under a binomial model. If n, the number of parameters in the saturated model (which, recall, is the number of conditions considered and not the total number of trials which is given by N), is fixed, then under the hypothesized model that produced $\widehat{\boldsymbol{p}}$, $D \to_d \chi^2_{n-q}$. The important emphasis here is on *fixed n*. The outcome after head injury dataset provides an example in which this assumption is valid since there are $n = 2 \times 2 \times 2 \times 3 = 24$ binomial trials being carried out at each combination of the levels of coma score, pupils, hematoma, and age.

When n is not fixed, the above result on the *absolute fit* is not relevant, but the *relative fit* may be assessed by comparing the difference in deviances. Specifically, consider nested models with q_j parameters under H_j, $j = 0, 1$. Further, the estimated probabilities and fitted values under hypothesis H_j will be denoted $\widehat{\boldsymbol{p}}_j$ and $\widehat{y}^{(j)}$, $j = 0, 1$, respectively. Then the reduction in deviance is

$$D_0 - D_1 = 2\{l(\widetilde{p}) - l(\widehat{p}_0) - [l(\widetilde{p}) - l(\widehat{p}_1)]\}$$
$$= 2[l(\widehat{p}_1) - l(\widehat{p}_0)]$$
$$= 2\sum_{i=1}^{n}\left[y_i \log\left(\frac{\widehat{y}_i^{(1)}}{\widehat{y}_i^{(0)}}\right) + (N_i - y_i)\log\left(\frac{N_i - \widehat{y}_i^{(1)}}{N_i - \widehat{y}_i^{(0)}}\right)\right].$$

Under H_0, $D_0 - D_1 \to_d \chi^2_{q_1 - q_0}$.

When the denominators N_i are small, the deviance should not be used, as we now illustrate in the case of $N_i = 1$. Suppose that $Y_i \mid p_i \sim_{ind} \text{Bernoulli}(p_i)$, with a logistic model, $\text{logit}(p_i) = x_i\beta$, for $i = 1, \ldots, n$. We fit this model using maximum likelihood, resulting in estimates $\widehat{\beta}$ and fitted probabilities $\widehat{p}$. In this case, (7.14) becomes

$$D = -2\sum_{i=1}^{n} y_i \log\left(\frac{\widehat{p}_i}{1 - \widehat{p}_i}\right) - 2\sum_{i=1}^{n} y_i \log(1 - \widehat{p}_i)$$
$$= -2y^\mathsf{T} x\widehat{\beta} - 2\sum_{i=1}^{n} \log(1 - \widehat{p}_i)$$
$$= -2\widehat{\beta}^\mathsf{T} x^\mathsf{T} y - 2\sum_{i=1}^{n} \log(1 - \widehat{p}_i)$$

since $y \log y = (1 - y)\log(1 - y) = 0$. At the MLE, $x^\mathsf{T} y = x^\mathsf{T} \widehat{p}$ so that

$$D = -2\widehat{\beta}^\mathsf{T} x^\mathsf{T} \widehat{p} - 2\sum_{i=1}^{n} \log(1 - \widehat{p}_i)$$

and the deviance is a function only of $\widehat{\beta}$. In other words, D is a deterministic function of $\widehat{\beta}$ only and cannot be used as a goodness of fit statistic. With small N_i, this is a problem for any link function.

An alternative goodness of fit measure for a model with q parameters is the Pearson statistic, as introduced in Sect. 6.5.3:

$$X^2 = \sum_{i=1}^{n} \frac{(Y_i - N_i\widehat{p}_i)^2}{N_i\widehat{p}_i(1 - \widehat{p}_i)}, \tag{7.15}$$

with $X^2 \to_d \chi^2_{n-q}$ under the null and under the assumption of fixed n. The Pearson statistic also has problems with small N_i. For example, for the model $Y_i \mid p \sim_{ind} \text{Bernoulli}(p)$, $\widehat{p} = \overline{y}$ and

$$X^2 = \sum_{i=1}^{n} \frac{(y_i - \overline{y})^2}{\overline{y}(1 - \overline{y})} = n,$$

which is not a useful goodness of fit measure (McCullagh and Nelder 1989, Sect. 4.4.5). The deviance also has problems under this Bernoulli model (Exercise 7.5).

7.6.3 Quasi-likelihood Inference for Logistic Regression Models

As we saw in Sect. 6.6, an extremely simple and appealing manner of dealing with overdispersion is to assume the model

$$E[Y_i \mid \beta] = N_i p_i(\beta)$$
$$\text{var}(Y_i \mid \beta) = \alpha N_i p_i(\beta) \left[1 - p_i(\beta)\right],$$

with $\text{cov}(Y_i, Y_j \mid \beta) = 0$, for $i \neq j$. Under this model, due to the proportionality of the variance model, the maximum quasi-likelihood estimator satisfies the score function (7.12), since the value of α is irrelevant to finding the root of the estimating equation. Hence, the quasi-likelihood estimator $\widehat{\beta}$ corresponds to the MLE. Interval estimates and tests are altered, however. In particular, asymptotic confidence intervals are derived from the variance–covariance $\widehat{\alpha}(\boldsymbol{D}^\text{T}\boldsymbol{V}^{-1}\boldsymbol{D})^{-1}$. An obvious estimator of α is provided by the method of moments, which corresponds to the Pearson statistic (7.15) divided by $n - k - 1$. This estimator is consistent if the first two moments are correctly specified. The reference χ^2 distribution under the null is also perturbed, as in (6.27).

7.6.4 Bayesian Inference for Logistic Regression Models

A Bayesian approach to inference combines the likelihood $L(\beta)$ with a prior $\pi(\beta)$, with a multivariate normal distribution being the obvious choice. For the binomial model there is no conjugate distribution for general regression models. In simple situations with a small number of discrete covariates, one could specify beta priors with known parameters for each combination of levels and obtain analytic posteriors, but there would be no linkage between the different groups, that is, no transfer of information. With multivariate normal priors, computation may be carried out using INLA (Sect. 3.7.4), though this approximation strategy may be inaccurate if the binomial denominators are small (Fong et al. 2010). An alternative is provided by MCMC (Sect. 3.8).

As discussed in Sect. 7.6.3, it is common to encounter excess-binomial variation. This may be dealt with in a Bayesian context via the introduction of random effects. The beta-binomial described in Sect. 7.5 provides one possibility. An alternative, more flexible formulation would assume the two-stage model:

Stage One: The likelihood:

$$Y_i \mid \boldsymbol{\beta}, b_i \sim_{ind} \text{Binomial}\,[N, p(\boldsymbol{x}_i)]$$

$$\log\left[\frac{p(\boldsymbol{x}_i)}{1 - p(\boldsymbol{x}_i)}\right] = \boldsymbol{x}_i \boldsymbol{\beta} + b_i$$

Stage Two: The random effects distribution:

$$b_i \mid \sigma_0^2 \sim_{iid} \text{N}(0, \sigma_0^2).$$

The parameter σ_0^2 controls the amount of overdispersion, though not in a simple fashion. A Bayesian approach adds priors on $\boldsymbol{\beta}$ and σ_0^2. This model is discussed further in Sect. 9.13.

Example: Outcome After Head Injury

Parameter estimation, whether via likelihood or Bayes, is straightforward for these data *given* a particular model. The difficult task in this problem is deciding upon a model. If prediction is all that is required, then Bayesian model averaging provides one possibility, and this is explored for these data in Chap. 12.

In exploratory mode, we illustrate some approaches to model selection. In Sect. 4.8, approaches to variable selection were reviewed and critiqued. In particular, the hierarchy principle, in which all interactions are accompanied by their constituent main effect, was discussed. Even applying the hierarchy principle here, there are still 167 models with $k = 4$ variables.

We begin by applying forward selection (obeying the hierarchy principle), beginning with the null model and using AIC as the selection criteria. This leads to a model with all main effects and the three two-way interactions H.P, H.A, and P.A. Since there are $n = 24$ fixed cells here we can assess the overall fit. The deviance associated with the model selected via forward selection is 13.6 on 13 degrees of freedom which indicates a good fit. Applying backward elimination produces a model with all main effects and five two-way interactions, the three selected using forward selection and, in addition, H.C and C.A. This model has a deviance of 7.0 on 10 degrees of freedom, so the overall fit is good.

Carrying out an exhaustive search over all 167 models using AIC as the criterion leads to the model selected with backward selection (i.e., main effects plus five two-way interactions). Using BIC as the criteria leads to a far simpler model with the main effects H, C, and A only. It is often found that BIC picks simpler models.

We consider inference for the model:

$$1+H+P+C+A2+A3+H.P+H.A2+H.A3+P.A2+P.A3, \qquad (7.16)$$

7.6 Logistic Regression Models

Table 7.2 Likelihood and Bayesian estimates and uncertainty measures for model (7.16) applied to the head injury data

	MLE	Std. err.	Post. mean	Post S.D.
1	−1.39	0.26	−1.37	0.26
H	1.03	0.35	1.02	0.35
P	2.05	0.30	2.04	0.29
C	−1.52	0.17	−1.53	0.17
A2	1.20	0.33	1.18	0.32
A3	3.69	0.48	3.68	0.47
H.P	−0.55	0.34	−0.55	0.34
H.A2	−0.39	0.36	−0.38	0.36
H.A3	−1.32	0.53	−1.29	0.52
P.A2	−0.57	0.37	−0.56	0.36
P.A3	−1.35	0.49	−1.33	0.48

that is, the model with main effects for hematoma (H), pupils (P), coma score (C), and age (with A2 and A3 representing the second and third levels) and with interactions between hematoma and pupils (H.P), hematoma and age (H.A2 and H.A3), and pupils and age (P.A2 and P.A3).

The MLEs and standard errors are given in Table 7.2, along with Bayesian posterior means and standard deviations. The prior on the intercept was taken as flat, and for the 10 log odds ratios, independent normal priors N(0, 4.70²) were taken, which correspond to 95% intervals for the odds ratios of [0.0001, 10000], that is, very weak prior information was incorporated. The INLA method was used for computation. The original scale of the parameters is given in the table, which is not ideal for interpretation, but makes sense for comparison of results since the sampling distributions and posteriors are close to normal. The first thing to note is that inference from the two approaches is virtually identical. This is not surprising, given the relatively large counts and weak priors.

The pupil and age variables and their interaction at the highest age level are clearly very important. The high coma score parameter is large and negative, and since the coma variable is not involved in any interactions, we can say that having a high coma score reduces the odds of death by $\exp(-1.52) = 0.22$.

The observed and fitted probabilities are displayed in Fig. 7.3 with different line types joining the observed probabilities (as in Fig. 7.1). The vertical lines join the fitted to the observed probabilities, with the same line type as the observed probabilities with which they are associated. There are no clear badly fitting cells.

Example: Aircraft Fasteners

Let Y_i be the number of fasteners failing at pressure x_i, and assume $Y_i \mid p_i \sim_{ind}$ Binomial(n_i, p_i), $i = 1, \ldots, n$, with the logistic model logit$(p_i) = \beta_0 + \beta_1 x_i$. This specification yields likelihood

Fig. 7.3 Probability of death after head injury as a function of age, hematoma score, and pupils. Panels (**a**) and (**b**) are for low and high coma scores, respectively. The *open circles* are the fitted values. The observed values are joined by different line types. The residuals $y/n - \widehat{p}$ are shown as *vertical lines* of the same line type

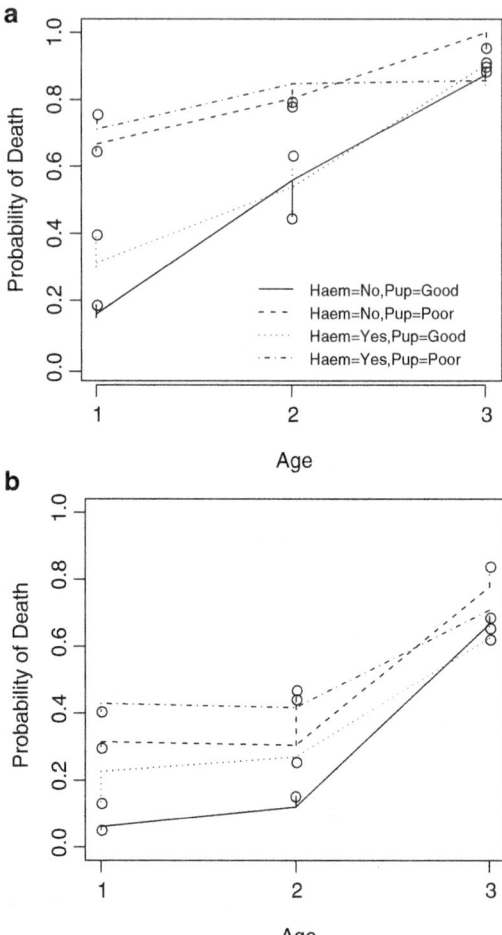

$$L(\boldsymbol{\beta}) = \exp\left(\beta_0 \sum_{i=1}^{n} y_i + \beta_1 \sum_{i=1}^{n} x_i y_i - \sum_{i=1}^{n} n_i \log\left[1 + \exp(\beta_0 + \beta_1 x_i)\right]\right)$$

(7.17)

where $\boldsymbol{\beta} = [\beta_0, \beta_1]^{\text{T}}$. The MLEs and variance–covariance matrix are

$$\widehat{\boldsymbol{\beta}} = \begin{bmatrix} -5.34 \\ 0.0015 \end{bmatrix}, \ \widehat{\text{var}}(\widehat{\boldsymbol{\beta}}) = \begin{bmatrix} 2.98 \times 10^{-1} & -8.50 \times 10^{-5} \\ -8.50 \times 10^{-5} & 2.48 \times 10^{-8} \end{bmatrix}. \quad (7.18)$$

The solid line in Fig. 7.4 is the fitted curve $\widehat{p}(x)$ corresponding to the MLE. The fit appears good. For comparison we also fit models with complementary log–log and log–log link functions, as described in Sect. 7.4.2. Figure 7.4 shows the fit from

7.6 Logistic Regression Models

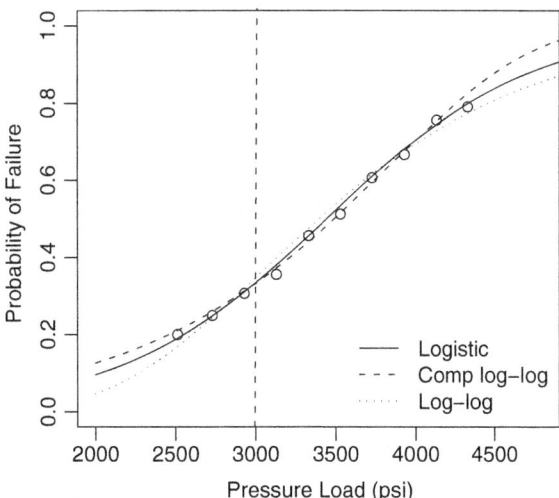

Fig. 7.4 Fitted curves for the aircraft fasteners data under three different link functions

these models. The residual deviance from logistic, complementary log–log, and log–log links are 0.37, 0.69, and 1.7, respectively. These values are not comparable via likelihood ratio tests since the models are not nested. AIC (Sect. 4.8.2) can be used for such comparisons, but the approximations inherent in the derivation are more accurate for nested models (Ripley 2004 and Sect. 10.6.4). The differences are so small here that we would not make any conclusions on the basis of these numbers. Since the number of x categories is not fixed in this example, we cannot formally examine the absolute fit of the models. In Fig. 7.6, we see that residual plots for these three models indicate that the logistic fit is preferable.

A 95% confidence interval for the odds ratio corresponding to a 500 psi increase in pressure load is

$$\exp\left[500 \times \widehat{\beta}_1 \pm 1.96 \times 500\sqrt{\text{var}(\widehat{\beta}_1)}\right] = [1.86, 2.53]. \quad (7.19)$$

We now present a Bayesian analysis. For these abundant data and without any available prior information, the improper uniform prior $\pi(\beta) \propto 1$ is assumed. The posterior is therefore proportional to (7.17). We use a bivariate Metropolis–Hastings random walk MCMC algorithm (Sect. 3.8.2) to explore the posterior. A bivariate normal proposal was used, with variance–covariance matrix proportional to the asymptotic variance–covariance matrix, $\widehat{\text{var}}(\widehat{\beta})$, (7.18). This matrix was multiplied by four to give an acceptance ratio of around 30%. Panels (a) and (b) of Fig. 7.5 show histograms of the dependent samples from the posterior $\beta_0^{(s)}$ and $\beta_1^{(s)}$, $s = 1, \ldots, S = 500$, and panel (c) the bivariate posterior. The posterior median for β is $[-5.36, 0.0015]$, and a 95% posterior interval for the odds ratio corresponding to a 500 psi increase in pressure is identical to the asymptotic likelihood interval (7.19).

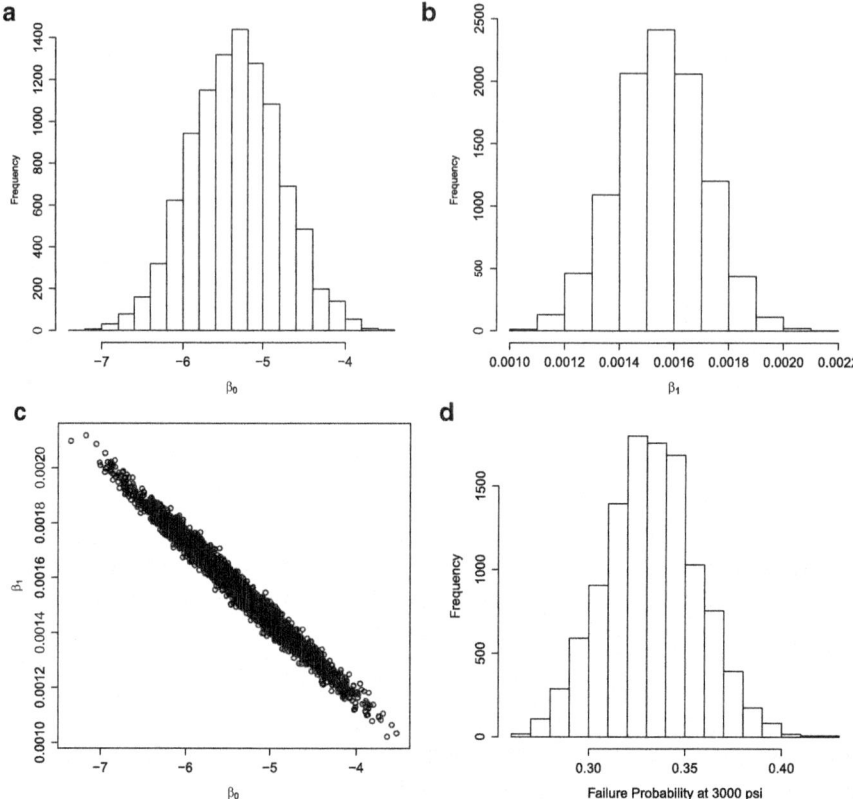

Fig. 7.5 Posterior summaries for the aircraft fasteners data: (a) $p(\beta_0|\mathbf{y})$, (b) $p(\beta_1|\mathbf{y})$, (c) $p(\beta_0, \beta_1|\mathbf{y})$, (d) $p(\exp(\theta)/[1+\exp(\theta)]|\mathbf{y})$, where $\theta = \beta_0 + \beta_1\tilde{x}$, that is, the posterior for the probability of failure at a load of $\tilde{x} = 3{,}000$ psi

We now imagine that it is of interest to give an interval estimate for the probability of failure at $\tilde{x} = 3{,}000$ psi (which is indicated as a dashed vertical line on Fig. 7.4). An asymptotic 95% confidence interval for $\theta = \beta_0 + \beta_1\tilde{x}$ is

$$\widehat{\theta} \pm 1.96 \times \sqrt{\operatorname{var}(\widehat{\theta})},$$

where

$$\widehat{\theta} = \widehat{\beta}_0 + \tilde{x}\widehat{\beta}_1$$
$$\operatorname{var}(\widehat{\theta}) = \operatorname{var}(\widehat{\beta}_0) + 2\tilde{x}\operatorname{cov}(\widehat{\beta}_0, \widehat{\beta}_1) + \tilde{x}^2\operatorname{var}(\widehat{\beta}_1).$$

Taking the expit transform of the endpoints of the confidence interval on the linear predictor scale leads to a 95% interval of [0.29, 0.38]. Substitution of the posterior

7.7 Conditional Likelihood Inference

Table 7.3 A generic 2 × 2 table

	$Y=0$	$Y=1$	
$X=0$	y_{00}	y_{01}	$y_{0\cdot}$
$X=1$	y_{10}	y_{11}	$y_{1\cdot}$
	$y_{\cdot 0}$	$y_{\cdot 1}$	$y_{\cdot\cdot}$

samples $\beta^{(s)}$ to give $\text{expit}(\theta^{(s)})$, $s=1,\ldots,S$ results in a 95% interval which is again identical to the frequentist interval.

7.7 Conditional Likelihood Inference

In Sect. 2.4.2, conditional likelihood was introduced as a procedure that could be used for eliminating nuisance parameters. In this chapter, conditional likelihood will be used for discrete data, which we denote y. Suppose the distribution for the data can be represented as,

$$p(y \mid \lambda, \phi) \propto p(t_1 \mid t_2, \lambda) p(t_2 \mid \lambda, \phi), \qquad (7.20)$$

where λ is a parameter of interest and ϕ is a nuisance parameter. Then inference for λ may be based on the *conditional likelihood*

$$L_c(\lambda) = p(t_1 \mid t_2, \lambda).$$

Perhaps the most popular use of conditional likelihood leads to Fisher's exact test. Consider the 2 × 2 layout of data shown in Table 7.3 with

$$y_{01} \mid p_0 \sim \text{Binomial}(y_{0\cdot}, p_0)$$
$$y_{11} \mid p_1 \sim \text{Binomial}(y_{1\cdot}, p_1),$$

which we combine with the logistic regression model:

$$\log\left(\frac{p_0}{1-p_0}\right) = \beta_0$$
$$\log\left(\frac{p_1}{1-p_1}\right) = \beta_0 + \beta_1.$$

Here,

$$\exp(\beta_1) = \frac{p_1/(1-p_1)}{p_0/(1-p_0)}$$

is the odds of a positive response in the $X = 1$ group, divided by the odds of a positive response in the $X = 0$ group, that is, the odds ratio. This setup gives likelihood

$$\Pr(y_{01}, y_{11} \mid \beta_0, \beta_1) = \binom{y_{0\cdot}}{y_{01}} \binom{y_{1\cdot}}{y_{11}} \frac{e^{y_{11}\beta_1}}{(1+e^{\beta_0+\beta_1})^{y_{1\cdot}}} \frac{e^{y_{\cdot 1}\beta_0}}{(1+e^{\beta_0})^{y_{0\cdot}}}.$$
(7.21)

Now $[y_{01}, y_{11}]$ implies the distribution of $[y_{11}, y_{\cdot 1}]$, so we can write

$$\Pr(y_{11}, y_{\cdot 1} \mid \beta_0, \beta_1) = \binom{y_{0\cdot}}{y_{\cdot 1}-y_{11}} \binom{y_{1\cdot}}{y_{11}} \frac{e^{y_{11}\beta_1}}{(1+e^{\beta_0+\beta_1})^{y_{1\cdot}}} \frac{e^{y_{\cdot 1}\beta_0}}{(1+e^{\beta_0})^{y_{0\cdot}}}.$$

We now show that by conditioning on the column totals, in addition to the row totals, we obtain a distribution that depends only on the parameter of interest β_1. Consider

$$\Pr(y_{11} \mid y_{\cdot 1}, \beta_0, \beta_1) = \frac{\Pr(y_{11}, y_{\cdot 1} \mid \beta_0, \beta_1)}{\Pr(y_{\cdot 1} \mid \beta_0, \beta_1)},$$

where the marginal distribution is obtained by summing over the possible values that y_{11} can take, that is,

$$\Pr(y_{\cdot 1} \mid \beta_0, \beta_1) = \sum_{u=u_0}^{u_1} \Pr(u, y_{\cdot 1} \mid \beta_0, \beta_1)$$

$$= \sum_{u=u_0}^{u_1} \binom{y_{0\cdot}}{y_{\cdot 1}-u} \binom{y_{1\cdot}}{u} \frac{e^{u\beta_1}}{(1+e^{\beta_0+\beta_1})^{y_{1\cdot}}} \frac{e^{y_{\cdot 1}\beta_0}}{(1+e^{\beta_0})^{y_{0\cdot}}}.$$

where $u_0 = \max(0, y_{\cdot 1} - y_{0\cdot})$ and $u_1 = \min(y_{1\cdot}, y_{\cdot 1})$ ensure that the marginals are preserved. With respect to (7.20), $\lambda \equiv \beta_1$, $\phi \equiv \beta_0$, $t_1 \equiv y_{11}$, and $t_2 \equiv y_{\cdot 1}$. Accordingly, the conditional distribution takes the form

$$\Pr(y_{11} \mid y_{\cdot 1}, \beta_1) = \frac{\binom{y_{0\cdot}}{y_{\cdot 1}-y_{11}} \binom{y_{1\cdot}}{y_{11}} e^{y_{11}\beta_1}}{\sum_{u=u_0}^{u_1} \binom{y_{0\cdot}}{y_{\cdot 1}-u} \binom{y_{1\cdot}}{u} e^{u\beta_1}},$$
(7.22)

an *extended hypergeometric* distribution. We have removed the conditioning on β_0 since this distribution depends on β_1 only (which was the point of this derivation). Inference for β_1 may be based on the conditional likelihood (7.22). In particular, the conditional MLE may be determined, though unfortunately no closed form exists.

Conventionally, estimates of β_0 and β_1 would be determined from the product of binomial likelihoods, (7.21). Unless the samples are small, the conditional and

7.7 Conditional Likelihood Inference

unconditional MLEs (and associated variances) will be in close agreement, but for small samples, the conditional MLE is preferred due to the following informal argument. Consider the original 2×2 data in Table 7.3. If we knew $y_{\cdot 1}$, then this alone would not help us to estimate β_1, *but* the precision of conclusions about β_1 will depend on this column total, and we should therefore condition on the observed value. This is to ensure that we attach to the conclusions the precision actually achieved and not that to be achieved hypothetically in a particular situation that has in fact not occurred. For further discussion, see Cox and Snell (1989, p. 27–29).

To derive the conditional MLE, first consider the conditional likelihood

$$L_c(\beta_1) = \frac{c(y_{11}) e^{y_{11} \beta_1}}{\sum_{u=u_0}^{u_1} c(u) e^{u \beta_1}}$$

where

$$c(u) = \binom{y_{0\cdot}}{y_{\cdot 1} - u} \binom{y_{1\cdot}}{u}.$$

The (conditional) score is

$$S_c(\beta_1) = \frac{\partial}{\partial \beta_1} \log L_c(\beta_1) = y_{11} - \frac{\sum_{u=u_0}^{u_1} c(u) u e^{\widehat{\beta}_1 u}}{\sum_{u=u_0}^{u_1} c(u) e^{\widehat{\beta}_1 u}}. \quad (7.23)$$

The extended hypergeometric distribution is a member of the exponential family (Exercise 7.6) and

$$\mathrm{E}[S_c(\beta_1)] = \left. \frac{\partial}{\partial \beta_1} \log L_c(\beta_1) \right|_{\widehat{\beta}_1} = 0,$$

at the MLE. Consequently, from (7.23), we can use the equation $\mathrm{E}[Y_{11} \mid \widehat{\beta}_1] = y_{11}$ to solve for $\widehat{\beta}_1$. Asymptotic inference is based on

$$I_c(\beta_1)^{1/2} \left(\widehat{\beta}_1 - \beta_1 \right) \to_d N(0, 1), \quad (7.24)$$

where the (conditional) information is

$$I_c(\beta_1) = -\frac{\partial^2}{\partial \beta_1^2} \log L_c(\beta_1) = \frac{\sum_{u=u_0}^{u_1} c(u) u^2 e^{\widehat{\beta}_1 u}}{\sum_{u=u_0}^{u_1} c(u) e^{\widehat{\beta}_1 u}} - \left(\frac{\sum_{u=u_0}^{u_1} c(u) u e^{\widehat{\beta}_1 u}}{\sum_{u=u_0}^{u_1} c(u) e^{\widehat{\beta}_1 u}} \right)^2$$

$$= \mathrm{var}(Y_{11} \mid \beta_1).$$

It is straightforward to test the null hypothesis $H_0 : \beta_1 = 0$ using the conditional likelihood. When $\beta_1 = 0$, the distribution (7.22) is hypergeometric, and so

Table 7.4 Data on tumor appearance within rats

		Tumor		
		Absent $Y=0$	Present $Y=1$	
Control	$X=0$	13	19	32
Treated	$X=1$	2	21	23
		15	40	55

$$\Pr(y_{11} \mid y_{\cdot 1}, \beta_1 = 0) = \frac{\binom{y_{0\cdot}}{y_{\cdot 1} - y_{11}} \binom{y_{1\cdot}}{y_{11}}}{\binom{y_{\cdot\cdot}}{y_{\cdot 1}}}. \tag{7.25}$$

The comparison of the observed y_{11} with the tail of this distribution is known as *Fisher's exact test* (Fisher 1935). Various possibilities are available to obtain a two-sided significance level, the simplest being to double the one-sided p-value. An alternative is provided by summing all probabilities less than the observed table. Confidence intervals for β_1 may be obtained from (7.24), or by inverting the test. See Agresti (1990, Sects. 3.5 and 3.6) for further discussion; in particular, the problems of the discreteness of the sampling distribution are discussed.

Example: Tumor Appearance Within Mice

We illustrate the application of conditional likelihood using data reported by Essenberg (1952) and presented in Table 7.4. To examine the carcinogenic effects of tobacco, 36 albino mice were placed in an enclosed chamber which was filled with the smoke of one cigarette every 12 h per day. Another group of mice were kept in an alternative chamber without smoke. After 1 year, autopsies were carried out on those mice that had survived for at least the first 2 months of the experiment. The data in Table 7.4 give the numbers of mice with and without tumors in the "control" and "treated" groups.

For these data, the permissible values of y_{11} lie between $u_0 = \max(0, 40 - 32) = 8$ and $u_1 = \min(23, 40) = 23$. Under $H_0 : \beta_1 = 0$, the probabilities of $y_{11} = 21, 22, 23$, from (7.25), are 0.00739, 0.00091, and 0.00005, which sum to 0.00834, the one-sided p-value. The simplest version of the two-sided p-value is therefore 0.0167, which would lead to rejection of H_0 under the usual threshold of 0.05. Summing the probabilities of more extreme tables gives a p-value of 0.0130.

Denoting by $\widehat{\beta}_1^u$ the (unconditional) MLE of the log odds ratio, we have

$$\widehat{\beta}_1^u = \log\left(\frac{21 \times 13}{2 \times 19}\right) = \log(7.18) = 1.97,$$

with asymptotic standard error

$$\sqrt{\widehat{\text{var}}(\widehat{\beta}_1^u)} = \sqrt{\frac{1}{2} + \frac{1}{21} + \frac{1}{13} + \frac{1}{19}} = 0.82,$$

to give asymptotic 95% confidence interval for the odds ratio of

$$\exp(1.97 \pm 1.96 \times 0.82) = [1.44, 35.8].$$

The Wald test p-value of 0.0166 is very close to that obtained from Fisher's exact test. The conditional MLE is

$$\widehat{\beta}_1 = \log(6.95) = 1.93$$

with conditional standard error

$$\sqrt{\text{var}(\widehat{\beta}_1)} = 0.61,$$

illustrating the extra precision gained by conditioning on $y_{\cdot 1}$. The conditional asymptotic 95% confidence interval for the odds ratio based on (7.24) is

$$\exp(1.93 \pm 1.96 \times 0.61) = [2.11, 22.9].$$

7.8 Assessment of Assumptions

In general, residual analysis is subjective, and though one might be able to conclude that a model is inadequate, concluding adequacy is much more difficult. Unfortunately, for logistic regression models with binary data, the assessment is even more tentative. Even when the model is true, little can be said about the moments and distribution of the residuals.

We briefly review Pearson and deviance residuals as defined for GLMs in Sect. 6.9. Pearson residuals are defined as $e_i^\star = (Y_i - \widehat{\mu}_i)/\sqrt{\widehat{\text{var}}(Y_i)}$, and for $Y_i \mid p_i \sim \text{Binomial}(n_i, p_i)$, we obtain

$$e_i^\star = \frac{y_i - n_i \widehat{p}_i}{[n_i \widehat{p}_i (1 - \widehat{p}_i)]^{1/2}},$$

$i = 1, \ldots, n$. Pearson's statistic is

$$X^2 = \sum_{i=1}^n \frac{(Y_i - n_i \widehat{p}_i)^2}{n_i \widehat{p}_i (1 - \widehat{p}_i)} = \sum_{i=1}^n (e_i^\star)^2,$$

showing the link between the measures of local and absolute fit.

Deviance residuals are defined as

$$e_i^\star = \text{sign}(y_i - \widehat{\mu}_i)\sqrt{D_i},$$

$i = 1, \ldots, N$. Note that the deviance $D = \sum_{i=1}^n (e_i^\star)^2$ where D is given by (7.14). For binary Y_i and a particular value of $\widehat{p}_i$, the residuals can only take one of two possible values, which is clearly a problem (this is illustrated later, in Fig. 7.8).

Few analytical results are available for the case of a binomial model, but, if the model is correct, both the Pearson and deviance residuals are asymptotically normally distributed. Hence, they may be put to many of the same uses as residual defined with respect to the normal linear regression model (as described in Sect. 5.11.3). For example, residuals may be plotted against covariates x and examined for outlying values. Interpretation is more difficult, however, as one must examine the appropriateness of the link function as well as the linearity assumption. A normal QQ plot of residuals can indicate outlying observations.

Empirical logits $\log[(y_i + 0.5)/(N_i - y_i + 0.5)]$ are useful for examining the adequacy of the logistic linear model. The addition of 0.5 removes problems when $y_i = 0$ or N_i. This adjustment is optimal; see Cox and Snell (1989, Sect. 2.1.6) for details. The mean–variance relationship can be examined by plotting residuals versus fitted values. In particular, different overdispersion models may be compared, as discussed in Sect. 7.5.

Example: Aircraft Fasteners

In this example, the denominators are relatively large (ranging between 40 and 100 for each of the 10 trials), and so the residuals are informative. Figure 7.6 shows Pearson residuals plotted against pressure load for each of three different link functions. On the basis of these plots, the logistic model looks the most reasonable since there are runs of positive and negative residuals associated with the other two link functions, signifying mean model misspecification.

Example: Outcome After Head Injury

The binary response in this example is cross-classified with respect to factors with 2 or 3 levels. We saw in Fig. 7.3 that the fit of model (7.16) appeared reasonable, though the distances $\frac{y_i}{n_i} - \widehat{p}_i$ that are displayed as vertical lines are not standardized, making interpretation difficult. Figure 7.7 gives a normal QQ plot of the Pearson residuals, and there are no obvious causes for concern with no outlying points.

7.8 Assessment of Assumptions

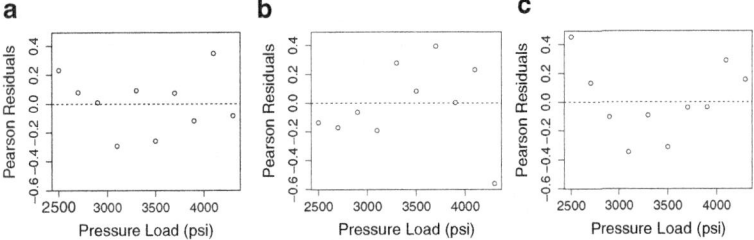

Fig. 7.6 Pearson residuals versus pressure load for the aircraft fasteners data for (**a**) logistic link model, (**b**) complementary log–log link model, and (**c**) log–log link model

Fig. 7.7 QQ plot of Pearson residuals for the head injury data

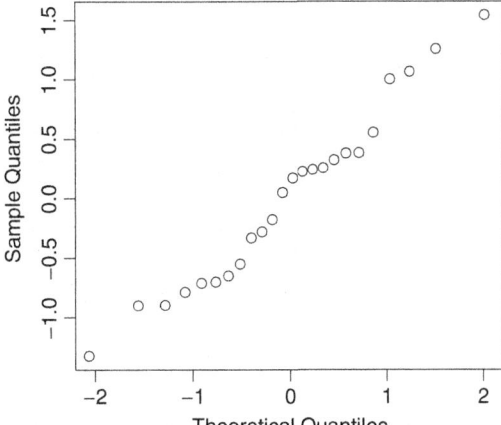

Example: BPD and Birth Weight

We fit a logistic regression model

$$\Pr(Y = 1 \mid x) = \frac{\exp(\beta_0 + \beta_1 x)}{1 + \exp(\beta_0 + \beta_1 x)}, \qquad (7.26)$$

with $Y = 0/1$ corresponding to absence/presence of BPD and x to birth weight. The curve arising from fitting this model is shown in Fig. 7.2, along with the curve from the use of the complementary log–log link. We might question whether either of these curves is adequate, since they are relatively inflexible, with forms determined by two parameters only. The Pearson residuals from the two models are plotted versus birth weight in Fig. 7.8. The binary nature of the response is evident in these plots, and assessing whether the models are adequate is not possible from this plot. In Chap. 11, we return to these data and fit flexible nonlinear models.

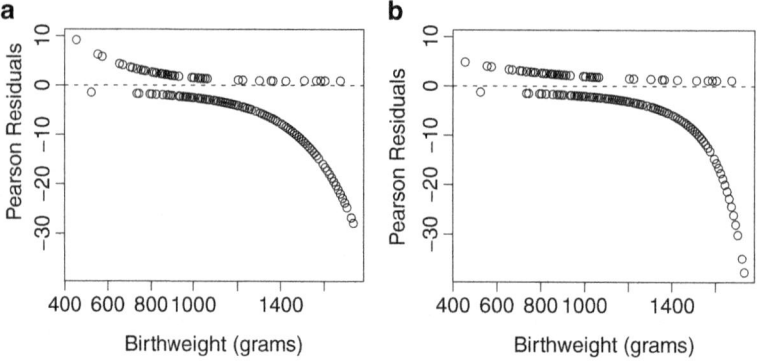

Fig. 7.8 Pearson residuals versus birth weight for the BPD data: (**a**) logistic model, (**b**) complementary log–log model

7.9 Bias, Variance, and Collapsibility

We begin by summarizing some of the results of Sect. 5.9 in which the bias and variance of estimators were examined for the linear model. Consider the models:

$$E[Y \mid x, z] = \beta_0 + \beta_1 x + \beta_2 z \tag{7.27}$$

$$E[Y \mid x] = \beta_0^\star + \beta_1^\star x. \tag{7.28}$$

First, suppose that x and z are orthogonal. Roughly speaking, if z is related to Y, then fitting model (7.27) will lead to a reduction in the variance of $\widehat{\beta}_1$, and $E[\widehat{\beta}_1] = E[\widehat{\beta}_1^\star]$ so that bias is not an issue. When x and z are not orthogonal, then fitting model (7.28) will lead to bias in the estimation of β_1 since $\beta_1^\star$ reflects not only x but also the effect of z through its association with x.

In this section we discuss these issues with respect to logistic regression models. To this end, consider the *logistic models*:

$$E[Y \mid x, z] = \frac{\exp(\beta_0 + \beta_1 x + \beta_2 z)}{1 + \exp(\beta_0 + \beta_1 x + \beta_2 z)} \tag{7.29}$$

$$E[Y \mid x] = \frac{\exp(\beta_0^* + \beta_1^* x)}{1 + \exp(\beta_0^* + \beta_1^* x)} = E_{z \mid x}\left[\frac{\exp(\beta_0 + \beta_1 x + \beta_2 z)}{1 + \exp(\beta_0 + \beta_1 x + \beta_2 z)}\right]. \tag{7.30}$$

The last equation indicates that determining the effects of omission of z will be very hard to determine due to the nonlinearity of the logistic function. As we illustrate shortly though, even if x and z are orthogonal, $E[\beta_1] \neq E[\beta_1^\star]$. Linear models for the probabilities are more straightforward to understand, but, as discussed previously, since probabilities are constrained $[0, 1]$, such models are rarely appropriate for binary data.

7.9 Bias, Variance, and Collapsibility

Table 7.5 Illustration of Simpson's paradox for the case of non-orthogonal x and z

		$z=0$		$z=1$		Marginal	
		$Y=0$	$Y=1$	$Y=0$	$Y=1$	$Y=0$	$Y=1$
Control	$x=0$	8	2	9	21	17	23
Treatment	$x=1$	18	12	2	8	20	20
	Odds Ratio	1.6		1.7		0.7	

We now discuss the marginalization of effect measures. Roughly speaking, if an effect measure is constant across strata (subtables) and equal to the measure calculated from the marginal table, it is known as *collapsible*. Non-collapsibility is sometimes referred to as Simpson's paradox (Simpson 1951) in the statistics literature. As in Greenland et al. (1999), we include the case of orthogonal x and z in Simpson's paradox, though first illustrate with a case in which x and z are non-orthogonal.

Consider the data in Table 7.5 in which $x = 0/1$ represents a control/treatment which is applied in two strata $z = 0/1$, with a binary response $Y = 0/1$ being recorded. In both z strata, the treatment appears beneficial with odds ratios of 1.6 and 1.7. However, when the data are collapsed over strata, the marginal association is reversed to give an odds ratio of 0.7 so that the treatment appears detrimental.

Mathematically, the paradox is relatively simple to understand. Let

$$p_{xz} = \Pr(Y=1 \mid X=x, Z=z)$$
$$p_x^* = \Pr(Y=1 \mid X=x)$$

be the conditional and marginal probabilities of a response and $q_x = \Pr(Z=1 \mid X=x)$ summarize the relationship between x and z, for $x, z = 0, 1$. The "paradox" reflects the fact that it is possible to have

$$p_{00} < p_{10} \text{ and } p_{01} < p_{11},$$

that is, the probability of a positive response being greater under $X = 1$ for both strata, but

$$p_{00}(1-q_0) + p_{01}q_0 = p_0^* > p_1^* = p_{10}(1-q_1) + p_{11}q_1$$

so that the marginal probability of a positive response is greater under $x = 0$ than under $x = 1$. For the data of Table 7.5,

$$p_{00} = \frac{2}{10} = 0.20, \quad p_{10} = \frac{13}{30} = 0.43, \quad p_{01} = \frac{21}{30} = 0.7, \quad p_{11} = \frac{8}{10} = 0.8,$$

and

$$p_0^* = \frac{23}{40} = 0.58, \quad p_1^* = \frac{20}{40} = 0.50,$$

Table 7.6 Illustration of Simpson's paradox for the case of orthogonal x and z

		$z = 0$		$z = 1$		Marginal	
		$Y = 0$	$Y = 1$	$Y = 0$	$Y = 1$	$Y = 0$	$Y = 1$
Control	$x = 0$	95	5	10	90	105	95
Treatment	$x = 1$	90	10	5	95	95	105
	Odds ratio	2.1		2.1		1.2	

with

$$q_0 = \frac{30}{40}, \quad q_1 = \frac{10}{40}.$$

It is important to realize that the paradox has nothing to do with the absolute values of the counts. Reversal of the association (as measured by the odds ratio) cannot occur if $q_0 = q_1$ (i.e., if there is no confounding), but the odds ratio is still non-collapsible, as the next example illustrates.

We now consider the situation in which $q_0 = q_1$. Such a balanced situation would occur, by construction, in a randomized clinical trial in which (say) equal numbers of $x = 0$ and $x = 1$ groups receive the treatment. We illustrate in Table 7.6 in which there are 100 patients in each of the four combinations of x and z. In each of the z stratum, we see an odds ratio for the treatment as compared to the control of 2.1. We do not see a reversal in the direction of the association but rather an attenuation toward the null, with the marginal association being 1.2.

We emphasize that the marginal estimator is not a biased estimate, but is rather estimating a different quantity, the averaged or marginal association. A second point to emphasize is that, as we have just illustrated, collapsibility and confounding are different issues and should not be confused. In particular, it is possible to have confounding present without non-collapsibility, as discussed in Greenland et al. (1999).

Another issue that we briefly discuss is the effect of stratification on the variance of an estimator. As discussed at the start of this section, if x and z are orthogonal but z is associated with y, then including z in a linear model will increase the precision of the estimator of the association between y and x. We illustrate numerically that this is not the case in the logistic regression context, again referring to the data in Table 7.6. Let p_{xz} represent the probability of disease for treatment group x and strata z. In the conditional analysis we fit the model

$$\log\left(\frac{p_{xz}}{1 - p_{xz}}\right) = \begin{cases} \beta_0 & \text{for } x = 0, z = 0 \\ \beta_0 + \beta_x & \text{for } x = 1, z = 0 \\ \beta_0 + \beta_z & \text{for } x = 0, z = 1 \\ \beta_0 + \beta_x + \beta_z & \text{for } x = 1, z = 1, \end{cases}$$

where we have not included an interaction between x and z. This results in $\exp(\beta_x) = \exp(0.75) = 2.1$, as expected from Table 7.6, with standard error 0.40.

Now suppose we ignore the stratum information and let $p_x^\star$ be the probability of disease for treatment group x. We fit the model

$$\log\left(\frac{p_x^\star}{1-p_x^\star}\right) = \begin{cases} \beta_0^\star & \text{for } x=0 \\ \beta_0^\star + \beta_x^\star & \text{for } x=1 \end{cases}$$

This gives $\exp(\beta_x^\star) = \exp(0.20) = 1.2$, again as expected from Table 7.6, but with standard error 0.20 which is a reduction from the conditional model and is in stark contrast to the behavior we saw with the linear model.

In any cross-classified table the summary we observe is an "averaged" measure, where the average is with respect to the population underlying that table. Consider the right-hand 2×2 set of counts in Table 7.6, in which we had equal numbers in each strata (which mimics a randomized trial). The odds ratio comparing treatment to control is 1.2 here and is the effect averaged across strata (and any other variables that were unobserved). Such measures are relevant to what are sometimes referred to as *population* contrasts. Depending on the context, we will often wish to include additional covariates in order to obtain effect measures most relevant to particular subgroups (or subpopulations). The issues here have much in common with marginal and conditional modeling as discussed in the context of dependent data in Chaps. 8 and 9.

We emphasize that, as mentioned above, the difference between population and subpopulation-specific estimates should not be referred to as "bias" since different quantities are being estimated. As a final note, the discussion in this section has centered on logistic regression models, but the same issues hold for other nonlinear summary measures.

7.10 Case-Control Studies

In this section we discuss a very popular design in epidemiology, the case-control study. In the econometrics literature, this design is known as *choice-based sampling*.

7.10.1 The Epidemiological Context

Cohort (prospective) studies investigate the causes of disease by proceeding in the natural way from cause to effect. Specifically, individuals in different exposure groups of interest are enrolled, and then one observes whether they develop the disease or not over some time period. In contrast, case-control (retrospective) studies proceed from effect to cause. Cases and disease-free controls are identified, and then the exposure status of these individuals is determined. Table 7.7 demonstrates the simplest example in which there is a single binary exposure, with y_{ij} representing

Table 7.7 Generic 2×2 table for a binary exposure and binary disease outcome

		Not diseased $Y=0$	Diseased $Y=1$	
Unexposed	$X=0$	y_{00}	y_{01}	n_0
Exposed	$X=1$	y_{10}	y_{11}	n_1
		m_0	m_1	n

the number of individuals in exposure group i, $i = 0, 1$ and disease group j, $j = 0, 1$. In a cohort study, n_0 and n_1, the numbers of unexposed and exposed individuals, are fixed by design, and the random variables are the number of unexposed cases y_{01} and the number of exposed cases y_{11}.

There are a number of strong motivations for carrying out a case-control study. Since many diseases are rare, a cohort study has to generally contain a large number of participants to demonstrate an association between a risk factor and disease because few individuals will develop the disease (unless the effect of the exposure of interest is very strong). It may be difficult to assemble a full picture of the disease across subgroups (as defined by covariates) within a cohort study because the cohort is assembled at a particular time, the start of the study. As the study proceeds, certain subgroups, for example, the young, disappear. In this case it will not be possible to investigate a calendar time/age interaction, that is, the effect of calendar time at different age groups. Finally, the disease may take a long time to develop (this is true, for example, for most cancers), and so the study may need to run for a long period.

The case-control study provides a way of overcoming these difficulties. With reference to Table 7.7, m_0 and m_1, the numbers of controls and case, are fixed by design, and the random variables are the number of exposed controls y_{10} and the number of exposed cases, y_{11}.

A case-control study is not without its drawbacks. Probabilities of disease given exposure status are no longer directly estimable without external information, as we will discuss in more detail shortly. Most importantly, the study participants must be selected very carefully. The probability of selection for the study, for both cases and controls, must not depend on exposure status; otherwise, *selection bias* will be introduced; this bias can arise in many subtle ways. The great benefit of case-control studies is that we can still estimate the strength of the relationship between exposure and disease, a topic we discuss in-depth in the next section.

7.10.2 Estimation for a Case-Control Study

Consider the situation in which we have a binary response Y taking the values 0/1 corresponding to disease-free/diseased and exposures contained in a $(k + 1) \times 1$ vector x. The exposures can be a mix of continuous and discrete variables. In the case-control scenario, we select individuals on the basis of their disease status y, and the random variables are the exposures X.

7.10 Case-Control Studies

In a cohort study with a binary endpoint, a logistic regression disease model is the most common choice for analysis, with form

$$\Pr(Y = 1 \mid \boldsymbol{x}) = p(\boldsymbol{x}) = \frac{\exp\left(\beta_0 + \sum_{j=1}^{k} x_j \beta_j\right)}{1 + \exp\left(\beta_0 + \sum_{j=1}^{k} x_j \beta_j\right)}. \quad (7.31)$$

The *relative risk* of individuals having exposures $\boldsymbol{x}$ and $\boldsymbol{x}^\star$ is defined as

$$\text{Relative risk} = \frac{\Pr(Y = 1 \mid \boldsymbol{x})}{\Pr(Y = 1 \mid \boldsymbol{x}^\star)}$$

and is an easily interpretable quantity that epidemiologists are familiar with. As already mentioned in Sect. 7.6.1, for rare diseases, the relative risk is well approximated by the odds ratio

$$\frac{\Pr(Y = 1 \mid \boldsymbol{x})/\Pr(Y = 0 \mid \boldsymbol{x})}{\Pr(Y = 1 \mid \boldsymbol{x}^\star)/\Pr(Y = 0 \mid \boldsymbol{x}^\star)}.$$

With respect to the logistic regression model (7.31),

$$\frac{p(\boldsymbol{x})/[1 - p(\boldsymbol{x})]}{p(\boldsymbol{x}^\star)/[1 - p(\boldsymbol{x}^\star)]} = \exp\left[\sum_{j=1}^{k} \beta_j (x_j - x_j^\star)\right],$$

so that, in particular, $\exp(\beta_j)$ represents the increase in the odds of disease associated with a unit increase in x_j, with all other covariates held fixed (Sect. 7.6.1). The parameter β_0 represents the baseline log odds of disease, corresponding to the odds when all of the exposures are set equal to zero.

We now turn to interpretation in a case-control study. We first introduce an indicator variable Z which represents the event that an individual was selected for the study ($Z = 1$) or not ($Z = 0$). Let $\pi_y = \Pr(Z = 1 \mid Y = y)$ denote the probabilities of selection, given response y, $y = 0, 1$. Typically, π_1 is much greater than π_0, since cases are rarer than non-cases. Now consider the probability that a person is diseased, given exposures $\boldsymbol{x}$ and selection for the study:

$$\Pr(Y = 1 \mid Z = 1, \boldsymbol{x}) = \frac{\Pr(Z = 1 \mid Y = 1, \boldsymbol{x}) \Pr(Y = 1 \mid \boldsymbol{x})}{\Pr(Z = 1 \mid \boldsymbol{x})}. \quad (7.32)$$

The denominator may be simplified to

$$\Pr(Z = 1 \mid \boldsymbol{x}) = \sum_{y=0}^{1} \Pr(Z = 1 \mid Y = y, \boldsymbol{x}) \Pr(Y = y \mid \boldsymbol{x})$$

$$= \sum_{y=0}^{1} \Pr(Z = 1 \mid Y = y) \Pr(Y = y \mid \boldsymbol{x}),$$

where we have made the crucial assumption that

$$\Pr(Z=1 \mid Y=y, \boldsymbol{x}) = \Pr(Z=1 \mid Y=y) = \pi_y,$$

for $y = 0, 1$, that is, the selection probabilities depend only on the disease status and *not* on the exposures (i.e., there is no selection bias). If we take a random sample of cases and controls, this assumption is valid. Substitution in (7.32), and assuming a logistic regression model, gives

$$\Pr(Y=1 \mid Z=1, \boldsymbol{x}) = \frac{\pi_1 \exp(\boldsymbol{x}\boldsymbol{\beta})/[1+\exp(\boldsymbol{x}\boldsymbol{\beta})]}{\pi_1 \exp(\boldsymbol{x}\boldsymbol{\beta})/[1+\exp(\boldsymbol{x}\boldsymbol{\beta})] + \pi_0/[1+\exp(\boldsymbol{x}\boldsymbol{\beta})]}$$

$$= \frac{\pi_1 \exp\left(\beta_0 + \sum_{j=1}^k x_j \beta_j\right)}{\pi_0 + \pi_1 \exp\left(\beta_0 + \sum_{j=1}^k x_j \beta_j\right)}$$

$$= \frac{\exp\left(\beta_0^\star + \sum_{j=1}^k x_j \beta_j\right)}{1 + \exp\left(\beta_0^\star + \sum_{j=1}^k x_j \beta_j\right)},$$

where $\beta_0^\star = \beta_0 + \log \pi_1/\pi_0$. Hence, we see that the probabilities of disease in a case-control study also follow a logistic model but with an altered intercept. In the usual case, $\pi_1 > \pi_0$ so that the intercept is increased to account for the over-sampling of cases. Unless information on π_0 and π_1 is available, we cannot obtain estimates of $\Pr(Y=1 \mid \boldsymbol{x})$ (the incidence for different exposure groups).

This derivation shows that assuming a logistic model in the cohort context implies that the disease frequency within the case-control sample also follows a logistic model, but does not illuminate how inference may be carried out. Suppose there are m_0 controls and m_1 cases. Since the exposures are random in a case-control context, the likelihood is of the form

$$L(\boldsymbol{\theta}) = \prod_{y=0}^{1} \prod_{j=1}^{m_y} p(\boldsymbol{x}_{yj} \mid y, \boldsymbol{\theta}),$$

where $\boldsymbol{x}_{yj}$ is the set of covariates for individual j in disease group y, and it appears that we are faced with the unenviable task of specifying forms, depending on parameters $\boldsymbol{\theta}$, for the distribution of covariates in the control and case populations. In a seminal paper, Prentice and Pyke (1979) showed that asymptotic likelihood inference for the odds ratio parameters was identical irrespective of whether the data are collected prospectively or retrospectively. The proof of this result hinges on assuming a logistic disease model, depending on parameters $\boldsymbol{\beta}$, with additional nuisance parameters being estimated via nonparametric maximum likelihood. Great care is required in this context because unless the sample space for $\boldsymbol{x}$ is finite (i.e., the covariates are all discrete with a fixed number of categories), the dimension of the nuisance parameter increases with the sample size.

7.10 Case-Control Studies

To summarize, when data are collected from a case-control study, a likelihood-based analysis with a logistic regression model may proceed with asymptotic inference, acting as if the data were collected in a cohort fashion, except that the intercept is no longer interpretable as the baseline log odds of disease.

7.10.3 Estimation for a Matched Case-Control Study

A common approach in epidemiological studies is to "match" the controls to the cases on the basis of known confounders. By choosing controls to be similar to cases, one "controls" for the confounding variables. This provides efficiency gains since the controls are more similar to the cases with respect to confounders, which increases power. It also removes the need to model the disease-confounder relationship.

In a *frequency-matched* design, the cases are grouped into broad strata (e.g., 10-year age bands), and controls are matched on the basis of these variables. In an *individually matched* study, controls are matched exactly, usually upon multiple variables, for example, age, gender, time of diagnosis, and area of residence. For both forms of matching, the nonrandom selection of controls must be acknowledged in the analysis by including a parameter for each matching set in the logistic model.

For matched data, let $j = 1, \ldots, J$ index the matched sets, and Y_{ij} and $\boldsymbol{x}_{ij}$ denote the responses and covariate vector of additional variables (i.e., beyond the matching variables) for individual i, with $i = 1, \ldots, m_{1j}$ representing the cases and $i = m_{1j} + 1, \ldots, m_{1j} + m_{0j}$ the controls. Hence, for $j = 1, \ldots, J$,

$$y_{ij} = 1 \quad \text{for} \quad i = 1, \ldots, m_{1j}$$

$$y_{ij} = 0 \quad \text{for} \quad i = m_{1j} + 1, \ldots, m_{1j} + m_{0j},$$

and there are $m_1 = \sum_{j=1}^{J} m_{1j}$ cases and $m_0 = \sum_{j=1}^{J} m_{0j}$ controls in total.
The disease model is

$$\log \left[\frac{p_j(\boldsymbol{x}_{ij})}{1 - p_j(\boldsymbol{x}_{ij})} \right] = \alpha_j + \boldsymbol{x}_{ij} \boldsymbol{\beta} \tag{7.33}$$

where

$$p_j(\boldsymbol{x}_{ij}) = \Pr(Y_{ij} = 1 \mid \boldsymbol{x}_{ij}, \text{stratum } j)$$

for $i = 1, \ldots, m_{0j} + m_{1j}, j = 1, \ldots, J$. In terms of inference, the key distinction between the two matching situations is that in the frequency matching situation, the number of matching strata J is fixed. In this case, the result outlined in Sect. 7.10.2 can be extended so that the matched data can be analyzed as if they were gathered prospectively, though the intercept parameters α_j are no longer interpretable as log odds ratios describing the association between disease and the variables defining stratum j. For the same reason, it is not possible to estimate interactions between stratum variables and exposures of interest. Calculations in Breslow and Day (1980) show that, in terms of efficiency gains, it is usually not worth exceeding 5 controls per case and 3 will often be sufficient. Exercise 7.8 considers the analysis of a particular set of data to illustrate the benefits of case-control sampling and matching.

For individually matched data, for simplicity, suppose there are M controls for each case so that $m_{1j} = 1$ and $m_{0j} = M$ for all j. Hence, $m_1 = J$ and $m_0 = MJ = Mm_1$. Also let $n = m_1$ represent the number of cases so that $m_0 = Mn$ is the number of controls. The likelihood contribution of the jth stratum is

$$p(\boldsymbol{x}_{1j} \mid Y_{1j} = 1) \prod_{i=2}^{M+1} p(\boldsymbol{x}_{ij} \mid Y_{ij} = 0), \tag{7.34}$$

but care is required for inference because the number of nuisance parameters, $\alpha_1, \ldots, \alpha_n$, is equal to the number of cases/matching sets, n, and so increases with sample size.

To overcome this violation of the usual regularity conditions, a conditional likelihood may be constructed. Specifically, for each j, one conditions on the collection of $M + 1$ covariate vectors within each matching set. The conditional contribution is the probability that subject $i = 1$ is the case, given it could have been any of the $M + 1$ subjects within that matching set. The numerator is (7.34), and the denominator is this expression but evaluated under the possibility that each of the $i = 1, \ldots, M + 1$ individuals could have been the case. Hence, the jth contribution to the conditional likelihood is

$$\frac{p(\boldsymbol{x}_{1j} \mid Y_{1j} = 1) \prod_{i=2}^{M+1} p(\boldsymbol{x}_{ij} \mid Y_{ij} = 0)}{\sum_{R_j} p(\boldsymbol{x}_{\pi(1),j} \mid Y_{1j} = 1) \prod_{i=2}^{M+1} p(\boldsymbol{x}_{\pi(i),j} \mid Y_{ij} = 0)}$$

where R_j is the set of $M + 1$ permutations, $[\boldsymbol{x}_{\pi(1),j}, \ldots, \boldsymbol{x}_{\pi(M+1),j}]$ of $[\boldsymbol{x}_{1j}, \ldots, \boldsymbol{x}_{M+1,j}]$. Applying Bayes theorem to each term,

$$p(\boldsymbol{x}_{ij} \mid Y = y) = \frac{p(Y = y \mid \boldsymbol{x}_{ij}) p(\boldsymbol{x}_{ij})}{p(Y = y)},$$

and taking the product across matching sets, we obtain

$$L_c(\boldsymbol{\beta}) = \prod_{j=1}^{n} \frac{p(Y_{1j} = 1 \mid \boldsymbol{x}_{1j}) \prod_{i=2}^{M+1} p(Y_{ij} = 0 \mid \boldsymbol{x}_{ij})}{\sum_{R_j} p(Y_{1j} = 1 \mid \boldsymbol{x}_{\pi(1),j}) \prod_{i=2}^{M+1} p(Y_{ij} = 0 \mid \boldsymbol{x}_{\pi(i),j})}.$$

Substitution of the logistic disease model (7.33) yields the conditional likelihood

$$L_c(\boldsymbol{\beta}) = \prod_{j=1}^{n} \frac{\exp(\boldsymbol{x}_{1j}\boldsymbol{\beta})}{\sum_{i=1}^{M+1} \exp(\boldsymbol{x}_{ij}\boldsymbol{\beta})}$$

$$= \prod_{j=1}^{n} \left(1 + \sum_{i=2}^{M+1} \exp\left[(\boldsymbol{x}_{ij} - \boldsymbol{x}_{1j})\boldsymbol{\beta}\right]\right)^{-1}$$

7.11 Concluding Remarks

Table 7.8 Notation for a matched-pair case-control study with n controls and n cases and a single exposure

		Not diseased $Y=0$	Diseased $Y=1$
Unexposed	$X=0$	m_{00}	m_{01}
Exposed	$X=1$	m_{10}	m_{11}
		n	n

with the α_j terms having canceled out, as was required. For further details, see Cox and Snell (1989) and Prentice and Pyke (1979, Sect. 6). As an example, if $M = 2$ (two controls per case), the conditional likelihood is

$$L_c(\boldsymbol{\beta}) = \prod_{j=1}^{n} \frac{\exp(\boldsymbol{x}_{1j}\boldsymbol{\beta})}{\exp(\boldsymbol{x}_{1j}\boldsymbol{\beta}) + \exp(\boldsymbol{x}_{2j}\boldsymbol{\beta}) + \exp(\boldsymbol{x}_{3j}\boldsymbol{\beta})}$$

$$= \prod_{j=1}^{n} \left(1 + \sum_{i=2}^{3} \exp\left[(\boldsymbol{x}_{ij} - \boldsymbol{x}_{1j})\boldsymbol{\beta}\right]\right)^{-1}.$$

The importance of the use of conditional likelihood can be clearly demonstrated in the matched-pairs situation, in which there is one control per case. Suppose that the data are as summarized in Table 7.8 so that there is a single exposure only. There are m_{00} concordant pairs in which neither case nor control is exposed and m_{11} concordant pairs in which both are exposed. Exercise 7.12 shows that the unconditional MLE of the odds ratio is $(m_{10}/m_{01})^2$, the square of the ratio of discordant pairs. In contrast, the estimate based on the appropriate conditional likelihood is m_{10}/m_{01}. Hence, the unconditional estimator is the square of the correct conditional estimator.

A further caveat to the use of individually matched case-control data is that it is more difficult to generalize inference to a specific population under this design because the manner of selection is far from that of a random sample.

7.11 Concluding Remarks

The analysis of binomial data is difficult unless the denominators are large because there is so little information in a single Bernoulli outcome. In addition, the models for probabilities are typically nonlinear. Logistic regression models are the obvious candidate for analysis, but the interpretation of odds ratios is not straightforward, unless the outcome of interest is rare. The effect of omitting variables is also nonobvious. The fact that the linear logistic model is a GLM does offer advantages in terms of consistency, however, and the logit being the canonical link gives simplifications in terms of computation.

The use of conditional likelihood in individually matched case-control studies in practice is uncontroversial, but its theoretical underpinning is not completely convincing (since the conditioning statistic is not ancillary). Fisher's exact test is historically popular, but, as discussed in Sect. 4.2, frequentist hypothesis testing can be difficult to implement in practice since p-values need to be interpreted in the context of the sample size. For Fisher's exact, the discreteness of the test statistic can also be problematic. Exercise 7.11 provides an alternative approach based on Bayes factors. The latter do not suffer from the discreteness of the sampling distribution (since one only uses the observed data and not other hypothetical realizations).

7.12 Bibliographic Notes

Robinson and Jewell (1991) examine the effects of omission of variables in logistic regression models and contrast the implications with the linear model case. Greenland et al. (1999) is a wide-ranging discussion on collapsibility and confounding. A seminal book on the design and analysis of case-control studies is Breslow and Day (1980). There is no Bayesian analog of the Prentice and Pyke (1979) result showing the equivalence of odds ratio estimation for prospective and retrospective sampling, though Seaman and Richardson (2004) show the equivalence in restricted circumstances. Simplified estimation based on nonparametric maximum likelihood has also been established for other outcome-dependent sampling schemes such as two-phase sampling; see, for example, White (1982) and Breslow and Chatterjee (1999). Again, no equivalent Bayesian approaches are available. A fully Bayesian approach in a case-control setting would require the modeling of the covariate distributions for each of the cases and controls, which is, in general, a difficult process and seems unnecessary given that there is no direct interest in these distributions. Hence, the nonparametric maximum likelihood procedure seems preferable, though a hybrid approach in which one simply combines the prospective likelihood with a prior would seem practically reasonable if one has prior information and/or one is worried about asymptotic inference.

Rice (2008) shows the equivalence between conditional likelihood and random effects approaches to the analysis of matched-pairs case-control data. In general, conditional likelihood does not have a Bayesian interpretation, though Bayesian analyses have been carried out in the individually matched case-control situation by combining a prior with the conditional likelihood. This approach avoids the difficulty of specifying priors over nuisance parameters with dimension equal to the number of matching sets (Diggle et al. 2000).

Fisher's exact test has been discussed extensively in the statistics literature; see, for example, Yates (1984). Altham (1969) published an intriguing result showing that Fisher's exact test is equivalent to a Bayesian analysis. Specifically, let $p_{00}, p_{10}, p_{01}, p_{11}$ denote the underlying probabilities in a 2×2 table with entries $\boldsymbol{y} = [y_{00}, y_{10}, y_{01}, y_{11}]^\mathrm{T}$ (see Table 7.3), and suppose the prior on these probabilities is (improper) Dirichlet with parameters (0,1,1,0). Then the posterior probability

$\Pr(p_{11}p_{22}/p_{12}p_{21} < 1 \mid \boldsymbol{y})$ equals the Fisher's exact test p-value for testing $H_0 : p_{00}p_{11} = p_{10}p_{01}$ versus $H_1 : p_{00}p_{11} < p_{10}p_{01}$. Hence, the prior (slightly) favors a negative association between rows and columns, which is related to the fact that conditioning on the margins (as is done in Fisher's exact test) does lead to a small loss of information.

7.13 Exercises

7.1 Suppose $Z \mid p \sim \text{Bernoulli}(p)$.

(a) Show that the moment-generating function (Appendix D) of Z is $M_Z(t) = 1 - p + p\exp(t)$. Hence, show that the moment-generating function of $Y = \sum_{i=1}^{n} Z_i$ is
$$M_Y = [1 - p + p\exp(t)]^n,$$
which is the moment-generating function of a binomial random variable.

(b) Suppose $Y \mid \lambda \sim \text{Poisson}(\lambda)$. Show that the cumulant-generating function (Appendix D) of Y is
$$\lambda[\exp(t) - 1].$$

(c) From part (a), obtain the form of the cumulant-generating function of Y. Suppose that $p \to 0$ and $n \to \infty$ in such a way that $\mu = np$ remains fixed. By considering the limiting form of the cumulant-generating function of Y, show that in this situation, the limiting distribution of Y is Poisson with mean μ.

7.2 Before the advent of GLMs, the arc sine variance stabilizing transformation was used for the analysis of binomial data. Suppose that $Y \mid p \sim \text{Binomial}(N, p)$ with N large. Using a Taylor series expansion, show that the random variable
$$W = \arcsin\left[\sqrt{(Y/N)}\right]$$
has approximate first two moments:
$$E[W] \approx \arcsin(\sqrt{p}) - \frac{1 - 2p}{8\sqrt{Np(1-p)}}$$
$$\text{var}(W) \approx \frac{1}{4N}.$$

7.3 Suppose $Z_j \mid \lambda_j \sim_{ind} \text{Poisson}(\lambda_j)$, $j = 1, 2$ are independent Poisson random variables with rates λ_j. Show that
$$Z_1 \mid Z_1 + Z_2, p \sim \text{Binomial}(Z_1 + Z_2, p),$$
with $p = \lambda_1/(\lambda_1 + \lambda_2)$.

7.4 Consider n Bernoulli trials with Z_{ij}, $j = 1, \ldots, N_i$ the outcomes within-trial i with $Y_i = \sum_{j=1}^{N_i} Z_{ij}$, $i = 1, \ldots, n$. By writing

$$\text{var}(Y_i) = \sum_{j=1}^{N_i} \text{var}(Z_{ij}) + \sum_{j=1}^{N_i}\sum_{j \neq k} \text{cov}(Z_{ij}, Z_{ik}),$$

show that

$$\text{var}(Y_i) = N_i p_i (1 - p_i) \times [1 + (N_i - 1)\tau_i^2].$$

7.5 With respect to Sect. 7.6.2, show that for Bernoulli data the Pearson statistic is $X^2 = n$. Find the deviance in this situation and comments on its usefulness as a test of goodness of fit.

7.6 Show that the extended hypergeometric distribution (7.22) is a member of the exponential family (Sect. 6.3), that is, the distribution can be written in the form

$$\Pr(y_{11} \mid \theta, \alpha) = \exp\left(\frac{y_{11}\theta - b(\theta)}{\alpha} + c(y, \alpha)\right)$$

for suitable choices of α, $b(\cdot)$, and $c(\cdot, \cdot)$.

7.7 In this question, a simulation study to investigate the impact on inference of omitting covariates in logistic regression will be performed, in the situation in which the covariates are independent of the exposure of interest. Let x be the covariate of interest and z another covariate. Suppose the true (adjusted) model is $Y_i \mid x_i, z_i \sim_{iid} \text{Bernoulli}(p_i)$, with

$$\log\left(\frac{p_i}{1 - p_i}\right) = \beta_0 + \beta_1 x_i + \beta_2 z_i. \tag{7.35}$$

A comparison with the unadjusted model $Y_i \mid x_i \sim_{iid} \text{Bernoulli}(p_i^\star)$, where

$$\log\left(\frac{p_i^\star}{1 - p_i^\star}\right) = \beta_0^\star + \beta_1^\star x_i, \tag{7.36}$$

for $i = 1, \ldots, n = 1{,}000$ will be made. Suppose x is binary with $\Pr(X=1) = 0.5$ and $Z \sim_{iid} N(0,1)$ with x and z independent. Combinations of the parameters $\beta_1 = 0.5, 1.0$ and $\beta_2 = 0.5, 1.0, 2.0, 3.0$, with $\beta_0 = -2$ in all cases, will be considered.

For each combination of parameters, compare the results from the two models, (7.35) and (7.36), with respect to:

(a) $\text{E}[\widehat{\beta}_1]$ and $\text{E}[\widehat{\beta}_1^\star]$, as compared to β_1
(b) The standard errors of $\widehat{\beta}_1$ and $\widehat{\beta}_1^\star$
(c) The coverage of 95% confidence intervals for β_1 and $\beta_1^\star$
(d) The probability of rejecting $H_0 : \beta_1 = 0$ in model (7.35) and the probability of rejecting $H_0 : \beta_1^\star = 0$ in model (7.36). These probabilities correspond to the powers of the tests. Calculate these probabilities using Wald tests.

7.13 Exercises

Table 7.9 *Left table*: leprosy cases and non-cases versus presence/absence of BCG scar. *Right table*: leprosy cases and controls versus presence/absence of BCG scar

BCG scar	Cases	Non-cases	BCG scar	Cases	Controls
Present	101	46,028	Present	101	554
Absent	159	34,594	Absent	159	446

Based on the results, summarize the effect of omitting a covariate that is independent of the exposure of interest, in particular in comparison with the linear model case (as discussed in Sect. 5.9).

7.8 This question illustrates the benefits of case-control and matched case-control sampling, taking data from Fine et al. (1986) and following loosely the presentation of Clayton and Hills (1993). Table 7.9 gives data from a cross-sectional survey carried out in Northern Malawi. The aim of this study was to investigate whether receiving a bacillus Calmette-Guérin (BCG) vaccination in early childhood (which protects against tuberculosis) gives any protection against leprosy. Let $X = 0/1$ denote absence/presence of BCG scar, $Y = 0/1$ denote leprosy-free/leprosy, and $p_x = \Pr(Y = 1 \mid X = x)$, $x = 0, 1$:

(a) Fit the logistic model

$$\log\left(\frac{p_x}{1 - p_x}\right) = \beta_0 + \beta_1 x$$

to the case/non-case data in the left half of Table 7.9. Report your findings in terms of an estimate of the odds ratio $\exp(\beta_1)$ along with an associated standard error.

(b) Now consider the case/control data in the right half of Table 7.9 (these data were simulated from the full dataset). Fit the logistic model

$$\log\left(\frac{p_x}{1 - p_x}\right) = \beta_0^\star + \beta_1 x$$

to the case/control data, and again report your findings in terms of the odds ratio $\exp(\beta_1)$ along with an associated standard error. Hence, use this example to describe the benefits, in terms of efficiency, of a case-control study.

(c) In this example, the population data are known and consequently the sampling fractions of cases and controls are also known. Hence, reconstruct an estimate of β_0, using the results from the case-control analysis.

(d) Next the benefits of matching will be illustrated. BCG vaccination was gradually introduced into the study region, and so older people are less likely to have been vaccinated but also more likely to have developed leprosy. Therefore, age is a potential confounder in this study.

Let $z = 0, 1, \ldots, 6$ denote age represented as a factor and $p_{xz} = \Pr(Y = 1 \mid X = x,)$, for $x = 0, 1$, denote the probability of leprosy

Table 7.10 *Left table*: leprosy cases and non-cases as a function of presence/absence of BCG scar and age. *Right table*: leprosy cases and matched controls as a function of presence/absence of BCG scar and age

| | BCG scar | | | | | BCG scar | | | |
| | Cases | | Non-cases | | | Cases | | Controls | |
Age	Absent	Present	Absent	Present	Age	Absent	Present	Absent	Present
0–4	1	1	7,593	11,719	0–4	1	1	3	5
5–9	11	14	7,143	10,184	5–9	11	14	48	52
10–14	28	22	5,611	7,561	10–14	28	22	67	133
15–19	16	28	2,208	8,117	15–19	16	28	46	130
20–24	20	19	2,438	5,588	20–24	20	19	50	106
25–29	36	11	4,356	1,625	25–29	36	11	126	62
30–34	47	6	5,245	1,234	30–34	47	6	174	38

for an individual with BCG status x and in age strata z. To adjust for age, fit the logistic model

$$\log\left(\frac{p_{xz}}{1-p_{xz}}\right) = \beta_0 + \beta_1 x + \beta_z z$$

to the data in the left half of Table 7.10. This model assumes a common odds ratio across age strata. Report your findings in terms of the odds ratio $\exp(\beta_1)$ and associated standard error.

(e) If it were possible to sample controls from the non-cases in the left half of Table 7.10, the age distribution would be highly skewed toward the young, which would lead to an inefficient analysis. As an alternative, the right half of Table 7.10 gives a simulated frequency-matched case-control study with 4 controls per case within each age strata. Analyze these data using the logistic model

$$\log\left(\frac{p_{xz}}{1-p_{xz}}\right) = \beta_0^\star + \beta_1 x + \beta_z^\star z,$$

and report your findings in terms of $\exp(\beta_1)$ and its associated standard error. Comment on the accuracy of inference as compared to the analysis using the complete data.

7.9 Table 7.11 gives data from a toxicological experiment in which the number of beetles that died after 5 h exposure to gaseous carbon disulphide at various doses.

(a) Fit complementary log–log, probit, and logit link models to these data using likelihood methods.
(b) Summarize the association for each model in simple terms.
(c) Examine residuals and report the model that you believe provides the best fit to these data, along with your reasoning.

7.13 Exercises

Table 7.11 Number of beetle deaths as a function of log dose, from Bliss (1935)

Log dose	No. beetles	No. killed
1.691	59	6
1.724	60	13
1.755	62	18
1.784	56	28
1.811	63	52
1.837	59	53
1.861	62	61
1.884	60	60

Table 7.12 Death penalty verdict by race of victim and defendant

Defendant's race	Victim's race	Death penalty Yes	No
White	White	19	132
	Black	0	9
Black	White	11	52
	Black	06	97

(d) Fit your favored model with a Bayesian approach using (improper) flat priors. Is there a substantive difference in the conclusions, as compared to the likelihood analysis?

7.10 Table 7.12 contains data from Radelet (1981) on death penalty verdict, cross-classified by defendant's race and victim's race.

(a) Fit a logistic regression model that includes factors for both defendant's race and victim's race. Estimate the odds ratios associated with receiving the death penalty if Black as compared to if White, for the situations in which the victim was White and in which the victim was Black.

(b) Fit a logistic regression model to the marginal 2×2 table that collapses across victim's race, and hence, estimate the odds ratio associated with receiving the death penalty if Black versus if White.

(c) Discuss the results of the two parts, in relation to Simpson's paradox. In particular, discuss the paradox in terms understandable to a layperson.

7.11 Suppose $Y_i \mid p_i \sim \text{Binomial}(N_i, p_i)$ for $i = 0, 1$ and that interest focuses on $H_0 : p_0 = p_1 = p$ versus $H_1 : p_0 \neq p_1$:

(a) Consider the Bayes factor (Sect. 3.10)

$$\text{BF} = \frac{\Pr(y_0, y_1 \mid H_0)}{\Pr(y_0, y_1 \mid H_1)}$$

with the priors: $p \sim \text{Be}(a_0, b_0)$ under H_0 and $p_i \sim \text{Be}(a_1, b_1)$, for $i = 0, 1$, under H_1. Obtain a closed-form expression for the Bayes factor.

(b) Calculate the Bayes factor for the tumor data given in Table 7.4 using uniform priors, that is, $a_0 = a_1 = b_0 = b_1 = 1$.

(c) Based on the Bayes factor, would you reject H_0? Why?

(d) Using the same priors as in the previous part, evaluate the posterior probability that $\Pr(p_0 > p_1 \mid y_0, y_1)$. Based on this probability, what would you conclude about equality of p_0 and p_1? Is your conclusion in agreement with the previous part?

[Hint: Obtaining samples from the posteriors $p(p_i \mid y_i)$, for $i = 0, 1$, is a simple way of obtaining the posterior of interest in the final part.]

7.12 This question derives unconditional and conditional estimators for the case of a matched-pairs case-control design with n pairs and a binary exposure. The notation is given in Table 7.8, and the logistic model in the jth matching set is

$$\Pr(Y = 1 \mid x, j) = \frac{\exp(\alpha_j + x\beta)}{1 + \exp(\alpha_j + x\beta)},$$

for $x = 0, 1$ and $j = 1, \ldots, n$.

(a) Show that the unconditional maximum likelihood estimator of β is the square of the ratio of the discordant pairs, $(m_{10}/m_{01})^2$.
(b) Show, by considering the distribution of m_{10} given the total $m_{10} + m_{01}$, that the estimate based on the appropriate conditional likelihood is m_{10}/m_{01}.

Part III
Dependent Data

Chapter 8
Linear Models

8.1 Introduction

In Part III of the book the conditional independence assumptions of Part II are relaxed as we consider models for dependent data. Such data occur in many contexts, with three common situations being when sampling is over time, space, or within families. We do not discuss pure time series applications in which data are collected over a single (usually long) series; this is a vast topic with many specialized texts. Generically, we consider regression modeling situations in which there are a set of units ("clusters") upon which multiple measurements have been collected. For example, when data are available over time for a group of units, we have *longitudinal* (also known as *repeated measures*) data, and each unit forms a cluster. We will often refer to the units as individuals. The methods described in Part II for calculating uncertainty measures (such as standard errors) are not applicable in situations in which the data are dependent.

Throughout Part III we distinguish approaches that specify a full probability model for the data (with likelihood or Bayesian approaches to inference) and those that specify first, and possibly second, moments only (with an estimating function being constructed for inference). As in Part II we believe it will often be advantageous to carry out inference from both standpoints in a complimentary fashion. In some instances the form of the question of interest may be best served by a particular approach, however, and this will be stressed at relevant points.

In this chapter we consider linear regression models. Such models are widely applicable with growth curves, such as the dental data of Sect. 1.3.5, providing a specific example. As another example, in the so-called *split-plot* design, fields are planted with different crops and within each field (unit), different subunits are treated with different fertilizers. We expect crop yields in the same field to be more similar than those in different fields, and yields may be modeled as a linear function of crop and fertilizer effects. With clustered data, we expect measurements on the same unit to exhibit *residual* dependence due to shared unmeasured variables, where the qualifier acknowledges that we have controlled for known regressors.

The structure of this chapter is as follows. We begin, in Sect. 8.2, with a brief overview of approaches to inference for dependent data, in the context of the dental data of Sect. 1.3.5. Section 8.3 provides a description of the efficiency gains that can be achieved with data collected over time in a longitudinal design. In Figure 8.1(a), linear mixed effects models, in which full probability models are specified for the data, are introduced. In Sects. 8.5 and 8.6, likelihood and Bayesian approaches to inference for these models are described. Section 8.7 discusses the generalized estimating equations (GEE) approach which is based on a marginal mean specification and empirical sandwich estimation of standard errors. We describe how the assumptions required for valid inference may be assessed in Sect. 8.8 and discuss the estimation of longitudinal and cohort effects in Sect. 8.9. Concluding remarks appear in Sect. 8.10 with bibliographic notes in Sect. 8.11.

8.2 Motivating Example: Dental Growth Curves

In Table 1.3 dental measurements of the distance in millimeters from the center of the pituitary gland to the pteryo-maxillary fissure are given for 11 girls and 16 boys, recorded at the ages of 8, 10, 12, and 14 years. In this section we concentrate on the data from the girls only. Figure 8.1(a) plots the dental measurements for each girl versus age. The slopes look quite similar, though there is clearly between-girl variability in the intercepts.

There are various potential aims for the analysis of data such as these:

1. Population inference, in which we describe the average growth as a function of age, for the population from which the sample of children were selected.
2. Assessment of the within- to between-child variability in growth measurements.

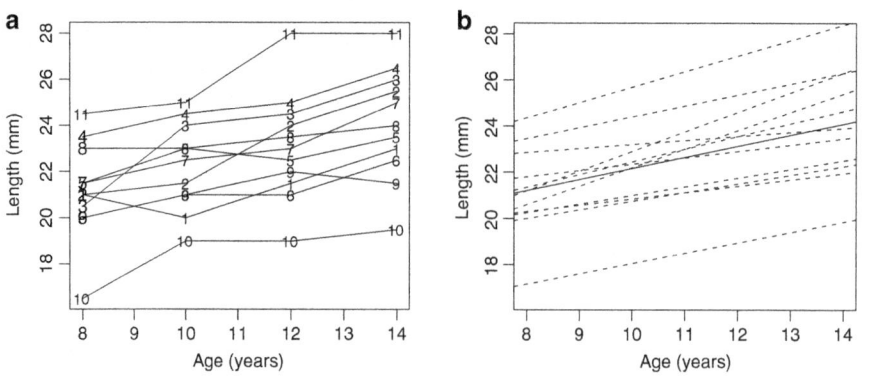

Fig. 8.1 Dental data for girls only: (**a**) individual observed data (with the girl index taken as plotting symbol), (**b**) individual *fitted curves* (*dashed*) and overall *fitted curve* (*solid*)

3. Individual-level inference, either for a child in the sample, or for a new unobserved child (from the same population). The latter could be used to construct a "growth chart" in which the percentile points of children's measurements at different ages are presented.

Part III of the book will provide extensive discussion of *mixed effects* models which contain both *fixed* effects that are shared by all individuals and *random effects* that are unique to particular individuals and are assumed to arise from a distribution. For longitudinal data there are two extreme fixed effects approaches. Proceeding naively, we could assume a single "marginal" curve for *all* of the girls data and carry out a standard analysis assuming independent data. Marginal here refers to *averaging* over girls in the population. At the other extreme we could assume a distinct curve for each girl. Figure 8.1(b) displays the least squares fitted lines corresponding to each of these fixed effects approaches.

Continuing with the marginal approach, let Y_{ij} denote the jth measurement, taken at time t_j on the ith child, $i = 1, \ldots, m = 11, j = 1, \ldots, n_i = 4$. Consider the model

$$E[Y_{ij}] = \beta_0^M + \beta_1^M t_j \tag{8.1}$$

where β_0^M and β_1^M represent *marginal* intercept and slope parameters. Then,

$$e_{ij}^M = Y_{ij} - \beta_0^M - \beta_1^M t_j,$$

$i = 1, \ldots, 11; j = 1, \ldots, 4$, denote marginal residuals. In Part II of the book, we emphasized conditional independence, so that observations were independent *given* a set of parameters; due to dependence of observations on the same girl, we would not expect the marginal residuals to be independent.

We fit the marginal model (8.1) to the data from all girls and let

$$\begin{bmatrix} \sigma_1 & & & \\ \rho_{12} & \sigma_2 & & \\ \rho_{13} & \rho_{23} & \sigma_3 & \\ \rho_{14} & \rho_{24} & \rho_{34} & \sigma_4 \end{bmatrix} \tag{8.2}$$

represent the standard deviation/correlation matrix of the residuals. Here,

$$\sigma_j = \sqrt{\text{var}(e_{ij}^M)},$$

is the standard deviation of the dental length at time t_j and

$$\rho_{jk} = \frac{\text{cov}(e_{ij}^M, e_{ik}^M)}{\sqrt{\text{var}(e_{ij}^M)\text{var}(e_{ik}^M)}},$$

is the correlation between residual measurements taken at times t_j and t_k on the same girl, $j \neq k, j, k = 1, \ldots, 4$. We assume four distinct standard deviations at each of the ages, and distinct correlations between measurements at each of the six combinations of pairs of ages, but assume that these standard deviations and correlations are constant across all girls. We empirically estimate the entries of (8.2) as

$$\begin{bmatrix} 2.12 & & & \\ 0.83 & 1.90 & & \\ 0.86 & 0.90 & 2.36 & \\ 0.84 & 0.88 & 0.95 & 2.44 \end{bmatrix} \tag{8.3}$$

illustrating that, not surprisingly, there is clear correlation between residuals at different ages on the same girl. Fitting a single curve to the totality of the data and using methods for independent data that assume within-girl correlations are zero will clearly give inappropriate standard errors/uncertainty estimates for $\widehat{\beta}_0^M$ and $\widehat{\beta}_1^M$. Fitting such a marginal model is appealing, however, since it allows the *direct* calculation of the average responses at different ages. Fitting a marginal model forms the basis of the GEE approach described in Sect. 8.7.

The alternative fixed effects approach is to assume a fixed curve for each child and analyze each set of measurements separately. However, while providing valid inference for each curve, there is no "borrowing of strength" across children, so that each girl's fit is based solely on her data only and not on the data of other children. We might expect that there is *similarity* between the curves, and therefore, it is reasonable to believe that the totality of data will enhance estimation for each child. In some instances, using the totality of data will be vital. For example, estimating the growth curve for a girl with just a single observation is clearly not possible using the observed data on that girl only. Suppose we are interested in making formal inference for the population of girls from which the $m = 11$ girls are viewed as a random sample; this is not formally possible using the collection of fixed effects estimates from each girl. The basis of the mixed effects model approach described in Sect. 8.4 is to assume a girl-specific set of *random effect* parameters that are assumed to arise from a population. In different contexts, random effects may have a direct interpretation as arising from a population of effects, or may simply be viewed as a convenient modeling tool, in situations in which there is no hypothetical population of effects to appeal to.

Throughout Part III, we will describe mixed effects and GEE approaches to analysis. The mixed effects approach can be seen as having a greater *contextual* basis, since it builds up a model from the level of the unit. In contrast, with a marginal model, as specified in GEE, the emphasis is on population inference based on *minimal* assumptions and on obtaining a reliable standard error via sandwich estimation.

8.3 The Efficiency of Longitudinal Designs

While making inference for dependent data is in general more difficult than for independent data, designs that collect dependent data can be very efficient. For example, in a longitudinal data setting, applying different treatments to the same

8.3 The Efficiency of Longitudinal Designs

patient over time can be very beneficial, since each patient acts as his/her own control. To illustrate, we provide a comparison between longitudinal and cross-sectional studies (in which data are collected at a single time point); this section follows the development of Diggle et al. (2002, Sect. 2.3).

We consider a very simple situation in which we wish to compare two treatments, coded as -1 and $+1$, and take four measurements in total. In the cross-sectional study a single measurement is taken on each of four individuals with

$$Y_{i1} = \beta_0 + \beta_1 x_{i1} + \epsilon_{i1}, \tag{8.4}$$

for $i = 1, \ldots, m = 4$. The error terms ϵ_{i1} are independent with $E[\epsilon_{i1}] = 0$ and $\text{var}(\epsilon_{i1}) = \sigma^2$. The design is such that $x_{11} = -1, x_{21} = -1, x_{31} = 1, x_{41} = 1$, so that individuals 1 and 2 (3 and 4) receive treatment -1 ($+1$). With this coding, the treatment effect is

$$E[Y_1 \mid x = 1] - E[Y_1 \mid x = -1] = 2\beta_1.$$

The (unbiased) ordinary least squares (OLS) estimators are

$$\widehat{\beta}_0^c = \frac{\sum_{i=1}^{4} Y_{i1}}{4}, \quad \widehat{\beta}_1^c = \frac{Y_{31} + Y_{41} - (Y_{11} + Y_{21})}{4},$$

and, more importantly for our purposes, the variance of the treatment estimator is

$$\text{var}(\widehat{\beta}_1^c) = \frac{\sigma^2}{4}.$$

The subscript here labels the relevant quantities as arising from the cross-sectional design.

For the longitudinal study we assume the model

$$Y_{ij} = \beta_0 + \beta_1 x_{ij} + b_i + \delta_{ij},$$

with b_i and δ_{ij} independent and $E[\delta_{i1}] = 0$, $\text{var}(\delta_{ij}) = \sigma_\delta^2$, $E[b_i] = 0$, $\text{var}(b_i) = \sigma_0^2$, for $i = 1, 2, j = 1, 2$, so that we record two observations on each of two individuals. The b_i represent random individual-specific parameters and ϵ_{ij} measurement error. Marginally, that is, averaging over individuals, and with $\boldsymbol{Y} = [Y_{11}, Y_{12}, Y_{21}, Y_{22}]^T$, we have $\text{var}(\boldsymbol{Y}) = \sigma^2 \boldsymbol{R}$ with

$$\boldsymbol{R} = \begin{bmatrix} 1 & \rho & 0 & 0 \\ \rho & 1 & 0 & 0 \\ 0 & 0 & 1 & \rho \\ 0 & 0 & \rho & 1 \end{bmatrix} \tag{8.5}$$

where $\sigma^2 = \sigma_0^2 + \sigma_\delta^2$ is the sum of the between- and within-individual variances and $\rho = \sigma_0^2/\sigma^2$ is the correlation between observations on the same individual. Notice that the cross-sectional variance model is a special case of (8.4) with $\epsilon_{i1} = b_i + \delta_{i1}$. We consider two designs. In the first, the treatment is constant over time for each individual: $x_{11} = x_{12} = -1, x_{21} = x_{22} = 1$, while in the second each individual receives both treatments: $x_{11} = x_{22} = 1, x_{12} = x_{21} = -1$. Generalized least squares gives unbiased estimator

$$\widehat{\boldsymbol{\beta}}^{\text{L}} = (\boldsymbol{x}^{\text{T}}\boldsymbol{R}^{-1}\boldsymbol{x})^{-1}\boldsymbol{x}^{\text{T}}\boldsymbol{R}^{-1}\boldsymbol{Y}, \tag{8.6}$$

with

$$\text{var}(\widehat{\boldsymbol{\beta}}^{\text{L}}) = (\boldsymbol{x}^{\text{T}}\boldsymbol{R}^{-1}\boldsymbol{x})^{-1}\sigma^2,$$

and where $\boldsymbol{R}$ is given by (8.5). The variance of the "slope" estimator is

$$\text{var}(\widehat{\beta}_1^{\text{L}}) = \frac{\sigma^2(1-\rho^2)}{4 - 2\rho(x_{11}x_{12} + x_{21}x_{22})}.$$

The *efficiency* of the longitudinal design, as compared to the cross-sectional design, is therefore

$$\frac{\text{var}(\widehat{\beta}_1^{\text{L}})}{\text{var}(\widehat{\beta}_1^{\text{C}})} = \frac{(1-\rho^2)}{1 - \rho(x_{11}x_{12} + x_{21}x_{22})/2}.$$

The efficiency of the longitudinal study with *constant* treatments across individuals is

$$1 + \rho,$$

so that in this case, the cross-sectional study is preferable in the usual situation in which observations on the same individual display positive correlation, that is, $\rho > 0$. When the treatment is constant within individuals, the treatment estimate is based on between-individual comparisons only, and so, it is more beneficial to obtain measurements on additional individuals.

The efficiency of the longitudinal study with treatments *changing* within individuals is

$$1 - \rho,$$

so that the longitudinal study is more efficient when $\rho > 0$, because each individual is acting as his/her own control. That is, we are making within-individual comparisons. If $\rho = 0$, the designs have the same efficiency. In practice, collecting two measurements on different individuals will often be logistically more straightforward than collecting two measurements on the same individual (e.g., with the possibility of missing data at the second time point), but in pure efficiency terms, the longitudinal design with changing treatment can be very efficient. Clearly, this discussion extends to other longitudinal situations in which covariates are changing over time (and more general situations with covariate variation within clusters).

8.4 Linear Mixed Models

8.4.1 The General Framework

The basic idea behind mixed effects models is to assume that each unit has a regression model characterized by both fixed effects, that are common to all units in the population, and unit-specific perturbations, or random effects. "Mixed" effects refers to the combination of both fixed and random effects. The frequentist interpretation of the random effects is that the units can be viewed as a random sample from a hypothetical super-population of units. A Bayesian interpretation arises through considerations of exchangeability (Sect. 3.9), as we discuss further in Sect. 8.6.2.

Let the multiple responses for the ith unit be $Y_i = [Y_{i1}, \ldots, Y_{in_i}]^T$, $i = 1, \ldots, m$. We assume that responses on different units are independent but that there is dependence between observations on the same unit. Let β represent a $(k+1) \times 1$ vector of fixed effects and b_i a $(q+1) \times 1$ vector of random effects, with $q \leq k$. In this chapter, we assume the mean for Y_{ij} is linear in the fixed and random effects. Let $x_{ij} = [1, x_{ij1}, \ldots, x_{ijk}]$ be a $(k+1) \times 1$ vector of covariates measured at occasion j, so that $x_i = [x_{i1}, \ldots, x_{in_i}]$ is the design matrix for the fixed effects for unit i. Similarly, let $z_{ij} = [1, z_{ij1}, \ldots, z_{ijq}]^T$ be a $(q+1) \times 1$ vector of variables that are a subset of x_{ij}, so that $z_i = [z_{i1}, \ldots, z_{in_i}]^T$ is the design matrix for the random effects.

We describe a two-stage linear mixed model (LMM).

Stage One: The response model, *conditional* on random effects b_i is

$$y_i = x_i\beta + z_i b_i + \epsilon_i, \tag{8.7}$$

where ϵ_i is an $n_i \times 1$ zero-mean vector of error terms, $i = 1, \ldots, m$.

Stage Two: The random terms in (8.7) satisfy

$$\mathrm{E}[\epsilon_i] = \mathbf{0}, \quad \mathrm{var}(\epsilon_i) = E_i(\alpha),$$
$$\mathrm{E}[b_i] = \mathbf{0}, \quad \mathrm{var}(b_i) = D(\alpha),$$
$$\mathrm{cov}(b_i, \epsilon_{i'}) = \mathbf{0}, \quad i, i' = 1, \ldots, m,$$

where α is an $r \times 1$ vector containing the collection of variance–covariance parameters. Further, $\mathrm{cov}(\epsilon_i, \epsilon_{i'}) = \mathbf{0}$ and $\mathrm{cov}(b_i, b_{i'}) = \mathbf{0}$, for $i \neq i'$.

The two stages may be collapsed, by averaging over the random effects, to give the marginal model:

$$\mathrm{E}[Y_i] = x_i\beta$$
$$\mathrm{var}(Y_i) = V_i(\alpha)$$
$$= z_i D(\alpha) z_i^T + E_i(\alpha) \tag{8.8}$$

for $i = 1, \ldots, m$, so that $V_i(\alpha)$ is an $n_i \times n_i$ matrix. The random effects have therefore induced dependence on an individual through the first term in (8.8). However, responses on individuals i and i', $i \neq i'$, are independent:

$$\text{cov}(Y_i, Y_{i'}) = 0$$

where $\text{cov}(Y_i, Y_{i'})$ is the $n_i \times n_{i'}$ matrix with element (j, j') corresponding to $\text{cov}(Y_{ij}, Y_{i'j'})$, $j = 1, \ldots, n_i$, $j' = 1, \ldots, n_{i'}$.

8.4.2 Covariance Models for Clustered Data

With respect to model (8.7), a common assumption is that $b_i \sim_{iid} N_{q+1}(0, D)$ and $\epsilon_i \sim_{ind} N_{n_i}(0, E_i)$. A common variance for all individuals and at all measurement occasions, along with uncorrelated errors, gives the simplified form $E_i = \sigma_\epsilon^2 I_{n_i}$. We will refer to σ_ϵ^2 as the measurement error variance, but as usual, the error terms may include contributions from model misspecification, such as departures from linearity, and data recording errors. The inclusion of random effects *induces* a marginal covariance model for the data. This may be contrasted with the direct specification of a marginal variance model. In this section we begin by deriving the marginal variance structure that arises from two simple random effects models, before describing more general covariance structures. It is important to examine the *marginal* variances and covariances, since these may be directly assessed from the observed data.

We first consider the random intercepts only model $z_i b_i = 1_{n_i} b_i$ with $\text{var}(b_i) = \sigma_0^2$, along with $E_i = \sigma_\epsilon^2 I_{n_i}$. From (8.8), it is straightforward to show that this stipulation gives the *exchangeable* or *compound symmetry* marginal variance model:

$$\text{var}(Y_i) = \sigma^2 \begin{bmatrix} 1 & \rho & \rho & \cdots & \rho \\ \rho & 1 & \rho & \cdots & \rho \\ \rho & \rho & 1 & \cdots & \rho \\ \vdots & \vdots & \vdots & \ddots & \vdots \\ \rho & \rho & \rho & \cdots & 1 \end{bmatrix}$$

where $\sigma^2 = \sigma_\epsilon^2 + \sigma_0^2$ and $\rho = \sigma_0^2/\sigma^2$. In this case we have two variance parameters so that $\alpha = [\sigma_\epsilon^2, \sigma_0^2]$. A consequence of between-individual variability in intercepts is therefore constant marginal within-individual correlation. The latter must be nonnegative under this model (since $\sigma_0^2 \geq 0$) which would seem reasonable in most situations.

The exchangeable model is particularly appropriate for clustered data with no time ordering as may arise, for example, in a split-plot design, or for multiple measurements within a family. It may be useful for longitudinal data also, particularly over short time scales. If we think of residual variability as being due to unmeasured variables, then the exchangeable structure is most appropriate when we believe such variables are relatively constant across responses within an individual.

We now consider a model with both random intercepts and random slopes. Such a model is a common choice in longitudinal studies. With respect to (8.7) and for $i = 1, \ldots, m$, the first stage model is

$$\begin{bmatrix} y_{i1} \\ y_{i2} \\ \vdots \\ y_{in_i} \end{bmatrix} = \begin{bmatrix} 1 & t_{i1} \\ 1 & t_{i2} \\ \vdots & \vdots \\ 1 & t_{in_i} \end{bmatrix} \begin{bmatrix} \beta_0 \\ \beta_1 \end{bmatrix} + \begin{bmatrix} 1 & t_{i1} \\ 1 & t_{i2} \\ \vdots & \vdots \\ 1 & t_{in_i} \end{bmatrix} \begin{bmatrix} b_{i0} \\ b_{i1} \end{bmatrix} + \begin{bmatrix} \epsilon_{i1} \\ \epsilon_{i2} \\ \vdots \\ \epsilon_{in_i} \end{bmatrix}$$

with $\boldsymbol{b}_i = [b_{i0}, b_{i1}]^T$ and $\text{var}(\boldsymbol{b}_i) = \boldsymbol{D}$ where

$$\boldsymbol{D} = \begin{bmatrix} \sigma_0^2 & \sigma_{01} \\ \sigma_{01} & \sigma_1^2 \end{bmatrix}.$$

Therefore, σ_0 is the standard deviation of the intercepts, σ_1 is the standard deviation of the slopes, and σ_{01} is the covariance between the two. This model induces a marginal variance at time t_{ij} which is quadratic in time:

$$\text{var}(Y_{ij}) = \sigma_\epsilon^2 + \sigma_0^2 + 2\sigma_{01} t_{ij} + \sigma_1^2 t_{ij}^2. \tag{8.9}$$

The marginal correlation between observations at times t_{ij} and t_{ik} is

$$\rho_{jk} = \frac{\sigma_0^2 + (t_{ij} + t_{ik})\sigma_{01} + t_{ij} t_{ik} \sigma_1^2}{(\sigma_\epsilon^2 + \sigma_0^2 + 2t_{ij}\sigma_{01} + t_{ij}^2 \sigma_1^2)^{1/2} (\sigma_\epsilon^2 + \sigma_0^2 + 2t_{ik}\sigma_{01} + t_{ik}^2 \sigma_1^2)^{1/2}} \tag{8.10}$$

for $j, k = 1, \ldots, n_i, j \neq k$. Therefore, the assumption of random slopes has induced marginal correlations that vary as a function of the timings of the measurements. After a model is fitted, the variances (8.9) and correlations (8.10) can be evaluated at the estimated variance components and compared to the empirical marginal variance and correlations.

In a longitudinal setting, an obvious extension to model (8.7) is provided by

$$\boldsymbol{y}_i = \boldsymbol{x}_i \boldsymbol{\beta} + \boldsymbol{z}_i \boldsymbol{b}_i + \boldsymbol{\delta}_i + \boldsymbol{\epsilon}_i, \tag{8.11}$$

with the error vectors $\boldsymbol{b}_i$, $\boldsymbol{\delta}_i$, and $\boldsymbol{\epsilon}_i$ representing individual-specific random effects, serial dependence, and measurement error. We assume

$$\text{E}[\boldsymbol{\epsilon}_i] = \boldsymbol{0}, \quad \text{var}(\boldsymbol{\epsilon}_i) = \sigma_\epsilon^2 \boldsymbol{I}_{n_i}$$

$$\text{E}[\boldsymbol{b}_i] = \boldsymbol{0}, \quad \text{var}(\boldsymbol{b}_i) = \boldsymbol{D}$$

$$\text{E}[\boldsymbol{\delta}_i] = \boldsymbol{0}, \quad \text{var}(\boldsymbol{\delta}_i) = \sigma_\delta^2 \boldsymbol{R}_i$$

$$\text{cov}(\boldsymbol{b}_i, \boldsymbol{\epsilon}_{i'}) = \boldsymbol{0}, \quad i, i' = 1, \ldots, m,$$

$$\operatorname{cov}(\boldsymbol{b}_i, \boldsymbol{\delta}_{i'}) = \mathbf{0}, \qquad i, i' = 1, \ldots, m,$$
$$\operatorname{cov}(\boldsymbol{\delta}_i, \boldsymbol{\epsilon}_{i'}) = \mathbf{0}, \qquad i, i' = 1, \ldots, m,$$

with $\operatorname{cov}(\boldsymbol{\epsilon}_i, \boldsymbol{\epsilon}_{i'}) = \mathbf{0}$, $\operatorname{cov}(\boldsymbol{\delta}_i, \boldsymbol{\delta}_{i'}) = \mathbf{0}$, and $\operatorname{cov}(\boldsymbol{b}_i, \boldsymbol{b}_{i'}) = \mathbf{0}$, for $i \neq i'$. Here, $\boldsymbol{R}_i$ is an $n_i \times n_i$ correlation matrix with elements R_{ijk}, for $j, k = 1, \ldots, n_i$ which correspond to within individual correlations.

In general, it is difficult to identify/estimate all three sources of variability, but this formulation provides a useful conceptual model.

We now discuss specific choices of $\boldsymbol{R}_i$, beginning with a widely-used time series model, the first-order autoregressive, or AR(1), process. We assume initially that responses are observed at equally spaced times. For $j \geq 2$ and $|\rho| < 1$ suppose

$$\delta_{ij} = \rho \delta_{i,j-1} + u_{ij}, \tag{8.12}$$

with $\boldsymbol{u}_i = [u_{i1}, \ldots, u_{in_i}]^\mathrm{T}$, $\mathrm{E}[\boldsymbol{u}_i] = \mathbf{0}$, $\operatorname{var}(\boldsymbol{u}_i) = \sigma_u^2 \mathbf{I}_{n_i}$, and with u_{ij} independent of all other error terms in the model. We first derive the marginal moments corresponding to this model. Repeated application of (8.12) gives, for $k > 0$,

$$\delta_{ij} = u_{ij} + \rho u_{i,j-1} + \rho^2 u_{i,j-2} + \ldots + \rho^{k-1} u_{i,j-k+1} + \rho^k \delta_{i,j-k} \tag{8.13}$$

so that

$$\operatorname{var}(\delta_{ij}) = \sigma_u^2 (1 + \rho^2 + \rho^4 + \ldots + \rho^{2(k-1)}) + \rho^{2k} \operatorname{var}(\delta_{i,j-k}).$$

Taking the limit as $k \to \infty$, and using $\sum_{l=1}^{\infty} x^{l-1} = (1-x)^{-1}$ for $|x| < 1$, gives

$$\operatorname{var}(\delta_{ij}) = \frac{\sigma_u^2}{(1-\rho^2)} = \sigma_\delta^2,$$

which is the marginal variance of all of the δ error terms. Using (8.13),

$$\operatorname{cov}(\delta_{ij}, \delta_{i,j-k}) = \mathrm{E}[\delta_{ij} \delta_{i,j-k}] = \rho^k \mathrm{E}[\delta_{i,j-k}^2] = \rho^k \operatorname{var}(\delta_{i,j-k}^2)$$
$$= \rho^k \sigma_\delta^2,$$

so that the correlations decline as observations become further apart in time. Under this model, the correlation matrix of $\boldsymbol{\delta}_i$ is

$$\boldsymbol{R}_i = \begin{bmatrix} 1 & \rho & \rho^2 & \cdots & \rho^{n_i-1} \\ \rho & 1 & \rho & \cdots & \rho^{n_i-2} \\ \rho^2 & \rho & 1 & \cdots & \rho^{n_i-3} \\ \vdots & \vdots & \vdots & \ddots & \vdots \\ \rho^{n_i-1} & \rho^{n_i-2} & \rho^{n_i-3} & \cdots & 1 \end{bmatrix}.$$

8.4 Linear Mixed Models

The autoregressive model is appealing in longitudinal settings and contains just two parameters, σ_δ^2 and ρ. The model can be extended to unequally spaced times to give covariance

$$\text{cov}(\delta_{ij}, \delta_{ik}) = \sigma_\delta^2 \rho^{|t_{ij}-t_{ik}|}. \tag{8.14}$$

A *Toeplitz* model assumes the variance is constant across time and that responses that are an equal distance apart in time have the same correlation.[1] For equally spaced responses in time:

$$\text{var}(\boldsymbol{\delta}_i) = \sigma_\delta^2 \begin{bmatrix} 1 & \rho_1 & \rho_2 & \cdots & \rho_{n_i-1} \\ \rho_1 & 1 & \rho_1 & \cdots & \rho_{n_i-2} \\ \rho_2 & \rho_1 & 1 & \cdots & \rho_{n_i-3} \\ \vdots & \vdots & \vdots & \ddots & \vdots \\ \rho_{n_i-1} & \rho_{n_i-2} & \rho_{n_i-3} & \cdots & 1 \end{bmatrix}.$$

This model may be useful in situations in which there is a common design across individuals, which allows estimation of the $n_i = n$ parameters ($n - 1$ correlations and a variance). The AR(1) model is a special case in which $\rho_k = \rho^k$.

An *unstructured* covariance structure allows for different variances at each occasion $\sigma_{\delta 1}^2, \ldots, \sigma_{\delta n_i}^2$ and distinct correlations for each pair of responses, that is,

$$\text{corr}(\boldsymbol{\delta}_i) = \begin{bmatrix} 1 & \rho_{12} & \rho_{13} & \cdots & \rho_{1n_i} \\ \rho_{21} & 1 & \rho_{23} & \cdots & \rho_{2n_i} \\ \rho_{31} & \rho_{32} & 1 & \cdots & \rho_{3n_i} \\ \vdots & \vdots & \vdots & \ddots & \vdots \\ \rho_{n_i 1} & \rho_{n_i 2} & \rho_{n_i 3} & \cdots & 1 \end{bmatrix}$$

with $\rho_{jk} = \rho_{kj}$, for $j, k = 1, \ldots, n_i$. This model contains $n_i(n_i + 1)/2$ parameters per individual, which is a large number if n_i is large. If one has a common design across individuals, it may be plausible to fit this model, but one would still need a large number of individuals m, in order for inference to be reliable. As usual, there is a trade-off between flexibility and parsimony.

8.4.3 Parameter Interpretation for Linear Mixed Models

In this section we discuss how β and b may be interpreted in the LMM; this interpretation requires care, as we illustrate in the context of a longitudinal study

---

[1] In linear algebra, a Toeplitz matrix is a matrix in which each descending diagonal, from left to right, is constant.

with both random intercepts and random slopes. For a generic individual at time t, suppose the model is

$$\mathrm{E}[Y \mid \boldsymbol{b}, t] = (\beta_0 + b_0) + (\beta_1 + b_1)(t - \bar{t})$$

with $\boldsymbol{b} = [b_0, b_1]^{\mathrm{T}}$. The marginal model is

$$\mathrm{E}[Y \mid t] = \beta_0 + \beta_1(t - \bar{t}).$$

so that β_0 is the expected response at $t = \bar{t}$ and the slope parameter β_1 is the expected change in response for a unit increase in time. These expectations are with respect to the distribution of random effects and are averages across the population of individuals.

For a generic individual, $\beta_0 + b_0$ is the expected response at $t = \bar{t}$, and $\beta_1 + b_1$ is the expected change in response for a unit increase in time. In a linear model, β_1 is also the average of the individual slopes, $\beta_1 + b_1$. Consequently, *since the model is linear*, β_1 is both the expected change in the average response in unit time (across individuals) and the average of the individual expected changes in unit time. An alternative interpretation is that β_1 is the change in response for a unit change in t for a "typical" individual, that is, an individual with $b_1 = 0$. In Chap. 9 we will illustrate how the interpretation of parameters in mixed models becomes far more complex when the model is nonlinear in the parameters, and we will see that the consideration of a typical individual is particularly useful in this case.

8.5 Likelihood Inference for Linear Mixed Models

We now turn to inference and first consider likelihood methods for the LMM

$$\boldsymbol{y}_i = \boldsymbol{x}_i \boldsymbol{\beta} + \boldsymbol{z}_i \boldsymbol{b}_i + \boldsymbol{\epsilon}_i.$$

To implement a likelihood approach, we need to specify a complete probability distribution for the data, and this follows by specifying distributions for $\boldsymbol{\epsilon}_i$ and $\boldsymbol{b}_i$, $i = 1, \ldots, m$. A common choice is $\boldsymbol{\epsilon}_i \mid \sigma_\epsilon^2 \sim_{iid} \mathrm{N}_{n_i}(\boldsymbol{0}, \sigma_\epsilon^2 \mathbf{I}_{n_i})$ and $\boldsymbol{b}_i \mid \boldsymbol{D} \sim_{iid} \mathrm{N}_{q+1}(\boldsymbol{0}, \boldsymbol{D})$ where

$$\boldsymbol{D} = \begin{bmatrix} \sigma_0^2 & \sigma_{01} & \sigma_{02} & \cdots & \sigma_{0q} \\ \sigma_{10} & \sigma_1^2 & \sigma_{12} & \cdots & \sigma_{1q} \\ \sigma_{20} & \sigma_{21} & \sigma_2^2 & \cdots & \sigma_{2q} \\ \vdots & \vdots & \vdots & \ddots & \vdots \\ \sigma_{q0} & \sigma_{q1} & \sigma_{q2} & \cdots & \sigma_q^2 \end{bmatrix},$$

8.5 Likelihood Inference for Linear Mixed Models

so that the vector of variance–covariance parameters is $\alpha = [\sigma_\epsilon^2, D]$. The marginal mean and variance are

$$\mathrm{E}[Y_i \mid \beta] = \mu_i(\beta) = x_i\beta \tag{8.15}$$

$$\mathrm{var}(Y_i \mid \alpha) = V_i(\alpha) = z_i D z_i^\mathrm{T} + \sigma_\epsilon^2 I_{n_i}. \tag{8.16}$$

We have refined notation in this section to explicitly condition on the relevant parameters. In general, inference may be required for the fixed effects regression parameters β, the variance components α, or the random effects, $b = [b_1, \ldots, b_m]^\mathrm{T}$. We consider each of these possibilities in turn.

8.5.1 Inference for Fixed Effects

Likelihood methods have traditionally been applied to nonrandom parameters, and so, we integrate over the random effects in the two-stage model to give

$$p(y \mid \beta, \alpha) = \int_b p(y \mid b, \beta, \alpha) \times p(b \mid \beta, \alpha)\, db.$$

Exploiting conditional independencies, we obtain the simplified form

$$p(y \mid \beta, \alpha) = \prod_{i=1}^m \int_{b_i} p(y_i \mid b_i, \beta, \sigma_\epsilon^2) \times p(b_i \mid D)\, db_i$$

and since a convolution of normals is normal, we obtain

$$y_i \mid \beta, \alpha \sim \mathrm{N}_{n_i}[\mu_i(\beta), V_i(\alpha)],$$

where the marginal mean $\mu_i(\beta)$ and variance $V_i(\alpha)$ correspond to (8.15) and (8.16), respectively. The log-likelihood is

$$l(\beta, \alpha) = -\frac{1}{2} \sum_{i=1}^m \log |V_i(\alpha)| - \frac{1}{2} \sum_{i=1}^m (y_i - x_i\beta)^\mathrm{T} V_i(\alpha)^{-1} (y_i - x_i\beta). \tag{8.17}$$

The MLEs for β and α are obtained via maximization of (8.17). The score equations for β are

$$\frac{\partial l}{\partial \beta} = \sum_{i=1}^m x_i^\mathrm{T} V_i^{-1} Y_i - \sum_{i=1}^m x_i^\mathrm{T} V_i^{-1} x_i \beta$$

$$= \sum_{i=1}^m x_i^\mathrm{T} V_i^{-1} (Y_i - x_i\beta) \tag{8.18}$$

and yield the MLE

$$\widehat{\boldsymbol{\beta}} = \left(\sum_{i=1}^m \boldsymbol{x}_i^\mathrm{T} \boldsymbol{V}_i(\widehat{\boldsymbol{\alpha}})^{-1} \boldsymbol{x}_i\right)^{-1} \left(\sum_{i=1}^m \boldsymbol{x}_i^\mathrm{T} \boldsymbol{V}_i(\widehat{\boldsymbol{\alpha}})^{-1} \boldsymbol{y}_i\right), \qquad (8.19)$$

which is a generalized least squares estimator. If $\boldsymbol{D} = \boldsymbol{0}$, then $\boldsymbol{V} = \sigma_\epsilon^2 \boldsymbol{I}_N$ (where $N = \sum_{i=1}^m n_i$), and $\widehat{\boldsymbol{\beta}}$ corresponds to the ordinary least squares estimator, as we would expect. The variance of $\widehat{\boldsymbol{\beta}}$ may be obtained either directly from (8.19), since the estimator is linear in $\boldsymbol{y}_i$, or from the second derivative of the log-likelihood.

The expected information matrix is block diagonal:

$$\boldsymbol{I}(\boldsymbol{\beta}, \boldsymbol{\alpha}) = \begin{bmatrix} \boldsymbol{I}_{\beta\beta} & \boldsymbol{0} \\ \boldsymbol{0} & \boldsymbol{I}_{\alpha\alpha} \end{bmatrix} \qquad (8.20)$$

so there is asymptotic independence between $\widehat{\boldsymbol{\beta}}$ and $\widehat{\boldsymbol{\alpha}}$ and any consistent estimator of $\boldsymbol{\alpha}$ will give an asymptotically efficient estimator for $\boldsymbol{\beta}$ (likelihood-based estimation of $\boldsymbol{\alpha}$ is considered in Sects. 8.5.2 and 8.5.3). Since

$$\boldsymbol{I}_{\beta\beta} = -\mathrm{E}\left[\frac{\partial^2 l}{\partial \boldsymbol{\beta} \partial \boldsymbol{\beta}^\mathrm{T}}\right] = \sum_{i=1}^m \boldsymbol{x}_i^\mathrm{T} \boldsymbol{V}_i^{-1} \boldsymbol{x}_i = -\frac{\partial^2 l}{\partial \boldsymbol{\beta} \partial \boldsymbol{\beta}^\mathrm{T}}, \qquad (8.21)$$

the observed and expected information matrices coincide. The estimator $\widehat{\boldsymbol{\beta}}$ is linear in the data $\boldsymbol{Y}_i$, and so under normality of the data, $\widehat{\boldsymbol{\beta}}$ is normal also. Under correct specification of the variance model, and with a consistent estimator $\widehat{\boldsymbol{\alpha}}$,

$$\left(\sum_{i=1}^m \boldsymbol{x}_i \boldsymbol{V}_i(\widehat{\boldsymbol{\alpha}})^{-1} \boldsymbol{x}_i\right)^{1/2} (\widehat{\boldsymbol{\beta}}_m - \boldsymbol{\beta}) \to_d \mathrm{N}_{k+1}(\boldsymbol{0}, \boldsymbol{I}),$$

as $m \to \infty$. Since $\widehat{\boldsymbol{\beta}}$ is linear in $\boldsymbol{Y}$, it follows immediately that this asymptotic distribution is also appropriate when the data and random effects are not normal. We require the second moments of the data to be correctly specified, however. In Sect. 8.7 we describe how a consistent variance estimator may be obtained when $\mathrm{cov}(\boldsymbol{Y}_i, \boldsymbol{Y}_{i'} \mid \boldsymbol{\alpha}) = \boldsymbol{0}$, but $\mathrm{var}(\boldsymbol{Y}_i \mid \boldsymbol{\alpha}) = \boldsymbol{V}_i(\boldsymbol{\alpha})$ is not necessarily correctly specified.

In terms of the asymptotics it is not sufficient to have m fixed and $n_i \to \infty$ for $i = 1, \ldots, m$. We illustrate for the LMM with $\boldsymbol{z}_i = \boldsymbol{x}_i$, in which case $\boldsymbol{V}_i = \boldsymbol{x}_i \boldsymbol{D} \boldsymbol{x}_i^\mathrm{T} + \sigma_\epsilon^2 \boldsymbol{I}_{n_i}$. Under this setup,

$$\mathrm{var}(\widehat{\boldsymbol{\beta}}) = \left(\sum_{i=1}^m \boldsymbol{x}_i^\mathrm{T} \boldsymbol{V}^{-1} \boldsymbol{x}_i\right)^{-1}$$

$$= \left(\sum_{i=1}^m \left[(\boldsymbol{x}_i^\mathrm{T} \boldsymbol{x}_i)^{-1} \sigma_\epsilon^{-2} + \boldsymbol{D}\right]^{-1}\right)^{-1},$$

8.5 Likelihood Inference for Linear Mixed Models

where we have used the matrix identity $\boldsymbol{x}_i^{\mathrm{T}}\boldsymbol{V}_i^{-1}\boldsymbol{x}_i = \left[(\boldsymbol{x}_i^{\mathrm{T}}\boldsymbol{x})^{-1}\sigma_\epsilon^2 + \boldsymbol{D}\right]^{-1}$ (which may be derived from (B.3) of Appendix B). When $n_i \to \infty$,

$$(\boldsymbol{x}_i^{\mathrm{T}}\boldsymbol{x}_i)^{-1} = O(n_i^{-1}) \to \boldsymbol{0},$$

and if m is fixed,

$$\mathrm{var}(\widehat{\boldsymbol{\beta}}) \to \frac{\boldsymbol{D}}{m},$$

showing that we require $m \to \infty$ for consistency of $\widehat{\boldsymbol{\beta}}$.

Likelihood ratio tests can be used to test hypotheses concerning elements of $\boldsymbol{\beta}$, for fixed $\boldsymbol{\alpha}$ or, in practice, the substitution of an estimate $\boldsymbol{\alpha}$. Various t and F-like approaches have been suggested for correcting for the estimation of $\boldsymbol{\alpha}$, see Verbeeke and Molenberghs (2000, Chap. 6), but if the sample size m is not sufficiently large for reliable estimation of $\boldsymbol{\alpha}$, we recommend resampling methods, or following a Bayesian approach to inference, since this produces inference for $\boldsymbol{\beta}$ that averages over the uncertainty in the estimation of $\boldsymbol{\alpha}$.

For more complex linear models, inference may not be so straightforward. For example, consider the model

$$Y_{ij} = \boldsymbol{x}_{ij}\boldsymbol{\beta} + \boldsymbol{z}_i\boldsymbol{b}_i + \epsilon_{ij} = \mu_{ij} + \epsilon_{ij}$$

but with nonconstant measurement error variance. A common model is $\mathrm{var}(Y_{ij}) = \sigma_\epsilon^2 \mu_{ij}^\gamma$, for known $\gamma > 0$. In this case the MLE for $\boldsymbol{\beta}$ is not available in closed form, and we do not have a diagonal information matrix as in (8.20). An example of the fitting of such a model in a nonlinear setting is given at the end of Sect. 9.20.

Maximum likelihood estimation is also theoretically straightforward for the extended model (8.11) in which we have a richer variance model, but identifiability issues may arise due to the complexity of the error structure.

8.5.2 Inference for Variance Components via Maximum Likelihood

The MLE $\widehat{\boldsymbol{\alpha}}$ is obtained from maximization of (8.17), but in general, there is no closed-form solution. However, the expectation-maximization (EM, Dempster et al. 1977) or Newton–Raphson algorithm may be applied to the profile likelihood:

$$l_p(\boldsymbol{\alpha}) = \max_{\boldsymbol{\beta}} l(\boldsymbol{\beta}, \boldsymbol{\alpha}) = -\frac{1}{2}\log|\boldsymbol{V}(\boldsymbol{\alpha})| - \frac{1}{2}(\boldsymbol{y} - \boldsymbol{x}\widehat{\boldsymbol{\beta}})^{\mathrm{T}}\boldsymbol{V}(\boldsymbol{\alpha})^{-1}(\boldsymbol{y} - \boldsymbol{x}\widehat{\boldsymbol{\beta}}),$$

since recall from Sect. 2.4.2 that the MLE for $\boldsymbol{\alpha}$ is identical to the estimate obtained from the profile likelihood. Under standard likelihood theory,

$$\boldsymbol{I}_{\alpha\alpha}^{1/2}(\widehat{\boldsymbol{\alpha}} - \boldsymbol{\alpha}) \to_d \mathrm{N}_r(\boldsymbol{0}, \boldsymbol{I}_r),$$

where r is the number of distinct elements of $\boldsymbol{\alpha}$. This distribution provides asymptotic confidence intervals for elements of $\boldsymbol{\alpha}$.

Testing whether random effect variances are zero requires care since the null hypothesis lies on the boundary, and so, the usual regularity conditions are not satisfied. We illustrate by considering the model

$$Y_{ij} = \beta_0 + \boldsymbol{x}_{ij}\boldsymbol{\beta} + b_i + \epsilon_{ij}$$

with $b_i \mid \sigma_0^2 \sim N(0, \sigma_0^2)$. Suppose we wish to test whether the random effects variance is zero, that is, $H_0 : \sigma_0^2 = 0$ versus $H_1 : \sigma_0^2 > 0$. In this case, the asymptotic null distribution is a 50:50 mixture of χ_0^2 and χ_1^2 distributions, where the former is the distribution that gives probability mass 1 to the value 0. For example, the 95% points of a χ_1^2 and the 50:50 mixture are 3.84 and 2.71, respectively. Consequently, if the usual χ_1^2 distribution is used, the null will be accepted too often, leading to a variance component structure that is too simple.

The intuition behind the form of the null distribution is the following. Estimating σ_0^2 is equivalent to estimating $\rho = \sigma_0^2/\sigma^2$ and setting $\widehat{\rho} = 0$ if the estimated correlation is negative, and under the null, this will happen half the time. If $\widehat{\rho} = 0$, then we recover the null for the distribution of the data, and so, the likelihood ratio will be 1. This gives the mass at the value 0, and combining with the usual χ_1^2 distribution gives the 50:50 mixture.

If H_0 and H_1 correspond to models with k and $k+1$ random effects, respectively, each with general covariance structures, then the asymptotic distribution is a 50:50 mixture of χ_k^2 and χ_{k+1}^2 distributions. Hence, for example, if we wish to test random intercepts only versus correlated random intercepts and random slopes (with $\boldsymbol{D}$ having elements $\sigma_0^2, \sigma_{01}, \sigma_1^2$), then the distribution of the likelihood ratio statistic is a 50:50 mixture of χ_1^2 and χ_2^2 distributions. Similar asymptotic results are available for more complex models/hypotheses; see, for example, Verbeeke and Molenberghs (2000).

8.5.3 Inference for Variance Components via Restricted Maximum Likelihood

While MLE for variance components yields consistent estimates under correct model specification, the estimation of $\boldsymbol{\beta}$ is not acknowledged, in the sense that inference proceeds as if $\boldsymbol{\beta}$ were known. We have already encountered this in Sect. 2.4.2 for the simple linear model where it was shown that the MLE of σ^2 is RSS/n, while the unbiased version is $RSS/(n - k - 1)$, where RSS is the residual sum of squares and k is the number of covariates. An alternative, and often preferable, method that acknowledges estimation of $\boldsymbol{\beta}$ is provided by restricted (or residual) maximum likelihood (REML). We provide a Bayesian justification for REML in Sect. 8.6 and here provide another derivation based on marginal likelihood.

8.5 Likelihood Inference for Linear Mixed Models

Recall the definition of marginal likelihood from Sect. 2.4.2. Let S_1, S_2, be minimal sufficient statistics and suppose

$$p(y \mid \lambda, \phi) \propto p(s_1, s_2 \mid \lambda, \phi) = p(s_1 \mid \lambda) p(s_2 \mid s_1, \lambda, \phi) \qquad (8.22)$$

where λ represents the parameters of interest and ϕ the remaining (nuisance) parameters. Inference for λ may be based on the marginal likelihood $L_m(\lambda) = p(s_1 \mid \lambda)$. We discuss how marginal likelihoods may be derived for general LMMs.

To derive a marginal likelihood, we need to find a function of the data, $U = f(Y)$, whose distribution does not depend upon β. We briefly digress to discuss an *error contrast*, $C^\mathsf{T} Y$, which is defined by the property that $\mathrm{E}[C^\mathsf{T} Y] = 0$ for all values of β, with C an N-dimensional vector. For the LMM

$$\mathrm{E}[C^\mathsf{T} Y] = 0 \text{ for all } \beta \text{ if and only if } C^\mathsf{T} x = 0.$$

When $C^\mathsf{T} x = 0$,

$$C^\mathsf{T} Y = C^\mathsf{T} z b + C^\mathsf{T} \epsilon,$$

which does not depend on β, suggesting that the marginal likelihood could be based on error contrasts. If x is of full rank, that is, is of rank $k + 1$, there are exactly $N - k - 1$ linearly independent error contrasts (since $k + 1$ fixed effects have been estimated, which induces dependencies in the error contrasts). Let $B = [C_1, \ldots, C_{N-k-1}]$ denote an error contrast matrix. Given two error contrast matrices B_1 and B_2, it can be shown that there exists a full rank, $(N - k - 1) \times (N - k - 1)$ matrix A such that $A B_1^\mathsf{T} = A B_2^\mathsf{T}$. Therefore, likelihoods based on $B_1^\mathsf{T} Y$ or on $B_2^\mathsf{T} Y$ will be proportional, and estimators based on either will be identical. Let $H = x(x^\mathsf{T} x)^{-1} x^\mathsf{T}$, and choose B such that $\mathbf{I} - H = B B^\mathsf{T}$ and $\mathbf{I} = B^\mathsf{T} B$. It is easily shown that B is an error contrast matrix since

$$B^\mathsf{T} x = B^\mathsf{T} B B^\mathsf{T} x = B^\mathsf{T} (\mathbf{I} - H) x = 0.$$

The function of the data we consider is therefore $U = B^\mathsf{T} Y$ which may be written as

$$U = B^\mathsf{T} Y = B^\mathsf{T} B B^\mathsf{T} Y = B^\mathsf{T} (\mathbf{I} - H) Y = B^\mathsf{T} r,$$

where $r = Y - x \widehat{\beta}_o$, and $\widehat{\beta}_o = (x^\mathsf{T} x)^{-1} x^\mathsf{T} Y$ is the OLS estimator, showing that $B^\mathsf{T} Y$ is a linear combination of residuals (hence the name "residual" maximum likelihood). Since $B^\mathsf{T} x = 0$, we can confirm that

$$U = B^\mathsf{T} Y = B^\mathsf{T} z b + B^\mathsf{T} \epsilon,$$

with $\mathrm{E}[U] = 0$. Further, the distribution of U does not depend upon β, as required for a marginal likelihood.

We now derive the distribution of U by considering the transformation from $Y \to [U, \widehat{\boldsymbol{\beta}}_G] = [B^T Y, G^T Y]$, where

$$\widehat{\boldsymbol{\beta}}_G = G^T Y = (x^T V^{-1} x)^{-1} x^T V^{-1} Y$$

is the generalized least squares (GLS) estimator. We derive the Jacobian of the transformation, using (B.1) and (B.2) in Appendix B:

$$|J| = \left|\frac{\partial(U, \widehat{\boldsymbol{\beta}}_G)}{\partial Y}\right| = |B\ G| = \left|\begin{bmatrix} B^T \\ G^T \end{bmatrix} [B\ G]\right|^{1/2}$$

$$= |B^T B|^{1/2} |G^T G - G^T B (B^T B)^{-1} B^T G|^{1/2}$$

$$= 1 \times |G^T G - G^T (I - H) G|^{1/2}$$

$$= G^T H G = |x^T x|^{-1/2} \neq 0$$

which implies that $[U, \widehat{\boldsymbol{\beta}}_G]$ is of full rank (and equal to N). The vector $[U, \widehat{\boldsymbol{\beta}}_G]$ is a linear combination of normals and so is normal, and

$$\mathrm{cov}(U, \widehat{\boldsymbol{\beta}}_G) = \mathrm{E}[U(\widehat{\boldsymbol{\beta}}_G - \boldsymbol{\beta})^T]$$

$$= \mathrm{E}[B^T Y Y^T G] - \mathrm{E}[B^T Y - \boldsymbol{\beta}^T]$$

$$= B^T [\mathrm{var}(Y) + \mathrm{E}(Y)\mathrm{E}(Y^T)] G + B^T x \boldsymbol{\beta} - \boldsymbol{\beta}^T$$

$$= B^T V G^T + B^T x \boldsymbol{\beta} (x \boldsymbol{\beta})^T$$

$$= \mathbf{0},$$

where we have repeatedly used $B^T x = \mathbf{0}$ and $V = \mathrm{var}(Y)$. So U and $\widehat{\boldsymbol{\beta}}_G$ are uncorrelated and, since they are normal, independent also. Consequently,

$$p(Y \mid \boldsymbol{\alpha}, \boldsymbol{\beta}) = p(U, \widehat{\boldsymbol{\beta}}_G \mid \boldsymbol{\alpha}, \boldsymbol{\beta}) \mid J \mid$$

$$= p(U \mid \widehat{\boldsymbol{\beta}}_G, \boldsymbol{\beta}) p(\widehat{\boldsymbol{\beta}}_G \mid \boldsymbol{\alpha}, \boldsymbol{\beta}) \mid J \mid$$

$$= p(U \mid \boldsymbol{\alpha}) p(\widehat{\boldsymbol{\beta}}_G \mid \boldsymbol{\alpha}, \boldsymbol{\beta}) \mid J \mid. \tag{8.23}$$

By comparison with (8.22), we have $s_1 = U$, $s_2 = \widehat{\boldsymbol{\beta}}_G$, $\lambda = \boldsymbol{\alpha}$, and $\phi = \boldsymbol{\beta}$, and $p(U \mid \boldsymbol{\alpha})$ is a marginal likelihood. Rearrangement of (8.23) gives

$$p(U \mid \boldsymbol{\alpha}) = \frac{p(y \mid \boldsymbol{\alpha}, \boldsymbol{\beta})}{p(\widehat{\boldsymbol{\beta}}_G \mid \boldsymbol{\alpha}, \boldsymbol{\beta})} |J|^{-1}.$$

Since

$$p(y \mid \boldsymbol{\alpha}, \boldsymbol{\beta}) = (2\pi)^{-N/2} |V|^{-1/2} \exp\left[-\frac{1}{2}(y - x\boldsymbol{\beta})^T V^{-1} (y - x\boldsymbol{\beta})\right],$$

8.5 Likelihood Inference for Linear Mixed Models

and

$$p(\widehat{\boldsymbol{\beta}}_G \mid \boldsymbol{\alpha}, \boldsymbol{\beta}) = (2\pi)^{-(k+1)/2} |\boldsymbol{x}^T \boldsymbol{V}^{-1} \boldsymbol{x}|^{1/2} \exp\left[-\frac{1}{2}(\widehat{\boldsymbol{\beta}}_G - \boldsymbol{\beta})^T \boldsymbol{x}^T \boldsymbol{V}^{-1} \boldsymbol{x} (\widehat{\boldsymbol{\beta}}_G - \boldsymbol{\beta})\right]$$

we obtain the marginal likelihood

$$p(U \mid \boldsymbol{\alpha}) = c \frac{|\boldsymbol{x}^T \boldsymbol{x}|^{1/2} |\boldsymbol{V}|^{-1/2}}{|\boldsymbol{x}^T \boldsymbol{V}^{-1} \boldsymbol{x}|^{1/2}} \exp\left[-\frac{1}{2}(\boldsymbol{y} - \boldsymbol{x}\widehat{\boldsymbol{\beta}}_G)^T \boldsymbol{V}^{-1} (\boldsymbol{y} - \boldsymbol{x}\widehat{\boldsymbol{\beta}}_G)\right]$$

with $c = (2\pi)^{-(N-k-1)/2}$, which (as already mentioned) does not depend upon B. Hence, we can choose any linearly independent combination of the residuals.

The restricted log-likelihood upon which inference for $\boldsymbol{\alpha}$ may be based is

$$l_R(\boldsymbol{\alpha}) = -\frac{1}{2} \log |\boldsymbol{x}^T \boldsymbol{V}(\boldsymbol{\alpha})^{-1} \boldsymbol{x}| - \frac{1}{2} \log |\boldsymbol{V}(\boldsymbol{\alpha})| - \frac{1}{2}(\boldsymbol{y} - \boldsymbol{x}\widehat{\boldsymbol{\beta}}_G)^T \boldsymbol{V}(\boldsymbol{\alpha})^{-1} (\boldsymbol{y} - \boldsymbol{x}\widehat{\boldsymbol{\beta}}_G).$$

Comparison with the profile log-likelihood for $\boldsymbol{\alpha}$,

$$l_P(\boldsymbol{\alpha}) = -\frac{1}{2} \log |\boldsymbol{V}(\boldsymbol{\alpha})| - \frac{1}{2}(\boldsymbol{y} - \boldsymbol{x}\widehat{\boldsymbol{\beta}}_G)^T \boldsymbol{V}(\boldsymbol{\alpha})^{-1} (\boldsymbol{y} - \boldsymbol{x}\widehat{\boldsymbol{\beta}}_G),$$

shows that we have an additional term, $-\frac{1}{2} \log |\boldsymbol{x}^T \boldsymbol{V}(\boldsymbol{\alpha})^{-1} \boldsymbol{x}|$, that may be viewed as accounting for the degrees of freedom lost in estimation of $\boldsymbol{\beta}$. Computationally, finding REML estimators is as straightforward as their ML counterparts, as the objective functions differ simply by a single term. Both ML and REML estimates may be obtained using EM or Newton–Raphson algorithms; see Pinheiro and Bates (2000) for details.

In general, REML estimators have finite sample bias, but they are less biased than ML estimators, particularly for small samples. So far, as estimation of the variance components are concerned, the asymptotic distribution of the REML estimator is normal, with variance given by the inverse of the Fisher's information matrix, where the latter is based on $l_R(\boldsymbol{\alpha})$.

REML is effectively based on a likelihood with data constructed from the distribution of the residuals $\boldsymbol{y} - \boldsymbol{x}\widehat{\boldsymbol{\beta}}_G$. Therefore, when two regression models are to be compared, the data under the two models are different; hence, REML likelihood ratio tests for elements of $\boldsymbol{\beta}$ cannot be performed. Consequently, when a likelihood ratio test is required to formally compare two nested regression models, maximum likelihood must be used to fit the models. Likelihood ratio tests for variance components are valid under restricted maximum likelihood, however, since the covariates, and hence residuals, are constant in both models.

Example: One-Way ANOVA

The simplest example of a LMM is the balanced one-way random effects ANOVA model:

$$Y_{ij} = \beta_0 + b_i + \epsilon_{ij},$$

with b_i and ϵ_{ij} independent and distributed as $b_i \mid \sigma_0^2 \sim_{iid} \mathrm{N}(0, \sigma_0^2)$ and $\epsilon_{ij} \mid \sigma_\epsilon^2 \sim_{iid} \mathrm{N}(0, \sigma_\epsilon^2)$, with n observations on each unit and $i = 1, \ldots, m$ to give $N = nm$ observations in total. In this example, $\boldsymbol{\beta} = \beta_0$ and $\boldsymbol{\alpha} = [\sigma_\epsilon^2, \sigma_0^2]$. This model was considered briefly in Sect. 5.8.4.

The model can be written in the form of (8.7) as

$$\boldsymbol{y}_i = \mathbf{1}_n \beta_0 + \mathbf{1}_n b_i + \boldsymbol{\epsilon}_i$$

where $\boldsymbol{y}_i = [y_{i1}, \ldots, y_{in}]^\mathrm{T}$ and $\boldsymbol{\epsilon}_i = [\epsilon_{i1}, \ldots, \epsilon_{in}]^\mathrm{T}$. Marginally, this specification implies that the data are normal with $\mathrm{E}[\boldsymbol{Y} \mid \boldsymbol{\beta}] = \mathbf{1}_N \beta_0$ and $\mathrm{var}(\boldsymbol{Y} \mid \boldsymbol{\alpha}) = \mathrm{diag}(\boldsymbol{V}_1, \ldots, \boldsymbol{V}_m)$ where

$$\boldsymbol{V}_i = \mathbf{1}_n \mathbf{1}_n^\mathrm{T} \sigma_0^2 + \mathbf{I}_n \sigma_\epsilon^2,$$

for $i = 1, \ldots, m$. In the case of $n = 3$ observations per unit, this yields the $N \times N$ marginal variance

$$\boldsymbol{V} = \sigma^2 \begin{bmatrix} 1 & \rho & \rho & 0 & 0 & 0 & \cdots & 0 & 0 & 0 \\ \rho & 1 & \rho & 0 & 0 & 0 & \cdots & 0 & 0 & 0 \\ \rho & \rho & 1 & 0 & 0 & 0 & \cdots & 0 & 0 & 0 \\ 0 & 0 & 0 & 1 & \rho & \rho & \cdots & 0 & 0 & 0 \\ 0 & 0 & 0 & \rho & 1 & \rho & \cdots & 0 & 0 & 0 \\ 0 & 0 & 0 & \rho & \rho & 1 & \cdots & 0 & 0 & 0 \\ \vdots & \vdots & \vdots & \vdots & \vdots & \vdots & \ddots & \vdots & \vdots & \vdots \\ 0 & 0 & 0 & 0 & 0 & 0 & \cdots & 1 & \rho & \rho \\ 0 & 0 & 0 & 0 & 0 & 0 & \cdots & \rho & 1 & \rho \\ 0 & 0 & 0 & 0 & 0 & 0 & \cdots & \rho & \rho & 1 \end{bmatrix},$$

where $\sigma^2 = \sigma_\epsilon^2 + \sigma_0^2$ is the marginal variance of each observation, and

$$\rho = \frac{\sigma_0^2}{\sigma^2} = \frac{\sigma_0^2}{\sigma_\epsilon^2 + \sigma_0^2}$$

is the marginal correlation between two observations on the same unit. The correlation, ρ, is induced by the shared random effect and is referred to as the *intra-class correlation coefficient*.

For some data/mixed effects model combinations, there are more combined fixed and random effects than data points, which is at first sight disconcerting, but the random effects have a special status since they are tied together through a common distribution. In the above ANOVA model, we have $m + 3$ unknown quantities if we include the random effects, but these random effects may be integrated from the model so that the distribution of the data may be written in terms of the three parameters, $[\beta_0, \sigma_0^2, \sigma_\epsilon^2]$ only, without reference to the random effects, that is,

$$\boldsymbol{Y} \mid \beta_0, \sigma_0^2, \sigma_\epsilon^2 \sim \mathrm{N}_N \left[\mathbf{1}\beta_0, \boldsymbol{V}(\sigma_0^2, \sigma_\epsilon^2) \right].$$

8.5 Likelihood Inference for Linear Mixed Models

A fixed effects model with a separate parameter for each group has $m+1$ parameters, which shows that the mixed effects model can offer a parsimonious description.

The MLE for β_0 is given by the grand mean, i.e., $\widehat{\beta}_0 = \overline{Y}_{..}$. With balanced data the ML and REML estimators for the variance components are available in closed form (see Exercise 8.2). We define the between- and within-group mean squares as

$$\text{MSA} = \frac{n\sum_{i=1}^{m}(\overline{y}_{i.} - \overline{y}_{..})^2}{m-1}, \quad \text{MSE} = \frac{\sum_{i=1}^{m}\sum_{j=1}^{n}(\overline{y}_{ij} - \overline{y}_{i.})^2}{m(n-1)}.$$

The MLEs of the variance components are

$$\widehat{\sigma}_\epsilon^2 = \text{MSE},$$

$$\widehat{\sigma}_0^2 = \max\left(0, \frac{(1-1/m)\text{MSA} - \text{MSE}}{n}\right).$$

The REML for $\widehat{\sigma}_\epsilon^2$ is the same as the MLE, but the REML estimate for σ_0^2 is

$$\widehat{\sigma}_0^2 = \max\left(0, \frac{\text{MSA} - \text{MSE}}{n}\right),$$

which is slightly larger than the ML estimate, having accounted for the estimation of β_0. Notice that the ML and REML estimators for σ_0^2 may be zero.

Example: Dental Growth Curves

We consider the full data and fit a model with distinct fixed effects (intercepts and slopes) for boys and girls and with random intercepts and slopes but with a common random effects distribution for boys and girls. Specifically, at stage one,

$$Y_{ij} = (\beta_0 + b_{i0}) + (\beta_1 + b_{i1})t_j + \epsilon_{ij}$$

for boys, $i = 1, \ldots, 16$, and

$$Y_{ij} = (\beta_0 + \beta_2 + b_{i0}) + (\beta_1 + \beta_4 + b_{i1})t_j + \epsilon_{ij}$$

for girls, $i = 17, \ldots, 27$. At stage two,

$$\boldsymbol{b}_i = \begin{bmatrix} b_{i0} \\ b_{i1} \end{bmatrix} \mid \boldsymbol{D} \sim_{iid} N_2(\boldsymbol{0}, \boldsymbol{D}), \quad \boldsymbol{D} = \begin{bmatrix} \sigma_0^2 & \sigma_{01} \\ \sigma_{01}^2 & \sigma_1^2 \end{bmatrix}$$

for $i = 1, \ldots, 27$. We take $[t_1, t_2, t_3, t_4] = [-2, -1, 1, 2]$ so that we have centered by the average age of 11 years. In the generic notation introduced in Sect. 8.4, the above model translates to

$$Y_{ij} = \boldsymbol{x}_{ij}\boldsymbol{\beta} + \boldsymbol{z}_{ij}\boldsymbol{b}_i + \epsilon_{ij}$$

where $\boldsymbol{\beta} = [\beta_0, \beta_1, \beta_2, \beta_3]^{\text{T}}$, and the design matrices for the fixed and random effects are

$$\boldsymbol{x}_{ij} = \begin{cases} [1, t_j, 0, 0] & \text{for } i = 1, \ldots, 16 \\ [1, t_j, 1, t_j] & \text{for } i = 17, \ldots, 27, \end{cases}$$

and $\boldsymbol{z}_{ij} = [1, t_j]$, where $j = 1, 2, 3, 4$. Therefore, β_0 is the average tooth length at 11 years for boys, β_1 is the slope for boys (specifically the average change in tooth length between two populations of boys whose ages differ by 1 year), β_2 is the difference between the average tooth lengths of girls and boys at 11 years, and β_3 is the average difference in slopes between girls and boys. The intercept random effects b_{i0} may be viewed as the accumulation of all unmeasured variables that contribute to the tooth length for child i differing from the relevant (boy or girl) population average length (measured at 11 years). The slope random effects b_{i1} are the child by time interaction terms and summarize all of the unmeasured variables for child i that lead to the rate of change in growth for this child differing from the relevant (boy or girl) population average.

Fitting this model via REML yields

$$\widehat{\boldsymbol{\beta}} = [25, 0.78, -2.3, -0.31]^{\text{T}}$$

with standard errors

$$[0.49, 0.086, 0.76, 0.14].$$

The asymptotic 95% confidence interval for the average difference in tooth lengths at 11 years is $[-3.8, -0.83]$, from which we conclude that the average tooth lengths at 11 years is greater for boys than for girls. The 95% interval for the slope difference is $[-0.57, -0.04]$ suggesting that the average rate of growth is greater for boys also.

There are a number of options to test whether gender-specific slopes are required, that is, to decide on whether $\beta_4 = 0$. A Wald test using the REML estimates gives a p-value of 0.026 (so that one endpoint of a 97.4% confidence interval is zero), which conventionally would suggest a difference in slopes. To perform a likelihood ratio test, we need to carry out a fit using ML, since REML is not valid, as explained in Sect. 8.5.2. Fitting the models with and without distinct slopes gives a change in twice the log-likelihood of 5.03, with an associated p-value of 0.036, which is consistent with the Wald test. Hence, there is reason to believe that the slopes for boys and girls are unequal, with the increase in the average growth over 1 year being estimated as 0.3 mm greater for boys than for girls.

The estimated variance–covariance matrices of the random effects, $\widehat{\boldsymbol{D}}$, under REML and ML are

$$\begin{bmatrix} 1.84^2 & 0.21 \times 1.84 \times 0.18 \\ 0.21 \times 1.84 \times 0.18 & 0.18^2 \end{bmatrix},$$

and

$$\begin{bmatrix} 1.75^2 & 0.23 \times 1.75 \times 0.15 \\ 0.23 \times 1.75 \times 0.15 & 0.15^2 \end{bmatrix}$$

8.5 Likelihood Inference for Linear Mixed Models

so that, as expected, the REML estimates are slightly larger. Although $\widehat{\boldsymbol{\beta}}$ depends on $\widehat{\boldsymbol{D}}$, the point estimates of $\boldsymbol{\beta}$ are identical under ML and REML here, because of the balanced design. The standard errors for elements of $\boldsymbol{\beta}$ are slightly larger under REML, due to the differences in $\widehat{\boldsymbol{V}}$.

Under REML, the estimated standard deviations of the distributions of the intercepts and slopes are $\widehat{\sigma}_0 = 1.84$ and $\widehat{\sigma}_1 = 0.18$, respectively. Whether these are "small" or "not small" relates to the scale of the variables with which they are associated. Interpretation of elements of $\boldsymbol{D}$ depends, in general, on how we parameterize the time variable. For example, if we changed the time scale via a location shift, we would change the definition of the intercept. As parameterized above, the off-diagonal term D_{01} describes the covariance between the child-specific responses at 11 years and the child-specific slopes (the REML estimates of the correlation between these quantities is 0.23).

Suppose we reparameterize stage one of the model as

$$\mathrm{E}[Y_{ij} \mid \boldsymbol{b}_i^\star] = (\beta_0^\star + b_{i0}^\star) + (\beta_1 + b_{i1})t_j^\star$$

with $[t_1^\star, t_2^\star, t_3^\star, t_4^\star] = [8, 10, 12, 14]$ and $\boldsymbol{b}_i^\star = [b_{i0}^\star, b_{i1}]^{\mathrm{T}}$. Then $\beta_0^\star = \beta_0 - \beta_1 \bar{t}$, $b_{i0}^\star = b_{i0} - b_{i1}\bar{t}$, and

$$D_{00}^\star = D_{00} - 2\bar{t}D_{01} + \bar{t}^2 D_{11}$$
$$D_{01}^\star = D_{01} - \bar{t}D_{11}$$
$$D_{11}^\star = D_{11}.$$

Consequently, only the interpretation of the variance of the slopes remains unchanged, when compared with the previous parameterization.

We return to the original parameterization and examine further the fitting of this model. Since we have assumed a common measurement error variance σ_ϵ^2, and common random effects variances $\boldsymbol{D}$ for boys and girls, the implied marginal standard deviations and correlations are the same for boys and girls and may be estimated from (8.9) and (8.10). Under REML, $\widehat{\sigma}_\epsilon = 1.31$ and the standard deviations (on the diagonal) and correlations (on the off-diagonal) are

$$\begin{bmatrix} 2.23 & & & \\ 0.65 & 2.23 & & \\ 0.64 & 0.65 & 2.30 & \\ 0.62 & 0.65 & 0.68 & 2.35 \end{bmatrix}. \tag{8.24}$$

We see that the standard deviations increases slightly over time, and the correlations decrease only slightly for observations further apart in time, suggesting that the random slopes are not contributing greatly to the fit. Fitting a random-intercepts-only model to these data produced a marginal variance estimate of 2.28^2 and common within-child correlations of 0.63.

The empirical standard deviations and correlations for boys and girls are given, respectively, by

$$\begin{bmatrix} 2.45 & & & \\ 0.44 & 2.14 & & \\ 0.56 & 0.39 & 2.65 & \\ 0.32 & 0.63 & 0.59 & 2.09 \end{bmatrix}, \quad \begin{bmatrix} 2.12 & & & \\ 0.83 & 1.90 & & \\ 0.86 & 0.90 & 2.36 & \\ 0.84 & 0.88 & 0.95 & 2.44 \end{bmatrix}$$

which suggests that our model needs refinement, since clearly the correlations for girls are greater than for boys.

8.5.4 Inference for Random Effects

In some situations, interest will focus on inference for the random effects. For example, for the dental data, we may be interested in the growth curve of a particular child. Estimates of random effects are also important for model checking.

Various approaches to inference for random effects have been proposed. The simplest, which we describe first, is to take an empirical Bayes approach. From a Bayesian standpoint, there is no distinction inferentially between fixed and random effects (the distinction is in the priors that are assigned). Consequently, inference is simply based on the posterior distribution $p(b_i \mid y)$. Consider the LMM

$$y_i = x_i\beta + z_i b_i + \epsilon_i,$$

and assume b_i and ϵ_i are independent with $b_i \mid D \sim_{iid} N_{q+1}(0, D)$ and $\epsilon_i \mid \sigma_\epsilon^2 \sim_{ind} N_{n_i}(0, \sigma_\epsilon^2 I)$, so that $\alpha = [\sigma_\epsilon^2, D]$. We begin by considering the simple, albeit unrealistic, situation, in which β and α are known. Letting $y_i^\star = y_i - x_i\beta$, we have

$$p(b_i \mid y_i, \beta, \alpha) \propto p(y_i \mid b_i, \beta, \alpha) \times \pi(b_i \mid \alpha)$$

$$\propto \exp\left[-\frac{1}{2\sigma_\epsilon^2}(y_i^\star - z_i b_i)^T(y_i^\star - z_i b_i) - \frac{1}{2}b_i^T D^{-1} b_i\right]$$

which we recognize as a multiple linear regression with a zero-centered normal prior on the parameters b_i (this model is closely linked to that used in ridge regression, see Sect. 10.5.1). Using a standard derivation, (5.7),

$$b_i \mid y_i, \beta, \alpha \sim N_{q+1}\left[E(b_i \mid y_i, \beta, \alpha), \operatorname{var}(b_i \mid y_i, \beta, \alpha)\right]$$

with mean and variance

$$E[b_i \mid y_i, \beta, \alpha] = \left(\frac{z_i^T z_i}{\sigma_\epsilon^2} + D^{-1}\right)^{-1} \frac{z_i^T}{\sigma_\epsilon^2}(y_i - x_i\beta)$$

$$= D z_i^T V_i^{-1}(y_i - x_i\beta) \qquad (8.25)$$

8.5 Likelihood Inference for Linear Mixed Models

$$\text{var}(\boldsymbol{b}_i \mid \boldsymbol{y}_i, \boldsymbol{\beta}, \boldsymbol{\alpha}) = \left(\frac{\boldsymbol{z}_i^{\text{T}} \boldsymbol{z}_i}{\sigma_\epsilon^2} + \boldsymbol{D}^{-1}\right)^{-1}$$

$$= \boldsymbol{D} - \boldsymbol{D}\boldsymbol{z}_i^{\text{T}}\boldsymbol{V}_i^{-1}\boldsymbol{z}_i\boldsymbol{D}, \tag{8.26}$$

see Exercise 8.4. As we will see in this section, the estimate (8.25) may be derived under a number of different formulations.

A fully Bayesian approach would consider

$$p(\boldsymbol{b} \mid \boldsymbol{y}) = \int \int p(\boldsymbol{b}, \boldsymbol{\beta}, \boldsymbol{\alpha} \mid \boldsymbol{y}) \, d\boldsymbol{\beta} d\boldsymbol{\alpha},$$

which emphasizes that the uncertainty in $\boldsymbol{\beta}, \boldsymbol{\alpha}$ is not acknowledged in the derivation of (8.25) and (8.26).

We now demonstrate how we may account for estimation of $\boldsymbol{\beta}$ with a flat prior on $\boldsymbol{\beta}$ and assuming $\boldsymbol{\alpha}$ known. The posterior mean and variance of $\boldsymbol{\beta}$ are

$$\text{E}[\boldsymbol{\beta} \mid \boldsymbol{y}, \boldsymbol{\alpha}] = \widehat{\boldsymbol{\beta}}_{\text{G}}$$

$$\text{var}(\boldsymbol{\beta} \mid \boldsymbol{y}, \boldsymbol{\alpha}) = (\boldsymbol{x}^{\text{T}}\boldsymbol{V}^{-1}\boldsymbol{x})^{-1}$$

where $\widehat{\boldsymbol{\beta}}_{\text{G}}$ is the GLS estimator (these forms are derived for more general priors later, see (8.35) and (8.36)). Consequently,

$$\text{E}[\boldsymbol{b}_i \mid \boldsymbol{y}, \boldsymbol{\alpha}] = \text{E}_{\beta \mid y, \alpha}\left[\text{E}(\boldsymbol{b}_i \mid \boldsymbol{y}, \boldsymbol{\alpha})\right]$$

$$= \boldsymbol{D}\boldsymbol{z}_i^{\text{T}}\boldsymbol{V}_i^{-1}(\boldsymbol{y}_i - \boldsymbol{x}_i\widehat{\boldsymbol{\beta}}_{\text{G}}) \tag{8.27}$$

$$\text{var}(\boldsymbol{b}_i \mid \boldsymbol{y}, \boldsymbol{\alpha}) = \text{E}_{\beta \mid y, \alpha}[\text{var}(\boldsymbol{b}_i \mid \boldsymbol{\beta}, \boldsymbol{y}, \boldsymbol{\alpha})] + \text{var}_{\beta \mid y, \alpha}(\text{E}[\boldsymbol{b}_i \mid \boldsymbol{y}, \boldsymbol{\alpha}])$$

$$= \text{E}_{\beta \mid y, \alpha}[\boldsymbol{D} - \boldsymbol{D}\boldsymbol{z}_i^{\text{T}}\boldsymbol{V}_i^{-1}\boldsymbol{z}_i\boldsymbol{D}] + \text{var}_{\beta \mid y, \alpha}(\boldsymbol{D}\boldsymbol{z}_i^{\text{T}}\boldsymbol{V}_i^{-1}(\boldsymbol{y}_i - \boldsymbol{x}_i\boldsymbol{\beta}))$$

$$= \boldsymbol{D} - \boldsymbol{D}\boldsymbol{z}_i^{\text{T}}\boldsymbol{V}_i^{-1}\boldsymbol{z}_i\boldsymbol{D} + \boldsymbol{D}\boldsymbol{z}_i^{\text{T}}\boldsymbol{V}_i^{-1}\boldsymbol{x}_i(\boldsymbol{x}^{\text{T}}\boldsymbol{V}^{-1}\boldsymbol{x})^{-1}\boldsymbol{x}_i^{\text{T}}\boldsymbol{V}_i^{-1}\boldsymbol{z}_i\boldsymbol{D}. \tag{8.28}$$

Therefore, we can easily account for the estimation of $\boldsymbol{\beta}$, but no such simple development is available to account for estimation of $\boldsymbol{\alpha}$.

From a frequentist perspective, inference for random effects is often viewed as *prediction* rather than estimation, since $[\boldsymbol{b}_1, \ldots, \boldsymbol{b}_m]$ are random variables and not unknown constants. Many different criteria may be used to find a predictor $\widehat{\boldsymbol{b}} = f(\boldsymbol{Y})$ of $\boldsymbol{b}$, for a generic unit.

We begin by defining the optimum predictor as that which minimizes the mean squared error (MSE). Let $\boldsymbol{b}^\star$ represent a general predictor and consider the MSE:

$$\text{MSE}(\boldsymbol{b}^\star) = \text{E}_{\boldsymbol{y}, \boldsymbol{b}}[(\boldsymbol{b}^\star - \boldsymbol{b})^{\text{T}} \boldsymbol{A} (\boldsymbol{b}^\star - \boldsymbol{b})],$$

where we emphasize that the expectation is with respect to both y and b, and A is any positive definite symmetric matrix. We show that the MSE is minimized by $\widehat{b} = \mathrm{E}[b \mid y]$. For the moment, we suppress the dependence on any additional parameters. We can express the MSE in terms of $b^\star$ and $\widehat{b}$:

$$\begin{aligned}
\mathrm{MSE}(b^\star) &= \mathrm{E}_{Y,b}[(b^\star - b)^{\mathrm{T}} A(b^\star - b)] \\
&= \mathrm{E}_{Y,b}[(b^\star - \widehat{b} + \widehat{b} - b)^{\mathrm{T}} A(b^\star - \widehat{b} + \widehat{b} - b)] \\
&= \mathrm{E}_{Y,b}[(b^\star - \widehat{b})^{\mathrm{T}} A(b^\star - \widehat{b})] + 2 \times \mathrm{E}_{Y,b}[(b^\star - \widehat{b}) A(\widehat{b} - b)] \\
&\quad + \mathrm{E}_{Y,b}[(\widehat{b} - b)^{\mathrm{T}} A(\widehat{b} - b)].
\end{aligned} \qquad (8.29)$$

The third term does not involve $b^\star$, and we may write the second expectation as

$$\begin{aligned}
\mathrm{E}_{Y,b}[(b^\star - \widehat{b}) A(\widehat{b} - b)] &= \mathrm{E}_Y \{\mathrm{E}_{b|y}[(b^\star - \widehat{b}) A(\widehat{b} - b) \mid y]\} \\
&= \mathrm{E}_Y[(b^\star - \widehat{b}) A(\widehat{b} - \widehat{b})] = \mathbf{0}
\end{aligned}$$

and so, minimizing MSE corresponds to minimizing the first term in (8.29). This quantity must be nonnegative, and so, the solution is to take $b^\star = \widehat{b}$. The latter is the solution irrespective of A. So the best prediction is that which estimates the random variable b by its conditional mean. We now examine properties of $\widehat{b}$.

The usual frequentist optimality criteria for a fixed effect θ concentrate upon unbiasedness and upon the variance of the estimator, $\mathrm{var}(\widehat{\theta})$, see Sect. 2.2. When inference is required for a random effect b, these criteria need adjustment. Specifically, an unbiased predictor $\widehat{b}$ is such that

$$\mathrm{E}[\widehat{b} - b] = \mathbf{0},$$

to give

$$\mathrm{E}[\widehat{b}] = \mathrm{E}[b]$$

so that the expectation of the predictor is equal to the expectation of the random variable that it is predicting. For $\widehat{b} = \mathrm{E}[b \mid y]$,

$$\mathrm{E}_Y[\widehat{b}] = \mathrm{E}_Y[\mathrm{E}_{b|y}(b \mid y)] = \mathrm{E}_b[b]$$

where the first step follows on substitution of $\widehat{b}$ and the second from iterated expectation; therefore, we have an unbiased predictor. We emphasize that we do not have an unbiased estimator in the usual sense, and in general, $\widehat{b}$ will display *shrinkage* toward zero, as we illustrate in later examples.

The variance of a random variable is defined with respect to a fixed number, the mean. In the context of prediction of a random variable, a more relevant summary of the variability is

$$\mathrm{var}(\widehat{b} - b) = \mathrm{var}(\widehat{b}) + \mathrm{var}(b) - 2 \times \mathrm{cov}(\widehat{b}, b).$$

8.5 Likelihood Inference for Linear Mixed Models

If this quantity is small, then the predictor and the random variable are moving in a stochastically similar way. We have

$$\text{cov}_{\widehat{b},b}(\widehat{b},b) = E_Y[\text{cov}(\widehat{b},b \mid y)] + \text{cov}_Y(E[\widehat{b} \mid y], E[b \mid y])$$
$$= E_Y[\text{cov}(\widehat{b},b \mid y)] + \text{cov}_Y(\widehat{b},\widehat{b})$$
$$= \text{var}(\widehat{b}), \tag{8.30}$$

since the first term in (8.30) is the covariance between the constant $E[\widehat{b} \mid y]$ (since y is conditioned upon) and $\widehat{b}$, and so is zero. To obtain the form of the second term in (8.30), we have used $E[\widehat{b} \mid y] = E[E[b \mid y] \mid y] = \widehat{b}$. Hence,

$$\text{var}(\widehat{b} - b) = \text{var}(b) - \text{var}(\widehat{b}) = D - \text{var}(\widehat{b}).$$

In order to determine the form of $\widehat{b} = E[b \mid y]$ and evaluate $\text{var}(\widehat{b} - b)$, we need to provide more information on the model that is to be used, so that the form of $p(b \mid y)$ can be determined.

For the LMM,

$$\begin{bmatrix} b_i \\ Y_i \end{bmatrix} \sim N_{q+1+n_i} \left(\begin{bmatrix} 0 \\ x_i\beta \end{bmatrix}, \begin{bmatrix} D & Dz_i^T \\ z_iD & V_i \end{bmatrix} \right)$$

since

$$\text{cov}(b_i, Y_i) = \text{cov}(b_i, x_i\beta + z_ib_i + \epsilon_i) = \text{cov}(b_i, z_ib_i) = Dz_i^T,$$

(Appendix B), and similarly, $\text{cov}(Y_i, b_i) = z_iD$. The conditional distribution of a multivariate normal distribution is normal also (Appendix D) with mean

$$\widehat{b}_i = E[b_i \mid y_i] = Dz_i^T V_i^{-1}(y_i - x_i\beta) \tag{8.31}$$

which coincides with the Bayesian derivation earlier, (8.25). From a frequentist perspective, (8.25) is known as the best linear unbiased predictor (BLUP), where unbiased refers to it satisfying $E[\widehat{b}_i] = E[b_i]$.

The form (8.31) is not of practical use since it depends on the unknown β and α; instead, we use

$$\widehat{b}_i = E[b_i \mid y_i, \widehat{\beta}, \widehat{\alpha}] = \widehat{D} z_i^T \widehat{V}_i^{-1}(y_i - x_i\widehat{\beta}_G) \tag{8.32}$$

where $\widehat{D} = D(\widehat{\alpha})$ and $\widehat{V} = V(\widehat{\alpha})$. The implications of the substitution of $\widehat{\beta}_G$ are not great, since it is an unbiased estimator and appears in (8.31) in a linear fashion, but the use of $\widehat{\alpha}$ is more problematic. In particular the predictor $\widehat{b}_i$ is no longer linear in the data, so that exact properties can no longer be derived.

The uncertainty in the prediction, accounting for the estimation of β, is

$$\mathrm{var}(\widehat{\boldsymbol{b}}_i - \boldsymbol{b}_i) = \boldsymbol{D} - \mathrm{var}(\widehat{\boldsymbol{b}}_i)$$
$$= \boldsymbol{D} - \boldsymbol{D}\boldsymbol{z}_i^{\mathrm{T}}\boldsymbol{V}_i^{-1}\boldsymbol{z}_i\boldsymbol{D} + \boldsymbol{D}\boldsymbol{z}_i^{\mathrm{T}}\boldsymbol{V}_i^{-1}\boldsymbol{x}_i(\boldsymbol{x}^{\mathrm{T}}\boldsymbol{V}^{-1}\boldsymbol{x})^{-1}\boldsymbol{x}_i^{\mathrm{T}}\boldsymbol{V}_i^{-1}\boldsymbol{z}_i\boldsymbol{D}$$

after tedious algebra (Exercise 8.5), so that (8.27) is recovered. We again emphasize that this estimate of variability of prediction does not acknowledge the uncertainty in $\widehat{\boldsymbol{\alpha}}$. Given correct specification of the marginal variance model, $\mathrm{var}(Y \mid \boldsymbol{\alpha}) = \boldsymbol{V}(\boldsymbol{\alpha})$, and a consistent estimator of $\boldsymbol{\alpha}$, $\widehat{\boldsymbol{b}}_i$ is asymptotically normal with a known distribution, which can be used to form interval estimates. As an alternative to the use of (8.25), we can implement a fully Bayesian approach (Sect. 8.6), though no closed-form solution emerges.

As a final derivation, rather than assume normality, we could consider estimators that are *linear* in $\boldsymbol{y}$. Exercise 8.6 shows that this again leads to

$$\widehat{\boldsymbol{b}}_i = \boldsymbol{D}\boldsymbol{z}_i^{\mathrm{T}}\boldsymbol{V}_i^{-1}(\boldsymbol{y}_i - \boldsymbol{x}_i\boldsymbol{\beta}).$$

The best linear predictor is therefore identical to the best predictor under normality. For general distributions, $\mathrm{E}[\boldsymbol{b}_i \mid \boldsymbol{y}_i]$ will not necessarily be linear in $\boldsymbol{y}_i$.

Since we now have a method for predicting $\boldsymbol{b}_i$, we can examine fitted values:

$$\widehat{Y}_i = \boldsymbol{x}_i\widehat{\boldsymbol{\beta}} + \boldsymbol{z}_i\widehat{\boldsymbol{b}}_i$$
$$= \boldsymbol{x}_i\widehat{\boldsymbol{\beta}} + \boldsymbol{z}_i\left[\boldsymbol{D}\boldsymbol{z}_i^{\mathrm{T}}\boldsymbol{V}_i^{-1}(\boldsymbol{y}_i - \boldsymbol{x}_i\widehat{\boldsymbol{\beta}})\right]$$
$$= (\boldsymbol{I}_{n_i} - \boldsymbol{W}_i)\boldsymbol{x}_i\widehat{\boldsymbol{\beta}} + \boldsymbol{W}_i\boldsymbol{y}_i,$$

with $\boldsymbol{W}_i = \boldsymbol{z}_i\boldsymbol{D}\boldsymbol{z}_i^{\mathrm{T}}\boldsymbol{V}_i^{-1}$, so that we have a weighted combination of the population profile and the unit's data. If $\boldsymbol{D} = \boldsymbol{0}$, we obtain $\widehat{Y}_i = \boldsymbol{x}_i\widehat{\boldsymbol{\beta}}$, and if $\boldsymbol{D}$ is "small," the fitted values are close to the population curve, which is reasonable if there is little between-unit variability. If elements of $\boldsymbol{D}$ are large, the fitted values are closer to the observed data.

Example: One-Way ANOVA

For the simple balanced ANOVA model previously considered, the calculation of $\mathrm{E}[b_i \mid \boldsymbol{y}_i, \widehat{\boldsymbol{\beta}}, \widehat{\boldsymbol{\alpha}}]$ results in

$$\widehat{b}_i = \frac{n\widehat{\sigma}_0^2}{\widehat{\sigma}_\epsilon^2 + n\widehat{\sigma}_0^2}(\overline{y}_i - \widehat{\beta}_0)$$

to give a predictor that is a weighted combination of the "residual" $\overline{y}_i - \widehat{\beta}_0$ and zero. For finite n, the predictor is biased towards zero. As $n \to \infty$, $\widehat{b}_i \to \overline{y}_i - \widehat{\beta}_0$, so that $\widehat{\beta}_0 + \widehat{b}_i \to \overline{y}_i$, illustrating that the shrinkage disappears as the number of observations on a unit n increases, as we would hope.

8.6 Bayesian Inference for Linear Mixed Models

8.6.1 A Three-Stage Hierarchical Model

We consider the LMM

$$y_i = x_i\beta + z_i b_i + \epsilon_i,$$

with b_i and ϵ_i independent and distributed as $b_i \mid D \sim_{iid} N_{q+1}(\mathbf{0}, D)$, and $\epsilon_i \mid \sigma_\epsilon^2 \sim_{ind} N_{n_i}(\mathbf{0}, \sigma_\epsilon^2 \mathbf{I})$, $i = 1, \ldots, m$.

The second stage assumption for b_i can be motivated using the concept of exchangeability that we encountered in Sect. 3.9. If we believe a priori that $b_1, \ldots, b_m$ are exchangeable (and are considered within a hypothetical infinite sequence of such random variables), then it can be shown using representation theorems (Sect. 3.9) that the prior has the form

$$p(b_1, \ldots, b_m) = \int \prod_{i=1}^{m} p(b_i \mid \phi) \pi(\phi) \, d\phi,$$

so that the collection $[b_1, \ldots, b_m]$ are conditionally independent, given *hyperparameters* ϕ, with the hyperparameters having a distribution known as a *hyperprior*.

Hence, we have a two-stage (hierarchical) prior:

$$b_i \mid \phi \sim_{iid} p(\cdot \mid \phi), \quad i = 1, \ldots, m$$

$$\phi \sim_{iid} \pi(\cdot).$$

Parametric choices for $p(\cdot \mid \phi)$ and $\pi(\cdot)$ are based on the application, though computational convenience may also be a consideration (as we discuss in Sect. 8.6.3). We initially consider the multivariate normal prior $N_{q+1}(\mathbf{0}, D)$ so that $\phi = D$. The practical importance of this representation is that under exchangeability the beliefs about each of the unit-specific parameters must be identical. For example, for the dental data, if we do not believe that the individual-specific deviations from the average intercepts and slopes for boys and girls are exchangeable, then we should consider separate prior specifications for each gender. In general, if collections of units cluster due to an observed covariate that we believe will influence b_i, then our prior should reflect this. This framework contrasts with the sampling theory approach in which the random effects are assumed to be a random sample from a hypothetical infinite population.

The three-stage model is

Stage One: Likelihood:
$$p(\boldsymbol{y}_i \mid \boldsymbol{\beta}, \boldsymbol{b}_i, \sigma_\epsilon^2), \quad i = 1, \ldots, m.$$

Stage Two: Random effects prior:
$$p(\boldsymbol{b}_i \mid \boldsymbol{D}), \quad i = 1, \ldots, m.$$

Stage Three: Hyperprior:
$$p(\boldsymbol{\beta}, \boldsymbol{D}, \sigma_\epsilon^2).$$

8.6.2 Hyperpriors

It is common to assume independent priors:
$$\pi(\boldsymbol{\beta}, \boldsymbol{D}, \sigma_\epsilon^2) = \pi(\boldsymbol{\beta})\pi(\boldsymbol{D})\pi(\sigma_\epsilon^2).$$

A multivariate normal distribution or $\boldsymbol{\beta}$ and an inverse gamma distribution for σ_ϵ^2 are often reasonable choices, since they are flexible enough to reflect a range of prior information. The data are typically informative on $\boldsymbol{\beta}$ and σ_ϵ^2 also. These choices also lead to conditional distributions that have convenient forms for Gibbs sampling (Sect. 3.8.4). The prior specification for $\boldsymbol{D}$ is less straightforward.

If $\boldsymbol{D}$ is a diagonal matrix with elements σ_k^2, $k = 0, 1, \ldots, q$, then an obvious choice is
$$\pi(\sigma_0^2, \ldots, \sigma_q^2) = \prod_{k=0}^{q} \text{IGa}(a_k, b_k),$$

where $\text{IGa}(a_k, b_k)$ denotes the inverse gamma distribution with prespecified parameters a_k, b_k, $k = 0, \ldots, q$. These choices also lead to conjugate conditional distributions for Gibbs sampling. Other choices are certainly possible, however, for example, those contained in Gelman (2006). A prior for non-diagonal $\boldsymbol{D}$ is more troublesome; there are $(q+2)(q+1)/2$ elements, with the restriction that the matrix of elements is positive definite. The inverse Wishart distribution is the conjugate choice and is the only distribution for which any great practical experience has been gathered.

We digress to describe how the Wishart distribution can be motivated. Suppose $\boldsymbol{Z}_1, \ldots, \boldsymbol{Z}_r \sim_{iid} \text{N}_p(\boldsymbol{0}, \boldsymbol{S})$, with $\boldsymbol{S}$ a non-singular variance–covariance matrix, and let
$$\boldsymbol{W} = \sum_{j=1}^{r} \boldsymbol{Z}_j \boldsymbol{Z}_j^{\text{T}}. \tag{8.33}$$

8.6 Bayesian Inference for Linear Mixed Models

Then W follows a Wishart distribution, denoted $\text{Wish}_p(r, S)$, with probability density function

$$p(w) = c^{-1} \mid w \mid^{(r-p-1)/2} \exp\left[-\frac{1}{2}\text{tr}(wS^{-1})\right]$$

where

$$c = 2^{rp/2} \Gamma_p(r/2) \mid S \mid^{r/2}, \qquad (8.34)$$

with

$$\Gamma_p(r/2) = \pi^{p(p-1)/4} \prod_{j=1}^{p} \Gamma[(r+1-j)/2]$$

the generalized gamma function. We require $r > p - 1$ for a proper density. The mean is

$$\text{E}[W] = rS.$$

Taking $p = 1$ yields

$$p(w) = \frac{(2S)^{-r/2}}{\Gamma(r/2)} w^{r/2-1} \exp(-w/2S),$$

for $w > 0$, revealing that the Wishart distribution is a multivariate version of the gamma distribution, parameterized as $\text{Ga}[r/2, 1/(2S)]$. Further, taking $S = 1$ gives a χ_r^2 random variable, which is clear from (8.33).

If $W \sim \text{Wish}_p(r, S)$, the distribution of $D = W^{-1}$ is known as the inverse Wishart distribution, denoted $\text{InvWish}_p(r, S)$, with density

$$p(d) = c^{-1} \mid d \mid^{-(r+p+1)/2} \exp\left[-\frac{1}{2}\text{tr}(d^{-1}S)\right],$$

where c is again given by (8.34). We denote this random variable by D in anticipation of subsequently specifying an inverse Wishart distribution as prior for the variance–covariance matrix of the random effects D. The mean is

$$\text{E}[D] = \frac{S^{-1}}{r - p - 1}$$

and is defined for $r > p + 1$. If $p = 1$, we recover the inverse gamma distribution $\text{IGa}(r/2, 1/2S)$ with

$$\text{E}[D] = \frac{1}{S(r-2)}$$

$$\text{var}(D) = \frac{1}{S^2(r-2)(r-4)},$$

so that small r gives a more dispersed distribution (which is true for general p). One way of thinking about prior specification is to imagine that the prior data for the precision consists of observing r multivariate normal random variables with empirical variance–covariance matrices $\boldsymbol{R} = \boldsymbol{S}^{-1}$. See Appendix D for further properties of the Wishart and inverse Wishart distributions.

Returning from our digression, within the LMM, we specify $\boldsymbol{W} = \boldsymbol{D}^{-1} \sim \mathrm{W}_{q+1}(r, \boldsymbol{R}^{-1})$ where we have taken $\boldsymbol{S} = \boldsymbol{R}^{-1}$ to aid in prior specification. We require choices for r and $\boldsymbol{R}$. Since

$$\mathrm{E}[\boldsymbol{D}] = \frac{\boldsymbol{R}}{r - q - 2},$$

$\boldsymbol{R}$ may be scaled to be a prior estimate of $\boldsymbol{D}$, with r acting as a strength of belief in the prior, with large r placing more mass close to the mean.

One method of specification that attempts to minimize the influence of the prior is to take $r = q + 3$ the smallest integer that gives a proper prior to give $\mathrm{E}[\boldsymbol{D}] = \boldsymbol{R}$, as the prior guess for $\boldsymbol{D}$. We now describe another way of specifying a Wishart prior, based on Wakefield (2009b). Marginalization over $\boldsymbol{D}$ gives $\boldsymbol{b}_i$ as multivariate Student's t with location $\boldsymbol{0}$, scale matrix $\boldsymbol{R}/(r - p + 1)$, and degrees of freedom $d = r - q + 2$. The margins of a multivariate Student's t are t also, which allows r and $\boldsymbol{R}$ to be chosen via specification of an interval for the jth element of $\boldsymbol{b}_i$, b_{ij}. Specifically, b_{ij} follows a univariate Student's t distribution with location 0, scale $R_{jj}/(r-q+2)$, and degrees of freedom $d = r - q$. For a required range of $[-V, V]$ with probability 0.95, we use the relationship $\pm t_{0.025}^d \sqrt{D_{jj}} = \pm V$, where t_p^d is the $100 \times p$th quantile of a Student's t random variable with d degrees of freedom. Picking the smallest integer that results in a proper prior gives $r = q + 1$ so that $d = 1$ and $R_{jj} = V^2 d / 2(t_{1-(1-p)/2}^d)^2$.

As an example of this procedure, consider a single random effect ($q = 0$). We specify a $\mathrm{Ga}[r/2, 1/(2S)]$ prior for σ_0^{-2}, so that marginally, b_i is a Student's t distribution with location 0, scale r/S, and degrees of freedom r. The above prescription gives $r = 1$ and $S = (t_{1-(1-p)/2}^d)^2 / V^2$. In the more conventional $\mathrm{Ga}(a,b)$ parameterization, we obtain $a = 0.5$ and $b = V^2 / [2(t_{1-(1-p)/2}^d)^2]$. For example, for the dental data, if we believe that a 95% range for the intercepts, about the population intercept, is $\pm V = \pm 0.2$, we obtain the choice $\mathrm{Ga}(0.5, 0.000124)$ for σ_0^{-2}. This translates into a prior for σ_0 (which is more interpretable) with 5%, 50%, and 95% points of [0.008, 0.023, and 0.25]. An important point to emphasize is that within the LMM, a proper prior is required for $\boldsymbol{D}$ to ensure propriety of the posterior distribution.

A weakness with the Wishart distribution is that it is deficient in second moment parameters, since there is only a single degrees of freedom parameter r. So, for example, it is not possible to have differing levels of certainty in the tightness of the prior distribution for different elements of $\boldsymbol{D}$. This contrasts with the situation in which $\boldsymbol{D}$ is diagonal, and we specify independent inverse gamma priors, which gives separate precision parameters for each variance.

8.6 Bayesian Inference for Linear Mixed Models

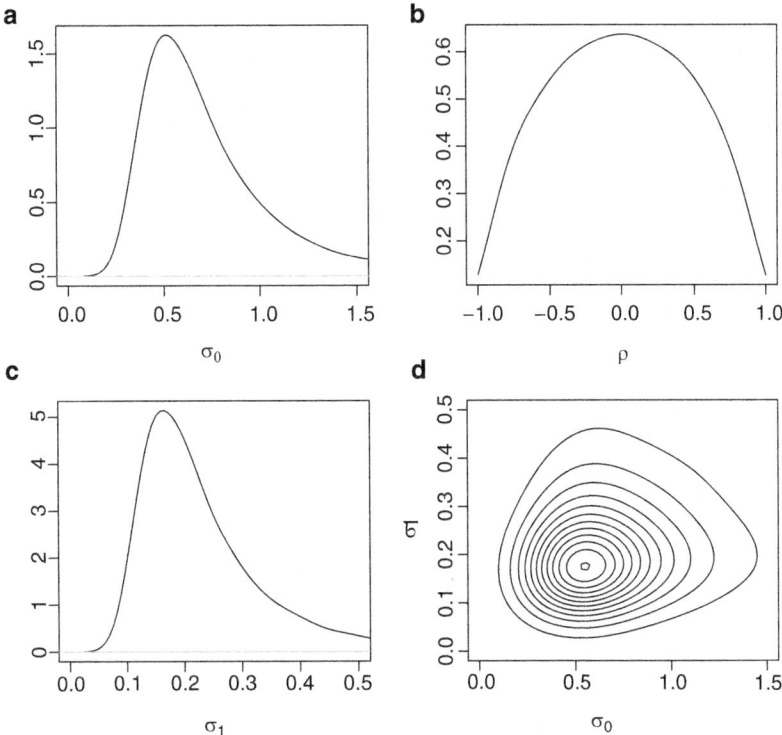

Fig. 8.2 Prior summaries for the prior $D^{-1} \sim W_2(r, R^{-1})$ with $r = 4$ and R containing elements $[1.0, 0, 0, 0.1]$. Univariate marginal densities for (**a**) σ_0, (**b**) ρ, (**c**) σ_1, and the bivariate density for (**d**) (σ_0, σ_1)

Figure 8.2 displays summaries for an example with a 2×2 variance–covariance matrix (so that $q = 1$). We assume $D^{-1} \sim W_2(r, R^{-1})$ with $r = 4$ and $E[D] = \frac{R}{4-1-2} = R$ with $R = \begin{bmatrix} 1.0 & 0 \\ 0 & 0.1 \end{bmatrix}$. We summarize samples from the Wishart via marginal distributions for σ_0, σ_1, and ρ since these are more interpretable. These plots were obtained by simulating samples for D^{-1} from the Wishart prior and then converting these samples to the required functions of interest. Finally, we smooth the sample histograms and scatter plots to produce Fig. 8.2. As we would expect, the prior on the correlation is symmetric about 0. Examination of intervals for σ_0, σ_1 can inform on whether we believe the prior is suitable for any given application. Going one step further, we could then simulate random effects from the zero mean normal with variance D, the latter being a draw from the prior; we might also continue to simulate data, though this would require draws from the other priors too.

8.6.3 Implementation

For simplicity, we suppose that $x_i = z_i$. It is convenient in what follows to reparameterize in terms of the set $[\boldsymbol{\beta}_1, \ldots, \boldsymbol{\beta}_m, \tau, \boldsymbol{\beta}, \boldsymbol{W}]$ where $\boldsymbol{\beta}_i = \boldsymbol{\beta} + \boldsymbol{b}_i$, $\tau = \sigma_\epsilon^{-2}$, and $\boldsymbol{W} = \boldsymbol{D}^{-1}$. The joint posterior is

$$p(\boldsymbol{\beta}_1, \ldots, \boldsymbol{\beta}_m, \tau, \boldsymbol{\beta}, \boldsymbol{W} \mid \boldsymbol{y}) \propto \prod_{i=1}^{m} \left[p(\boldsymbol{y}_i \mid \boldsymbol{\beta}_i, \tau) p(\boldsymbol{\beta}_i \mid \boldsymbol{\beta}, \boldsymbol{W}) \right] \pi(\boldsymbol{\beta}) \pi(\tau) \pi(\boldsymbol{W}),$$

with priors:

$$\boldsymbol{\beta} \sim \text{N}_{q+1}(\boldsymbol{\beta}_0, \boldsymbol{V}_0), \quad \tau \sim \text{Ga}(a_0, b_0), \quad \boldsymbol{W} \sim \text{W}_{q+1}(r, \boldsymbol{R}^{-1}).$$

Marginal distributions, and summaries of these distributions, are not available in closed form. Various approaches to obtaining quantities of interest are available. The INLA procedure described in Sect. 3.7.4 is ideally suited to the LMM. As an alternative, we describe an MCMC strategy using Gibbs sampling (Sect. 3.8.4). The required conditional distributions are

- $p(\boldsymbol{\beta} \mid \tau, \boldsymbol{W}, \boldsymbol{\beta}_1, \ldots, \boldsymbol{\beta}_m, \boldsymbol{y})$.
- $p(\tau \mid \boldsymbol{\beta}, \boldsymbol{W}, \boldsymbol{\beta}_1, \ldots, \boldsymbol{\beta}_m, \boldsymbol{y})$.
- $p(\boldsymbol{\beta}_i \mid \boldsymbol{\beta}, \tau, \boldsymbol{W}, \boldsymbol{y}), i = 1, \ldots, m$.
- $p(\boldsymbol{W} \mid \boldsymbol{\beta}, \tau, \boldsymbol{\beta}_1, \ldots, \boldsymbol{\beta}_m, \boldsymbol{y})$.

where we block update $\boldsymbol{\beta}$, $\boldsymbol{W}$, and $\boldsymbol{\beta}_i$ to reduce dependence in the Markov chain.

The conditional distributions for $\boldsymbol{\beta}$, τ, and $\boldsymbol{\beta}_i$ are straightforward to derive (Exercise 8.10) and are given, respectively, by

$$\boldsymbol{\beta} \mid \boldsymbol{\beta}_1, \ldots, \boldsymbol{\beta}_m, \boldsymbol{W} \propto \prod_{i=1}^{m} p(\boldsymbol{\beta}_i \mid \boldsymbol{\beta}, \boldsymbol{W}) \pi(\boldsymbol{\beta})$$

$$\sim \text{N}_{q+1} \left[(m\boldsymbol{W} + \boldsymbol{V}_0^{-1})^{-1} \left(\boldsymbol{W} \sum_{i=1}^{m} \boldsymbol{\beta}_i + \boldsymbol{V}_0^{-1} \boldsymbol{\beta}_0 \right), \right.$$

$$\left. (m\boldsymbol{W} + \boldsymbol{V}_0^{-1})^{-1} \right]$$

$$\tau \mid \boldsymbol{\beta}_i, \boldsymbol{y} \propto \prod_{i=1}^{m} p(\boldsymbol{y}_i \mid \boldsymbol{\beta}_i, \tau) \pi(\tau)$$

$$\sim \text{Ga} \left[a_0 + \frac{\sum_{i=1}^{m} n_i}{2}, b_0 + \frac{1}{2} \sum_{i=1}^{m} (\boldsymbol{y}_i - \boldsymbol{x}_i \boldsymbol{\beta}_i)^{\text{T}} (\boldsymbol{y}_i - \boldsymbol{x}_i \boldsymbol{\beta}_i) \right]$$

$$\boldsymbol{\beta}_i \mid \tau, \boldsymbol{W}, \boldsymbol{y} \propto \prod_{i=1}^{m} p(\boldsymbol{y}_i \mid \boldsymbol{\beta}_i, \tau) p(\boldsymbol{\beta}_i \mid \boldsymbol{\beta}, \boldsymbol{W})$$

$$\sim \text{N}_{q+1} \left[(\tau \boldsymbol{x}_i^{\text{T}} \boldsymbol{x}_i + \boldsymbol{W})^{-1} (\tau \boldsymbol{x}_i^{\text{T}} \boldsymbol{y}_i + \boldsymbol{W} \boldsymbol{\beta}), (\tau \boldsymbol{x}_i^{\text{T}} \boldsymbol{x}_i + \boldsymbol{W})^{-1} \right].$$

8.6 Bayesian Inference for Linear Mixed Models

Conditional independencies have been exploited, and in each case, the notation explicitly conditions on only those parameters on which the conditional distribution depends. For example, to derive the conditional distribution for β, we only require $[\beta_1, \ldots, \beta_m]$ and W. The conditional for β_i is, once we reparameterize, identical to the empirical Bayes estimates derived for the random effects in Sect. 8.5.4 (Exercise 8.11). This comparison illustrates how the uncertainty in β and $\alpha = [\tau, W]$ is accounted for across iterates of the Gibbs sampler.

Deriving the conditional distribution for W is a little more involved. First, note that

$$(\beta_i - \beta)^T W(\beta_i - \beta) = \text{tr}[(\beta_i - \beta)^T W(\beta_i - \beta)] = \text{tr}[W(\beta_i - \beta)(\beta_i - \beta)^T].$$

Then

$$W \mid \beta_i, \beta \propto \prod_{i=1}^{m} p(\beta_i \mid W) \times \pi(W)$$

$$\propto \mid W \mid^{(m+r-q-1-1)/2} \exp\left\{-\frac{1}{2}\left[\sum_{i=1}^{m}(\beta_i - \beta)^T W(\beta_i - \beta) + \text{tr}(WR)\right]\right\}$$

$$= \mid W \mid^{(m+r-q-1-1)/2} \exp\left\{-\frac{1}{2}\text{tr}\left(W\left[\sum_{i=1}^{m}(\beta_i - \beta)(\beta_i - \beta)^T + R\right]\right)\right\}$$

to give the conditional distribution

$$W \mid \beta_1, \ldots, \beta_m, \beta \sim W_{q+1}\left[r+m, \left(R + \sum_{i=1}^{m}(\beta_i - \beta)(\beta_i - \beta)^T\right)^{-1}\right].$$

This illustrates how r and R are comparable to m and the between-unit sum of squares, respectively, which aids in prior specification. Since

$$E[D \mid \beta_1, \ldots, \beta_m, \beta] = \frac{R + \sum_{i=1}^{m}(\beta_i - \beta)(\beta_i - \beta)^T}{r + m - q - 2}$$

the form of the conditional distribution suggests that it is better to err on the side of picking R too small, since a large R will always dominate the sum of squares. If m is small, the prior is always influential.

If we collapse over β_i, $i = 1, \ldots, m$, we obtain the two-stage model with

Stage One: Marginal likelihood:

$$y \mid \beta, \tau, W \sim N_N(x\beta, V),$$

where $V = V(W, \tau)$.

Stage Two: Priors:

$$\pi(\beta)\pi(W)\pi(\tau).$$

An MCMC algorithm iterates between

- $p(\beta \mid y, W, \tau)$
- $p(\tau \mid y, \beta, W)$
- $p(W \mid y, \beta)$

This approach is appealing since it is over a reduced parameter space, but the form of $p(W \mid y, \beta, \tau)$ is extremely awkward. The conditional for β offers some intuition on the Bayesian approach, however. Specifically, writing $\alpha = [\tau, W]$, we obtain the conditional distribution:

$$\beta \mid y, \alpha \sim N_{q+1}\left[E(\beta \mid y, \alpha), \text{var}(\beta \mid y, \alpha)\right]$$

where the mean and variance can be written in the weighted forms

$$E[\beta \mid y, \alpha] = w \times \widehat{\beta}_G + (I - w) \times \beta_0 \tag{8.35}$$

$$\text{var}(\beta \mid y, \alpha) = w \times \text{var}(\widehat{\beta}_G). \tag{8.36}$$

Here, $\widehat{\beta}_G = (x^T V^{-1} x)^{-1} x^T V^{-1} y$ is the GLS estimator with variance $\text{var}(\widehat{\beta}_G) = (x^T V^{-1} x)^{-1}$, and the $(q+1) \times (q+1)$ weight matrix is

$$w = (x^T V^{-1} x + V_0^{-1})^{-1} x^T V^{-1} x.$$

As the prior becomes more diffuse, that is, as $V_0^{-1} \to 0$, the weight $w \to I$, the conditional posterior mean approaches the GLS estimator $\widehat{\beta}_G$, and the conditional posterior variance approaches $\text{var}(\widehat{\beta}_G)$. In contrast, as $V^{-1} \to 0$, so that the prior becomes more concentrated about β_0, $w \to 0$ and the conditional posterior moments approach the prior distribution. Since

$$E[\beta \mid y] = E_{\alpha \mid y}\left[E(\beta \mid y, \alpha)\right],$$

the posterior mean is the conditional posterior mean averaged over $\alpha \mid y$. As is typical, the Bayesian estimate integrates over α, while the GLS estimator conditions on $\widehat{\alpha}$ for evaluation of V. We would expect likelihood and Bayesian point and interval estimates to be similar for large samples because the posterior $\alpha \mid y$ will become increasingly concentrated about $\widehat{\alpha}$.

8.6.4 *Extensions*

Computationally, under a Bayesian approach via MCMC, it is relatively straightforward to extend the basic LMM. The conditional distributions may not be of

8.6 Bayesian Inference for Linear Mixed Models

conjugate form, but Metropolis–Hastings steps can be substituted (Sect. 3.8.2). For example, great flexibility in the distributional assumptions and error models is available, though prior specification will usually require greater care. To automatically protect against outlying measurements/individuals, Student's t errors may be specified at stage one/stage two of the hierarchy, though when regression is the focus of the analysis, the greatest effort should be concentrated upon specifying appropriate mean–variance relationships at the two stages.

With the advent of MCMC, there is a temptation to fit complex models that attempt to reflect every possible nuance of the data. However, the statistical properties of complex models (such as consistency of estimation under incorrect model specification) are difficult to determine, as are the implied marginal distributions for the data (which can aid in model assessment). Overfitting is also always a hazard. Consequently, caution should be exercised in model refinement. One of the arts of statistical analysis is deciding on when model refinement is warranted.

Example: Dental Growth Curves

We analyze the data from the $m = 11$ girls only and adopt the following three-stage hierarchical model:

Stage One: As likelihood, we assume

$$y_{ij} = \beta_{i0} + \beta_{i1} t_j + \epsilon_{ij},$$

with $\epsilon_{ij} \mid \tau \sim_{iid} \mathrm{N}(0, \tau^{-1})$, $j = 1, \ldots, 4$, $i = 1, \ldots, 11$.

Stage Two: Let

$$\boldsymbol{\beta}_i = \begin{bmatrix} \beta_{i0} \\ \beta_{i1} \end{bmatrix} \quad \boldsymbol{\beta} = \begin{bmatrix} \beta_0 \\ \beta_1 \end{bmatrix} \quad \boldsymbol{D} = \begin{bmatrix} \sigma_0^2 & \sigma_{01} \\ \sigma_{10} & \sigma_1^2 \end{bmatrix},$$

with random effects prior

$$\boldsymbol{\beta}_i \mid \boldsymbol{\beta}, \boldsymbol{D} \sim \mathrm{N}_2(\boldsymbol{\beta}, \boldsymbol{D}), \quad i = 1, \ldots, m.$$

Stage Three: As hyperprior, we assume

$$\pi(\tau, \boldsymbol{\beta}, \boldsymbol{D}^{-1}) = \pi(\tau) \times \pi(\boldsymbol{\beta}) \times \pi(\boldsymbol{D}^{-1})$$

with improper priors on τ and $\boldsymbol{\beta}$:

$$\pi(\tau) \propto \tau^{-1}, \quad \pi(\boldsymbol{\beta}) \propto 1$$

and

$$\boldsymbol{D}^{-1} \sim W_2(r, \boldsymbol{R}^{-1}).$$

In the LMM, there is typically abundant information in the data with respect to τ and $\boldsymbol{\beta}$. By placing a flat prior on $\boldsymbol{\beta}$ (which are often the parameters of interest), we are also basing inference on the data alone (in nonlinear models, more care is required since a proper prior is often required to ensure propriety of the posterior).

With just 11 girls, we would expect inference for $\boldsymbol{D}$ to be sensitive to the prior, and so, we consider three choices of r and $\boldsymbol{R}$. Each prior has the same mean of

$$\mathrm{E}[\boldsymbol{D}] = \begin{bmatrix} 1.0 & 0 \\ 0 & 0.1 \end{bmatrix} = \frac{\boldsymbol{R}}{r - q - 2} \qquad (8.37)$$

with $q = 1$ here. The above specification corresponds to an a priori belief that the spread of the expected response at 11 years across girls is

$$\pm 1.96 \mathrm{E}[\sigma_0] \approx \pm 1.96 \sqrt{R_{11}} = \pm 1.96$$

and the variability in slopes across girls is expected to be

$$\pm 1.96 \mathrm{E}[\sigma_1] \approx \pm 1.96 \sqrt{R_{22}} = \pm 0.62.$$

The exact intervals can be evaluated in an obvious fashion using simulation. The off-diagonal of $\boldsymbol{R}$ is 0 as we assume there is no reason to believe the correlation between intercepts and slopes will be positive or negative.

The degrees of freedom r is on the same scale as m and may be viewed as a prior sample size. We pick $r = 4, 7, 28$, and to obtain the same prior mean, (8.37), $\boldsymbol{R}$ is specified as

$$\begin{bmatrix} 1.0 & 0 \\ 0 & 0.1 \end{bmatrix}, \quad \begin{bmatrix} 4.0 & 0 \\ 0 & 0.4 \end{bmatrix}, \quad \begin{bmatrix} 25 & 0 \\ 0 & 2.5 \end{bmatrix},$$

for each of $r = 4, 7, 28$, respectively. To obtain a proper posterior, we require $r > 1$. We pick $r = 4$ as our smallest choice since the mean exists for this value. Samples from this prior are displayed in Fig. 8.2.

We present the results in terms of elements of $\boldsymbol{D}$, for direct comparison with the prior. If we were reporting substantive conclusions, we would choose σ_0, σ_1, ρ, or interval estimates for $\boldsymbol{\beta}_{i*} = [\beta_{i*0}, \beta_{i*1}]$, the parameters of a new girl who is exchangeable with those in the study. Table 8.1 gives posterior medians and 95% interval estimates for the fixed effects and variance components. We see sensitivity to the prior with respect to inference for $\boldsymbol{D}$. As r increases, the posterior medians draw closer to the prior means of 1.0 and 0.1. For β_0 and β_1, the medians are robust to the prior specification, while the width of the intervals for β_0 and β_1 change in proportion to the behavior of σ_0^2 and σ_1^2, respectively. The interval estimates for β_0 narrow, while those for β_1 widen, though the changes are modest. With only 11 subjects, we would expect sensitivity to the prior on $\boldsymbol{D}$. For $r = 7$, the "total degrees of freedom" is 18 with a prior contribution of 7 and a data contribution of 11.

Table 8.1 Posterior medians and 95% intervals for fixed effects and variance components, under three priors for the dental growth data for girls

	Prior	$r=4$		$r=7$		$r=28$	
β_0	–	22.6	[21.4,23.8]	22.6	[21.5,23.7]	22.6	[21.8,23.5]
β_1	–	0.48	[0.33,0.63]	0.48	[0.31,0.65]	0.48	[0.28,0.67]
σ_0^2	1.0	3.48	[1.66,8.75]	2.97	[1.51,6.63]	1.78	[1.14,2.97]
σ_{01}	0.0	0.13	[−0.10,0.54]	0.10	[−0.14,0.46]	0.04	[−0.10,0.20]
σ_1^2	0.1	0.03	[0.01,0.10]	0.05	[0.02,0.12]	0.08	[0.05,0.14]

The population intercept is β_0 and the population slope is β_1. The variances of the random intercepts and random slopes are σ_0^2 and σ_1^2, respectively, and the covariance between the two is σ_{01}

8.7 Generalized Estimating Equations

8.7.1 Motivation

We now describe the GEE approach to modeling/inference. GEE attempts to make minimal assumptions about the data-generating process and is constructed to answer population-level, rather than individual-level, questions. There are some links with the quasi-likelihood approach described in Sect. 2.5 in that, rather than specify a full probability model for the data, only the first two moments are specified. GEE is motivated by dependent data situations, however, and exploits replication across units to empirically estimate standard errors through sandwich estimation. GEE uses a "working" second moment assumption; "working" refers to the choice of a variance model that may not necessarily correspond to exactly the form we believe to be true but rather to be a choice that is statistically convenient (we elaborate on this point subsequently). Any discrepancies from the truth are corrected using sandwich estimation to give a procedure that gives a consistent estimator of both the regression parameters and the standard errors (so long as we have independence between individuals).

We assume the marginal mean model

$$\mathrm{E}[\boldsymbol{Y}_i] = \boldsymbol{x}_i \boldsymbol{\beta},$$

and consider the $n_i \times n_i$ *working* variance–covariance matrix:

$$\mathrm{var}(\boldsymbol{Y}_i) = \boldsymbol{W}_i \tag{8.38}$$

with $\mathrm{cov}(\boldsymbol{Y}_i, \boldsymbol{Y}_{i'}) = \boldsymbol{0}$ for $i \neq i'$, so that observations on different individuals are assumed uncorrelated. To motivate GEE, we begin by assuming that $\boldsymbol{W}_i$ is known and does not depend on unknown parameters. In this case the GLS estimator minimizes

$$\sum_{i=1}^{m} (\boldsymbol{Y}_i - \boldsymbol{x}_i\boldsymbol{\beta})^\mathrm{T} \boldsymbol{W}_i^{-1} (\boldsymbol{Y}_i - \boldsymbol{x}_i\boldsymbol{\beta}),$$

and is given by the solution to the estimating equation

$$\sum_{i=1}^{m} x_i^\mathrm{T} W_i^{-1}(Y_i - x_i \widehat{\beta}) = 0,$$

which is

$$\widehat{\beta} = \left(\sum_{i=1}^{m} x_i^\mathrm{T} W_i^{-1} x_i\right)^{-1} \sum_{i=1}^{m} x_i^\mathrm{T} W_i^{-1} Y_i.$$

We have $\mathrm{E}[\widehat{\beta}] = \beta$, and if the information about β grows with increasing m, then $\widehat{\beta}$ is consistent. The vital observation is that $\widehat{\beta}$ is a consistent estimator for *any* fixed $W = \mathrm{diag}(W_1, \ldots, W_m)$. The weighting of observations by the latter dictates the efficiency of the estimator but not its consistency. The variance, $\mathrm{var}(\widehat{\beta})$, is

$$\left(\sum_{i=1}^{m} x_i^\mathrm{T} W_i^{-1} x_i\right)^{-1} \left(\sum_{i=1}^{m} x_i^\mathrm{T} W_i^{-1} \mathrm{var}(Y_i) W_i^{-1} x_i\right) \left(\sum_{i=1}^{m} x_i^\mathrm{T} W_i^{-1} x_i\right)^{-1}.$$

(8.39)

If the assumed variance–covariance matrix is substituted, that is, $\mathrm{var}(Y_i) = W_i$, then we obtain the model-based variance

$$\mathrm{var}(\widehat{\beta}) = \left(\sum_{i=1}^{m} x_i^\mathrm{T} W_i^{-1} x_i\right)^{-1}.$$

A Gauss–Markov theorem shows that, in this case, the estimator is efficient amongst linear estimators *if* the variance model (8.38) is correct (Exercise 8.6). The novelty of GEE is that rather than depend on a correctly specified variance model, sandwich estimation, via (8.39), is used to repair any deficiency in the working variance model.

8.7.2 The GEE Algorithm

We now suppose that $\mathrm{var}(Y_i) = W_i(\alpha)$ where α are unknown parameters in the variance–covariance model. A common approach is to assume

$$W_i(\alpha) = \alpha_1 R_i(\alpha_2),$$

where $\alpha_1 = \mathrm{var}(Y_{ij})$ is the variance of the response, for all i and j, and $R_i(\alpha_2)$ is a working correlation matrix that depends on parameters α_2. There are a number of choices for R_i, including independence, exchangeable and AR(1) models (as described in Sect. 8.4.2). For known α, $\widehat{\beta}$ is the root of the estimating equation

8.7 Generalized Estimating Equations

$$G(\beta) = \sum_{i=1}^{m} x_i^{\mathsf{T}} W_i^{-1}(\alpha)(Y_i - x_i\beta) = 0. \tag{8.40}$$

When α is unknown, we require an estimator $\widehat{\alpha}$ that converges to "something" so that, informally speaking, we have a stable weighting matrix $W(\widehat{\alpha})$ in the estimating equation.

The sandwich variance estimator is

$$\widehat{\text{var}}(\widehat{\beta}) = \left(\sum_{i=1}^{m} x_i^{\mathsf{T}} \widehat{W}_i^{-1} x_i\right)^{-1} \left(\sum_{i=1}^{m} x_i^{\mathsf{T}} \widehat{W}_i^{-1} \text{var}(Y_i) \widehat{W}_i^{-1} x_i\right) \left(\sum_{i=1}^{m} x_i^{\mathsf{T}} \widehat{W}_i^{-1} x_i\right)^{-1} \tag{8.41}$$

where $\widehat{W}_i = W_i(\widehat{\alpha})$ and $\text{var}(Y_i)$ is estimated by the variance–covariance matrix of the residuals:

$$(Y_i - x_i\widehat{\beta})(Y_i - x_i\widehat{\beta})^{\mathsf{T}}. \tag{8.42}$$

This produces a consistent estimate of $\text{var}(\widehat{\beta})$, so long as we have independence between units, that is, $\text{cov}(Y_i, Y_{i'}) = 0$ for $i \neq i'$. It is the replication across units that produces consistency, and so, the approach cannot succeed if we have no replication. Exercise 8.12 shows that we cannot estimate $\text{var}(Y)$ using the analog of (8.42) when there is dependence between units.

For inference, we may use the asymptotic distribution

$$\widehat{\text{var}}(\widehat{\beta})^{-1/2}(\widehat{\beta} - \beta) \sim N_{k+1}(0, I),$$

where we emphasize that the asymptotics are in the number of units, m. The variance estimator is sometimes referred to as *robust*, but *empirical* is a more appropriate description since the form can be highly unstable for small m.

In the most general case of working variance model specification, we may allow the working variance model to depend on β also, so that we have $W_i(\alpha, \beta)$ to allow mean–variance relationships. For example, in a longitudinal setting, the variance may depend on the square of the marginal mean μ_{ij} with an autoregressive covariance model:

$$\text{var}(Y_{ij}) = \alpha_1 \mu_{ij}^2$$
$$\text{cov}(Y_{ij}, Y_{ik}) = \alpha_1 \alpha_2^{|t_{ij}-t_{ik}|} \mu_{ij} \mu_{ik}$$
$$\text{cov}(Y_{ij}, Y_{i'k}) = 0, \quad i \neq i'$$

with $j = 1, \ldots, n_i$, $k, k' = 1, \ldots, n_{i'}$ and where t_{ij} is the time associated with response Y_{ij}. In this model, α_1 is the component of the variance that does not depend on the mean (and is assumed constant across time and across individuals), α_2 is the correlation between responses on the same individual which are one unit of time apart and $\alpha = [\alpha_1, \alpha_2]$. In general the roots of the estimating equation

$$\sum_{i=1}^{m} x_i^{\mathrm{T}} W_i^{-1}(\alpha, \beta)(Y_i - x_i \beta) = 0 \tag{8.43}$$

are not available in closed form when β appears in W.

We can write the $(k+1) \times 1$ estimating function in a variety of forms, for example:

$$x^{\mathrm{T}} W^{-1}(Y - x\beta)$$

$$\sum_{i=1}^{m} x_i^{\mathrm{T}} W_i^{-1}(Y_i - x_i \beta)$$

$$\sum_{i=1}^{m} \sum_{j=1}^{n_i} \sum_{k=1}^{n_i} x_{ij} W_i^{jk}(Y_{ik} - x_{ik}\beta)$$

where W_i^{ij} denotes entry (i,j) of W_i^{-1}. We will often use the middle form, since this emphasizes that the basic unit of replication (upon which the asymptotic properties depend) is indexed by i.

The GEE approach is constructed to carry out marginal inference, and so we cannot perform individual-level inference. For a linear model, marginalizing a LMM produces a marginal model identical to that used in a GEE approach. As a consequence, parameter interpretation, as discussed in Sect. 8.4.3 in the marginal setting, is identical in the LMM and in GEE. When nonlinear models are considered in Chap. 9 there is no equivalence and the differences between the conditional and marginal approaches to inference becomes more pronounced. For the linear model, sandwich estimation may be applied to the MLE of β.

So far, as the choice of "working" correlation structure is concerned, we encounter the classic efficiency/robustness trade-off. If we choose a simple structure, there are few elements in α to estimate, but there is a potential loss of efficiency. A more complex model may provide greater efficiency if the variance model is closer to the true data-generating mechanism but more instability in estimation of α. Clearly, this choice should be based on the sample size, with relatively sparse data encouraging the use of a simple model.

We summarize the GEE approach to modeling/estimation when the working variance model depends on α and not on β. The steps of the approach are:

1. Specification of a mean model, $\mathrm{E}[Y_i] = x_i \beta$.
2. Specification of a working variance model, $\mathrm{var}(Y_i) = W_i(\alpha)$.
3. From (1) and (2), an estimating function is constructed, and sandwich estimation is applied to the variance of the resultant estimator.

In general, iteration is needed to simultaneously estimate β and α. Let $\widehat{\alpha}^{(0)}$ be an initial estimate, set $t = 0$, and iterate between:

1. Solve $G(\widehat{\beta}, \widehat{\alpha}^{(t)}) = 0$, with G given by (8.40), to give $\widehat{\beta}^{(t+1)}$.
2. Estimate $\widehat{\alpha}^{(t+1)}$ based on $\widehat{\beta}^{(t+1)}$.

Set $t \to t+1$, and return to 1.

Example: Linear Regression

We illustrate the use of a working variance assumption in an independent data situation. Suppose

$$E[Y_i] = x_i\beta,$$

for $i = 1, \ldots, n$. Under the working *independence* variance model, $\text{var}(Y) = \alpha I$, the OLS estimator

$$\widehat{\beta} = (x^T x)^{-1} x^T Y$$

is recovered. The sandwich form of variance estimate is

$$\text{var}(\widehat{\beta}) = (x^T x)^{-1} x^T \text{var}(Y) x (x^T x)^{-1}. \tag{8.44}$$

Assuming the working variance is "true" gives the model-based estimate

$$\text{var}(\widehat{\beta}) = (x^T x)^{-1} \alpha,$$

and α may be estimated by

$$\widehat{\alpha} = \frac{1}{n-k-1} \sum_{i=1}^{n}(Y_i - x_i\widehat{\beta})^2,$$

which is formerly equivalent to quasi-likelihood. If we replace $\text{var}(Y)$ in (8.44) by a diagonal matrix with diagonal elements $(Y_i - x_i\widehat{\beta})^2$, then we obtain a variance estimator that protects (asymptotically) against errors with nonconstant variance. We cannot protect against correlated outcomes, however, since there is no replication.

8.7.3 Estimation of Variance Parameters

To formalize the estimation of α, we may introduce a second estimating equation. In the context of data with $\mu_{ij} = E[Y_{ij}]$ and $\text{var}(Y_{ij}) \propto v(\mu_{ij})$, we define residuals $R_{ij} = Y_{ij} - x_{ij}\beta$. Recall that β is a $(k+1) \times 1$ vector of parameters, and suppose α is an $r \times 1$ vector of variance parameters. We then consider the pair of estimating equations:

$$G_1(\beta, \alpha) = \sum_{i=1}^{m} x_i^T W_i^{-1}(Y_i - x_i\beta) \tag{8.45}$$

$$G_2(\beta, \alpha) = \sum_{i=1}^{m} E_i^T H_i^{-1}[T_i - \Sigma_i(\alpha)] \tag{8.46}$$

where the "data" in the second estimating equation are

$$T_i^{\mathrm{T}} = [R_{i1}R_{i2}, \ldots, R_{in_i-1}R_{in_i}, R_{i1}^2, \ldots, R_{in_i}^2],$$

an $[n_i + n_i(n_i - 1)/2]$-dimensional vector with

$$\Sigma_i(\alpha) = \mathrm{E}[T_i]$$

a model for the variances of, and correlations between, the residuals. In (8.46), $E_i = \partial \Sigma_i / \partial \alpha$ is the $[n_i + n_i(n_i-1)/2] \times r$ vector of derivatives, and $H_i = \mathrm{cov}(T_i)$ is the $[n_i + n_i(n_i-1)/2] \times [n_i + n_i(n_i-1)/2]$ working covariance model for the squared and cross residual terms. If G_2 is correctly specified, then there will be efficiency gains. A further advantage of this approach is that it is straightforward to incorporate a regression model for the variance–covariance parameters, that is, $\alpha = g(x)$, for some link function $g(\cdot)$. For general H, we will require the estimation of fourth order statistics, that is, $\mathrm{var}(T)$, which is a highly unstable endeavor unless we have an abundance of data. For this reason, working independence, $H_i = I$, is often used.

If $\mathrm{E}[T] \neq \Sigma$, then we will not achieve consistent estimation of the true variance model but, crucially, consistency of β through G_1 is guaranteed, so long as $\widehat{\alpha}$ converges to "something." We reiterate that a consistent estimate of $\mathrm{var}(\widehat{\beta})$ is guaranteed through the use of sandwich estimation, so long as units are independent.

As an illustration of the approach, assume for simplicity $n_i = n = 3$ so that

$$T_i^{\mathrm{T}} = [R_{i1}R_{i2}, R_{i1}R_{i3}, R_{i2}R_{i3}, R_{i1}^2, R_{i2}^2, R_{i3}^2].$$

With an exchangeable variance model:

$$\Sigma_i(\alpha)^{\mathrm{T}} = \mathrm{E}[T_i^{\mathrm{T}}] = [\alpha_1\alpha_2, \alpha_1\alpha_2, \alpha_1\alpha_2, \alpha_1, \alpha_1, \alpha_1]$$

so that α_1 is the marginal variance, and α_2 is the correlation between observations on the same unit. With $H_i = I$, that is, a working independence model for the variance parameters, the estimating function for α is

$$G_2(\widehat{\beta}, \alpha) = \sum_{i=1}^m \begin{bmatrix} \alpha_2 & \alpha_2 & \alpha_2 & 1 & 1 & 1 \\ \alpha_1 & \alpha_1 & \alpha_1 & 0 & 0 & 0 \end{bmatrix} \left(\begin{bmatrix} R_{i1}R_{i2} \\ R_{i1}R_{i3} \\ R_{i2}R_{i3} \\ R_{i1}^2 \\ R_{i2}^2 \\ R_{i3}^2 \end{bmatrix} - \begin{bmatrix} \alpha_1\alpha_2 \\ \alpha_1\alpha_2 \\ \alpha_1\alpha_2 \\ \alpha_1 \\ \alpha_1 \\ \alpha_1 \end{bmatrix} \right).$$

8.7 Generalized Estimating Equations

We therefore need to simultaneously solve the two equations:

$$\sum_{i=1}^{m} \widehat{\alpha}_2 \left[\sum_{j<k} R_{ij} R_{ik} - \widehat{\alpha}_1 \widehat{\alpha}_2 \right] + \sum_{j=1}^{3} (R_{ij}^2 - \widehat{\alpha}_1) = 0$$

$$\sum_{i=1}^{m} \widehat{\alpha}_1 \left[\sum_{j<k} R_{ij} R_{ik} - \widehat{\alpha}_1 \widehat{\alpha}_2 \right] = 0.$$

Dividing the second of these by $\widehat{\alpha}_1$ shows that

$$\widehat{\alpha}_1 \widehat{\alpha}_2 = \frac{1}{3m} \sum_{i=1}^{m} \sum_{j<k} R_{ij} R_{ik}$$

and substituting this into the first equation gives

$$\widehat{\alpha}_1 = \frac{1}{3m} \sum_{i=1}^{m} \sum_{j<k} R_{ij}^2,$$

to yield a pair of method of moments estimators.

Example: Dental Growth Curve

We use a GEE approach with the marginal model:

$$\mathrm{E}[Y_{ij}] = \boldsymbol{x}_{ij} \boldsymbol{\beta},$$

and interactions so that

$$\boldsymbol{x}_{ij} = \begin{cases} [1, t_j, 0, 0] & \text{for } i = 1, \ldots, 16 \\ [1, t_j, 1, t_j] & \text{for } i = 17, \ldots, 27, \end{cases}$$

where $j = 1, 2, 3, 4$ and $[t_1, t_2, t_3, t_4] = [-2, -1, 1, 2]$. Table 8.2 summarizes analyses with independence and exchangeable working correlation models, including standard errors under the assumption that the working model is correct (the "model" standard errors) and under sandwich estimation.

The point estimates and model-based standard errors under working independence always correspond to those from an OLS fit. The point estimates under the two working models are also identical here due to the balanced design. This agreement will not hold in general. The marginal variance is estimated as 2.26, and the correlation parameter under the exchangeable model as 0.61. These are in very close agreement with the equivalent values of 2.28 and 0.63 obtained from the random intercepts LMM. As we would expect for these data, the model-based and sandwich

Table 8.2 Summaries for the dental growth data of fixed effects from GEE analyses, under independence and exchangeable working correlation matrices; β_0 and β_1 are the population intercept and population slope for boys and $\beta_0 + \beta_2$ and $\beta_1 + \beta_3$ are the population intercept and population slope for girls

	Independence			Exchangeable		
		Standard error			Standard error	
	Estimate	Model	Sandwich	Estimate	Model	Sandwich
β_0	25.0	0.28	0.44	25.0	0.47	0.44
β_1	0.78	0.13	0.098	0.78	0.079	0.098
β_2	−2.32	0.44	0.75	−2.32	0.74	0.75
β_3	−0.31	0.20	0.12	−0.31	0.12	0.12

standard errors are quite similar under the exchangeable working model, because we have seen that the empirical estimates of the second moments are close to those of an exchangeable correlation structure. In contrast, the working independence standard errors change quite considerably. The sandwich standard errors are larger for the time static intercepts and smaller for the parameters associated with time (the two slopes).

Likelihood inference for a LMM with random intercepts and slopes produced identical point estimates to those in Table 8.2 and standard errors of [0.49, 0.086, 0.76, 0.14], which are in reasonable agreement with the sandwich standard errors reported in the table.

Example: Dental Data, Reduced Dataset

In the dental example the balanced design and relative abundance of data leads to summaries that might suggest that the alternative methods we have described are always in complete agreement. To correct this illusion, we now report summaries from an artificially created dental growth curve data set in which it is assumed that children randomly drop out of the study at some point after the first measurement. This yielded the data in Fig. 8.3 with 39 measurements on boys (previously there were 64) and 25 on girls (previously there were 44).

We analyze these data using GEE and LMMs, the latter via likelihood and Bayesian approaches to inference. For GEE, we implement independence and exchangeable working correlation structures. Table 8.3 gives point estimates along with uncertainty measures. For GEE, we report sandwich standard errors, for the likelihood LMM model-based standard errors and for the Bayes LMM posterior (model-based) standard deviations. The posterior distributions for the regression parameters were close to normal, with interval estimates based on a normal approximation virtually identical to those based directly on samples from the posterior. For the Bayesian analysis, we used a flat prior on β, and the Wishart prior for D^{-1} had prior mean (8.37), with $r = 4$.

8.7 Generalized Estimating Equations

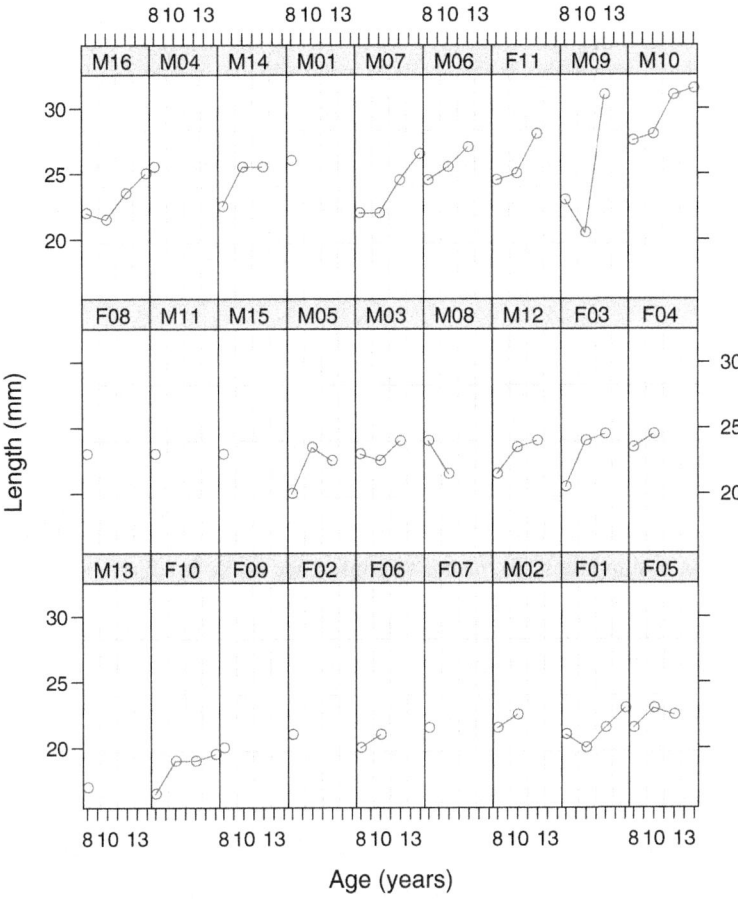

Fig. 8.3 Distance versus age for reduced dental data

Table 8.3 Summaries for the reduced dental growth data of fixed effects from GEE under independent and exchangeable working correlation matrices and likelihood and Bayesian LMMs; β_0 and β_1 are the population intercept and population slope for boys and $\beta_0 + \beta_2$ and $\beta_1 + \beta_3$ are the population intercept and population slope for girls

	GEE independence		GEE exchangeable		LMM likelihood		LMM Bayesian	
	Est.	s.e.	Est.	s.e.	Est.	s.e.	Est.	s.d.
β_0	24.9	0.75	24.8	0.63	24.7	0.65	24.8	0.63
β_1	0.77	0.20	0.71	0.11	0.70	0.14	0.70	0.16
β_2	−2.70	1.23	−2.01	0.97	−1.92	1.04	−1.98	1.02
β_3	−0.53	0.27	−0.21	0.15	−0.17	0.23	−0.19	0.26

For these data, none of the analyses are completely satisfactory since the small number of observations does not give confidence in the sandwich standard errors, nor are the data sufficiently abundant to allow any reliable evaluation of assumptions for the LMM analyses. The exchangeable standard errors appear too small for the slope parameters, though the point estimates are in reasonable agreement with their LMM counterparts. The GEE independence standard errors are more in line with the LMM analyses, though the point estimates are quite different for β_2 and β_3. As expected under these priors, the likelihood and Bayes analyses are in reasonable agreement.

8.8 Assessment of Assumptions

8.8.1 Review of Assumptions

Each of the approaches to modeling that we have described depend, to a varying degree, upon assumptions. To ensure that inference is accurate, we need to check that these assumptions are at least approximately valid. We begin by reviewing the assumptions, starting with GEE (since it depends on the fewest assumptions).

For GEE, we have the marginal mean model:

$$Y_i = x_i\beta + e_i,$$

and working covariance var$(e_i) = W_i(\alpha)$, $i = 1, \ldots, m$. The first consideration is whether the marginal model E$[Y_i] = x_i\beta$ is appropriate. In particular, one must check whether the model requires refinement by, for example, the addition of quadratic terms or interactions. We may also examine whether additional variables, such as confounders, are required in the model. These considerations are common to all approaches. If the mean model is inadequate, but all other assumptions are satisfied, then we will still have a consistent estimator of the assumed form, but the relevance of inference is open to question. For example, suppose the true relationship is quadratic, but we incorrectly assume a linear model. The linear association will still be consistently estimated but may be very misleading. Deciding on a course of action if the mean model is inadequate depends on the nature of the analysis. If we are in exploratory mode, then fitting different models is not problematic. But if we are in confirmatory mode, then we would want to minimize changes to the model, though knowing of inadequacies is important for interpretation.

The use of a sandwich estimate for the standard errors is reliable in the sense of giving consistent estimates regardless of whether the working covariance model mimics the truth, but a working model that is far from the truth will lead to a loss of efficiency (so that the standard errors are bigger than they need to be), which suggests one should examine whether the assumed working model is far from that suggested by the data. In addition, if the number of units m is not large, then the estimate of the sandwich standard errors could be very unstable, and asymptotic inference may be inappropriate. As usual, there is no easy recipe for deciding whether m is "sufficiently large", since this depends on the design across individuals in the sample. The decision may be based on simulation, though experience with similar datasets is beneficial.

For the LMM, the usual model is

$$Y_i = x_i\beta + z_i b_i + \epsilon_i,$$

with $b_i \mid D \sim_{iid} N_{q+1}(\mathbf{0}, D)$, $\epsilon_i \mid \sigma_\epsilon^2 \sim_{ind} N_{n_i}(\mathbf{0}, \sigma_\epsilon^2 I)$, and b_i, ϵ_i independent, $i = 1, \ldots, m$. This leads to the marginal model $Y_i \mid \beta, \alpha \sim N_{n_i}(x_i\beta, V_i)$ and estimator

$$\widehat{\beta} = \left(x^\mathsf{T} V^{-1} x\right)^{-1} x^\mathsf{T} V^{-1} Y \tag{8.47}$$

with

$$\left(x^\mathsf{T} V^{-1} x\right)^{1/2} (\widehat{\beta} - \beta) \sim N_{k+1}(\mathbf{0}, I).$$

Therefore, if m is large, we do not require the data or the random effects to be normally distributed since the estimator is linear in the data, and so we can appeal to a central limit theorem. For an accurate standard error, we require the model-based form of the variance to be close to the truth, however. It is particularly important that there are no unmodeled mean–variance relationships. Another key requirement is that the random effects arise from a common distribution. Often, unit-specific covariates will be available, and these may define subpopulations that have different distributions (e.g., differing variance–covariance matrices D) in covariate-defined subpopulations. If m is small we require, in addition, the data to be "close to normal" for valid inference. Sandwich estimation can be easily applied to obtain an empirical standard error, keeping in mind the caveats expressed above with regard to the need for sufficiently large m.

For prediction of the random effects, we have seen that the BLUP estimator is optimal under a number of different criteria. Normality of the random effects or the errors is not required, though an appropriate variance model is again important.

A Bayesian analysis of the LMM adds hyperpriors for β and α to the two-stage likelihood model. Each of the modeling assumptions required for likelihood-based inference are needed for a Bayesian analysis. However, asymptotic inference is not needed if, for example, MCMC is used. Accurate inference requires checking of the first and second stage assumptions because inference relies on the model being correct (or in practice, close to correct). Also, thought is required when priors are specified because inference may well be sensitive to the choices made. In particular, care is called for in the specification for D. We emphasize that normality of the data and the random effects is not needed for a valid analysis if the sample size is large. For example, for inference with respect to β, the posterior for β will be accurate so long as the asymptotic distribution of the estimator, (8.47), is faithful. Essentially, the asymptotic distribution replaces the likelihood contribution to the posterior.

8.8.2 Approaches to Assessment

For those individuals with sufficient data, individual-specific models may be fitted to allow examination of the appropriateness of initially hypothesized models in terms of the linear component and assumptions about the errors, such as constant variance, serial correlation, and normality if m is small. Following the fitting of marginal or mixed models, the assumptions may then be assessed further, with examination of residuals a useful exercise.

Residuals may be defined with respect to different levels. With respect to the usual LMM, a vector of unstandardized *population-level* (marginal) residuals is

$$e_i = Y_i - x_i\beta$$

and these are most useful for analyses based on the marginal (GEE) approach. A vector of unstandardized *unit-level* (stage one) residuals is

$$\epsilon_i = Y_i - x_i\beta - z_i b_i.$$

The vector of random effects b_i is also a form of (stage two) residual. Estimated versions of these residuals are

$$\widehat{e}_i = Y_i - x_i\widehat{\beta} \qquad (8.48)$$

$$\widehat{\epsilon}_i = Y_i - x_i\widehat{\beta} - z_i\widehat{b}_i \qquad (8.49)$$

and $\widehat{b}_i$, $i = 1, \ldots, m$.

We first discuss the population residuals (8.48). Recall, from consideration of the ordinary linear model (Sect. 5.11), that estimated residuals have dependencies induced by replacement of parameters by their estimates. The situation is far worse for dependent data because we would expect the population residuals to be dependent, even if the true parameter values were known. If $V_i(\alpha)$ is the true error structure, then

$$\text{var}(e_i) = V_i \text{ and } \text{var}(\widehat{e}_i) \approx V_i(\widehat{\alpha}),$$

showing the dependence of the residuals under the model. This means that, when working with e_i, it is difficult to check whether the covariance model is correctly specified. Plotting $\widehat{e}_{ij}$ versus the lth covariate x_{ijl}, $l = 1, \ldots, k$ may also be misleading due to the dependence within the residuals. Therefore, standardization is essential to remove the dependence.

Let $\widehat{V}_i = L_i L_i^{\text{T}}$ be the Cholesky decomposition of $\widehat{V}_i = V_i(\widehat{\alpha})$. We can use this decomposition to form

$$\widehat{e}_i^\star = L_i^{-1}\widehat{e}_i = L_i^{-1}(Y_i - x_i\widehat{\beta})$$

so that $\text{var}(\widehat{e}_i^\star) \approx I_{n_i}$. We may then work with the model

$$Y_i^\star = x_i^\star \beta + e_i^\star$$

8.8 Assessment of Assumptions

where $Y_i^\star = L_i^{-1} Y_i$, $x_i^\star = L_i^{-1} x_i$, and $e_i^\star = L_i^{-1} e_i$. Plots of $\widehat{e}_{ij}^\star$ against $x_{ijl}^\star$, $l = 1, \ldots, k$ should not show systematic patterns if the assumed linear form is correct.

QQ plots of $\widehat{e}_{ij}^\star$ versus the expected residuals from a normal distribution can be used to assess normality (unless m is small, normal errors are not required for accurate inference, but the closer to normality are the data, the smaller the m required for the asymptotics to have practically "kicked in"). If e_{ij} are normal, then standardized residuals will be normally distributed also, since $e_{ij}^\star$ is a linear combination of elements of e_i.

The correctness of the mean–variance relationship can be assessed by plotting $\widehat{e}_{ij}^{\star 2}$ (or $|\widehat{e}_{ij}^\star|$) against fitted values $\widehat{\mu}_{ij}^\star = x_{ij}^\star \widehat{\beta}$. Any systematic (non-horizontal) trends suggest problems. Local smoothers (as described in Chap. 11) can be added to plots to aid interpretation and plotting symbols such as unit or observation number can also be useful to identify collections of observations for which the model is not adequate.

For the LMM with $\epsilon_i \mid \sigma_\epsilon^2 \sim_{iid} N_{n_i}(\,0, \sigma_\epsilon^2 I\,)$, the stage one residuals (8.49) may be formed. Standardized versions are $\widehat{e}_{ij}^\star = \widehat{e}_{ij}/\widehat{\sigma}_\epsilon$. As usual, these residuals may be plotted against covariates. One may construct normal QQ plots, though a correct mean–variance relationship is more influential than lack of normality (so long as the sample size is not small). The constant variance assumption may be examined via a plot of $\widehat{e}_{ij}^{\star 2}$ (or $|\widehat{e}_{ij}^\star|$) versus $\widehat{\mu}_{ij} = x_{ij}\widehat{\beta} + z_{ij}\widehat{b}_i$.

Recall the model
$$y_i = x_i \beta + z_i b_i + \delta_i + \epsilon_i, \tag{8.50}$$

introduced in Sect. 8.4.2, with $b_i \mid D \sim_{iid} N_{q+1}(\,0, D\,)$ and $\epsilon_i \mid \sigma_\epsilon^2 \sim_{iid} N_{n_i}(\,0, \sigma_\epsilon^2 I\,)$ representing random effects and measurement error and δ_{ij} being zero-mean normal error terms with serial dependence in time. A simple and commonly used form for serial dependence is the AR(1) model (also described in Sect. 8.4.2) which gives covariances

$$\text{cov}(\delta_{ij}, \delta_{ik}) = \sigma_\delta^2 \rho^{|t_{ij} - t_{ik}|} = \sigma_\delta^2 R_{ijk}.$$

Conditional on b_i, this leads to the variance–covariance for responses on unit i:

$$\text{var}(Y_i \mid b_i) = V_i = \sigma_\delta^2 R_i + \sigma_\epsilon^2 I_{n_i}. \tag{8.51}$$

If model (8.50) is fitted, then residuals of the form (8.49) may be formed, but these should be standardized in the same way as just described for population residuals (i.e., using the decomposition $\widehat{V}_i = L_i L_i^\mathsf{T}$) since they will have marginal variance (8.51).

In a temporal setting, one may want to detect whether serial correlation is present in the residuals. Two tools for such detection are the *autocorrelation function* and the *semi-variogram*. We describe the autocorrelation function and the semi-variogram generically with respect to the model

$$Y_t = \mu_t + \epsilon_t,$$

for $t = 1, \ldots, n$. We assume the error terms ϵ_t are *second-order stationary*, which means that $\mathrm{E}[\epsilon_t] = \mu$ is constant, and $\mathrm{cov}(\epsilon_t, \epsilon_{t+d}) = C(d)$, where $d \geq 0$, that is, the covariance only depends on the temporal spacing between the variables. This implies that the variance of ϵ_t is constant, and equal to $C(0)$, for all t. The autocorrelation function (ACF) is defined, for time points $d \geq 0$ apart, as

$$\rho(d) = \frac{\mathrm{cov}(\epsilon_t, \epsilon_{t+d})}{\sqrt{\mathrm{var}(\epsilon_t)\mathrm{var}(\epsilon_{t+d})}} = \frac{C(d)}{C(0)},$$

for all t. Now, suppose we have estimates of the errors $\widehat{\epsilon}_t$ for responses equally spaced over time, which we label as $t = 1, \ldots, n$. The *empirical* ACF is defined as

$$\widehat{\rho}(d) = \frac{\widehat{C}(d)}{\widehat{C}(0)} = \frac{\sum_{t=1}^{n-d} \widehat{\epsilon}_t \widehat{\epsilon}_{t+d}/(n-d)}{\sum_{t=1}^{n} \widehat{\epsilon}_t^2/n},$$

for $d = 0, 1, \ldots, n-1$. A *correlogram* plots $\widehat{\rho}(d)$ versus d for $d = 0, 1, 2, \ldots, n-1$. If the residuals are a white noise process (i.e., uncorrelated), then asymptotically

$$\sqrt{n}\, \widehat{\rho}(d) \to_d \mathrm{N}(0, 1),$$

for $d = 1, 2, \ldots$, to give, for example, 95% confidence bands of $\pm 1.96/\sqrt{n}$.

We now turn to a description of the semi-variogram, a tool which was introduced by Matheron (1971) in the context of spatial analysis (more specifically, geostatistics) and is described in the context of longitudinal data by Diggle et al. (2002, Chap. 3.4). Define the semi-variogram of the residuals ϵ_t, as

$$\gamma(d) = \frac{1}{2}\mathrm{var}(\epsilon_t - \epsilon_{t+d}) = \frac{1}{2}\mathrm{E}\left[(\epsilon_t - \epsilon_{t+d})^2\right]$$

for $d \geq 0$. The reason for the $1/2$ term will soon become apparent. The semi-variogram exists under weaker conditions than the ACF, specifically under *intrinsic stationarity*, which means that ϵ_t has constant mean and $\mathrm{var}(\epsilon_t - \epsilon_{t+d})$ only depends on d (so that the covariance need not be defined). For zero-mean error terms and under second-order stationarity,

$$\gamma(d) = \frac{1}{2}\mathrm{var}(\epsilon_t) + \frac{1}{2}\mathrm{var}(\epsilon_{t+d}) - \mathrm{cov}(\epsilon_t, \epsilon_{t+d})$$
$$= C(0) - C(d)$$
$$= C(0)[1 - \rho(d)].$$

Suppose we now have estimated errors $\widehat{\epsilon}_l$, along with associated times t_l, $l = 1, \ldots, n$. The sample semi-variogram uses the empirical halved squared differences between pairs of residuals

$$v_{ll'} = \frac{1}{2}(\widehat{\epsilon}_l - \widehat{\epsilon}_{l'})^2,$$

8.8 Assessment of Assumptions

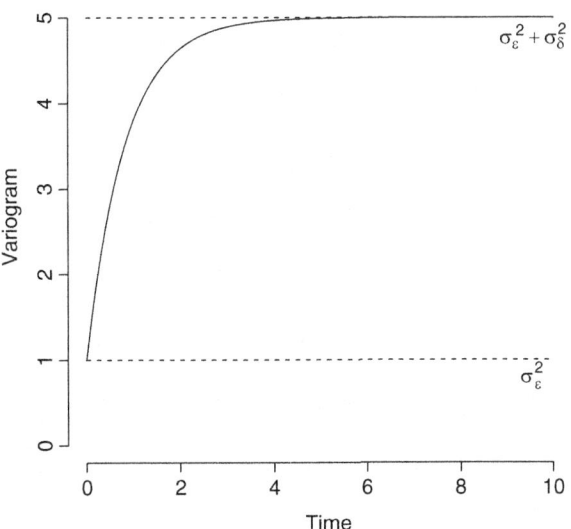

Fig. 8.4 Theoretical (semi-)variogram corresponding to (8.53) with $\sigma_\epsilon^2 = 1$, $\sigma_\delta^2 = 4$ and $\rho = 0.3$

along with the spacings $d_{ll'} = |t_l - t_{l'}|$ for $l = 1, \ldots, n$ and $l < l' = 1, \ldots, n$. With irregular sampling times, the variogram can be estimated from the pairs $(d_{ll'}, v_{ll'})$, with the resultant plot being smoothed.[2] An example of such a plot is given in Fig. 8.9. Under normality of the data, the marginal distribution of each $v_{ll'}$ is $C(0)\chi_1^2$, and this large variability can make the variogram difficult to interpret. In addition, because each residual contributes to $n - 1$ terms in the empirical cloud of points, the points are not independent, and a single outlying point can influence the plot at different time lags.

Suppose now we are in a longitudinal setting, in which response y_{ij} is observed at time t_{ij}, and we fit the LMM

$$\boldsymbol{y}_i = \boldsymbol{x}_i \boldsymbol{\beta} + \boldsymbol{z}_i \boldsymbol{b}_i + \boldsymbol{\epsilon}_i, \qquad (8.52)$$

with the usual forms for $\boldsymbol{b}_i$ and $\boldsymbol{\epsilon}_i$. After fitting, we form the stage one residuals (8.49), that is, $\widehat{\epsilon}_{ij} = y_{ij} - \boldsymbol{x}_{ij}\widehat{\boldsymbol{\beta}} - \boldsymbol{z}_{ij}\widehat{\boldsymbol{b}}_i$. We might believe the serial dependence takes the same form across individuals. For equally spaced times, we can examine the empirical ACF of the residuals where, for simplicity, we assume that there are n responses on each of the m individuals,

$$\widehat{\rho}(d) = \frac{\sum_{i=1}^m \sum_{j=1}^{n-d} \widehat{\epsilon}_{ij} \widehat{\epsilon}_{i,j+d}/(n-d)}{\sum_{i=1}^m \sum_{j=1}^n \widehat{\epsilon}_{ij}^2/n},$$

for $d = 0, 1, \ldots, n - 1$.

---

[2] For unequally spaced times, the longitudinal data literature often recommends the construction of the empirical semi-variogram (Diggle et al. 2002, Sect. 3.4; Fitzmaurice et al. 2004, Sect. 9.4), though one could construct and smooth the empirical covariance function in a similar fashion.

Now suppose that we again fit model (8.52), and we have n_i responses for individual i with sampling times t_{ij}. We then define the semi-variogram for the ith individual as

$$\gamma_i(d_{ijk}) = \frac{1}{2}\text{E}\left[(\epsilon_{ij} - \epsilon_{ik})^2\right]$$

where $d_{ijk} = |t_{ij} - t_{ik}|$. We now form

$$v_{ijk} = \frac{1}{2}(\widehat{\epsilon}_{ij} - \widehat{\epsilon}_{ik})^2$$

and the semi-variogram can then be estimated by plotting the pairs (d_{ijk}, v_{ijk}) for $i = 1, \ldots, m$ and $j < k = 1, \ldots, n_i$ and smoothing. If no serial dependence is present, the smoother should be roughly horizontal.

Consider the interpretation of the variogram when model (8.50) is the "truth," but suppose we fit a LMM without the autocorrelated terms. We consider stage one residuals, which under (8.50) will take the form

$$\epsilon'_{ij} = Y_{ij} - \boldsymbol{x}_{ij}\boldsymbol{\beta} - \boldsymbol{z}_{ij}\boldsymbol{b}_i = \delta_{ij} + \epsilon_{ij}.$$

For differences in residuals on the same individual,

$$\epsilon'_{ij} - \epsilon'_{ik} = \delta_{ij} + \epsilon_{ij} - \delta_{ik} - \epsilon_{ik}$$
$$= (\delta_{ij} - \delta_{ik}) + (\epsilon_{ij} - \epsilon_{ik}),$$

and so the semi-variogram takes the form

$$\gamma_i(d_{ijk}) = \frac{1}{2}\text{E}\left[(\epsilon'_{ij} - \epsilon'_{ik})^2\right]$$
$$= \frac{1}{2}\text{E}\left[(\delta_{ij} - \delta_{ik})^2 + (\epsilon_{ij} - \epsilon_{ik})^2\right]$$
$$= \sigma_\delta^2[1 - \rho(d_{ijk})] + \sigma_\epsilon^2. \quad (8.53)$$

As $d_{ijk} \to 0$, $\gamma_i(d_{ijk}) \to \sigma_\epsilon^2$. The rate at which asymptote $\sigma_\delta^2 + \sigma_\epsilon^2$ is reached as $d_{ijk} \to \infty$ is determined by ρ. This variogram is illustrated in Fig. 8.4.

We now briefly consider the use of population residuals, starting with the random intercepts model:

$$Y_{ij} = \boldsymbol{x}_{ij}\boldsymbol{\beta} + b_i + \delta_{ij} + \epsilon_{ij},$$

with $b_i \mid \sigma_0^2 \sim_{iid} \text{N}(0, \sigma_0^2)$ and the AR(1) model for δ_{ij}. The population residuals under this model are

$$e_{ij} = Y_{ij} - \boldsymbol{x}_{ij}\boldsymbol{\beta} = b_i + \delta_{ij} + \epsilon_{ij},$$

$i = 1, \ldots, m; j = 1, \ldots n_i$. For differences in residuals on the same individual,

8.8 Assessment of Assumptions

$$e_{ij} - e_{ik} = b_i + \delta_{ij} + \epsilon_{ij} - b_i - \delta_{ik} - \epsilon_{ik}$$
$$= (\delta_{ij} - \delta_{ik}) + (\epsilon_{ij} - \epsilon_{ik}),$$

and so we obtain the same semi-variogram, (8.53), as before. Since b_i is constant for individual i, its variance does not appear.

In general, the variogram is limited in its use for population residuals for the LMM, as we now illustrate. Consider the LMM with random intercepts and independent random slopes:

$$b_{i0} \mid D_0 \sim N(0, D_0), \quad b_{i1} \mid D_1 \sim N(0, D_1).$$

This leads to marginal variance

$$\text{var}(Y_{ij}) = \sigma_\epsilon^2 + D_0 + D_1 t_{ij}^2,$$

which is not constant over time. Therefore, a semi-variogram of population residuals should not be constructed, because we do not have second-order stationarity.

Predictions of the random effects $\widehat{b}_i$ may be used to assess assumptions associated with the random effects distribution, though since these have undergone shrinkage, they may be deceptive. One may instead carry out individual fitting and then use the resultant estimates to assess the normality assumption. The latter may be assessed via QQ plots, but the interpretation of plots requires care since estimates and not observed quantities are being plotted; see Lange and Ryan (1989). We may also assess whether the variance of the random effects is independent of covariates x_i. If the spread of the random effects distribution depends on the levels of covariates, and this is missed, then inaccurate inference can result (Heagerty and Kurland 2001). For the LMM, it is better to examine stage one and stage two residuals separately, rather than population residuals, since the latter are a mixture of the two, and so, if something appears amiss, it is difficult to determine the stage at which the inadequacy is occurring. As usual, as discussed in Sect. 4.9, the implications of changing the model should be carefully considered, and one should avoid the temptation to model every nuance of the data.

Example: FEV1 Over Time

The dental data that have formed our running illustration are balanced, and there are few individuals and time points, and so, these data are not ideal for illustrating model checking. Hence, we introduce data from an epidemiological study described by van der Lende et al. (1981). We analyze a sample of 133 men and women, initially aged 15–44, from the rural area of Vlagtwedde in the Netherlands. Study participants were followed over time to obtain information on the prevalence of, and risk factors for, chronic obstructive lung diseases. These data were previously

Table 8.4 Mean FEV1 (and sample size) by smoking status and time

Time	Former smoker	Current smoker
0	3.52 (23)	3.23 (85)
3	3.58 (27)	3.12 (95)
6	3.26 (28)	3.09 (89)
9	3.17 (30)	2.87 (85)
12	3.14 (29)	2.80 (81)
15	2.87 (24)	2.68 (73)
19	2.91 (28)	2.50 (74)

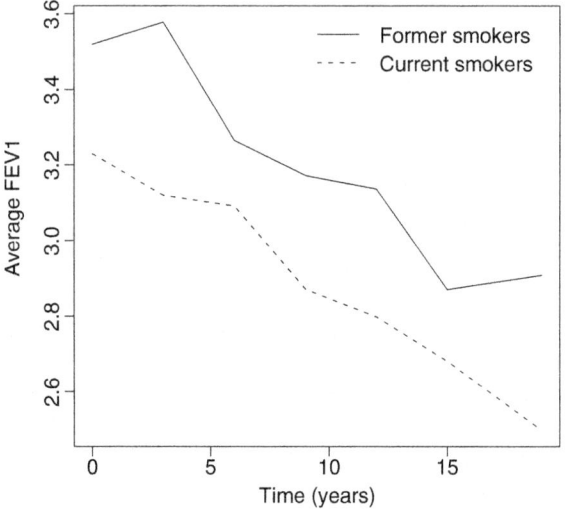

Fig. 8.5 Mean FEV1 profiles versus time for two smoking groups

analyzed by Fitzmaurice et al. (2004). Follow-up surveys provided information on respiratory symptoms and smoking status. Pulmonary function was measured by spirometry, and a measure of forced expiratory volume (FEV1) was obtained every 3 years for the first 15 years of the study and also at year 19. Each study participant was either a current or a former smoker, with current smoking defined as smoking at least one cigarette per day. In this dataset, FEV1 was not recorded for every subject at each of the planned measurement occasions so that the number of measurements of FEV1 on each subject varied between 1 and 7. Table 8.4 shows the numbers of observations available at each time point. There are 32 former smokers and 101 current smokers in total, and we see that the numbers with missing observations at each time point are not drastically different.

Figure 8.5 plots the mean FEV1 profiles versus time for former smokers (solid line) and current smokers (dashed line). It is clear that there is a difference in the overall level, with former smokers having higher responses. Whether the rate of decline in FEV1 is different in the two groups is not so obvious. Figure 8.6 plots the individual trajectories versus time for former smokers (solid lines) and current smokers (dashed lines). There is clearly large between-individual variability in levels so that observations on the same individual will be correlated.

8.8 Assessment of Assumptions

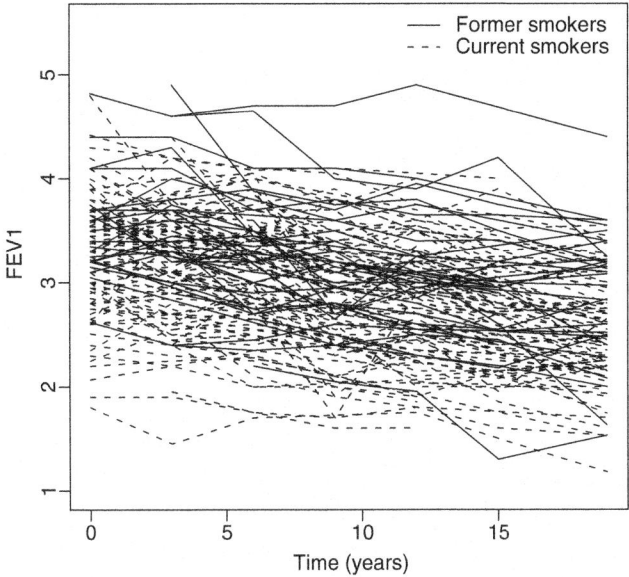

Fig. 8.6 FEV1 versus time for 133 individuals, former and current smokers are indicated by *solid* and *dashed lines* respectively

Let Y_{ij} represent the FEV1 on individual i at time (from baseline) t_{ij} (in years), with $S_i = 0/1$ indicating former/current smoker. We treat this example as illustrative only and therefore fit various models to examine the effects on inference and to demonstrate model assessment and comparison. We initially fit the following three models using REML:

$$Y_{ij} = \beta_0 + \beta_1 t_{ij} + b_i + \epsilon_{ij} \tag{8.54}$$

$$Y_{ij} = \beta_0 + \beta_1 t_{ij} + \beta_2 S_i + b_i + \epsilon_{ij} \tag{8.55}$$

$$Y_{ij} = \beta_0 + \beta_1 t_{ij} + \beta_2 S_i + \beta_3 S_i \times t_{ij} + b_i + \epsilon_{ij} \tag{8.56}$$

with $b_i \mid \sigma_0^2 \sim_{iid} N(0, \sigma_0^2)$ and $\epsilon_{ij} \mid \sigma_\epsilon^2 \sim_{iid} N(0, \sigma_\epsilon^2)$ and with ϵ_{ij} and b_i independent, $i = 1, \ldots, m$, $j = 1, \ldots, n_i$. We emphasize that the random effect distribution is assumed common to both former and current smokers. Estimates and standard errors for β_1, β_2, and β_3 are given in Table 8.5. We include an ordinary least squares (OLS) fit of the model $E[Y_{ij}] = \beta_0 + \beta_1 t_{ij} + \beta_2 S_i + \beta_3 S_i \times t_{ij}$. This model is clearly inappropriate since it assumes independent observations but, when compared to the equivalent LMM, (8.56), illustrates that the standard errors of the estimates corresponding to time-varying covariates (time β_1 and the interaction β_3) are reduced under the LMM. This behavior occurs because within-individual comparisons are more efficient in a longitudinal study (as discussed in Sect. 8.3).

Table 8.5 Results of various LMM analyses and an ordinary least squares (OLS) fit to the FEV1 data

Model	β_1 (Time)	s.e.	β_2 (Smoke)	s.e.	β_3 (Inter)	s.e.
LMM TIME	−0.037	0.0013	–	–	–	–
LMM TIME + SMOKE	−0.037	0.0013	−0.31	0.11	–	–
LMM TIME × SMOKE	−0.034	0.0026	−0.27	0.11	−0.0046	0.0030
OLS TIME × SMOKE	−0.038	0.0067	−0.31	0.085	−0.00041	0.0077

Table 8.6 Results of LMM (likelihood and Bayesian) and GEE analyses for the FEV1 data

Model	β_1 (Time)	s.e.	β_2 (Smoke)	s.e.	σ_0
Likelihood LMM	−0.037	0.0013	−0.31	0.11	0.53
Bayes LMM	−0.037	0.0013	−0.31	0.12	0.53
GEE	−0.037	0.0015	−0.31	0.11	–
Likelihood LMM AR(1)	−0.037	0.0013	−0.31	0.11	0.53

σ_0 is the standard deviation of the random intercepts

To compare the three LMMs in Table 8.5, we must use MLE for likelihood ratio tests, since the data are not constant under the different models under REML (due to different $\widehat{\beta}_G$, Sect. 8.5.3). For

$$H_0 : \text{Model (8.54) versus } H_1 : \text{Model (8.55)}$$

we have a likelihood ratio statistic of 8.22 on 1 degree of freedom and a p-value of 0.0042. Hence, there is strong evidence to reject the null, and we conclude that there are differences in intercepts for former and current smokers (as we suspected from Fig. 8.5). For

$$H_0 : \text{Model (8.55) versus } H_1 : \text{Model (8.56)}$$

we have a likelihood ratio statistic of 2.29 on 1 degree of freedom and a p-value of 0.13. Hence, under conventional levels of significance, there is no reason to reject the null, and we conclude that the interaction is not needed, so that the decline in FEV1 with time is the same for both former and current smokers.

We now report a Bayesian analysis of model (8.55) with improper flat priors on $\beta_0, \beta_1, \beta_2$, the improper prior $\sigma_\epsilon^2 \propto \frac{1}{\sigma_\epsilon^2}$ and $\sigma_0^{-2} \sim \text{Ga}(0.5, 0.02)$. The latter prior gives 95% of its mass for σ_0, the standard deviation of the between-individual intercepts, between 0.09 and 6.5. Table 8.6 gives the results, which are very similar to those of the likelihood-based approach, which is reassuring.

We now fit the marginal model version of (8.55) using GEE. We use an exchangeable correlation structure, since clearly we have dependence between measurements on the same individual at different times, but the exact form of the correlation is not clear. The results are given in Table 8.6 and again show good agreement for the regression coefficients. In the exchangeable correlation structure, there are two components to α, a marginal variance, α_1, and a common marginal correlation, α_2. The model may be compared to the random intercepts

8.8 Assessment of Assumptions

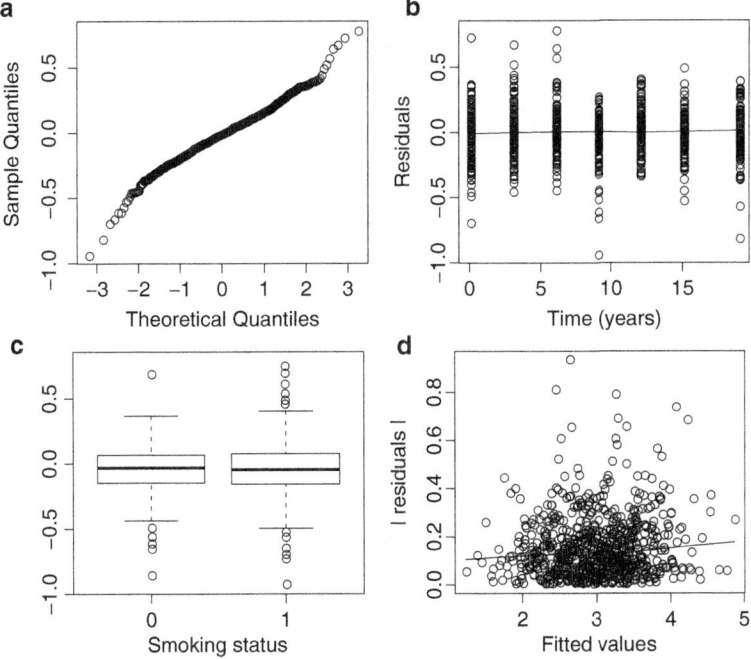

Fig. 8.7 Stage one residual plots for the FEV1 data: (**a**) normal QQ plot, (**b**) residuals versus time, (**c**) residuals as a function of smoking status (0 = former smoker, 1 = current smoker), (**d**) absolute value of residuals versus fitted values

model in which we have marginal variance $\alpha_1 = \sigma_0^2 + \sigma_\epsilon^2$ and marginal correlation $\alpha_2 = \sigma_0^2/(\sigma_0^2 + \sigma_\epsilon^2)$. From the GEE analysis, $\widehat{\alpha}_1 = 0.31$ and $\widehat{\alpha}_2 = 0.82$ to give $\sqrt{\widehat{\alpha}_1 \times \widehat{\alpha}_2} = 0.50$, which is comparable to the estimates of $\widehat{\sigma}_0 = 0.53$ in Table 8.6.

We now examine the assumptions of the various approaches. We focus on the linear model that includes time and smoking (but no interaction). Figure 8.7 summarizes the stage one residuals:

$$\widehat{\epsilon}_{ij} = y_{ij} - x_{ij}\widehat{\beta} - z_i\widehat{b}_i.$$

Panel (a) shows that the distribution of the errors is symmetric but heavier tailed than normal. With such a large sample, there is nothing troubling in this plot, and, there are no outlying points. Panels (b) and (c) plot the residuals against time and smoking status. We see no nonlinear behavior in the time plot and no great divergence from constant variance in either plot. A very important assumption in mixed effects modeling is that a common random effects distribution across covariates is appropriate. To examine this assumption, separate analyses were carried out for former and current smokers. The estimates of the variance components for former smokers were $\widehat{\sigma}_\epsilon = 0.22$ and $\widehat{\sigma}_0 = 0.58$ and for current smokers, $\widehat{\sigma}_\epsilon = 0.21$ and $\widehat{\sigma}_0 = 0.51$. The differences between estimates in the two groups are small, and we

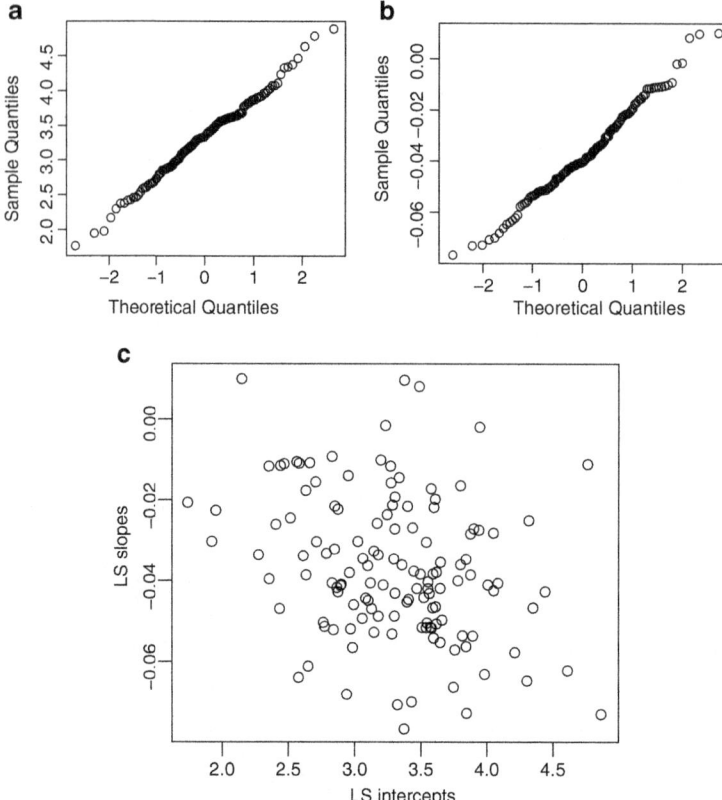

Fig. 8.8 Normal QQ plots of OLS estimates for the FEV1 data for (**a**) intercepts, (**b**) slopes, and (**c**) scatterplot of pairs of least square estimates

conclude that a common random effects distribution is reasonable. Panel (d) plots the absolute value of the residuals versus the fitted values $x_{ij}\widehat{\beta} + \widehat{b}_i$, along with a smoother. If the variance function is correctly specified, then we should see no systematic pattern. Here, there is nothing to be too concerned about since there is only a slight increase in variability as the mean increases. These residual plots are based on residuals from the likelihood analysis (the Bayesian versions are similar).

For the 132 individuals with more than a single response, individual OLS fits were performed. Figure 8.8 shows normal QQ plots of the intercept and slope parameter estimates in panels (a) and (b) and a bivariate scatter plot of the pairs of estimates in panel (c). The estimates look remarkably normal, at least in (a) and (b), and there are no outlying individuals.

Finally, we examine the residuals for serial correlation. Figure 8.9 gives the semi-variogram of the stage one residuals along with a smoother and indicates some evidence of dependence. In panel (a), the pattern is not apparent, but in panel (b), the semi-variance axis is reduced for clarity, which allows the trend to be more

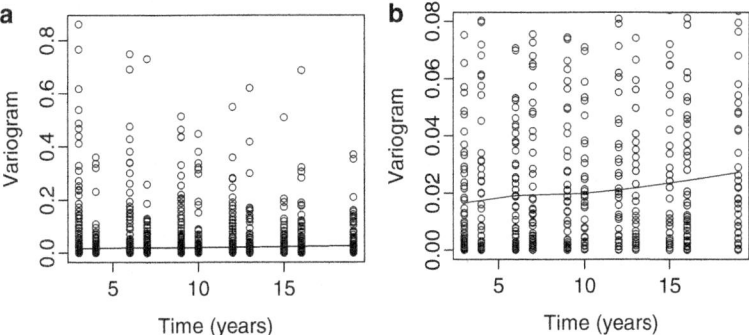

Fig. 8.9 For the FEV1 data: (**a**) the (semi)-variogram of stage one residuals, (**b**) on a truncated semi-variogram scale

clearly seen. Consequently, we fit an AR(1) model to the residuals (Sect. 8.4.2), using restricted maximum likelihood, and obtain the parameter estimates in the last row of Table 8.6. This model is a significant improvement over the non-serial correlation model (as measured by a likelihood ratio test, $p = 0.0002$). However, there is virtually no change in the estimates/standard errors, since the AR correlation parameter is just 0.20, with an asymptotic 95% confidence interval of [0.087, 0.30].

We may also examine whether random slopes are required. Fitting this model via restricted likelihood gave a standard deviation of $\widehat{\sigma}_1 = 0.0099$. The likelihood ratio statistic test for correlated random intercepts and slopes, versus random intercepts only, is 10.9 which is significant at around the 0.0025 level (where the distribution under the null is a mixture of χ_1^2 and χ_2^2 distributions, see Sect. 8.5.2).

In terms of the fixed effects, there is little sensitivity to the assumed random effects structure. Inference under the random intercepts and slopes models is similar to the random intercepts only model, since the between-individual variability in slopes is small (though statistically significant). The population change in FEV1 is a drop of 0.0371 per year, with a standard error of 0.0013–0.0015 depending on the model. The posterior median for the intraindividual correlation, $\sigma_0^2/(\sigma_\epsilon^2 + \sigma_0^2)$, is 0.84 with 95% interval [0.82, 0.89] suggesting that the majority of the variability in FEV1 is between individual.

8.9 Cohort and Longitudinal Effects

We now describe another benefit of longitudinal studies, the ability to estimate both longitudinal and cohort effects. We frame the discussion around the modeling of $Y = $ FEV1 as a function of age. We might envisage that FEV1 changes as age increases within an individual and that individuals may have different baseline levels of FEV_1 due to "cohort" effects. A birth cohort is a group of individuals who were

Fig. 8.10 Three population (cohort) trajectories over time

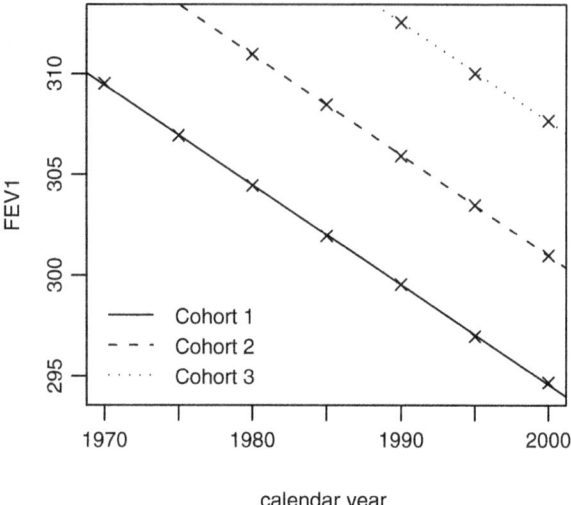

Fig. 8.11 Relationship between cross-sectional and longitudinal effects in a hypothetical example with three populations. The *dashed line* (which is the top line) represents the cross-sectional slope

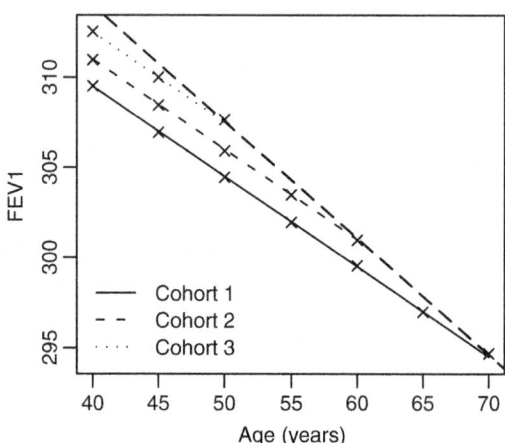

born in the same year. Cohort effects may include the effects of environmental pollutants and differences in lifestyle choices or medical treatment received. In a cross-sectional study, a group of individuals are measured at a single time point. A great advantage of longitudinal studies, as compared to cross-sectional studies, is that both cohort and aging (longitudinal) effects may be estimated.

As an illustration, Fig. 8.10 shows the trajectories of three hypothetical individuals as a function of calendar time. The starting positions are different due to cohort effects. Figure 8.11 shows the same individuals but with trajectories plotted versus age. The cross-sectional association, which would result from observing the final measurement only, is highlighted and displays a steeper decline than seen in the longitudinal slope.

8.9 Cohort and Longitudinal Effects

To examine in more detail the issues, consider the model

$$E[Y_{ij} \mid x_{ij}, x_{i1}] = \beta_0 + \beta_C x_{i1} + \beta_L(x_{ij} - x_{i1})$$

where Y_{ij} is the jth FEV1 measurement on individual i and x_{ij} is the age of the individual at occasion j, with x_{i1} being the age on a certain day (so that all the individuals are comparable). At the first occasion,

$$E[Y_{i1} \mid x_{i1}] = \beta_0 + \beta_C x_{i1},$$

so that β_C is the average change in response between two populations who differ by one unit in their baseline ages. Said another way, we are examining the differences in FEV1 between two birth cohorts a year apart, so that β_C is the *cohort effect*.
Since

$$E[Y_{ij} \mid x_{ij}, x_{i1}] - E[Y_{i1} \mid x_{i1}] = \beta_L(x_{ij} - x_{i1})$$

it is evident that β_L is the *longitudinal effect*, that is, the change in the average FEV1 between two populations who are in the same birth cohort and whose ages differ by 1 year. The usual cross-sectional model is

$$E[Y_{ij} \mid x_{ij}] = \beta_0 + \beta_1 x_{ij} \qquad (8.57)$$
$$= \beta_0 + \beta_1 x_{i1} + \beta_1(x_{ij} - x_{i1})$$

so that the model implicitly assumes equal longitudinal and cohort effects, that is, $\beta_1 = \beta_C = \beta_L$.

In the cross-sectional study with model (8.57),

$$\widehat{\beta}_1 = \frac{\sum_{i=1}^m \sum_{j=1}^{n_i}(x_{ij} - \overline{x})(Y_{ij} - \overline{Y})}{\sum_{i=1}^m \sum_{j=1}^{n_i}(x_{ij} - \overline{x})^2}$$

with $\overline{x} = \frac{1}{N}\sum_{i=1}^m \sum_{j=1}^{n_i} x_{ij}$ and $\overline{Y} = \frac{1}{N}\sum_{i=1}^m \sum_{j=1}^{n_i} Y_{ij}$ with $N = \sum_{i=1}^m n_i$. The expected value of this estimator is

$$E[\widehat{\beta}_1] = \beta_L + \frac{\sum_{i=1}^m n_i(x_{i1} - \overline{x}_1)(\overline{x}_i - \overline{x})}{\sum_{i=1}^m \sum_{j=1}^{n_i}(x_{ij} - \overline{x})^2}(\beta_C - \beta_L) \qquad (8.58)$$

(Exercise 8.15) so that the estimate is a combination of cohort and longitudinal effects. The cross-sectional regression model will give an unbiased estimate of the longitudinal association if $\beta_L = \beta_C$ or if $\{x_{i1}\}$ and $\{\overline{x}_i\}$ are orthogonal. To conclude, longitudinal studies can be powerfully employed to separate cohort and longitudinal effects.

8.10 Concluding Remarks

In this chapter we have described two approaches to fitting linear models to dependent data: LMMs and GEE. GEE has the fewest assumptions and is designed for population-level inference. Asymptotics are required for inference, and so, GEE is less appealing when the number of individuals m is small. A sufficiently large sample size is required for both normality of the estimator and reliability of the sandwich variance estimator. The use of the sandwich variance estimator makes GEE the most dependable method in large sample situations. However, there can be losses in efficiency if we choose a working correlation matrix that is far from reality. With GEE, it is not possible to make inference for individuals or incorporate prior information.

LMMs are more flexible than GEE in terms of the questions that can be addressed with the data, but this flexibility comes at the price of a greater number of assumptions. For likelihood inference, as with GEE, we require the number of units m to be sufficiently large for asymptotic inference. Prior information cannot be incorporated in a likelihood analysis; for that, we need a Bayesian approach. For a small number of individuals, a Bayesian approach fully captures the uncertainty, but inference is completely model-based, and with a small number of individuals, it is unlikely that we will be able to check the modeling assumptions.

8.11 Bibliographic Notes

For descriptions of linear mixed effects models, see Hand and Crowder (1996, Chap. 5), Diggle et al. (2002, Sects. 4.4 and 4.5), and Verbeeke and Molenberghs (2000). Covariance models are described in Verbeeke and Molenberghs (2000, Chap. 10), Pinheiro and Bates (2000, Chap. 5), and Diggle et al. (2002, Chap. 5). Demidenko (2004) provides theory for mixed models, including the linear case. Robinson (1991) provides an interesting discussion of BLUP estimates. Two early influential references on the LMM from Bayesian and likelihood perspectives, respectively, are Lindley and Smith (1972) and Laird and Ware (1982).

The name GEE was coined by Liang and Zeger (1986) and Zeger and Liang (1986). See also Gourieroux et al. (1984) who considered sandwich estimation for regression parameters with a consistent estimator of additional parameters. Prentice (1988) introduced a second estimating equation for estimation of α. Crowder (1995) points out that the existence of the α parameters in the working covariance matrix is not guaranteed, in which case the asymptotics break down. Fitzmaurice et al. (2004) is an excellent practical text on longitudinal modeling.

8.12 Exercises

8.1 *A Gauss–Markov Theorem for Dependent Data:* Suppose $E[Y] = x\beta$ and $\text{var}(Y) = V$, with $Y = [Y_1^T, \ldots, Y_m^T]^T$ and where $Y_i = [Y_{i1}, \ldots, Y_{in_i}]^T$ and $x = [x_1, \ldots, x_m]^T$ is $N \times (k+1)$ with $x_i = [x_{i1}, \ldots, x_{in_i}]$, $x_{ij} = [1, x_{ij1}, \ldots, x_{ijk}]^T$, $N = \sum_i n_i$ and β is the $(k+1) \times 1$ vector of regression coefficients.

Consider linear estimators of the form

$$\widetilde{\beta}_{\text{w}} = (x^T W^{-1} x)^{-1} x^T W^{-1} Y,$$

where W is symmetric and positive definite. Show that:

(a) $E[\widetilde{\beta}_{\text{w}}] = \beta$.
(b) $\text{var}(\widetilde{\beta}_{\text{v}}) \leq \text{var}(\widetilde{\beta}_{\text{w}})$.

[Hint: In (b), show that $\text{var}(\widetilde{\beta}_{\text{w}}) - \text{var}(\widetilde{\beta}_{\text{v}})$ is positive semi-definite.]

8.2 Consider the data in Table 5.4 (from Davies 1967) that were presented in Sect. 5.8.1. These data consist of the yield in grams from six randomly chosen batches of raw material, with five replicates each. The aim of this experiment was to find out to what extent batch-to-batch variation was responsible for variation in the final product yield.

One possibility for a model for these data is the one-way analysis of variance with

$$y_{ij} = \mu + b_i + \epsilon_{ij},$$

with $j = 1, \ldots, n$, replicates on $i = 1, \ldots, m$, batches, $b_i \mid \sigma_0^2 \sim_{iid} N(0, \sigma_0^2)$, $\epsilon_{ij} \mid \sigma_\epsilon^2 \sim_{iid} N(0, \sigma_\epsilon^2)$, with b_i and ϵ_{ij} independent.

In what follows the following identity is useful. Let I_n denote the $n \times n$ identity matrix and J_n the $n \times n$ matrix of 1's. Then

$$(aI_n + bJ_n)^{-1} = \frac{1}{a}\left(I_n - \frac{b}{a+nb}J_n\right), \quad a \neq 0, \ a \neq -nb,$$

and

$$|aI_n + bJ_n| = a^{n-1}(a+nb).$$

(a) Derive the log-likelihood for $\mu, \sigma_0^2, \sigma_\epsilon^2$.
(b) Differentiate the log-likelihood, and show that the MLEs are

$$\hat{\mu} = \bar{y}_{..},$$

$$\widehat{\sigma}_\epsilon^2 = \text{MSE},$$

$$\widehat{\sigma}_0^2 = \frac{(1 - 1/m)\text{MSA} - \text{MSE}}{n},$$

where MSA= $n\sum_{i=1}^{m}(\bar{y}_{i\cdot} - \bar{y}_{\cdot\cdot})^2/(m-1)$ and MSE= $\sum_{i=1}^{m}\sum_{j=1}^{n}(y_{ij} - \bar{y}_{i\cdot})^2/[m(n-1)]$.
[Hint: Life is easier if the model is parameterized in terms of $\lambda = \sigma_\epsilon^2 + n\sigma_0^2$.]

- (c) Obtain the form of var($\hat{\mu}$), and give an estimator of this quantity.
- (d) Find the REML estimators of σ_0^2 and σ_ϵ^2.
- (e) In the one-way random effects model with balanced data, it can be shown that

$$\frac{\text{MSA}/(n\sigma_0^2 + \sigma_\epsilon^2)}{\text{MSE}/\sigma_\epsilon^2} \sim F_{m-1, m(n-1)},$$

the F distribution on $m-1$ and $m(n-1)$ degrees of freedom. Use this result to explain why $F = \text{MSA}/\text{MSE}$ may be compared with an $F_{m-1,m(n-1)}$ distribution to provide a test of $H_0 : \sigma_0^2 = 0$.
- (f) Using the last part, show that the probability that the REML estimator $\hat{\sigma}_0^2$ is negative is the probability that an $F_{m(n-1),(m-1)}$ random variable is bigger than $1 + n\sigma_0^2/\sigma_\epsilon^2$.
- (g) Numerically obtain an MLE, with associated standard error, for μ. Additionally, find ML and REML estimates of σ_0^2 and σ_ϵ^2.
- (h) Confirm these estimates using a statistical package.

8.3 Consider the so-called Neymann–Scott problem (previously considered in Exercises 2.6 and 3.3) in which

$$Y_{ij} \mid \mu_i, \sigma^2 \sim_{ind} N(\mu_i, \sigma^2),$$

for $i = 1, \ldots, n$, $j = 1, 2$.

- (a) Obtain the MLE for σ^2, and show that it is inconsistent. Why are there problems here?
- (b) Consider a REML approach. Assign an improper uniform prior to $\mu_1, \ldots, \mu_n$, and integrate out these parameters. Obtain the REML of σ^2, and show that it is an unbiased estimator.

8.4 Derive (8.25) and (8.26).
[Hint: The identities

$$\left(\frac{z_i^T z_i}{\sigma_\epsilon^2} + D^{-1}\right)^{-1} \frac{z_i^T}{\sigma_\epsilon^2} = D z_i^T V_i^{-1}$$

$$\left(D^{-1} + \frac{z_i^T z_i}{\sigma_\epsilon^2}\right)^{-1} = D - D z_i^T V^{-1} z_i D$$

are useful. These follow from

8.12 Exercises

$$(E+F)^{-1}E = I - (E+F)^{-1}F$$
$$(G + EFE^{\text{T}})^{-1} = G^{-1} - G^{-1}E(E^{\text{T}}G^{-1}E + F^{-1})^{-1}E^{\text{T}}G^{-1},$$

respectively.]

8.5 Show that

$$\text{var}(\widehat{b}_i - b_i) = \text{var}(b_i) - \text{var}(\widehat{b}_i) = D - \text{var}(\widehat{b}_i)$$
$$= D - Dz_i^{\text{T}}V_i^{-1}z_iD + Dz_i^{\text{T}}V_i^{-1}x_i(x^{\text{T}}V^{-1}x)^{-1}x_i^{\text{T}}V_i^{-1}z_iD.$$

8.6 Consider the class of linear predictors $b^*(y) = a + By$, where a and B are constants of dimensions $(q+1) \times 1$ and $(q+1) \times n$. Let $W = b - By$, and show that

$$\text{E}[(b^* - b)^{\text{T}}A(b^* - b)] = [a - \text{E}(W)]^{\text{T}}A[a - \text{E}(W)] + \text{tr}[A\text{var}(W)].$$

Deduce that this expression is minimized by taking $a = -Bx\beta$ and $B = Dz^{\text{T}}V^{-1}$. Hence, show that

$$Dz^{\text{T}}V^{-1}(y - x\beta)$$

is the best linear predictor of b, whatever the distributions of b and y.

8.7 Prove that if the prior distribution for $\theta^{\text{T}} = [\theta_1, \ldots, \theta_m]$ can be written as

$$p(\theta) = \int \prod_{i=1}^{m} p(\theta_i \mid \phi) p(\phi) \, d\phi,$$

then the covariances $\text{cov}(\theta_i, \theta_j)$ are all nonnegative.

[Hint: You may assume that $\text{E}[\theta_i \mid \phi] = \text{E}[\theta_j \mid \phi]$ for $i \neq j$.]

8.8 We return to the yield data of Exercise 8.2.

(a) Numerically evaluate the formula

$$\widehat{b}_i = \text{E}[b_i \mid y_i] = \widehat{D}z_i^{\text{T}}\widehat{V}_i^{-1}(y_i - x_i\widehat{\beta})$$

in your favorite package, and obtain predictions for the yield data.
(b) Obtain measures of the variability of the prediction via $\text{var}(\widehat{b}_i - b_i)$.
(c) Confirm your predictions using LMM software.

8.9 A Bayesian analysis of the yield data of Exercise 8.2 will now be performed. In terms of the parameters $\beta_0, \sigma_\epsilon^2$, and $\lambda = \sigma_\epsilon^2 + n\sigma_0^2$, the likelihood is

$$p(y \mid \beta_0, \sigma_\epsilon^2, \lambda) = (2\pi)^{-nm/2}(\sigma_\epsilon^2)^{-m(n-1)/2}\lambda^{-m/2}$$
$$\times \exp\left\{-\frac{1}{2}\left[\frac{nm(y_{++} - \beta_0)^2}{\lambda} + \frac{\text{SS}_\text{B}}{\lambda} + \frac{\text{SS}_\text{W}}{\sigma_\epsilon^2}\right]\right\}$$

where

$$\text{SS}_\text{B} = n \sum_{i=1}^{m}(y_{i+} - y_{++})^2, \quad \text{SS}_\text{W} = \sum_{i=1}^{m}\sum_{i=1}^{n}(y_{ij} - y_{i+})^2.$$

Assume the improper prior

$$\pi(\beta_0, \sigma_\epsilon^2, \lambda) \propto \frac{1}{\sigma_\epsilon^2 \lambda}.$$

(a) Integrate β_0 from the joint posterior $p(\beta_0, \sigma_\epsilon^2, \lambda \mid y)$ to obtain $p(\sigma_\epsilon^2, \lambda \mid y)$. Show that this distribution has the form of a product of independent inverse gamma distributions with an additional term that is due to the constraint $\lambda > \sigma_\epsilon^2 > 0$.

(b) Obtain the distribution of $p(\beta_0 \mid \sigma_\epsilon^2, \lambda, y)$.

(c) Give details of a composition algorithm (as described in Sect. 3.8.4) for simulating from the posterior $p(\beta_0, \sigma_\epsilon^2, \lambda \mid y)$.

(d) Implement the algorithm for the yield data.

  (i) Give histograms and 5%, 50%, 95% quantile summaries of the univariate posterior distributions for $\beta_0, \sigma_\epsilon^2, \lambda, \sigma_0^2$, and $\rho = \sigma_0^2/(\sigma_0^2 + \sigma_\epsilon^2)$.

  (ii) Obtain a bivariate scatterplot representation of the posterior distribution $p(\sigma_\epsilon^2, \sigma_0^2 \mid y)$.

  (iii) Using samples from the distribution for ρ, answer the original question concerning the extent of batch-to-batch variability that is contributing to the total variability.

(e) Obtain the distribution of $p(b_i \mid \beta_0, \sigma_\epsilon^2, \sigma_0^2, y)$. Hence, describe an algorithm for simulating from the posterior $p(b_i \mid y)$. Implement this algorithm for the yield data, and give 5%, 50%, 95% quantile summaries for $p(b_i \mid y), i = 1, \ldots, m$.

(f) Now, consider an alternative computational approach assuming independent priors with an improper flat prior on μ, the improper prior $\pi(\sigma_\epsilon^2) \propto \sigma_\epsilon^{-2}$, and a $\text{Ga}(0.05, 0.01)$ prior for σ_0^{-2}. Implement a Gibbs sampling algorithm for sampling from the conditionals:

- $\mu \mid \sigma_\epsilon^2, \sigma_0^2, b, y$
- $\sigma_\epsilon^2 \mid \mu, \sigma_0^2, b, y$
- $\sigma_0^2 \mid \mu, \sigma_\epsilon^2, b, y$
- $b \mid \mu, \sigma_\epsilon^2, \sigma_0^2, y$,

where $b = [b_1, \ldots, b_m]^\text{T}$.

Report posterior medians and 90% credible intervals for $\mu, \sigma_0^2, \sigma_\epsilon^2, b$, and ρ, and compare your answers with those using the alternative priors derived in the earlier part of the question.

8.12 Exercises

8.10 Derive the conditional distributions, given in Sect. 8.6.3, that are required for Gibbs sampling in the LMM.

8.11 Show the equivalence of the BLUP predictor $\widehat{b}_i$ and the Gibbs conditional distribution $b_i \mid y, \beta, \alpha$.

8.12 Consider the tooth growth data that were analyzed in this chapter. These data are available in the R package nlme as Orthodont. Let Y_{ij} denote the growth (in mm) at occasion t_j (in years) for boy i, $i = 1, \ldots, m$, $j = 1, \ldots, 4$, with $t_1 = 8$, $t_2 = 10$, $t_3 = 12$, $t_4 = 14$.

(a) Code up a GEE algorithm with working independence in your favorite package, and report $\widehat{\beta}$ and var($\widehat{\beta}$).

(b) Using an available option in a statistical package such as R confirm the results of the previous part.

(c) Show that var(Y) = $(Y - x\beta)^T(Y - x\beta) = 0$ if we attempt to use sandwich estimation in the situation in which cov($Y_i, Y_{i'}) \neq 0$.

8.13 In this question the effect of using different correlation structures, designs, and sample sizes in the GEE approach will be examined. Let Y_{ij} represent the observed growth on individual i at time x_{ij}, $i = 1, \ldots, m$; $j = 1, \ldots, n_i$. Let $N = \sum_{i=1}^{m} n_i$.

Assume the marginal model is

$$E[Y_{ij}] = \beta_0 + \beta_1 x_{ij},$$

so that $E[Y] = x\beta$ where Y is of dimension $N \times 1$, x is $N \times 2$, and β is 2×1. Consider the estimating function

$$\sum_{i=1}^{m} x_i^T W_i^{-1} (Y_i - x_i \beta),$$

with working covariances W_i of dimension $n_i \times n_i$, $i = 1, \ldots, m$.

Assume throughout that $\beta_0 = 18$, $\beta_1 = 0.5$, $\alpha_1 = 1$, and α_2 is set to either 0.5 or 0.9. Simulate data from the multivariate normal distribution $Y_i \sim N_{n_i}(x_i \beta, V_i)$, with the form of V_i taken as either the exchangeable or the AR(1) matrices W_i that are given below, for $i = 1, \ldots, m$. Examine the efficiency of these working models as a function of:

- The number of individuals, with $m = 8, 20, 60$
- Two designs:
    - *Design I:* Balanced with $n_i = n = 4$, $i = 1, \ldots, m$ and $x_1 = 8$, $x_2 = 10$, $x_3 = 12$, $x_4 = 14$ for all individuals
    - *Design II:* Unbalanced with $n_i = n = 3$, $i = 1, \ldots, m$ and

$$x_{i1} = 8, \quad x_{i2} = 10, \quad x_{i3} = 12 \quad \text{for } i = 1, \ldots, m/4$$
$$x_{i1} = 8, \quad x_{i2} = 10, \quad x_{i3} = 14, \quad \text{for } i = m/4+1, \ldots, m/2$$
$$x_{i1} = 8, \quad x_{i3} = 12, \quad x_{i3} = 14, \quad \text{for } i = m/2+1, \ldots, 3m/4$$
$$x_{i2} = 10, \quad x_{i3} = 12, \quad x_{i4} = 14, \quad \text{for } i = 3m/4+1, \ldots, m$$

- The working covariance structure $\alpha_1 \boldsymbol{W}_i$ with:
  - *Independence:* $\boldsymbol{W}_i = \mathbf{I}_{n_i}$ where $\mathbf{I}_{n_i}$ is the $n_i \times n_i$ identity matrix.
  - *Exchangeable:* $\boldsymbol{W}_i$ has diagonal elements 1 and off-diagonal elements α_2.
  - *First-order autocorrelation:* $\boldsymbol{W}_i$ has diagonal elements 1 and off-diagonal elements $W_{ijk} = \alpha_2^{|x_{ij}-x_{ik}|}$, $j,k = 1, \ldots, n_i, j \neq k, i = 1, \ldots, m$.

In total, there are $3 \times 2 \times 4 = 24$ sets of simulations, and for each you should:

(a) Report the 95% confidence interval coverage for β_1.
(b) Report the standard errors and efficiencies. For each working covariance model, there are two standard error calculations; the "true" standard errors are obtained across simulations while $\widehat{\text{var}(\widehat{\beta}_1)}$ describes the average (across simulations) of the reported squared standard error, where the latter is calculated using the sandwich formula. To evaluate the efficiencies, the (sandwich) variance of the estimators under each of the working models should be calculated.

8.14 Crowder and Hand (1990) describe data on the body weight of rats measured over 64 days. These data are available in the R package nlme and are named BodyWeight. Body weight is measured (in grams) on day 1, and every 7 days subsequently until day 64, with an extra measurement on day 44. There are 3 groups of rats, each on a different diet; 8 rats are on a control diet, and two sets of 4 rats are each on a different treatment.

(a) Fit LMMs to these data using ML/REML, with the primary aim being to determine whether there are differences in intercepts and slopes for each of the diets. Repeat this procedure using GEE.
(b) Carefully describe the models that you fit, in particular the choice of random effects structure in the LMM, and summarize your findings in simple terms.
(c) Now, analyze the first group of rats using a Bayesian analysis. Specifically, suppose Y_{ij} is the body weight of rat i at time t_j, and consider the three-stage model:
Stage One:

$$Y_{ij} = \beta_0 + b_i + \beta_1 t_j + \epsilon_{ij}$$

with $\epsilon_{ij} \mid \tau \sim_{iid} N(0, \tau^{-1})$, $i = 1, \ldots, m, j = 1, \ldots, n$.

8.12 Exercises

Stage Two: $b_i \mid \tau_0 \sim_{iid} N(0, \tau_0^{-1})$, with b_i independent of the ϵ_{ij}, $i=1,\ldots,m, j=1,\ldots,n$.
Stage Three: Independent hyperpriors with:

$$\pi(\boldsymbol{\beta}) \propto 1,$$
$$\pi(\tau) \propto \tau^{-1},$$
$$\pi(\tau_0) \sim \text{Ga}(0.1, 0.5)$$

where $\boldsymbol{\beta} = [\beta_0, \beta_1]^\text{T}$.

(d) Find the form of the conditional distributions that are required for constructing a Gibbs sampling algorithm to explore the posterior distribution $p(\boldsymbol{\beta}, \tau, b_1, \ldots, b_m, \tau_0 \mid \boldsymbol{y})$:

- $p(\boldsymbol{\beta} \mid \tau, b_1, \ldots, b_m, \tau_0, \boldsymbol{y})$.
- $p(\tau \mid \boldsymbol{\beta}, b_1, \ldots, b_m, \tau_0, \boldsymbol{y})$.
- $p(\tau_0 \mid \boldsymbol{\beta}, \tau, b_1, \ldots, b_m, \boldsymbol{y})$.
- $p(b_i \mid \boldsymbol{\beta}, \tau, b_j, j \neq i, \tau_0, \boldsymbol{y}), i = 1, \ldots, m$.

(e) Implement this algorithm for the data on the 8 rats in the control group. Provide trace plots of selected parameters to provide evidence of convergence of the Markov chain. Report two sets of summaries, consisting of the 5%, 50%, 95% quantiles, from two chains started from different values.

(f) Check your answers using available software, such as INLA or WinBUGS.

8.15 Prove (8.58).

Chapter 9
General Regression Models

9.1 Introduction

In this chapter we consider dependent data but move from the linear models of Chap. 8 to general regression models. As in Chap. 6, we consider generalized linear models (GLMs) and, more briefly, nonlinear models. We first give an outline of this chapter. In Sect. 9.2 we describe three motivating datasets to which we return throughout the chapter. The GLMs discussed in Sect. 6.3 can be extended to incorporate dependences in observations on the same unit; as with the linear model, an obvious way to carry out modeling in this case is to introduce unit-specific random effects. Within a GLM a natural approach is for these random effects to be included on the linear predictor scale. The resultant *conditional* models are known as generalized linear mixed models (GLMMs), and these are introduced in Sect. 9.3. In Sects. 9.4 and 9.5 we describe likelihood and conditional likelihood methods of estimation, with Sect. 9.6 devoted to a Bayesian treatment. Section 9.7 illustrates some of the flexibility of GLMMs by describing and applying a particular model for spatial dependence. An alternative random effects specification, based on conjugacy, is described in Sect. 9.8. An important approach to the modeling and analysis of dependent data that is philosophically different from the random effects formulation is via marginal models and generalized estimating equations (GEE), and these are the subject of Sect. 9.9. In Sect. 9.10, a second GEE approach is described in which the estimating equations for the mean are supplemented with a second set for the variances/covariances. For GLMMs, extra care must be taken with parameter interpretation, and Sect. 9.11 discusses this issue, emphasizing how interpretation differs between conditional and marginal models. In Part II of the book, which focused on independent data, Chap. 7 was devoted to models for binary data. For dependent data, models binary data are less well developed, and so we do not devote a complete chapter to their description. However, Sect. 9.12 introduces the modeling of dependent binary data, and, subsequently, Sects. 9.13 and 9.14 describe conditional (mixed) and marginal models for binary data. Section 9.15 considers how nonlinear models, as defined in Sect. 6.10, can be extended to the dependent data case. For such models, many applications concentrate on inference for units, and so the introduction of random effects is again suggested. We refer to the resultant class of models as nonlinear mixed models (NLMMs). Section 9.16 considers issues related to the parameterization of the nonlinear model. Inference for nonlinear mixed models via likelihood and Bayes approaches is covered in Sects. 9.17 and 9.18, while GEE is briefly considered in Sect. 9.19. The assessment of assumptions for general regression models is described in Sect. 9.20, with concluding comments contained in Sect. 9.21. Additional references appear in Sect. 9.22.

9.2 Motivating Examples

In this chapter we will analyze the lung cancer and radon data introduced in Sect. 1.3.3 and three additional datasets.

9.2.1 Contraception Data

Fitzmaurice et al. (2004) reanalyze data originally appearing in Machin et al. (1988) concerning a randomized longitudinal contraception trial. Each of 1,151 women received injections of 100 or 150 mg of depot medroxyprogesterone acetate (DMPA) on the day of randomization and three additional injections at 90-day intervals. There was a final follow-up 3 months after the last injection (a year after the initial injection). The women completed a menstrual diary throughout the study, and the binary response is whether the woman had experienced amenorrhea, the absence of menstrual bleeding for a specified number of days, during each of the four 3-month intervals. There was dropout in this study, but we will not address this issue, important though it is. The sample sizes, across measurement occasions, in the low- and high-dose groups are [576, 477, 409, 361] and [575, 476, 389, 353], respectively. Plotting the individual-level 0/1 data is usually not informative for binary data, and so in Fig. 9.1, we plot the averages, that is, the probabilities of amenorrhea over time for the two treatment groups. We see increasing probabilities of amenorrhea in both groups, with the probabilities in the 150-mg dose group being greater than in the 100-mg dose group.

As we will discuss in Sect. 9.14, for binary data, there is no obvious natural measure of dependence, unlike normal data for which the correlation is routinely used. However, Table 9.1 gives the empirical correlations between responses at different measurement occasions in the low- and high-dose groups, respectively. In both groups there is appreciable correlation between observations on the same woman, with a suggestion that the correlations decrease on measurements taken further apart. To explicitly acknowledge the dependence over time in responses on the same woman, multivariate binary data models are required.

9.2 Motivating Examples

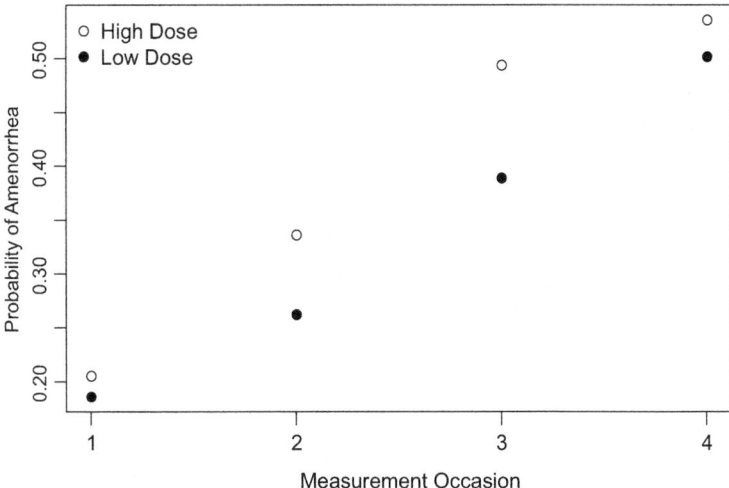

Fig. 9.1 Probability of amenorrhea over time in low- and high-dose groups, in the contraception data

Table 9.1 Empirical variances (on the *diagonal*) and correlations (on the *upper diagonal*), between measurements on the same woman at different observation occasions (1–4), in the low- (*left*) and high- (*right*) dose groups of the contraception data

	1	2	3	4		1	2	3	4
1	0.15	0.40	0.28	0.27	1	0.16	0.31	0.25	0.29
2		0.19	0.45	0.35	2		0.22	0.43	0.43
3			0.24	0.13	3			0.25	0.47
4				0.25	4				0.25

9.2.2 Seizure Data

Thall and Vail (1990) describe data on epileptic seizures in 59 individuals. For each patient, the number of epileptic seizures was recorded during a baseline period of 8 weeks, after which patients were randomized to one of two groups: treatment with either the antiepileptic drug progabide or with placebo. The numbers of individuals in the placebo and progabide groups were 28 and 31, respectively. The number of seizures was recorded in four consecutive 2-week periods. For these data, let Y_{ij} represent the number of counts for patient i, $i = 1, \ldots, 59$ at occasion j, with $j = 0$ the baseline period and $j = 1, \ldots, 4$ the subsequent set of four 2-week measurement periods. Also, let T_j be the length (in weeks) of the observation period (which is the same for all individuals), with $T_0 = 8$ and $T_j = 2$ for $j = 1, \ldots, 4$. We might consider the model

$$Y_{ij} \mid \mu_{ij} \sim \text{Poisson}(\mu_{ij})$$

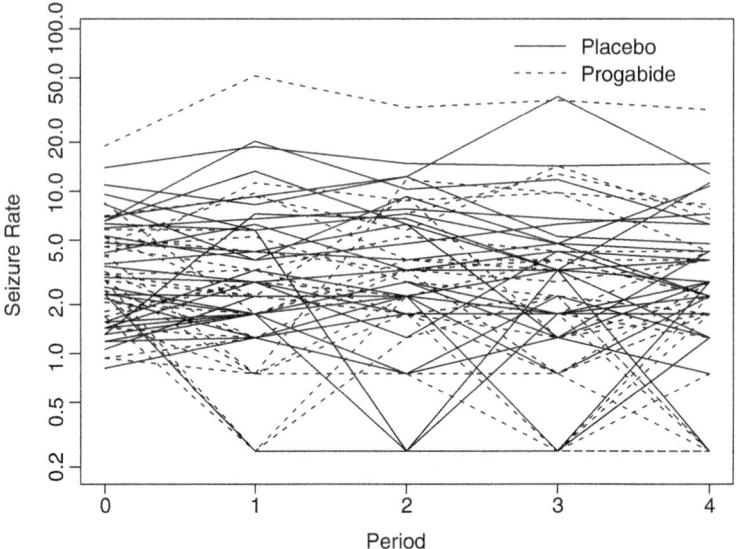

Fig. 9.2 Log of seizure rates by period for individuals on placebo and progabide

where $\mu_{ij} = T_j \exp(\boldsymbol{x}_{ij}\boldsymbol{\beta})$ and $\exp(\boldsymbol{x}_{ij}\boldsymbol{\beta})$ is a loglinear regression model. There are two immediate issues with this model: data on the same individual are unlikely to be independent and there may be excess-Poisson variation.

As a first look at the data, we plot the log seizure rate $\log[(Y_{ij}+0.5)/T_j]$ for each individual versus period j in Fig. 9.2. The 0.5 is added to avoid taking the log of zero. The line types distinguish the placebo and progabide groups. It is difficult to discern much pattern from this plot. In particular, it is not clear if progabide provides a drop in the rate of seizures, though there is clearly large between-individual variability in the rates. One individual's profile appears to be outlying and high, with the rate of seizures increasing after treatment with progabide.

Figure 9.3 displays the average seizure counts by period and by treatment group. In three out of the four post-baseline periods, the averages are lower in the progabide group. To assess the excess Poisson, we calculate the ratio of the variance of the counts to the mean, that is, $\mathrm{var}(Y_{ij})/\mathrm{E}[Y_{ij}]$, by period and treatment group. Table 9.2 gives these ratios and clearly shows that there is a great deal of excess-Poisson variability for these data.

9.2.3 Pharmacokinetics of Theophylline

Twelve subjects were given an oral dose of the antiasthmatic agent theophylline, with 11 concentration measurements obtained from each individual over 25 h. The doses ranged between 3.10 and 5.86 mg/kg. As is usual with experiments such as

9.2 Motivating Examples

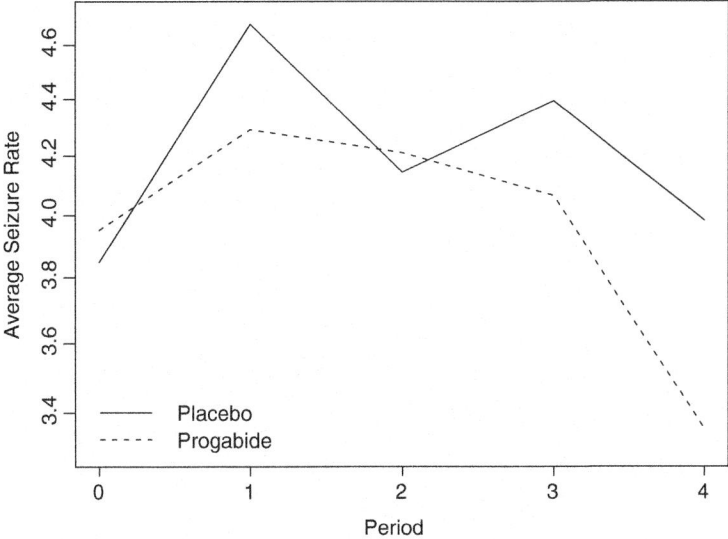

Fig. 9.3 Average seizure rates by period and treatment group

Table 9.2 Ratio of the variance of seizure counts to the mean of the seizure counts, by period and treatment group

Group	Period				
	0	1	2	3	4
Placebo	22.1	11.0	8.0	24.5	7.3
Progabide	24.8	38.8	16.7	23.7	18.9

this, there is abundant sampling at early times in an attempt to capture the absorption phase, which is rapid. Further background on pharmacokinetic modeling is given in Example 1.3.4 where the data for the first individual were presented. Section 6.2 introduced a mean model for these data (for a generic individual) as

$$\frac{Dk_a}{V(k_a - k_e)}[\exp(-k_e x) - \exp(-k_a x)]$$

where x is the sampling time, $k_a > 0$ is the absorption rate constant, $k_e > 0$ is the elimination rate constant, and $V > 0$ is the (apparent) volume of distribution (that converts total amount of drug into concentration). Figure 9.4 shows the concentration–time data. The curves follow a similar pattern, but there is clearly between-subject variability.

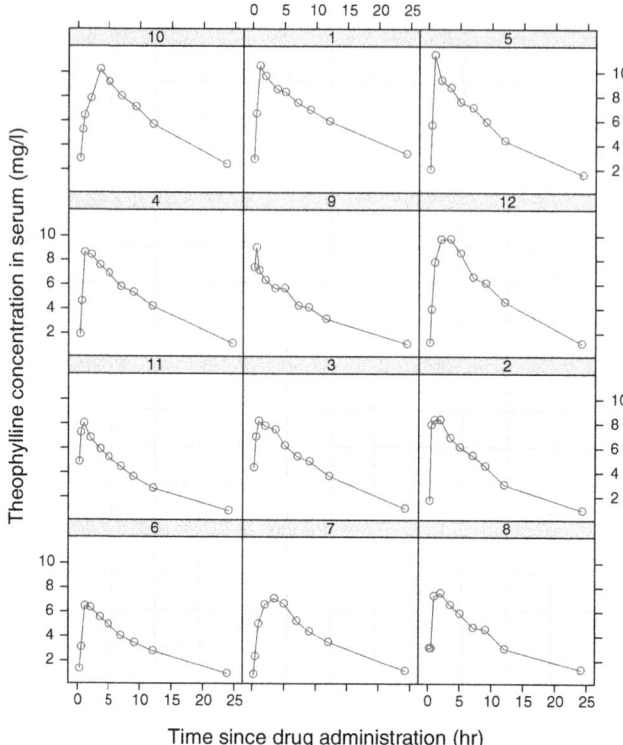

Fig. 9.4 Concentrations versus time for 12 individuals who received the drug theophylline

9.3 Generalized Linear Mixed Models

In this section we describe a modeling framework that allows the introduction of random effects into GLMs; these models induce dependence between responses on the same unit. Adding normal random effects on the linear predictor scale gives a GLMM. The paper of Breslow and Clayton (1993) popularized these models, by discussing implementation and providing a number of cases studies.

We first describe notation. Let Y_{ij} be the jth observation on the ith unit for $i = 1, \ldots, m$, $j = 1, \ldots, n_i$. The responses for the ith unit will be denoted $\boldsymbol{Y}_i = [Y_{i1}, \ldots, Y_{in_i}]^\mathrm{T}$, $i = 1, \ldots, m$. Responses on different units will be assumed independent. Let $\boldsymbol{\beta}$ represent a $(k+1) \times 1$ vector of fixed effects and $\boldsymbol{b}_i$ a $(q+1) \times 1$ vector of random effects, with $q \leq k$. Let $\boldsymbol{x}_{ij} = [1, x_{ij1}, \ldots, x_{ijk}]$ be a $(k+1) \times 1$ vector of covariates, so that $\boldsymbol{x}_i = [\boldsymbol{x}_{i1}, \ldots, \boldsymbol{x}_{in_i}]$ is the design matrix for the fixed effects of unit i, and let $\boldsymbol{z}_{ij} = [1, z_{ij1}, \ldots, z_{ijq}]^\mathrm{T}$ be a $(q+1) \times 1$ vector of variables that are a subset of $\boldsymbol{x}_{ij}$, so that $\boldsymbol{z}_i = [\boldsymbol{z}_{i1}, \ldots, \boldsymbol{z}_{in_i}]^\mathrm{T}$ is the design matrix for the random effects of unit i.

9.3 Generalized Linear Mixed Models

A GLMM is defined by the following two-stage model:

Stage One: The distribution of the data is $Y_{ij} \mid \theta_{ij}, \alpha \sim p(\cdot)$ where $p(\cdot)$ is a member of the exponential family, that is

$$p(y_{ij} \mid \theta_{ij}, \alpha) = \exp\{[y_{ij}\theta_{ij} - b(\theta_{ij})]/a(\alpha) + c(y_{ij}, \alpha)\}, \quad (9.1)$$

for $i = 1, \ldots, m$ units and $j = 1, \ldots, n_i$, measurements per unit. The variance is

$$\text{var}(Y_{ij} \mid \theta_{ij}, \alpha) = \alpha v(\mu_{ij}).$$

Let $\mu_{ij} = \text{E}[Y_{ij} \mid \theta_{ij}, \alpha]$ and, for a link function $g(\cdot)$, suppose

$$g(\mu_{ij}) = \boldsymbol{x}_{ij}\boldsymbol{\beta} + \boldsymbol{z}_{ij}\boldsymbol{b}_i,$$

so that random effects are introduced on the scale of the linear predictor. This defines the *conditional* part of the model.

Stage Two: The random effects are assigned a normal distribution:

$$\boldsymbol{b}_i \mid \boldsymbol{D} \sim_{iid} \text{N}_{q+1}(\boldsymbol{0}, \boldsymbol{D}).$$

For a number of reasons, including parameter interpretation, it is important to investigate the marginal moments that are induced by the random effects. Since marginal summaries may be calculated for the observed data, comparison with the theoretical forms is useful for model checking. The marginal mean is

$$\text{E}[Y_{ij}] = \text{E}[\text{E}_{b_i}(Y_{ij} \mid \boldsymbol{b}_i)]$$
$$= \text{E}[\mu_{ij}] = \text{E}_{b_i}[g^{-1}(\boldsymbol{x}_{ij}\boldsymbol{\beta} + \boldsymbol{z}_{ij}\boldsymbol{b}_i)].$$

The variance is

$$\text{var}(Y_{ij}) = \text{E}[\text{var}(Y_{ij} \mid \boldsymbol{b}_i)] + \text{var}(\text{E}[Y_{ij} \mid \boldsymbol{b}_i])$$
$$= \alpha \text{E}_{b_i}[v\{g^{-1}(\boldsymbol{x}_{ij}\boldsymbol{\beta} + \boldsymbol{z}_{ij}\boldsymbol{b}_i)\}] + \text{var}_{b_i}[g^{-1}(\boldsymbol{x}_{ij}\boldsymbol{\beta} + \boldsymbol{z}_{ij}\boldsymbol{b}_i)].$$

The covariances between outcomes on the same unit are

$$\text{cov}(Y_{ij}, Y_{ik}) = \text{E}[\text{cov}(Y_{ij}, Y_{ik} \mid \boldsymbol{b}_i)] + \text{cov}[\text{E}(Y_{ij} \mid \boldsymbol{b}_i), \text{E}[Y_{ik} \mid \boldsymbol{b}_i)]$$
$$= \text{cov}_{b_i}[g^{-1}(\boldsymbol{x}_{ij}\boldsymbol{\beta} + \boldsymbol{z}_{ij}\boldsymbol{b}_i), g^{-1}(\boldsymbol{x}_{ik}\boldsymbol{\beta} + \boldsymbol{z}_{ik}\boldsymbol{b}_i)]$$
$$\neq 0,$$

for $j \neq k$ due to shared random effects, and

$$\text{cov}(Y_{ij}, Y_{i'k}) = 0,$$

for $i \neq i'$, as there are no random effects in the model that are shared by different units. Explicit forms of the moments are available for some choices of exponential family, as we see later in the chapter, though the marginal distribution of the data is not typically available (outside of the normal case discussed in Chap. 8).

9.4 Likelihood Inference for Generalized Linear Mixed Models

As discussed in Sect. 8.5, there are three distinct sets of parameters for which inference may be required: fixed effects β, variance components α, and random effects $b = [b_1, \ldots, b_m]^T$. As with the linear mixed model (LMM), we maximize the likelihood $L(\beta, \alpha)$, where α denote the variance components in D and the scale parameter α (if present). The likelihood is obtained by integrating $[b_1, \ldots, b_m]$ from the model:

$$L(\beta, \alpha) = \prod_{i=1}^{m} \int p(y_i \mid \beta, b_i) \times p(b_i \mid \alpha) \, db_i.$$

There are m integrals to evaluate, each of dimension equal to the number of random effects, $q + 1$. For non-Gaussian GLMMs, these integrals are not available in closed form, and so some sort of analytical, numerical, or simulation-based approximation is required (Sect. 3.7). Common approaches include analytic approximations such as the Laplace approximation (Sect. 3.7.2) or the use of adaptive Gauss–Hermite numerical integration rules (Sect. 3.7.3). There are two difficulties with inference for GLMMs: carrying out the required integrations and maximizing the resultant (approximated) likelihood function. The likelihood function can be unwieldy, and, in particular, the second derivatives may be difficult to determine, so the Newton–Raphson method cannot be directly used. An alternative is provided by the quasi-Newton approach in which the derivatives are approximated (Dennis and Schnabel 1996).

One approach to the integration/maximization difficulties is the following. In Sect. 6.5.2 the iteratively reweighted least squares (IRLS) algorithm was described as a method for finding MLEs in a GLM. The penalized-IRLS (P-IRLS) algorithm is a variant in which the working likelihood is augmented with a penalization term corresponding to the (log of the) random effects distribution. This algorithm may be used in a GLMM context in order to obtain, conditional on α, estimates of β and b, with α being estimated via a profile log-likelihood (Sect. 2.4.2); see Bates (2011). The P-IRLS is also used for nonparametric regression and is described in this context in Sect. 11.5.1.

The method of penalized quasi-likelihood (PQL) was historically popular (Breslow and Clayton 1993) but can be unacceptably inaccurate, in particular, for binary outcomes. See Breslow (2005) for a recent perspective.

9.4 Likelihood Inference for Generalized Linear Mixed Models

Approximate inference for $[\beta, \alpha]$ is carried out via the usual asymptotic normality of the MLE which is, with sloppy notation,

$$\begin{bmatrix} \widehat{\beta} \\ \widehat{\alpha} \end{bmatrix} \sim N\left(\begin{bmatrix} \beta \\ \alpha \end{bmatrix}, \begin{bmatrix} I_{\beta\beta} & I_{\beta\alpha} \\ I_{\alpha\beta} & I_{\alpha\alpha} \end{bmatrix}^{-1} \right) \tag{9.2}$$

where $I_{\beta\beta}$, $I_{\beta\alpha}$, $I_{\alpha\beta}$, and $I_{\alpha\alpha}$ are the relevant information matrices. An important observation is that in general $I_{\beta\alpha} \neq 0$, and so we cannot separately estimate the regression and variance parameters, so consistency requires correct specification of both mean and variance models. Likelihood ratio tests are available for fixed effects though it requires experience or simulation to determine whether the sample size m is large enough for the null χ^2 distribution to be accurate.

In terms of the random effects, one estimator is

$$E[b_i \mid y] = \frac{\int_{b_i} b_i p(y \mid b_i) p(b_i \mid D) \, db_i}{\int_{b_i} p(y \mid b_i) p(b_i \mid D) \, db_i}.$$

Unless the first stage is normal, the integrals in numerator and denominator will not be analytically tractable, though Laplace approximations or adaptive Gauss–Hermite may be used. In practice, empirical Bayes estimators, $E[b_i \mid y, \widehat{\beta}, \widehat{\alpha}]$, are used.

Example: Seizure Data

Recall that Y_{ij} is the number of seizures on patient i during period j, $j = 0, 1, 2, 3, 4$, and T_j is the observation period during period j, $j = 0, 1, 2, 3, 4$ with $T_0 = 8$ weeks and $T_j = 2$ weeks for $j = 1, \ldots, 4$. It is clear from Fig. 9.2 that there is considerable between-patient variability in the level of seizures, which suggests that a random effects model should include at least random intercepts. A random intercepts GLMM for the seizure data is:

Stage One: $Y_{ij} \mid \beta, b_i \sim_{ind} \text{Poisson}(\mu_{ij})$, with

$$g(\mu_{ij}) = \log \mu_{ij} = \log T_{ij} + x_{ij}\beta + b_i,$$

and where x_{ij} is the design matrix for individual i at period j, with associated fixed effect β. A particular model will be discussed shortly. The first two-conditional moments are

$$E[Y_{ij} \mid b_i] = \mu_{ij} = T_{ij} \exp(x_{ij}\beta + b_i),$$
$$\text{var}(Y_{ij} \mid b_i) = \mu_{ij}.$$

Table 9.3 Parameter interpretation for the model defined by (9.3)

Group	Period 0	Period 1,2,3,4
Placebo	$\exp(\beta_0)$	$\exp(\beta_0 + \beta_2)$
Progabide	$\exp(\beta_0 + \beta_1)$	$\exp(\beta_0 + \beta_1 + \beta_2 + \beta_3)$

Stage Two: $b_i \mid \sigma_0^2 \sim_{iid} N(0, \sigma_0^2)$.

Writing $\alpha = \sigma_0^2$, the likelihood is

$$L(\boldsymbol{\beta}, \boldsymbol{\alpha}) = \prod_{i=1}^{m} \int \prod_{j=1}^{n_i} \frac{\exp[-\mu_{ij}(b_i)]\mu_{ij}(b_i)^{y_{ij}}}{y_{ij}!}$$

$$\times (2\pi\sigma_0^2)^{-1/2} \exp\left(-\frac{b_i^2}{2\sigma_0^2}\right) db_i$$

$$= (2\pi\sigma_0^2)^{-m/2} \prod_{i=1}^{m} \exp\left(\sum_{j=1}^{n_i} y_{ij} x_{ij}\boldsymbol{\beta}\right)$$

$$\int \exp\left(-e^{b_i}\sum_{j=1}^{n_i} e^{x_{ij}\boldsymbol{\beta}} + \sum_{j=1}^{n_i} y_{ij} b_i - \frac{b_i^2}{2\sigma_0^2}\right) db_i.$$

The latter integral is analytically intractable. A Laplace approximation would expand each of the m integrands about the maximizing value of b_i, or, alternatively, numerical integration can be used, for example, using adaptive Gauss–Hermite.

Let $x_{1i} = 0/1$ if patient i was assigned placebo/progabide, $x_{2j} = 0/1$ if $j = 0/1, 2, 3, 4$, and $x_{ij3} = x_{1i} \times x_{2j}$ for $j = 0, 1, 2, 3, 4$. Therefore, x_1 is a treatment indicator, x_2 is an indicator of pre-/post-baseline, and x_3 takes the value 1 for progabide individuals who are post-baseline and is zero otherwise. The first model we fit is

$$\boldsymbol{x}_{ij}\boldsymbol{\beta} = \beta_0 + \beta_1 x_{1i} + \beta_2 x_{2j} + \beta_3 x_{3ij}, \tag{9.3}$$

so that $\boldsymbol{x}_{ij}$ is 1×4 and $\boldsymbol{\beta}$ is 4×1. Table 9.3 summarizes the form of the model across groups and periods.

We first provide an interpretation from a conditional perspective. In the following interpretation, a "typical" patient corresponds to a patient whose random effect is zero, that is, $b = 0$. On the more interpretable rate scale:

- $\exp(\beta_0)$ is the rate of seizures for a typical individual under placebo in time period 0.
- $\exp(\beta_1)$ is the ratio of the seizure rate of a typical individual under progabide to a typical individual under placebo, in time period 0. If the groups are comparable

at the time of treatment assignment and there are no other corrupting factors, we would expect this parameter to be estimated as close to 1.
- $\exp(\beta_2)$ is the ratio of the seizure rate post-baseline (T_j, $j = 1, 2, 3, 4$) as compared to baseline (T_0), for a typical individual in the placebo group.
- $\exp(\beta_3)$ is the ratio of the seizure rate for a typical individual in the progabide group post-baseline, as compared to a typical individual in the placebo group in the same period. Hence, $\exp(\beta_3)$ is the rate ratio parameter of interest.

Alternatively, we may interpret these rates and ratios of rates as being between two individuals with the same baseline rate of seizures (i.e., the same random effect b) prior to treatment assignment.

We now evaluate the implied *marginal model*. We recap the first two moments of a lognormal random variable. If $Z \sim \text{LogNorm}(\mu, \sigma^2)$, then

$$E[Z] = \exp(\mu + \sigma^2/2)$$
$$\text{var}(Z) = \exp(2\mu + \sigma^2) \times [\exp(\sigma^2) - 1]$$
$$= E[Z]^2 \times [\exp(\sigma^2) - 1].$$

Therefore, since $\exp(b_i) \sim \text{LN}(0, \sigma_0^2)$, the marginal mean is

$$E[Y_{ij}] = E_{b_i}\left[E(Y_{ij} \mid b_i)\right]$$
$$= T_{ij} \exp(\boldsymbol{x}_{ij}\boldsymbol{\beta}) E_{b_i}\left[\exp(b_i)\right]$$
$$= T_{ij} \exp(\boldsymbol{x}_{ij}\boldsymbol{\beta} + \sigma_0^2/2).$$

Consequently, for this model, relative rates $\exp(\beta_k)$, $k = 1, 2, 3$ (which, recall, are ratios) have a marginal interpretation, since the $\exp(\sigma_0^2/2)$ terms cancel in numerator and denominator (under the model). For example, $\exp(\beta_1)$ is the ratio of the average seizure rate in the progabide group to the average rate in the placebo group, in time period 0. Further discussion of parameter interpretation in marginal and conditional models is provided in Sect. 9.11. The marginal variance is

$$\text{var}(Y_{ij}) = E_{b_i}\left[\text{var}(Y_{ij} \mid b_i)\right] + \text{var}_{b_i}\left[E(Y_{ij} \mid b_i)\right]$$
$$= E_{b_i}\left[T_{ij} \exp(\boldsymbol{x}_{ij}\boldsymbol{\beta} + b_i)\right] + \text{var}_{b_i}\left[T_{ij} \exp(\boldsymbol{x}_{ij}\boldsymbol{\beta} + b_i)\right]$$
$$= E[Y_{ij}][1 + E(Y_{ij})][\exp(\sigma_0^2) - 1)]$$
$$= E[Y_{ij}][1 + E(Y_{ij}) \times \kappa]$$

where

$$\kappa = \exp(\sigma_0^2) - 1 > 0$$

illustrating quadratic excess-Poisson variation which increases as σ_0^2 increases.

The marginal covariance between observations on the same individual is

$$\text{cov}(Y_{ij}, Y_{ik}) = \text{cov}\left[T_{ij}\exp(\boldsymbol{x}_{ij}\boldsymbol{\beta} + b_i), T_{ij}\exp(\boldsymbol{x}_{ik}\boldsymbol{\beta} + b_i)\right]$$
$$= T_{ij}T_{ik}\exp(\boldsymbol{x}_{ij}\boldsymbol{\beta} + \boldsymbol{x}_{ik}\boldsymbol{\beta}) \times \exp(\sigma_0^2)[\exp(\sigma_0^2) - 1]$$
$$= \text{E}[Y_{ij}]\text{E}[Y_{ik}]\kappa.$$

To summarize, for individual i, the variance–covariance matrix is

$$\begin{bmatrix} \mu_{i1} + \mu_{i1}^2\kappa & \mu_{i1}\mu_{i2}\kappa & \mu_{i1}\mu_{i3}\kappa & \mu_{i1}\mu_{i4}\kappa \\ \mu_{i2}\mu_{i1}\kappa & \mu_{i2} + \mu_{i2}^2\kappa & \mu_{i2}\mu_{i3}\kappa & \mu_{i2}\mu_{i4}\kappa \\ \mu_{i3}\mu_{i1}\kappa & \mu_{i3}\mu_{i2}\kappa & \mu_{i3} + \mu_{i3}^2\kappa & \mu_{i3}\mu_{i4}\kappa \\ \mu_{i4}\mu_{i1}\kappa & \mu_{i4}\mu_{i2}\kappa & \mu_{i4}\mu_{i2}\kappa & \mu_{i4} + \mu_{i4}^2\kappa \end{bmatrix}.$$

For a random intercepts only LMM the marginal correlation is constant within a unit with correlation $\sigma_0^2/(\sigma_0^2 + \sigma_\epsilon^2)$, regardless of μ_{ij}, μ_{ik}. In contrast, for the Poisson random intercepts mixed model, the marginal correlation is

$$\text{corr}(Y_{ij}, Y_{ik}) = \frac{\kappa\sqrt{\mu_{ij}\mu_{ik}}}{\sqrt{(1 + \kappa\mu_{ij})(1 + \kappa\mu_{ik})}}$$
$$= \left[1 + \frac{1}{\kappa}\left(\frac{1}{\mu_{ij}} + \frac{1}{\mu_{ik}}\right) + \frac{1}{\kappa^2}\frac{1}{\mu_{ij}\mu_{ik}}\right]^{-1/2} \quad (9.4)$$

so that correlations vary in a complicated fashion as a function of the mean responses. However, the correlations increase as κ increases and as the means increase. A deficiency of this model is that we have only a single parameter (σ_0^2) to control both excess-Poisson variability and the strength of dependence over time. We address this in Sect. 9.6 by adding a second random effect to the model. For observations on different individuals, $\text{cov}(Y_{ij}, Y_{i'k}) = 0$ for $i \neq i'$.

Using a Laplace approximation to evaluate the integrals that define the likelihood, we obtain the estimates and standard errors given in Table 9.4. An alternative approach using Gauss–Hermite with 50 points to evaluate the integrals gave the same answers, so we conclude that the Laplace approximation is accurate in this example.

In terms of the parameter of interest β_3, there is an estimated drop in the seizure rates of 10% in the progabide group as compared to placebo, but this drop is not significant when assessed using a likelihood ratio test under conventional significance levels. The estimated value of β_1 indicates that the placebo and progabide groups are comparable at baseline, though the value of β_2 and its standard error suggest there is some evidence that the rate of seizures increased in the placebo group after randomization.

The random intercepts standard deviation is estimated as $\hat{\sigma}_0^2 = 0.61$ to give $\hat{\kappa} = 0.84$. For an individual whose rate of seizures is constant over the study period

9.5 Conditional Likelihood Inference for Generalized Linear Mixed Models

Table 9.4 MLEs and standard errors from a generalized linear mixed model fit to the seizure data

	Estimate	Std. err.
β_0	1.03	0.15
β_1	-0.024	0.21
β_2	0.11	0.047
β_3	-0.10	0.065
σ_0	0.78	–

Parameter meaning: β_0 is the log baseline seizure rate in the placebo group for a typical individual; β_1 is the log of the ratio of seizure rates between typical individuals in the progabide and placebo groups, at baseline; β_2 is the log of the ratio of seizure rates of typical individuals in the post-baseline and baseline placebo groups; β_3 is the log of the ratio of the seizure rate for a typical individual in the progabide group as compared to a typical individual in the placebo group, post-baseline; σ_0 is the standard deviation of the random intercepts

at levels $\mu_{ij} = \mu_{ik} = 1, 2, 5$, the correlations between responses on this individual, from (9.4), are estimated as $0.46, 0.63, 0.81$. We conclude that the correlations are appreciable.

9.5 Conditional Likelihood Inference for Generalized Linear Mixed Models

An alternative approach to estimation in the GLMM is provided by conditional likelihood (Sect. 2.4.2). The basic idea is to split the data into components t_1 and t_2 in such a way that t_1 contains information on parameter of interests, while t_2 contains information primarily on nuisance parameters. In a GLMM setting, the aim is to condition on a part of the data that eliminates the random effects, hence avoiding both the need for their estimation and the need to specify their distribution. A consequence of the conditioning is that we also eliminate all regression coefficients in the model that are associated with covariates that are constant within an individual.

We now work through the details and assume a discrete GLMM and a canonical link function so that

$$g(\mu_{ij}) = \theta_{ij} = \boldsymbol{\beta}^T \boldsymbol{x}_{ij}^T + \boldsymbol{b}_i^T \boldsymbol{z}_{ij}^T.$$

We further assume $\alpha = 1$, as is true for Poisson and binomial models. Viewing both $\boldsymbol{\beta}$ and $\boldsymbol{b}$ as fixed effects gives, from (9.1),

$$\Pr(\boldsymbol{y} \mid \boldsymbol{\beta}, \boldsymbol{b}) \propto \exp\left[\boldsymbol{\beta}^T \sum_{i=1}^{m}\sum_{j=1}^{n_i} \boldsymbol{x}_{ij}^T y_{ij} + \sum_{i=1}^{m} \boldsymbol{b}_i^T \sum_{j=1}^{n_i} \boldsymbol{z}_{ij}^T y_{ij} - \sum_{i=1}^{m}\sum_{j=1}^{n_i} b(\theta_{ij})\right],$$

(9.5)

where $b'(\theta_{ij}) = \mathrm{E}[Y_{ij} \mid b_i]$. Define

$$t_{1i} = \sum_{j=1}^{n_i} x_{ij}^\mathrm{T} y_{ij}$$

$$t_{2i} = \sum_{j=1}^{n_i} z_{ij}^\mathrm{T} y_{ij},$$

and let $t_1 = [t_{11}, \ldots, t_{1m}]^\mathrm{T}$ and $t_2 = [t_{21}, \ldots, t_{2m}]^\mathrm{T}$ so that t_1 and t_{2i} are sufficient statistics for β and b_i, respectively. Conditioning on the sufficient statistics for b_i, we obtain

$$\Pr\left(y_i \mid \sum_{j=1}^{n_i} z_{ij}^\mathrm{T} Y_{ij} = t_{2i}, \beta\right) = \frac{\Pr\left(y_i, \sum_{j=1}^{n_i} z_{ij}^\mathrm{T} Y_{ij} = t_{2i} \mid \beta, b_i\right)}{\Pr\left(\sum_{j=1}^{n_i} z_{ij}^\mathrm{T} Y_{ij} = t_{2i} \mid \beta, b_i\right)}$$

$$= \frac{\sum_{S_{1i}} \exp(\beta^\mathrm{T} t_{1i} + b_i t_{2i})}{\sum_{S_{2i}} \exp(\beta^\mathrm{T} x_{ij}^\mathrm{T} y_{ij} + b_i t_{2i})},$$

so that the conditional likelihood is

$$L_c(\beta) = \frac{\sum_{S_{1i}} \exp(\beta^\mathrm{T} t_{1i})}{\sum_{S_{2i}} \exp(\beta^\mathrm{T} x_{ij}^\mathrm{T} y_{ij})},$$

where

$$S_{1i} = \left\{ y_i \mid \sum_{j=1}^{n_i} x_{ij}^\mathrm{T} y_{ij} = t_{1i}, \sum_{j=1}^{n_i} z_{ij}^\mathrm{T} y_{ij} = t_{2i} \right\}$$

$$S_{2i} = \left\{ y_i \mid \sum_{j=1}^{n_i} z_{ij}^\mathrm{T} y_{ij} = t_{2i} \right\}.$$

The set S_{1i} denotes the possible outcomes $Y_{ij}, j = 1, \ldots, n_i$ that are consistent with t_{1i} and t_{2i}, given x_i and z_i. The conditional MLE has the usual properties of an MLE. In particular, under regularity conditions, it is consistent and asymptotically normally distributed with the variance–covariance matrix determined from the second derivatives of the conditional log-likelihood.

The conditional likelihood approach allows the specification of a model (via the parameters b_i) to acknowledge dependence but eliminates these parameters from the model. We emphasize that no distribution has been specified for the b_i, as they have been viewed as fixed effects. Depending on the structure of x_{ij} and z_{ij}, some of the β parameters may be eliminated from the model. For example, if $x_{ij} = z_{ij}$, the collections S_{1i} and S_{2i} coincide and the complete β vector would be conditioned away.

Example: Seizure Data

We derive the conditional likelihood in this example, using the random intercepts only model so that $z_{ij}b_i = b_i$. The loglinear random intercept model is

$$\log \mathrm{E}[Y_{ij} \mid \boldsymbol{\beta}^\star, \lambda_i] = \log T_{ij} + \lambda_i + \beta_2 x_{2j} + \beta_3 x_{3ij}$$
$$= \log T_{ij} + \lambda_i + \boldsymbol{x}_{ij}\boldsymbol{\beta}^\star$$
$$= \log \mu_{ij}$$

where $\boldsymbol{\beta}^\star = [\beta_2, \beta_3]^\mathrm{T}$ represents the regression coefficients that are not conditioned from the model (since they are associated with covariates that change within an individual), $\boldsymbol{x}_{ij} = [x_{2j}, x_{3ij}]$, and $\lambda_i = \beta_0 + \beta_1 x_{1i} + b_i$. We cannot estimate β_1 because the associated covariate x_{1i} is a treatment indicator and constant within an individual in this study; hence, it is eliminated from the model by the conditioning, along with b_i and β_0. This parameter is not a parameter of primary interest, however.

To derive the conditional likelihood, we first write $c_{1i}^{-1} = \prod_{j=0}^{4} y_{ij}!$, and then the joint distribution of the data for the ith individual is

$$p(\boldsymbol{y}_i \mid \boldsymbol{\beta}^\star, \lambda_i) = c_{1i} \exp\left(\sum_{j=0}^{4} y_{ij} \log \mu_{ij} - \sum_{j=0}^{4} \mu_{ij}\right)$$
$$= c_{1i} \exp\left[\lambda_i y_{i+} + \sum_{j=0}^{4} y_{ij}(\log T_{ij} + \boldsymbol{x}_{ij}\boldsymbol{\beta}^\star) - \mu_{i+}\right].$$

In this case, the conditioning statistic is y_{i+}, and its distribution is straightforward to derive

$$Y_{i+} \mid \boldsymbol{\beta}^\star, \lambda_i \sim \mathrm{Poisson}(\mu_{i+}).$$

Letting $c_{2i}^{-1} = y_{i+}!$, and recognizing that $\mu_{i+} = \exp(\lambda_i) \sum_{j=0}^{4} T_{ij} \exp(\boldsymbol{x}_{ij}\boldsymbol{\beta}^\star)$, gives

$$p(y_{i+} \mid \boldsymbol{\beta}^\star, \lambda_i) = c_{2i} \prod_{i=1}^{m} \exp\left(-\mu_{i+} + y_{i+} \log \mu_{i+}\right)$$
$$= c_{2i} \prod_{i=1}^{m} \exp\left[\lambda_i y_{i+} + y_{i+} \log\left(\sum_{j=0}^{4} T_{ij} \exp(\boldsymbol{x}_{ij}\boldsymbol{\beta}^\star)\right) - \mu_{i+}\right].$$

Hence,

$$p(\boldsymbol{y}_i \mid y_{i+}, \boldsymbol{\beta}^\star) = \frac{p(\boldsymbol{y}_i \mid \boldsymbol{\beta}^\star, \lambda_i)}{p(y_{i+} \mid \boldsymbol{\beta}^\star, \lambda_i)}$$

simplifies to

$$p(\boldsymbol{y}_i \mid y_{i+}, \boldsymbol{\beta}^\star) = \frac{c_{1i}}{c_{2i}} \exp\left[\sum_{j=0}^{4} y_{ij}(\log T_{ij} + \boldsymbol{x}_{ij}\boldsymbol{\beta}^\star) - y_{i+} \log\left(\sum_{j=0}^{4} T_{ij} \exp(\boldsymbol{x}_{ij}\boldsymbol{\beta}^\star)\right)\right]$$

$$= \frac{c_{1i}}{c_{2i}} \prod_{j=0}^{4} \exp\left\{y_{ij}\left[\log T_{ij} + \boldsymbol{x}_{ij}\boldsymbol{\beta}^\star - \log\left(\sum_{j=0}^{4} T_{ij} \exp(\boldsymbol{x}_{ij}\boldsymbol{\beta}^\star)\right)\right]\right\}$$

$$= \frac{y_{i+}!}{\prod_{j=0}^{4} y_{ij}!} \prod_{j=0}^{4} \left(\frac{T_{ij} \exp(\boldsymbol{x}_{ij}\boldsymbol{\beta}^\star)}{\sum_{l=0}^{4} T_{il} \exp(\boldsymbol{x}_{il}\boldsymbol{\beta}^\star)}\right)^{y_{ij}}$$

which is a multinomial likelihood (we have conditioned a set of Poisson counts on their total so this is no surprise). More transparently,

$$y_{ij} \mid y_{i+}, \boldsymbol{\beta}^\star \sim \text{Multinomial}_4(y_{i+}, \boldsymbol{\pi}_i)$$

where $\boldsymbol{\pi}_i = [\pi_{i0}, \ldots, \pi_{i4}]^\text{T}$ and

$$\pi_{ij} = \frac{T_{ij} \exp(\boldsymbol{x}_{ij}\boldsymbol{\beta}^\star)}{\sum_{l=0}^{4} T_{il} \exp(\boldsymbol{x}_{il}\boldsymbol{\beta}^\star)}.$$

Since $\boldsymbol{x}_{i1} = \boldsymbol{x}_{i2} = \boldsymbol{x}_{i3} = \boldsymbol{x}_{i4}$ and $T_{i0} = 8 = \sum_{j=1}^{4} T_{ij}$, we effectively have two observation periods of equal length. Letting $Y_i^\star = \sum_{j=1}^{4} Y_{ij}$,

$$Y_i^\star \mid y_{i+}, \boldsymbol{\beta}^\star \sim_{ind} \text{Binomial}(y_{i+}, \pi_i^\star)$$

where the odds are such that

$$\frac{\pi_i^\star}{1 - \pi_i^\star} = \begin{cases} \exp(\beta_2) & i = 1, \ldots, 28, \quad \text{placebo group} \\ \exp(\beta_2 + \beta_3) & i = 29, \ldots, 59, \quad \text{progabide group.} \end{cases}$$

Hence, fitting can be simply performed using logistic regression. For the seizure data, the sum of the denominators are 1,825 and 1,967 for placebo and progabide with 963 and 987 total seizures in the post-treatment period. These values result in estimates (standard errors) of $\widehat{\beta}_2 = 0.11$ (0.047) and $\widehat{\beta}_3 = -0.10$ (0.065). The estimate suggests a positive effect of progabide, but the difference from zero is not significant. Performing Fisher's exact test (Sect. 7.7) makes little difference for these data since the counts are large.

The conditional likelihood approach is quite intuitive in this example and results in a two-period design in which each person is acting as their own control. Conditioning on the sum of the two counts results in a single outcome per patient and removes the need to confront the dependency issue.

9.6 Bayesian Inference for Generalized Linear Mixed Models

9.6.1 Model Formulation

A Bayesian approach to inference for a GLMM requires a prior distribution for β, α. As with the linear mixed model (Sect. 8.6), a proper prior is required for the matrix D. A proper prior is not always necessary for β, but care is required. The exponential family and canonical link lead to a likelihood that is well behaved (in particular, with respect to tail behavior), though it is safer to specify a proper prior since impropriety of the posterior can occur in some cases (e.g., with noncanonical links or when counts are either equal to zero or to the denominator; see Sect. 6.8.1). As with the LMM, closed-form inference is unavailable, but MCMC (Sect. 3.8) is almost as straightforward as in the LMM, and the integrated nested Laplace approximation approach (Sect. 3.7.4) is also available though the approximation is not always accurate for the GLMM (Fong et al. 2010).

Let $W = D^{-1}$, and assume that there are no unknown scale parameters at stage one of the model (i.e., $\alpha = 1$), as is the case for binomial and Poisson models. The joint posterior is

$$p(\beta, W, b \mid y) \propto \prod_{i=1}^{m} \left[p(y_i \mid \beta, b_i) p(b_i \mid W) \right] \pi(\beta, W).$$

We assume independent hyperpriors:

$$\beta \sim N_{q+1}(\beta_0, V_0)$$
$$W \sim \text{Wish}_{q+1}(r, R^{-1})$$

where $\text{Wish}_{q+1}(r, R^{-1})$ denotes a Wishart distribution of dimension $q+1$ with degrees of freedom r and scale matrix R^{-1}; see Sect. 8.6.2 for further discussion. The conditional distribution for W is unchanged from the LMM case. There are no closed-form conditional distributions for β, or for b_i, but if an MCMC approach is followed, Metropolis–Hastings steps can be used.

9.6.2 Hyperpriors

In a GLMM we can often specify priors for more meaningful parameters than the original elements of β. For example, $\exp(\beta)$ is the relative risk/rate in a loglinear model and is the odds ratio in a logistic model. It is convenient to specify lognormal priors for a generic parameter $\theta > 0$, since one may specify two quantiles of the distribution, and directly solve for the two parameters of the prior. Denote by LogNorm(μ, σ) the lognormal prior distribution for θ with E[$\log \theta$] = μ and var($\log \theta$) = σ^2, and let θ_1 and θ_2 be the q_1 and q_2 quantiles of this prior. Then, (3.15) and (3.16) give the lognormal parameters. As an example, in a Poisson model, suppose we believe there is a 50% chance that the relative risk is less than 1 and a 95% chance that it is less than 5. With $q_1 = 0.5, \theta_1 = 1.0$ and $q_2 = 0.95, \theta_2 = 5.0$, we obtain lognormal parameters $\mu = 0$ and $\sigma = \log(5/1.96) = 0.98$.

Consider the random intercepts model with $b_i \mid \sigma_0^2 \sim_{iid} N(0, \sigma_0^2)$. It is not straightforward to specify a prior for σ_0, which represents the standard deviation of the residuals on the linear predictor scale and is consequently not easy to interpret. We specify a gamma prior Ga(a, b) for the precision $\tau_0 = 1/\sigma_0^2$, with parameters a, b specified a priori. The choice of a gamma distribution is convenient since it produces a marginal distribution for the "residuals" in closed form. As discussed in Sect. 8.6.2, the marginal distribution for b_i is $t_d(0, \lambda^2)$, a Student's t distribution with $d = 2a$ degrees of freedom, location zero, and scale $\lambda^2 = b/a$. These summaries allow prior specification based on beliefs concerning the residuals on a natural scale.

As an example, consider a log link, in which case the above prior specification is equivalent to the residual relative risks following a log Student's t distribution. We specify the range $\exp(\pm V)$ within which we expect the residual relative risks to lie with probability q and use the relationship $\pm t_{q/2}^d \lambda = \pm V$, where t_q^d is the qth quantile of a Student's t random variable with d degrees of freedom, to give $a = d/2, b = V^2 d/2(t_{q/2}^d)^2$. For example, if we assume a priori that the residual relative risks follow a log Student's t distribution with 2 degrees of freedom and that 95% of these risks fall in the interval [0.5,2.0], then we obtain the prior, Ga(1, 0.0260). In terms of σ_0, this results in [2.5%,97.5%] quantiles of [0.084,1.01] with posterior median 0.19.

It is important to assess whether the prior allows all reasonable levels of variability in the residual relative risks, in particular, small values should not be excluded. The prior Ga(0.001,0.001), which has been widely used under the guise of being relatively non-informative, should be avoided for this reason. This prior corresponds to the relative risks following a log Student's t distribution with 0.002 degrees of freedom, so that the spread is enormous. For example, the 0.01 quantile for σ_0 is 6.4 so that it is unlikely a priori that the standard deviation is small.

Example: Seizure Data

For illustration, we consider three models for the seizure data:

Model 1: The conditional mean model we start with has stages one and two given by:

$$Y_{ij} \mid b_i \sim_{ind} \text{Poisson}[T_{ij} \exp(\boldsymbol{x}_{ij} \boldsymbol{\beta} + b_i)]$$
$$b_i \mid \sigma_0^2 \sim_{iid} N(0, \sigma_0^2). \tag{9.6}$$

9.6 Bayesian Inference for Generalized Linear Mixed Models

For a Bayesian analysis, we require priors for $\boldsymbol{\beta}$ and σ_0^2. In this and the following two models, we take the improper prior $\pi(\boldsymbol{\beta}) \propto 1$. We assume $\sigma_0^{-2} = \tau_0 \sim$ Ga(1, 0.260). This prior corresponds to a Student's t_2 distribution for the residual rates with a 95% prior interval of [0.5, 2.0].

Model 2: We assume the same first and second stages as model 1 but address the sensitivity to the prior on τ_0. Specifically, we perturb the prior to $\tau_0 \sim$ Ga(2, 1.376), which corresponds to a Student's t_4 distribution for the residual rates with a 95% interval [0.1, 10.0].

Model 3: As pointed out in Sect. 9.4, a Poisson mixed model with a single random effect has a single parameter σ_0 only to model excess-Poisson variability *and* within-individual dependence. Therefore, we introduce "measurement error" into the model via the introduction of an additional random effect in the linear predictor. To motivate this model, consider the random intercepts only LMM model:

$$E[Y_{ij} \mid b_i] = \boldsymbol{x}_{ij}\boldsymbol{\beta} + b_i + \epsilon_{ij}$$
$$b_i \mid \sigma_0^2 \sim N(0, \sigma_0^2)$$
$$\epsilon_{ij} \mid \sigma_\epsilon^2 \sim N(0, \sigma_\epsilon^2),$$

with b_i and ϵ_{ij} independent. By analogy, consider the model:

$$Y_{ij} \mid b_i, \epsilon_{ij} \sim \text{Poisson}[T_{ij}\exp(\boldsymbol{x}_{ij}\boldsymbol{\beta} + b_i + \epsilon_{ij})]$$
$$b_i \mid \sigma_0^2 \sim N(0, \sigma_0^2)$$
$$\epsilon_{ij} \mid \sigma_\epsilon^2 \sim N(0, \sigma_\epsilon^2)$$

with b_i and ϵ_{ij} independent. There are now two parameters to allow for between-individual variability, σ_0^2, and within-individual variability, σ_ϵ^2 (with both producing excess-Poisson variability). Unfortunately, there is no simple marginal interpretation of σ_0^2 and σ_ϵ^2 since

$$E[Y_{ij}] = \mu_{ij} = T_{ij}\exp(\boldsymbol{x}_{ij}\boldsymbol{\beta} + \sigma_0^2/2 + \sigma_\epsilon^2/2)$$
$$\text{var}(Y_{ij}) = \mu_{ij}\{1 + \mu_{ij}[\exp(\sigma_0^2) - 1][\exp(\sigma_\epsilon^2) - 1]\}$$
$$\text{cov}(Y_{ij}, Y_{ik}) = T_{ij}T_{ik}\exp[(\boldsymbol{x}_{ij} + \boldsymbol{x}_{ik})\boldsymbol{\beta}]\exp(\sigma_0^2)[\exp(\sigma_0^2) - 1].$$

The expression for the marginal covariance shows that σ_0^2 is controlling the within-individual dependence in the model, with large values giving high dependence. The expression for the marginal variance is quadratic in the mean and is controlled by both σ_0^2 and σ_ϵ^2, with large values corresponding to greater excess-Poisson variability. We assign independent priors $\sigma_0^{-2} \sim$ Ga(1, 0.260), $\sigma_\epsilon^{-2} \sim$ Ga(1, 0.260).

All three models were implemented using MCMC. Table 9.5 gives summaries for the three models. Model 1 gives very similar inference to the likelihood approach described in Sect. 9.4 (specifically, the result presented in Table 9.4), which is not surprising given the relatively large sample size and weak priors. Model 2 shows little sensitivity to the prior distribution on σ_0 which is again not surprising given

Table 9.5 Posterior means and standard deviations for Bayesian analyses of the seizure data

	Model 1		Model 2		Model 3	
	Estimate	Std. err.	Estimate	Std. err.	Estimate	Std. err.
β_0	1.03	0.16	1.04	0.16	1.04	0.18
β_1	−0.036	0.21	−0.030	0.22	0.062	0.25
β_2	0.11	0.047	0.11	0.047	0.0064	0.10
β_3	−0.10	0.065	−0.10	0.065	−0.29	0.14
σ_0	0.80	0.078	0.81	0.077	0.82	0.084
σ_ϵ	–	–	–	–	0.39	0.033

See the caption of Table 9.4 for details on parameter interpretation. Models 1 and 2 are standard GLMMs and differ only in the priors placed on σ_0 which is the standard deviation of the random intercepts. Model 3 adds an additional measurement error random effect, with standard deviation σ_ϵ

the number of individuals. Model 3 shows substantive differences, however. The parameter of interest β_3 is now greatly reduced, with a 95% credible interval for the rate being [0.56,0.99]. The reason for the change is that in the progabide group, there is a single individual (as seen in Fig. 9.2) who is very influential; this individual has counts of 151, 102, 65, 72 and 63 in the five time periods. The introduction of measurement error accommodates this individual. The posterior medians of ϵ_{ij} for this individual show a negative error term at baseline, followed by a run of positive terms post-baseline: −0.61, 0.61, 0.17, 0.27, 0.14. The difference in signs explains why the between-individual random effect cannot accommodate this individual's data. Notice also that β_2 (the log ratio of seizure rates in the post-baseline period relative to the baseline period, for typical individuals in the placebo group) is now close to zero, whereas in models 1 and 2, it is 0.11. This shows that the aberrant individual's measurements were responsible for the high value of β_2 in the first two models. The estimate for σ_ϵ is less than half the estimate for σ_0 so that between-individual variability is greater than within-individual variability for these data.

In analyses presented in Diggle et al. (2002), the influential individual was dropped, and in their Table 9.7, the single random effect analysis produced an estimate (standard error) of −0.30 (0.070), which is very similar to that for model 3. We would always prefer to not remove individuals from the analysis, however, unless there are substantive reasons to do so.

Another possibility for modeling excess-Poisson variability, by combining the Poisson likelihood with a gamma random effects distribution, is considered in Sect. 9.8. □

In the last example we saw that the introduction of normal random effects accounted for both measurement error and between-individual variability. This flexibility is a great benefit of the GLMM framework. One way of approaching modeling is to first imagine that the response is continuous and then decide upon a model that would be considered in this case. The same structure can then be assumed for the data at hand but on the linear predictor scale. In the next example, the versatility is further illustrated with a model for spatial dependence.

9.7 Generalized Linear Mixed Models with Spatial Dependence

9.7.1 A Markov Random Field Prior

The topic of modeling residual spatial dependence is vast and here we only scratch the surface and present a model that is popular in the spatial epidemiology literature, and fits within the GLMM framework. We first describe the model and then illustrate its use on the lung cancer and radon data of Sect. 1.3.3.

The following three-stage model was introduced by Besag et al. (1991) in the context of disease mapping:

Stage One: The distribution of the response in area i is

$$Y_i \mid \mu_i, \epsilon_i, S_i \sim_{ind} \text{Poisson}[E_i \mu_i \exp(\epsilon_i + S_i)]$$

with loglinear mean model

$$\log \mu_i = \beta_0 + \beta_i x_i, \tag{9.7}$$

where x_i is the radon level in area i. The random effects ϵ_i and S_i represent error terms without and with spatial structure, respectively. We have already encountered the nonspatial version when a Poisson-Gamma model was described for these data in Chap. 6. There are many models one might envision for the spatial terms S_i, $i = 1, \ldots, m$. An obvious isotropic form would be $\boldsymbol{S} = [S_1, \ldots, S_m]^\text{T} \sim \text{N}_m(\boldsymbol{0}, \sigma_s^2 \boldsymbol{R})$ with $\boldsymbol{R}$ a correlation matrix with $R_{ii'}$ describing the correlation between areas i and i', $i, i' = 1, \ldots, m$. A common form is $R_{ii'} = \rho^{d_{ii'}}$ where $d_{ii'}$ is the distance between the centroids of areas i and i'. We have already seen this form of correlation in the context of longitudinal data; see in particular (8.14).

Marginally, this model gives

$$\text{E}[Y_i] = E_i \mu_i \exp(\sigma_\epsilon^2/2 + \sigma_s^2/2)$$
$$\text{var}(Y_i) = \text{E}[Y_i] \left\{ 1 + \text{E}[Y_i][\exp(\sigma_\epsilon^2) - 1][\exp(\sigma_s^2) - 1)] \right\}$$
$$\text{cov}(Y_i, Y_{i'}) = E_i \mu_i E_{i'} \mu_{i'} \exp(\sigma_s^2)[\exp(\sigma_s^2) - 1].$$

This *isotropic* model is computationally expensive within an MCMC scheme because we need to invert $\boldsymbol{R}$ at each iteration to obtain the conditional distribution. We describe an alternative which is both computationally feasible and statistically appealing.

Stage Two: The random effects distributions are

$$\epsilon_i \mid \sigma_\epsilon^2 \sim_{iid} \mathrm{N}(0, \sigma_\epsilon^2) \tag{9.8}$$

$$S_i \mid S_{i'}, i' \in \mathrm{ne}(i), \sigma_s^2 \sim_{ind} \mathrm{N}\left(\overline{S}_i, \frac{\sigma_s^2}{n_i}\right) \tag{9.9}$$

where $\overline{S}_i = \frac{1}{n_i} \sum_{i' \in \mathrm{ne}(i)} S_{i'}$ is the mean of the "neighbors" of area i, with $\mathrm{ne}(i)$ defining the set of, and n_i the number of, such neighbors. This *intrinsic conditional autoregressive* (ICAR) model is very appealing since it provides local spatial smoothing and may be viewed as providing stochastic interpolation (Besag and Kooperberg 1995). A common definition (which we adopt in the example at the end of this section) is that two areas are neighbors if they share a common boundary. In non-lattice systems, this is clearly ad hoc.

An interesting aspect of this model is that the joint distribution is undefined. The form of the joint "density" is

$$p(\boldsymbol{s} \mid \sigma_s^2) \propto \sigma_s^{-(m-r)} \exp\left[-\frac{1}{2\sigma_s^2} \sum_{i<i'} W_{ii'} (s_i - s_{i'})^2\right], \tag{9.10}$$

where $W_{ii'} = 1$ if areas i and i' are neighbors and $W_{ii'} = 0$ otherwise. In the spatial context, r is the number of connected regions. So if $r = 1$, there are no collection of areas that are not neighbors of the remaining areas, which means that we cannot break the study region into collections of areas that are unconnected. One way of thinking about this model is that it specifies a prior on the differences between levels in different areas but not on the overall level.

There are two equivalent representations of model (9.10) that are commonly used. In one approach, the intercept β_0 is removed from the mean model (9.7), while in the other, we allow an intercept β_0, along with an improper uniform prior for this parameter, and then constrain $\overline{S} = 0$. In the following we assume that the intercept has been excluded from the model. See Besag and Kooperberg (1995) and Rue and Held (2005) for further discussion of this model.

Stage Three: Hyperpriors:

$$\beta_1 \sim \mathrm{N}(\mu_\beta, \Sigma_\beta)$$
$$\sigma_\epsilon^{-2} \sim \mathrm{Gamma}(a_\epsilon, b_\epsilon)$$
$$\sigma_s^{-2} \sim \mathrm{Gamma}(a_s, b_s).$$

9.7.2 Hyperpriors

Picking a prior for σ_s is not straightforward because of its interpretation as the *conditional* standard deviation. In particular, σ_s and σ_ϵ are not directly comparable since the latter has a marginal interpretation (on the log relative risk scale).

We describe how to simulate realizations from (9.10) to examine candidate prior distributions. As already noted, due to the rank deficiency, (9.10) does not define a probability density, and so we cannot directly simulate from this prior. We need to define some new notation in order to describe the method of simulation. The model can be written in the form

$$p(\boldsymbol{s} \mid \sigma_s^2) = (2\pi)^{-(m-r)/2} |\boldsymbol{Q}^\star|^{1/2} \sigma_s^{-(m-r)} \exp\left(-\frac{1}{2\sigma_s^2} \boldsymbol{s}^\mathrm{T} \boldsymbol{Q} \boldsymbol{s}\right) \quad (9.11)$$

where $\boldsymbol{s} = [s_1, \ldots, s_m]$ is the collection of random effects, $\boldsymbol{Q}$ is a (scaled) "precision" matrix of rank $m - r$, with

$$Q_{ij} = \sigma_s^{-2} \begin{cases} n_i & \text{if } i = j \\ -1 & \text{if } i \text{ and } j \text{ are neighbors} \\ 0 & \text{otherwise} \end{cases}$$

and $|\boldsymbol{Q}^\star|$ is a generalized determinant which is the product over the $m - r$ nonzero eigenvalues of $\boldsymbol{Q}$.

Rue and Held (2005) give the following algorithm for generating samples from (9.11):

1. Simulate $z_j \sim \mathrm{N}(0, \lambda_j^{-1})$, for $j = m-r+1, \ldots, m$, where λ_j are the eigenvalues of $\boldsymbol{Q}$ (recall there are $m - r$ nonzero eigenvalues as $\boldsymbol{Q}$ has rank $m - r$).
2. Return $\boldsymbol{s} = z_{m-r+1} \boldsymbol{e}_{n-r+1} + z_3 \boldsymbol{e}_3 + \ldots + z_n \boldsymbol{e}_m = \boldsymbol{E} \boldsymbol{z}$ where $\boldsymbol{e}_j$ are the corresponding eigenvectors of $\boldsymbol{Q}$, $\boldsymbol{E}$ is the $m \times (m - r)$ matrix with these eigenvectors as columns, and $\boldsymbol{z}$ is the $(m - r) \times 1$ vector containing z_j, $j = m - r + 1, \ldots, m$.

The simulation algorithm is conditioned so that samples are zero in the null-space of $\boldsymbol{Q}$. If $\boldsymbol{s}$ is a sample and the null-space is spanned by $\boldsymbol{v}_1$ and $\boldsymbol{v}_2$, then $\boldsymbol{s}^\mathrm{T} \boldsymbol{v}_1 = \boldsymbol{s}^\mathrm{T} \boldsymbol{v}_2 = \boldsymbol{0}$. For example, suppose $\boldsymbol{Q} \boldsymbol{1} = \boldsymbol{0}$ so that the null-space is spanned by $\boldsymbol{1}$ and the rank deficiency is 1. Then $\boldsymbol{Q}$ is of rank $m - 1$, since the eigenvalue corresponding to $\boldsymbol{1}$ is zero, and samples $\boldsymbol{s}$ produced by the algorithm are such that $\boldsymbol{s}^\mathrm{T} \boldsymbol{1} = 0$. It is also useful to note that if we wish to compute the marginal variances, only then simulation is not required, as they are available as the diagonal elements of the matrix $\sum_j \lambda_j^{-1} \boldsymbol{e}_j \boldsymbol{e}_j^T$.

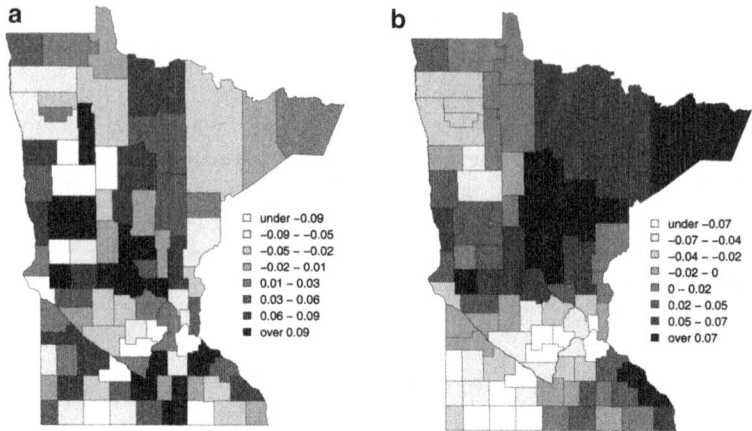

Fig. 9.5 (a) Nonspatial and (b) spatial random effects for the Minnesota lung cancer data

Example: Lung Cancer and Radon

We apply the Poisson model with nonspatial and spatial normal random effects, that is, the model given by (9.8) and (9.9). We note that model (9.7) does not aggregate correctly from a plausible individual-level model; see Wakefield (2007b) and the discussion leading to model (6.19). The prior on β_1 is $N(0, 1.17^2)$ which gives a 95% interval for the relative risk of [0.1,10].

The priors on σ_ϵ^2 and σ_s^2 require more care, but we would like to specify priors in such a way that the nonspatial and spatial contributions are approximately equal. This is complicated by σ_s^2 having a conditional interpretation, as just discussed. We specify gamma priors for each of the precisions, σ_ϵ^{-2} and σ_s^{-2}. To make the priors compatible, we first specify a prior for σ_ϵ^{-2} and evaluate the average of the marginal variances over the 87 areas, when $\sigma_s^2 = 1$, as described at the end of Sect. 9.7.2. We then match up the means of the gamma distributions. Following the development of Sect. 9.6.2 for the unstructured variability, we assume that the unstructured residual relative risks lie in the interval [0.2, 5] with probability 0.95 and assume $d = 2$ to give the exponential prior distribution Ga(1,0.140) for σ_ϵ^{-2}. The average of the marginal variances over the study region for the spatial random effects is 0.21; hence, the average of the marginal precisions is approximately 1/0.21. The prior for σ_s^{-2} is therefore Ga(0.21,0.140), to give $E[\sigma_s^{-2}] = 0.21 \times E[\sigma_\epsilon^{-2}]$.

The fitting of this model (using INLA) results in the posterior mean estimates $\widehat{\epsilon}_i$ and $\widehat{S}_i$ mapped in Fig. 9.5(a) and (b) respectively. Notice that the scale is narrower in panel (b), since the spatial contribution to the residuals is relatively small here, though the spatial pattern in these residuals is apparent. As we discussed with respect to prior specification, the variances σ_ϵ^2 and σ_s^2 are not directly comparable,

9.7 Generalized Linear Mixed Models with Spatial Dependence

Table 9.6 Parameter estimates for β_1, the area-level log relative risk corresponding to radon, and measures of uncertainty (standard errors and posterior standard deviations) under various models, for the Minnesota lung cancer data

Model	Estimate ($\times 10^2$)	Uncertainty ($\times 10^3$)
Poisson	−3.6	5.4
Quasi-likelihood	−3.6	8.8
Negative binomial	−2.9	8.2
Nonspatial random effects	−2.8	9.1
Nonspatial and ICAR random effects	−2.8	9.7

and so we calculate an approximate proportion of the total residual variance that is spatial by comparing σ_ϵ^2 with an empirical estimate of the marginal variance of the collection of random effects $\{\widehat{S}_i, i = 1, \ldots, m\}$. Specifically, we calculate

$$\frac{\text{var}(\widehat{S}_i)}{\text{var}(\widehat{S}_i) + \widehat{\sigma}_\epsilon^2}$$

where $\text{var}(\widehat{S}_i)$ is the empirical variance of the random effects and $\widehat{\sigma}_\epsilon^2$ is the posterior median. From this calculation, the fraction of the total residual variability that is attributed to the spatial component is 0.13.

Table 9.6 provides estimates and standard error/posterior standard deviations for the log relative risk associated with a unit increase in radon, for a variety of models. We include a model with nonspatial normal random effects only. The Poisson and quasi-likelihood methods assume the same form of (proportional) mean–variance relationship, while the negative binomial and nonspatial normal random effects approaches imply a variance that is quadratic in the mean. The marginal variance does not exist under the improper spatial model, but here the spatial contributions are small. We might therefore expect to see similar conclusions to the negative binomial and nonspatial normal random effects models. This is borne out in the table, with the last three models giving similar estimates that are closer to zero than the first two models. The standard error from the spatial model does increase a little over the nonspatial random effects model.

In general, if strong spatial effects are present and the exposure surface has spatial structure, then when spatial random effects are added to a model, large changes may be seen in the regression coefficient associated with exposure. This phenomenon, which is sometimes known as *confounding by location*, is a big practical headache since it is difficult to decide on whether to attribute spatial variability in risk to the exposure or to the spatial random effects (which may be acting as surrogates for unmeasured confounders). Wakefield (2007b) and Hodges and Reich (2010) provide further discussion.

9.8 Conjugate Random Effects Models

An obvious approach to extending models for independent data is to assume a random effects distribution that is conjugate to the likelihood. We illustrate this approach, and its shortcomings, through two examples.

Example: Lung Cancer and Radon

A Poisson-Gamma conjugate model was fitted to the lung cancer/radon data in Sect. 6.9 with:

Stage One: $Y_i \mid \mu_i, \delta_i \sim_{ind} \text{Poisson}(E_i \delta_i)$, with $\log \mu_i = \beta_0 + \beta_1 x_i$, for $i = 1, \ldots, m$.

Stage Two: $\delta_i \mid b \sim_{iid} \text{Gamma}(b, b)$ for $i = 1, \ldots, m$.

The advantage of this model is that the random effects can be analytically integrated from the model to give $Y_i \mid \mu_i, b \sim_{ind} \text{NegBin}(\mu_i, b)$, $i = 1, \ldots, m$. However, the extension to allow spatial dependence is not obvious, unless one introduces normal random effects, as in the last section.

Example: Seizure Data

Letting $\mu_{ij} = T_{ij} \exp(x_{ij} \beta)$, consider the two-stage model:

$$Y_{ij} \mid \mu_{ij}, \xi_{ij} \sim_{ind} \text{Poisson}(\mu_{ij} \xi_{ij})$$

$$\xi_{ij} \mid b \sim_{iid} \text{Ga}(b, b).$$

This results in $Y_{ij} \mid \mu_i, b \sim_{ind} \text{NegBin}(\mu_i, b)$ with $\text{E}[Y_i] = \mu_{ij}$ and $\text{var}(Y_{ij}) = \mu_{ij}(1 + \mu_{ij}/b)$. This model allows for excess-Poisson variability but not for dependence of observations on the same patient. The introduction of patient-specific random effects allows for the latter but loses the analytical tractability. Specifically, the two-stage model

$$Y_{ij} \mid \mu_{ij}, \delta_i \sim_{ind} \text{Poisson}(\mu_{ij} \delta_i)$$

$$\delta_i \mid b \sim_{iid} \text{Ga}(b, b)$$

leads to a marginal model for the data of the ith individual of

$$\Pr(y_{i0}, \ldots, y_{i4} \mid \mu_{ij}, b) = \left(\prod_{j=0}^{4} \frac{\mu_{ij}^{y_{ij}}}{y_{ij}!} \right) \frac{b^b}{\Gamma(b)} \frac{\Gamma(b + y_{i+})}{(b + \mu_{i+})^{b+y_{i+}}},$$

which is not of negative binomial form.

9.9 Generalized Estimating Equations for Generalized Linear Models

The GEE approach was described in Sect. 8.7 for linear models. The extension to GLM mean models is conceptually straightforward, since all that is required is specification of a mean model and a working covariance model. The mean is

$$g(\mu_{ij}) = \boldsymbol{x}\boldsymbol{\gamma}$$

where $\mu_{ij} = \mathrm{E}[Y_{ij}]$, $g(\cdot)$ is a link function, $\boldsymbol{x}$ is a $n \times (k+1)$ design matrix, and $\boldsymbol{\gamma}$ is a $(k+1) \times 1$ vector. We use $\boldsymbol{\gamma}$ to denote the parameters in the *marginal* mean model to distinguish them from the parameters $\boldsymbol{\beta}$ which have been used to represent the mixed model *conditional* parameters. The working covariance matrix is

$$\mathrm{var}(\boldsymbol{Y}) = \boldsymbol{W}.$$

and in a GLM setting, $\boldsymbol{W}$ will usually depend on $\boldsymbol{\gamma}$ and on additional parameters $\boldsymbol{\alpha}$ so that $\boldsymbol{W} = \boldsymbol{W}(\boldsymbol{\gamma}, \boldsymbol{\alpha})$. Suppose $\widehat{\boldsymbol{\alpha}}$ is a consistent estimator of $\boldsymbol{\alpha}$. Then, GEE takes the estimator $\widehat{\boldsymbol{\gamma}}$ that satisfies

$$\boldsymbol{G}(\widehat{\boldsymbol{\gamma}}, \widehat{\boldsymbol{\alpha}}) = \sum_{i=1}^{m} \boldsymbol{D}_i^\mathrm{T} \boldsymbol{W}_i^{-1} (\boldsymbol{Y}_i - \widehat{\boldsymbol{\mu}}_i) = \boldsymbol{0},$$

where $\boldsymbol{D}_i = \partial \boldsymbol{\mu}_i / \partial \boldsymbol{\gamma}$ is $n_i \times (k+1)$ and $\boldsymbol{W}_i = \boldsymbol{W}_i(\boldsymbol{\gamma}, \widehat{\boldsymbol{\alpha}})$ is the $n_i \times n_i$ working covariance model for unit i, $i = 1, \ldots, m$. The estimator $\widehat{\boldsymbol{\gamma}}$ will not be of closed form, unless the link is linear. Under mild regularity conditions,

$$\boldsymbol{V}_\gamma^{-1/2}(\widehat{\boldsymbol{\gamma}} - \boldsymbol{\gamma}) \to_d \mathrm{N}_{k+1}(\boldsymbol{0}, \boldsymbol{I}_{k+1}),$$

where $\boldsymbol{V}_\gamma$ takes the sandwich form

$$\left(\sum_{i=1}^{m} \boldsymbol{D}_i^\mathrm{T} \boldsymbol{W}_i^{-1} \boldsymbol{D}_i \right)^{-1} \left[\sum_{i=1}^{m} \boldsymbol{D}_i^\mathrm{T} \boldsymbol{W}_i^{-1} \mathrm{cov}(\boldsymbol{Y}_i) \boldsymbol{W}_i^{-1} \boldsymbol{D}_i \right] \left(\sum_{i=1}^{m} \boldsymbol{D}_i^\mathrm{T} \boldsymbol{W}_i^{-1} \boldsymbol{D}_i \right)^{-1}.$$

(9.12)

In practice, an empirical estimator of $\mathrm{cov}(\boldsymbol{Y}_i)$ is substituted to obtain $\widehat{\boldsymbol{V}}_\gamma$. This produces a consistent estimator of the standard error of $\widehat{\boldsymbol{\gamma}}$, so long as we have independence between units $i \neq i'$, $i, i' = 1, \ldots, m$. For small m, the variance estimator may be unstable, however.

As in the linear case, various assumptions about the form of the working covariance are available. We write

$$\boldsymbol{W}_i = \boldsymbol{\Delta}_i^{1/2} \boldsymbol{R}_i(\boldsymbol{\alpha}) \boldsymbol{\Delta}_i^{1/2},$$

where $\boldsymbol{\Delta}_i = \text{diag}[\text{var}(Y_{i1}), \ldots, \text{var}(Y_{in_i})]^\text{T}$ and $\boldsymbol{R}_i$ is a working correlation model. Common choices include independence, exchangeable, AR(1), and unstructured. For discrete data, there is often no natural choice since, in this setting, the correlation is not an intuitive measure of dependence.

For small m, the sandwich estimator will have high variability, and so model-based variance estimators may be preferable (and we would probably not rely on asymptotic normality if m were small anyway). Model-based estimators are more efficient if the model is correct and efficiency will be improved if we can pick a working correlation matrix that is close to the true structure.

Published comments on whether to assume working independence or a more complex form are a little in conflict: Liang and Zeger (1986) state that there is "little difference when correlation is moderate," in agreement with McDonald (1993) who states "the independence estimator may be recommended for practical purposes." On the other hand, Zhao et al. (1992) assert that assuming independence "can lead to important losses of efficiency," in line with Fitzmaurice et al. (1993) who state that it is "important to obtain a close approximation to $\text{cov}(\boldsymbol{Y}_i)$ in order to achieve high efficiency." The issue is complex since it depends on, among other things, the design and whether the covariates corresponding to the parameters are constant within an individual or not.

9.10 GEE2: Connected Estimating Equations

In an approach coined by Liang et al. (1992) as GEE2, there is a *connected* set of joint estimating equations for $\boldsymbol{\gamma}$ and $\boldsymbol{\alpha}$. This approach is particularly appealing if the variance–covariance model is of interest. To motivate a pair of estimating equations, consider the following model for a single individual with n *independent* observations:

$$Y_i \mid \boldsymbol{\gamma}, \boldsymbol{\alpha} \sim_{ind} \text{N}\left[\mu_i(\boldsymbol{\gamma}), \Sigma_i(\boldsymbol{\gamma}, \boldsymbol{\alpha})\right].$$

For example, we may have $\Sigma_i(\boldsymbol{\gamma}, \boldsymbol{\alpha}) = \alpha \mu_i(\boldsymbol{\gamma})^2, i = 1, \ldots, n$. The log-likelihood is

$$l(\boldsymbol{\gamma}, \boldsymbol{\alpha}) = -\frac{1}{2} \sum_{i=1}^{n} \log(\Sigma_i) - \frac{1}{2} \sum_{i=1}^{n} \frac{(Y_i - \mu_i)^2}{\Sigma_i}.$$

Differentiation gives the score equations as

$$\begin{aligned}
\frac{\partial l}{\partial \boldsymbol{\gamma}} &= -\frac{1}{2} \sum_{i=1}^{n} \left(\frac{\partial \Sigma_i}{\partial \boldsymbol{\gamma}}\right)^\text{T} \frac{1}{\Sigma_i} + \sum_{i=1}^{n} \left(\frac{\partial \mu_i}{\partial \boldsymbol{\gamma}}\right)^\text{T} \frac{(Y_i - \mu_i)}{\Sigma_i} + \frac{1}{2} \sum_{i=1}^{n} \left(\frac{\partial \Sigma_i}{\partial \boldsymbol{\gamma}}\right)^\text{T} \frac{(Y_i - \mu_i)^2}{\Sigma_i^2} \\
&= \sum_{i=1}^{n} \left(\frac{\partial \mu_i}{\partial \boldsymbol{\gamma}}\right)^\text{T} \frac{(Y_i - \mu_i)}{\Sigma_i} + \sum_{i=1}^{n} \left(\frac{\partial \Sigma_i}{\partial \boldsymbol{\gamma}}\right)^\text{T} \frac{[(Y_i - \mu_i)^2 - \Sigma_i]}{2\Sigma_i^2} \quad (9.13)
\end{aligned}$$

9.10 GEE2: Connected Estimating Equations

and

$$\frac{\partial l}{\partial \alpha} = -\frac{1}{2}\sum_{i=1}^{n}\left(\frac{\partial \Sigma_i}{\partial \alpha}\right)^{\mathrm{T}}\frac{1}{\Sigma_i} + \frac{1}{2}\sum_{i=1}^{n}\left(\frac{\partial \Sigma_i}{\partial \alpha}\right)^{\mathrm{T}}\frac{(Y_i - \mu_i)^2}{\Sigma_i^2}$$

$$= \sum_{i=1}^{n}\left(\frac{\partial \Sigma_i}{\partial \alpha}\right)^{\mathrm{T}}\frac{[(Y_i - \mu_i)^2 - \Sigma_i]}{2\Sigma_i^2}. \quad (9.14)$$

This pair of quadratic estimating functions is unbiased given correct specification of the first two moments; to emphasize, normality of the data is not required. A disadvantage of the use of these functions, compared to the original GEE method (which is sometimes referred to as GEE1), is that if the variance model is wrong, we are no longer guaranteed a consistent estimator of γ. If the model is correct, however, there will be a gain in efficiency.

Let

$$S_i = (Y_i - \mu_i)^2$$

with $\mathrm{E}[S_i] = \Sigma_i$. Under normality,

$$\mathrm{var}(S_i) = \mathrm{E}[S_i^2] - \mathrm{E}[S_i]^2 = 3\Sigma_i^2 - \Sigma_i^2 = 2\Sigma_i^2$$

Hence, (9.13) and (9.14) can be written

$$\frac{\partial l}{\partial \gamma} = \sum_{i=1}^{n} D_i^{\mathrm{T}} V_i^{-1}(Y_i - \mu_i) + \sum_{i=1}^{n} E_i W_i^{-1}(S_i - \Sigma_i) \quad (9.15)$$

$$\frac{\partial l}{\partial \alpha} = \sum_{i=1}^{n} F_i W_i^{-1}(S_i - \Sigma_i) \quad (9.16)$$

where $D_i = \partial \mu_i/\partial \beta$, $E_i = \partial \Sigma_i/\partial \beta$, $F_i = \partial \Sigma_i/\partial \alpha$, $V_i = \Sigma_i$, and $W_i = 2\Sigma_i^2$. This pair of estimating equations can be compared with the usual estimating equation specification

$$\frac{\partial l}{\partial \beta} = \sum_{i=1}^{n} D_i^{\mathrm{T}} V_i^{-1}(Y_i - \mu_i).$$

The additional term is the information about γ in the variance.

We turn to the dependent data situation and let $\boldsymbol{\mu}_i$ denote the $n_i \times 1$ mean vector and $\boldsymbol{\Sigma}_i$ the $n_i \times n_i$ covariance matrix. The general form of estimating equations is

$$\sum_{i=1}^{m}\begin{bmatrix} D_i & 0 \\ E_i & F_i \end{bmatrix}^{\mathrm{T}}\begin{bmatrix} V_i & C_i \\ C_i^{\mathrm{T}} & W_i \end{bmatrix}^{-1}\begin{bmatrix} Y_i - \mu_i \\ S_i - \Sigma_i \end{bmatrix} = \begin{bmatrix} 0 \\ 0 \end{bmatrix}$$

where $D_i = \partial \boldsymbol{\mu}_i/\partial \boldsymbol{\beta}$, $E_i = \partial \boldsymbol{\Sigma}_i/\partial \boldsymbol{\beta}$, and $F_i = \partial \boldsymbol{\Sigma}_i/\partial \boldsymbol{\alpha}$ and we have "working" variance–covariance structure

$$V_i = \text{var}(Y_i)$$
$$C_i = \text{cov}(Y_i, S_i)$$
$$W_i = \text{var}(S_i).$$

When $C_i = 0$, we obtain

$$G_1(\boldsymbol{\gamma}, \boldsymbol{\alpha}) = \sum_{i=1}^{m} D_i^{\text{T}} V_i^{-1}(Y_i - \boldsymbol{\mu}_i) + \sum_{i=1}^{m} E_i W_i^{-1}(S_i - \boldsymbol{\Sigma}_i) \quad (9.17)$$

$$G_2(\boldsymbol{\gamma}, \boldsymbol{\alpha}) = \sum_{i=1}^{m} F_i W_i^{-1}(S_i - \boldsymbol{\Sigma}_i) \quad (9.18)$$

which are the dependent data version of the normal score equations we obtained earlier, that is, (9.15) and (9.16). In the dependent data pair of equations, we have freedom in choosing V_i and W_i. In particular, the latter need not be chosen to coincide with that under a multivariate normal model, and, since this choice is difficult, we could instead choose working independence.

It can be shown (Prentice and Zhao 1991, Appendix 2) that (9.17) and (9.18) arise from the *quadratic exponential model*

$$p(\boldsymbol{y}_i \mid \boldsymbol{\theta}_i, \boldsymbol{\lambda}_i) = \Delta_i^{-1} \exp[\boldsymbol{y}_i^{\text{T}} \boldsymbol{\theta}_i + \boldsymbol{w}_i^{\text{T}} \boldsymbol{\lambda}_i + c_i(\boldsymbol{y}_i)], \quad (9.19)$$

where $\boldsymbol{\theta}_i = [\theta_{i1}, \ldots, \theta_{in_i}]^{\text{T}}$ is the canonical parameter,

$$\boldsymbol{w}_i = [y_{i1}^2, y_{i1}y_{i2}, \ldots, y_{i2}^2, y_{i2}y_{i3}, \ldots]^{\text{T}}$$

is the vector of squared responses, $c_i(\cdot)$ is a function that defines the "shape," $\Delta_i = \Delta_i(\boldsymbol{\theta}_i, \boldsymbol{\lambda}_i, c_i)$ is a normalization constant, and $\boldsymbol{\lambda}_i = [\lambda_{i11}, \lambda_{i12}, \ldots, \lambda_{i22}, \lambda_{i23}, \ldots]^{\text{T}}$. As an example of this form, if all the responses are continuous on the whole real line and $c_i = 0$, the multivariate normal is recovered (Exercise 9.2). Gourieroux et al. (1984) showed that the quadratic exponential family is unique in giving consistent estimates of the mean and covariance parameters, even in the situation in which the data actually arise from outside this family. So, as the exponential family produces desirable consistency properties for mean parameters, the quadratic exponential family has the same properties when mean and variance parameters are of interest.

To emphasize: For consistency of $\widehat{\boldsymbol{\gamma}}$, we require the models for both Y_i and S_i to be correct, and there is increased efficiency over the single estimating equation version (GEE1) if this is the case. This approach is useful if the variance–covariance parameters are of primary interest as, for example, in some breeding and genetic applications. Otherwise, if can, be prudent to stick with GEE1.

9.11 Interpretation of Marginal and Conditional Regression Coefficients

To illustrate the differences in interpretation of marginal and conditional coefficients, we examine the meaning of parameters for a loglinear model. In a marginal model, such as is considered under GEE, we have

$$E[Y \mid x] = \exp(\gamma_0 + \gamma_1 x),$$

in which case $\exp(\gamma_1)$ is the multiplicative change in the average response over two populations of individuals whose x values differ by one unit. Under the conditional mixed model, the interpretation of regression coefficients is conditional on the value of the random effect. For the model

$$E[Y \mid x, b] = \exp(\beta_0 + \beta_1 x + b),$$

with $b \mid \sigma_0^2 \sim_{iid} N(0, \sigma^2)$, $\exp(\beta_1)$ is therefore the change in the expected response for two individuals with identical random effects. Sometimes, the comparison is described as between two *typical* (i.e., $b = 0$) individuals who differ in x by one unit. The marginal mean corresponding to this model follows from the variance of a lognormal distribution:

$$E[Y \mid x] = E_b[E(Y \mid x, b)] = \exp(\beta_0 + \sigma^2/2 + \beta_1 x).$$

Therefore, for the random intercepts, loglinear model $\exp(\beta_1)$ has the same marginal interpretation to $\exp(\gamma_1)$ and the marginal intercept is $\gamma_0 = \beta_0 + \sigma^2/2$.

We now consider the random intercepts and slopes model

$$E[Y \mid x, \boldsymbol{b}] = \exp\left[(\beta_0 + b_0) + (\beta_1 + b_1)x\right]$$

where $\boldsymbol{b} = [b_0, b_1]$ and

$$\begin{bmatrix} b_0 \\ b_1 \end{bmatrix} \sim N_2 \left(\begin{bmatrix} 0 \\ 0 \end{bmatrix}, \begin{bmatrix} D_{00} & D_{01} \\ D_{10} & D_{11} \end{bmatrix} \right).$$

In this model $\exp(\beta_1)$ is the relative risk between two individuals with the same $\boldsymbol{b}$ but with x values that differ by one unit. That is,

$$\exp(\beta_1) = \frac{E[Y \mid x, \boldsymbol{b}]}{E[Y \mid x - 1, \boldsymbol{b}]}.$$

An alternative interpretation is to say that it is the expected change between two "typical individuals," that is, individuals with specific values of the random effects, $\boldsymbol{b} = \boldsymbol{0}$. Under this model, the marginal mean is

$$E[Y \mid x] = \exp\left[\beta_0 + D_{00}/2 + x(\beta_1 + D_{01}) + x^2 D_{11}/2\right]$$

so that a quadratic loglinear marginal model has been induced by the conditional formulation. The marginal *median* is $\exp(\beta_0 + \beta_1 x)$ so that $\exp(\beta_1)$ is the ratio of median responses between two populations whose x values differ by one unit. There is no such simple interpretation in terms of marginal means.

Hence, marginal inference is possible under a mixed model formulation, though care must be taken to derive the exact form of the marginal model. Estimation of marginal parameters via GEE produces a consistent estimator in more general circumstances than mixed model estimation, though there is an efficiency loss if the random effects model is correct.

Example: Seizure Data

The marginal mean version of the conditional model fitted previously in this chapter is
$$\mathrm{E}[Y_{ij}] = T_{ij} \exp(\gamma_0 + \gamma_1 x_{i1} + \gamma_2 x_{ij2} + \gamma_3 x_{i1} x_{ij2}).$$

The parameters are interpreted as follows:

- $\exp(\gamma_0)$ is the expected rate of seizures in the placebo group during the baseline period, $j = 0$ (this expectation is over the hypothetical population of individuals who were assigned to the placebo group).
- $\exp(\gamma_1)$ is the ratio of the expected seizure rate in the progabide group, compared to the placebo group, during the baseline period.
- $\exp(\gamma_2)$ is the ratio of the expected seizure rate post-baseline as compared to baseline, in the placebo group.
- $\exp(\gamma_3)$ is the ratio of the expected seizure rates in the progabide group in the post-baseline period, as compared to the placebo group, in the same period. Hence, $\exp(\gamma_3)$ is a period by treatment effect and is the parameter of interest.

The loglinear mean model suggests the variance model $\mathrm{var}(Y_{ij}) = \alpha_1 \mu_{ij}$. We consider various forms for the working correlation. Table 9.7 gives estimates and standard errors under various models. The Poisson, quasi-likelihood, and working independence GEE models have estimating equation

$$\boldsymbol{G}(\widehat{\boldsymbol{\gamma}}, \widehat{\boldsymbol{\alpha}}) = \sum_{i=1}^{m} \boldsymbol{x}_i^{\mathrm{T}}(\boldsymbol{Y}_i - \widehat{\boldsymbol{\mu}}_i) = \boldsymbol{0}.$$

Consequently, the point estimates coincide but the models differ in the manner by which the standard errors are calculated. The Poisson standard errors are clearly much too small. The coincidence of the estimates and standard errors for independence and exchangeable working correlations is a consequence of the balanced design. The quasi-likelihood standard errors are increased by $\sqrt{19.7} = 4.4$ (in line with the empirical estimates in Table 9.2) but do not acknowledge dependence of observations on the same individual (so estimation is carried out

Table 9.7 Parameter estimates and standard errors under various models for the seizure data

	Estimates and standard errors									
	Poisson		Quasi-Lhd		GEE independence		GEE exchangeable		GEE AR(1)	
γ_0	1.35	0.034	1.35	0.15	1.35	0.16	1.35	0.16	1.31	0.16
γ_1	0.027	0.047	0.027	0.21	0.027	0.22	0.027	0.22	0.015	0.21
γ_2	0.11	0.047	0.11	0.21	0.11	0.12	0.11	0.12	0.16	0.11
γ_3	−0.10	0.065	−0.10	0.29	−0.10	0.22	−0.10	0.22	−0.13	0.27
α_1, α_2	1.0	0	19.7	0	19.4	0	19.4	0.78	20.0	0.89

Parameter meaning: γ_0 is the log baseline seizure rate in the placebo group; γ_1 is the log of the ratio of seizure rates between the progabide and placebo groups, at baseline; γ_2 is the log of the ratio of seizure rates in the post-baseline and baseline placebo groups; γ_3 is the log of the ratio of the seizure rate in the progabide group as compared to the placebo group, post-baseline; α_1 and α_2 are variance and correlation parameters, respectively

as if we have 59×5 independent observations). The standard errors of estimated parameters that are associated with time-varying covariates (γ_2 and γ_3) are reduced under GEE, since within-person comparisons are being made and a longitudinal design can be very efficient in such a study, if there is strong within-individual dependence (as discussed in Sect. 8.3). In none of the analyses would the treatment effect of interest be judged significantly different from zero, under conventional levels.

9.12 Introduction to Modeling Dependent Binary Data

Binary outcomes are the simplest form of data but are, ironically, one of the most challenging to model. For a single binary variable Y all moments are determined by $p = \mathrm{E}[Y]$. Specifically, $\mathrm{E}[Y^r] = p$ for $r \geq 1$, so that Bernoulli random variables cannot be overdispersed. Before turning to observations on multiple units, we initially adopt a simplified notation and consider n binary observations $\boldsymbol{Y} = [Y_1, \ldots, Y_n]^\mathrm{T}$. Under conditional independence and with probabilities $p_j = \mathrm{E}[Y_j]$,

$$\Pr(\boldsymbol{Y} = \boldsymbol{y} \mid \boldsymbol{p}) = \prod_{j=1}^n p_j^{y_j}(1-p_j)^{1-y_j},$$

with $\boldsymbol{p} = [p_1, \ldots, p_n]^\mathrm{T}$. In Chap. 7, we saw that a common mean form is the logistic regression model with $\log[p_j/(1-p_j)] = \boldsymbol{x}_j\boldsymbol{\beta}$. In this chapter we wish to formulate models that allow for dependence between binary outcomes, with a starting point being the specification of a multivariate binary distribution. Such a joint distribution can be used with a likelihood-based approach, or one can use the first one or two moments only within a GEE approach. The difficulty with multivariate binary data is that there is no natural way to characterize dependence between pairs, triples, etc., of binary responses. In the dependent binary data situation, we will show that correlation parameters are tied to the means, making estimation from a model based on means and correlations unattractive.

To specify the joint distribution of n binary responses requires 2^n probabilities so that the saturated model has $2^n - 1$ parameters. This may be contrasted with a saturated multivariate normal model which has n means, n variances, and $n(n-1)/2$ correlations. As n becomes large, the number of parameters in the binary saturated model is very large. With $n = 10$, for example, there are $2^{10} - 1 = 1,023$ parameters in the binary model as compared to 65 in the normal model. Our aim is to reduce the $2^n - 1$ distinct probabilities to give formulations that allow both a parsimonious description and the interpretable specification of a regression model.

We begin our description of models for multivariate binary data in Sect. 9.13 with a discussion of mixed models, with likelihood, Bayesian and conditional likelihood approaches to inference. Next, in Sect. 9.14, marginal models are described.

9.13 Mixed Models for Binary Data

9.13.1 Generalized Linear Mixed Models for Binary Data

In Sect. 7.5, we discussed a beta-binomial model for overdispersed data. This form is not very flexible, for the reasons described in Sect. 9.8, and so we describe an alternative mixed model with normal random effects. Let Y_{ij} be the binary "success" indicator with $j = 1, \ldots, n_i$ trials on each of $i = 1, \ldots, m$ units.

Consider the GLMM with logistic link:

Stage One: Likelihood: $Y_{ij} \mid p_{ij} \sim_{ind} \text{Bernoulli}(n_{ij}, p_{ij})$ with the linear logistic model

$$\log\left(\frac{p_{ij}}{1-p_{ij}}\right) = \boldsymbol{x}_{ij}\boldsymbol{\beta} + \boldsymbol{z}_{ij}\boldsymbol{b}_i.$$

In this model, $\boldsymbol{\beta}$ represents a $(k+1) \times 1$ vector of fixed effects and $\boldsymbol{b}_i$ a $(q+1) \times 1$ vector of random effects, with $q \leq k$. Let $\boldsymbol{x}_{ij} = [1, x_{ij1}, \ldots, x_{ijk}]$ be a $(k+1) \times 1$ vector of covariates, so that $\boldsymbol{x}_i = [\boldsymbol{x}_{i1}, \ldots, \boldsymbol{x}_{in_i}]$ is the design matrix for the fixed effects, and let $\boldsymbol{z}_{ij} = [1, z_{ij1}, \ldots, z_{ijq}]^T$ be a $(k+1) \times 1$ vector of variables that are a subset of $\boldsymbol{x}_{ij}$, so that $\boldsymbol{z}_i = [\boldsymbol{z}_{i1}, \ldots, \boldsymbol{z}_{in_i}]^T$ is the design matrix for the random effects.

Stage Two: Random effects distribution: $\boldsymbol{b}_i \mid \boldsymbol{D} \sim_{iid} N_{q+1}(\boldsymbol{0}, \boldsymbol{D})$ for $i = 1, \ldots, m$.

As we have repeatedly stressed, the conditional parameters $\boldsymbol{\beta}$ and the marginal parameters $\boldsymbol{\gamma}$ have different interpretations in nonlinear situations, and for a logistic model, there is no exact analytical relationship between the two. However, we

9.13 Mixed Models for Binary Data

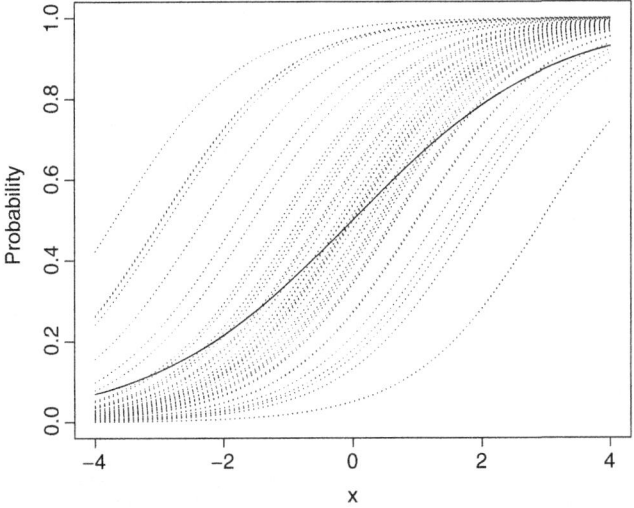

Fig. 9.6 Individual-level curves (*dotted lines*) from a random intercept logistic GLMM, along with marginal curve (*solid line*). The specific model is logit (E[Y | b]) = $\beta_0 + \beta_1 x$, with $\beta_0 = 0, \beta_1 = 1$ and $b \sim_{iid} N(0, 2^2)$. The approximate attenuation factor of the marginal curve, which is given by the denominator of (9.21), is 1.54

may approximate the relationship. For the random intercepts model $b_i \mid \sigma_0^2 \sim_{iid}$ $N(0, \sigma_0^2)$, we have, for a generic Bernoulli response Y with associated random effect b,

$$E[Y] = \frac{\exp(\boldsymbol{x\gamma})}{1 + \exp(\boldsymbol{x\gamma})} = E_b[E(Y \mid b)]$$

$$= E_b\left[\frac{\exp(\boldsymbol{x\beta} + b)}{1 + \exp(\boldsymbol{x\beta} + b)}\right] \approx \frac{\exp(\boldsymbol{x\beta}/[c^2\sigma_0^2 + 1]^{1/2})}{1 + \exp(\boldsymbol{x\beta}/[c^2\sigma_0^2 + 1]^{1/2})} \quad (9.20)$$

where $c = 16\sqrt{3}/(15\pi)$ (Exercise 9.1), so that

$$\gamma \approx \frac{\beta}{[c^2\sigma_0^2 + 1]^{1/2}}. \quad (9.21)$$

Consequently, the marginal coefficients are attenuated toward zero. Figure 9.6 illustrates this phenomena for particular values of $\beta_0, \beta_1, \sigma_0^2$. We observe that the averaging of the conditional curves results in a flattened marginal curve. This attenuation was first encountered in Sect. 7.9 when the lack of collapsibility of the odds ratio was discussed. We emphasize that one should not view the difference in marginal and conditional parameter estimates as bias. If $\sigma_0 > 0$ and $\beta_1 \neq 0$, the parameters will differ, but they are estimating different quantities. In practice, if we fit marginal and conditional models and we do not see attenuation, then the approximation could be poor (e.g., if σ_0^2 is large) or some of the assumptions of the conditional model could be inaccurate.

For the general logistic mixed model

$$\log\left(\frac{\mathrm{E}[Y \mid b]}{1 - \mathrm{E}[Y \mid b]}\right) = x\beta + zb$$

with $b \mid D \sim_{iid} \mathrm{N}_{q+1}(0, D)$, we obtain

$$\mathrm{E}[Y] = \frac{\exp(x\gamma)}{1 + \exp(x\gamma)} \approx \frac{\exp\left(x\beta/\mid c^2 Dzz^\mathrm{T} + I_{q+1} \mid^{(q+1)/2}\right)}{1 + \exp\left(x\beta/\mid c^2 Dzz^\mathrm{T} + I_{q+1} \mid^{(q+1)/2}\right)}$$

so that

$$\gamma \approx \frac{\beta}{\mid c^2 Dzz^\mathrm{T} + I_{q+1} \mid^{(q+1)/2}}.$$

With random slopes or more complicated random effects structures, it is therefore far more difficult to understand the relationship between conditional and marginal parameters.

Marginal inference is possible with mixed models, but one needs to do a little work. Specifically, if one requires marginal inference, then the above approximations may be invoked, or one may directly calculate the required integrals using a Monte Carlo estimate. For example, the marginal probability at x is

$$\widehat{\mathrm{E}}[Y \mid x] = \frac{1}{S} \sum_{s=1}^{S} \frac{\exp(x\widehat{\beta} + b^{(s)})}{1 + \exp(x\widehat{\beta} + b^{(s)})} \tag{9.22}$$

where the random effects are simulated as $b^{(s)} \mid \widehat{D} \sim \mathrm{N}_{q+1}(0, \widehat{D})$, $s = 1, \ldots, S$. A more refined Bayesian approach would replace $\widehat{D}$ by samples from the posterior $p(D \mid y)$.

An important distinction between conditional and marginal modeling through GEE is that the latter is likely to be more robust to model misspecification, since it directly models marginal associations.

Recall that the logistic regression model for binary data can be derived by consideration of an unobserved (latent) continuous logistic random variable (Sect. 7.6.1). This latent formulation can be extended to the mixed model. In particular, assume $U_{ij} = \mu_{ij} + b_i$, where $b_i \mid \sigma_0^2 \sim \mathrm{N}(0, \sigma_0^2)$ and U_{ij} follows the standard logistic distribution, that is, $U_{ij} \mid b_i \sim_{ind} \mathrm{Logistic}(\mu_{ij} + b_i, 1)$. Without loss of generality set, $Y_{ij} = 1$ if $U_{ij} > c$ and 0 otherwise. Then

$$\mathrm{Pr}(Y_{ij} = 1 \mid b_i) = \mathrm{Pr}(U_{ij} > c \mid b_i) = \frac{\exp(\mu_{ij} + b_i - c)}{1 + \exp(\mu_{ij} + b_i - c)}$$

and taking $\mu_{ij} = x_{ij}\beta + c$ produces the random effects logistic model.

9.13 Mixed Models for Binary Data

An interpretation of σ_0^2 is obtained by comparing its magnitude to $\pi^2/3$ (the variance of the logistic distribution, which can be viewed as the within-person variability) via the intra-class correlation:

$$\rho = \text{corr}(U_{ij}, U_{ik}) = \frac{\sigma_0^2}{\sigma_0^2 + \pi^3/3}.$$

Note that ρ is the marginal correlation (averaged over the random effects) among the unobserved latent variables U_{ij} and not the marginal correlation among the Y_{ij}'s. See Fitzmaurice et al. (2004, Sect. 12.5) for further discussion.

We examine the marginal moments further. The marginal mean is $\text{E}[Y_{ij}] = \text{Pr}(Y_{ij} = 1) = \text{E}_{b_i}[p_{ij}]$ where we continue to consider the random intercepts only model

$$p_{ij} = \frac{\exp(\boldsymbol{x}_{ij}\boldsymbol{\beta} + b_i)}{1 + \exp(\boldsymbol{x}_{ij}\boldsymbol{\beta} + b_i)}.$$

The expectation is over the distribution of the random effect. We have already derived the approximate marginal mean (9.20), which we write as

$$p_{ij}^\star = \frac{\exp[\boldsymbol{x}_{ij}\boldsymbol{\beta}/(c^2\sigma_0^2 + 1)^{1/2}]}{1 + \exp[\boldsymbol{x}_{ij}\boldsymbol{\beta}/(c^2\sigma_0^2 + 1)^{1/2}]}.$$

The variance is

$$\begin{aligned}\text{var}(Y_{ij}) &= \text{E}[\text{var}(Y_{ij} \mid b_i)] + \text{var}[\text{E}(Y_{ij} \mid b_i)] \\ &= \text{E}[p_{ij} - p_{ij}^2] + \text{E}[p_{ij}^2] - \text{E}[p_{ij}]^2 \\ &= p_{ij}^\star(1 - p_{ij}^\star),\end{aligned}$$

illustrating again that there is no overdispersion for a Bernoulli random variable. This gives the diagonal elements of the marginal variance–covariance matrix. The covariances between responses on the same unit i are

$$\begin{aligned}\text{cov}(Y_{ij}, Y_{ik}) &= \text{cov}\left(\frac{\exp(\boldsymbol{x}_{ij}\boldsymbol{\beta} + b_i)}{1 + \exp(\boldsymbol{x}_{ij}\boldsymbol{\beta} + b_i)}, \frac{\exp(\boldsymbol{x}_{ik}\boldsymbol{\beta} + b_i)}{1 + \exp(\boldsymbol{x}_{ik}\boldsymbol{\beta} + b_i)}\right) \\ &= \text{E}\left[\left(\frac{\exp(\boldsymbol{x}_{ij}\boldsymbol{\beta} + b_i)}{1 + \exp(\boldsymbol{x}_{ij}\boldsymbol{\beta} + b_i)}\right)\left(\frac{\exp(\boldsymbol{x}_{ik}\boldsymbol{\beta} + b_i)}{1 + \exp(\boldsymbol{x}_{ik}\boldsymbol{\beta} + b_i)}\right)\right] - p_{ij}^\star p_{ik}^\star,\end{aligned}$$

so note that the marginal covariance is not constant and not of easily interpretable form. With a single random effect, the correlations are all determined by the single parameter σ_0.

9.13.2 Likelihood Inference for the Binary Mixed Model

As with the GLMMs described in Sect. 9.4, the integrals required to evaluate the likelihood for the fixed effects β and variance components $\alpha = D$ are analytically intractable. Unfortunately the Laplace approximation method may not be reliable for binary GLMMs, particularly if the random effects variances are large. For this reason adaptive Gauss–Hermite quadrature methods are often resorted to, though care in implementation is required to ensure that sufficient points are used to obtain an accurate approximation. When maximization routines encounter convergence problems, it may be an indication that either the model being fitted is not supported by the data or that the data do not contain sufficient data to estimate all of the parameters.

9.13.3 Bayesian Inference for the Binary Mixed Model

A Bayesian approach to binary GLMMs requires priors to be specified for β and D. As in Sect. 9.6.2, the priors may be specified in terms of interpretable quantities, for example, the residual odds of success. The information in binary data is limited, and so sensitivity to the priors may be encountered, particularly the prior on **D**. As with likelihood-based approaches, greater care is required in computation with binary data. Fong et al. (2010) report that the INLA method is relatively inaccurate for binary GLMMs so that MCMC is the more reliable method if the binomial denominators are small.

Example: Contraception Data

We illustrate likelihood inference for a binary GLMM using the contraception data introduced in Sect. 9.2.1. Let $Y_{ij} = 0/1$ denote the absence/presence of amenorrhea in the ith woman at time t_{ij}, where the latter takes the values 1, 2, 3, or 4. Also, let $d_i = 0/1$ represent the randomization indicators to doses of 100 mg/150 mg, for $i = 1, \ldots, 1151$ women (576 and 575 women received the low and high doses, respectively). There are n_i observations per woman, up to a maximum of 4. We consider the following two-stage model:

Stage One: The response model is $Y_{ij} \mid p_{ij} \sim_{ind}$ Bernoulli (p_{ij}) with

$$\log\left(\frac{p_{ij}}{1-p_{ij}}\right) = \beta_0 + \beta_1 t_{ij} + \beta_2 t_{ij}^2 + \beta_3 d_i t_{ij} + \beta_4 d_i t_{ij}^2 + b_i, \qquad (9.23)$$

so that we have separate quadratic models in time for each of the two-dose levels.

9.13 Mixed Models for Binary Data

Table 9.8 Mixed effects model parameter estimates for the contraception data

Parameter	Likelihood Laplace		Likelihood G–H[a]		Bayesian MCMC	
	Est.	Std. err.	Est.	Std. err.	Est.	Std. err.
Intercept	−3.8	0.27	−3.8	0.30	−3.6	0.27
Low-dose time	1.1	0.25	1.1	0.27	0.99	0.25
Low-dose time^2	−0.044	0.052	−0.042	0.055	−0.015	0.052
High-dose time	0.55	0.18	0.56	0.21	0.55	0.18
High-dose time^2	−0.11	0.051	−0.11	0.050	−0.11	0.058
σ_0	2.1	–	2.3	0.11	2.2	0.13

[a] Adaptive Gauss–Hermite with 50 points

Stage Two: The random effects model is $b_i \mid \sigma_0^2 \sim_{iid} N(0, \sigma_0^2)$.

We do not include a term for the main effect of dose, since we assume that randomization has ensured that the two-dose groups are balanced at baseline ($t = 0$). The conditional odds ratios $\exp(\beta_1)$ and $\exp(\beta_2)$ represent linear and quadratic terms in time for a typical individual ($b_i = 0$) in the low-dose group. Similarly, $\exp(\beta_1 + \beta_3)$ and $\exp(\beta_2 + \beta_4)$ represent linear and quadratic terms in time for a typical individual ($b_i = 0$) in the high-dose group.

Table 9.8 gives parameter estimates and standard errors for a number of analyses, including Laplace and adaptive Gauss–Hermite rules for likelihood calculation. We initially concentrate on the Gauss–Hermite results which are more reliable than those based on the Laplace implementation. Informally, comparing the estimates with the standard errors, the linear terms in time are clearly needed, while it is not so obvious that the quadratic terms are required.

In terms of substantive conclusions, a woman assigned the high dose, when compared to a woman assigned the low dose, both with the same baseline risk of amenorrhea (i.e., with the same random effect) will have increased odds at time t of

$$\exp(\widehat{\beta}_3 t + \widehat{\beta}_4 t^2)$$

giving increases of 1.6, 2.0, 2.0, 1.6 at times 1, 2, 3, 4, respectively. Hence, the difference between the groups increases and then decreases as a function of time, though it is always greater than zero.

The standard deviation of the random effects $\widehat{\sigma} = 2.3$ is substantial here. An estimate of a 95% interval for the risk of amenorrhea in the low-dose group at occasion 1 is

$$\frac{\exp(-3.8 + 1.1 - 0.042 \pm 1.96 \times 2.3)}{1 + \exp(-3.8 + 1.1 - 0.042 \pm 1.96 \times 2.3)} = [0.0007, 0.85],$$

so that we have very large between-woman variability in risk. The marginal intraclass correlation coefficient is estimated as $\rho = 0.61$ (recall this is the correlation for the latent variable and not for the marginal responses).

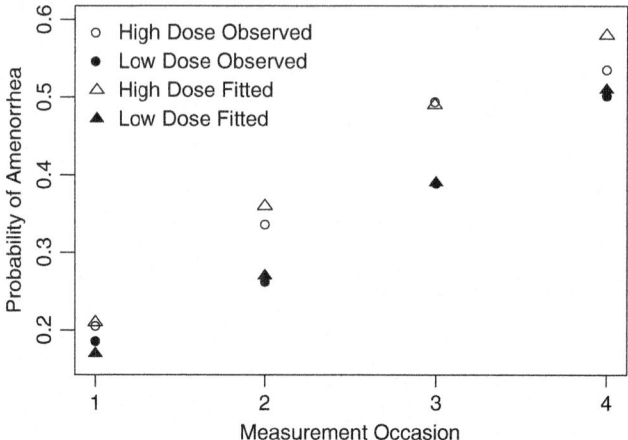

Fig. 9.7 Probability of amenorrhea over time in low- and high-dose groups in the contraception data, along with fitted probabilities. The latter are calculated via Monte Carlo simulation, with likelihood estimation in the mixed model, implemented with Gauss–Hermite quadrature

Table 9.9 Monte Carlo estimated variances (on the *diagonal*) and correlations (*upper diagonal*), between measurements on the same woman, at different observations occasions (1–4), in the low- (*left*) and high- (*right*) dose groups

	1	2	3	4		1	2	3	4
1	0.14	0.38	0.36	0.33	1	0.17	0.39	0.36	0.33
2		0.20	0.41	0.39	2		0.23	0.42	0.40
3			0.24	0.43	3			0.25	0.43
4				0.25	4				0.24

These estimates are based on likelihood estimation in the mixed model, implemented with Gauss–Hermite quadrature

Allowing the random effects variance to vary by covariate groups is important to investigate since missing such dependence can lead to serious inaccuracies (Heagerty and Kurland 2001). The assumption of a common σ_0 in the two groups is important for accurate inference in this example. We fit separate logistic GLMMs to the two-dose groups and obtain estimates of 2.3 and 2.2, illustrating that a common σ_0 is supported by the data.

We evaluate the marginal means calculation using Monte Carlo integration. These means are shown, along with the observed proportions, in Fig. 9.7. We see that the overall fit is good, apart from the last time point (for which there is reduced data due to dropout).

In Table 9.9, we estimate the marginal variance–covariance and correlation matrices for the two-dose groups using Monte Carlo integration. As we have already discussed in Sect. 9.13.1 a random intercepts only model does not lead to correlations that are constant across time (unlike the linear model). In general, the estimates are in reasonable agreement with the empirical variances and correlations reported in Table 9.1.

9.13 Mixed Models for Binary Data

For the Bayesian analysis, the prior for the intercept was relatively flat, $\beta_0 \sim$ N$(0, 2.38^2)$. If there was no effect of time (i.e., if $\beta_1 = \beta_2 = \beta_3 = \beta_4 = 0$) then a 95% interval for the probabilities for a typical individual would be $\exp(\pm 1.96 \times 2.38) = [0.009, 0.99]$. For the regression coefficients, we specify $\beta_k \sim$ N$(0, 0.98^2)$ which gives a 95% interval for the odds ratios of $\exp(\pm 1.96 \times 0.98) = [0.15, 6.8]$. Finally, for σ_0^{-2}, we assume a Gamma(0.5,0.1) prior which gives a 95% interval for σ_0 of [0.06,4.5]. More informatively, a 95% interval for the residual odds is [0.17,6.0]. These priors are not uninformative but correspond to ranges for probabilities and odds ratios that are consistent with the application.

The posterior means and standard deviations are given in Table 9.8, and we see broad agreement with the MLEs and standard errors found using Gauss–Hermite. The intra-class correlation coefficient is estimated as 0.60 with 95% credible interval [0.55, 0.67].

9.13.4 Conditional Likelihood Inference for Binary Mixed Models

Recall that conditional likelihood is a technique for eliminating nuisance parameters, in this case the random effects in the mixed model. Following from Sect. 9.5, we outline the approach as applied to the binary mixed model with random intercepts. Consider individual i with binary observations $y_{i1}, \ldots, y_{in_i}$ and assume the random intercepts model $Y_{ij} \mid \lambda_i, \boldsymbol{\beta}^\star \sim$ Bernoulli(p_{ij}), where

$$\log\left(\frac{p_{ij}}{1 - p_{ij}}\right) = \boldsymbol{x}_{ij}\boldsymbol{\beta}^\star + \lambda_i$$

and $\lambda_i = \boldsymbol{x}_i\boldsymbol{\beta}^\dagger + b_i$ so that $\boldsymbol{\beta}^\dagger$ represents those parameters associated with covariates that are constant within an individual and $\boldsymbol{\beta}^\star$ those that vary. Mimicking the development in Sect. 9.5, the joint distribution for the responses of the ith unit is

$$\Pr(y_{i1}, \ldots, y_{in_i} \mid \lambda_i, \boldsymbol{\beta}^\star) = \prod_{j=1}^{n_i} \frac{\exp\left(\lambda_i y_{ij} + \boldsymbol{\beta}^{\star\mathrm{T}} \boldsymbol{x}_{ij}^\mathrm{T} y_{ij}\right)}{1 + \exp\left(\lambda_i + \boldsymbol{\beta}^{\star\mathrm{T}} \boldsymbol{x}_{ij}^\mathrm{T}\right)}$$

$$= \frac{\exp\left(\lambda_i \sum_{j=1}^{n_i} y_{ij} + \boldsymbol{\beta}^{\star\mathrm{T}} \sum_{j=1}^{n_i} \boldsymbol{x}_{ij}^\mathrm{T} y_{ij}\right)}{\prod_{j=1}^{n_i} \left[1 + \exp\left(\lambda_i + \boldsymbol{\beta}^{\star\mathrm{T}} \boldsymbol{x}_{ij}^\mathrm{T}\right)\right]}$$

$$= \frac{\exp\left(\lambda_i t_{2i} + \boldsymbol{\beta}^{\star\mathrm{T}} \boldsymbol{t}_{1i}\right)}{\prod_{j=1}^{n_i} \left[1 + \exp\left(\lambda_i + \boldsymbol{\beta}^{\star\mathrm{T}} \boldsymbol{x}_{ij}^\mathrm{T}\right)\right]}$$

$$= \frac{\exp\left(\lambda_i t_{2i} + \boldsymbol{\beta}^{\star\mathrm{T}} \boldsymbol{t}_{1i}\right)}{k(\lambda_i, \boldsymbol{\beta})}$$

$$= p(\boldsymbol{t}_{1i}, t_{2i} \mid \lambda_i, \boldsymbol{\beta}^\star)$$

where

$$t_{1i} = \sum_{j=1}^{n_i} \boldsymbol{x}_{ij}^{\mathrm{T}} y_{ij}, \quad t_{2i} = \sum_{j=1}^{n_i} y_{ij} = y_{i+}$$

and

$$k(\lambda_i, \boldsymbol{\beta}^\star) = \prod_{j=1}^{n_i} \left[1 + \exp\left(\lambda_i + \boldsymbol{\beta}^{\star\mathrm{T}} \boldsymbol{x}_{ij}^{\mathrm{T}}\right)\right].$$

Therefore, the conditioning statistic is the number of successes on the ith unit. We have conditional likelihood

$$L_c(\boldsymbol{\beta}) = \prod_{i=1}^{m} p(t_{1i} \mid t_{2i}, \boldsymbol{\beta}^\star) = \prod_{i=1}^{m} \frac{p(t_{1i}, t_{2i} \mid \lambda_i, \boldsymbol{\beta}^\star)}{p(t_{2i} \mid \lambda_i, \boldsymbol{\beta}^\star)}$$

where

$$p(t_{2i} \mid \lambda_i, \boldsymbol{\beta}^\star) = \frac{\sum_{l=1}^{\binom{n_i}{y_{i+}}} \exp\left(\lambda_i y_{i+} + \boldsymbol{\beta}^{\star\mathrm{T}} \sum_{k=1}^{n_i} \boldsymbol{x}_{ik}^{\mathrm{T}} y_{ik}^{(l)}\right)}{k(\lambda_i, \boldsymbol{\beta}^\star)},$$

and the summation is over the $\binom{n_i}{y_{i+}}$ ways of choosing y_{i+} ones out of n_i and $\boldsymbol{y}_i^{(l)} = [y_{i1}^{(l)}, \ldots, y_{in_i}^{(l)}]$, $l = 1, \ldots, \binom{n_i}{y_{i+}}$ is the collection of these ways. Inference may be based on the conditional likelihood

$$L_c(\boldsymbol{\beta}^\star) = \prod_{i=1}^{m} \frac{\exp\left(\lambda_i y_{i+} + \boldsymbol{\beta}^{\star\mathrm{T}} \sum_{j=1}^{n_i} \boldsymbol{x}_{ij}^{\mathrm{T}} y_{ij}\right)}{\sum_{l=1}^{\binom{n_i}{y_{i+}}} \exp\left(\lambda_i y_{i+} + \boldsymbol{\beta}^{\star\mathrm{T}} \sum_{k=1}^{n_i} \boldsymbol{x}_{ik}^{\mathrm{T}} y_{ik}^{(l)}\right)}$$

$$= \prod_{i=1}^{m} \frac{\exp\left(\boldsymbol{\beta}^{\star\mathrm{T}} \sum_{j=1}^{n_i} \boldsymbol{x}_{ij}^{\mathrm{T}} y_{ij}\right)}{\sum_{l=1}^{\binom{n_i}{y_{i+}}} \exp\left(\boldsymbol{\beta}^{\star\mathrm{T}} \sum_{k=1}^{n_i} \boldsymbol{x}_{ik}^{\mathrm{T}} y_{ik}^{(l)}\right)}.$$

Hence, there is no need to specify a distribution for the unit-specific parameters that allow for within-unit dependence, as they are eliminated by the conditioning argument.

As an example, if $n_i = 3$ and $\boldsymbol{y}_i = [0, 0, 1]$ so that $y_{i+} = 1$, then

$$\boldsymbol{y}_i^{(1)} = [1, 0, 0], \quad \boldsymbol{y}_i^{(2)} = [0, 1, 0], \quad \boldsymbol{y}_i^{(3)} = [0, 0, 1]$$

and the contribution to the conditional likelihood is

$$\frac{\exp(\boldsymbol{\beta}^{\star\mathrm{T}} \boldsymbol{x}_{i3}^{\mathrm{T}})}{\exp(\boldsymbol{\beta}^{\star\mathrm{T}} \boldsymbol{x}_{i1}^{\mathrm{T}}) + \exp(\boldsymbol{\beta}^{\star\mathrm{T}} \boldsymbol{x}_{i2}^{\mathrm{T}}) + \exp(\boldsymbol{\beta}^{\star\mathrm{T}} \boldsymbol{x}_{i3}^{\mathrm{T}})}.$$

As a second example, if $n_i = 3$ and $\boldsymbol{y}_i = [1,0,1]$ so that $y_{i+} = 2$, then

$$\boldsymbol{y}_i^{(1)} = [1,1,0], \quad \boldsymbol{y}_i^{(2)} = [1,0,1], \quad \boldsymbol{y}_i^{(3)} = [0,1,1],$$

and the contribution to the conditional likelihood is

$$\frac{\exp(\boldsymbol{\beta}^{\star\mathrm{T}}\boldsymbol{x}_{i1}^{\mathrm{T}} + \boldsymbol{\beta}^{\star\mathrm{T}}\boldsymbol{x}_{i3}^{\mathrm{T}})}{\exp(\boldsymbol{\beta}^{\star\mathrm{T}}\boldsymbol{x}_{i1}^{\mathrm{T}} + \boldsymbol{\beta}^{\star\mathrm{T}}\boldsymbol{x}_{i2}^{\mathrm{T}}) + \exp(\boldsymbol{\beta}^{\star\mathrm{T}}\boldsymbol{x}_{i1}^{\mathrm{T}} + \boldsymbol{\beta}^{\star\mathrm{T}}\boldsymbol{x}_{i3}^{\mathrm{T}}) + \exp(\boldsymbol{\beta}^{\star\mathrm{T}}\boldsymbol{x}_{i2}^{\mathrm{T}} + \boldsymbol{\beta}^{\star\mathrm{T}}\boldsymbol{x}_{i3}^{\mathrm{T}})}.$$

There is no contribution to the conditional likelihood from individuals with $n_i = 1$ or $y_{i+} = 0$ or $y_{i+} = n_i$. The conditional likelihood can be computationally expensive to evaluate if n_i is large, for example, if $n_i = 20$ and $y_{i+} = 10$ there are $\binom{n_i}{y_{i+}} = 184{,}756$ variations. The similarity to Cox's partial likelihood (e.g., Kalbfleisch and Prentice 2002, Chap. 4) may be exploited to carry out computation, however.

We reiterate that the conditional likelihood estimates those elements of $\boldsymbol{\beta}^\star$ that are associated with covariates that vary within individuals. If a covariate only varies between individuals, then its effect cannot be estimated using conditional likelihood. For covariates that vary both between and within individuals, only the within-individual contrasts are used.

9.14 Marginal Models for Dependent Binary Data

We now consider the marginal modeling of dependent binary data. We begin by describing how the GEE approach of Sect. 9.9 can be used for binary data and then describe alternative approaches.

9.14.1 Generalized Estimating Equations

For the marginal Bernoulli outcome $Y_{ij} \mid \mu_{ij} \sim \text{Bernoulli}(\mu_{ij})$ and with a logistic regression model, we have the exponential family representation

$$\Pr(Y_{ij} = y_{ij} \mid \boldsymbol{x}_{ij}) = \mu_{ij}^{y_{ij}}(1 - \mu_{ij})^{1-y_{ij}}$$
$$= \exp\left\{y_{ij}\theta_{ij} - \log[1 + \exp(\theta_{ij})]\right\},$$

where

$$\theta_{ij} = \log\left(\frac{\mu_{ij}}{1 - \mu_{ij}}\right) = \boldsymbol{x}_{ij}\boldsymbol{\gamma}.$$

For independent responses, the likelihood is

$$\Pr(Y \mid x) = \exp\left\{\sum_{i=1}^{m}\sum_{j=1}^{n_i} y_{ij}\theta_{ij} - \sum_{i=1}^{m}\sum_{j=1}^{n_i} \log[1 + \exp(\theta_{ij})]\right\}$$

$$= \exp\left(\sum_{i=1}^{m}\sum_{j=1}^{n_i} l_{ij}\right).$$

To find the MLEs, we consider the score equation

$$G(\gamma) = \frac{\partial l}{\partial \gamma} = \sum_{i=1}^{m}\sum_{j=1}^{n_i} \frac{\partial l_{ij}}{\partial \theta_{ij}}\frac{\partial \theta_{ij}}{\partial \gamma}$$

$$= \sum_{i=1}^{m}\sum_{j=1}^{n_i} x_{ij}(y_{ij} - \mu_{ij}) = \sum_{i=1}^{m} x_i^T(y_i - \mu_i)$$

with $\mu_i = [\mu_{i1}, \ldots, \mu_{in_i}]^T$. This form is identical to the use of GEE with working independence and so can be implemented with standard software, though we need to "fix up" the standard errors via sandwich estimation. Hence, the above estimating equation construction offers a very simple approach to inference which may be adequate if the dependence between observations on the same unit is small. If the correlations are not small, then efficiency considerations suggest that nonindependence working covariance models should be entertained.

As with other types of data (Sect. 9.9), we can model the correlation structure (Liang and Zeger 1986) and assume $\text{var}(Y_i) = W_i$ with $W_i = \Delta_i^{1/2} R_i(\alpha) \Delta_i^{1/2}$ with Δ_i a diagonal matrix with jth diagonal entry $\text{var}(Y_{ij}) = \mu_{ij}(1 - \mu_{ij})$ and $R_i(\alpha)$ a working correlation model depending on parameters α. In this case, the estimating function is

$$G(\gamma, \alpha) = \sum_{i=1}^{m} D_i^T W_i^{-1}(y_i - \mu_i), \qquad (9.24)$$

where $D_i = \partial \mu_i / \partial \gamma$. As usual, an estimate of α is required, with an obvious choice being a method of moments estimator. The variance of the estimator takes the usual sandwich form (9.12).

9.14.2 Loglinear Models

We now consider another approach to constructing models for dependent binary data that may form the basis for likelihood or GEE procedures. Loglinear models are a

9.14 Marginal Models for Dependent Binary Data

Table 9.10 Probabilities of the four possible outcomes for two binary variables via a loglinear representation

y_1	y_2	$\Pr(Y_1 = y_1, Y_2 = y_2)$
0	0	$c(\boldsymbol{\theta})$
1	0	$c(\boldsymbol{\theta})\exp(\theta_1^{(1)})$
0	1	$c(\boldsymbol{\theta})\exp(\theta_2^{(1)})$
1	1	$c(\boldsymbol{\theta})\exp(\theta_1^{(1)} + \theta_2^{(1)} + \theta_{12}^{(2)})$

popular choice for cross-classified discrete data (Cox 1972; Bishop et al. 1975). We begin by returning to the situation in which we have n responses on a single unit, $y_j, j = 1, \ldots, n$. A saturated loglinear model is

$$\Pr(\boldsymbol{Y} = \boldsymbol{y}) = c(\boldsymbol{\theta})\exp\left(\sum_{j=1}^n \theta_j^{(1)} y_j + \sum_{j_1 < j_2} \theta_{j_1 j_2}^{(2)} y_{j_1} y_{j_2} + \ldots + \theta_{12\ldots n}^{(n)} y_1 \ldots y_n\right),$$

with $2^n - 1$ parameters $\boldsymbol{\theta} = [\theta_1^{(1)}, \ldots, \theta_n^{(1)}, \theta_{12}^{(2)}, \ldots, \theta_{n-1,n}^{(2)}, \ldots, \theta_{12\ldots n}^{(n)}]^\mathrm{T}$, and normalizing constant $c(\boldsymbol{\theta})$. To provide an interpretation of the parameters, consider the case of $n = 2$ trials for which

$$\Pr(Y_1 = y_1, Y_2 = y_2) = c(\boldsymbol{\theta})\exp\left(\theta_1^{(1)} y_1 + \theta_2^{(1)} y_2 + \theta_{12}^{(2)} y_1 y_2\right),$$

where $\boldsymbol{\theta} = [\theta_1^{(1)}, \theta_2^{(1)}, \theta_{12}^{(2)}]^\mathrm{T}$ and

$$c(\boldsymbol{\theta})^{-1} = \sum_{y_1=0}^1 \sum_{y_2=0}^1 \exp\left(\theta_1^{(1)} y_1 + \theta_2^{(1)} y_2 + \theta_{12}^{(2)} y_1 y_2\right).$$

Table 9.10 gives the forms of the probabilities for the loglinear representation, from which we can determine the interpretation of the three parameters:

$$\exp(\theta_1^{(1)}) = \frac{\Pr(Y_1 = 1 \mid y_2 = 0)}{\Pr(Y_1 = 0 \mid y_2 = 0)}$$

is the odds of an event at trial 1, given no event at trial 2,

$$\exp(\theta_2^{(1)}) = \frac{\Pr(Y_2 = 1 \mid y_1 = 0)}{\Pr(Y_2 = 0 \mid y_1 = 0)}$$

is the odds of an event at trial 2, given no event at trial 1, and

$$\exp(\theta_{12}^{(12)}) = \frac{\Pr(Y_2 = 1 \mid y_1 = 1)/\Pr(Y_2 = 0 \mid y_1 = 1)}{\Pr(Y_2 = 1 \mid y_1 = 0)/\Pr(Y_2 = 0 \mid y_1 = 0)}$$

is the ratio of the odds of an event at trial 2 given an event at trial 1, divided by the odds of an event at trial 2 given no event at trial 1. Consequently, if this parameter is larger than 1, there is positive dependence between Y_1 and Y_2.

For general n, a simplified version of the loglinear model is provided when third- and higher-order terms are set to zero, so that

$$\Pr(\boldsymbol{Y} = \boldsymbol{y}) = c(\boldsymbol{\theta}) \exp\left(\sum_{j=1}^{n} \theta_j^{(1)} y_j + \sum_{j<k} \theta_{jk}^{(2)} y_j y_k\right). \quad (9.25)$$

For this model,

$$\frac{\Pr(Y_j = 1 \mid Y_k = y_k, Y_l = 0, l \neq j, k)}{\Pr(Y_j = 0 \mid Y_k = y_k, Y_l = 0, l \neq j, k)} = \exp(\theta_j^{(1)} + \theta_{jk}^{(2)} y_k).$$

so that $\exp(\theta_j^{(1)})$ is the (conditional) odds of an event at trial j, given all other responses are zero. Further, $\exp(\theta_{jk}^{(2)})$ is the odds ratio describing the association between Y_j and Y_k, given all other responses are set equal to zero, that is,

$$\frac{\Pr(Y_j = 1, Y_k = 1 \mid Y_l = 0, l \neq j, k) \Pr(Y_j = 0, Y_k = 0 \mid Y_l = 0, l \neq j, k)}{\Pr(Y_j = 1, Y_k = 0 \mid Y_l = 0, l \neq j, k) \Pr(Y_j = 0, Y_k = 1 \mid Y_l = 0, l \neq j, k)}$$
$$= \exp(\theta_{jk}^{(2)}).$$

The quadratic model (9.25) was described in Sect. 9.10 and was suggested for the analysis of binary data by Zhao and Prentice (1990). Recall that this model has the appealing property of consistency so long as the first two moments are correctly specified. The quadratic exponential model is unique in this respect.

Unfortunately, parameterizing in terms of the $\boldsymbol{\theta}$ parameters is unappealing for regression modeling where the primary aim is to model the response as a function of $\boldsymbol{x}$. To illustrate, consider binary longitudinal data with a binary covariate x and suppose we let the parameters $\boldsymbol{\theta}$ depend on x. The difference between the log odds $\theta_j^{(1)}(x=1)$ and $\theta_j^{(1)}(x=0)$ represents the effect of x on the *conditional* log odds of an event at period j, *given* that there were no events at any other trials, which is difficult to interpret. We would rather model the *marginal* means $\boldsymbol{\mu}$, and these are a function of both $\boldsymbol{\theta}^{(1)}$ and $\boldsymbol{\theta}^{(2)}$. For example, for the $n=2$ case presented in Table 9.10, the marginal means are

$$E[Y_1] = c(\boldsymbol{\theta}) \exp(\theta_1^{(1)})[1 + \exp(\theta_1^{(1)} + \theta_{12}^{(2)})]$$
$$E[Y_2] = c(\boldsymbol{\theta}) \exp(\theta_2^{(1)})[1 + \exp(\theta_2^{(1)} + \theta_{12}^{(2)})],$$

and these forms do not lend themselves to straightforward incorporation of covariates. Hence, alternative approaches have been proposed as we now discuss.

9.14.3 Further Multivariate Binary Models

A number of approaches are based on assuming a marginal mean model, to overcome the problems described in the previous section, along with a second set of parameters to model the dependence.

First, we may reparameterize the model via the mean vector $\boldsymbol{\mu}$ and second- and higher-order loglinear parameters. For example, we may consider second-order parameters only and work with $\boldsymbol{\mu}$ and the loglinear parameters $\boldsymbol{\theta}^{(2)}$, as suggested by Fitzmaurice and Laird (1993). The latter used maximum likelihood for estimation. There are two disadvantages to this approach. First, the interpretation of the $\boldsymbol{\theta}^{(2)}$ parameters depends on the number of responses n. This is particularly a problem in a longitudinal setting with differing n_i. Hence, this approach is most useful for data that have $n_i = n$ for all i. Second, if interest lies in understanding the structure of the dependence, the conditional odds ratio parameters do not have the attractive simple interpretation of *marginal* odds ratios.

A second approach is based on modeling the correlations in addition to the means. Let

$$e^{\star}_{ijk} = \frac{Y_{ij} - \mu_{ij}}{[\mu_{ij}(1-\mu_{ij})]^{1/2}}$$

$$\rho_{ijk} = \text{corr}(Y_{ij}, Y_{ik}) = \text{E}[e^{\star}_{ij}e^{\star}_{ik}]$$

$$\rho_{ijkl} = \text{E}[e^{\star}_{ij}e^{\star}_{ik}e^{\star}_{il}]$$

$$\cdots \quad \cdots$$

$$\rho_{i1\ldots n_i} = \text{E}[e^{\star}_{i1}e^{\star}_{i2}\ldots e^{\star}_{in_i}].$$

The correlations have marginal interpretations. For example, ρ_{ijkl} is a three-way association parameter. Bahadur (1961) defined a multivariate binary model based on the marginal means and these correlations. The probability for the set of outcomes on unit i is

$$\Pr(\boldsymbol{Y}_i = \boldsymbol{y}_i) = \prod_{j=1}^{n_i} \mu_{ij}^{y_{ij}}(1-\mu_{ij})^{1-y_{ij}} \times$$

$$\left(1 + \sum_{j<k} \rho_{ijk}e^{\star}_{ij}e^{\star}_{ik} + \sum_{j<k<l} \rho_{ijkl}e^{\star}_{ij}e^{\star}_{ik}e^{\star}_{il} + \ldots + \rho_{i1\ldots n}e^{\star}_{i1}e^{\star}_{i2}\ldots e^{\star}_{in_i}\right).$$

Unfortunately, the correlations are constrained in complicated ways by the marginal means. As an example, consider two measurements on a single individual, Y_{i1} and Y_{i2}, with means μ_{i1} and μ_{i2}. The correlation is

$$\text{corr}(Y_{i1}, Y_{i2}) = \frac{\Pr(Y_{i1}=1, Y_{i2}=1) - \mu_{i1}\mu_{i2}}{[\mu_{i1}(1-\mu_{i1})\mu_{i2}(1-\mu_{i2})]^{1/2}}$$

Table 9.11 Notation in the case of $n_i = 2$ binary responses on individual i

		Y_{i2}		
		0	1	
Y_{i1}	0	$1 - \mu_{i1} - \mu_{i2} + \mu_{i12}$	$\mu_{i2} - \mu_{i12}$	$1 - \mu_{i1}$
	1	$\mu_{i1} - \mu_{i12}$	μ_{i12}	μ_{i1}
		$1 - \mu_{i2}$	μ_{i2}	

and

$$\max(0, \mu_{i1} + \mu_{i2} - 1) \leq \Pr(Y_{i1} = 1, Y_{i2} = 1) \leq \min(\mu_{i1}, \mu_{i2}),$$

which implies complicated constraints on the correlation. For example, if $\mu_{i1} = 0.8$ and $\mu_{i2} = 0.2$, then $0 \leq \text{corr}(Y_{i1}, Y_{i2}) \leq 0.25$. The message here is that correlations are not a natural measure of dependence for binary data so that the Bahadur representation is not appealing.

A third approach (Lipsitz et al. 1991; Liang et al. 1992) is to parameterize in terms of the marginal means and the marginal odds ratios defined by. Let

$$\delta_{ijk} = \frac{\Pr(Y_{ij} = 1, Y_{ik} = 1) \Pr(Y_{ij} = 0, Y_{ik} = 0)}{\Pr(Y_{ij} = 1, Y_{ik} = 0) \Pr(Y_{ij} = 0, Y_{ik} = 1)}$$

$$= \frac{\Pr(Y_{ij} = 1 \mid Y_{ik} = 1) / \Pr(Y_{ij} = 0 \mid Y_{ik} = 1)}{\Pr(Y_{ij} = 1 \mid Y_{ik} = 0) / \Pr(Y_{ij} = 0 \mid Y_{ik} = 0)},$$

which is the odds (for individual i) that the jth observation is a 1, given the kth observation is a 1, divided by the odds that the jth observation is a 1, given the kth observation is a 0. Therefore, we have a set of marginal odds ratios, and if $\delta_{ijk} > 1$, we have positive dependence between outcomes j and k. It is then possible to obtain the joint distribution in terms of the means $\boldsymbol{\mu}$, where $\mu_{ij} = \Pr(Y_{ij} = 1)$, the odds ratios $\boldsymbol{\delta}_i = [\delta_{i12}, \ldots, \delta_{i,n_i-1,n_i}]$ and contrasts of odds ratios. To determine the probability distribution of the data, we need to find

$$\mu_{ijk} = \text{E}[Y_{ij} Y_{ik}] = \Pr(Y_{ij} = 1, Y_{ik} = 1),$$

so that we can write down either the likelihood function or an estimating function.

For the case of $n_i = 2$ (see Table 9.11), we have

$$\delta_{i12} = \frac{\Pr(Y_{i1} = 1, Y_{i2} = 1) \Pr(Y_{i1} = 0, Y_{i2} = 0)}{\Pr(Y_{i1} = 1, Y_{i2} = 0) \Pr(Y_{i1} = 0, Y_{i2} = 1)} = \frac{\mu_{i12}(1 - \mu_{i1} - \mu_{i2} + \mu_{i12})}{(\mu_{i1} - \mu_{i12})(\mu_{i2} - \mu_{i12})},$$

and so

$$\mu_{i12}^2(\delta_{i12} - 1) + \mu_{i12} b_i + \delta_{i12} \mu_{i1} \mu_{i2} = 0,$$

where $b_i = (\mu_{i1} + \mu_{i2})(1 - \delta_{i12}) - 1$, to give

$$\mu_{i12} = \frac{-b_i \pm \sqrt{b_i^2 - 4(\delta_{i12} - 1)\mu_{i1}\mu_{i2}\delta_{i12}}}{2(\delta_{i12} - 1)}$$

9.14 Marginal Models for Dependent Binary Data

if $\delta_{i12} \neq 1$ and $\mu_{i12} = \mu_{ij}\mu_{ik}$ if $\delta_{i12} = 1$. The likelihood is

$$\mu_{i1}^{y_{i1}}(1-\mu_{i1})^{1-y_{i1}}\mu_{i2}^{y_{i2}}(1-\mu_{i2})^{1-y_{i2}} + (-1)^{(y_{i1}-y_{i2})}(\mu_{i12} - \mu_{i1}\mu_{i2}) \quad (9.26)$$

(Exercise 9.3).

As the number of binary responses increases so does the complexity of solving for the μ_{ijk}'s; see Liang et al. (1992) for further details. In the case of large n_i, there are a large numbers of nuisance odds ratios, and assumptions such as $\delta_{ijk} = \delta$ for $i = 1, \ldots, m, j, k = 1, \ldots, n_i$ may be made.

In a longitudinal setting, another possibility is to take

$$\log \delta_{ijk} = \alpha_0 + \alpha_1 |t_{ij} - t_{ik}|^{-1},$$

so that the degree of association is inversely proportional to the time between observations. Computation may be carried out by setting up an estimating equation for y_i and a method of moments estimator for estimation of the covariance parameters. As an alternative, GEE2 may be used with a pair of linked estimating equations (Sect. 9.10).

Letting $\alpha_{ijk} = \log \delta_{ijk}$, Carey et al. (1993) suggest the following approach for estimating β and α. It is easy to show that

$$\frac{\Pr(Y_{ij}=1 \mid Y_{ik}=y_{ik})}{\Pr(Y_{ij}=0 \mid Y_{ik}=y_{ik})} = \exp(y_{ik}\alpha_{ijk})\frac{\Pr(Y_{ij}=1, Y_{ik}=0)}{\Pr(Y_{ij}=0, Y_{ik}=0)}$$

$$= \exp(y_{ik}\alpha_{ijk})\left(\frac{\mu_{ij} - \mu_{ijk}}{1 - \mu_{ij} - \mu_{ik} + \mu_{ijk}}\right),$$

which can be written as a logistic regression model for the conditional probabilities $E[Y_{ij} \mid Y_{ik}]$:

$$\text{logit}(E[Y_{ij} \mid Y_{ik}]) = \log\left(\frac{\Pr(Y_{ij}=1 \mid Y_{ik}=y_{ik})}{\Pr(Y_{ij}=0 \mid Y_{ik}=y_{ik})}\right)$$

$$= y_{ik}\alpha_{ijk} + \log\left(\frac{\mu_{ij} - \mu_{ijk}}{1 - \mu_{ij} - \mu_{ik} + \mu_{ijk}}\right)$$

where the term on the right is an offset (given estimates of the means). Suppose, for simplicity, that $\alpha_{ijk} = \alpha$. Then, given current estimates of β, α, we can fit a logistic regression model by regressing Y_{ij} on Y_{ik} for $1 \leq j < k \leq n_i$, to reestimate α. The offset is a function of α and β so iteration is required. Consequently, Carey et al. (1993) named this approach *alternating logistic regressions*. Once the α parameters are estimated, one may solve for $\text{var}(Y_i)$ in order to use the estimating function (9.24).

In some situations, interest may focus on estimating/modeling the within-unit dependence. Basing a model on correlation parameters is not appealing, but using marginal log odds ratios suggests the model $\alpha_{ijk} = x^*_{ijk}\Psi$ for a set of covariates of interest x^*_{ijk} with associated regression coefficients Ψ.

Table 9.12 GEE parameter estimates for the contraception data

Parameter	GEE independence		GEE exchangeable		GEE ALR[a]	
	Est.	Std. err.	Est.	Std. err.	Est.	Std. err.
Intercept	−2.2	0.18	−2.2	0.18	−2.3	0.16
Low-dose time	0.67	0.16	0.70	0.16	0.70	0.15
Low-dose time^2	−0.030	0.033	−0.033	0.032	−0.033	0.031
High-dose time	0.30	0.11	0.33	0.11	0.34	0.11
High-dose time^2	−0.062	0.030	−0.064	0.029	−0.067	0.028

[a] Alternating logistic regression

Example: Contraception Data

Table 9.12 gives parameter estimates and standard errors for various implementations of GEE, for the marginal model

$$\log\left(\frac{p_{ij}}{1-p_{ij}}\right) = \gamma_0 + \gamma_1 t_{ij} + \gamma_2 t_{ij}^2 + \gamma_3 d_i t_{ij} + \gamma_4 d_i t_{ij}^2, \qquad (9.27)$$

where the γ notation emphasizes that we are estimating marginal parameters. We initially implement GEE with working independence; in general, this is not to be recommended unless it is thought that the outcomes within a cluster are close to independent. We also allow a working exchangeable structure, with the latter parameterized in terms of correlations. Finally, we assume a working exchangeable model parameterized in terms of a common (marginal) log odds ratio. For these data, there are few substantive differences between the approaches. Under the exchangeable models, the common correlation is estimated as 0.36 (0.024) (which is in line with the correlations in Table 9.1), while the common log odds ratio is estimated as 2.0 (0.11). The latter is log of the ratio of the the odds of amenorrhea at time t, given amenorrhea at time s, to the odds of amenorrhea at time t, given no amenorrhea at time s, $s \neq t$.

We may compare these results with a random intercept GLMM. The Bayesian marginal estimates obtained by dividing the posterior means and the posterior standard deviations by $(c^2\widehat{\sigma}_0^2 + 1)^{1/2}$ result in the estimates (standard errors): −2.3 (0.17), 0.68 (0.15), −0.019 (0.032), 0.34 (0.11), and −0.066 (0.035), which are in close agreement with the point and interval estimates in Table 9.12. The marginal probabilities from the GEE exchangeable model were identical to those obtained via Monte Carlo integration in the mixed model (and displayed on Fig. 9.7).

As we have already mentioned, model checking is very difficult with binary data. For data with replication across common x variables, one may obtain empirical probabilities and/or logits (as in Fig. 9.1), which may suggest model forms in an exploratory model building exercise or may be compared with fitted summaries. Similarly, the dependence structure may be examined across covariate groups, via empirical correlations or odds.

9.15 Nonlinear Mixed Models

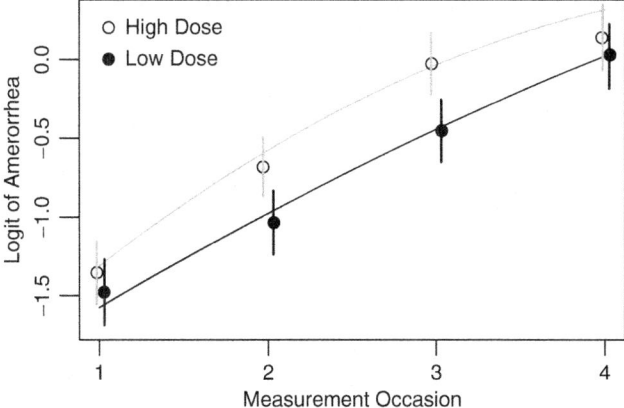

Fig. 9.8 Logit of probability of amenorrhea over time in high- and low- dose groups with marginal fits from exchangeable GEE model

Figure 9.8 shows the fitted logistic curves in each dose group versus time along with the logits of the probabilities of amenorrhea. The vertical lines represent 95% confidence intervals for the logits. These intervals increase slightly in width over time as dropout occurs. Here, we would conclude that the model fit is reasonable.

9.15 Nonlinear Mixed Models

We now turn attention to the nonlinear mixed model (NLMM). Our development will be much shorter for this class of models. One reason for this is that the nonlinearity results in very little analytical theory being available. Also, traditionally, dependent nonlinear data have been analyzed with mixed models and not GEE because the emphasis is often on unit-level inference. The fitting, inferential summarization and assessment of assumptions will be illustrated using the theophylline data described in Sect. 9.2.3.

In a nonlinear mixed model (NLMM), the first stage of a linear mixed model is replaced by a nonlinear form. We describe a specific two-stage form that is useful in many longitudinal situations. The response at time t_{ij} is y_{ij}, and $\boldsymbol{x}_{ij}$ are covariates measured at these times, $i = 1, \ldots, m$, $j = 1, \ldots, n_i$. Let $N = \sum_{i=1}^{m} n_i$:

Stage One: Conditional on random effects, $\boldsymbol{b}_i$, the response model is

$$y_{ij} = f(\eta_{ij}, t_{ij}) + \epsilon_{ij}, \tag{9.28}$$

where $f(\cdot, \cdot)$ is a nonlinear function and

$$\eta_{ij} = \boldsymbol{x}_{ij}\boldsymbol{\beta} + \boldsymbol{z}_{ij}\boldsymbol{b}_i,$$

with a $(k+1) \times 1$ vector of fixed effects $\boldsymbol{\beta}$, a $(q+1) \times 1$ vector of random effects, $\boldsymbol{b}_i$, with $q \leq k$, $\boldsymbol{x}_i = [\boldsymbol{x}_{i1}, \ldots, \boldsymbol{x}_{in_i}]^T$ the design matrix for the fixed effect with $\boldsymbol{x}_{ij} = [1, x_{ij1}, \ldots, x_{ijk}]^T$ and $\boldsymbol{z}_i = [\boldsymbol{z}_{i1}, \ldots, \boldsymbol{z}_{in_i}]^T$ the design matrix for the random effects with $\boldsymbol{z}_{ij} = [1, z_{ij1}, \ldots, z_{ijq}]^T$.

Stage Two: Random terms:

$$E[\boldsymbol{\epsilon}_i] = \boldsymbol{0}, \quad \text{var}(\boldsymbol{\epsilon}_i) = \boldsymbol{E}_i(\boldsymbol{\alpha}),$$
$$E[\boldsymbol{b}_i] = \boldsymbol{0}, \quad \text{var}(\boldsymbol{b}_i) = \boldsymbol{D}(\boldsymbol{\alpha}),$$
$$\text{cov}(\boldsymbol{b}_i, \boldsymbol{\epsilon}_i) = \boldsymbol{0}$$

where $\boldsymbol{\alpha}$ is the vector of variance–covariance parameters. A common model assumes

$$\boldsymbol{\epsilon}_i \sim_{ind} N(\boldsymbol{0}, \sigma_\epsilon^2 \boldsymbol{I}_{n_i}),$$
$$\boldsymbol{b}_i \sim_{iid} N(\boldsymbol{0}, \boldsymbol{D}).$$

For this model, $\boldsymbol{\alpha} = [\sigma_\epsilon^2, \boldsymbol{D}]$.

For nonlinear models even the first two moments are not available in closed form. In general:

$$E[Y_{ij}] = E_{\boldsymbol{b}_i}[f(\boldsymbol{x}_{ij}\boldsymbol{\beta} + \boldsymbol{z}_{ij}\boldsymbol{b}_i, t_{ij})] \neq f(\boldsymbol{x}_{ij}\boldsymbol{\beta}, t_{ij})$$

where $f(\boldsymbol{x}_{ij}\boldsymbol{\beta}, t_{ij})$ is the nonlinear curve evaluated at $\boldsymbol{b}_i = \boldsymbol{0}$. Hence, unlike the LMM, the nonlinear curve at a time point averaged across individuals is not equal to the nonlinear curve at that time for an average individual (i.e., one with $\boldsymbol{b}_i = \boldsymbol{0}$). The variance is

$$\text{var}(Y_{ij}) = \sigma_\epsilon^2 + \text{var}_{\boldsymbol{b}_i}[f(\boldsymbol{x}_{ij}\boldsymbol{\beta} + \boldsymbol{z}_{ij}\boldsymbol{b}_i, t_{ij})]$$

so that the marginal variance of the response is not constant across time, even when we have a random intercepts only model (unlike the LMM). For responses on the same individual, dependence is induced through the common random effects:

$$\text{cov}(Y_{ij}, Y_{ij'}) = \text{cov}_{\boldsymbol{b}_i}[f(\boldsymbol{x}_{ij}\boldsymbol{\beta} + \boldsymbol{z}_{ij}\boldsymbol{b}_i, t_{ij}), f(\boldsymbol{x}_{ij'}\boldsymbol{\beta} + \boldsymbol{z}_{ij'}\boldsymbol{b}_i, t_{ij'})]$$

but, as with the GLMM, there is no closed form for the covariance. Finally, for observations on different individuals:

$$\text{cov}(Y_{ij}, Y_{i'j'}) = 0$$

for $i \neq i'$. The data do not have a closed-form marginal distribution. These forms illustrate that picking particular random effect structures cannot be based on specific requirements in terms of the marginal variance and covariance. Rather, this choice should be based on the context and on data availability.

In a NLMM, the interpretation of parameters is usually tied to the particular model. In a GLMM, one can make use of linearity on the linear predictor scale to have an interpretation in terms of unit changes in covariates (as we have illustrated for loglinear and logistic linear models). In a NLMM, this will not be possible, however (since the model is *nonlinear*!).

We next briefly consider parameterization of the model, before considering likelihood and Bayesian inference in Sects. 9.17 and 9.18, respectively. A GEE approach is briefly considered in Sect. 9.19, but as previously mentioned, this is not as popular as likelihood and Bayes approaches, and so this section is short. The nonlinearity of the model means there is no sufficient statistic for β, and so conditional likelihood cannot be used.

9.16 Parameterization of the Nonlinear Model

In contrast to LMMs and GLMMs, there is no obvious way to parameterize a NLMM, and the way one proceeds is an art form. Given the normal random effects distribution, one usually parameterizes to quantities on the whole real line. This issue relates to the discussion of the solution locus and the parameterization of nonlinear models given in Sect. 6.15.

Example: A Simple Pharmacokinetic Model

The simplest pharmacokinetic model is

$$\mathrm{E}[Y \mid V, k_e] = \frac{D}{V} \exp(-k_e t)$$

where D is the known dose, $V > 0$ is the volume of distribution, and $k_e > 0$ is the elimination rate constant. The obvious parameterization is $\beta_0 = \log V, \beta_1 = \log k_e$. A key parameter of interest is the clearance, defined as $Cl = V \times k_e$, and so one may alternatively take $\beta_1^\star = \log Cl$ with $\beta_0^\star = \beta_0$ as before. This parameterization has a number of advantages. A first advantage is that the clearance for individual i is often modeled as a function of covariates, for example, via a loglinear model of the form

$$\log Cl = \alpha_0 + \alpha_1 x_i \qquad (9.29)$$

where x_i is a covariate of interest such as weight. A second advantage is that the clearance is a very stable parameter to estimate. The clearance is the dose D divided by the area under the concentration–time curve, and this area tends to be very well estimated (unless there are few sample points at large times) and hence so does the clearance, Cl.

If a Bayesian approach is adopted, then the prior must clearly be specific to the parameterization. For example, for $\boldsymbol{\beta} = [\beta_0, \beta_1]^{\mathrm{T}}$ and $\boldsymbol{\beta}^* = [\beta_0^*, \beta_1^*]^{\mathrm{T}}$ the prior $\boldsymbol{\beta} \sim \mathrm{N}_2(\boldsymbol{\mu}_0, \boldsymbol{\Sigma}_0)$ with fixed $\boldsymbol{\mu}_0, \boldsymbol{\Sigma}_0$, will clearly give different inference to assuming $\boldsymbol{\beta}^* \sim \mathrm{N}_2(\boldsymbol{\mu}_0, \boldsymbol{\Sigma}_0)$. $\square$

There is some theoretical work on choosing parameterizations (Bates and Watts 1980), but good parameterizations are often found through experience with particular models. The accuracy of asymptotic approximations is also crucially dependent on the choice of parameterization, with stable parameters likely to display good asymptotic properties. The examination of likelihood contours (as was done in Sect. 6.12) can indicate whether asymptotic distributions are likely to be accurate or not.

With many nonlinear models, care must be taken to ensure the model is identifiable in the sense that if $\boldsymbol{\theta} \neq \boldsymbol{\theta}'$, $f(\boldsymbol{\theta}) \neq f(\boldsymbol{\theta}')$. If there is non-identifiability, then one may either reparameterize the model or enforce identifiability through the prior. The latter can be messy, however.

Unfortunately, preserving identifiability and retaining an interpretable parameter cannot usually be simultaneously achieved. We illustrate the problems with an example.

Example: Pharmacokinetics of Theophylline

As discussed in Sect. 6.2, the one-compartment open model is non-identifiable. We illustrate by parameterizing as $[k_e, k_a, Cl]$ to give the mean model, for a generic individual, as

$$\mathrm{E}[Y] = \frac{D k_e k_a}{Cl(k_a - k_e)} \left[\exp(-k_e t) - \exp(-k_a t)\right]. \tag{9.30}$$

This form is known as the "flip-flop" model because the parameters $[k_e, k_a, Cl]$ give the same curve as the parameters $[k_a, k_e, Cl]$. To enforce identifiability, it is typical to assume that $k_a > k_e > 0$, since for many drugs, absorption is faster than elimination. This suggests the parameterization $[\log k_e, \log(k_a - k_e), \log Cl]$.

9.17 Likelihood Inference for the Nonlinear Mixed Model

As with the linear mixed and generalized linear mixed models already considered, the likelihood is defined with respect to fixed effects β and variance components α:

$$p(\boldsymbol{y} \mid \boldsymbol{\beta}, \boldsymbol{\alpha}) = \prod_{i=1}^{m} \int_{\boldsymbol{b}_i} p(\boldsymbol{y}_i \mid \boldsymbol{b}_i, \boldsymbol{\beta}, \sigma_\epsilon^2) \times p(\boldsymbol{b}_i \mid \boldsymbol{D}) \, d\boldsymbol{b}_i, \qquad (9.31)$$

with $\boldsymbol{\alpha} = [\boldsymbol{D}, \sigma_\varepsilon^2]$.

The first difficulty to overcome is how to calculate the required integrals, which for nonlinear models are analytically intractable (recall for the LMM they were available in closed form). As with the GLMM, two obvious approaches are to resort to Laplace approximations or adaptive Gauss–Hermite. Pinheiro and Bates (2000, Chap. 7) contains extensive details on these approaches (see also Bates 2011). We wish to evaluate

$$p(\boldsymbol{y}_i \mid \boldsymbol{\beta}, \boldsymbol{\alpha}) = (2\pi\sigma_\epsilon^2)^{-n_i/2}(2\pi)^{-(q+1)/2}|\boldsymbol{D}|^{-1/2} \int \exp[\,n_i g(\boldsymbol{b}_i)\,]\, d\boldsymbol{b}_i,$$

where

$$-2n_i g(\boldsymbol{b}_i) = [\boldsymbol{y}_i - \boldsymbol{f}_i(\boldsymbol{\beta}, \boldsymbol{b}_i, \boldsymbol{x}_i)]^{\mathrm{T}}[\boldsymbol{y}_i - \boldsymbol{f}_i(\boldsymbol{\beta}, \boldsymbol{b}_i, \boldsymbol{x}_i)]/\sigma_\epsilon^2 + \boldsymbol{b}_i^{\mathrm{T}} \boldsymbol{D}^{-1} \boldsymbol{b}_i \qquad (9.32)$$

and

$$\boldsymbol{f}_i(\boldsymbol{\beta}, \boldsymbol{b}_i) = [f(\boldsymbol{x}_{i1}\boldsymbol{\beta} + \boldsymbol{z}_{i1}\boldsymbol{b}_i, t_{i1}), \ldots, f(\boldsymbol{x}_{in_i}\boldsymbol{\beta} + \boldsymbol{z}_{in_i}\boldsymbol{b}_i, t_{in_i})]^{\mathrm{T}}.$$

The Laplace approximation (Sect. 3.7.2) is a second-order Taylor series expansion of $g(\cdot)$ about

$$\widehat{\boldsymbol{b}}_i = \arg\min_{\boldsymbol{b}_i} \; [-g(\boldsymbol{b}_i)]$$

where this minimization constitutes a penalized least squares problem. For a nonlinear model, numerical methods are required for this minimization, but the dimensionality, $q+1$, is typically small. With respect to (9.31), the second difficulty is how to maximize the likelihood as a function of $\boldsymbol{\beta}$ and $\boldsymbol{\alpha}$; again see Pinheiro and Bates (2000) and Bates (2011) for details.

In terms of the random effects, empirical Bayes estimates may be calculated, as with the GLMM. In the example that follows, we evaluate the MLEs using the procedure described in Lindstrom and Bates (1990) in which estimates of $\boldsymbol{b}_i$ are $\boldsymbol{\beta}$ are first obtained by minimizing the penalized least squares criteria (9.32), given estimates of $\boldsymbol{D}$ and σ_ϵ^2. Then a first-order Taylor series expansion of $\boldsymbol{f}_i$ about the current estimates of $\boldsymbol{\beta}$ and $\boldsymbol{b}_i$ is carried out, which results in a LMM. For such a model, the random effects may be integrated out analytically, and the subsequent (approximate) likelihood can be maximized with respect to $\boldsymbol{D}$ and σ_ϵ^2. This procedure is then iterated until convergence.

Approximate inference for $[\beta, \alpha]$ is carried out via asymptotic normality of the MLE:

$$\begin{bmatrix} \widehat{\beta} \\ \widehat{\alpha} \end{bmatrix} \sim N\left(\begin{bmatrix} \beta \\ \alpha \end{bmatrix}, \begin{bmatrix} I_{\beta\beta} & I_{\beta\alpha} \\ I_{\alpha\beta} & I_{\alpha\alpha} \end{bmatrix}^{-1} \right)$$

where $I_{\beta\beta}$, $I_{\beta\alpha}$, $I_{\alpha\beta}$, and $I_{\alpha\alpha}$ are the relevant information matrices.

Many approximation strategies have been suggested for nonlinear hierarchical models, but care is required since validity of the asymptotic distribution depends on the approximation used. For example, a historically popular approach (Beal and Sheiner 1982) was to carry out a first-order Taylor series about $E[b_i] = 0$ to give

$$y_{ij} = f(x_{ij}\beta_i + z_{ij}b_i, t_{ij}) + \epsilon_{ij}$$

$$\approx f(x_{ij}\beta_i, t_{ij}) + b_i^T \left.\frac{\partial f}{\partial b_i}\right|_{b_i=0} + \epsilon_{ij}.$$

This first-order estimator is inconsistent, however, and has bias even if n_i and m both go to infinity; see Demidenko (2004, Chap. 8).

Example: Pharmacokinetics of Theophylline

For these data, the one-compartment model with first-order absorption and elimination is a good starting point for analysis. This model was described in some detail in Sect. 6.16.3. The mean concentration at time point t_{ij} for subject i is

$$\frac{D_i k_{ai} k_{ei}}{Cl_i(k_{ai} - k_{ei})} \left[\exp(-k_{ei}t_{ij}) - \exp(k_{ai}t_{ij})\right], \tag{9.33}$$

where we have parameterized in terms of $[Cl_i, k_{ai}, k_{ei}]$ and D_i is the initial dose.

We first fit the above model to each individual, using nonlinear least squares; Fig. 9.9 gives the resultant 95% asymptotic confidence intervals. The between-individual variability is evident, particularly for $\log k_a$. Figure 9.10 displays the data along with the fitted curves. The general shape of the curve seems reasonable, but the peak is missed for a number of individuals (e.g., numbers 10, 1, 5, and 9).

Turning now to a NLMM, we assume that each of the parameters is treated as a random effect so that

$$\log k_{ei} = \beta_1 + b_{1i} \tag{9.34}$$

$$\log k_{ai} = \beta_2 + b_{2i} \tag{9.35}$$

$$\log Cl_i = \beta_3 + b_{3i} \tag{9.36}$$

9.17 Likelihood Inference for the Nonlinear Mixed Model

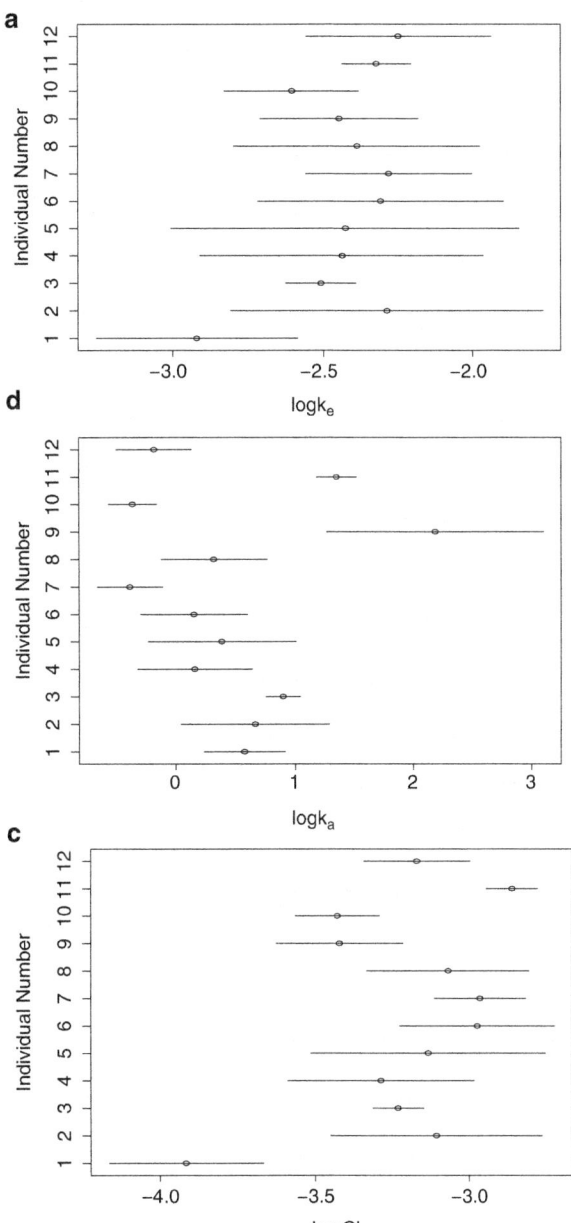

Fig. 9.9 95% confidence intervals for each of the three parameters and 12 individuals in the theophylline data. Obtained via individual fitting

with $b_i \mid D \sim N_3(0, D)$ where $b_i = [b_{i1}, b_{i2}, b_{i3}]^{\text{T}}$. The estimates resulting from the Lindstrom and Bates (1990) method described in the previous section are given in Table 9.13. The standard deviation of the random effects for $\log k_a$ is large, as we anticipated from examination of Fig. 9.9.

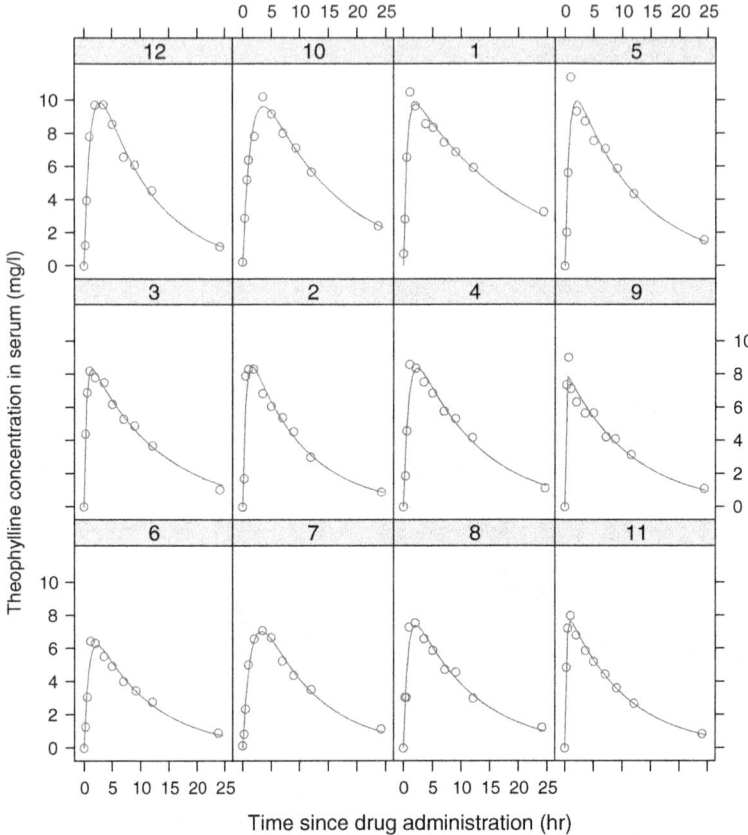

Fig. 9.10 Concentrations versus time for 12 individuals given the drug theophylline, along with individual nonlinear least squares fits

9.18 Bayesian Inference for the Nonlinear Mixed Model

The first two stages of the model are as in the likelihood formulation. We first discuss how hyperpriors may be specified, before discussing inference for functions of interest.

9.18.1 Hyperpriors

A Bayesian approach requires a prior distribution for β, α. As with the LMM, a proper prior is required for the matrix D. In contrast to the LMM, a proper prior is required for β also, to ensure the propriety of the posterior distribution.

9.18 Bayesian Inference for the Nonlinear Mixed Model

Table 9.13 Comparison of likelihood and Bayesian NLMM estimation techniques for the theophylline data

PK label	Parameter	Likelihood		Bayes normal		Bayes lognorm		Bayes power	
		Est.	(s.e.)	Est.	(s.d.)	Est.	(s.d.)	Est.	(s.d.)
log k_e	β_1	−2.43	(0.063)	−2.46	(0.077)	−2.43	(0.075)	−2.25	(0.083)
log k_a	β_2	0.45	(0.20)	0.47	(0.19)	0.26	(0.23)	0.45	(0.22)
log Cl	β_3	−3.21	(0.081)	−3.23	(0.082)	−3.22	(0.090)	−3.22	(0.092)
log k_e	$\sqrt{D_{11}}$	0.13	(−)	0.19	(0.049)	0.22	(0.059)	0.23	(0.061)
log k_a	$\sqrt{D_{22}}$	0.64	(−)	0.62	(0.15)	0.72	(0.19)	0.69	(0.18)
log Cl	$\sqrt{D_{33}}$	0.25	(−)	0.25	(0.051)	0.30	(0.071)	0.29	(0.072)

For the likelihood summaries, we report the MLEs and the asymptotic standard errors, while for the Bayesian analysis, we report the mean and standard deviation of the posterior distribution. The three Bayesian models differ in the error models assumed at the first stage with normal, lognormal, and power models being considered

If parameters occur linearly, then proper priors are not required, but, as usual, the safest strategy is to specify proper priors.

For simplicity, we assume that random effects are associated with all parameters and as, in Sect. 8.6.3, parameterize the model as $\tau = \sigma_\epsilon^{-2}$, $\boldsymbol{W} = \boldsymbol{D}^{-1}$, and $\boldsymbol{\beta}_i = \boldsymbol{\beta} + \boldsymbol{b}_i$ for $i = 1, \ldots, m$, with the dimensionality of $\boldsymbol{\beta}_i$ being $k + 1$. The joint posterior is

$$p(\boldsymbol{\beta}_1, \ldots, \boldsymbol{\beta}_m, \tau, \boldsymbol{\beta}, \boldsymbol{W} \mid \boldsymbol{y}) \propto \prod_{i=1}^{m} [p(\boldsymbol{y}_i \mid \boldsymbol{\beta}_i, \tau) p(\boldsymbol{\beta}_i \mid \boldsymbol{\beta}, \boldsymbol{W})] \pi(\boldsymbol{\beta}) \pi(\tau) \pi(\boldsymbol{W}).$$

We assume the priors

$$\boldsymbol{\beta} \sim \text{N}_{k+1}(\boldsymbol{\beta}_0, \boldsymbol{V}_0), \quad \tau \sim \text{Ga}(a_0, b_0), \quad \boldsymbol{W} \sim \text{Wish}_{k+1}(r, \boldsymbol{R}^{-1}),$$

for further discussion of this specification, see Sect. 8.6.2. Closed-form inference is unavailable, but MCMC is almost as straightforward as in the LMM case. The INLA approach is not (at time of writing) available for the Bayesian analysis of nonlinear models. With respect to MCMC, the conditional distributions for $\boldsymbol{\beta}, \tau, \boldsymbol{W}$ are unchanged from the linear case. There is no closed-form conditional distribution for $\boldsymbol{\beta}_i$, which is given by

$$p(\boldsymbol{\beta}_i \mid \boldsymbol{\beta}, \tau, \boldsymbol{W}, \boldsymbol{y}) \propto p(\boldsymbol{y}_i \mid \boldsymbol{\beta}_i, \tau) \times p(\boldsymbol{\beta}_i \mid \boldsymbol{\beta}, \boldsymbol{W})$$

but a Metropolis–Hastings step can be used (to give a Metropolis within Gibbs algorithm, as described in Sect. 3.8.5).

9.18.2 Inference for Functions of Interest

We discuss prior choice and inferential summaries in the context of fitting a NLMM to the theophylline data. For these data, the parameterization

$$\boldsymbol{\beta}_i = [\log k_{ei}, \log k_{ai}, \log Cl_i]$$

was initially adopted, with random effects normal prior $\boldsymbol{\beta}_i \mid \boldsymbol{\beta}, \boldsymbol{D} \sim_{iid} \mathrm{N}_3(\boldsymbol{\beta}, \boldsymbol{D})$. We assume independent normal priors for the elements of $\boldsymbol{\beta}$, centered at 0 and with large variances (recall that we need proper priors). For $\boldsymbol{D}^{-1}$, we assume a Wishart$(r, \boldsymbol{R}^{-1})$ distribution with diagonal $\boldsymbol{R}$ (see Sect. 8.6.2 and Appendix D for discussion of the Wishart distribution). We describe the procedure that is followed in order to choose the diagonal elements.

Consider a generic univariate "natural" parameter θ (e.g., k_e, k_a, or Cl) for which we assume the lognormal prior LogNorm(β, D). Pharmacokineticists have insight into the coefficient of variation for θ, that is, $\mathrm{CV}(\theta) = \mathrm{sd}(\theta)/\mathrm{E}[\theta]$. Recall the first two moments of a lognormal

$$\mathrm{E}[\theta] = \exp(\beta + D/2)$$
$$\mathrm{var}(\theta) = \mathrm{E}[\theta]^2[\exp(D) - 1]$$
$$\mathrm{sd}(\theta) = \mathrm{E}[\theta]\sqrt{\exp(D) - 1}$$
$$\approx \mathrm{E}[\theta]\sqrt{D}$$

so that

$$\mathrm{CV}(\theta) \approx \sqrt{D}.$$

We can therefore assign a prior for D by providing a prior estimate of $\sqrt{D}$. Under the Wishart parameterization, we have adopted $\mathrm{E}[\boldsymbol{D}^{-1}] = r\boldsymbol{R}^{-1}$. We take $r = 3$ (which is the smallest integer that gives a proper prior) and $\boldsymbol{R} = \mathrm{diag}(1/5, 1/5, 1/5)$ which gives $\mathrm{E}[D_{kk}^{-1}] = 15$ so that, for $k = 1, 2, 3$, $\mathrm{E}[\sqrt{D_{kk}}] \approx 1/\sqrt{15} = 0.26$, or an approximate prior expectation of the coefficient of variation of 26%, which is reasonable in this context (Wakefield et al. 1999).

For inference, again consider a generic parameter θ with prior LogNorm(β, D). The mode, median, and mean of the *population distribution* of θ are

$$\exp(\beta - \sqrt{D}), \qquad \exp(\beta), \qquad \exp(\beta + D/2),$$

respectively. Further, $\exp(\beta \pm 1.96\sqrt{D})$ is a 95% interval for θ in the population. Consequently, given samples from the posterior $p(\beta, D \mid \boldsymbol{y})$, one may simply convert to samples for any of these summaries.

In a pharmacokinetic context, interest often focuses on various functions of the natural parameters. As a first example, consider the terminal half-life which is given by $t_{1/2} = k_e^{-1} \log 2$. In the parameterization adopted in the theophylline study, $\log k_e \sim \text{N}(\beta_1, D_{11})$, and so the distribution of the log half-life is normal also:

$$\log t_{1/2} \sim \text{N}[\log(\log 2) - \beta_1, D_{11}]$$

which simplifies inference since one can summarize the population distribution in the same way as was just described for a generic parameter θ. Other parameters of interest are not simple linear combinations, however. For example, the time to maximum is

$$t_{\max} = \frac{1}{k_a - k_e} \log\left(\frac{k_a}{k_e}\right)$$

and the maximum concentration is

$$\text{E}[Y \mid t_{\max}] = \frac{Dk_a}{V(k_a - k_e)} [\exp(-k_e t_{\max}) - \exp(-k_a t_{\max})]$$

$$= \frac{D}{V} \left(\frac{k_a}{k_e}\right)^{k_a/(k_a - k_e)}.$$

For such summaries, the population distribution may be examined by simulating parameter sets $[\log k_e, \log k_a, \log Cl]$ for new individuals from the population distribution, and then converting to the functions of interest.

As noted in Sect. 9.16, the parameterization $[\log k_e, \log k_a, \log Cl]$ that we have adopted is non-identifiable since the same likelihood values are achieved with the set $[\log k_a, \log k_e, \log Cl]$. For the theophylline data, we performed MCMC with two chains, and one of the chains "flipped" between the two non-identifiable regions in the parameter space, as illustrated in Fig. 9.11 (note that in panels (a) and (b), the vertical axes have the same scale). In this plot the three population parameters β_1, β_2, β_3 are plotted in the three rows. Here, the labeling of β_1 and β_2 is arbitrary. The parameter β_3 is unaffected by the flip-flop behavior because the mean log clearance is the same under each nonidentifiable set. In Fig. 9.11(a), the chain represented by the solid line corresponds to the smaller of the two rate constants and, after a small period of burn-in, remains in the region of the parameter space corresponding to the smaller constant. In contrast, the chain represented by the dotted line flips to the region corresponding to the larger rate constant at around (thinned) iteration number 200. In panel (b), we see that the dotted chain flips the other way, as it is required to do.

We now constrain the parameters by enforcing the known ordering on the rates: $k_{ai} > k_{ei} > 0$. To avoid the flip-flop problem, we use the parameterization

$$\theta_{1i} = \log k_{ei} = \beta_1 + b_{1i} \tag{9.37}$$

$$\theta_{2i} = \log(k_{ai} - k_{ei}) = \beta_2 + b_{2i} \tag{9.38}$$

$$\theta_{3i} = \log Cl_i = \beta_3 + b_{3i} \tag{9.39}$$

Fig. 9.11 Demonstration of flip-flop behavior for the theophylline data and the unconstrained parameterization given by (9.34)–(9.36): (**a**) β_1, (**b**) β_2, (**c**) β_3. Thinned realizations from two chains appear in each plot

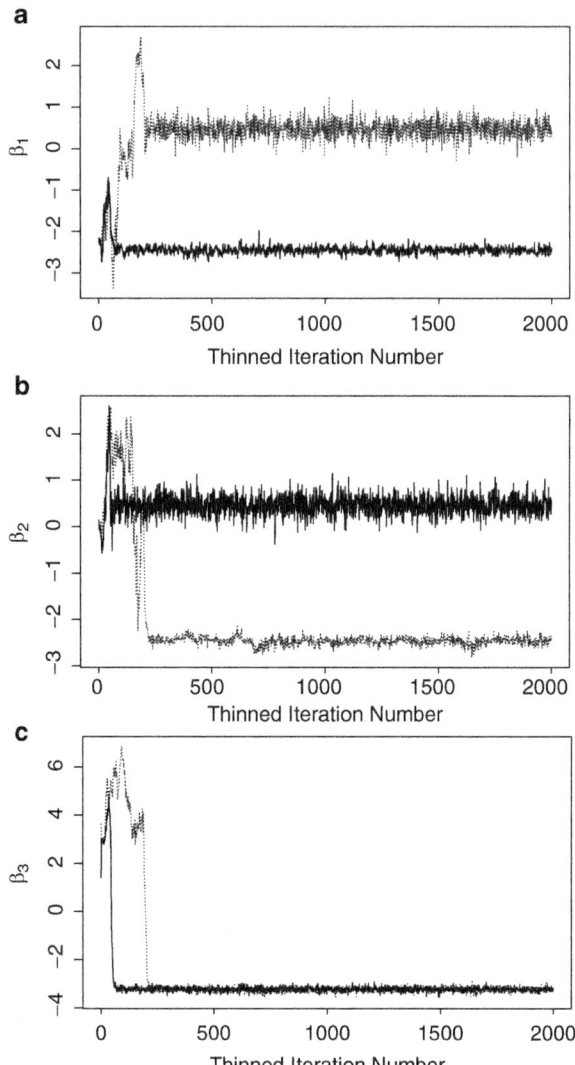

with $\boldsymbol{b}_i = [b_{1i}, b_{2i}, b_{3i}]^{\mathrm{T}} \sim \mathrm{N}_3(\boldsymbol{0}, \boldsymbol{D})$. This is a different model to the model that does not prevent flip-flop since the prior inputs are different. In this case, we keep the same priors which correspond to assuming that the coefficient of variation for $k_a - k_e$ is around 26% which is clearly less meaningful, but in this example, k_a is considerably larger than k_e.

We can convert to the original parameters via

$$k_{ei} = \exp(\theta_{1i})$$
$$k_{ai} = \exp(\theta_{1i}) + \exp(\theta_{2i})$$
$$Cl_i = \exp(\theta_{3i}).$$

Inference for the population distribution of k_{ei} and Cl_i is straightforward, but for k_{ai}, more work is required. However, the expectation of the population absorption rate is

$$\mathrm{E}[k_{ai}] = \mathrm{E}[\exp(\theta_{1i}) + \exp(\theta_{2i})]$$
$$= \exp\left(\beta_1 + \sqrt{D_{11}}/2\right) + \exp(\beta_1 + \sqrt{D_{11}}/2).$$

A full Bayesian analysis is postponed until later in the chapter (at the end of Sect. 9.20).

9.19 Generalized Estimating Equations

If interest lies in population parameters, then we may use the estimator $\widehat{\gamma}$ that satisfies

$$G(\gamma, \widehat{\alpha}) = \sum_{i=1}^{m} D_i^{\mathrm{T}} W_i^{-1}(Y_i - f_i) = 0, \tag{9.40}$$

where $D_i = \partial f_i/\partial \gamma$, $W_i = W_i(\gamma, \widehat{\alpha})$ is the working covariance model, $f_i = f_i(\gamma)$, and $\widehat{\alpha}$ is a consistent estimator of α. Sandwich estimation may be used to obtain an empirical estimate of the variance V_γ:

$$\left(\sum_{i=1}^{m} D_i^{\mathrm{T}} W_i^{-1} D_i\right)^{-1} \left[\sum_{i=1}^{m} D_i^{\mathrm{T}} W_i^{-1} \mathrm{cov}(Y_i) W_i^{-1} D_i\right] \left(\sum_{i=1}^{m} D_i^{\mathrm{T}} W_i^{-1} D_i\right)^{-1}. \tag{9.41}$$

We then have the usual asymptotic result: $V_\gamma^{-1/2}(\widehat{\gamma} - \gamma) \to_d \mathrm{N}(0, I)$.

GEE has not been extensively used in a nonlinear (non-GLM) setting. This is partly because in many settings (e.g., pharmacokinetics/pharmacodynamics), interest focuses on understanding between-individual variability, and explaining this in terms of individual-specific covariates, or making predictions for particular individuals. The interpretation of the parameters within a GEE implementation is also not straightforward. For a marginal GLM, there is a link function and a *linear predictor* which allows interpretation in terms of differences in averages between

Table 9.14 GEE estimates of marginal parameters for the theophylline data

PK label	Parameter	Est.	(s.e.)
log k_e	γ_1	−2.52	(0.068)
log k_a	γ_2	0.40	(0.17)
log Cl	γ_3	−3.25	(0.076)

populations defined by covariates; see Sect. 9.11. Consider a nonlinear model over time. In a mixed model, the population mean parameters are averages of individual-level parameters. A marginal approach models the average response as a nonlinear function of time, and the parameters do not, in general, have interpretations as averages of parameters. Rather, parameters within a marginal nonlinear model determine a population-averaged curve. The parameters can be made a function of covariates such as age and gender, but the interpretation is less clear when compared to a mixed model formulation. For example, in (9.29), we model the individual-level log clearance as a function of a covariate x_i. We could include covariates in the marginal model in an analogous fashion, but it is not individual clearance we are modeling, and the subsequent analysis cannot be used in the same way to derive optimal doses as a function of x, for example. Obviously, GEE cannot provide estimates of between-individual variability or obtain predictions for individuals.

Example: Pharmacokinetics of Theophylline

GEE was implemented with mean model

$$\text{E}[Y_{ij}] = f_i(\boldsymbol{\gamma}) = \frac{D_i \exp(\gamma_1 + \gamma_2)}{\exp(\gamma_3)[\exp(\gamma_2) - \exp(\gamma_1)]} \left[\exp(-e^{\gamma_1} t_{ij}) - \exp(-e^{\gamma_2} t_{ij})\right]. \tag{9.42}$$

As just discussed, the interpretation of the parameters for this model is not straightforward since we are simply modeling a population-averaged curve. So, for example, $k_e = \exp(\gamma_1)$ is the rate of elimination that defines the population-averaged curve and is *not* the average elimination rate in the population.

We use working independence ($\boldsymbol{W}_i = \boldsymbol{I}_{n_i}$) so that (9.40) is equivalent to a nonlinear least squares criteria, which allows the estimates to be found using standard software. The variance estimate (9.41) simplifies under working independence, and the most tedious part is evaluating the $n_i \times 3$ matrix of partial derivatives $\boldsymbol{D}_i = \partial \boldsymbol{f}_i / \partial \boldsymbol{\gamma}$. The estimates and standard errors are given in Table 9.14. It is not possible to directly compare these estimates with those obtained from a mixed model formulation.

9.20 Assessment of Assumptions for General Regression Models

Model checking proceeds as with the linear model with dependent data (Sect. 8.8) except that interpretation is not as straightforward since the properties of residuals are difficult to determine even when the model is correct. We focus on generalized and nonlinear mixed models. For both of these classes Pearson (stage one) residuals,

$$e_{ij} = \frac{Y_{ij} - \mathrm{E}[Y_{ij} \mid \boldsymbol{b}_i]}{\sqrt{\mathrm{var}(Y_{ij} \mid \boldsymbol{b}_i)}}$$

are straightforward to calculate.

With respect to mixed models, as with the LMM, there are assumptions at each of the stages, and one should endeavor to provide checks at each stage. If we are in the situation in which there are individuals with sufficient data to reliably estimate the parameters from these data alone, we should use the resultant estimates to provide checks. Residuals from individual fits can be used to assess whether the nonlinear model is appropriate and if the assumed variance model is appropriate. One may also construct normal QQ plots and bivariate plots of the estimated individual-level parameters to see if the second-stage normality assumption appears reasonable. In a nonlinear setting, there are few results availability on consistency of estimates, unless the model is correct, and so it is far more important to have random effects distributions that are approximately correctly specified.

If individual-level covariates are available, then the estimated parameters may be plotted against these to determine whether a second-stage regression model is appropriate (if we are in exploratory mode). In the pharmacokinetic context, one may model clearance as a function of weight, for example, via a loglinear model as in (9.29). Examining whether the spread of the random effects estimates changes with covariates is also an important step.

All of the above checks can be carried out based on the (shrunken) estimates obtained from random effects modeling, but caution is required as these estimates may be strongly influenced by the assumption of normality. If n_i is large, then this will be less problematic.

Example: Pharmacokinetics of Theophylline

We present some diagnostics for the theophylline data. We first carry out individual fitting using nonlinear least squares (which is possible here since $n_i = 11$), and Fig. 9.12 gives normal QQ plots of the $\log k_e$, $\log k_a$, and $\log Cl$ parameters. There is at least one outlying individual here, but there is nothing too worrying in these plots.

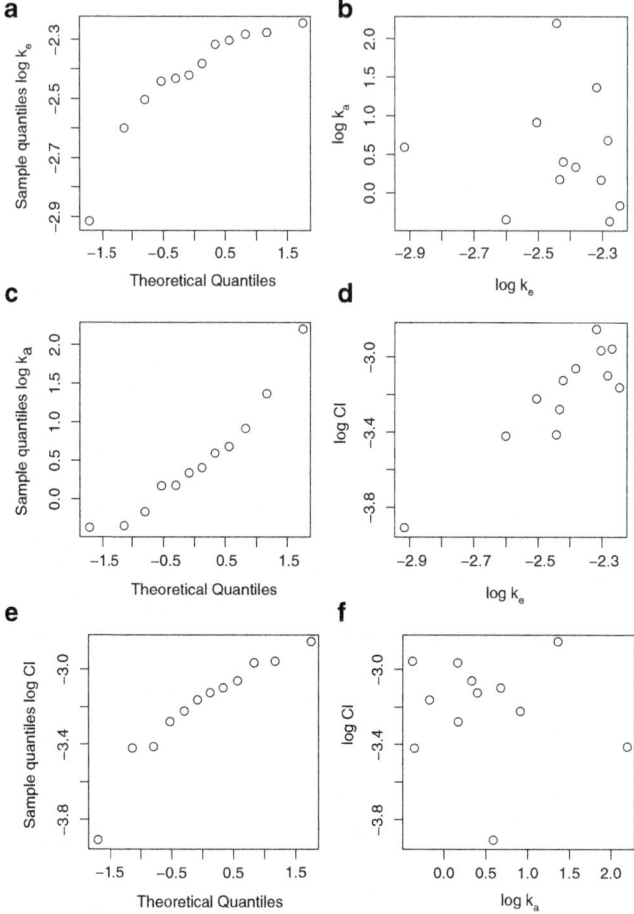

Fig. 9.12 Normal QQ plots (*left column*) and scatterplots (*right column*) of the parameter estimates from individual nonlinear least square fits for the theophylline data. (**a**) QQ plot for $\log k_e$, (**b**) $\log k_a$ versus $\log k_e$, (**c**) QQ plot for $\log k_a$, (**d**) $\log Cl$ versus $\log k_e$, (**e**) QQ plot for $\log Cl$, (**f**) $\log Cl$ versus $\log k_a$

In the following, a number of mixed models are fitted in an exploratory fashion in order to demonstrate some of the flexibility of NLMMs. We first fit a mixed model using MLE and the nonlinear form (9.33). The error terms were assumed to be normal on the concentration scale, with constant variance. Plots of the Pearson residuals versus fitted value and versus time are displayed in Figs. 9.13(a) and (b).

Figure 9.13(b) suggests the variance changes with time (or that the model is inadequate for time points close to 0), and we carry out another analysis with the model

$$\log y_{ij} = \log(\mu_{ij}) + \delta_{ij} \tag{9.43}$$

9.20 Assessment of Assumptions for General Regression Models

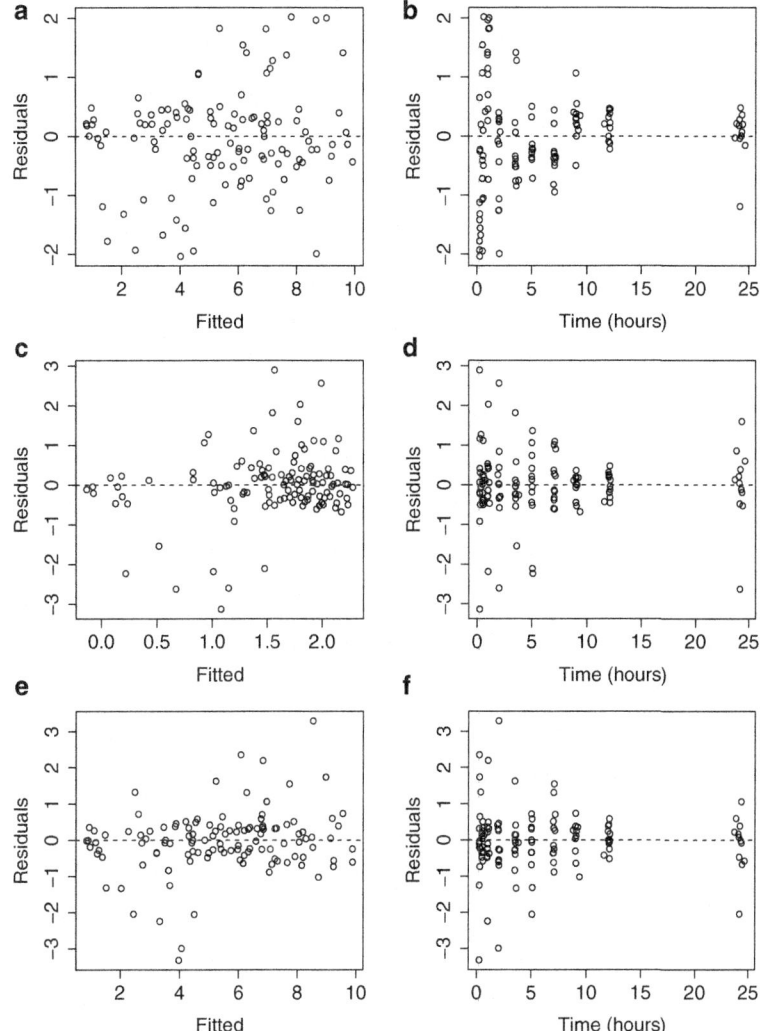

Fig. 9.13 Residuals obtained from various NLMM fits to the theophylline data: (**a**) normal model: residuals against fitted values (**b**) normal model: residuals against time, (**c**) lognormal model: residuals against fitted values (**d**) lognormal model: residuals against time, (**e**) power model: residuals against fitted values (**f**) power model: residuals against time

with μ_{ij} again given by (9.33) and $\delta_{ij} \mid \sigma_\delta^2 \sim_{iid} N(0, \sigma_\delta^2)$. This lognormal model has (approximately) a constant coefficient of variation. To fit this model, the responses at time 0 were removed since $\mu_{ij} = 0$ for $t_{ij} = 0$. This time, we adopt the parameterization that prevents flip-flop, that is, the model with (9.37)–(9.39). This model produced the Bayesian summaries given in Table 9.13 which are reasonably consistent with those in the normal model. Unfortunately, the residual plot in Fig. 9.13(d) shows only a slight improvement over the normal model in panel (b).

The next model considered was $y_{ij} = \mu_{ij} + \epsilon_{ij}$ with the *power* model

$$\epsilon_{ij} \mid \mu_{ij}, \sigma_0^2, \sigma_1^2, \gamma \sim_{ind} \mathrm{N}\left(0, \sigma_0^2 + \sigma_1^2 \mu_{ij}^\gamma\right) \tag{9.44}$$

with μ_{ij} given by (9.33) and $0 < \gamma \leq 2$. This model has two components of variance and may be used when an assay method displays constant measurement at low concentrations with the variance increasing with the mean for larger concentrations. See Davidian and Giltinan (1995, Sect. 2.2.3) for further discussion of variance models.

The joint prior on $[\sigma_0, \sigma_1, \gamma]$ can be difficult to specify since there is dependence between σ_1 and γ in particular. For simplicity, uniform priors on the range [0,2] were placed on σ_0 and σ_1. The parameter γ controls the strength of the mean–variance relationship, and, considering the second component only, the constant coefficient of variation model corresponds to $\gamma = 2$. In the pharmacokinetics literature, fixing $\gamma = 1$ or 2 is not uncommon. A uniform prior on [0,2] was specified for γ also. Figures 9.13(e) and (f) show the residual plots for this model, and we see some improvement over the other two error models, though there is still some misspecification evident at low time points in panel (f). Further analyses for these data might examine other absorption models (since the kinetics may be nonlinear, which could explain the poor fit at low times).

Posterior summaries for the power variance model are given in Fig. 9.14. The strong dependence between σ_1 and γ is evident in panel (f). There is a reasonable amount of uncertainty in the posterior for γ, but the median is 0.71. The parameter estimates for β and D are given in Table 9.13 and are similar to those from the normal and lognormal error models. Following the procedure described in Sect. 9.18.2, samples for the population medians for k_e, k_a, and Cl were generated, and these are displayed in Fig. 9.15, with notable skewness in the posteriors for k_a and Cl.

9.21 Concluding Remarks

The modeling of generalized linear and nonlinear dependent data is inherently more difficult than the modeling of linear dependent data due to mathematical tractability, the required computations to perform inference and parameter interpretation. Conceptually, however, the adaption of mixed (conditional) and GEE (marginal) models to the generalized linear and nonlinear scenarios is straightforward. With respect to parameter interpretation, the clear distinction between marginal and conditional models is critical and needs to be recognized.

There is little theory on the consistency of estimators in the face of model misspecification for GLMMs and NLMMs. This suggests that one should be more

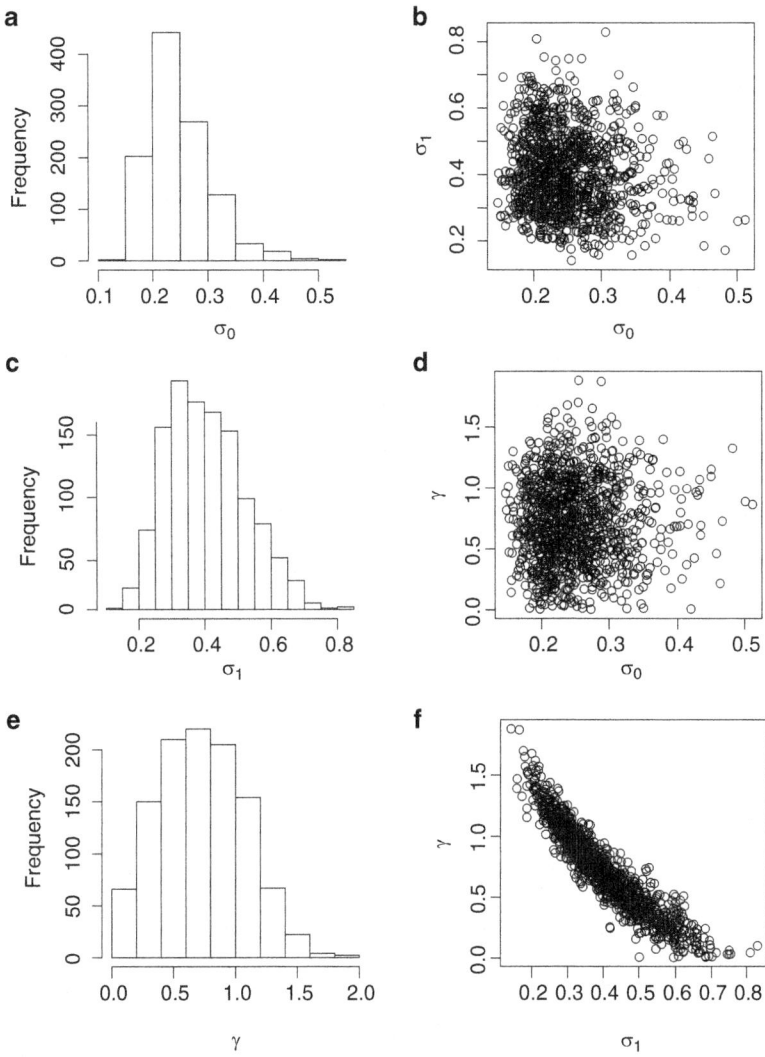

Fig. 9.14 Posterior summaries for the two-component power error model (9.44) fitted to the theophylline data. Posterior marginals for σ_0, σ_1, γ in the left common and bivariate plots in the right column

cautious in interpretation of the results from GLMMs and NLMMs, when compared to LMMs, and model checking should be carefully carried out. The effects of model misspecification with mixed models have attracted a lot of interest. Heagerty and Kurland (2001) illustrate the bias that is introduced when the random effects variances are a function of covariates. McCulloch and Neuhaus (2011) show that misspecification of the assumed random effects distribution has less impact on prediction of random effects. Sensitivity analyses, with respect to the random effects

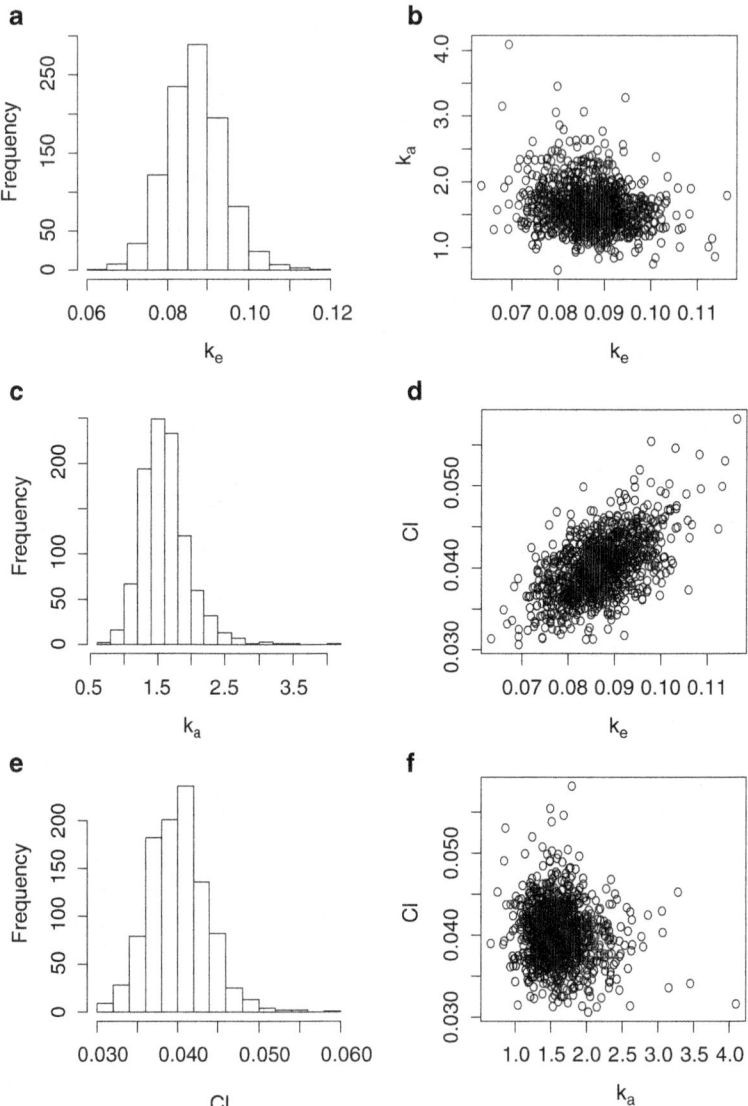

Fig. 9.15 Posterior distributions from the power model (9.44) fitted to the theophylline data: (**a**) population median k_e, (**b**) population median k_a versus population median k_e, (**c**) population median k_a, (**d**) population median k_e versus population median Cl, (**e**) population median Cl, (**f**) population median Cl versus population median k_a

distribution, for example, can be useful. The Bayesian approach, with computation via MCMC, is ideally suited to this endeavor. If the number of observations per unit, or the number of units, is small, then the MCMC route is appealing because one does not have to rely on asymptotic inference. Model checking is difficult in this situation, however.

We have not discussed REML in the context of GLMs; Smyth and Verbyla (1996) show how REML may be derived from a conditional likelihood approach in the context of GLMs with dispersion parameters and canonical link functions.

The modeling of dependent binary data is a difficult enterprise since binary observations contain little information, and there is no obvious choice of multivariate binary distribution. Logistic mixed models are intuitively appealing but are restrictive in the dependence structure they impose on the data. Care in computation is required, and the use of adaptive Gauss–Hermite for MLE, or MCMC for Bayes, is recommended. As always, GEE has desirable robustness properties for large numbers of clusters. In the GLM context, we emphasize the fitting of both types of model in a complimentary fashion. We have illustrated how marginal inference may be carried out with the logistic mixed model, which allows direct comparison of results with GEE.

9.22 Bibliographic Notes

Liang and Zeger (1986) and Zeger and Liang (1986) popularized GEE by considering GLMs with dependence within units (in the context of longitudinal data). Prentice (1988) proposed using a second set of estimating equations for α. Gourieroux et al. (1984) considered the quadratic exponential model. Zhao and Prentice (1990) discussed the use of this model for multivariate binary data and Prentice and Zhao (1991) for general responses (to give the approach labelled GEE2). Qaqish and Ivanova (2006) describe an algorithm for detecting when an arbitrary set of logistic contrasts correspond to a valid set of joint probabilities and for computing them if they provide a legal set. Fitzmaurice et al. (2004) is a very readable account of the modeling of longitudinal data with GLMs, from a frequentist (GEE and mixed model) perspective.

An extensive treatment of Bayesian multilevel modeling is described in Gelman and Hill (2007). We have concentrated on inverse gamma priors for random effects variances, but a popular alternative is the half-normal prior; see Gelman (2006) for further details. Fong et al. (2010) describe how the INLA computational approach may be used for GLMMs, including a description of its shortcomings, in terms of accuracy, for the analysis of binary data. Models and methods of analysis for spatial data are reviewed in Gelfand et al. (2010).

Davidian and Giltinan (1995) is an extensive and excellent treatment of nonlinear modeling with dependent responses, mostly from a non-Bayesian perspective. Pinheiro and Bates (2000) is also excellent and covers mixed models (again primarily from a likelihood perspective) and is particularly good on computation.

9.23 Exercises

9.1 Consider the model

$$\mathrm{E}[Y \mid b] = \frac{\exp(\boldsymbol{\beta}\boldsymbol{x} + b)}{1 + \exp(\boldsymbol{\beta}\boldsymbol{x} + b)},$$

with $b \mid \sigma_0^2 \sim_{iid} \mathrm{N}(0, \sigma_0^2)$. Prove that

$$\mathrm{E}[Y] \approx \frac{\exp\left[\boldsymbol{\beta}\boldsymbol{x}/(c^2\sigma_0^2 + 1)^{1/2}\right]}{1 + \exp\left[\boldsymbol{\beta}\boldsymbol{x}/(c^2\sigma_0^2 + 1)^{1/2}\right]}$$

where $c = 16\sqrt{3}/(15\pi)$.
[Hint: $G(z) \approx \Phi(cz)$ where $G(z) = (1 + e^{-z})^{-1}$ is the CDF of a logistic random variable, and $\Phi(\cdot)$ is the CDF of a normal random variable.]

9.2 Show that if each response is on the whole real line, then the density (9.19), with $c_i = 0$, corresponds to the multivariate normal model.

9.3 With respect to Table 9.11, show that, for a model for two binary responses parameterized in terms of the marginal means and marginal odds ratio, the likelihood is given by (9.26).

9.4 Sommer (1982) contains details of a study on 275 children in Indonesia. This study examined, among other things, the association between the risk of respiratory infection and xerophthalmia (dry eye syndrome), which may be caused by vitamin A deficiency. These data are available in the R package epicalc and are named Xerop.

Consider the marginal model for the jth observation on the ith child

$$\log\left(\frac{\mathrm{E}[Y_{ij}]}{1 - \mathrm{E}[Y_{ij}]}\right) = \gamma_0 + \gamma_1 \texttt{gender}_{ij} + \gamma_2 \texttt{hfora}_{ij} + \gamma_3 \texttt{cos}_{ij} +$$

$$\gamma_4 \texttt{sin}_{ij} + \gamma_5 \texttt{xero}_{ij} + \gamma_6 \texttt{age}_{ij} + \gamma_7 \texttt{age}_{ij}^2 \qquad (9.45)$$

where:

- Y_{ij} is the absence/presence of respiratory infection.
- $\texttt{gender}_{ij}$ is the gender (0 = male, 1 = female).
- $\texttt{hfora}_{ij}$ is the height-for-age.
- $\texttt{cos}_{ij}$ is the cosine of time of measurement i, j (time is in number of quarters).
- $\texttt{sin}_{ij}$ is the sine of time of measurement i, j (time is in number of quarters).
- $\texttt{xero}_{ij}$ is the absence/presence (0/1) of xerophthalmia.
- $\texttt{age}_{ij}$ is the age.

See Example 9.4 of Diggle et al. (2002) for more details on this model.

(a) Interpret each of the coefficients in (9.45).
(b) Provide parameter estimates and standard errors from a GEE analysis.

9.23 Exercises

(c) Consider a GLMM logistic analysis with a normally distributed random intercept and the conditional version of the regression model (9.45). Interpret the coefficients of this model.

(d) Provide parameter estimates and standard errors from the GLMM analysis.

(e) Summarize the association between respiratory infection and xeropthalmia and age.

9.5 On the book website, you will find data on illiteracy and race collected during the US 1930 census. Wakefield (2009b) provides more information on these data. *Illiterate* is defined as being unable to read and over 10 years of age. For each of the $i = 1, \ldots, 49$ states that existed in 1930, the data consist of the number of illiterate individuals Y_{ij} and the total population aged 10 years and older N_{ij} by race, coded as native-born White ($j = 1$), foreign-born White ($j = 2$), and Black ($j = 3$). Let p_{ij} be the probability of being illiterate for an individual residing in state i and of race j. An additional binary state-level variable $x_i = 0/1$ describes whether Jim Crow laws were absent/present in state $i = 1, \ldots, 49$. These laws enforced racial segregation in all public facilities.

The association between illiteracy and race, state, and Jim Crow laws will be examined using logistic regression models. In particular, interest focuses on whether illiteracy in 1930 varied by race, varied across states, and was associated with the presence/absence of Jim Crow laws:

(a) Calculate the empirical logits of the p_{ij}'s, and provide informative plots that graphically display the association between illiteracy and state, race, and Jim Crow laws.

(b) First consider the native-born White data only (Y_{i1}, N_{i1}), $i = 1, \ldots, 49$, with the following models:

- *Binomial:* $Y_{i1} \mid p_{i1} \sim \text{Binomial}(N_{i1}, p_{i1})$, with the logistic model

$$\log\left(\frac{p_{i1}}{1 - p_{i1}}\right) = \gamma_1 \qquad (9.46)$$

for $i = 1, \ldots, 49$.

- *Quasi-Likelihood:* Model (9.46) with

$$\text{E}[Y_{i1}] = N_{i1} p_{i1}, \qquad \text{var}(Y_{i1}) = \kappa \times N_{i1} p_{i1}(1 - p_{i1}).$$

- *GEE:* Model (9.46) with $\text{E}[Y_{i1}] = N_{i1} p_{i1}$ and working independence.
- *GLMM*

$$\log\left(\frac{p_{i1}}{1 - p_{i1}}\right) = \beta_1 + b_{i1} \qquad (9.47)$$

with $b_{i1} \mid \sigma_1^2 \sim_{iid} \text{N}(0, \sigma_1^2)$, $i = 1, \ldots, 49$.

(i) Give careful definitions of $\exp(\gamma_1)$ in the GEE model and $\exp(\beta_1)$ in the GLMM.

(ii) Fit the binomial model to the native-born White data and give a 95% confidence interval for the odds of native-born White illiteracy. Is this model appropriate?

(iii) Fit the quasi-likelihood and GEE models to the native-born White data and give 95% confidence interval for the odds of native-born White illiteracy in each case. How does the GEE approach differ from quasi-likelihood here? Which do you prefer?

(iv) Fit the GLMM model to the data using a likelihood approach and give a 95% confidence interval for the odds of native-born White illiteracy along with an estimate of the between-state variability in logits. Are the results consistent with the GEE analysis?

(c) Now consider data on all three races. Using GEE fit, *separate* models to the data of each race. Give a 95% confidence interval for the odds ratios comparing illiteracy between foreign-born Whites and native-born Whites, and comparing Blacks with native-born Whites. Is there any problem with this analysis?

(d) Use GEE to fit a model to all three races simultaneously and compare your answer with the previous part. Which analysis is the most appropriate and why?

(e) Fit the GLMM

$$\log\left(\frac{p_{ij}}{1-p_{ij}}\right) = \beta_j + b_{ij} \qquad (9.48)$$

with $b_{ij} \mid \sigma_j^2 \sim_{ind} N(0, \sigma_j^2)$, $j = 1, 2, 3$, using likelihood-based methods. Give 95% confidence intervals for the odds ratios comparing illiteracy between foreign-born Whites and native-born Whites, and comparing Blacks with native-born Whites. Are your conclusions the same as with the GEE analysis? Does this model require refinement?

(f) The state-level Jim Crow law indicator will now be added to the analysis. Consider the model

$$\log\left(\frac{p_{ij}}{1-p_{ij}}\right) = \gamma_{0j} + \gamma_{1j} x_i \qquad (9.49)$$

Give interpretations of each of $\exp(\gamma_{0j}), \exp(\gamma_{1j})$ for $j = 1, 2, 3$. Fit this model using GEE and interpret and summarize the results in a clear fashion.

(g) Consider Bayesian fitting of the GLMM:

$$\log\left(\frac{p_{ij}}{1-p_{ij}}\right) = \beta_{0j} + \beta_{1j} x_i + b_{ij} \qquad (9.50)$$

where $b_i \mid D \sim_{iid} N_3(0, D)$ with $b_i = [b_{i1}, b_{i2}, b_{i3}]^T$ and

$$D = \begin{bmatrix} \sigma_1^2 & \rho_{12}\sigma_1\sigma_2 & \rho_{13}\sigma_1\sigma_3 \\ \rho_{12}\sigma_2\sigma_1 & \sigma_2^2 & \rho_{23}\sigma_2\sigma_3 \\ \rho_{13}\sigma_3\sigma_1 & \rho_{23}\sigma_3\sigma_2 & \sigma_3^2 \end{bmatrix}$$

is a 3×3 variance–covaraince matrix for the random effects b_i. Assume improper flat priors for $\beta_{0j}, \beta_{1j}, j = 1, 2, 3$, and the Wishart prior $W = D^{-1} \sim$ Wishart(r, S) parameterized so that $E[W] = rS$, with $r = 3$ and

$$S = \begin{bmatrix} 30.45 & 0 & 0 \\ 0 & 30.45 & 0 \\ 0 & 0 & 30.45 \end{bmatrix}.$$

Carry out a Bayesian analysis using this model and interpret and summarize the results in a clear fashion.

(h) Write a short summary of what you have found, concentrating on the particular substantive questions of interest stated in the introduction.

9.6 For the theophylline data considered in this chapter, reproduce the results in Table 9.14 by coding up the nonlinear GEE model with working independence. These data are available as Theoph in the R package.

9.7 Throughout this chapter, mixed models with clustering induced by normally distributed random effects have been considered. In this question, a non-normal random effects distribution will be considered. Suppose, for paired binary observations, that the data-generating mechanism is the following:

$$Y_{ij} \mid \mu_{ij} \sim_{ind} \text{Bernoulli}(\mu_{ij}),$$

for $i = 1, \ldots, n, j = 1, 2$, with

$$\mu_{ij} = \frac{\exp(\beta_0 + \beta_1 x_{ij} + b_i)}{1 + \exp(\beta_0 + \beta_1 x_{ij} + b_i)}$$

$$b_i = \begin{cases} -\gamma & \text{with probability } 1/2 \\ \gamma & \text{with probability } 1/2. \end{cases}$$

and $X_{ij} \sim_{iid}$ Unif$(-10, 10)$. The parameters $\beta_1 \in \mathbb{R}$ and $\gamma > 0$ are unknown, and all b_i are independent and identically distributed. For simplicity, assume $\beta_0 = 0$ throughout:

(a) For $0 \leq \beta_1 \leq 1$ and $0 \leq \gamma \leq 5$, calculate the correlation between the outcomes Y_{ij} and $Y_{ij'}$ within cluster i, averaged over the distribution of clusters.

(b) For $\beta_1 = 1$ and $0 \leq \gamma \leq 5$, calculate the numerical value of the true slope parameter estimated by a GEE logistic regression model of y on x, with working independence within clusters. Compare this value to the true β_1.

(c) Consider a study with paired observations and binary outcomes (e.g., a matched-pairs case-control study as described in Sect. 7.10.3). The true data-generating mechanism is as above with $\beta_1 = 1, \gamma = 5$. First plot y versus x for all observations and add a smoother. This plot seems to indicate that there are low-, medium-, and high-risk subjects, depending on levels of x.

(d) In truth, of course, there are not three levels of risk. For some example data, give a plot that illustrates this and write an explanation of what your plot shows. The plot should use only observed, and not latent, variables.

Part IV
Nonparametric Modeling

Chapter 10
Preliminaries for Nonparametric Regression

10.1 Introduction

In all other chapters we assume that the regression model, $f(x)$, takes an a priori specified, usually simple, parametric form. Such models have a number of advantages: If the assumed parametric form is approximately correct, then efficient estimation will result; having a specific linear or nonlinear form allows concise summarization of an association; inference for parametric models is often relatively straightforward. Further, a particular model may be justifiable from the context.

In this and the following two chapters, we consider situations in which a greater degree of flexibility is desired, at least when modeling some components of the covariate vector x. Nonparametric modeling is particularly useful when one has little previous experience with the specific data-generating context. Typically, one may desire $f(\cdot)$ to arise from a class of functions with restrictions on smoothness and continuity. Although the models of this and the next two chapters are referred to as nonparametric,[1] they often assume parametric forms but depend on a large number of parameters which are constrained in some way, in order to prevent overfitting of the data. For some approaches, for example, the regression tree models described in Sect. 12.7, the model is specified implicitly through an algorithm, with the specific form (including the number of parameters) being selected adaptively.

There are a number of contexts in which flexible modeling is required. The simplest is when a graphical description of a set of data is needed, which is often referred to as *scatterplot smoothing*. Formal inference is also possible within a nonparametric framework, however. In some circumstances, estimation of a parametric relationship between a response and an x variable may be of interest, while requiring flexible nonparametric modeling of other nuisance variables (including confounders). The example described in Sect. 1.3.6 is of this form, with the association between spinal bone mineral density and ethnicity being of primary

---
[1] Some authors prefer the label *semiparametric*.

interest, but with a flexible model for age being desired. Finally, an important and common use of nonparametric modeling is prediction. In this case, the focus is on the accuracy of the final prediction, with little interest in the values of the parameters in the model. Prediction with a discrete outcome is often referred to as *classification*.

Much of the development of nonparametric methods, in particular those associated with classification, has occurred in computer science and, more specifically, machine learning, with a terminology that is quite different to that encountered in the statistics literature. The data with which the model is fitted constitute the *training sample*; nonparametric regression is referred to as *learning a function*; the covariates are called *features*; and adding a penalty term to an objective function (e.g., a residual sum of squares) is called *regularization*. In *supervised learning* problems, there is an outcome variable that we typically wish to predict, while in *unsupervised learning* there is no single outcome to predict, rather the aim is to explore how the data are organized or clustered. Only supervised learning is considered here.

The layout of this chapter is as follows. In Sect. 10.2, we discuss a number of motivating examples. Section 10.3 examines what response summary should be reported in a prediction setting using a decision theory framework, while in Sect. 10.4 various measures of predictive accuracy are reviewed. A recurring theme will be the bias-variance trade-off encountered when fitting flexible models containing a large number of parameters. To avoid excess variance of the prediction, various techniques that reduce model complexity will be described; a popular approach is to penalize large values of the parameters. This concept is illustrated in Sect. 10.5 with descriptions of *ridge regression* and the *lasso*. These *shrinkage* methods are introduced in the context of multiple linear regression.[2] Controlling the complexity of a model is a key element of nonparametric regression and is usually carried out using smoothing (or tuning) parameters. In Sect. 10.6, smoothing parameter estimation is considered. Concluding comments appear in Sect. 10.7. There is a huge and rapidly growing literature on nonparametric modeling, and the surface is only scratched here; Sect. 10.8 gives references to broader treatments and to more detailed accounts of specific techniques.

The next two chapters also consider nonparametric modeling. In Chap. 11, two popular approaches to smoothing are described: Those based on splines and those based on kernels; the focus of the latter is local regression. Chapter 11 only considers situations with a single covariate, with multiple predictors considered in Chap. 12, along with methods for classification.

10.2 Motivating Examples

Three examples that have been previously introduced will be used for illustrating nonparametric modeling: The prostate cancer data described in Sect. 1.3.1 are used for illustration in this chapter and in Chap. 12; the spinal bone marrow data of

---

[2]Ridge regression is also briefly encountered in Sect. 5.12.

10.2 Motivating Examples

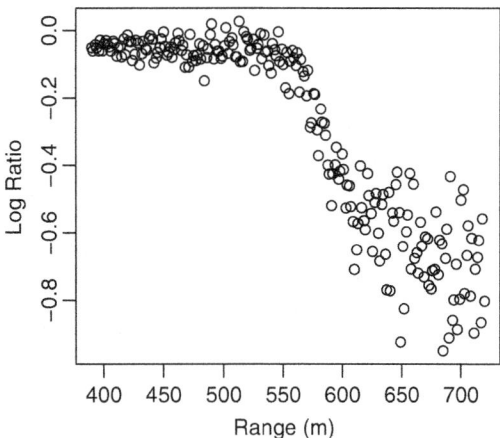

Fig. 10.1 Log ratio of two laser sources, as a function of the range, in the LIDAR data

Sect. 1.3.6 will be analyzed in Chap. 11; and the bronchopulmonary dysplasia data described in Sect. 7.2.3 will be examined in Chaps. 11 and 12. In this section, two additional datasets are described.

10.2.1 Light Detection and Ranging

Figure 10.1 shows data, taken from Holst et al. (1996), from a light detection and ranging (LIDAR) experiment. The LIDAR technique (which is similar to radar technology) uses the reflection of laser-emitted light to monitor the distribution of atmospheric pollutants. The data we consider concern mercury. The x-axis measures distance traveled before light is reflected back to its source (and is referred to as the range), and the y-axis is the logarithm of the ratio of distance measured for two laser sources: One source has a frequency equal to the resonant frequency of mercury, and the other has a frequency off this resonant frequency. For these data, point and interval estimates for the association between the log ratio and range are of interest. Figure 10.1 shows a clear nonlinear relationship between the log ratio and range, with greater variability at larger ranges.

10.2.2 Ethanol Data

This example concerns data collected in a study reported by Brinkman (1981). The data consist of $n = 88$ measurements on three variables: NOx, the concentration of nitric oxide (NO) and nitrogen dioxide (NO_2) in the engine exhaust, with normalization by the work done by the engine; C, the compression ratio of the engine; and E, the equivalence ratio at which the engine was run, a measure of

Fig. 10.2 A three-dimensional display of the ethanol data, showing the normalized concentration of nitric oxide and nitrogen dioxide (NOx) as a function of the equivalance ratio at which the engine was run (E) and the compression ratio of the engine (C)

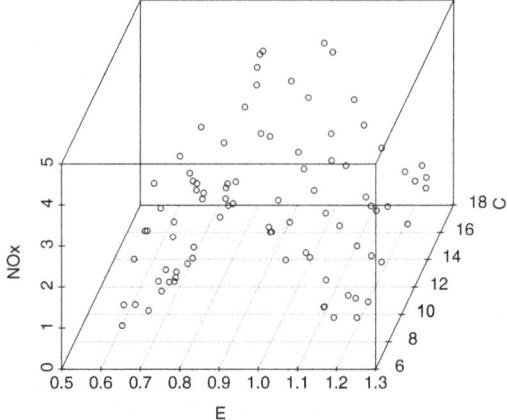

the air/ethanol mix. Figure 10.2 gives a three-dimensional display of these data. The aim is to build a predictive model, and a simple linear model is clearly inadequate since there is a strong nonlinear (inverse U-shaped) association between NOx and E. The form of the association between NOx and C is less clear.

10.3 The Optimal Prediction

Before considering model specification and describing methods for fitting, we use a decision theory framework to decide on which summary of the distribution of $Y \mid x$ we should report if the aim of analysis is prediction, where x is a $1 \times (k+1)$ design vector corresponding to the intercept and k covariates. Throughout this section, we will suppose we are in an idealized situation in which all aspects of the data-generating mechanism are known, and we need only decide on which quantity to report.

The specific decision problem we consider is the following. Imagine we are involved in a game in which the aim is to predict a new observation y, using a function of covariates x, $f(x)$. Further, we know that our predictions will be penalized via a loss function $L[y, f(x)]$ that is the penalty incurred when predicting y by $f(x)$. The optimal prediction is that which minimizes the expected loss defined as

$$\mathrm{E}_{X,Y}\left\{L\left[Y, f(X)\right]\right\}, \tag{10.1}$$

where the expectation is with respect to the joint distribution of the random variables Y and X.

10.3.1 Continuous Responses

The most common choice of loss function is squared error loss, with $f(x)$ chosen to minimize the expected (squared) prediction error:

$$E_{X,Y}\left\{[Y - f(X)]^2\right\}, \tag{10.2}$$

that is, the quadratic loss. Writing (10.2) as

$$E_X\left[E_{Y|X=x}\left\{[Y - f(x)]^2 \mid X = x\right\}\right]$$

indicates that we may minimize pointwise, with solution

$$\widehat{f}(x) = E[Y \mid x],$$

that is, the conditional expectation (Exercise 10.4). Hence, the best prediction, $\widehat{f}(x)$, is the usual regression function.

As an alternative, with absolute loss, $E_{X,Y}[\,|Y - f(X)|\,]$, the solution is the conditional median

$$\widehat{f}(x) = \mathrm{median}(Y \mid x)$$

(Exercise 10.4). Modeling via the median, rather than the mean, provides greater robustness to outliers but with an increase in computational complexity. Exercise 10.4 also considers a generalization of absolute loss.

Other choices have also been suggested for specific situations. For example, the scaled quadratic loss function

$$L[y, f(x)] = \left(\frac{y - f(x)}{y}\right)^2 \tag{10.3}$$

has been advocated for random variables $y > 0$ (e.g., Bernardo and Smith 1994, p. 301). This loss function is scaling departures $y - f(x)$ by y, so that discrepancies in the predictions of the same magnitude are penalized more heavily for small y than for large y. Taking the expectation of (10.3) with respect to $Y \mid x$ leads to

$$\widehat{f}(x) = \frac{E[Y^{-1} \mid x]}{E[Y^{-2} \mid x]}. \tag{10.4}$$

For details, see Exercise 10.5. As an example, suppose the data are gamma distributed as

$$Y \mid \mu(x), \alpha \sim_{iid} \mathrm{Ga}\left\{\alpha^{-1}, [\mu(x)\alpha]^{-1}\right\},$$

where $E[Y \mid x] = \mu(x)$, and $\alpha^{-1/2}$ is the coefficient of variation. Then Exercise 10.5 shows that (10.4) is equal to

$$\widehat{f}(x) = (1 - 2\alpha)\mu(x), \tag{10.5}$$

for $\alpha < 0.5$. Hence, under the scaled quadratic loss function, we should scale the mean function by $1 - 2\alpha$ when reporting.

10.3.2 Discrete Responses with K Categories

Now suppose the response is categorical, with $Y \in \{0, 1, \ldots, K - 1\}$. Again, we must decide on which summary measure to report. One approach is to assign a class in $\{0, 1, \ldots, K - 1\}$ to a new case via a classification rule $g(x)$. Alternatively, a probability distribution over the classes may be reported.[3]

Suppose the distributions of x given $Y = k$, $p(x \mid Y = k)$, are known along with prior probabilities on the classes, $\Pr(Y = k) = \pi_k$. Then, via Bayes theorem, the posterior classifications may be obtained:

$$\Pr(Y = k \mid x) = \frac{p(x \mid Y = k)\pi_k}{\sum_{l=0}^{K-1} p(x \mid Y = l)\pi_l}. \tag{10.6}$$

Choosing the k that maximizes these probabilities gives a *Bayes classifier*.

For the situation in which we wish to assign a class label, the loss function is a $K \times K$ matrix L with element $L(j, k)$ representing the loss incurred when the truth is $Y = j$, and the classification is $g(x) = k$, with $j, k \in \{0, 1, \ldots, K - 1\}$. A sensible loss function is

$$L(j, k) = \begin{cases} 0 & \text{if } j = k \\ \geq 0 & \text{if } j \neq k. \end{cases} \tag{10.7}$$

In most cases, we will assign $L(j, k) > 0$ for $j \neq k$ but in some contexts incorrect classifications will not be penalized if they are of no consequence. We emphasize that the *class* predictor $g(x)$ takes a value from the set $\{0, 1, \ldots, K - 1\}$ and is a function of $\Pr(Y = k \mid x)$. The expected loss is

$$E_{X,Y}\{L[Y, g(X)]\} = E_X\left[E_{Y \mid x}\{L[Y, g(x)] \mid X = x\}\right]$$

$$= E_X\left[\sum_{k=0}^{K-1} L[Y = k, g(x)]\Pr(Y = k \mid x)\right]. \tag{10.8}$$

---

[3]It is possible to also have a "doubt" category that is assigned if there is sufficient ambiguity but we do not consider this possibility. See Ripley (1996) for further discussion.

10.3 The Optimal Prediction

Table 10.1 Loss table for a binary decision problem

		Predicted Class	
		$g(x) = 0$	$g(x) = 1$
True Class	$Y = 0$	0	$L(0, 1)$
	$Y = 1$	$L(1, 0)$	0

where we are assuming the form of $g(x)$ is known. The inner expectation of (10.8) is known as the *Bayes risk* (e.g., Ripley 1996), with minimum

$$\widehat{g}(x) = \mathrm{argmin}_{g(x) \in \{0,\ldots,K-1\}} \sum_{k=0}^{K-1} L[Y = k, g(x)] \Pr(Y = k \mid x).$$

The $K = 2$ situation will now be considered in greater detail. Table 10.1 gives the table of losses for this case. The Bayes risk is minimized by the choice

$$\widehat{g}(x) =$$
$$\mathrm{argmin}_{g(x) \in \{0,1\}} \{L[Y = 0, g(x)] \Pr(Y = 0 \mid x) + L[Y = 1, g(x)] \Pr(Y = 1 \mid x)\}.$$

Hence,

$$g(x) = 0 \text{ gives Bayes risk } = L(1, 0) \times \Pr(Y = 1 \mid x)$$
$$g(x) = 1 \text{ gives Bayes risk } = L(0, 1) \times [1 - \Pr(Y = 1 \mid x)]$$

and so the Bayes risk is minimized by $g(x) = 1$ if

$$L(1, 0) \times \Pr(Y = 1 \mid x) > L(0, 1) \times [1 - \Pr(Y = 1 \mid x)]$$

or equivalently if

$$\frac{\Pr(Y = 1 \mid x)}{1 - \Pr(Y = 1 \mid x)} > \frac{L(0, 1)}{L(1, 0)} = R \qquad (10.9)$$

with the consequence that only the ratio of losses R requires specification. A final restatement is to classify a new case with covariates x as $g(x) = 1$ if

$$\Pr(Y = 1 \mid x) > \frac{L(0, 1)}{L(0, 1) + L(1, 0)} = \frac{R}{1 + R}.$$

If classifying as $g(x) = 1$ when $Y = 0$ is much worse than classifying as $g(x) = 0$ when $Y = 1$, then R should be given a value greater than 1. In this case, if $\Pr(Y = 1 \mid x) > 0.5$ then we assign $g(x) = 1$. For example, if $R = 4$, we set $g(x) = 1$ only if $\Pr(Y = 1 \mid x) > 0.8$.

Returning to the case of general K, in the most straightforward case of all errors being equal, we simply assign an observation to the most likely class, using the probabilities $\Pr(Y = k \mid x)$, $k = 0, 1, \ldots, K - 1$.

We now turn to the second situation in which a classification is not required, but rather a set of probabilities over $\{0, 1, \ldots, K-1\}$, that is, we require $\boldsymbol{f}(\boldsymbol{x}) = [f_0(\boldsymbol{x}), \ldots, f_{K-1}(\boldsymbol{x})]$. First, consider the $K = 2$ (binary) case. In this case we simplify notation and write $\boldsymbol{f}(\boldsymbol{x}) = [f(\boldsymbol{x}), 1 - f(\boldsymbol{x})]$. We may specify a loss function which is proportional to the negative Bernoulli log-likelihood

$$L[y, f(\boldsymbol{x})] = -2y \log[f(\boldsymbol{x})] - 2(1-y) \log[1 - f(\boldsymbol{x})] \qquad (10.10)$$

where $f(\boldsymbol{x})$ is the function that we will report. Therefore, if the log-likelihood is high the loss is low. The expectation of (10.10) is

$$-2 \Pr(Y = 1 \mid \boldsymbol{x}) \log[f(\boldsymbol{x})] - 2[1 - \Pr(Y = 1 \mid \boldsymbol{x})] \log[1 - f(\boldsymbol{x})]$$

where $E[Y \mid \boldsymbol{x}] = \Pr(Y = 1 \mid \boldsymbol{x})$ are the true probabilities, given covariates $\boldsymbol{x}$. The solution is $\widehat{f}(\boldsymbol{x}) = \Pr(Y = 1 \mid \boldsymbol{x})$. Hence, to minimize the expected deviance-type loss function, the true probabilities should be reported, which is not a great surprise.

In the general case of K classes and a multinomial likelihood with one trial and probabilities $\boldsymbol{f}(\boldsymbol{x}) = [f_0(\boldsymbol{x}), \ldots, f_{K-1}(\boldsymbol{x})]$, the deviance loss function is

$$L[y, \boldsymbol{f}(\boldsymbol{x})] = -2 \sum_{k=0}^{K-1} I(Y = k) \log f_k(\boldsymbol{x}), \qquad (10.11)$$

where $I(\cdot)$ is the indicator function that equals 1 if its argument is true and 0 otherwise. The expected loss is

$$-2 \sum_{k=0}^{K-1} \Pr(Y = k \mid \boldsymbol{x}) \log f_k(\boldsymbol{x})$$

which is minimized by $\widehat{f_k}(\boldsymbol{x}) = \Pr(Y = k \mid \boldsymbol{x})$.

10.3.3 General Responses

In general, if we are willing to speculate on a distribution for the data, we may take the loss function as

$$L[y, f(\boldsymbol{x})] = -2 \log p_f(y \mid \boldsymbol{x}), \qquad (10.12)$$

which is the deviance (Sect. 6.5.3), up to an additive constant not depending on f. The notation p_f emphasizes that the distribution of the data depends on f. The previous section gave examples of this loss function for binomial, (10.10), and multinomial, (10.11), data. The loss function (10.12) is an obvious measure of the closeness of y to the predictor function $f(\boldsymbol{x})$ since it is a general measure of the *discrepancy* between the data y and $f(\boldsymbol{x})$. When $Y \mid \boldsymbol{x} \sim N[f(\boldsymbol{x}), \sigma^2]$, we obtain

$$L[y, f(\boldsymbol{x})] = \log(2\pi\sigma^2) + [y - f(\boldsymbol{x})]^2 / \sigma^2,$$

which produces $\widehat{f}(x) = \mathrm{E}[Y \mid x]$, as with quadratic loss. Similarly, choosing a Laplacian distribution, that is, $Y \mid x \sim \mathrm{Lap}[f(x), \phi]$ (Appendix D) leads to the posterior median as the optimal choice.

10.3.4 *In Practice*

Sections 10.3.1 and 10.3.2 describe which summary should be reported, if one is willing to specify a loss function. Such a loss function will often have been based on an implicit model for the distribution of the data or upon an estimation method.

For example, a quadratic loss function is consistent with a model for *continuous responses with additive errors* which is of the form

$$Y = f(x) + \epsilon \tag{10.13}$$

with $\mathrm{E}[\epsilon] = 0$, $\mathrm{var}(\epsilon) = \sigma^2$ and errors on different responses being uncorrelated. This form may be supplemented with the assumption of normal errors or one may simply proceed with least squares estimation. Modeling proceeds by assuming some particular form for $f(x)$. A simple approach is to assume that the conditional mean, $f(x)$, is approximated by the linear model $x\beta$, as in Chap. 5. Alternative nonlinear models are described in Chap. 6.

Relaxing the constant variance assumption, one may consider generalized linear model (GLM) type situations, to allow for more flexible mean-variance modeling. GLMs are also described in Chap. 6. An assumption of a particular distributional form may be combined with the deviance-type loss function (10.12).

In Sect. 10.3.2 discrete responses were considered, and we saw that with equal losses, one may classify on the basis of the probabilities $\Pr(Y = k \mid x)$. As described in Chap. 12, there are two broad approaches to classification. The first approach directly models the probabilities $\Pr(Y = k \mid x)$. For example, in the case of binary ($K = 2$) responses, logistic modeling provides an obvious approach (as described in Sect. 7.6). Chap. 12 describes a number of additional methods to model the probabilities as a function of x. The second approach is to assume forms for the distributions of x given $Y = k$, $p(x \mid Y = k)$ and then combine these with prior probabilities on the classes, $\Pr(Y = k) = \pi_k$, to form posterior classifications, via (10.6); Chapter 12 also considers this situation.

10.4 Measures of Predictive Accuracy

As already noted, nonparametric modeling is often used for *prediction*, and so the conventional criteria by which methods of parameter estimation are compared (as discussed in Sect. 2.2) are not directly relevant. In a prediction context, there is less concern about the values of the constituent parts of the prediction equation, rather interest is on the total contribution. In Sect. 10.3, loss functions were introduced in order to determine how to report the prediction. In this section, loss functions are used to provide an overall measure of the "error" of a procedure.

The *generalization error* is defined as

$$\text{GE}(\widehat{f}) = \text{E}_{\boldsymbol{X},Y}\left\{L\left[Y,\widehat{f}(\boldsymbol{X})\right]\right\}, \quad (10.14)$$

where $\widehat{f}(\boldsymbol{X})$ is the prediction for Y at a point $\boldsymbol{X}$, with $\boldsymbol{X}, Y$ drawn from their joint distribution. Hence, we are in the so-called *X-random*, as opposed to *X-fixed*, case (Breiman and Spector 1992). The terminology with respect to different measures of accuracy can be confusing and is also inconsistent in the literature; the notation used here is summarized in Table 10.2.

Hastie et al. (2009, Sect. 7.2) describe how one would ideally split the data into three portions with one part being used to fit (or train) models, a second (validation) part to choose a model (which includes both choosing between different classes of models and selecting smoothing parameters within model classes), and a third part to estimate the generalization error of the final model on a test dataset. Unfortunately, there are often insufficient data for division into three parts. Consequently, when prediction methods are to be compared, a common approach is to separate the data into *training* and *test* datasets. The training data are used to train the model and then approximate the validation step using methods to be described in Sect. 10.6. The *test* data are used to estimate the generalization error (10.14) using the function $\widehat{f}(\boldsymbol{x})$ estimated from the training data. We now discuss the form of the generalization error for different data types.

10.4.1 Continuous Responses

To gain flexibility and so minimize bias, predictive models $f(\boldsymbol{x})$ that contain many parameters are appealing. However, if the parameters are not constrained in some way, such models produce wide predictive intervals because a set of data only contains a limited amount of information. In general, as the number of parameters increases, the uncertainty in the estimation of each increases in tandem, which results in greater uncertainty in the prediction also. Consequently, throughout this and the next two chapters, we will repeatedly encounter the bias-variance trade-off. Section 5.9 provides a discussion of this trade-off in the linear model context.

The *expected squared prediction error* is a special case of the generalization error with squared error loss:

$$\text{ESPE}(\widehat{f}) = \text{E}_{\boldsymbol{X},Y}\left\{\left[Y - \widehat{f}(\boldsymbol{X})\right]^2\right\}, \quad (10.15)$$

where $\widehat{f}(\boldsymbol{X})$ is again the prediction for Y at a point $\boldsymbol{X}$, with $\boldsymbol{X}, Y$ drawn from their joint distribution.

Estimators $\widehat{f}$ with small $\text{ESPE}(\widehat{f})$ are sought, but balancing the bias in estimation with the variance will be a constant challenge. To illustrate, suppose we wish to

10.4 Measures of Predictive Accuracy

Table 10.2 Summary of predictive accuracy measures

Name	Short-hand	Definition
Generalization error	$\text{GE}(\widehat{f})$	$\text{E}_{\boldsymbol{X},Y}\left\{L[Y,\widehat{f}(\boldsymbol{X})]\right\}$
Expected squared prediction error	$\text{ESPE}(\widehat{f})$	$\text{E}_{\boldsymbol{X},Y}\left\{[Y-\widehat{f}(\boldsymbol{X})]^2\right\}$
Mean squared error (or risk)	$\text{MSE}\left[\widehat{f}(\boldsymbol{x}_0)\right]$	$\text{E}_{\boldsymbol{Y}_n}\left\{[\widehat{f}(\boldsymbol{x}_0)-f(\boldsymbol{x}_0)]^2\right\}$
Predictive risk	$\text{PR}\left[\widehat{f}(\boldsymbol{x}_0)\right]$	$\text{E}_{\boldsymbol{Y}_n,Y_0}\left\{[Y_0-\widehat{f}(\boldsymbol{x}_0)]^2\right\}$
		$= \sigma^2 + \text{MSE}\left[\widehat{f}(\boldsymbol{x}_0)\right]$
Integrated mean squared error	$\text{IMSE}\left(\widehat{f}\right)$	$\int \text{E}_{\boldsymbol{Y}_n}\left\{[\widehat{f}(\boldsymbol{x})-f(\boldsymbol{x})]^2\right\} p(\boldsymbol{x})\,d\boldsymbol{x}$
		$= \int \text{MSE}\left(\widehat{f}(\boldsymbol{x})\right) p(\boldsymbol{x})\,d\boldsymbol{x}$
Average mean squared error	$\text{AMSE}\left(\widehat{f}\right)$	$n^{-1}\sum_{i=1}^n \text{e}_{\boldsymbol{Y}_n}\left\{[\widehat{f}(\boldsymbol{x}_i)-f(\boldsymbol{x}_i)]^2\right\}$
		$= \sum_{i=1}^n \text{MSE}\left[\widehat{f}(\boldsymbol{x}_i)\right]$
Average predictive risk	$\text{APR}\left(\widehat{f}\right)$	$n^{-1}\sum_{i=1}^n \text{e}_{\boldsymbol{Y}_n,\boldsymbol{Y}_n^*}\left\{[Y_i^*-\widehat{f}(\boldsymbol{x}_i)]^2\right\}$
		$= \sigma^2 + \text{AMSE}\left(\widehat{f}\right)$
Residual sum of squares	$\text{RSS}\left(\widehat{f}\right)$	$n^{-1}\sum_{i=1}^n [y_i-\widehat{f}(\boldsymbol{x}_i)]^2$
Leave-one-out (ordinary) CV score	$\text{OCV}\left(\widehat{f}\right)$	$n^{-1}\sum_{i=1}^n [y_i-\widehat{f}_{-i}(\boldsymbol{x}_i)]^2$
Generalized CV score	$\text{GCV}\left[\widehat{f}\right]$	$[n-\text{tr}(\boldsymbol{S})]^{-1}\sum_{i=1}^n [y_i-\widehat{f}(\boldsymbol{x}_i)]^2$

All rows of the table but the first are based on integrated or summed *squared* quantities and, hence, are appropriate for a model of the form $y = f(\boldsymbol{x}) + \epsilon$ with the error terms ϵ having zero mean, constant variance σ^2, and with error terms at different $\boldsymbol{x}$ being uncorrelated. CV is short for cross-validation, with OCV and GCV being described in Sects. 10.6.2 and 10.6.3, respectively. Notation: The predictive model evaluated at covariates $\boldsymbol{x}$ is $f(\boldsymbol{x})$, with prediction $\widehat{f}(\boldsymbol{x})$ based on the observed data $\boldsymbol{Y}_n = [Y_1, \ldots, Y_n]$; Y_0 is a new response with associated covariates $\boldsymbol{x}_0$; the observed data are $[y_i, \boldsymbol{x}_i]$, $i = 1, \ldots, n$; $\boldsymbol{Y}_n^* = [Y_1^*, \ldots, Y_n^*]$ are a set of new observations with covariates $\boldsymbol{x}_1, \ldots, \boldsymbol{x}_n$ that we would like to predict; $p(\boldsymbol{x})$ is the distribution of the covariates; $\widehat{f}_{-i}(\boldsymbol{x}_i)$ is the prediction at the point $\boldsymbol{x}_i$ based on the observed data with the i-th case, $[y_i, \boldsymbol{x}_i]$, removed; $\boldsymbol{S}$ is the "smoother" hat matrix and is described in Sect. 10.6.1. The entries in the last three lines are all estimates of $\text{ESPE}(\widehat{f})$

predict a response Y_0 with associated covariates $\boldsymbol{x}_0$. We calculate the expected squared distance between the response Y_0 and the fitted function $\widehat{f}(\boldsymbol{x}_0)$. The expectation is with respect to both Y_0 and repeat (training) data $\boldsymbol{Y}_n = [Y_1, \ldots, Y_n]$ with Y_0 and $\boldsymbol{Y}_n$ being independent. The resultant measure is known as the *predictive risk* and may be decomposed as

$$\text{E}_{\boldsymbol{Y}_n,Y_0}\left\{\left[Y_0 - \widehat{f}(\boldsymbol{x}_0)\right]^2\right\} = \text{E}_{\boldsymbol{Y}_n,Y_0}\left\{\left[Y_0 - f(\boldsymbol{x}_0) + f(\boldsymbol{x}_0) - \widehat{f}(\boldsymbol{x}_0)\right]^2\right\}$$

$$= \text{E}_{Y_0}\left\{[Y_0 - f(\boldsymbol{x}_0)]^2\right\} + \text{E}_{\boldsymbol{Y}_n}\left\{\left[\widehat{f}(\boldsymbol{x}_0) - f(\boldsymbol{x}_0)\right]^2\right\}$$

$$+ 2 \times \text{E}_{Y_0}\{[Y_0 - f(\boldsymbol{x}_0)]\}\, \text{E}_{\boldsymbol{Y}_n}\left\{\left[\widehat{f}(\boldsymbol{x}_0) - f(\boldsymbol{x}_0)\right]\right\}$$

$$= \sigma^2 + \mathrm{E}_{\boldsymbol{Y}_n} \left\{ \left[\widehat{f}(\boldsymbol{x}_0) - f(\boldsymbol{x}_0) \right]^2 \right\}$$

$$= \sigma^2 + \mathrm{MSE} \left[\widehat{f}(\boldsymbol{x}_0) \right].$$

Writing

$$\mathrm{MSE} \left[\widehat{f}(\boldsymbol{x}_0) \right] = \mathrm{E}_{\boldsymbol{Y}_n} \left\{ \left[f(\boldsymbol{x}_0) - \mathrm{E}_{\boldsymbol{Y}_n} \left(\widehat{f}(\boldsymbol{x}_0) \right) + \mathrm{E}_{\boldsymbol{Y}_n} \left(\widehat{f}(\boldsymbol{x}_0) \right) - \widehat{f}(\boldsymbol{x}_0) \right]^2 \right\}$$

we have

$$\mathrm{E}_{\boldsymbol{Y}_n, Y_0} \left\{ \left[Y_0 - \widehat{f}(\boldsymbol{x}_0) \right]^2 \right\} = \sigma^2 + \mathrm{E}_{\boldsymbol{Y}_n} \left\{ \left[\mathrm{E}_{\boldsymbol{Y}_n} \left(\widehat{f}(\boldsymbol{x}_0) \right) - f(\boldsymbol{x}_0) \right]^2 \right\}$$

$$+ \mathrm{E}_{\boldsymbol{Y}_n} \left\{ \left[\widehat{f}(\boldsymbol{x}_0) - \mathrm{E}_{\boldsymbol{Y}_n} \left(\widehat{f}(\boldsymbol{x}_0) \right) \right]^2 \right\}$$

$$= \sigma^2 + \mathrm{bias} \left[\widehat{f}(\boldsymbol{x}_0) \right]^2 + \mathrm{var}_{\boldsymbol{Y}_n} \left[\widehat{f}(\boldsymbol{x}_0) \right].$$

In terms of the prediction error we can achieve given a particular model, nothing can be done about σ^2, which is referred to as the *irreducible error*. Therefore, we concentrate on the MSE of the estimator $\widehat{f}(\boldsymbol{x}_0)$:

$$\mathrm{MSE} \left[\widehat{f}(\boldsymbol{x}_0) \right] = \mathrm{E}_{\boldsymbol{Y}_n} \left\{ \left[\widehat{f}(\boldsymbol{x}_0) - f(\boldsymbol{x}_0) \right]^2 \right\} = \mathrm{bias} \left[\widehat{f}(\boldsymbol{x}_0) \right]^2 + \mathrm{var} \left[\widehat{f}(\boldsymbol{x}_0) \right]$$

where we emphasize that the MSE is calculated at the point $\boldsymbol{x}_0$, with the expectation over training samples. As we discuss subsequently, the estimators $\widehat{f}$ we consider are indexed by a smoothing parameter, and selection of this parameter influences the characteristics of $\widehat{f}$. Little smoothing produces a wiggly $\widehat{f}$, with low bias and high variance. More extensive smoothing produces $\widehat{f}$ with greater bias but reduced variance.

To summarize the MSE over the range of x, we may consider the *integrated mean squared error* (IMSE). For univariate x, over an interval $[a, b]$, and with density $p(x)$:

$$\mathrm{IMSE} \left(\widehat{f} \right) = \int_a^b \mathrm{E}_{\boldsymbol{Y}_n} \left\{ \left[\widehat{f}(x) - f(x) \right]^2 \right\} p(x) \, dx$$

$$= \int_a^b \mathrm{bias} \left[\widehat{f}(x) \right]^2 p(x) \, dx + \int_a^b \mathrm{var} \left[\widehat{f}(x) \right] p(x) \, dx.$$

(10.16)

This summary will be encountered in Sect. 11.3.2.

10.4 Measures of Predictive Accuracy

An alternative to the IMSE, that may be more convenient to use, is the *average mean squared error* (AMSE), which only considers the errors at the observations:

$$\text{AMSE}\left(\widehat{f}\right) = \frac{1}{n}\sum_{i=1}^{n} E_{Y_n}\left\{\left[\widehat{f}(x_i) - f(x_i)\right]^2\right\}$$

$$= \frac{1}{n}\sum_{i=1}^{n} \text{bias}\left[\widehat{f}(x_i)\right]^2 + \frac{1}{n}\sum_{i=1}^{n} \text{var}\left[\widehat{f}(x_i)\right]. \quad (10.17)$$

For the additive errors model (10.13), the *average predictive risk* (APR) is

$$\text{APR}\left(\widehat{f}\right) = \frac{1}{n}\sum_{i=1}^{n} E_{Y_n, Y_n^\star}\left\{\left[Y_i^\star - \widehat{f}(x_i)\right]^2\right\}$$

$$= \sigma^2 + \text{AMSE}\left(\widehat{f}\right).$$

where $Y_n^\star = [Y_1^\star, \ldots, Y_n^\star]$ are the new set of observations which we would like to predict at $x_1, \ldots, x_n$, and are independent of Y_n. In Sect. 10.6.1, a procedure for estimating the APR will be described in the context of smoothing parameter choice.

We denote the test data by $[y_i^\star, x_i^\star]$, $i = 1, \ldots, m$. For continuous data and quadratic loss, we may evaluate an estimate of the expected squared prediction error (10.15):

$$\frac{1}{m}\sum_{i=1}^{m}\left[y_i^\star - \widehat{f}(x_i^\star)\right]^2, \quad (10.18)$$

where $\widehat{f}(x_i^\star)$ is the estimator based on the training data.

10.4.2 Discrete Responses with K Categories

With the loss function (10.7), and with equal losses, the generalization error is

$$\Pr_{X,Y}[\widehat{g}(X) \neq Y], \quad (10.19)$$

which is also known as the *misclassification probability*. Given test data $[y_i^\star, x_i^\star]$, $i = 1, \ldots, m$, the empirical estimate is

$$\frac{1}{m}\sum_{i=1}^{m} I\left[\widehat{g}(x_i^\star) \neq y_i^\star\right],$$

which is simply the proportion of misclassified observations.

We now consider the binary case and introduce terminology that is common in a medical context, before describing additional measures that are useful summaries of a procedure in this case. Suppose we wish to predict disease status given covariates (symptoms) x. Define

$$Y = \begin{cases} 0 & \text{if true state is no disease} \\ 1 & \text{if true state is disease.} \end{cases}$$

A classification rule $g(x)$ is

$$g(x) = \begin{cases} 0 & \text{if prediction is no disease} \\ 1 & \text{if prediction is disease.} \end{cases}$$

The *sensitivity* of a rule is the probability of predicting disease for a diseased individual:

$$\text{Sensitivity} = \Pr\left[g(x) = 1 \mid Y = 1\right].$$

The *specificity* is the probability of predicting disease-free for an individual without disease:

$$\text{Specificity} = \Pr\left[g(x) = 0 \mid Y = 0\right].$$

With respect to Table 10.1, recall that $L(0,1)$ is the loss for predicting $g(x) = 1$ when in reality $Y = 0$ (so we predict disease for a healthy individual) and $L(1,0)$ is the loss associated with predicting healthy for a diseased individual. Consequently, if we increase the former loss $L(0,1)$ while holding $L(1,0)$ constant, we will be more conservative in declaring a patient as diseased, which will increase the specificity and decrease the sensitivity.[4] An alternative, closely related, pair of summaries are the *false-positive fraction* (FPF) and *true-positive fraction* (TPF) defined, respectively, as

$$\text{FPF} = \Pr\left[g(X) = 1 \mid Y = 0\right]$$

and

$$\text{TPF} = \Pr\left[g(X) = 1 \mid Y = 1\right].$$

The sensitivity is the TPF, and the specificity is $(1 - \text{FPF})$. Two additional measures are the *positive predictive value* (PPV) and the *negative predictive value* (NPV), defined as

$$\text{PPV} = \Pr\left[Y = 1 \mid g(X) = 1\right] = \frac{\Pr\left[g(X) = 1 \mid Y = 1\right] \Pr(Y = 1)}{\Pr\left[g(X) = 1\right]}$$

$$\text{NPV} = \Pr\left[Y = 0 \mid g(X) = 0\right] = \frac{\Pr\left[g(X) = 0 \mid Y = 0\right] \Pr(Y = 0)}{\Pr\left[g(X) = 0\right]},$$

which give the probabilities of correct assignments, given classification.

---

[4] We note that the decision problem considered here has many elements in common with that in which we choose between two hypotheses, as discussed in Sect. 4.3.1. The sensitivity is analogous to the power of a test, while 1−specificity is analogous to the type I error.

Now define a classification rule that, based on a model $g(\boldsymbol{x})$ (whose parameters will be estimated from the data), assigns $g(\boldsymbol{x}) = 1$ if the odds of disease

$$\frac{\Pr(Y = 1 \mid \boldsymbol{x})}{\Pr(Y = 0 \mid \boldsymbol{x})} > \frac{L(0,1)}{L(1,0)} = R,$$

as discussed in more detail in relation to (10.9). Plotting TPF(R) versus FPF(R) produces a *receiver-operating characteristic* (ROC) curve. The ROC curve gives the complete behavior of FPF and TPF over the range of R. Pepe (2003) provides an in-depth discussion of the above summary measures.

10.4.3 *General Responses*

For general data types we may evaluate the deviance-like loss function (10.12) over the test data $[y_i^\star, \boldsymbol{x}_i^\star], i = 1, \ldots, m$:

$$-\frac{2}{m} \sum_{i=1}^{m} \log p_{\hat{f}}(y_i^\star \mid \boldsymbol{x}_i^\star),$$

to measure the error of a procedure.

10.5 A First Look at Shrinkage Methods

We describe two penalization methods that are used in the context of multiple linear regression, ridge regression and the lasso.

10.5.1 *Ridge Regression*

We first assume that $\boldsymbol{y}$ has been centered and that each covariate has been standardized, that is,

$$\sum_{i=1}^{n} y_i = 0, \quad \frac{1}{n} \sum_{i=1}^{n} x_{ij} = 0, \quad \frac{1}{n} \sum_{i=1}^{n} x_{ij}^2 = 1.$$

Consider the linear model

$$\boldsymbol{y} = \boldsymbol{x}\boldsymbol{\beta} + \boldsymbol{\epsilon}$$

with $\boldsymbol{x}$ the $n \times k$ design matrix, $\boldsymbol{\beta} = [\beta_1, \ldots, \beta_k]^\mathsf{T}$ the $k \times 1$ vector of parameters, and $\mathrm{E}[\boldsymbol{\epsilon}] = \boldsymbol{0}$, $\mathrm{var}(\boldsymbol{\epsilon}) = \sigma^2 \mathbf{I}$. Note that there is no intercept in the model due to the centering of $y_1, \ldots, y_n$.

We saw in Chap. 5 that linear models are an analytically and computationally appealing class but, with many predictors, fitting the full model without penalization may result in large predictive intervals, unless the sample size is very large relative to k. Ridge regression is an approach to modeling that addresses this deficiency by placing a particular form of constraint on the parameters. Specifically, $\widehat{\boldsymbol{\beta}}^{\mathrm{RIDGE}}$ is chosen to minimize the *penalized sum of squares*:

$$\sum_{i=1}^{n} \left(y_i - \sum_{j=1}^{k} x_{ij} \beta_j \right)^2 + \lambda \sum_{j=1}^{k} \beta_j^2, \tag{10.20}$$

for some $\lambda > 0$. Using a Lagrange multiplier argument (Exercise 10.6), minimization of (10.20) is equivalent to minimization of

$$\sum_{i=1}^{n} \left(y_i - \sum_{j=1}^{k} x_{ij} \beta_j \right)^2$$

subject to, for some $s \geq 0$,

$$\sum_{j=1}^{k} \beta_j^2 \leq s, \tag{10.21}$$

so that the size of the sum of the squared coefficients is constrained (which is known as an L_2 penalty). The intuition behind ridge regression is that, with many parameters to estimate, the estimator can be highly variable, but by constraining the sum of the squared coefficients, this shortcoming can be alleviated.

Figure 10.3 shows the effect of ridge regression with two parameters, β_1 and β_2. The elliptical contours in the top right of the figure correspond to the sum of squares. In ridge regression this sum of squares is minimized subject to the constraint (10.21), and for $k = 2$, this constraint corresponds to a circle, centered at zero. The estimate is given by the point at which the ellipse and the circle touch.

Writing the penalized sum of squares (10.20) as

$$(\boldsymbol{y} - \boldsymbol{x}\boldsymbol{\beta})^\mathsf{T}(\boldsymbol{y} - \boldsymbol{x}\boldsymbol{\beta}) + \lambda \boldsymbol{\beta}^\mathsf{T} \boldsymbol{\beta} \tag{10.22}$$

it is easy to see that the minimizing solution is

$$\widehat{\boldsymbol{\beta}}^{\mathrm{RIDGE}} = (\boldsymbol{x}^\mathsf{T} \boldsymbol{x} + \lambda \mathbf{I}_k)^{-1} \boldsymbol{x}^\mathsf{T} \boldsymbol{Y}. \tag{10.23}$$

Since the estimator (10.23) is linear, it is straightforward to calculate the variance–covariance matrix, for a given λ, as

$$\mathrm{var}\left(\widehat{\boldsymbol{\beta}}^{\mathrm{RIDGE}}\right) = \sigma^2 (\boldsymbol{x}^\mathsf{T} \boldsymbol{x} + \lambda \mathbf{I}_k)^{-1} \boldsymbol{x}^\mathsf{T} \boldsymbol{x} (\boldsymbol{x}^\mathsf{T} \boldsymbol{x} + \lambda \mathbf{I}_k)^{-1}. \tag{10.24}$$

10.5 A First Look at Shrinkage Methods

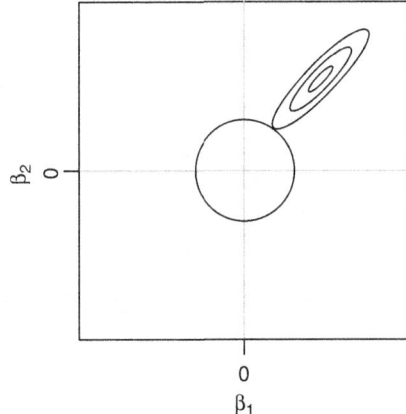

Fig. 10.3 Pictorial representation of ridge regression, for two covariates. The *elliptical contours* represent the sum of squares, and the *circle* represents the constraint corresponding to the L_2 penalty

Beginning with a normal likelihood $y \mid \beta \sim N_n(x\beta, \sigma^2 I_n)$ and adding the penalty term $\lambda \beta^T \beta$ to the log-likelihood also leads to minimization of (10.22). The resultant estimator (10.23) is therefore sometimes referred to as a *maximum penalized likelihood estimator* (MPLE).

It is well known that the least squares estimator $\widehat{\beta}^{LS}$ is an unbiased estimator, with variance $(x^T x)^{-1} \sigma^2$ (under correct second moment specification). If we write $R = (x^T x)^{-1}$, then the ridge regression estimator may be written as (Exercise 10.7)

$$\widehat{\beta}^{RIDGE} = (I_k + \lambda R)^{-1} \widehat{\beta}^{LS}, \qquad (10.25)$$

showing that it is clearly biased. Turning now to a consideration of the variance, let $x = UDV^T$ be the singular value decomposition (SVD) of x. In the SVD U is $n \times n$, V is $k \times k$ and D is an $n \times k$ diagonal matrix with diagonal elements $d_1, \ldots, d_k$. Then, the variance of the ridge estimator (10.24) may be written as

$$\mathrm{var}\left(\widehat{\beta}^{RIDGE}\right) = \sigma^2 (x^T x + \lambda I_k)^{-1} x^T x (x^T x + \lambda I_k)^{-1} = \sigma^2 V A V^T, \qquad (10.26)$$

where A is a diagonal matrix whose elements are $d_i^2 / (d_i^2 + \lambda)^2$. The variance of the least squares estimator is

$$\mathrm{var}\left(\widehat{\beta}^{LS}\right) = \sigma^2 V W V^T, \qquad (10.27)$$

where W is a diagonal matrix whose elements are $1/d_i^2$. Hence, the reduction in variance of the ridge regression estimator is apparent. The derivations of (10.26) and (10.27) are left as Exercise 10.7.

With respect to the frequentist methods described in Chap. 2, penalized least squares correspond to a method that produces an estimating function with finite sample bias but with potentially lower mean squared error as a consequence of the penalization term, which reduces the variance.

For *orthogonal* covariates $x^{\mathsf{T}}x = n \times \mathbf{I}_k$, the ridge regression estimator is

$$\widehat{\beta}^{\text{RIDGE}} = \frac{n}{n+\lambda}\widehat{\beta}^{\text{LS}}.$$

Hence, in this case, the ridge estimator always produces shrinkage towards 0. Figure 10.4(a) illustrates the shrinkage (towards zero) performed by ridge regression for a single parameter in the case of orthogonal covariates. For non-orthogonal covariates, the collection of estimators undergoes shrinkage, though individual components of $\widehat{\beta}^{\text{RIDGE}}$ may increase in absolute value.

The fitted value at a particular value $\widetilde{x}$ is

$$\widehat{f}(\widetilde{x}) = \widetilde{x}\,\widehat{\beta}^{\text{RIDGE}} \qquad (10.28)$$

$$= \widetilde{x}(x^{\mathsf{T}}x + \lambda \mathbf{I}_k)^{-1}x^{\mathsf{T}}Y \qquad (10.29)$$

with

$$\text{var}\left[\widehat{f}(\widetilde{x})\right] = \sigma^2 \widetilde{x}(x^{\mathsf{T}}x + \lambda\mathbf{I}_k)^{-1}x^{\mathsf{T}}x(x^{\mathsf{T}}x + \lambda\mathbf{I}_k)^{-1}\widetilde{x}^{\mathsf{T}}. \qquad (10.30)$$

An important concept in shrinkage is the "effective" degrees of freedom associated with a set of parameters. In a ridge regression setting, if we choose $\lambda = 0$, we have k parameters, while for $\lambda > 0$ the parameters are constrained and the degrees of freedom will effectively be lower, tending to 0 as $\lambda \to \infty$. Many smoothers are linear in the sense that $\widehat{y} = S^{(\lambda)}y$, with ridge regression being one example, as can be seen from (10.29). For linear smoothers, the *effective (or equivalent) degrees of freedom* may be defined as

$$p^{(\lambda)} = \text{df}(\lambda) = \text{tr}\left(S^{(\lambda)}\right), \qquad (10.31)$$

where the notation $p^{(\lambda)}$ emphasizes the dependence on the smoothing parameter. For the ridge estimator, the effective degrees of freedom associated with estimation of $\beta_1, \ldots, \beta_k$ is defined as

$$\text{df}(\lambda) = \text{tr}\left[x(x^{\mathsf{T}}x + \lambda\mathbf{I}_k)^{-1}x^{\mathsf{T}}\right]. \qquad (10.32)$$

Notice that $\lambda = 0$, which corresponds to no shrinkage, gives $\text{df}(\lambda) = k$ (so long as $x^{\mathsf{T}}x$ is non-singular), as we would expect.

There is a one-to-one mapping between λ and the degrees of freedom, so in practice, one may simply pick the effective degrees of freedom that one would like associated with the fit and solve for λ. As an alternative to a user-chosen λ, a number of automated methods for choosing λ are described in Sect. 10.6.

Insight into the ridge estimator can be gleaned from the following Bayesian formulation. Consider the model with likelihood

$$y \mid \beta, \sigma^2 \sim \text{N}_n(x\beta, \sigma^2 \mathbf{I}_n), \qquad (10.33)$$

10.5 A First Look at Shrinkage Methods

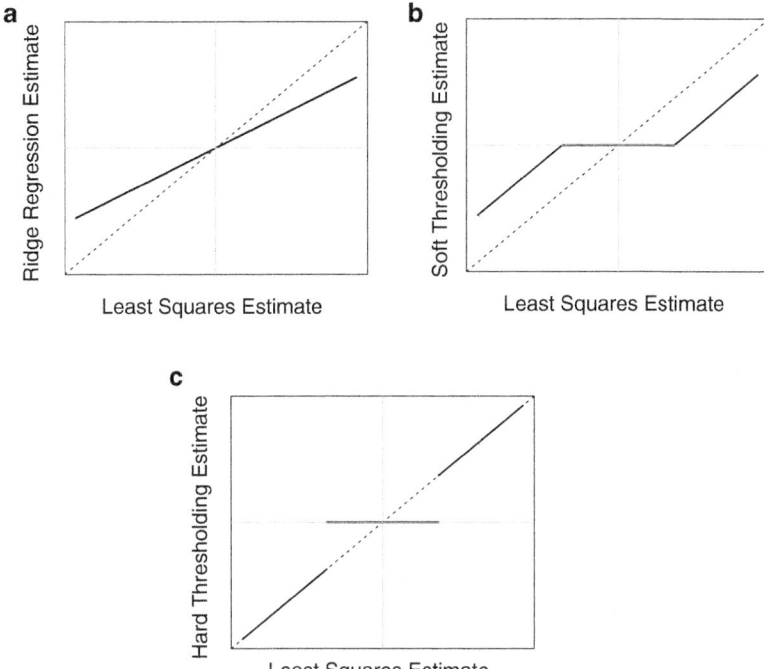

Fig. 10.4 The comparison for single estimate of different forms of shrinkage, with alternative estimates plotted against the least squares estimate $\widehat{\beta}^{\text{LS}}$ and in the case of orthogonal covariates: (**a**) ridge regression, (**b**) *soft thresholding* as carried out by the lasso, and (**c**) *hard thresholding* as carried out by conventional variable selection. On all plots, the line of equality, representing the unrestricted estimate, is drawn as *dashed*

with σ^2 known, and prior

$$\beta \mid \sigma^2 \sim \text{N}_k\left(\mathbf{0}, \frac{\sigma^2}{\lambda}\mathbf{I}_k\right).$$

The latter form shows that a large value of λ corresponds to a prior that is more tightly concentrated around zero and so leads to greater *shrinkage* of the collection of coefficients towards zero. A common λ for each β_j makes it clear that we need to standardize each of the covariates in order for them to be comparable.

Using derivations similar to those of Sect. 5.7, the posterior is

$$\beta \mid y \sim \text{N}_k\left[\widehat{\beta}^{\text{RIDGE}}, \sigma^2(\boldsymbol{x}^{\text{T}}\boldsymbol{x} + \lambda\mathbf{I}_k)^{-1}\right],$$

where $\widehat{\beta}^{\text{RIDGE}}$ corresponds to (10.23), confirming that the posterior mean and mode coincide with the ridge regression estimator, (10.23). Interestingly, the posterior variance $\text{var}(\beta \mid y)$ differs from $\text{var}\left(\widehat{\beta}^{\text{RIDGE}}\right)$, as given in (10.24).

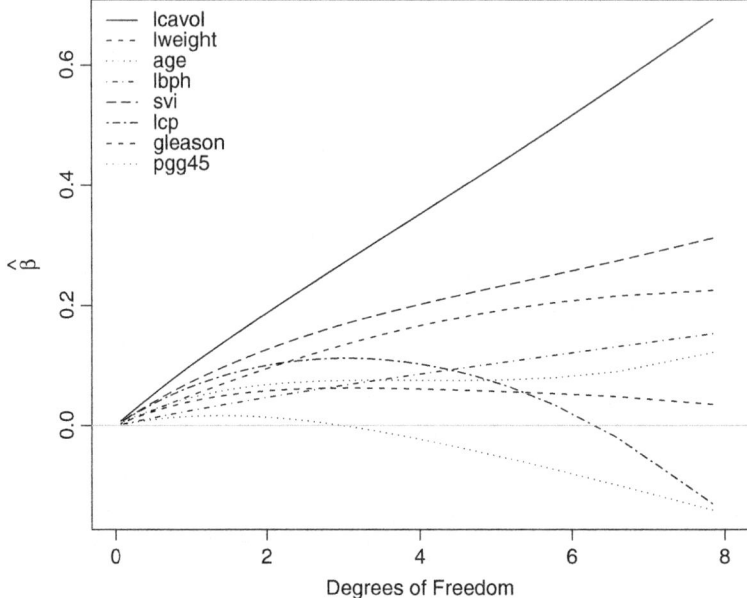

Fig. 10.5 Ridge estimates for the prostate data, as a function of the effective degrees of freedom

Example: Prostate Cancer

As described in Sect. 1.3.1, the response in this dataset is log (PSA), and there are eight covariates. In this chapter, we take the aim of the analysis as prediction of log PSA. In Chap. 5, we analyzed these data using a Bayesian approach with normal priors for each of the eight standardized coefficients, as summarized in (5.66). In that case, the standard deviation of the normal prior was chosen on substantive grounds. Here, we illustrate the behavior of the estimates as a function of the smoothing parameter.

Figure 10.5 shows the eight ridge estimates as a function of the effective degrees of freedom (which ranges between 0 and 8, because there is no intercept in the model). For small values of λ, the effective degrees of freedom is close to 8, and estimates show little shrinkage. In contrast, large values of λ give effective degrees of freedom close to 0 and strong shrinkage. Notice that the curves do not display monotonic shrinkage due to the non-orthogonality of the covariates.

10.5.2 The Lasso

The *least absolute shrinkage and selection operator*, or *lasso*, as described in Tibshirani (1996),[5] is a technique that has received a great deal of interest. As with ridge regression, we assume that the covariates are standardized to have mean zero and standard deviation 1. The lasso estimate minimizes the penalized sum of squares

$$\sum_{i=1}^{n}\left(y_i - \beta_0 - \sum_{j=1}^{k} x_{ij}\beta_j\right)^2 + \lambda \sum_{j=1}^{k} |\beta_j|, \qquad (10.34)$$

with respect to β. The L_2 penalty of ridge regression is therefore being replaced by an L_1 penalty. As with ridge regression, the minimization of (10.34) can be shown to be equivalent to minimization of

$$\sum_{i=1}^{n}\left(y_i - \beta_0 - \sum_{j=1}^{k} x_{ij}\beta_j\right)^2 \qquad (10.35)$$

subject to

$$\sum_{j=1}^{k} |\beta_j| \leq s, \qquad (10.36)$$

for some $s \geq 0$.

Let $\widehat{\beta}^{\text{LS}}$ and $\widehat{\beta}^{\text{LASSO}}$ denote the least squares and lasso estimates, respectively, and define $s_0 = \sum_{j=1}^{k} |\widehat{\beta}_j^{\text{LS}}|$ as the L_1 norm of the least squares estimate. Values of $s < s_0$ cause shrinkage of $\sum_{j=1}^{k} |\widehat{\beta}_j^{\text{LASSO}}|$ towards zero. If, for example, $s = s_0/2$, then the average absolute shrinkage of the least squares coefficients is 50%, though individual coefficients may increase rather than decrease in absolute value.

A key characteristic of the lasso is that individual parameter estimates may be set to zero, a phenomenon that does not occur with ridge regression. Figure 10.6 gives the intuition behind this behavior in the case of two coefficients β_1 and β_2. The lasso performs L_1 shrinkage so that there are "corners" in the constraint; the diamond represents constraint (10.36) for $k = 2$. If the ellipse (10.35) "hits" one of these corners, then the coefficient corresponding to the axis that is touched is shrunk to zero. In the example in Fig. 10.6, neither of the coefficients would be set to zero, because the ellipse does not touch a corner. As k increases, the multidimensional diamond has an increasing number of corners, and so there is an increasing chance of coefficients being set to zero. Consequently, the lasso effectively produces a form of *continuous* subset (or feature) selection. The lasso is sometimes referred to as offering a *sparse solution* due to this property of setting coefficients to zero.

---

[5]The method was also introduced into the signal-processing literature, under the name *basis pursuit*, by Chen et al. (1998).

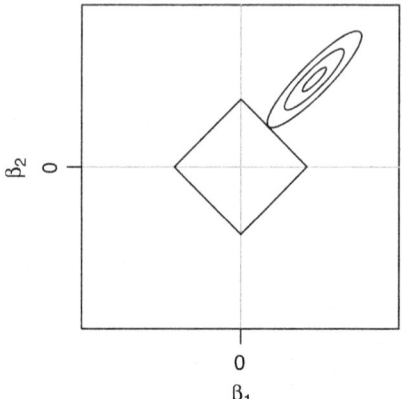

Fig. 10.6 Pictorial representation of the lasso for two covariates. The *elliptical contours* represent the sum of squares, and the *diamond* indicates the constraint corresponding to the L_1 penalty

In the case of orthonormal covariates, for which $x^\mathsf{T} x = \mathbf{I}_k$, the lasso performs so-called *soft thresholding*. Specifically, for component j of the lasso estimator:

$$\widehat{\beta}_j^{\text{LASSO}} = \text{sign}\left(\widehat{\beta}_j^{\text{LS}}\right) \left(|\widehat{\beta}_j^{\text{LS}}| - \frac{\lambda}{2}\right)_+,$$

where "sign" denotes the sign of its argument (± 1), and z_+ represents the positive part of z. As the smoothing parameter is varied, the sample path of the estimates moves continuously to zero, as displayed in Fig. 10.4(b). In contrast, conventional hypothesis testing performs *hard thresholding*, as illustrated in Fig. 10.4(c), since the coefficient is set equal to zero when the absolute value of the estimate drops below some critical value, giving discontinuities in the graph.

The lasso solution is nonlinear in y. Efficient algorithms exist for computation based on coordinate descent; however, see Meier et al. (2008) and Wu and Lange (2008). Tibshirani (2011) gives a brief history of the computation of the lasso solution. Due to the nonlinearity of the solution and the subset selection nature of estimation, inference is not straightforward and remains an open problem. Standard errors for elements of $\widehat{\beta}^{\text{LASSO}}$ are not immediately available, though they may be calculated via the bootstrap. Since the lasso estimator is not linear, the effective degrees of freedom cannot be defined as in (10.31); an alternative definition exists as

$$\text{df} = \frac{1}{\sigma^2} \sum_{i=1}^{n} \text{cov}(\widehat{y}_i, y_i),$$

see Hastie et al. (2009) equation (3.60).

More generally, penalties of the form

$$\lambda \sum_{j=1}^{k} |\beta_j|^q$$

10.5 A First Look at Shrinkage Methods

may be considered, for $q \geq 0$. Ridge regression and the lasso correspond to $q = 2$ and $q = 1$, respectively. For $q < 1$, the constraint is non-convex, which makes optimization more difficult. Convex penalties occur for $q \geq 1$ and feature selection for $q \leq 1$, so that the lasso (with $q = 1$) achieves both.

Many variants of the lasso have appeared since its introduction (Tibshirani 2011). In some contexts, we may wish to treat a set of regressors as a group, for example, when we have a categorical covariate with more than two levels. The *grouped lasso* (Yuan and Lin 2007) addresses this problem by considering the simultaneous shrinkage of (pre-defined) groups of coefficients.

In the case in which $k > n$, the lasso cannot select more than n variables. Furthermore, the lasso will typically assign only one nonzero coefficient to a set of highly correlated covariates (Zou and Hastie 2005), which is an obvious disadvantage and was a motivation for the group lasso (Yuan and Lin 2007). Empirical observation indicates that the lasso produces inferior performance to ridge regression when there are a large number of small effects (Tibshirani 1996). These deficiencies motivated the *elastic net* (Zou and Hastie 2005) which attempts to combine the desirable properties of ridge regression and the lasso via a penalty of the form

$$\lambda_1 \sum_{j=1}^{k} |\beta_j| + \lambda_2 \sum_{j=1}^{k} \beta_j^2.$$

The lasso estimate is equivalent to the mode of the posterior distribution under a normal likelihood, (10.33), and independent Laplace (double exponential) priors on elements of β:

$$\pi(\beta_j) = \frac{\lambda}{2} \exp\left(-\lambda |\beta_j|\right)$$

for $j = 1, \ldots, k$ (the variance of this distribution is $2/\lambda^2$, Appendix D). Under this prior, the posterior is not available in closed form, but the posterior mean will not equal the posterior mode. Hence, if used as a summary, the posterior means will not produce the same lasso shrinkage of coefficients to zero. Thus, regardless of the value of λ, all k covariates are retained in a Bayesian analysis, even though the posterior mode may lie at zero. Markov chain Monte Carlo allows inference under the normal/Laplace model but without the subset selection aspect, which lessens the appeal of this Bayesian version of the lasso.

Example: Prostate Cancer

We illustrate the use of the lasso for the prostate cancer data. Figure 10.7 shows the lasso estimates as a function of the shrinkage factor:

$$\frac{\sum_{j=1}^{k} |\widehat{\beta}_j^{\text{LASSO}}|}{\sum_{j=1}^{k} |\widehat{\beta}_j^{\text{LS}}|}.$$

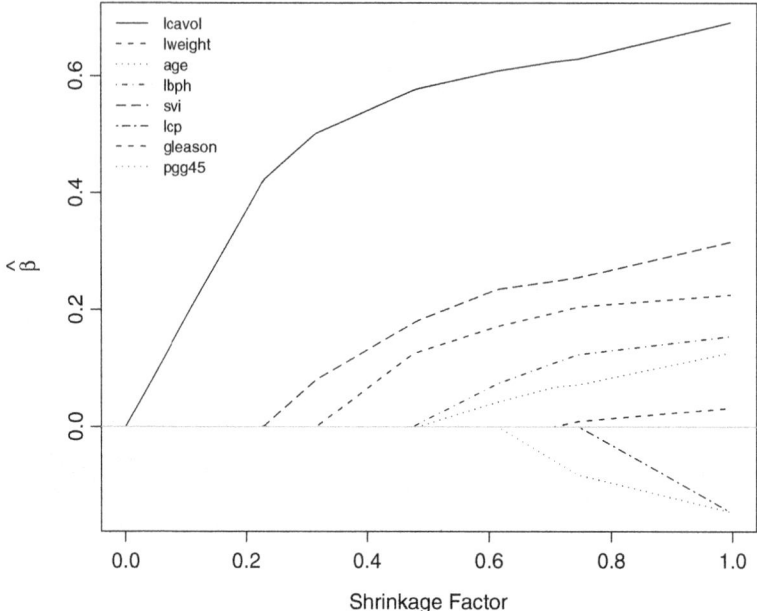

Fig. 10.7 Lasso estimates for the prostate data, as a function of the shrinkage factor, $\sum_{j=1}^{k} |\hat{\beta}_j^{\text{LASSO}}| / \sum_{j=1}^{k} |\hat{\beta}_j^{\text{LS}}|$

When the shrinkage factor is 1, the lasso estimates are the same as the least squares estimates. Beginning with the coefficient associated with log capsular penetration and ending with that associated with log cancer volume each of the coefficients is absorbed at zero, as the coefficient trajectories are traced out. For example, at a shrinkage factor of 0.4, only 3 coefficients are nonzero, those associated with log cancer volume, log weight and Gleason. In this example, the curves decrease monotonically to zero, but this phenomenon will not occur in all examples. The piecewise linear nature of the solution is apparent.

10.6 Smoothing Parameter Selection

For both ridge regression and the lasso, as well as a number of methods to be described in Chaps. 11 and 12, a key element of implementation is smoothing parameter selection.[6] We denote a generic smoothing parameter by λ and the estimated function at this λ, for a particular covariate value x, by $\widehat{f}^{(\lambda)}(x)$.

---

[6]We use the name "smoothing" parameter because we concentrate on nonparametric regression smoothers in this and the next two chapters, but in the context of ridge regression and the lasso, the label "tuning" parameter is often used.

10.6 Smoothing Parameter Selection

In this section, the overall strategy is to derive methods for minimizing, with respect to λ, estimates of the generalization error, or related measures. We initially assume a quadratic loss function before describing smoothing parameter selection in generalized linear model situations.

In Sect. 10.6.1, an analytic method of minimizing the AMSE (Table 10.2) is described and shown to be equivalent to Mallows C_P (Sect. 4.8.2). Two popular approaches for smoothing parameter selection, ordinary and generalized cross-validation, are described in Sects. 10.6.2 and 10.6.3, and in Sect. 10.6.4, we describe the AIC model selection statistic, which extends Mallows C_P to general data types. Finally, Sect. 10.6.5 briefly describes cross-validation for generalized linear models.

Bayesian approaches include choosing λ on substantive grounds (as carried out in Sect. 5.12) or treating λ as an unknown parameter. In the latter case, a prior is specified for λ, which is then estimated in the usual way. Section 11.2.8 adopts a mixed model formulation and describes a frequentist approach to smoothing parameter estimation, with restricted maximum likelihood (REML, see Sect. 8.5.3) being emphasized. Section 11.2.9 takes the same formulation but describes a Bayesian approaches to estimation.

Smoothing parameter choice is an inherently difficult problem because, in many situations, the data do not indicate a clear "optimal" λ. Therefore, there is no universally reliable method for smoothing parameter selection. Consequently, in practice, one should not blindly accept the solution provided by any method. Rather, one should treat the solution as a starting point for further exploration, including the use of alternative methods.

10.6.1 Mallows C_P

In this section we assume that the smoothing method produces a *linear smoother* of the form $\widehat{y} = S^{(\lambda)} y$. Ridge regression provides an example with $S^{(\lambda)} = x(x^T x + \lambda \mathbf{I}_k)^{-1} x^T$; the lasso does not fall within this class. Many methods that we describe in Chap. 11 produce smoothers of linear form.

Recall, from Sect. 5.11.2, that in linear regression $\widehat{y} = Sy$ where $S = x(x^T x)^{-1} x^T$ is the *hat* matrix, and tr(S) is both the number of regression parameters in the model and the degrees of freedom. Equation (10.31) defined the effective degrees of freedom for linear smoothers as $p^{(\lambda)} = \text{df}(\lambda) = \text{tr}\left(S^{(\lambda)}\right)$. One approach to smoothing parameter choice is to simply pick λ to produce the desired effective degrees of freedom $p^{(\lambda)}$, if we have some a priori sense of the degrees of freedom that is desirable. This allows a direct comparison with parametric models. For example, one may pick $p^{(\lambda)} = 4$ to provide a fit with effective degrees of freedom equal to the number of parameters in a cubic polynomial regression model.

An appealing approach is to choose the smoothing parameter to minimize the average mean squared error, (10.17):

$$\text{AMSE}^{(\lambda)} = \text{AMSE}(\widehat{\boldsymbol{f}}^{(\lambda)}) = \frac{1}{n} \sum_{i=1}^{n} \text{E}\left\{ \left[f(\boldsymbol{x}_i) - \widehat{f}^{(\lambda)}(\boldsymbol{x}_i) \right]^2 \right\}$$

$$= \frac{1}{n} \text{E}\left[\left(\boldsymbol{f} - \widehat{\boldsymbol{f}}^{(\lambda)} \right)^{\text{T}} \left(\boldsymbol{f} - \widehat{\boldsymbol{f}}^{(\lambda)} \right) \right], \quad (10.37)$$

where $\boldsymbol{f} = [f(\boldsymbol{x}_1), \ldots, f(\boldsymbol{x}_n)]^{\text{T}}$ and $\widehat{\boldsymbol{f}}^{(\lambda)} = \left[\widehat{f}^{(\lambda)}(\boldsymbol{x}_1), \ldots, \widehat{f}^{(\lambda)}(\boldsymbol{x}_n) \right]^{\text{T}}$. The AMSE depends on the unknown $\boldsymbol{f}$ and so is not directly of use. A more applicable version is obtained by replacing $\boldsymbol{f}$ by $\boldsymbol{Y} - \boldsymbol{\epsilon}$ (with $\text{E}[\boldsymbol{\epsilon}] = 0$) and taking $\widehat{\boldsymbol{f}}^{(\lambda)} = \boldsymbol{S}^{(\lambda)} \boldsymbol{Y}$ to give

$$\text{AMSE}^{(\lambda)} = \frac{1}{n} \text{E}\left[\left(\boldsymbol{Y} - \boldsymbol{\epsilon} - \boldsymbol{S}^{(\lambda)} \boldsymbol{Y} \right)^{\text{T}} \left(\boldsymbol{Y} - \boldsymbol{\epsilon} - \boldsymbol{S}^{(\lambda)} \boldsymbol{Y} \right) \right]$$

$$= \frac{1}{n} \text{E}\left[\left(\boldsymbol{Y} - \boldsymbol{S}^{(\lambda)} \boldsymbol{Y} \right)^{\text{T}} \left(\boldsymbol{Y} - \boldsymbol{S}^{(\lambda)} \boldsymbol{Y} \right) \right]$$

$$+ \frac{1}{n} \text{E}[\boldsymbol{\epsilon}^{\text{T}} \boldsymbol{\epsilon}] - \frac{1}{n} \text{E}\left[2\boldsymbol{\epsilon}^{\text{T}} (\boldsymbol{I} - \boldsymbol{S}^{(\lambda)}) \boldsymbol{Y} \right].$$

Replacing $\boldsymbol{Y}$ by $\boldsymbol{f} + \boldsymbol{\epsilon}$ in the final term and rearranging gives

$$\text{AMSE}^{(\lambda)} = \frac{1}{n} \text{E}\left[\left(\boldsymbol{Y} - \boldsymbol{S}^{(\lambda)} \boldsymbol{Y} \right)^{\text{T}} \left(\boldsymbol{Y} - \boldsymbol{S}^{(\lambda)} \boldsymbol{Y} \right) \right]$$

$$- \frac{1}{n} \text{E}\left[\boldsymbol{\epsilon}^{\text{T}} \boldsymbol{\epsilon} + 2\boldsymbol{\epsilon}^{\text{T}} \boldsymbol{f} - 2\boldsymbol{\epsilon}^{\text{T}} \boldsymbol{S}^{(\lambda)} \boldsymbol{f} - 2\boldsymbol{\epsilon}^{\text{T}} \boldsymbol{S}^{(\lambda)} \boldsymbol{\epsilon} \right].$$

Since

$$\text{E}\left[\boldsymbol{\epsilon}^{\text{T}} \boldsymbol{S}^{(\lambda)} \boldsymbol{\epsilon} \right] = \text{E}\left[\text{tr}\left(\boldsymbol{\epsilon}^{\text{T}} \boldsymbol{S}^{(\lambda)} \boldsymbol{\epsilon} \right) \right] = \text{E}\left[\text{tr}\left(\boldsymbol{S}^{(\lambda)} \boldsymbol{\epsilon} \boldsymbol{\epsilon}^{\text{T}} \right) \right] = \text{tr}\left(\boldsymbol{S}^{(\lambda)} \boldsymbol{I} \sigma^2 \right) = \sigma^2 \text{tr}\left(\boldsymbol{S}^{(\lambda)} \right)$$

$$= \sigma^2 p^{(\lambda)},$$

and $\text{E}[2\boldsymbol{\epsilon}^{\text{T}} \boldsymbol{f}] = \text{E}[2\boldsymbol{\epsilon}^{\text{T}} \boldsymbol{S}^{(\lambda)} \boldsymbol{f}] = 0$, we obtain

$$\text{AMSE}^{(\lambda)} = \frac{1}{n} \text{E}\left[\left(\boldsymbol{Y} - \boldsymbol{S}^{(\lambda)} \boldsymbol{Y} \right)^{\text{T}} \left(\boldsymbol{Y} - \boldsymbol{S}^{(\lambda)} \boldsymbol{Y} \right) \right] - \sigma^2 + \frac{2}{n} \text{E}\left[\boldsymbol{\epsilon}^{\text{T}} \boldsymbol{S}^{(\lambda)} \boldsymbol{\epsilon} \right]$$

$$= \frac{1}{n} \text{E}\left[\left(\boldsymbol{Y} - \boldsymbol{S}^{(\lambda)} \boldsymbol{Y} \right)^{\text{T}} \left(\boldsymbol{Y} - \boldsymbol{S}^{(\lambda)} \boldsymbol{Y} \right) \right] - \sigma^2 + \frac{2}{n} p^{(\lambda)} \sigma^2. \quad (10.38)$$

The natural estimator of (10.38) is

$$\widehat{\text{AMSE}}^{(\lambda)} = \frac{1}{n} \left(\boldsymbol{Y} - \boldsymbol{S}^{(\lambda)} \boldsymbol{Y} \right)^{\text{T}} \left(\boldsymbol{Y} - \boldsymbol{S}^{(\lambda)} \boldsymbol{Y} \right) - \widehat{\sigma}_{\max}^2 + \frac{2}{n} p^{(\lambda)} \widehat{\sigma}_{\max}^2$$

$$= \frac{\widehat{\sigma}_{\max}^2}{n} \left[\frac{\text{RSS}^{(\lambda)}}{\widehat{\sigma}_{\max}^2} - \left(n - 2 p^{(\lambda)} \right) \right]$$

where $\widehat{\sigma}^2_{\max} > 0$ is an estimate from a maximal model (e.g., the full model in a regression setting). Minimizing the estimated AMSE$^{(\lambda)}$ as a function of λ is therefore equivalent to minimization of Mallows C_P statistic, (4.25):

$$\frac{\text{RSS}^{(\lambda)}}{\widehat{\sigma}^2_{\max}} - \left(n - 2p^{(\lambda)}\right). \tag{10.39}$$

A useful quantity to evaluate is the *average predictive risk* (APR, Table 10.2), which is the predictive risk at the observed $\boldsymbol{x}_i$, $i = 1, \ldots, n$. Specifically,

$$\text{APR} = \sigma^2 + \text{AMSE}, \tag{10.40}$$

which can be estimated by

$$\widehat{\text{APR}}^{(\lambda)} = \widehat{\sigma}^2_{\max} + \frac{1}{n}\text{RSS}^{(\lambda)} - \frac{1}{n}\left(n - 2p^{(\lambda)}\right)\widehat{\sigma}^2_{\max}$$

$$= \frac{\text{RSS}^{(\lambda)}}{n} + \frac{2p^{(\lambda)}}{n}\widehat{\sigma}^2_{\max}. \tag{10.41}$$

Estimating APR by the average residual sum of squares (i.e. the first term in (10.41)) is clearly subject to *overfitting* (and hence will be an underestimate), but this is corrected for by the second term.

10.6.2 K-Fold Cross-Validation

A widely used and simple method for estimating prediction error, and hence smoothing parameters, is cross-validation. If we try to estimate the APR, as given by (10.40), from the data directly, that is, using

$$\frac{1}{n}\sum_{i=1}^{n}\left[y_i - \widehat{f}^{(\lambda)}(\boldsymbol{x}_i)\right]^2 = \frac{\text{RSS}^{(\lambda)}}{n},$$

we will obtain an optimistic estimate because the data have been used twice: once to fit the model and once to estimate the predictive risk, as we saw in (10.41). The problem is that the idiosyncrasies of the particular realization of the data will influence coefficient estimates so that the model will, in turn, predict the data "too well". As noted in Sect. 10.4, ideally one would split the data to produce a validation dataset, with estimation of the generalization error being performed using the validation data. Unfortunately there are frequently insufficient data to carry out this step. However, cross-validation provides an approach in the same spirit to estimate the APR.

In K-fold validation, a fraction $(K-1)/K$ of the data are used to fit the model. The remaining fraction, $1/K$, are predicted, and these data are used to produce an estimate of the predictive risk. Let $\boldsymbol{y} = [\boldsymbol{y}_1, \ldots, \boldsymbol{y}_K]$ represent a particular

K-fold split of the $n \times 1$ data vector $\boldsymbol{y}$. Further, let $J(k)$ be the set of elements of $\{1, 2, \ldots, n\}$ that correspond to the indices of data points within split k, with $n_k = |J(k)|$ representing the cardinality of set k. Let $\boldsymbol{y}_{-k}$ be the data with the portion $\boldsymbol{y}_k$ removed and $\widehat{f}_{-k}^{(\lambda)}(\boldsymbol{x}_i)$ represent the i-th fitted value, computed from fitting a model using $\boldsymbol{y}_{-k}$. Cross-validation proceeds by cycling over $k = 1, \ldots, K$ through the following two steps:

1. Fit the model using $\boldsymbol{y}_{-k}$.
2. Use the fitted model to obtain predictions for the removed data, $\boldsymbol{y}_k$, and estimate the error as

$$\mathrm{CV}_k^{(\lambda)} = \frac{1}{n_k} \sum_{i \in J(k)} \left[y_i - \widehat{f}_{-k}^{(\lambda)}(\boldsymbol{x}_i) \right]^2. \quad (10.42)$$

The K prediction errors are averaged to give

$$\mathrm{CV}^{(\lambda)} = \frac{1}{K} \sum_{k=1}^{K} \mathrm{CV}_k^{(\lambda)}.$$

This procedure is repeated for each potential value of the smoothing parameter, λ. We emphasize that the data are split into K pieces *once*, and so the resultant datasets are the same across all λ.

Typical choices for K include 5, 10, and n, the latter being known as *leave-one-out* or *ordinary* cross-validation (OCV). Picking $K = n$ produces an estimate of the expected prediction error with the least bias, but this estimate can have high variance because the n training sets are so similar to one another. The computational burden of OCV can be heavy, though for a large class of smoothers this burden can be sidestepped, as we describe shortly. For smaller values of K, the variance of the expected prediction error estimator is smaller but there is greater bias. Breiman and Spector (1992) provide some discussion on choice of K and recommend $K = 5$ based on simulations in which the aim was subset selection. A number of authors (e.g., Hastie et al. 2009) routinely create an estimate of the standard error of the cross-validation score, (10.42). This estimate assumes independence of $\mathrm{CV}_k^{(\lambda)}$, $k = 1, \ldots, K$, which is clearly not true since each pair of splits share a proportion $1 - 1/(K-1)$ of the data.

We consider leave-one-out cross-validation in more detail. It would appear that we need to fit the model n times, but we show that, for a particular class of smoothers (to be described below),

$$\frac{1}{n} \sum_{i=1}^{n} \left[y_i - \widehat{f}_{-i}^{(\lambda)}(\boldsymbol{x}_i) \right]^2 = \frac{1}{n} \sum_{i=1}^{n} \frac{\left[y_i - \widehat{f}^{(\lambda)}(\boldsymbol{x}_i) \right]^2}{\left(1 - S_{ii}^{(\lambda)} \right)^2} \quad (10.43)$$

where $S_{ii}^{(\lambda)}$ is the ith diagonal element of $\boldsymbol{S}^{(\lambda)}$, and $\widehat{f}_{-i}^{(\lambda)}(\boldsymbol{x}_i)$ is the ith fitted point, based on $\boldsymbol{y}_{-i}$.

10.6 Smoothing Parameter Selection

We prove (10.43), for a particular class of smoothers, based on the derivation in Wood (2006, Sect. 4.5.2). For many smoothing methods, including ridge regression, we can write the model as $f = h\beta$ where $h = [h_1, \ldots, h_n]^{\mathrm{T}}$ is an $n \times J$ design matrix with h_i a $1 \times J$ vector, and β is a $J \times 1$ vector of parameters. We prove the result (10.43) for a class of problems involving minimization of a sum of squares plus a quadratic penalty term:

$$\sum_{i=1}^{n}(y_i - h_i\beta)^2 + \lambda\beta^{\mathrm{T}}D\beta,$$

for a known matrix D. Section 11.2.5 gives further examples of smoothers that fall within this class. Fitting the model to the $n-1$ points contained in y_{-i} involves minimization of

$$\sum_{j=1, j\neq i}^{n}[y_j - f_{-i}(x_j)]^2 + \lambda\beta^{\mathrm{T}}D\beta = \sum_{j=1}^{n}[y_j^\star - f_{-i}(x_j)]^2 + \lambda\beta^{\mathrm{T}}D\beta$$

(10.44)

where

$$y_j^\star = \begin{cases} y_j & \text{if } j \neq i \\ y_i - y_i + f_{-i}(x_i) & \text{if } j = i. \end{cases}$$

Minimization of (10.44) yields

$$\widehat{f} = S^{(\lambda)}y^\star = h(h^{\mathrm{T}}h + \lambda D)^{-1}h^{\mathrm{T}}y^\star,$$

and $\widehat{f}_{-i}^{(\lambda)}(x_i) = S_i^{(\lambda)}y^\star$, where $S_i^{(\lambda)}$ is the ith row of $S^{(\lambda)}$ and $y^\star = [y_1^\star, \ldots, y_n^\star]$. Now

$$\widehat{f}_{-i}^{(\lambda)}(x_i) = S_i^{(\lambda)}y^\star$$
$$= S_i^{(\lambda)}y - S_{ii}^{(\lambda)}y_i + S_{ii}^{(\lambda)}\widehat{f}_{-i}^{(\lambda)}(x_i)$$
$$= \widehat{f}^{(\lambda)}(x_i) - S_{ii}^{(\lambda)}y_i + S_{ii}^{(\lambda)}\widehat{f}_{-i}^{(\lambda)}(x_i)$$

so that

$$\widehat{f}_{-i}^{(\lambda)}(x_i) = \frac{\widehat{f}^{(\lambda)}(x_i) - S_{ii}^{(\lambda)}y_i}{1 - S_{ii}}$$

and

$$y_i - \widehat{f}_{-i}^{(\lambda)}(x_i) = \frac{y_i(1 - S_{ii}^{(\lambda)}) - \widehat{f}^{(\lambda)}(x_i) + S_{ii}^{(\lambda)}y_i}{1 - S_{ii}^{(\lambda)}} = \frac{y_i - \widehat{f}^{(\lambda)}(x_i)}{1 - S_{ii}^{(\lambda)}}, \quad (10.45)$$

as required. To calculate the leave-one-out CV score, we therefore need only the residuals from the fit to the complete data and the diagonal elements of the smoother matrix. Note that the effect of $\left(1 - S_{ii}^{(\lambda)}\right)^2$ in the denominator of (10.43) is to inflate the residual at the i-th point, hence accounting for the underestimation of simply using the residual sum of squares. Formula (10.43) is true for all linear smoothers.

In practice, curves of the estimated prediction error against λ (the smoothing parameter) can be very flat, as shown for instance in Fig. 10.9. Therefore, as already noted, simply blindly using the value of λ that minimizes the cross-validation sum of squares is not a reliable strategy. In Hastie et al. (2009), it is recommended that λ be chosen such that the prediction error is no greater than one standard error above that with the lowest error. This approach results in a more parsimonious model being selected, though this recommendation is based on judgement and experience rather than theory.

10.6.3 Generalized Cross-Validation

So-called generalized cross-validation (GCV) provides an alternative to K-fold cross-validation. The GCV score is

$$\text{GCV}^{(\lambda)} = \frac{n}{\left[n - \text{tr}\left(S^{(\lambda)}\right)\right]^2} \sum_{i=1}^{n} \left(y_i - S_i^{(\lambda)} y\right)^2 \qquad (10.46)$$

for a linear smoother $\widehat{y} = S^{(\lambda)} y$. An important early reference on the use of GCV is Craven and Wabha (1979). Recall that $\text{tr}\left(S^{(\lambda)}\right)$ is the effective degrees of freedom of a linear smoother, (10.31), with larger values of λ corresponding to increased smoothing. Therefore, the denominator of (10.46) is the squared effective residual degrees of freedom and a measure of complexity: increasing λ decreases the effective number of parameters, that is, the complexity of the model, and this reduction produces lower variability. However, the numerator is the residual sum of squares and as such is a measure of squared bias with larger λ giving a poorer fit and increased bias. Consequently, we see that the GCV score is providing a trade-off between bias and variance. Unlike K-fold cross-validation, GCV does not require splitting of the data into cross-validation folds and repeatedly training and testing the model.

GCV may be justified/motivated in a number of different ways. On computational grounds, the GCV score is simpler to evaluate than the OCV score, since one only needs the trace of $S^{(\lambda)}$ and not the diagonal elements $S_{ii}^{(\lambda)}$. Recall from Sect. 5.11.2 that in the context of a linear model, the *leverage* of y_i is defined as $S_{ii}^{(\lambda)}$, and so the OCV score can be highly influenced by a small number of data points (due to the presence of $1 - S_{ii}^{(\lambda)}$ in the denominator of (10.43)), which can be undesirable. Therefore, one interpretation of GCV is that it is simply a robust alternative to OCV with $1 - S_{ii}^{(\lambda)}$ replaced by $1 - \text{tr}(S^{(\lambda)})/n$, which is clear if we rewrite (10.46) as

10.6 Smoothing Parameter Selection

$$\text{GCV}^{(\lambda)} = \frac{1}{n} \sum_{i=1}^{n} \frac{\left(y_i - \boldsymbol{S}_i^{(\lambda)} \boldsymbol{y}\right)^2}{\left(1 - S_{ii}^{(\lambda)}\right)^2} \left(\frac{1 - S_{ii}^{(\lambda)}}{1 - \text{tr}\left(\boldsymbol{S}^{(\lambda)}\right)/n}\right)^2$$

$$= \frac{1}{n} \sum_{i=1}^{n} \left[y_i - \widehat{f}_{-i}^{(\lambda)}(\boldsymbol{x}_i)\right]^2 \left(\frac{1 - S_{ii}^{(\lambda)}}{1 - \text{tr}\left(\boldsymbol{S}^{(\lambda)}\right)/n}\right)^2.$$

This representation illustrates that those observations with large leverage are being down-weighted, as compared to OCV.

A final justification for using GCV, which was emphasized by Golub et al. (1979), is an invariance property. Namely, GCV is invariant to certain transformations of the data whereas OCV is not. Suppose we transform $\boldsymbol{y}$ and $\boldsymbol{x}$ to $\boldsymbol{Qy}$ and $\boldsymbol{Qx}$, respectively, where $\boldsymbol{Q}$ is any $n \times n$ orthogonal matrix (i.e., $\boldsymbol{QQ}^\top = \boldsymbol{Q}^\top \boldsymbol{Q} = \boldsymbol{I}_n$). For fixed λ, minimization with respect to $\boldsymbol{\beta}$ of

$$(\boldsymbol{y} - \boldsymbol{x}\boldsymbol{\beta})^\top (\boldsymbol{y} - \boldsymbol{x}\boldsymbol{\beta}) + \lambda \boldsymbol{\beta}^\top \boldsymbol{\beta}$$

leads to inference that is identical to minimization of

$$(\boldsymbol{Qy} - \boldsymbol{Qx}\boldsymbol{\beta})^\top (\boldsymbol{Qy} - \boldsymbol{Qx}\boldsymbol{\beta}) + \lambda \boldsymbol{\beta}^\top \boldsymbol{\beta}.$$

However, for fixed λ, the OCV scores are not identical, so that $\widehat{\lambda}$ obtained via minimization of the OCV will differ depending on whether we work with $\boldsymbol{y}$ or $\boldsymbol{Qy}$.

If $\boldsymbol{S}^{(\lambda)}$ is the linear smoother for the original data, then

$$\boldsymbol{S}_Q^{(\lambda)} = \boldsymbol{Q}\boldsymbol{S}^{(\lambda)}\boldsymbol{Q}^\top$$

is the linear smoother for the rotated data. Note that

$$\text{tr}\left(\boldsymbol{S}_Q^{(\lambda)}\right) = \text{tr}\left(\boldsymbol{Q}\boldsymbol{S}^{(\lambda)}\boldsymbol{Q}^\top\right) = \text{tr}\left(\boldsymbol{S}^{(\lambda)}\boldsymbol{Q}^\top\boldsymbol{Q}\right) = \text{tr}\left(\boldsymbol{S}^{(\lambda)}\right),$$

and GCV is invariant to the choice of $\boldsymbol{Q}$ (Golub et al. 1979). It can be shown (e.g., Wood 2006, Sect. 4.5.3) that GCV corresponds to the rotation of the data that results in each of the diagonal elements of $\boldsymbol{S}_Q^{(\lambda)}$ being equal. Since the expected prediction error is invariant to the rotation used, the GCV score shares with the OCV score the interpretation as an estimate of the expected prediction error.

Using the approximation $(1 - x)^{-2} \approx 1 + 2x$ we obtain

$$\text{GCV}^{(\lambda)} \approx \frac{1}{n} \sum_{i=1}^{n} \left[y_i - \widehat{f}^{(\lambda)}(\boldsymbol{x}_i)\right]^2 + \frac{2\text{tr}\left(\boldsymbol{S}^{(\lambda)}\right)}{n} \frac{1}{n} \sum_{i=1}^{n} \left[y_i - \widehat{f}^{(\lambda)}(\boldsymbol{x}_i)\right]^2$$

$$= \frac{\text{RSS}^{(\lambda)}}{n} + \frac{2p^{(\lambda)}}{n} \widehat{\sigma}^2,$$

which is proportional to Mallows C_P if we replace $\widehat{\sigma}^2_{\max}$ in (10.39) with $\widehat{\sigma}^2$, up to a constant not depending on λ.

10.6.4 AIC for General Models

The AIC was introduced in Sect. 4.8.2; here we provide a derivation as a generalization of Mallows C_P. Consider the prediction of new observations $Y_1^\star, \ldots, Y_n^\star$ with model

$$Y_i^\star \mid \boldsymbol{\beta} \sim_{ind} \mathrm{N}\left[f_i(\boldsymbol{\beta}), \sigma^2\right],$$

for $i = 1, \ldots, n$. Suppose we fit a model using data $\boldsymbol{Y}_n = [Y_1, \ldots, Y_n]$ and obtain the MLE $\widehat{\boldsymbol{\beta}}$. The expected value of the negative maximized log-likelihood evaluated at $\widehat{\boldsymbol{\beta}}$ is

$$-\mathrm{E}\left[l_n(\widehat{\boldsymbol{\beta}})\right] = \frac{n}{2}\log 2\pi + n\log\sigma + \frac{1}{2\sigma^2}\sum_{i=1}^n \mathrm{E}\left\{\left[Y_i^\star - f_i(\widehat{\boldsymbol{\beta}})\right]^2\right\}.$$

Considering the last term only, we saw in Sect. 10.4.1 that

$$\sum_{i=1}^n \mathrm{E}\left\{\left[Y_i^\star - f_i(\widehat{\boldsymbol{\beta}})\right]^2\right\} = n\sigma^2 + \sum_{i=1}^n \mathrm{E}\left\{\left[f_i(\widehat{\boldsymbol{\beta}}) - f_i(\boldsymbol{\beta})\right]^2\right\}, \quad (10.47)$$

and Mallows C_P was derived as an approximation to the second term, with "good" models having a low C_p.

We now consider a general log-likelihood based on n observations $l_n(\boldsymbol{\beta})$, with our aim being to find a criterion to judge the "fits" of a collection of models, taking into account model complexity. The basis of AIC is to evaluate a model based on its ability to predict new data $Y_i^\star$, $i = 1, \ldots, n$. The prediction is based on the model $p(\boldsymbol{y}^\star \mid \widehat{\boldsymbol{\beta}})$ with $\widehat{\boldsymbol{\beta}}$ being the MLE based on an independent sample of size n, $\boldsymbol{Y}_n$.

The criterion that is used for discrimination, that is, to decide on whether the prediction is good, is the Kullback–Leibler distance (as discussed in Sect. 2.4.3) between the true model and the assumed model. The distance between the true (unknown) distribution $p_\mathrm{T}(\boldsymbol{y}^\star)$ and a model $p(\boldsymbol{y}^\star \mid \boldsymbol{\beta})$ is

$$\mathrm{KL}\left[p_\mathrm{T}(\boldsymbol{y}^\star), p(\boldsymbol{y}^\star \mid \boldsymbol{\beta})\right] = \int \log\left(\frac{p_\mathrm{T}(\boldsymbol{y}^\star)}{p(\boldsymbol{y}^\star \mid \boldsymbol{\beta})}\right) p_\mathrm{T}(\boldsymbol{y}^\star)\, d\boldsymbol{y}^\star \geq 0.$$

A good model with estimator $\widehat{\boldsymbol{\beta}}$ will produce a small value of

$$\mathrm{KL}\left[p_\mathrm{T}(\boldsymbol{y}^\star), p(\boldsymbol{y}^\star \mid \widehat{\boldsymbol{\beta}})\right]. \quad (10.48)$$

Unfortunately (10.48) cannot be directly used, since $p_\mathrm{T}(\boldsymbol{y}^\star)$ is unknown, but we show how it may be approximated, up to an additive constant.

Result: Let $\boldsymbol{Y}_n = [Y_1, \ldots, Y_n]$ be a random sample from $p_\mathrm{T}(y)$ and suppose a model $p(y \mid \boldsymbol{\beta})$ is fitted to these data and yields MLE $\widehat{\boldsymbol{\beta}}$, where $\boldsymbol{\beta}$ is a parameter vector of dimension p. For simplicity, we state and prove the result for independent and identically distributed data but the result is true in the nonidentically distributed case also. We wish to predict an independent sample, $Y_i^\star$, $i = 1, \ldots, n$, using $p(y^\star \mid \widehat{\boldsymbol{\beta}})$.

10.6 Smoothing Parameter Selection

Two times the expected distance between the true distribution and the assumed distribution, evaluated at the estimator $\widehat{\beta}$, is

$$D^\star = 2 \times E_{Y^\star}\left[\sum_{i=1}^n \log\left(\frac{p_{\mathrm{T}}(Y_i^\star)}{p(Y_i^\star \mid \widehat{\beta})}\right)\right]$$
$$= 2n \times \mathrm{KL}\left[p_{\mathrm{T}}(y^\star), p(y^\star \mid \widehat{\beta})\right]. \tag{10.49}$$

Then, we have the approximation

$$D^\star \approx 2n \times \mathrm{KL}\left[p_{\mathrm{T}}(y^\star), p(y^\star \mid \beta_{\mathrm{T}})\right] + p, \tag{10.50}$$

where β_{T} is the value of β that minimizes the Kullback–Leibler distance between $p_{\mathrm{T}}(y)$ and $p(y \mid \beta)$ (for discussion, see Sect. 2.4.3). The difference between (10.49) and (10.50) therefore gives the increase in the discrepancy when $p(y^\star \mid \beta_{\mathrm{T}})$ is replaced by $p(y^\star \mid \widehat{\beta})$.

An estimate of $D^\star$ is

$$\widehat{D}^\star = -2 \times l_n(\widehat{\beta}) + 2p + 2c_{\mathrm{T}}$$

where $c_{\mathrm{T}} = \int \log[p_{\mathrm{T}}(y^\star)]p_{\mathrm{T}}(y^\star)\,dy^\star$ is a constant that is common to all models under comparison. Ignoring this constant gives Akaike's *An Information Criterion*[7] (AIC, Akaike 1973):

$$\mathrm{AIC} = -2 \times l_n(\widehat{\beta}) + 2p.$$

Outline Derivation

The outline proof presented below is based on Davison (2003, Sect. 4.7). The distance measure $D^\star$ given in (10.49) is two times the expected difference between log-likelihoods:

$$D^\star = E\left[2n \log p_{\mathrm{T}}(Y^\star) - 2n \log p(Y^\star \mid \widehat{\beta})\right], \tag{10.51}$$

where the expectation is with respect to the true model $p_{\mathrm{T}}(y^\star)$. We proceed by first approximating the second term via a Taylor series expansion about β_{T}. Let

$$S_1(\beta) = \frac{\partial}{\partial \beta} \log p(Y \mid \beta), \quad I_1(\beta) = -E\left[\frac{\partial^2}{\partial \beta \partial \beta^{\mathrm{T}}} \log p(Y \mid \beta)\right]$$

denote the score and information in a sample of size one. Then

$$2n \log p(Y^\star \mid \widehat{\beta}) \approx 2n \log p(Y^\star \mid \beta_{\mathrm{T}}) + 2n(\widehat{\beta} - \beta_{\mathrm{T}})^{\mathrm{T}} S_1(\beta_{\mathrm{T}})$$
$$- n(\widehat{\beta} - \beta_{\mathrm{T}})^{\mathrm{T}} I_1(\beta_{\mathrm{T}})(\widehat{\beta} - \beta_{\mathrm{T}}).$$

---

[7] Commonly AIC is referred to as *Akaike's Information Criterion*.

Note that $\mathrm{E}[\boldsymbol{S}_1(\boldsymbol{\beta}_{\mathrm{T}})] = \boldsymbol{0}$ and $n(\widehat{\boldsymbol{\beta}} - \boldsymbol{\beta}_{\mathrm{T}})^{\mathrm{T}}\boldsymbol{I}_1(\boldsymbol{\beta}_{\mathrm{T}})(\widehat{\boldsymbol{\beta}} - \boldsymbol{\beta}_{\mathrm{T}})$ is asymptotically χ_p^2 (Sect. 2.9.4) so its expectation is p, the number of elements of $\boldsymbol{\beta}$. Hence, the second term in (10.51) may be approximated by

$$\mathrm{E}\left[2n \log p(Y^\star \mid \widehat{\boldsymbol{\beta}})\right] \approx \mathrm{E}\left[2n \log p(Y^\star \mid \boldsymbol{\beta}_{\mathrm{T}})\right] - p. \tag{10.52}$$

Therefore,

$$\begin{aligned}
D^\star &\approx 2n \times \mathrm{E}\left[\log p_{\mathrm{T}}(Y^\star) - \log p(Y^\star \mid \boldsymbol{\beta}_{\mathrm{T}})\right] + p \\
&= 2n \int \log\left(\frac{p_{\mathrm{T}}(y^\star)}{p(y^\star \mid \boldsymbol{\beta}_{\mathrm{T}})}\right) p_{\mathrm{T}}(y^\star) \, dy^\star + p \\
&= 2n \times \mathrm{KL}\left[p_{\mathrm{T}}(\boldsymbol{y}^\star), p(\boldsymbol{y}^\star \mid \boldsymbol{\beta}_{\mathrm{T}})\right] + p
\end{aligned} \tag{10.53}$$

proving (10.50).

This expression for $D^\star$ is not usable because $p_{\mathrm{T}}(\cdot)$ is unknown. An estimator of $\mathrm{KL}[p_{\mathrm{T}}(y^\star), p(y^\star \mid \boldsymbol{\beta}_{\mathrm{T}})]$ can be based on $\mathrm{E}\left[l_n(\widehat{\boldsymbol{\beta}})\right] = \mathrm{E}\left[\log p(Y \mid \widehat{\boldsymbol{\beta}})\right]$, however. We write

$$\begin{aligned}
-2 \times \mathrm{E}\left[l_n(\widehat{\boldsymbol{\beta}})\right] &= 2 \times \mathrm{E}\left[-l_n(\boldsymbol{\beta}_{\mathrm{T}}) - \left\{l_n(\widehat{\boldsymbol{\beta}}) - l_n(\boldsymbol{\beta}_{\mathrm{T}})\right\}\right] \\
&\approx 2n \times \mathrm{E}\left[-\log p(Y^\star \mid \boldsymbol{\beta}_{\mathrm{T}})\right] - p \\
&= 2n \times \mathrm{E}\left[-\log p(Y^\star \mid \boldsymbol{\beta}_{\mathrm{T}}) + \log p_{\mathrm{T}}(Y^\star) - \log p_{\mathrm{T}}(Y^\star)\right] - p \\
&= 2n \times \mathrm{KL}\left[p_{\mathrm{T}}(y^\star), p(\boldsymbol{y} \mid \boldsymbol{\beta}_{\mathrm{T}})\right] - 2c_{\mathrm{T}} - p
\end{aligned} \tag{10.54}$$

where

$$c_{\mathrm{T}} = \int \log[p_{\mathrm{T}}(\boldsymbol{y}^\star)] \, p_{\mathrm{T}}(\boldsymbol{y}^\star) \, d\boldsymbol{y}^\star,$$

and we have used the asymptotic result that

$$2\left[l_n(\widehat{\boldsymbol{\beta}}) - l_n(\boldsymbol{\beta}_{\mathrm{T}})\right] \to \chi_p^2, \tag{10.55}$$

as $n \to \infty$, see (2.55). It follows, by rearrangement of (10.54), that

$$2n \times \mathrm{KL}\left[p_{\mathrm{T}}(y^\star), p(y^\star \mid \boldsymbol{\beta}_{\mathrm{T}})\right] \approx -2 \times \mathrm{E}\left[l_n(\widehat{\boldsymbol{\beta}})\right] + p + 2c_T$$

which suggests an estimator of

$$2n \times \widehat{\mathrm{KL}}\left[p_{\mathrm{T}}(y^\star), p(y^\star \mid \boldsymbol{\beta}_{\mathrm{T}})\right] = -2 \times l_n(\widehat{\boldsymbol{\beta}}) + p + 2c_{\mathrm{T}}.$$

This estimate can be substituted into (10.53) to give the estimator

$$\begin{aligned}
\widehat{D}^\star &= -2 \times l_n(\widehat{\boldsymbol{\beta}}) + 2p + 2c_{\mathrm{T}} \\
&= \mathrm{AIC} + 2c_{\mathrm{T}}
\end{aligned}$$

where AIC $= -2 \times l_n(\widehat{\boldsymbol{\beta}}) + 2p$. Since the term on the right is common to all models, the AIC may be used to compare models, with relatively good models producing a small value of the AIC. Some authors suggest retaining all models whose AIC is within 2 of the minimum (e.g. Ripley 2004). □

The above derivation is based on a number of assumptions (Ripley 2004) including the model under consideration being true. The accuracy of the approximations is also much greater if the models under comparison are nested.

In a GLM smoothing setting, the AIC may be minimized as a function of λ, with the degrees of freedom p being replaced by $\operatorname{tr}\left(\boldsymbol{S}^{(\lambda)}\right)$. The AIC criteria in this case is

$$\mathrm{AIC}^{(\lambda)} = -2l(\widehat{\boldsymbol{\beta}}) + 2 \times \operatorname{tr}\left(\boldsymbol{S}^{(\lambda)}\right), \tag{10.56}$$

with the second term again measuring complexity.

An Aside

The derivation of AIC was carried out under the assumption of a correct model, which was required to obtain (10.52) and (10.55). If the model is wrong, then $\sqrt{n}(\widehat{\boldsymbol{\beta}} - \boldsymbol{\beta}_{\mathrm{T}})$ is asymptotically normal with zero mean and variance $\boldsymbol{I}^{-1}\boldsymbol{K}\boldsymbol{I}^{\mathrm{T}-1}$ where

$$\boldsymbol{K} = \boldsymbol{K}(\boldsymbol{\beta}_{\mathrm{T}}) = \mathrm{E}\left[\left(\frac{\partial}{\partial \boldsymbol{\beta}}\log p(Y \mid \boldsymbol{\beta}_{\mathrm{T}})\right)\left(\frac{\partial}{\partial \boldsymbol{\beta}}\log p(Y \mid \boldsymbol{\beta}_{\mathrm{T}})\right)^{\mathrm{T}}\right],$$

see Sect. 2.4.3. Hence, using identity (B.4) from Appendix B, the expectation of $n(\widehat{\boldsymbol{\beta}} - \boldsymbol{\beta}_{\mathrm{T}})^{\mathrm{T}}\boldsymbol{I}_1(\boldsymbol{\beta}_{\mathrm{T}})(\widehat{\boldsymbol{\beta}} - \boldsymbol{\beta}_{\mathrm{T}})$ is

$$\operatorname{tr}\left[\boldsymbol{I}_1(\boldsymbol{\beta}_{\mathrm{T}})\boldsymbol{I}_1(\boldsymbol{\beta}_{\mathrm{T}})^{-1}\boldsymbol{K}(\boldsymbol{\beta}_{\mathrm{T}})\boldsymbol{I}_1(\boldsymbol{\beta}_{\mathrm{T}})^{-1}\right] = \operatorname{tr}\left[\boldsymbol{K}(\boldsymbol{\beta}_{\mathrm{T}})\boldsymbol{I}_1(\boldsymbol{\beta}_{\mathrm{T}})^{-1}\right].$$

Similarly, under a wrong model, the likelihood ratio statistic $2\left[l_n(\widehat{\boldsymbol{\beta}}) - l_n(\boldsymbol{\beta}_{\mathrm{T}})\right]$ has an asymptotic distribution proportional to χ_p^2 but with mean $\operatorname{tr}\left[\boldsymbol{K}(\boldsymbol{\beta}_{\mathrm{T}})\boldsymbol{I}_1(\boldsymbol{\beta}_{\mathrm{T}})^{-1}\right]$. This follows since, via a Taylor series approximation,

$$2\left[l_n(\widehat{\boldsymbol{\beta}}) - l_n(\boldsymbol{\beta}_{\mathrm{T}})\right] \approx n(\widehat{\boldsymbol{\beta}} - \boldsymbol{\beta}_{\mathrm{T}})^{\mathrm{T}}\boldsymbol{I}_1(\boldsymbol{\beta}_{\mathrm{T}})(\widehat{\boldsymbol{\beta}} - \boldsymbol{\beta}_{\mathrm{T}}).$$

Replacing p by $\operatorname{tr}\left[\boldsymbol{K}(\boldsymbol{\beta}_{\mathrm{T}})\boldsymbol{I}_1(\boldsymbol{\beta}_{\mathrm{T}})^{-1}\right]$ in the above derivation gives the alternative *network information criterion* (NIC)

$$\mathrm{NIC} = -2l(\widehat{\boldsymbol{\beta}}) + 2 \times \operatorname{tr}\left[\boldsymbol{K}(\widehat{\boldsymbol{\beta}})\boldsymbol{I}_1(\widehat{\boldsymbol{\beta}})^{-1}\right],$$

as introduced by Stone (1977).

10.6.5 Cross-Validation for Generalized Linear Models

As discussed in Sect. 10.3.1, for general outcomes, a loss function for measuring the accuracy of a prediction is the negative log-likelihood. Hence, cross-validation can be extended to general data situations by replacing the sum of squares in (10.42) with a loss function to give

$$\text{CV}_k^{(\lambda)} = \frac{1}{n_k} \sum_{i \in J(k)} L\left[y_i, \widehat{f}_{-k}^{(\lambda)}(\boldsymbol{x}_i)\right].$$

In particular, the negative log-likelihood loss (10.12) produces

$$\text{CV}_k^{(\lambda)} = -\frac{2}{n_k} \sum_{i \in J(k)} \log p_{\widehat{f}_{-k}^{(\lambda)}}(y_i \mid \boldsymbol{x}_i),$$

where this notation emphasizes that the prediction at the point x_i is based upon the fitted value $\widehat{f}_{-k}^{(\lambda)}$. Similarly, a natural extension of (10.46) is the generalized cross-validation score based on the log-likelihood

$$\text{GCV}^{(\lambda)} = -\frac{2n}{\left[n - \text{tr}(\boldsymbol{S}^{(\lambda)})\right]^2} \sum_{i=1}^{n} \log p_{\widehat{f}_{-k}^{(\lambda)}}(y_i \mid \boldsymbol{x}_i).$$

Some authors (e.g., Ruppert et al. 2003, p. 220) replace the log-likelihood by the deviance, which adds a term that does not depend on λ.

Example: Prostate Cancer

We illustrate smoothing/tuning parameter choice and estimation of the prediction error using various approaches to modeling and a number of the methods described in Sect. 10.6 for smoothing parameter estimation. The modeling approaches we compare are fitting the full model using least squares, and picking the "best" subset of variables via an exhaustive search based on Mallows C_P, ridge regression, the lasso, and Bayesian model averaging (Sect. 3.6). We divide the prostate data into a training dataset of 67 randomly selected individuals and a test dataset of the remaining 30 individuals. Since the sample size is small, we repeat this splitting 500 times and then evaluate, for the different methods, the average error and its standard deviation over the train/test splits. An important point to emphasize is that we standardize the x variables in the training dataset and then apply the same standardization in the test dataset (and this procedure is repeated separately for each of the 500 splits).

10.6 Smoothing Parameter Selection

Table 10.3 Average test errors over 500 train/test splits of the prostate cancer data, along with the standard deviation over these splits

	Null	Full	Best subset	Ridge	Lasso	BMA
Mean	1.30	0.59	0.76	0.59	0.60	0.59
SD	0.32	0.15	0.35	0.14	0.14	0.14

Table 10.3 gives summaries of the test error, calculated via (10.18), for the five approaches. We also report the error that results from fitting the null (intercept only) model. The latter is a baseline reference, and gives an error of 1.30. The estimate of error corresponding to the full model fitted with least squares is 0.59, a reduction of 71%. The exhaustive search over model space (i.e., the $2^8 = 256$ combinations of 8 variables), using Mallows C_P as the model selection criterion, was significantly worse giving an error of 0.76 with a large standard deviation. Table 10.4 shows the variability across train/test splits in the model chosen by the exhaustive search procedure. For example, 34.2% of models contained only the variables log(can vol), log(weight), and SVI. The seven most frequently occurring models account for 73.8% of the total, with the remainder being spread over 27 other combinations of variables. The table illustrates the discreteness of the exhaustive search procedure (as discussed in Sect. 4.9) and explains the poor prediction performance. Ridge regression and the lasso were applied to each train/test split with λ chosen via minimization of the OCV score. The entries in Table 10.3 show that, for these data, the shrinkage methods provide prediction errors which are comparable to, and not an improvement on, the full model. The reason for this is that in this example the ratio of the sample size to the number of parameters is relatively large, and so there is little penalty for including all parameters in the model.

Figure 10.8 illustrates the variability across train/test splits of the optimal effective degrees of freedom, chosen via minimization of (a) the OCV score and (b) Mallows C_P, for the ridge regression analyses. The two measures are then plotted against each other in (c) and show reasonable agreement. There is a reasonable amount of variability in the optimal degrees of freedom across simulations.

The final approach included in this experiment was Bayesian model averaging. In this example, the performance of BMA matches that of ridge regression and the lasso. BMA is superior to exhaustive search because covariates are not excluded entirely, but rather every model is assigned a posterior weight so that all covariates contribute to the fit. A number of successful approaches to prediction, including boosting, bagging, and random forests (Hastie et al. 2009), gain success from averaging over models, since different models can pick up different aspects of the data, and the variance is reduced by averaging. Bagging and random forests are described in Sects. 12.8.5 and 12.8.6, respectively.

We now provide more detail on the ridge regression, lasso, and Bayesian model averaging approaches. We first consider in greater detail the application of ridge regression. Figure 10.9 shows estimates of the test error, evaluated via different methods, as a function of the effective degrees of freedom, for a single train/test split. The minimizing values are indicated as vertical lines. The dotted line shows

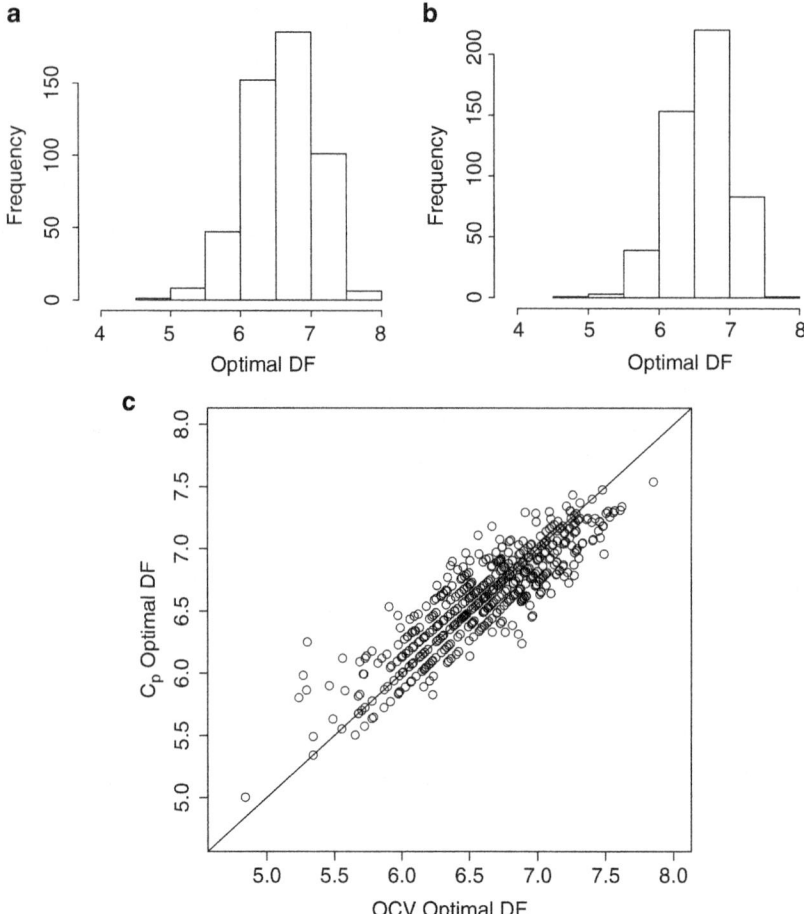

Fig. 10.8 Minimizing values of the effective degrees of freedom for ridge regression from 500 train/test splits of the prostate cancer data using: (**a**) OCV, (**b**) Mallows C_P as the minimizing criteria. Panel (**c**) plots the optimal degrees of freedom arising from each criteria against each other

the estimate as the AMSE plus the estimate of the error variance, (10.41). The minimizing value of AMSE (which is equivalent to minimizing Mallows C_P) is very similar to that obtained with the OCV criteria and is also virtually identical to that obtained from GCV. In all cases, the curves are flat close to the minimum, so one would not want to overinterpret specific numerical values. The effective degrees of freedom corresponding to the minimum OCV is 5.9, while under GCV and Mallows, the values are identical and equal to 5.7. The fivefold CV estimate is minimized for a slightly larger value than for OCV for this train/test split (effective degrees of freedom of 6.6); over all train/test splits, fivefold CV produced a comparable prediction error to OCV. Also included in the figure is the average residual sum of

10.6 Smoothing Parameter Selection

Table 10.4 Percentage of models selected in an exhaustive best subset search, over 500 train/test splits of the prostate cancer data

Variables selected								
lcavol	lweight	age	lbph	svi	lcp	gleason	pgg45	Percentage
1	1	0	0	1	0	0	0	34.2
1	1	1	1	1	0	0	0	11.4
1	1	0	1	1	0	0	0	11.0
1	0	0	1	1	0	0	0	5.8
1	0	1	1	1	0	0	0	4.8
1	1	0	0	1	0	0	1	3.4
1	1	1	0	1	0	0	0	3.2

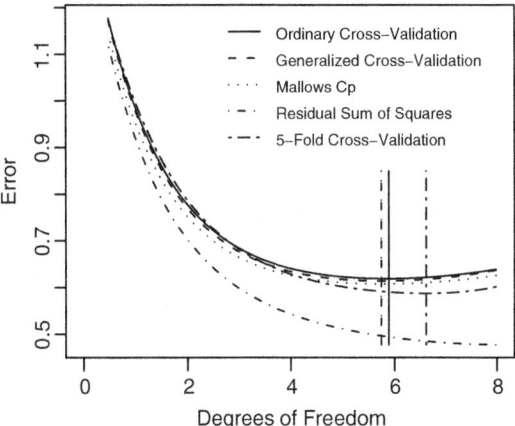

Fig. 10.9 Various estimates of error, as a function of the effective degrees of freedom, for ridge regression applied to the prostate cancer data. Minimizing values are shown as *vertical lines*. Also shown is the residual sum of squares (which has a minimum at 8 degrees of freedom)

squares, which is minimized at the most complex model (degrees of freedom equal to 8), as expected, and underestimates the predictive error, since the data are being used twice.

Turning now to the lasso, Figs. 10.10(a) and (b) show the OCV and GCV estimates of error versus the coefficient shrinkage factor, along with estimates of the standard error. As with ridge regression, the curves are relatively flat close to the minimum, indicating that we should not be wedded to the exact minimizing value of the smoothing parameter. For this train/test split, the minimizing value of the OCV function leads to three coefficients being set to zero.

Finally, for Bayesian model averaging, Fig. 10.11 provides a plot in which the horizontal axis orders the models in terms of decreasing posterior probability (going from left to right), with the variables indicated on the vertical axis. Black rectangles denote inclusion of that variable and gray, no inclusion. The posterior model percentages for the top five models are 23%, 17%, 8%, 7%, and 6%.

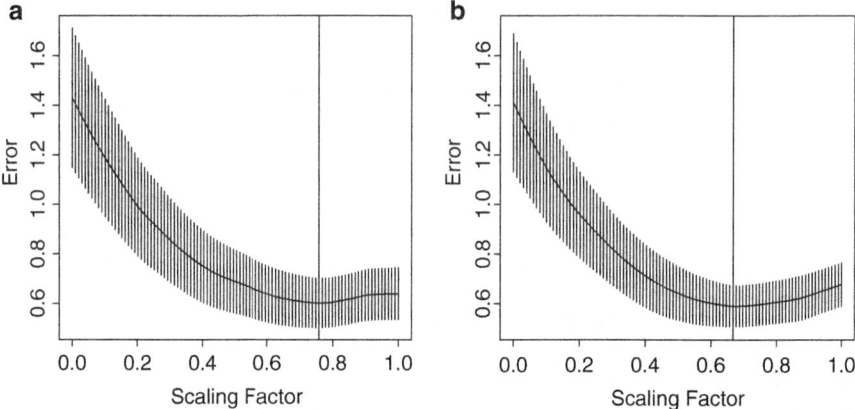

Fig. 10.10 (a) OCV and (b) fivefold CV estimates of error for the lasso, as a function of the scaling factor, $\sum_{j=1}^{k} |\widehat{\beta}_j| / \sum_{j=1}^{k} |\widehat{\beta}_j^{LS}|$, for the prostate cancer data. The minimizing value of the CV estimates of error is shown as a *solid vertical line*. Also shown are approximate standard error bands evaluated as if the CV estimates were independent (as discussed in Sect. 10.6.2)

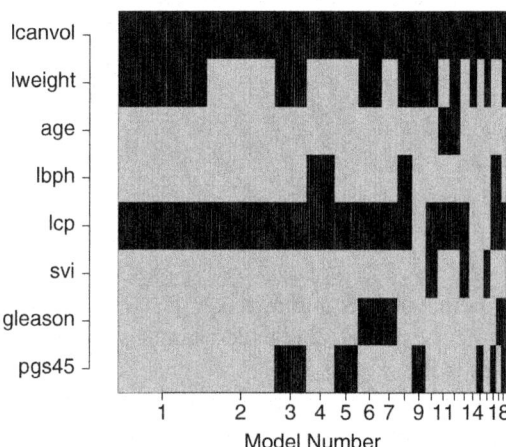

Fig. 10.11 From left to right this plot shows, for a particular split of the prostate cancer data, the models with the highest posterior probability, as evaluated via Bayesian model averaging

10.7 Concluding Comments

Whether parametric or nonparametric models are used, the bias-variance trade-off is a key consideration. In nonparametric modeling there are explicit smoothing parameters that determine this trade-off. We saw this with both ridge regression and the lasso, and this issue will return repeatedly in Chaps. 11 and 12. The choice of smoothing parameter is, therefore, crucial and a variety of approaches for selection, including cross-validation and the minimization of Mallows C_P have been described. Additional methods will be described in Chap. 11, but no single approach will work in all situations, and often subjective judgement is required. Härdle et al.

(1988) have shown that smoothing parameter methods such as Mallows C_P and GCV converge slowly to the optimum as the sample size increases. A number of simulation studies have been carried out and back up the above comments, see, for example, Ruppert et al. (2003, Sect. 5.4) and references therein.

10.8 Bibliographic Notes

There are many excellent texts on nonparametric regression, including Green and Silverman (1994), Simonoff (1997), Ruppert et al. (2003), Wood (2006), and, more recently and with a large range of topics, Hastie et al. (2009). Gneiting and Raftery (2007) provide an excellent review of scoring rules, which are closely related to the loss functions considered in Sect. 10.3. An important early reference on ridge regression is Hoerl and Kennard (1970). Since its introduction in Tibshirani (1996), the lasso has been the subject of much interest, see Tibshirani (2011) and the ensuing discussion for a summary. There is a considerable literature on the theoretical aspects of the lasso, for example, examining its properties with respect to prediction loss and model selection, see Meinshausen and Yu (2009) and references therein.

10.9 Exercises

10.1 For the LIDAR data described in Sect. 10.2.1 fit polynomials of increasing degree as a function of range and comment on the fit to the data. These data are available in the R package SemiPar and are named lidar. What degree of polynomial is required to obtain an adequate fit to these data? [Hint: One method of assessing the latter is to examine residuals.]

10.2 The BPD data described in Sect. 7.2.3 are available on the book website. Fit linear and quadratic logistic regression models to these data and interpret the parameters.

10.3 Carry out backwards elimination for the prostate cancer data, which are available in the R package lasso2 and are named Prostate. Comment on the standard errors of the estimates in the final model that you arrive at, as compared to the corresponding estimates in the full model.

10.4 With reference to Sect. 10.3.1:

a. Show that minimization of expected quadratic loss, $E_{X,Y}\left\{[Y - f(X)]^2\right\}$ leads to $\widehat{f}(x) = E[Y \mid x]$.

b. Show that minimization of expected absolute value loss, $E_{X,Y}[\ |Y - f(X)|\ ]$ leads to $\widehat{f}(x) = \text{median}(Y \mid x)$.

c. Consider the bilinear loss function

$$L[y, f(\boldsymbol{x})] = \begin{cases} a\,[y - f(\boldsymbol{x})] & \text{if } f(\boldsymbol{x}) \leq y \\ b\,[f(\boldsymbol{x}) - y] & \text{if } f(\boldsymbol{x}) \geq y. \end{cases}$$

Deduce that this leads to the optimal $f(\boldsymbol{x})$ being the $100 \times a/(a+b)\%$ point of the distribution function of Y.

10.5 a. Show that the expected value of scaled quadratic loss

$$\mathrm{E}_{Y \mid \boldsymbol{x}} \left\{ \frac{[Y - f(\boldsymbol{x})]^2}{Y^2} \right\}$$

is minimized by

$$\widehat{f}(\boldsymbol{x}) = \frac{\mathrm{E}[Y^{-1} \mid \boldsymbol{x}]}{\mathrm{E}[Y^{-2} \mid \boldsymbol{x}]}.$$

b. Suppose $Y \mid \mu(\boldsymbol{x}), \alpha \sim \mathrm{Ga}\{\alpha^{-1}, [\mu(\boldsymbol{x})\alpha]^{-1}\}$ and that prediction of Y using $f(\boldsymbol{x})$ is required, under scaled quadratic loss. Show that $\widehat{f}(\boldsymbol{x}) = \mathrm{E}[Y^{-1} \mid \boldsymbol{x}] = (1 - 2\alpha)\mu(\boldsymbol{x})$.
[Hint: If $Y \mid a, b \sim \mathrm{Ga}(a, b)$, then $Y^{-1} \mid a, b \sim \mathrm{InvGa}(a, b)$.]

10.6 From Sect. 10.5.1 show, using a Lagrange multiplier argument, that minimizing the penalized sum of squares:

$$\sum_{i=1}^{n} \left(y_i - \beta_0 - \sum_{j=1}^{k} x_{ij}\beta_j \right)^2 + \lambda \sum_{j=1}^{k} \beta_j^2,$$

is equivalent to minimization of

$$\sum_{i=1}^{n} \left(y_i - \beta_0 - \sum_{j=1}^{k} x_{ij}\beta_j \right)^2$$

subject to

$$\sum_{j=1}^{k} \beta_j^2 \leq s,$$

for some s.

10.7 Prove the alternative formulas (10.25)–(10.27) for ridge regression.

10.8 Show, using (10.45), that

$$\mid y_i - \widehat{f}_{-i}^{(\lambda)}(\boldsymbol{x}_i) \mid \; \geq \; \mid y_i - \widehat{f}^{(\lambda)}(\boldsymbol{x}_i) \mid .$$

Interpret this result.

10.9 Cross-validation can fail completely for some problems, as will now be illustrated.

(a) Suppose we smooth a response y_i, by minimizing, with respect to μ_i, $i = 1, \ldots, n$, the ridge regression sum of squares

$$\sum_{i=1}^{n}(y_i - \mu_i)^2 + \lambda \sum_{i=1}^{n}\mu_i^2,$$

where λ is the smoothing parameter. Show that for this problem, the OCV and GCV scores are identical and independent of λ.

(b) By considering the basic principle of OCV, explain what causes the failure of the previous part.

(c) Given the explanation of the failure of cross-validation for the ridge regression problem in part (a), it might be expected that the following modified approach will work better. Suppose a covariate x_i is observed for each y_i (and for convenience, assume $x_i < x_{i+1}$ for all i). Define $\mu(x)$ to be the piecewise linear function with $n-1$ linear segments between x_i and x_{i-1} for $i = 2, \ldots, n$. In this case μ_i could be estimated by minimizing the following penalized least squares objective:

$$\sum_{i=1}^{n}(y_i - \mu_i)^2 + \lambda \int \mu(x)^2 dx,$$

with respect to μ_i, $i = 1, \ldots, n$.
Now consider three equally spaced points x_1, x_2, x_3 with corresponding μ values μ_1, μ_2, μ_3. Suppose that $\mu_1 = \mu_3 = \mu^*$, but that μ_2 can be freely chosen. Show that in order to minimize $\int_{x_1}^{x_3} \mu(x)^2 dx$, μ_2 should be set to $-\mu^*/2$. What does this imply about trying to choose λ by cross-validation?
[Hint: think about what the penalty will do to μ_i if we "leave out" y_i.]

(d) Would the penalty

$$\int \mu'(x)^2 \, dx$$

suffer from the same problem as the penalty used in part (c)?

(e) Would you expect to encounter these sorts of problems with penalized regression smoothers? Explain your answer.

10.10 In this question data in the R package faraway that are named meatspec will be analyzed. Theses data concern the fat content, which is the response, measured in 215 samples of finely chopped meat, along with 100 covariates measuring the absorption at 100 wavelengths. Perform ridge regression on these data using OCV and GCV to choose the smoothing parameter. You should include a plot of how the estimates change as a function of the smoothing parameter and a plot displaying the cross-validation scores as a function of the smoothing parameter.

10.11 For the prostate cancer data considered throughout this chapter, reproduce the summaries in Table 10.3, coding up "by hand" the cross-validation procedures.

Chapter 11
Spline and Kernel Methods

11.1 Introduction

Spline models are based on *piecewise* polynomial fitting, while kernel regression models are based on *local* polynomial fitting. These two approaches to modeling are extremely popular, and so we dedicate a whole chapter to their description.

The layout of this chapter is as follows. In Sect. 11.2, a variety of approaches to spline modeling are described, while Sect. 11.3 discusses kernel-based methods. For inference, an estimate of the error variance is required; this topic is discussed in Sect. 11.4. In this chapter we concentrate on a single x variable only. However, we do consider general responses and, in particular, the class of generalized linear models. Approaches for these types of data are described in Sect. 11.5. Concluding comments appear in Sect. 11.6. There is an extensive literature on spline and kernel modeling; Sect. 11.7 gives references to key contributions and book-length treatments.

11.2 Spline Methods

11.2.1 Piecewise Polynomials and Splines

For *continuous responses*, splines are simply linear models, with an enhanced basis set that provides flexibility.[1] Let $h_j(x) : \mathbb{R} \to \mathbb{R}$ denote the jth function of x, for $j = 1, \ldots, J$. A generic linear model consists of the *linear basis expansion* in x:

$$f(x) = \sum_{j=1}^{J} \beta_j h_j(x).$$

---

[1] Appendix C gives a brief review of bases.

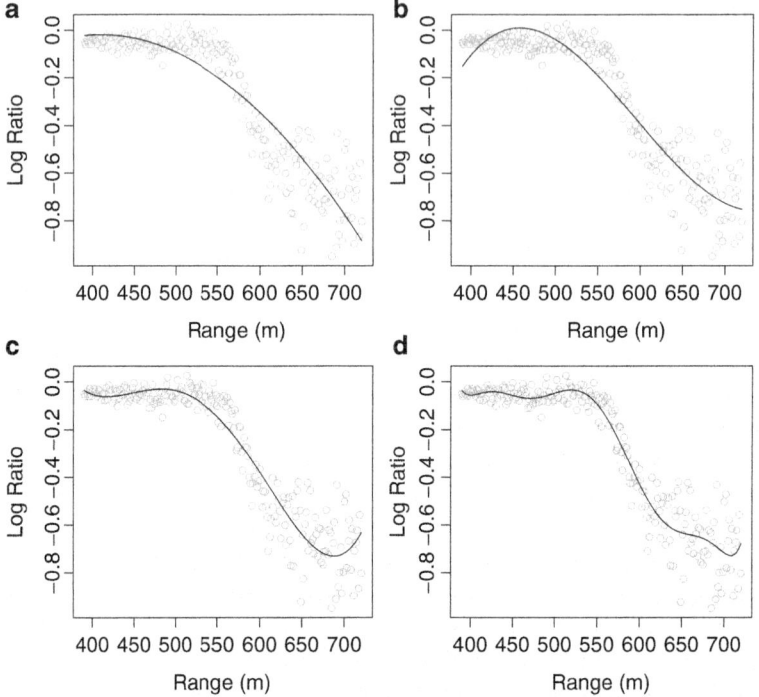

Fig. 11.1 Polynomial fits to the LIDAR data: (**a**) quadratic, (**b**) cubic, (**c**) quartic, and (**d**) degree-8 polynomial

An obvious choice of basis is a polynomial of degree $J - 1$, but the global behavior of such a choice can be poor in the sense that the polynomial will not provide a good fit over the complete range of x. However, *local* behavior can be well represented by relatively low-order polynomials.

Example: Light Detection and Ranging

Figure 11.1 shows degree 2, 3, 4, and 8 polynomial fits to the LIDAR data. The quadratic and cubic models fit very badly, while the quartic model produces a poor fit for ranges of 500–560 m. The degree-8 polynomial fit is also not completely satisfactory with wiggles at the extremes of the range variable due to the global nature of the fitting.

To motivate spline models, we fit piecewise-constant, linear, quadratic, and cubic models using least squares, with three pieces in each case. The fits are displayed in Fig. 11.2. We focus on the piecewise linear model, as shown in Fig. 11.2(b). By forcing the curve to be continuous but only allowing linear segments, we see that

11.2 Spline Methods

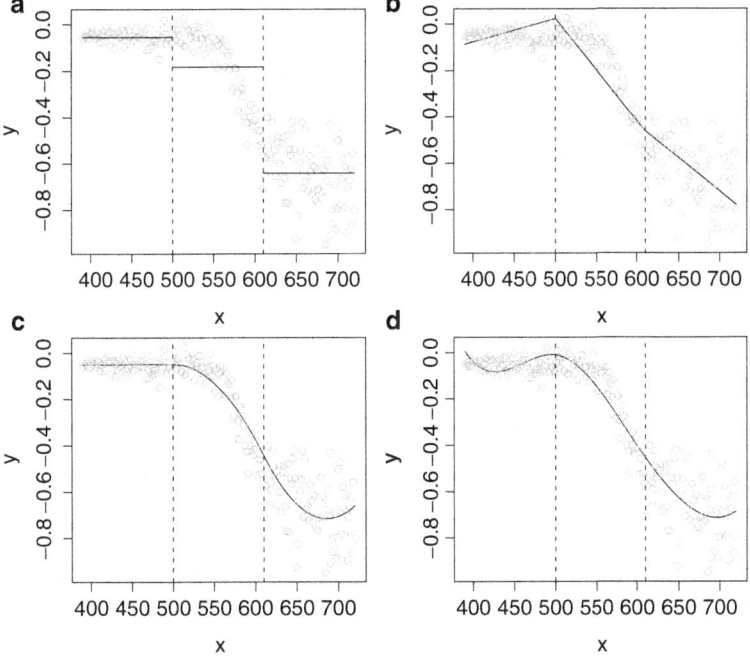

Fig. 11.2 Piecewise polynomials for the LIDAR data: (**a**) constant, (**b**) linear, (**c**) quadratic, and (**d**) cubic

the fit is not good (particularly in the first segment). The lack of smoothness is also undesirable. The quadratic and cubic fits in panels (c) and (d) are far more appealing visually, though neither provide satisfactory fits because we have only allowed three piecewise polynomials. In particular, in panel (d), the cubic fit is still poor at the left endpoint. □

We now start the description of spline models by introducing some notation. Let $\xi_1 < \xi_2 < \ldots < \xi_L$ be a set of ordered points, called *knots*, contained in some interval $[a, b]$. An *M-th order spline* is a piecewise $M - 1$ degree polynomial with $M - 2$ continuous derivatives at the knots.[2] Splines are very popular in nonparametric modeling though, as we shall see, care is required in choosing the degree of smoothing. The latter depends on a variety of factors including the order of the spline and the number and position of the knots.

We begin with a discussion on the order of the spline. The most basic piecewise polynomial is a piecewise-constant function, which is a first-order spline. With two knots, ξ_1 and ξ_2, one possible set of three basis functions is

---

[2]From the Oxford dictionary, a *spline* is a "flexible wood or rubber strip, for example, used in drawing large curves especially in railway work."

$$h_1(x) = I(x < \xi_1), \quad h_2(x) = I(\xi_1 \leq x < \xi_2), \quad h_3(x) = I(\xi_2 \leq x)$$

where $I(\cdot)$ is the indicator function. Note that there are no continuous derivatives at the knots; Fig. 11.2(a) clearly shows the undesirability of this aspect.

To obtain linear models in each of the intervals, we may introduce three additional bases

$$h_{3+j} = h_j(x)x, \quad j = 1, 2, 3,$$

to give the model

$$f(x) = I(x < \xi_1)(\beta_1 + \beta_4 x) + I(\xi_1 \leq x < \xi_2)(\beta_2 + \beta_5 x) + I(\xi_2 \leq x)(\beta_3 + \beta_6 x),$$

which contains six parameters. Lack of continuity is a problem with this model, but we can impose two constraints to enforce $f(\xi_1^-) = f(\xi_1^+)$ and $f(\xi_2^-) = f(\xi_2^+)$, which imply the two conditions

$$\beta_1 + \xi_1 \beta_4 = \beta_2 + \xi_1 \beta_5$$

$$\beta_2 + \xi_2 \beta_5 = \beta_3 + \xi_2 \beta_6,$$

to give four parameters in total. A neater way of incorporating these constraints is with the basis set:

$$h_1(x) = 1, \quad h_2(x) = x, \quad h_3(x) = (x - \xi_1)_+, \quad h_4(x) = (x - \xi_2)_+ \quad (11.1)$$

where t_+ denotes the positive part. We refer to the generic basis $(x - \xi)_+$ as a *truncated line*.[3] The resultant function

$$f(x) = \beta_0 + \beta_1 x + \beta_2 (x - \xi_1)_+ + \beta_3 (x - \xi_2)_+$$

is continuous at the knots since all prior basis functions are contributing to the fit up to any single x value. The model defined by the basis (11.1) is an order-2 spline, and the first derivative is discontinuous. Figure 11.3 shows the basis functions for this representation and Fig. 11.2(b) the fit of this model to the LIDAR data.

We now consider how the piecewise linear model may be extended. Naively, we might assume the quadratic form:

$$f(x) = \beta_0 + \beta_1 x + \beta_2 x^2 + \beta_3 (x - \xi_1)_+ + \beta_4 (x - \xi_1)_+^2 + \beta_5 (x - \xi_2)_+ + \beta_6 (x - \xi_2)_+^2,$$

(11.2)

which is continuous but has first derivative

$$f'(x) = \beta_1 + 2\beta_2 x + \beta_3 I(x > \xi_1) + 2\beta_4 (x - \xi_1)_+ + \beta_5 I(x > \xi_2) + 2\beta_6 (x - \xi_2)_+,$$

---

[3] It is conventional to define the truncated lines with respect to bases that take the positive part, but we could have defined the same model with respect to bases taking the negative part.

11.2 Spline Methods

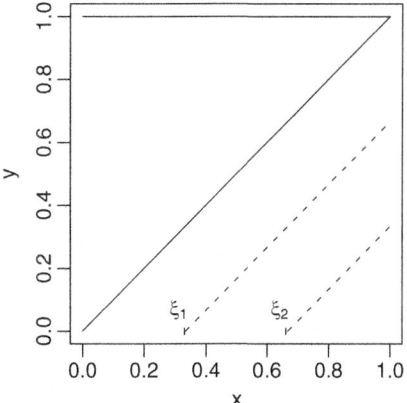

Fig. 11.3 Basis functions for a piecewise linear model with two knots at ξ_1 and ξ_2. The *solid lines* are the bases 1 and x, and the *dashed lines* are the bases $(x - \xi_1)_+$ and $(x - \xi_2)_+$

which is discontinuous at the knot points ξ_1 and ξ_2 due to the linear truncated bases associated with β_3 and β_5 in (11.2). This lack of smoothness at the knots is undesirable. Hence, we drop the truncated linear bases to give the regression model

$$f(x) = \beta_0 + \beta_1 x + \beta_2 x^2 + \beta_3 (x - \xi_1)_+^2 + \beta_4 (x - \xi_2)_+^2$$

which has continuous first derivative:

$$f'(x) = \beta_1 + 2\beta_2 x + 2\beta_3 (x - \xi_1)_+ + 2\beta_4 (x - \xi_2)_+.$$

The second derivative is discontinuous, however, which may also be undesirable. Consequently, a popular form (which we justify more rigorously shortly) is a cubic spline. We will concentrate on cubic splines in some detail, and so we introduce a slight change of notation with respect to the truncated cubic parameters. With two knots the function and first three derivatives are

$$f(x) = \beta_0 + \beta_1 x + \beta_2 x^2 + \beta_3 x^3 + b_1 (x - \xi_1)_+^3 + b_2 (x - \xi_2)_+^3$$
$$f'(x) = \beta_1 + 2\beta_2 x + 3\beta_3 x^2 + 3b_1 (x - \xi_1)_+^2 + 3b_2 (x - \xi_2)_+^2$$
$$f''(x) = 2\beta_2 + 6\beta_3 x + 6b_1 (x - \xi_1)_+ + 6b_2 (x - \xi_2)_+$$
$$f'''(x) = 6\beta_3 + 6b_1 I(x > \xi_1) + 6b_2 I(x > \xi_2).$$

The latter is discontinuous, with a jump at the knots. Figure 11.4 shows the basis functions for the cubic spline, with two knots, and Fig. 11.2(d) shows the fit to the LIDAR data.

For L knots, we write the cubic spline function as

$$f(x) = \beta_0 + \beta_1 x + \beta_2 x^2 + \beta_3 x^3 + \sum_{l=1}^{L} b_l (x - \xi_l)_+^3, \qquad (11.3)$$

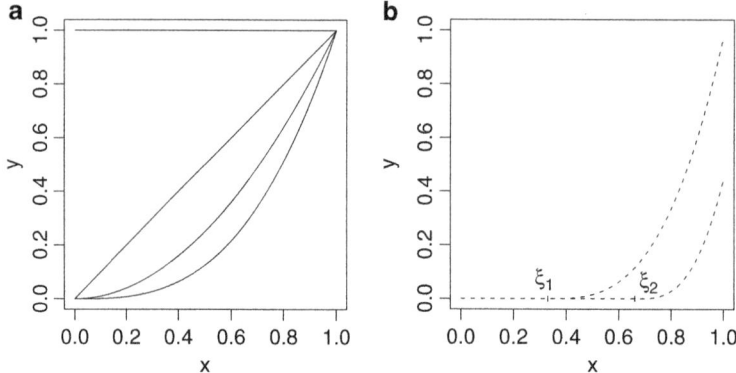

Fig. 11.4 Basis functions for a piecewise cubic spline model with two knots at ξ_1 and ξ_2. Panel (**a**) shows the bases 1, x, x^2, and x^3 and panel (**b**) the bases $(x-\xi_1)_+^3$ and $(x-\xi_2)_+^3$. Note that in (**b**) the bases have been scaled in the vertical direction for clarity

so that we have $L+4$ coefficients. The key to implementation is to recognize that we simply have a linear model, $f(x) = \mathrm{E}[\mathbf{Y} \mid \mathbf{z}] = \mathbf{z}\boldsymbol{\gamma}$, where $\mathbf{z} = \mathbf{z}(x)$ and

$$\mathbf{z} = \begin{bmatrix} 1 & x_1 & x_1^2 & x_1^3 & (x_1-\xi_1)_+^3 & \cdots & (x_1-\xi_L)_+^3 \\ 1 & x_2 & x_2^2 & x_2^3 & (x_2-\xi_1)_+^3 & \cdots & (x_2-\xi_L)_+^3 \\ \vdots & \vdots & \vdots & \vdots & \vdots & \ddots & \vdots \\ 1 & x_n & x_n^2 & x_n^3 & (x_n-\xi_1)_+^3 & \cdots & (x_n-\xi_L)_+^3 \end{bmatrix}, \quad \boldsymbol{\gamma} = \begin{bmatrix} \beta_0 \\ \beta_1 \\ \beta_2 \\ \beta_3 \\ b_1 \\ \vdots \\ b_L \end{bmatrix}.$$

The obvious estimator is therefore $\widehat{\boldsymbol{\gamma}} = (\mathbf{z}^\mathsf{T}\mathbf{z})^{-1}\mathbf{z}^\mathsf{T}\mathbf{Y}$, which gives the linear smoother $\widehat{\mathbf{Y}} = \mathbf{S}\mathbf{Y}$, where $\mathbf{S} = \mathbf{z}(\mathbf{z}^\mathsf{T}\mathbf{z})^{-1}\mathbf{z}^\mathsf{T}$.

11.2.2 Natural Cubic Splines

Spline models such as (11.3) can produce erratic behavior beyond the extreme knots. A *natural* spline enforces linearity beyond the boundary knots, that is,

$$f(x) = a_1 + a_2 x \quad \text{for} \quad x \leq \xi_1$$
$$f(x) = a_3 + a_4 x \quad \text{for} \quad x \geq \xi_L.$$

The first condition only considers values of x before the knots, and therefore, the b_l parameters in (11.3) are irrelevant. Consequently, it is straightforward to see that we require

$$\beta_2 = \beta_3 = 0. \tag{11.4}$$

11.2 Spline Methods

For $x \geq \xi_L$,

$$f(x) = \beta_0 + \beta_1 x + \sum_{l=1}^{L} b_l (x - \xi_l)^3$$

$$= \beta_0 + \beta_1 x + \sum_{l=1}^{L} b_l (x^3 - 3x^2 \xi_l + 3x \xi_l^2 - \xi_l^3),$$

and so, for linearity,

$$\sum_{l=1}^{L} b_l = \sum_{l=1}^{L} b_l \xi_l = 0. \tag{11.5}$$

Hence, we have four additional constraints in total, so that the basis for a natural cubic spline has L elements. Exercise 11.3 describes an alternative basis.

11.2.3 Cubic Smoothing Splines

So far we have examined splines in a heuristic way, as flexible functions with certain desirable properties in terms of the continuity of the function and the first and second derivatives at the knots. We now present a formal justification for the natural cubic spline.

Result. Consider the penalized least squares criterion

$$\sum_{i=1}^{n} [y_i - f(x_i)]^2 + \lambda \int f''(x)^2 dx, \tag{11.6}$$

where the second term penalizes the *roughness* of the curve and λ controls the degree of this roughness. It is clear that without the penalization term, we could choose an infinite number of curves that interpolate the data (in the case of unique x values, at least), with arbitrary behavior in between. Quite remarkably, the $f(\cdot)$ that minimizes (11.6) is the *natural cubic spline* with knots at the unique data points; we call this function $g(x)$.

Proof. The proof has two parts and is based on Green and Silverman (1994, Chap. 2). We begin by showing that a natural cubic spline minimizes (11.6) amongst all interpolating functions and then extend to non-interpolating functions. Assume that $x_1 < \ldots < x_n$. We consider all functions that are continuous in $[x_1, x_n]$ with continuous first and second derivatives and which interpolate $[x_i, y_i]$, $i = 1, \ldots, n$. Since the first term of (11.6) is zero, we need to show that the natural cubic spline, $g(x)$, minimizes

$$\int_{x_1}^{x_n} f''(x)^2 dx.$$

Let $\widetilde{g}(x)$ be another interpolant of (x_i, y_i), and define $h(x) = \widetilde{g}(x) - g(x)$. Then,

$$\int_{x_1}^{x_n} \widetilde{g}''(x)^2 \, dx = \int_{x_1}^{x_n} [g''(x) + h''(x)]^2 \, dx$$

$$= \int_{x_1}^{x_n} \left[g''(x)^2 + 2g''(x)h''(x) + h''(x)^2 \right] \, dx.$$

Applying integration by parts to the cross term,

$$\int_{x_1}^{x_n} g''(x)h''(x)dx = [g''(x)h'(x)]_{x_1}^{x_n} - \int_{x_1}^{x_n} g'''(x)h'(x) \, dx$$

$$= - \int_{x_1}^{x_n} g'''(x)h'(x) \, dx \text{ since } g''(x_1) = g''(x_n) = 0$$

$$= - \sum_{i=1}^{n-1} g'''(x_i^+) \int_{x_i}^{x_{i+1}} h'(x) \, dx$$

since $g'''(x)$ is constant in, and x_i^+ is a point in, $[x_i, x_{i+1}]$

$$= - \sum_{i=1}^{n-1} g'''(x_i^+) [h(x_{i+1}) - h(x_i)]$$

$$= 0$$

since $h(x_{i+1}) = \widetilde{g}(x_{i+1}) - g(x_{i+1})$ and both are interpolants (and similarly for $h(x_i)$). We have shown that

$$\int_{x_1}^{x_n} \widetilde{g}''(x)^2 dx = \int_{x_1}^{x_n} g''(x)^2 dx + \int_{x_1}^{x_n} h''(x)^2 \, dx$$

$$\geq \int_{x_1}^{x_n} g''(x)^2 \, dx$$

with equality if and only if $h''(x) = 0$ for $x_1 < x < x_n$. The latter implies $h(x) = a + bx$, but $h(x_1) = h(x_n) = 0$, and so $a = b = 0$. Consequently, any interpolant that is not identical to $g(x)$ will have a higher integrated squared second derivative. Therefore, the natural cubic spline with knots at the unique x values is the smoothest interpolant in the sense of minimizing $\int f''(x)^2 \, dx$. This is of use in, for example, numerical analysis, where interpolation of $[x_i, y_i]$ is of interest. But, in statistical applications, the data are measured with error, and we typically do not wish to restrict attention to interpolating functions.[4]

---

[4]There are some analogies here with bias, variance, and mean squared error. The penalized sum of squares (11.6) is analogous to the mean squared error, and interpolating functions are "unbiased"

11.2 Spline Methods

We have shown that a natural cubic spline minimizes (11.6) amongst all interpolating functions but the minimizing function need not necessarily be an interpolant since an interpolating function may have a large associated penalty contribution. The second part of the proof considers functions that do not necessarily interpolate the data but have n free parameters $g(x_i)$ with the aim being minimization of (11.6). The resulting $g(x)$ is known as a *smoothing spline*. Suppose some function $f^\star(x)$, other than the cubic smoothing spline, minimizes (11.6). Let $g(x)$ be the natural cubic spline that interpolates $[x_i, f^\star(x_i)]$, $i = 1, \ldots, n$. Obviously, $f^\star$ and g produce the same residual sum of squares in (11.6) since $f^\star(x_i) = g(x_i)$. But, by the first part of the proof,

$$\int f^{\star''}(x)^2 dx > \int g''(x)^2 dx.$$

Hence, the natural cubic spline is the function that minimizes (11.6); this spline is known as a *cubic smoothing spline*.

The above result has shown us that if we wish to minimize (11.6), we should take as model class the cubic smoothing splines. The coefficient estimates of the fit will depend on the value chosen for λ. We stress that the fitted natural cubic smoothing spline will not typically interpolate the data, and the level of smoothness will be determined by the value of λ chosen. Small values of λ, which correspond to a large effective degrees of freedom (Sect. 10.5.1), impose little smoothness and bring the fit closer to interpolation, while large values will result in the fit being close to linear in x (in the limit, a zero second derivative is required).

In terms of interpretation, if a thin piece of flexible wood (a mechanical spline) is placed over the points $[x_i, y_i]$, $i = 1, \ldots, n$, then the position taken up by the piece of wood will be of minimum energy and will describe a curve that approximately minimizes $\int f''^2$ over curves that interpolate the data.

Example: Light Detection and Ranging

We fit a natural cubic spline to the LIDAR data. Figure 11.5 shows the ordinary and generalized cross-validation scores (as described in Sects. 10.6.2 and 10.6.3, respectively) versus the effective degrees of freedom. The curves are very similar with well-defined minima since these data are abundant and the noise level is relatively low. The OCV and GCV scores are minimized at 9.3 and 9.4 effective degrees of freedom, respectively. Figure 11.6 shows the fit (using the GCV minimum corresponding to $\widehat{\lambda} = 959$), which appears good. In particular, we note that the boundary behavior is reasonable.

---

but may have large variability. However, we can obtain a better estimator if we are prepared to examine "biased" (i.e., non-interpolating) functions.

Fig. 11.5 Ordinary and generalized cross-validation scores versus effective degrees of freedom for the LIDAR data and a natural cubic spline model

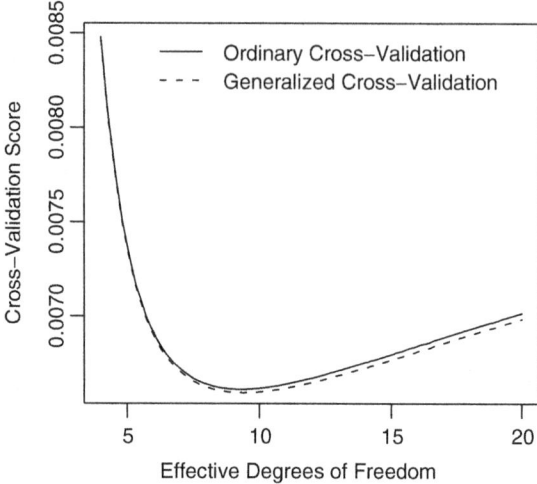

Fig. 11.6 Cubic spline fits to the LIDAR data. The natural cubic spline fit has smoothing parameter chosen by generalized cross-validation. The mixed model cubic spline has smoothing parameter chosen by REML

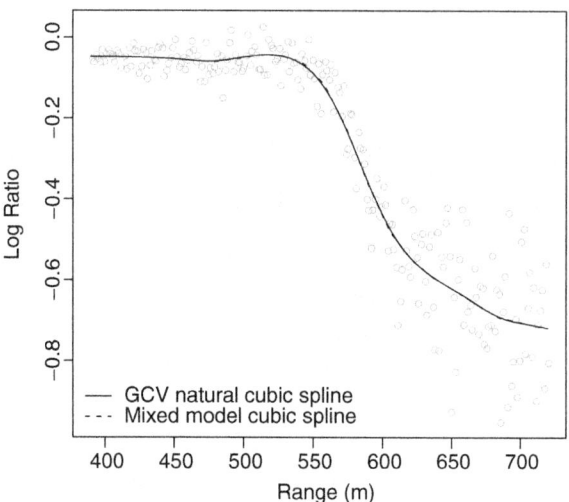

11.2.4 B-Splines

There are many ways of choosing a basis to represent a cubic spline; the so-called B-spline basis functions are popular, a primary reason being that they are nonzero over a limited range which aids in computation. B-splines also form the building blocks for other spline models as we describe in Sect. 11.2.5. The classic text on B-splines is de Boor (1978).

B-splines are available for splines of general order, which we again denote by M (so that for a cubic spline, $M = 4$). The number of basis functions is $L + M$ since we have an $M - 1$ degree polynomial (giving M bases) and one basis for each

11.2 Spline Methods

knot. The original set of knots are denoted ξ_l, $l = 1, \ldots, L$, and we let $\xi_0 < \xi_1$ and $\xi_L < \xi_{L+1}$ represent two boundary knots. We define an augmented set of knots, τ_j, $j = 1, \ldots, L + 2M$, with

$$\tau_1 \leq \tau_2 \leq \ldots \leq \tau_M \leq \xi_0$$

$$\tau_{j+M} = \xi_j, \; j = 1, \ldots, L$$

$$\xi_{L+1} \leq \tau_{L+M+1} \leq \tau_{L+M+2} \leq \ldots \leq \tau_{L+2M}$$

where the choice of the additional knots is arbitrary and so we may, for example, set $\tau_1 = \ldots = \tau_M = \xi_0$ and $\xi_{L+1} = \tau_{L+M+1} = \ldots = \tau_{L+2M}$. These additional knots ensure the basis functions detailed below are defined close to the boundaries. To construct the bases, first define

$$B_j^1(x) = \begin{cases} 1 & \text{if } \tau_j \leq x < \tau_{j+1} \\ 0 & \text{otherwise} \end{cases} \quad (11.7)$$

for $j = 2, \ldots, L + 2M - 1$. For $1 < m \leq M$, define

$$B_j^m(x) = \frac{x - \tau_j}{\tau_{j+m-1} - \tau_j} B_j^{m-1} + \frac{\tau_{j+m} - x}{\tau_{j+m} - \tau_{j+1}} B_{j+1}^{m-1} \quad (11.8)$$

for $j = 1, \ldots, L + 2M - m$. If we divide by zero, then we define the relevant basis element to be zero. The B-spline bases are nonzero over a domain spanned by at most $M + 1$ knots. For example, the support of cubic B-splines ($M = 4$) is at most five knots. At any x, M of the B-splines are nonzero.

The cubic B-spline model is

$$f(x) = \sum_{j=1}^{L+4} B_j^4(x) \beta_j. \quad (11.9)$$

For further details on computation, see Hastie et al. (2009, p. 186). Figure 11.7 shows the cubic B-spline basis (including the intercept) for $L = 9$ knots.

11.2.5 Penalized Regression Splines

Although the result of Sect. 11.2.3 is of theoretical interest, in general, we would like to have a functional form that has less parameters than data points. *Regresssion splines* are defined with respect to a reduced set of $L < n$ knots. Automatically deciding on the number and location of knots is difficult. For example, starting with n knots and then selecting via stepwise methods (Sect. 4.8.1) is fraught with difficulties since there are 2^n models to choose from (assuming the intercept

Fig. 11.7 *B*-spline basis functions corresponding to a cubic spline ($M = 4$) with $L = 9$ equally spaced knots (whose positions are shown as *open circles* on the x-axis). There are $L + M = 13$ bases in total. Note that six distinct line types are used so that, for example, there are three splines represented by *solid curves*: the leftmost, the central, and the rightmost

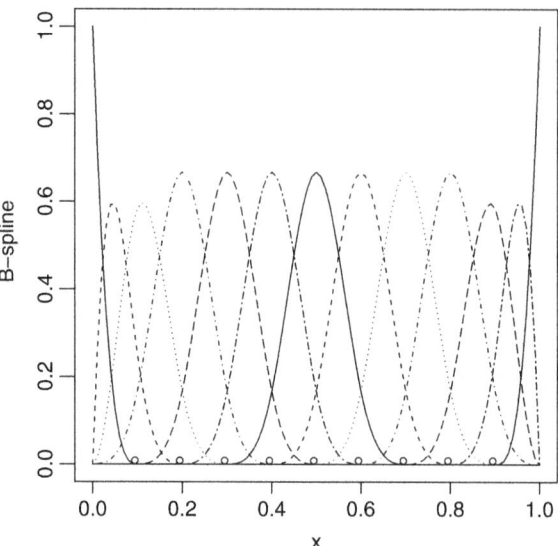

and linear terms are always present). An alternative *penalized regression spline* approach, with $L < n$ knots, is to choose sufficient knots for flexibility and then to penalize the parameters associated with the knot bases. If this approach is followed, the number and selection of knots is far less important than the choice of smoothing parameter. An obvious choice is to place an L_2 penalty on the coefficients, that is, to include the term $\lambda \sum_{l=1}^{L} b_l^2$ in a penalized least squares form. So-called *low-rank* smoothers use considerably fewer than n basis functions.

We now consider linear smoothers of the form:

$$f(x) = \sum_{j=1}^{J} h_j(x)\beta_j = \boldsymbol{h}(x)\boldsymbol{\beta},$$

where $\boldsymbol{h}(x)$ is a $1 \times J$ vector. A general *penalized* regression spline is $\widehat{\boldsymbol{\beta}}^{\mathrm{T}} \boldsymbol{h}(x)$, where $\widehat{\boldsymbol{\beta}}$ is the minimizer of

$$\sum_{i=1}^{n} (y_i - \boldsymbol{h}_i \boldsymbol{\beta})^2 + \lambda \boldsymbol{\beta}^{\mathrm{T}} \boldsymbol{D} \boldsymbol{\beta}, \tag{11.10}$$

with $\boldsymbol{h}_i = \boldsymbol{h}(x_i)$, $\boldsymbol{D}$ is a symmetric-positive semi-definite matrix, and $\lambda > 0$ is a scalar. If we let $\boldsymbol{h} = [\boldsymbol{h}_1, \ldots, \boldsymbol{h}_n]^{\mathrm{T}}$ represent the $n \times J$ design matrix, then

$$\widehat{\boldsymbol{\beta}} = (\boldsymbol{h}^{\mathrm{T}} \boldsymbol{h} + \lambda \boldsymbol{D})^{-1} \boldsymbol{h}^{\mathrm{T}} \boldsymbol{Y}. \tag{11.11}$$

The penalty

$$\lambda \int f''(x)^2 dx \tag{11.12}$$

11.2 Spline Methods

is of the form (11.10) since, for a linear smoother $f(x)$,

$$\int f''(x)^2 dx = \boldsymbol{\beta}^{\text{T}} \left[\int \boldsymbol{h}''(x)\boldsymbol{h}''(x)^{\text{T}} dx \right] \boldsymbol{\beta}$$
$$= \boldsymbol{\beta}^{\text{T}} \boldsymbol{D} \boldsymbol{\beta}$$

with $\boldsymbol{D}$ a matrix of known coefficients. The penalty is measuring complexity: For $\lambda = 0$, there is no cost to fitting a very complex function, while $\lambda = \infty$ gives the simple linear least squares line.

O'Sullivan splines (O'Sullivan 1986) use the cubic B-spline basis representation (11.9), combined with the penalty (11.12), which takes the form:

$$\lambda \int \left(\sum_{j=1}^{L+4} B_j^4(x)'' \beta_j \right)^2 dx.$$

Hence, the penalty matrix $\boldsymbol{D}$ has (j,k)-th element $\int B_j^4(x)'' B_k^4(x)'' \, dx$. O'Sullivan splines correspond to cubic smoothing splines for $L = n$ and distinct x_i (Green and Silverman 1994, Sect. 3.6).

The construction of P-splines is based on a different penalty in which a set of B-spline basis functions are used with a collection of equally spaced knots (Eilers and Marx 1996). The form of the penalty is

$$\lambda \sum_{j=k+1}^{J} (\Delta^k \beta_j)^2 \tag{11.13}$$

with $\Delta \beta_j = \beta_j - \beta_{j-1}$, the difference operator, and where k is a positive integer. For $k = 2$, the penalty is

$$\lambda \sum_{j=1}^{J-1} (\beta_{j+1} - \beta_j)^2 = \beta_1^2 - 2\beta_1\beta_2 + 2\beta_2^2 + \ldots + 2\beta_{J-1}^2 - 2\beta_{J-1}\beta_J + \beta_J^2,$$

which corresponds to the general penalty $\boldsymbol{\beta}^{\text{T}}\boldsymbol{D}\boldsymbol{\beta}$ with

$$\boldsymbol{D} = \begin{bmatrix} 1 & -1 & 0 & \cdot & \cdot \\ -1 & 2 & -1 & \cdot & \cdot \\ 0 & -1 & 2 & \cdot & \cdot \\ \cdot & \cdot & \cdot & \cdot & \cdot \end{bmatrix}.$$

This form penalizes large changes in adjacent coefficients, providing an alternative representation of smoothing. The P-spline approach was heavily influenced by the derivation of O'Sullivan splines (O'Sullivan 1986), and the P-spline penalty is an approximation to the integrated squared derivative penalty. See Eilers and Marx (1996) for a careful discussion of the two approaches. Wand and Ormerod (2008) also contrast O'Sullivan splines (which they refer to as *O-splines*) with P-splines and argue that O-splines are an attractive option for nonparametric regression.

With respect to penalized regression splines, a number of suggestions exist for the number and location of the knots. For example, Ruppert et al. (2003) take as default choice:

$$L = \min\left(\frac{1}{4} \times \text{number of unique } x_i, 35\right),$$

with knots ξ_l taken at the $(l+1)/(L+2)$th points of the unique x_i. These authors say that these choices "work well in most of the examples we come across" but urge against the unquestioning use of these rules.

11.2.6 A Brief Spline Summary

The terminology associated with splines can be confusing, so we provide a brief summary. For simplicity, we assume that the covariate x is univariate and that $x_1, \ldots, x_n$ are unique. A *smoothing spline* contains n knots, and a *cubic smoothing spline* is piecewise cubic. A *natural spline* is linear beyond the boundary knots. If there are $L < n$ knots, we have a *regression spline*. A *penalized regression spline* imposes a penalty on the coefficients associated with the piecewise polynomial. The penalty terms may take a variety of forms.

The number of basis functions that define the spline depends on the number of knots and the degree of the polynomial; natural splines have a reduced number of bases. Spline models may be parameterized in many different ways.

11.2.7 Inference for Linear Smoothers

Nonparametric regression may be used for a variety of purposes. The simplest use is as a scatterplot smoother for pure exploration. In such a context, a plot of $\widehat{f}(x)$ versus x is perhaps all that is required. In other instances, we may wish to produce interval estimates, either pointwise or simultaneous, in order to examine the uncertainty as a function of x.

We consider linear smoothers with J basis functions and write $f(x) = h(x)\beta$ for a prediction at x with β a $J \times 1$ vector and $h(x)$ the $J \times 1$ design matrix associated with x. Further, assume $Y(x) = f(x) + \epsilon(x)$, with the error terms $\epsilon(x)$ uncorrelated and with constant variance σ^2. We emphasize that J is not equal to the effective degrees of freedom, which is given by $p^{(\lambda)} = \text{tr}\left[h(h^\top h + \lambda D)^{-1} h^\top\right]$ where $h = [h(x_1), \ldots, h(x_n)]^\top$. Differentiation of (11.10) with respect to β and setting equal to zero gives

$$\widehat{\beta} = (h^\top h + \lambda D)^{-1} h^\top Y.$$

11.2 Spline Methods

Assuming a fixed λ, asymptotic inference for β is straightforward since

$$\left[(\mathbf{h}^\mathsf{T}\mathbf{h} + \lambda \mathbf{D})^{-1}\mathbf{h}^\mathsf{T}\mathbf{h}(\mathbf{h}^\mathsf{T}\mathbf{h} + \lambda \mathbf{D})^{-1}\right]^{-1/2}\left(\widehat{\beta} - \beta\right) \to \mathrm{N}_J(\mathbf{0}, \sigma^2 \mathbf{I}).$$

In a nonparametric regression context, interest often focuses on inference for the underlying function; we first consider inference at a single point x, $f(x)$.

Since the estimator is linear in the data,

$$\widehat{f}(x) = \mathbf{h}(x)\widehat{\beta} = \mathbf{S}(x)\mathbf{Y} = \sum_{i=1}^{n} S_i(x) Y_i \tag{11.14}$$

where $\mathbf{S}(x) = \mathbf{h}(x)(\mathbf{h}^\mathsf{T}\mathbf{h} + \lambda \mathbf{D})^{-1}\mathbf{h}^\mathsf{T}$ is the $1 \times n$ vector with elements $S_i(x)$, $i = 1, \ldots, n$. This estimator has mean

$$\mathrm{E}\left[\widehat{f}(x)\right] = \sum_{i=1}^{n} S_i(x) f(x_i)$$

and variance

$$\mathrm{var}\left(\widehat{f}(x)\right) = \sigma^2 \sum_{i=1}^{n} S_i(x)^2 = \sigma^2 \|\mathbf{S}(x)\|^2. \tag{11.15}$$

A major difficulty with (11.14) is that there will be bias $b(x)$ present in the estimator. If this bias were known, then

$$\frac{\widehat{f}(x) - f(x) - b(x)}{\sigma \|\mathbf{S}(x)\|} \to_d \mathrm{N}(0, 1), \tag{11.16}$$

via a central limit theorem. Note that it is "local" sample size that is relevant here, with a precise definition depending on the smoothing technique used (which defines $\mathbf{S}(x)$. Estimation of the bias is difficult since it involves estimation of $f''(x)$ (for a derivation in the context of density estimation, see Sect. 11.3.4).

Often the bias is just ignored. The interpretation of the resultant confidence intervals is that they are confidence intervals for $\overline{f}(x) = \mathrm{E}\left[\widehat{f}(x)\right]$, which may be thought of as a smoothed version of $f(x)$. We have

$$\frac{\widehat{f}(x) - f(x)}{\sigma \|\mathbf{S}(x)\|} = \frac{\widehat{f}(x) - \overline{f}(x)}{\sigma \|\mathbf{S}(x)\|} + \frac{\overline{f}(x) - f(x)}{\sigma \|\mathbf{S}(x)\|}$$

$$= Z_n(x) + \frac{b(x)}{\sigma \|\mathbf{S}(x)\|}, \tag{11.17}$$

which is a restatement of (11.16) and where $Z_n(x)$ converges to a standard normal. Hence, a $100(1-\alpha)\%$ asymptotic confidence interval for $\overline{f}(x)$ is $\widehat{f}(x) \pm c_\alpha \sigma \|\mathbf{S}(x)\|$, where c_α is the appropriate cutoff point of a standard normal distribution. In parametric inference, the bias is usually much smaller than the standard deviation of the estimator, so the bias term goes to zero as the sample size increases.[5] In a smoothing context, we have repeatedly seen that optimal smoothing corresponds to balancing bias and variance, and the second term does not disappear from (11.17), even for large sample sizes (recall that $S(x)$ will depend on λ, whose choice will depend on sample size).

We now turn to simultaneous confidence bands of the function $f(x)$ over an interval $x \in [a,b]$ with $a = \min(x_i)$ and $b = \max(x_i)$, $i = 1,\ldots,n$. In the following, we will assume that the confidence bands are for the smoothed function $\overline{f}(x) = \mathrm{E}\left[\widehat{f}(x)\right]$, thus sidestepping the bias issue. We again assume linear smoothers so that (11.14) holds.

One way to think about a simultaneous confidence band is to begin with a finite grid of x values: $x_j = a + j(b-a)/m$, $j = 1,\ldots,m$. Now suppose we wish to obtain a simultaneous confidence band for $\overline{f}(x_j)$, $j = 1,\ldots,m$. One way of approaching this problem is to consider the probability that each of the m estimated functions simultaneously lie within c standard errors of $\overline{f}$, that is,

$$\bigcap_{j=1}^{m} \left\{ \left| \frac{\widehat{f}(x_j) - \overline{f}(x_j)}{\sigma \|S(x_j)\|} \right| \le c \right\},$$

where c is chosen to correspond to the required $1-\alpha$ level of the confidence statement. Then

$$\Pr\left(\bigcap_{j=1}^{m} \left\{ \left| \frac{\widehat{f}(x_j) - \overline{f}(x_j)}{\sigma \|S(x_j)\|} \right| \le c \right\} \right) = \Pr\left(\max_{x_1,\ldots,x_m} \left| \frac{\widehat{f}(x_j) - \overline{f}(x_j)}{\sigma \|S(x_j)\|} \right| \le c \right).$$

(11.18)

Now suppose that $m \to \infty$ to give the limiting expression for (11.18) as

$$\Pr\left(\sup_{x \in [a,b]} \left| \frac{\widehat{f}(x) - \overline{f}(x)}{\sigma \|S(x)\|} \right| \le c \right) = \Pr(M \le c).$$

Sun and Loader (1994), following Knafl et al. (1985), considered approximating this probability in the present context. Let $T(x) = S(x)/\|S(x)\|$. Based on the theory of Gaussian processes,

$$\Pr(M \ge c) \approx 2\left[1 - \Phi(c)\right] + \frac{\kappa_0}{\pi} \exp(-c^2/2),$$

where

$$\kappa_0 = \int_a^b \|T'(x)\|\, dx,$$

---

[5]With parametric models, we are often interested in simple models with a fixed number of parameters, even if we know they are not "true". For example, when we carry out linear regression, we do not usually believe that the "true" underlying function is linear; rather, we simply wish to estimate the linear association.

11.2 Spline Methods

$T'(x) = [T'_1(x), \ldots, T'_n(x)]^{\mathsf{T}}$ and $T'_i(x) = \partial T_i(x)/\partial x$ for $i = 1, \ldots, n$. We choose c to solve

$$\alpha = 2\left[1 - \Phi(c)\right] + \frac{\kappa_0}{\pi}\exp(-c^2/2), \tag{11.19}$$

and κ_0 may be evaluated using numerical integration over a grid of x values. To summarize, once an α level is chosen, we obtain κ_0 and c and then form bands $\widehat{f}(x) \pm c\sigma\|\boldsymbol{S}(x)\|$.

In the case of nonconstant variance, we replace σ by $\sigma(x)$. Section 11.4 contains details on estimation of the error variance. Throughout this section, we have conditioned upon a λ value, which is usually estimated from the data. Hence, in practice, the uncertainty in λ is not accounted for in the construction of interval estimates. A Bayesian mixed model approach (Sect. 11.2.9) treats λ as a parameter, assigns a prior, and then averages over the uncertainty in λ in subsequent inference.

In some contexts, interest may focus on testing the adequacy of a parametric model, comparing nested smoothing models, or testing whether the relationship between the expected response and x is flat. In each of these cases, likelihood ratio or F tests can be performed (see, e.g., Wood 2006, Sect. 4.8.5), though the nonstandard context suggests that the significance of test statistics should be judged via simulation.

Example: Light Detection and Ranging

We fit a cubic penalized regression spline, with penalization $\lambda \sum_{l=1}^{L} b_l^2$ and λ estimated using generalized cross-validation. Figure 11.8(a) gives pointwise confidence intervals and simultaneous confidence bands under the assumption of constant variance. Figure 11.8(b) presents the more appropriate intervals with allowance for nonconstant variance (for details on how $\sigma(x)$ is estimated, see the example at the end of Sect. 11.4). The coverage probability is 0.95, and the critical value for c is 1.96 for the pointwise intervals and 3.11 for the simultaneous intervals, as calculated from (11.19), with κ_0 estimated as 15.4. Under a nonconstant variance assumption, the intervals are very tight for low ranges and increase in width as the range increases.

11.2.8 Linear Mixed Model Spline Representation: Likelihood Inference

In this section we describe an alternative mixed model framework for the representation of regression spline models. A benefit of this framework is that the smoothing parameter may be estimated using standard inference (e.g., likelihood or Bayesian) techniques. It is also possible to build complex mixed models that can model dependencies within the data using random effects, in addition to performing

Fig. 11.8 Pointwise confidence intervals and simultaneous confidence bands for $\overline{f}(x)$ for the LIDAR data under the assumption of (**a**) homoscedastic errors and (**b**) heteroscedastic errors

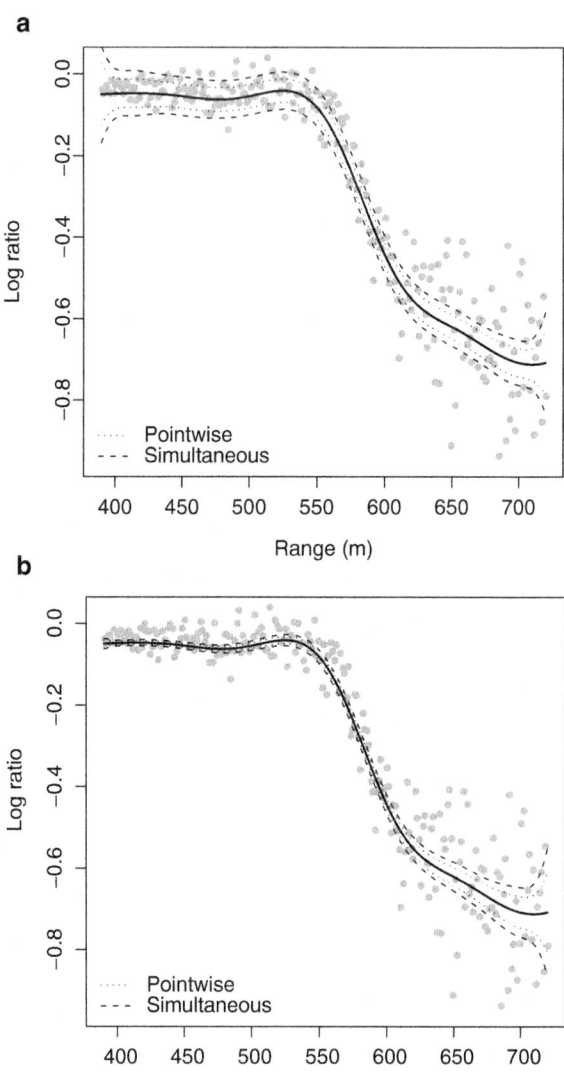

the required smoothing. In the following, we lean heavily on the material on linear random effects modeling contained in Chap. 8. Consider the $(p+1)$th-order (degree p polynomial) penalized regression spline with L knots, that is,

$$f(x) = \beta_0 + \beta_1 x + \ldots + \beta_p x^p + \sum_{l=1}^{L} b_l (x - \xi_l)_+^p.$$

A penalized least squares approach with L_2 penalization of the L truncated cubic coefficients leads to minimization of

11.2 Spline Methods

$$\sum_{i=1}^{n}(y_i - \boldsymbol{x}_i\boldsymbol{\beta} - \boldsymbol{z}_i\boldsymbol{b})^2 + \lambda \sum_{l=1}^{L} b_l^2, \qquad (11.20)$$

where

$$\boldsymbol{x}_i = [1, x_i, \ldots, x_i^p], \quad \boldsymbol{\beta} = \begin{bmatrix} \beta_0 \\ \beta_1 \\ \vdots \\ \beta_p \end{bmatrix}, \quad \boldsymbol{z}_i = [\,(x_i - \xi_1)_+^p, \ldots, (x_i - \xi_L)_+^p\,], \quad \boldsymbol{b} = \begin{bmatrix} b_1 \\ \vdots \\ b_L \end{bmatrix}.$$

Let $\boldsymbol{D} = \text{diag}(\boldsymbol{0}_{p+1}, \boldsymbol{1}_L)$ and $\boldsymbol{c}$ be the $n \times (p+1+L)$ matrix with ith row $\boldsymbol{c}_i = [1, x_i, \ldots, x_i^p, (x_i - \xi_1)_+^p, \ldots, (x_i - \xi_L)_+^p]$, so that $\boldsymbol{c} = [\boldsymbol{x}, \boldsymbol{z}]$, where $\boldsymbol{x} = [\boldsymbol{x}_1, \ldots, \boldsymbol{x}_n]^{\mathrm{T}}$ and $\boldsymbol{z} = [\boldsymbol{z}_1, \ldots, \boldsymbol{z}_n]^{\mathrm{T}}$. The penalized sum of squares (11.20) can be written as

$$(\boldsymbol{y} - \boldsymbol{c}\boldsymbol{\gamma})^{\mathrm{T}}(\boldsymbol{y} - \boldsymbol{c}\boldsymbol{\gamma}) + \lambda \boldsymbol{\gamma}^{\mathrm{T}} \boldsymbol{D} \boldsymbol{\gamma}, \qquad (11.21)$$

where $\boldsymbol{\gamma} = [\boldsymbol{\beta}, \boldsymbol{b}]^{\mathrm{T}}$.

We now reframe this approach in mixed model form with mean model

$$\begin{aligned} y_i &= f(x_i) + \epsilon_i \\ &= \boldsymbol{x}_i \boldsymbol{\beta} + \boldsymbol{z}_i \boldsymbol{b} + \epsilon_i, \end{aligned}$$

and covariance structure and distributional form determined by $\epsilon_i \mid \sigma_\epsilon^2 \sim_{iid} N(0, \sigma_\epsilon^2)$ and $b_l \mid \sigma_b^2 \sim_{iid} N(0, \sigma_b^2)$ with ϵ_i and b_l independent, $i = 1, \ldots, n$, $l = 1, \ldots, L$. This formulation sheds some light on the nature of the penalization. Since the distribution of b_l is independent of $b_{l'}$ for $l \neq l'$, we are assuming that the size of the contribution due to the lth basis is not influenced by any other contributions, in particular, the closest (in terms of x) basis. For example, knowing the sign of b_{l-1} does not imply we believe that b_l is of the same sign. This is in contrast to the P-spline difference penalty described in Sect. 11.2.5.

Minimization of (11.21) with respect to $\boldsymbol{\beta}$ and $\boldsymbol{b}$ is then equivalent to minimization of

$$\frac{1}{\sigma_\epsilon^2}\left[(\boldsymbol{y} - \boldsymbol{x}\boldsymbol{\beta} - \boldsymbol{z}\boldsymbol{b})^{\mathrm{T}}(\boldsymbol{y} - \boldsymbol{x}\boldsymbol{\beta} - \boldsymbol{z}\boldsymbol{b}) + \frac{\sigma_\epsilon^2}{\sigma_b^2}\boldsymbol{b}^{\mathrm{T}}\boldsymbol{b}\right]$$

so that $\lambda = \sigma_\epsilon^2/\sigma_b^2$. We summarize likelihood-based inference for this linear mixed model; Sect. 8.5 contains background details. The maximum likelihood (ML) estimate of $\boldsymbol{\beta}$ is

$$\widehat{\boldsymbol{\beta}} = \left(\boldsymbol{x}^{\mathrm{T}} \boldsymbol{V}^{-1} \boldsymbol{x}\right)^{-1} \boldsymbol{x}^{\mathrm{T}} \boldsymbol{V}^{-1} \boldsymbol{Y} \qquad (11.22)$$

where $\boldsymbol{V} = \sigma_b^2 \boldsymbol{z}\boldsymbol{z}^{\mathrm{T}} + \sigma_\epsilon^2 \boldsymbol{I}_n$, and the best linear unbiased predictor (BLUP) estimator/predictor of $\boldsymbol{b}$ is

$$\widehat{\boldsymbol{b}} = \sigma_b^2 \boldsymbol{z}^{\mathrm{T}} \boldsymbol{V}^{-1}(\boldsymbol{y} - \boldsymbol{x}\boldsymbol{\beta}) \qquad (11.23)$$

Let $\widehat{\sigma}_\epsilon^2$ and $\widehat{\sigma}_b^2$ be the restricted maximum likelihood (REML) estimators (see Sect. 8.5.3) of σ_ϵ^2 and σ_b^2 so that

$$\widehat{\lambda} = \left(\frac{\widehat{\sigma}_\epsilon^2}{\widehat{\sigma}_b^2}\right).$$

In practice, we use

$$\widehat{\boldsymbol{\beta}} = (\boldsymbol{x}^{\scriptscriptstyle\mathsf{T}}\widehat{\boldsymbol{V}}^{-1}\boldsymbol{x})^{-1}\boldsymbol{x}^{\scriptscriptstyle\mathsf{T}}\widehat{\boldsymbol{V}}^{-1}\boldsymbol{Y}$$
$$\widehat{\boldsymbol{b}} = \widehat{\sigma}_b^2 \boldsymbol{z}^{\scriptscriptstyle\mathsf{T}} \widehat{\boldsymbol{V}}^{-1} (\boldsymbol{y} - \boldsymbol{x}\widehat{\boldsymbol{\beta}}).$$

The (penalized) estimator of $\boldsymbol{\gamma} = [\boldsymbol{\beta}, \boldsymbol{b}]^{\scriptscriptstyle\mathsf{T}}$ can be written as

$$\widehat{\boldsymbol{\gamma}} = (\boldsymbol{c}^{\scriptscriptstyle\mathsf{T}}\boldsymbol{c} + \lambda \boldsymbol{D})^{-1}\boldsymbol{c}^{\scriptscriptstyle\mathsf{T}}\boldsymbol{Y} \tag{11.24}$$

(Exercise 11.2). Hence, we can write the fitted values as the linear smoother:

$$\widehat{\boldsymbol{f}} = \boldsymbol{c}\widehat{\boldsymbol{\gamma}} = \boldsymbol{S}^{(\lambda)}\boldsymbol{Y}$$
$$= \boldsymbol{c}(\boldsymbol{c}^{\scriptscriptstyle\mathsf{T}}\boldsymbol{c} + \lambda \boldsymbol{D})^{-1}\boldsymbol{c}^{\scriptscriptstyle\mathsf{T}}\boldsymbol{Y}.$$

The degrees of freedom of the model is defined as

$$\mathrm{df}(\lambda) = \mathrm{tr}\left(\boldsymbol{S}^{(\lambda)}\right)$$
$$= \mathrm{tr}\left[\boldsymbol{c}(\boldsymbol{c}^{\scriptscriptstyle\mathsf{T}}\boldsymbol{c} + \lambda\boldsymbol{D})^{-1}\boldsymbol{c}^{\scriptscriptstyle\mathsf{T}}\right]. \tag{11.25}$$

We consider inference for a particular value x:

$$\widehat{f}(x) = \boldsymbol{x}(x)\widehat{\boldsymbol{\beta}} + \boldsymbol{z}(x)\widehat{\boldsymbol{b}}$$
$$= \boldsymbol{c}(x)\widehat{\boldsymbol{\gamma}}$$
$$= \boldsymbol{c}(x)(\boldsymbol{c}^{\scriptscriptstyle\mathsf{T}}\boldsymbol{c} + \lambda\boldsymbol{D})^{-1}\boldsymbol{c}^{\scriptscriptstyle\mathsf{T}}\boldsymbol{Y}$$

where $\boldsymbol{x}(x) = [1, x, \ldots, x^p]$, $\boldsymbol{z}(x) = [(x - \xi_1)^p, \ldots, (x - \xi_L)^p]$ and $\boldsymbol{c}(x) = [\boldsymbol{x}(x), \boldsymbol{z}(x)]$.

The variance, conditional on $\boldsymbol{b}$, is

$$\mathrm{var}\left(\widehat{f}(x) \mid \boldsymbol{b}\right) = \sigma_\epsilon^2 \boldsymbol{c}(x)(\boldsymbol{c}^{\scriptscriptstyle\mathsf{T}}\boldsymbol{c} + \lambda\boldsymbol{D})^{-1}\boldsymbol{c}^{\scriptscriptstyle\mathsf{T}}\boldsymbol{c}(\boldsymbol{c}^{\scriptscriptstyle\mathsf{T}}\boldsymbol{c} + \lambda\boldsymbol{D})^{-1}\boldsymbol{c}(x)^{\scriptscriptstyle\mathsf{T}},$$

which is identical to the variance obtained from ridge regression (10.30). Ruppert et al. (2003, Sect. 6.4) argue for conditioning on $\boldsymbol{b}$ to give the appropriate measure of variability. Specifically, they state (in the notation used here): "Randomness of $\boldsymbol{b}$ is a device used to model curvature, while ϵ accounts for variability about the curve." Asymptotic 95% pointwise confidence intervals for $f(x)$ are

$$\widehat{f}(x) \pm 1.96 \times \sqrt{\mathrm{var}\left(\widehat{f}(x)\right)}.$$

Approximate or fully Bayesian approaches to confidence interval construction for the complete curve have been recently advocated and have shown to be accurate in

11.2 Spline Methods

simulation studies; see Chap. 17 of Ruppert et al. (2003) and the detailed account of Marra and Wood (2012). These accounts build upon the work of Wabha (1983); Silverman (1985), and Nychka (1988). The latter showed, for univariate x, that a Bayesian interval estimate of the curve, constructed using a cubic smoothing spline, has good frequentist coverage probabilities when the bias in curve estimation is a small contributor to the overall mean squared error. In this case, the average posterior variance is a good approximation to the mean squared error of the collection of predictions. Marra and Wood (2012) provide a far-ranging discussion of Bayesian confidence interval construction, in the context of generalized additive models, as described in Sect. 12.2; included is a discussion of when the coverage probability of the interval is likely to be poor, one instance being when a relatively large amount of bias occurs, for example, when one over-smooths.

Tests of the adequacy of a parametric model or of a null association via likelihood ratio and F tests are described in Ruppert et al. (2003, Sects. 6.6 and 6.7). We illustrate confidence interval construction with an example.

Example: Light Detection and Ranging

We fit a cubic spline with 20 equally spaced knots (so that we have 4 fixed effects and 20 random effects) with REML estimation of the smoothing parameter. The resultant fit is shown in Fig. 11.6 as a dashed line. The variance components are estimated as $\widehat{\sigma}_\epsilon^2 = 0.079^2$ and $\widehat{\sigma}_b^2 = 0.012^2$, to give smoothing parameter $\widehat{\lambda} = 45.8$, which equates to an effective degrees of freedom of 8.5. This is quite similar to the effective degrees of freedom of 9.4 that was chosen by GCV for the natural cubic spline fit, which is also shown in Fig. 11.6. The fits are virtually indistinguishable, which is reassuring. Again we point out that this analysis ignores the clear heteroscedasticity in these data. Within the linear mixed model framework, it would be natural to assume a parametric or nonparametric model for σ_ϵ^2 as a function of x.

In Fig. 11.9, we display the contributions $\widehat{b}_l(x - \xi_l)_+^3$ from the $l = 1, \ldots, 20$, truncated cubic segments. The contribution from the fixed effect cubic, $\widehat{\beta}_0 + \widehat{\beta}_1 x + \widehat{\beta}_2 x^2 + \widehat{\beta}_3 x^3$, is shown as the solid line in each of the plots in this figure. The 1st and 16th–20th cubic segments offer virtually no contribution to the fit, while the contribution of the 4th–14th segments is considerable, which reflects the strong rate of change in the response between ranges of 550 m and 650 m.

11.2.9 Linear Mixed Model Spline Representation: Bayesian Inference

We now discuss a Bayesian mixed model approach. The model is the same as in the last section, with carefully chosen priors. We will not discuss implementation in detail, but lean on the INLA method described in Sect. 3.7.4.

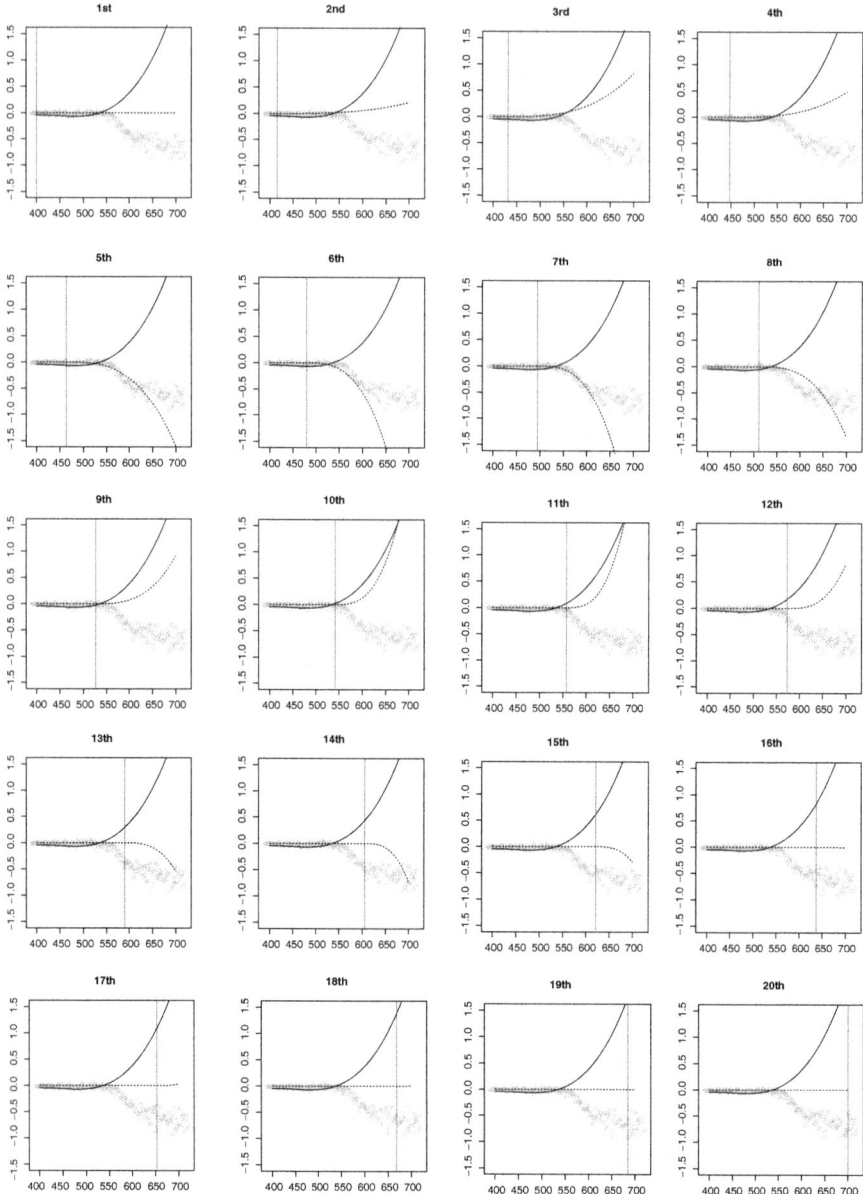

Fig. 11.9 Contributions of the 20 spline bases to the linear mixed model fit to the LIDAR data. The cubic fixed effects fitted line is drawn as the *solid line* on each plot, and the 20 contributions from each of the truncated cubic segments are drawn as *dotted lines* on each plot. The *dotted vertical line* on each plot indicates the knot location associated with the truncated line segment displayed in that plot

11.2 Spline Methods

Prior distributions on smoothing parameters have the potential to increase the stability of the fit, if the priors are carefully specified. An approach suggested by Fong et al. (2010) is to place a prior on σ_b^2 and examine the induced prior on the effective degrees of freedom, a more easily interpretable quantity. The idea is to experiment with prior choices on σ_b^2 until one settles on a prior on the effective degrees of freedom that one is comfortable with. The effective degrees of freedom is given by (11.25) and can be rewritten as

$$\mathrm{df}(\lambda) = \mathrm{tr}[(\boldsymbol{c}^\mathsf{T}\boldsymbol{c} + \lambda \boldsymbol{D})^{-1}\boldsymbol{c}^\mathsf{T}\boldsymbol{c}].$$

The total degrees of freedom can be decomposed into the degrees of freedom associated with $\boldsymbol{\beta}$ and $\boldsymbol{b}$. This decomposition can be extended easily to situations in which we have additional random effects beyond those associated with the spline basis. In each of these situations, the degrees of freedom associated with the respective parameter are obtained by summing the appropriate diagonal elements of $(\boldsymbol{c}^\mathsf{T}\boldsymbol{c} + \lambda\boldsymbol{D})^{-1}\boldsymbol{c}^\mathsf{T}\boldsymbol{c}$. Specifically, for d sets of parameters, let $\boldsymbol{E}_j$ be the $(p+1+L) \times (p+1+L)$ diagonal matrix with ones in the diagonal positions corresponding to set j, $j = 1, \ldots, d$. Then, the degrees of freedom associated with this set are

$$\mathrm{df}_j(\lambda) = \mathrm{tr}[\boldsymbol{E}_j(\boldsymbol{c}^\mathsf{T}\boldsymbol{c} + \lambda\boldsymbol{D})^{-1}\boldsymbol{c}^\mathsf{T}\boldsymbol{c}].$$

Note that the effective degrees of freedom change as a function of L, as expected. To evaluate λ, σ_ϵ^2 is required; Fong et al. (2010) recommend the substitution of an estimate of σ_ϵ^2. For example, one may use an estimate obtained from the fitting of a spline model in a likelihood implementation. For further discussion of prior choice for σ_b^2 in a spline context, see Crainiceanu et al. (2005). We first illustrate the steps in prior construction in a toy example, before presenting a more complex example.

Example: One-Way ANOVA Model

As a simple non-spline demonstration of the derived effective degrees of freedom, consider the one-way ANOVA model:

$$Y_{ij} = \beta_0 + b_i + \epsilon_{ij},$$

with $b_i \mid \sigma_b^2 \sim_{iid} \mathrm{N}(0, \sigma_b^2)$ and $\epsilon_{ij} \mid \sigma_\epsilon^2 \sim_{iid} \mathrm{N}(0, \sigma_\epsilon^2)$ for $i = 1, \ldots, m$ groups and $j = 1, \ldots, n$ observations per group. This model may be written as $\boldsymbol{y} = \boldsymbol{c\gamma} + \boldsymbol{\epsilon}$, where $\boldsymbol{c}$ is the $nm \times (m+1)$ design matrix

$$\boldsymbol{c} = \begin{bmatrix} \boldsymbol{1}_n & \boldsymbol{1}_n & \boldsymbol{0}_n & \cdots & \boldsymbol{0}_n \\ \boldsymbol{1}_n & \boldsymbol{0}_n & \boldsymbol{1}_n & \cdots & \boldsymbol{0}_n \\ \vdots & \vdots & \vdots & \ddots & \vdots \\ \boldsymbol{1}_n & \boldsymbol{0}_n & \boldsymbol{0}_n & \cdots & \boldsymbol{1}_n \end{bmatrix},$$

and $\boldsymbol{\gamma} = [\beta_0, b_1, \ldots, b_m]^\mathsf{T}$. The effective degrees of freedom are given by (11.25), with $\lambda = \sigma_\epsilon^2/\sigma_b^2$ and $\boldsymbol{D}$ a diagonal matrix with a single zero followed by m ones.

For illustration, assume $m = 10$ and $\sigma_b^{-2} \sim \mathrm{Ga}(0.5, 0.005)$. Figure 11.10 displays the prior distribution for σ_b, the implied prior distribution on the effective

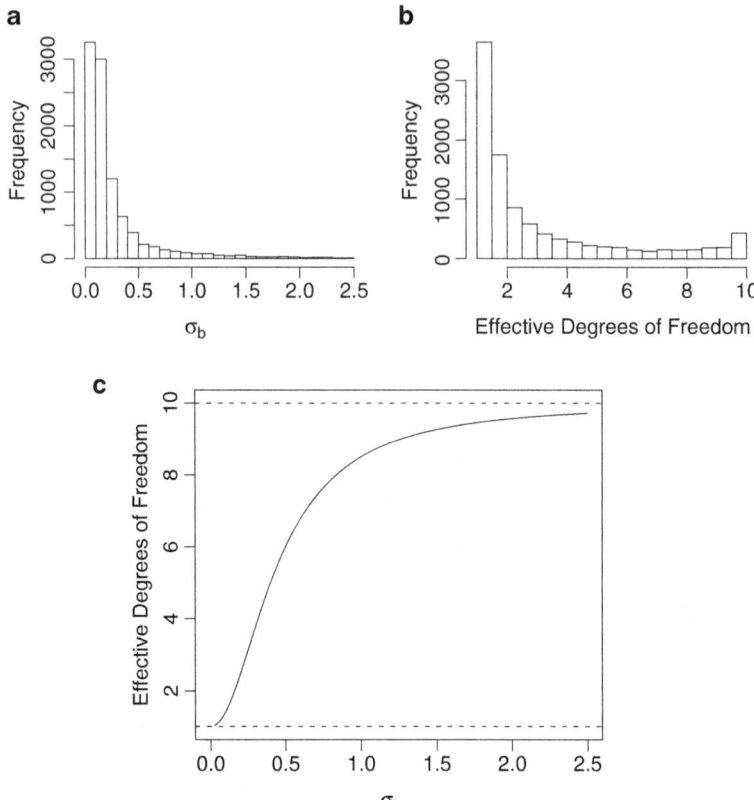

Fig. 11.10 Gamma prior for σ_b^{-2} with parameters 0.5 and 0.005, for the one-way ANOVA example. (**a**) Implied prior for σ_b, (**b**) implied prior for the effective degrees of freedom, and (**c**) effective degrees of freedom versus σ_b

degrees of freedom, and the bivariate plot of these quantities. For clarity, values of σ_b greater than 2.5 (corresponding to 4% of points) are excluded from the plots. In panel (c), we have placed horizontal lines at effective degrees of freedom equal to 1 (complete smoothing) and 10 (no smoothing). We also highlight the strong nonlinearity. From panel (b), we conclude that this prior choice favors quite strong smoothing.

Example: Spinal Bone Marrow Density

We demonstrate the use of the mixed model for nonparametric smoothing using O'Sullivan splines, which, as described in Sect. 11.2.5, are based on a B-spline basis, and using data introduced in Sect. 1.3.6. Recall that these data concern

11.2 Spline Methods

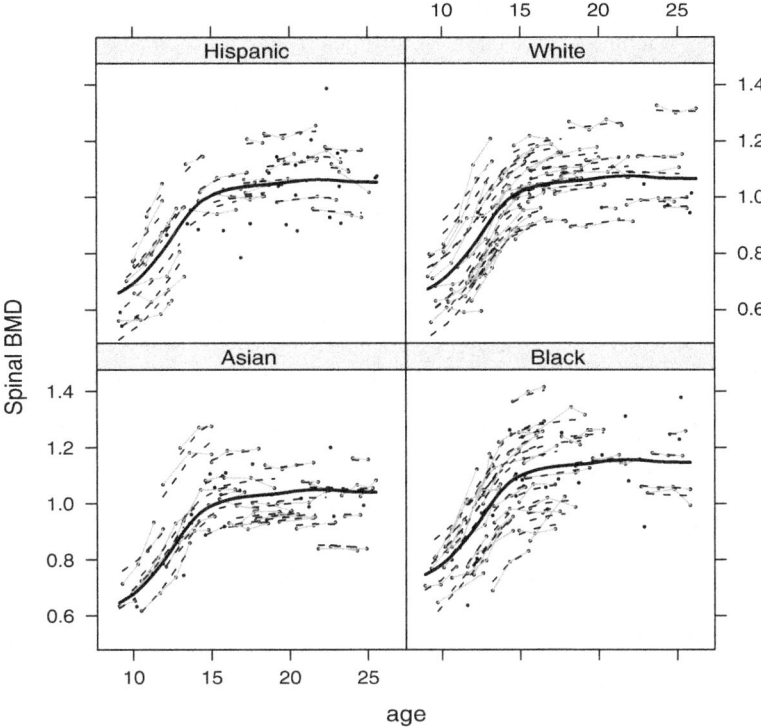

Fig. 11.11 Spinal bone mineral density measurements versus age by ethnicity. Measurements on the same woman are joined with *gray lines*. The *bold solid curve* corresponds to the fitted spline, and the *dashed lines* to the individual fits

longitudinal measurements of spinal bone mineral density (SBMD) on 230 female subjects aged between 8 and 27 years and of one of four ethnic groups: Asian, Black, Hispanic, and White. Let y_{ij} denote the SBMD measure for subject i at occasion j, for $i = 1, \ldots, m = 230$ and $j = 1, \ldots, n_i$ and with n_i ranging between 1 and 4. Let $N = \sum_{i=1}^{m} n_i$. Figure 11.11 shows these data with joined points indicating measurements on the same woman. For these data, we would like a model in which the response is a smooth function of age and in which between-woman variability in response is acknowledged. We therefore assume the model:

$$y_{ij} = \boldsymbol{x}_{ij}\boldsymbol{\beta}_1 + \text{age}_{ij} \times \beta_2 + \sum_{l=1}^{L} z_{ijl} b_{1l} + b_{2i} + \epsilon_{ij}$$

where $\boldsymbol{x}_{ij}$ is a 1×4 vector containing an indicator for the ethnicity of individual i, with $\boldsymbol{\beta}_1$ the associated 4×1 vector of fixed effects, z_{ijl} is the lth basis associated with age, with associated parameters $b_{1l} \mid \sigma_1^2 \sim N(0, \sigma_1^2)$ and $b_{2i} \mid \sigma_2^2 \sim N(0, \sigma_2^2)$ are the woman-specific random effects, and $\epsilon_{ij} \mid \sigma_\epsilon^2 \sim_{iid} N(0, \sigma_\epsilon^2)$ represent the residual errors. All random terms are assumed independent. Note that the spline model is assumed common to all ethnic groups and all women, though it would be straightforward to allow, for example, a different spline for each ethnicity. Let $\beta = [\beta_1, \beta_2]^T$ and x_i be the $n_i \times 5$ fixed effect design matrix with j-th row $[x_{ij}, \text{age}_{ij}]$, $j = 1, \ldots, n_i$ (each row is identical since age_{ij} is the initial age). Also, let z_{1i} be the $n_i \times L$ matrix of age basis functions, $b_1 = [b_1, \ldots, b_L]^T$ be the vector of associated coefficients, z_{2i} represent the $n_i \times 1$ vector of ones, and $\epsilon_i = [\epsilon_{i1}, \ldots, \epsilon_{in_i}]^T$. Then

$$y_i = x_i\beta + z_{1i}b_1 + z_{2i}b_i + \epsilon_i$$

and we may write:

$$y = x\beta + z_1 b_1 + z_2 b_2 + \epsilon$$
$$= c\gamma + \epsilon,$$

where $y = [y_1, \ldots, y_m]^T$, $x = [x_1, \ldots, x_m]^T$, $z_1 = [z_{11}, \ldots, z_{1m}]^T$, $z_2 = [z_{21}, \ldots, z_{2m}]^T$, and $b_2 = [b_{21}, \ldots, b_{2m}]^T$.

We examine two approaches to inference, one based on REML (Sect. 8.5.3) and the other Bayesian, using INLA for computation. In each case, to fit the model, we first construct the basis functions and from these, the required design matrices. Running the REML version of the model, we obtain $\hat{\sigma}_\epsilon = 0.033$, which we use to evaluate the effective degrees of freedom associated with the priors for each of σ_1^2 and σ_2^2. We assume the usual improper prior, $\pi(\sigma_\epsilon^2) \propto 1/\sigma_\epsilon^2$ for σ_ϵ^2. After some experimentation, we settled on the prior $\sigma_1^{-2} \sim \text{Ga}(0.5, 5 \times 10^{-6})$. For σ_2^2, we desire a 90% interval for b_{2i} of ± 0.3 which, with 1 degree of freedom for the marginal distribution, leads to $\sigma_2^{-2} \sim \text{Ga}(0.5, 0.00113)$. See Sect. 8.6.2 for details on the rationale for this approach. Figures 11.12(a) and (d) shows the priors for σ_1 and σ_2, with the priors on the implied effective degrees of freedom displayed in panels (b) and (e). For the spline component, the 90% prior interval on the effective degrees of freedom is $[2.4, 10]$. Figures 11.12(c) and (f) shows the relationship between the standard deviations and the effective degrees of freedom.

Table 11.1 compares estimates from REML and INLA implementations of the model, and we see close correspondence between the two. Figures 11.12(a) and (d) show the posterior medians for σ_1 and σ_2, which correspond to effective degrees of freedom of 8 and 214 for the spline model and random intercepts, respectively, as displayed on panels (b) and (e). The effective degrees of freedom of 214 associated with the random intercepts show that there is considerable variability between the 230 women here. This is confirmed in Fig. 11.11, where we observe large vertical differences between the profiles. This figure also shows the fitted spline, which appears to mimic the age trend in the data well.

11.3 Kernel Methods

We now turn to another class of smoothers that are based on kernels. Kernel methods are used in both density estimation and nonparametric regression, and it is the latter on which we concentrate (though we touch on the former in Sect. 11.3.2). The basic

11.3 Kernel Methods

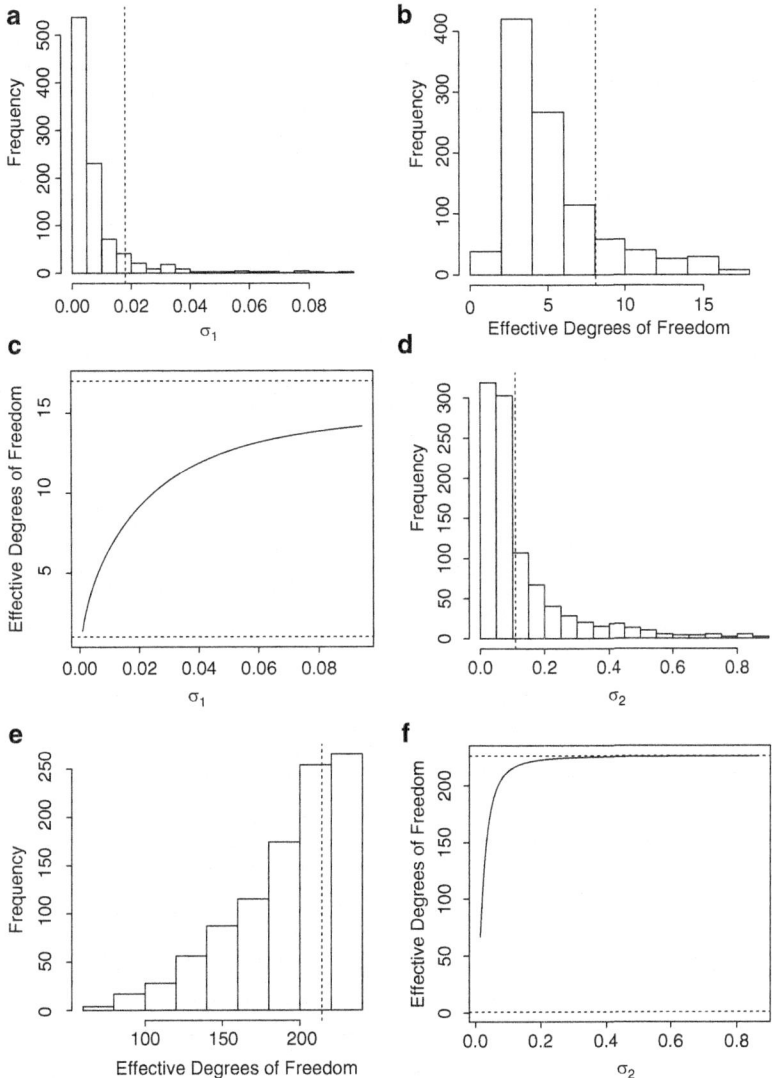

Fig. 11.12 Prior summaries for the spinal bone mineral density data. (**a**) σ_1, the standard deviation of the spline coefficients; (**b**) effective degrees of freedom associated with the prior for the spline coefficients; (**c**) effective degrees of freedom versus σ_1; (**d**) σ_2, the standard deviation of the between-individual random effects; (**e**) effective degrees of freedom associated with the individual random effects; and (**f**) effective degrees of freedom versus σ_2. The *lower and upper dashed horizontal lines* in panels (**c**) and (**f**) are the minimum and maximum attainable degrees of freedom, respectively. The *vertical dashed lines* on panels (**a**), (**b**), (**d**), and (**e**) correspond to the posterior medians

Table 11.1 REML and INLA summaries for the spinal bone data. The intercept corresponds to the Asian group. For the entries marked with a *, standard errors were unavailable

Variable	REML	INLA
Intercept	0.560 ± 0.029	0.563 ± 0.031
Black	0.106 ± 0.021	0.106 ± 0.021
Hispanic	0.013 ± 0.022	0.013 ± 0.022
White	0.026 ± 0.022	0.026 ± 0.022
Age	0.021 ± 0.002	0.021 ± 0.002
σ_1	$0.018^\star$	0.024 ± 0.006
σ_2	$0.109^\star$	0.109 ± 0.006
σ_ϵ	$0.033^\star$	0.033 ± 0.002

idea underlying kernel methods is to estimate the density/regression function *locally* with the kernel function weighting the data in an appropriate fashion. We begin by briefly defining, and giving examples of, kernels.

11.3.1 Kernels

A *kernel* is a smooth function $K(\cdot)$ such that $K(x) \geq 0$, with

$$\int K(u)\,du = 1, \quad \int uK(u)\,du = 0, \quad \sigma_K^2 = \int u^2 K(u)\,du < \infty. \quad (11.26)$$

In practice, a kernel is applied to a standardized variable, and so, in what follows, we do not include a scale parameter since the standardization has removed the dependence on scale.

We describe four common examples of kernel functions. The *Gaussian* kernel is

$$K(x) = (2\pi)^{-1/2} \exp\left(-\frac{x^2}{2}\right)$$

and is nonzero for all x, which makes this kernel relatively computationally expensive to work with since all points must be considered in calculations for a single x. We describe three alternatives but first define

$$I(x) = \begin{cases} 1 & \text{if } |x| \leq 1 \\ 0 & \text{if } |x| > 1. \end{cases}$$

The *Epanechnikov* kernel has the form

$$K(x) = \frac{3}{4}(1-x^2)I(x), \quad (11.27)$$

while the *tricube* kernel is

$$K(x) = \frac{70}{81}\left(1-|x|^3\right)^3 I(x). \quad (11.28)$$

11.3 Kernel Methods

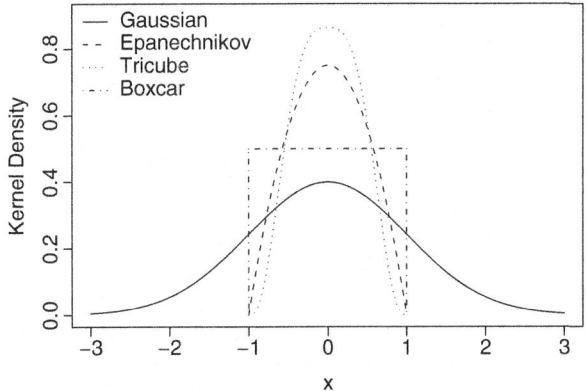

Fig. 11.13 Pictorial representation of four commonly used kernels

Finally, the *boxcar* kernel is

$$K(x) = \frac{1}{2}I(x). \tag{11.29}$$

All four kernels are displayed in Fig. 11.13. We first describe kernel density estimation, which is a simple technique used in a classification context (as described in Sect. 12.8.3).

11.3.2 Kernel Density Estimation

Consider a *random univariate sample* $x_1, \ldots, x_n$ from a density $p(\cdot)$. The kernel density estimate (KDE) of the unknown density, given a smoothing parameter λ, is

$$\widehat{p}^{(\lambda)}(x) = \frac{1}{n\lambda} \sum_{i=1}^{n} K\left(\frac{x - x_i}{\lambda}\right), \tag{11.30}$$

so that the estimate of the density at x is potentially built upon contributions from all n observed values, though for the finite range kernels (11.27)–(11.29), the sum will typically be over far fewer points. Choosing $K(\cdot)$ as a probability density function ensures that $\widehat{p}^{(\lambda)}(x)$ is also a density. We write $K_\lambda(u) = \lambda^{-1}K(u/\lambda)$ for a slightly more compact notation.

We now informally state a number of properties of the kernel density estimator. A number of regularity conditions are required, the most important of which is that the second derivative $p''(x)$ is absolutely continuous; Wand and Jones (1995, Chap. 2) contains more details. We also assume the conditions on $K(\cdot)$ given in (11.26).

Since $x_1, \ldots, x_n$ are a random sample from $p(\cdot)$, the expectation of the density estimator can be written as

$$\mathrm{E}\left[\widehat{p}^{(\lambda)}(x)\right] = \frac{1}{n\lambda} \sum_{i=1}^{n} \mathrm{E}_{X_i}\left[K\left(\frac{x - X_i}{\lambda}\right)\right]$$

$$= \mathrm{E}_T\left[K_\lambda(x - T)\right]$$

$$= \int K_\lambda(x - t) p(t) \, dt, \qquad (11.31)$$

which is a convolution of the true density with the kernel. Smoothing has, therefore, produced a biased estimator whose mean is a smoothed version of the true density. Clearly, we wish to have $\lambda \to 0$ as $n \to \infty$, so that the kernel concentrates more and more on x with increasing n, ensuring that the bias goes to zero.

We write λ_n to emphasize the dependence on n. It is straightforward to show that, as $n \to \infty$, with $\lambda_n \to 0$ and $n\lambda_n \to \infty$:

$$\mathrm{E}\left[\widehat{p}^{(\lambda_n)}(x)\right] = p(x) + \frac{1}{2}\lambda_n^2 p''(x) \sigma_K^2 + o(\lambda_n^2)$$

so that the estimator is *asymptotically unbiased*.

Proof. With $\widehat{p}^{(\lambda_n)}(x)$ given by (11.30),

$$\mathrm{E}[\widehat{p}^{(\lambda_n)}(x)] = \int K_{\lambda_n}(x - t) p(t) dt$$

$$= \int K(u) p(x - \lambda_n u) du$$

$$= \int K(u) \left[p(x) - \lambda_n u p'(x) + \frac{\lambda_n^2 u^2}{2} p''(x) + \ldots\right] du$$

$$= p(x) + \frac{\lambda_n^2}{2} p''(x) \sigma_K^2 + o(\lambda_n^2). \qquad \square$$

The bias is large whenever the absolute value of the second derivative is large. In peaks, $p''(x) < 0$, and the bias is negative since $\widehat{p}^{(\lambda_n)}(x)$ underestimates $p(x)$, and in troughs, the bias is positive as $\widehat{p}^{(\lambda_n)}(x)$ overestimates $p(x)$.

Via a similar calculation,

$$\mathrm{var}\left[\widehat{p}^{(\lambda_n)}(x)\right] = \frac{1}{n\lambda_n} p(x) K_2 + o\left(\frac{1}{n\lambda_n}\right),$$

where $K_2 = \int K(u)^2 \, du$ and $n\lambda_n$ is a "local sample size" (so that larger λ_n gives a larger effective sample size). The variance is also proportional to the height of the density. Overall, as λ_n decreases to zero, the bias diminishes, while the variance increases, with the opposite behavior occurring as λ_n increases. The combined

11.3 Kernel Methods

effect is that, in order to obtain an estimator which converges to the true density, we require both λ_n and $1/n\lambda_n$ to decrease as sample size increases.

As discussed in Sect. 10.4, the accuracy of an estimator may be assessed by evaluating the mean squared error (MSE). For $\widehat{p}^{(\lambda_n)}(x)$,

$$\begin{aligned}
\text{MSE}\left[\widehat{p}^{(\lambda_n)}(x)\right] &= \text{E}\left[\left(\widehat{p}^{(\lambda_n)}(x) - p(x)\right)^2\right] \\
&= \text{bias}\left[\widehat{p}^{(\lambda_n)}(x)\right]^2 + \text{var}\left[\widehat{p}^{(\lambda_n)}(x)\right] \\
&\approx \frac{\lambda_n^4}{4}p''(x)^2\sigma_K^4 + \frac{1}{n\lambda_n}p(x)K_2,
\end{aligned} \quad (11.32)$$

where the expectation in (11.32) is over the uncertainty in $\widehat{p}^{(\lambda_n)}(x)$, that is, over the sampling distribution of $X_1, \ldots, X_n$.

Averaging the MSE over x gives the *integrated mean squared error*

$$\begin{aligned}
\text{IMSE}\left[\widehat{p}^{(\lambda_n)}(x)\right] &= \int \text{MSE}\left[\widehat{p}^{(\lambda_n)}(x)\right] dx \\
&\approx \frac{1}{4}\lambda_n^4 \sigma_K^4 \int p''(x)^2 dx + \frac{1}{n\lambda_n}K_2.
\end{aligned} \quad (11.33)$$

If we differentiate (11.33) with respect to λ_n and set equal to zero, we obtain an asymptotic optimal bandwidth of

$$\lambda_n^{\star} = \left(\frac{K_2}{n\sigma_K^4 \int p''(x)^2 dx}\right)^{1/5}. \quad (11.34)$$

This formula is useful since it informs us that the optimal bandwidth decreases at rate $n^{-1/5}$. Then, substitution in (11.33) shows that the IMSE is of $O(n^{-4/5})$. It can be shown that there does not exist any estimator that converges faster than this rate, assuming only the existence of second derivatives, p''; for more details, see Chap. 24 of van der Vaart (1998).[6]

We turn now to a discussion of estimation of the amount of smoothing to carry out, that is, how to estimate the optimal λ_n. So-called "plug-in" estimators substitute estimates for unknown quantities (here the integrated squared second derivative in the denominator) in order to evaluate $\lambda_n^{\star}$. If we assume that $p(\cdot)$ is normal in (11.34), we obtain

$$\lambda_n^{\star} = (4/3)^{1/5} \times \sigma n^{-1/5}, \quad (11.35)$$

where σ is the standard deviation of the normal.

---

[6]The histogram estimator converges at rate $O(n^{-2/3})$; see, for example, Wand and Jones (1995, Sect. 2.5).

Leave-one-out cross-validation may be used to choose λ_n in order to minimize a measure of estimation accuracy. One convenient quantity that may be minimized is the *integrated squared error* (ISE), defined as

$$\text{ISE}\left[\widehat{p}^{(\lambda_n)}(x)\right] = \int \left[\widehat{p}^{(\lambda_n)}(x) - p(x)\right]^2 dx$$

$$= \int \widehat{p}^{(\lambda_n)}(x)^2 \, dx - 2\int p(x)\widehat{p}^{(\lambda_n)}(x) \, dx + \int p(x)^2 \, dx.$$

The last term does not involve λ_n, and the other terms can be approximated by

$$\frac{1}{n}\sum_{i=1}^{n}\left(\int \widehat{p}^{(\lambda_n)}_{-i}(x)^2 \, dx\right) - \frac{2}{n}\sum_{i=1}^{n}\widehat{p}^{(\lambda_n)}_{-i}(x_i),$$

where $\widehat{p}^{(\lambda_n)}_{-i}(x)$ is the estimator constructed from the data without observation x_i. The use of normal kernels gives a very convenient form for estimation, as described by Bowman and Azzalini (1997, p. 37).

11.3.3 The Nadaraya–Watson Kernel Estimator

We now turn to nonparametric regression and estimation of

$$f(x) = \text{E}[Y \mid x]$$
$$= \int y p(y \mid x) \, dy$$
$$= \frac{1}{p(x)} \int y p(x, y) \, dy. \qquad (11.36)$$

Suppose we estimate $p(x, y)$ by the product kernel

$$\widehat{p}^{(\lambda_x, \lambda_y)}(x, y) = \frac{1}{n\lambda_x \lambda_y} \sum_{i=1}^{n} K_x\left(\frac{x - x_i}{\lambda_x}\right) K_y\left(\frac{y - y_i}{\lambda_y}\right),$$

and $p(x)$ by

$$\widehat{p}^{(\lambda_x)}(x) = \frac{1}{n\lambda_x} \sum_{i=1}^{n} K_x\left(\frac{x - x_i}{\lambda_x}\right).$$

Substitution of these estimates in (11.36) gives the *Nadaraya–Watson* kernel regression estimator (Nadaraya 1964; Watson 1964):

11.3 Kernel Methods

$$\widehat{f}(x) = \frac{\frac{1}{n\lambda_x\lambda_y}\sum_{i=1}^{n}\int yK_x\left(\frac{x-x_i}{\lambda_x}\right)K_y\left(\frac{y-y_i}{\lambda_y}\right)dy}{\frac{1}{n\lambda_x}\sum_{i=1}^{n}K_x\left(\frac{x-x_i}{\lambda_x}\right)}$$

$$= \frac{\sum_{i=1}^{n}K_x\left(\frac{x-x_i}{\lambda_x}\right)\int(y_i+u\lambda_y)K_y(u)\,du}{\sum_{i=1}^{n}K_x\left(\frac{x-x_i}{\lambda_x}\right)}$$

$$= \frac{\sum_{i=1}^{n}K\left(\frac{x-x_i}{\lambda}\right)y_i}{\sum_{i=1}^{n}K\left(\frac{x-x_i}{\lambda}\right)} \tag{11.37}$$

where we have used $\int K_y(u)\,du = 1$ and $\int uK_y(u)\,du = 0$. We also write $\lambda = \lambda_x$ and $K_x = K$ in the final line. This estimator may be written as the linear smoother:

$$\widehat{f}^{(\lambda)}(x) = \sum_{i=1}^{n} S_i^{(\lambda)}(x) Y_i,$$

where the weights $S_i^{(\lambda)}(x)$ are defined as

$$S_i^{(\lambda)}(x) = \frac{K\left(\frac{x-x_i}{\lambda}\right)}{\sum_{i=1}^{n}K\left(\frac{x-x_i}{\lambda}\right)}.$$

As a special case, a rectangular window (i.e., the boxcar kernel) produces a smoother that is a simple moving average. As with spline models, the choice of the smoothing parameter λ is crucial for reasonable behavior of the estimator.

We now examine the asymptotic IMSE which, as usual, can be decomposed into contributions due to squared bias and variance. An advantage of local polynomial regression estimators is that the form of the bias and variance is relatively simple, thus enabling analytic study. For the subsequent calculations, and those that appear later in this chapter, we state results without regularity conditions. See Fan (1992, 1993) for a more rigorous treatment.

As $\lambda_n \to 0$ and $n\lambda_n \to \infty$, the bias of the Nadaraya–Watson estimator at the point x is

$$\text{bias}\left[\widehat{f}^{(\lambda_n)}(x)\right] \approx \frac{\lambda_n^2\sigma_K^2}{2}\left(f''(x) + 2f'(x)\frac{p'(x)}{p(x)}\right), \tag{11.38}$$

where $p(x)$ is the true but unknown density of x. The bias increases with increasing λ_n as we would expect. The bias also increases at points at which $f(\cdot)$ increases in "wiggliness" (i.e., large $f''(x)$) and where the derivative of the "design density," $p'(x)$, is large. The so-called *design bias* is defined as $2f'(x)p'(x)/p(x)$ and, as we will see in Sect. 11.3.4, may be removed if locally *linear* polynomial models are used.

The variance at the point x is

$$\text{var}\left[\widehat{f}^{(\lambda_n)}(x)\right] \approx \frac{K_2 \sigma^2}{n\lambda_n} \frac{1}{p(x)}, \qquad (11.39)$$

where we have assumed, for simplicity, that the variance $\sigma^2 = \text{var}(Y \mid x)$ is constant. The variance of the estimator decreases with decreasing measurement error, increasing density of x values, and increasing local sample size $n\lambda_n$. Consequently, we see the "usual" trade-off with small λ reducing the bias but increasing the variance. Combining the squared bias and variance and integrating over x gives the IMSE:

$$\text{IMSE}\left(\widehat{f}^{(\lambda_n)}\right) \approx \frac{\lambda_n^4 \sigma_K^4}{4} \int \left(f''(x) + 2f'(x)\frac{p'(x)}{p(x)}\right)^2 dx + \frac{K_2 \sigma^2}{n\lambda_n} \int \frac{1}{p(x)} dx.$$

(11.40)

If we differentiate this expression and set equal to zero, we obtain the optimal bandwidth as

$$\lambda_n^\star = \left(\frac{1}{n}\right)^{1/5} \left(\frac{\sigma^2 K_2 \int p(x)^{-1} \, dx}{\sigma_K^4 \int (f''(x) + 2f'(x)p'(x)/p(x))^2 \, dx}\right)^{1/5} \qquad (11.41)$$

so that $\lambda^\star = O(n^{-1/5})$. Plugging this expression into (11.40) shows that the IMSE is $O(n^{-4/5})$, which holds for many nonparametric estimators and is in contrast to most parametric estimators whose variance is $O(n^{-1})$. The loss in efficiency is the cost of the flexibility offered by nonparametric methods. Expression (11.41) depends on many unknown quantities, and while there are "plug-in" methods for estimating these terms, a popular approach is cross-validation.

11.3.4 Local Polynomial Regression

We now describe a generalization of the Nadaraya–Watson kernel estimator, *local polynomial regression*, with improved theoretical properties. Let $w_i(x) = K\left[(x_i - x)/\lambda\right]$ be a weight function and choose $\beta_{0x} = f(x)$ to minimize the weighted sum of squares

$$\sum_{i=1}^{n} w_i(x) (Y_i - \beta_{0x})^2$$

with solution

$$\widehat{f}(x) = \widehat{\beta}_{0x} = \frac{\sum_{i=1}^{n} w_i(x) Y_i}{\sum_{i=1}^{n} w_i(x)},$$

showing that the Nadaraya–Watson kernel regression estimator (11.37) corresponds to a locally constant model, estimated using weighted least squares. For notational

11.3 Kernel Methods

simplicity, we have not acknowledged that the weight $w_i(x)$ depends on the smoothing parameter λ. We emphasize that we carry out a separate weighted least squares fit for each prediction that we wish to obtain.

This formulation suggests an extension in which a local polynomial replaces the locally constant model of the Nadaraya–Watson kernel estimator. For values of u in a neighborhood of a fixed x, define the polynomial:

$$P_x(u; \boldsymbol{\beta}_x) = \beta_{0x} + \beta_{1x}(u - x) + \frac{\beta_{2x}}{2!}(u - x)^2 + \ldots + \frac{\beta_{px}}{p!}(u - x)^p,$$

with $\boldsymbol{\beta}_x = [\beta_{0x}, \ldots, \beta_{px}]$. The idea is to approximate f in a neighborhood of x by the polynomial $P_x(u; \boldsymbol{\beta}_x)$.[7] The parameter $\widehat{\boldsymbol{\beta}}_x$ is chosen to minimize the locally weighted sum of squares:

$$\sum_{i=1}^{n} w_i(x) \left[Y_i - P_x(x_i; \boldsymbol{\beta}_x)\right]^2. \tag{11.42}$$

The ensuing local estimate of f at u is

$$\widehat{f}(u) = P_x(u; \widehat{\boldsymbol{\beta}}_x).$$

We could use this estimate in a local neighborhood of x, but instead, we fit a new local polynomial for *every* target x value. At a target value $u = x$,

$$\widehat{f}(x) = P_x(x; \widehat{\boldsymbol{\beta}}_x) = \widehat{\beta}_{0x}.$$

The weight function is $w(x_i) = K[(x_i - x)/\lambda]$, so that the level of smoothing is controlled by the smoothing parameter λ, with $\lambda = 0$ resulting in $\widehat{f}(x_i) = y_i$ and $\lambda = \infty$ being equivalent to the fitting of a linear model. It is important to emphasize that $\widehat{f}(x)$ only depends on the intercept $\widehat{\beta}_{0x}$ of a local polynomial model, but should not be confused with the fitting of a locally constant model.

For estimating the function f at the point x, local regression is equivalent to applying weighted least squares to the model:

$$\boldsymbol{Y} = \boldsymbol{x}_x \boldsymbol{\beta}_x + \boldsymbol{\epsilon}_x, \tag{11.43}$$

with $E[\boldsymbol{\epsilon}_x] = \boldsymbol{0}$, $\text{var}(\boldsymbol{\epsilon}_x) = \sigma^2 \boldsymbol{W}_x^{-1}$,

$$\boldsymbol{x}_x = \begin{bmatrix} 1 & x_1 - x & \cdots & \frac{(x_1 - x)^p}{p!} \\ 1 & x_2 - x & \cdots & \frac{(x_2 - x)^p}{p!} \\ \vdots & \vdots & \ddots & \vdots \\ 1 & x_n - x & \cdots & \frac{(x_n - x)^p}{p!} \end{bmatrix}$$

representing the $n \times (p+1)$ design matrix and $\boldsymbol{W}_x$ the $n \times n$ diagonal matrix with elements $w_i(x)$, $i = 1, \ldots, n$. Large values of w_i correspond to $x - x_i$ being small,

---

[7] This approximation may be formally motivated via a Taylor series approximation argument.

so that data points x_i close to x are most influential. With the finite range kernels described in Sect. 11.3.1, some of the $w_i(x)$ elements will be zero, in which case we would only consider the data with nonzero elements within (11.43). Note that W_x depends on the kernel function, $K(\cdot)$, and therefore upon the bandwidth, λ. Minimization of
$$(Y - x_x\beta_x)^T W_x (Y - x_x\beta_x)$$
gives
$$\widehat{\beta}_x = (x_x^T W_x x_x)^{-1} x_x^T W_x Y. \tag{11.44}$$

Taking the inner product of the first row of $(x_x^T W_x x_x)^{-1} x_x^T W_x$ with Y gives $\widehat{f}(x) = \widehat{\beta}_{0x}$.

From (11.44), it is clear that this estimator is linear in the data:
$$\widehat{f}(x) = \sum_{i=1}^{n} S_i^{(\lambda)}(x) Y_i.$$

This estimator has mean
$$\mathrm{E}[\widehat{f}(x)] = \sum_{i=1}^{n} S_i^{(\lambda)}(x) f(x_i)$$

and variance
$$\mathrm{var}\left[\widehat{f}(x)\right] = \sigma^2 \sum_{i=1}^{n} S_i^{(\lambda)}(x)^2 = \sigma^2 \|S^{(\lambda)}(x)\|^2,$$

where we have again assumed the error variance is constant and that the observations are uncorrelated. The effective degrees of freedom can be defined as $p^{(\lambda)} = \mathrm{tr}(S^{(\lambda)})$ where $S^{(\lambda)}$ is the "hat" matrix determined from $\widehat{Y} = S^{(\lambda)} Y$.

Asymptotic analysis suggests that local polynomials of odd degree dominate those of even degree (Fan and Gijbels 1996), though Wand and Jones (1995) emphasize that the practical implications of this result should not be overinterpreted. Often $p = 1$ will be sufficient for estimating $f(\cdot)$. It can also be shown (Exercise 11.6) that with a linear local polynomial, we obtain

$$\widehat{f}(x) = \frac{\sum_{i=1}^{n} w_i(x) Y_i}{\sum_{i=1}^{n} w_i(x)} + (x - \overline{x}_w) \frac{\sum_{i=1}^{n} w_i(x)(x_i - \overline{x}_w) Y_i}{\sum_{i=1}^{n} w_i(x)(x_i - \overline{x}_w)^2},$$

where $\overline{x}_w = \sum_{i=1}^{n} w_i(x) x_i / \sum_{i=1}^{n} w_i(x)$ and $w_i(x) = K((x - x_i)/\lambda)$. Therefore, the estimator is the locally constant (Nadaraya–Watson) estimator plus a term that corrects for the local slope and skewness of the x_i.

For the linear local polynomial model, we have
$$\mathrm{E}\left[\widehat{f}^{(\lambda_n)}(x)\right] \approx f(x) + \frac{1}{2}\lambda_n^2 f''(x) \sigma_K^2$$

Fig. 11.14 Local linear polynomial fits to the LIDAR data, with three different kernels. The fits are indistinguishable

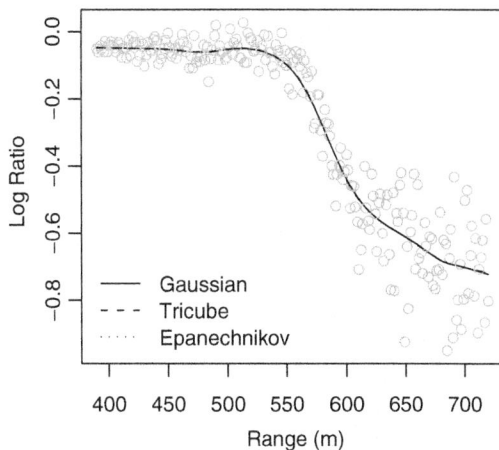

and

$$\text{var}\left[\widehat{f}^{(\lambda_n)}(x)\right] \approx \frac{1}{n\lambda_n}K_2\sigma^2\frac{1}{p(x)}.$$

Proofs of these expressions may be found in Wand and Jones (1995, Sect. 5.3). Notice that the bias is dominated by the second derivative, which is reflecting the error in the linear approximation. If f is linear in x, then $\widehat{f}$ is exactly unbiased.

For the local linear polynomial estimator,

$$\text{IMSE}\left(\widehat{f}^{(\lambda_n)}\right) = \text{bias}\left[\widehat{f}^{(\lambda_n)}\right]^2 + \text{var}\left[\widehat{f}^{(\lambda_n)}\right]$$

$$\approx \frac{\lambda_n^4 \sigma_K^4}{4}\left[\int f''(x)^2\,dx\right] + \frac{K_2\sigma^2}{n\lambda_n}\int \frac{1}{p(x)}\,dx.$$

In comparison with (11.40), the design bias is zero, showing a clear advantage of the linear polynomial over the Nadaraya–Watson estimator. The optimal λ is therefore

$$\lambda_n^\star = \left(\frac{1}{n}\right)^{1/5}\left(\frac{\sigma^2 K_2 \int p(x)^{-1}dx}{\sigma_K^4 \int f''(x)^2\,dx}\right)^{1/5}. \tag{11.45}$$

Each of the terms in expression (11.45) can be estimated to give a "plug-in" estimator of λ_n, or cross-validation may be used. Since the local polynomial regression estimator is a linear smoother, inference for this model follows as in Sect. 11.2.7.

Example: Light Detection and Ranging

Figure 11.14 shows scatterplot smoothing of the LIDAR data using local linear polynomials and Gaussian, tricube and Epanechnikov kernels. In each case the smoothing parameter is chosen via generalized cross-validation, as described in Sect. 10.6.3. The choice of kernel is clearly unimportant in this example.

11.4 Variance Estimation

Accurate inference, for example, confidence intervals for $f(x)$ at a particular x, depends on accurate estimation of the error variance, which may be nonconstant.

We begin by assuming that the model is

$$y_i = \mathrm{E}[Y_i \mid x_i] + \epsilon_i = f(x_i) + \epsilon_i,$$

with $\mathrm{var}(\epsilon_i \mid x_i) = \sigma^2$ and $\mathrm{cov}(\epsilon_i, \epsilon_j \mid x_i, x_j) = 0$. We have made the crucial, and strong, assumption that the errors have constant variance (i.e., are *homoscedastic*) and are uncorrelated. We assume a linear smoother so that $\widehat{f} = SY$ with $p = \mathrm{tr}(S)$ the effective degrees of freedom and suppressing the dependence on the smoothing parameter.

The expectation of the residual sum of squares is

$$\begin{aligned}
\mathrm{E}[(Y - \widehat{f})^{\mathrm{T}}(Y - \widehat{f})] &= \mathrm{E}[(Y - SY)^{\mathrm{T}}(Y - SY)] \\
&= \mathrm{E}[Y^{\mathrm{T}}(I - S)^{\mathrm{T}}(I - S)Y] \\
&= f^{\mathrm{T}}(I - S)^{\mathrm{T}}(I - S)f + \mathrm{tr}\left[(I - S)^{\mathrm{T}}(I - S)I\sigma^2\right] \\
&\quad \text{using identity (B.4) from Appendix B} \\
&= f^{\mathrm{T}}(I - S)^{\mathrm{T}}(I - S)f + \sigma^2 \mathrm{tr}(I - S^{\mathrm{T}} - S + S^{\mathrm{T}}S) \\
&= f^{\mathrm{T}}(I - S)^{\mathrm{T}}(I - S)f + \sigma^2 (n - 2p + \widetilde{p})
\end{aligned}$$

where

$$\widetilde{p} = \mathrm{tr}(S^{\mathrm{T}}S).$$

The bias is

$$f - \mathrm{E}[\widehat{f}] = f - S\mathrm{E}[Y] = f - Sf = (I - S)f.$$

Therefore,

$$\mathrm{E}\left[\frac{\mathrm{RSS}}{n - 2p + \widetilde{p}}\right] = \sigma^2 + \frac{f^{\mathrm{T}}(I - S)^{\mathrm{T}}(I - S)f}{n - 2p + \widetilde{p}}$$

11.4 Variance Estimation

with the second term being the sum of squared bias terms divided by a particular form of degrees of freedom. If the second term is small, it may be ignored to give the estimator:

$$\widehat{\sigma}^2 = \frac{\sum_{i=1}^{n} \left(Y_i - \widehat{f}(x_i)\right)^2}{n - 2p + \widetilde{p}}. \tag{11.46}$$

Notice that for idempotent $\mathbf{S}$, we have $\mathbf{S}^\mathsf{T}\mathbf{S} = \mathbf{S}$, $p = \widetilde{p}$, and (11.46) results in an estimator with a more familiar form, that is, with denominator $n - p$ with p the effective degrees of freedom.

We now derive an alternative *local differencing* (method of moments) estimator (Rice 1984). We begin by considering the expected differences:

$$\mathrm{E}\left[(Y_{i+1} - Y_i)^2\right] = \mathrm{E}\left[(f_{i+1} + \epsilon_{i+1} - f_i - \epsilon_i)^2\right]$$
$$= (f_{i+1} - f_i)^2 + \mathrm{E}\left[(\epsilon_{i+1} - \epsilon_i)^2\right] \tag{11.47}$$
$$= (f_{i+1} - f_i)^2 + 2\sigma^2 \tag{11.48}$$

for $i = 1, \ldots, n-1$. If $f_{i+1} \approx f_i$, then $\mathrm{E}[(Y_{i+1} - Y_i)^2] \approx 2\sigma^2$, leading to

$$\widehat{\sigma}^2 = \frac{1}{2(n-1)} \sum_{i=1}^{n-1} (y_{i+1} - y_i)^2. \tag{11.49}$$

This estimator will be inflated, as is clear from (11.48). An improved method of moments estimator, proposed by Gasser et al. (1986), is based on weighted second differences of the data. Specifically, first consider the line joining the points $[x_{i-1}, y_{i-1}]$ and $[x_{i+1}, y_{i+1}]$. This line is obtained by solving

$$y_{i+1} = \alpha_i + \beta_i x_{i+1}$$
$$y_{i-1} = \alpha_i + \beta_i x_{i-1},$$

to give

$$\widehat{\alpha}_i = \frac{y_{i-1} x_{i+1} - y_{i+1} x_{i-1}}{x_{i+1} - x_{i-1}}$$

$$\widehat{\beta}_i = \frac{y_{i+1} - y_{i-1}}{x_{i+1} - x_{i-1}}.$$

Define a pseudo-residual as

$$\widetilde{\epsilon}_i = \widehat{\alpha}_i + \widehat{\beta}_i x_i - y_i$$
$$= a_i y_{i-1} + b_i y_{i+1} - y_i,$$

where

$$a_i = \frac{x_{i+1} - x_i}{x_{i+1} - x_{i-1}}$$

$$b_i = \frac{x_i - x_{i-1}}{x_{i+1} - x_{i-1}}.$$

Gasser et al. (1986) show that $\text{var}(\widetilde{\epsilon}_i) = [a_i^2 + b_i^2 + 1]\sigma^2 + O(n^{-2})$ (the final term here is required because the pseudo-residuals do not have mean zero). We are therefore led to the estimator:

$$\widetilde{\sigma}^2 = \frac{1}{n-2} \sum_{i=2}^{n-1} c_i^2 \, \widetilde{\epsilon}_i^2 \qquad (11.50)$$

where $c_i^2 = (a_i^2 + b_i^2 + 1)^{-1}$, for $i = 2, \ldots, n$. Note that the variance estimators (11.49) and (11.50) depend only on (y_i, x_i), $i = 1, \ldots, n$ and not on the model that is fitted.

Now suppose we believe the data exhibit nonconstant variance (*heteroscedasticity*). If the variance depends on $f(x)$ via some known form, for example, $\sigma^2(x) = \sigma^2 f(x)$, then quasi-likelihood (Sect. 2.5) may be used. Otherwise, consider the model:

$$Y_i = f(x_i) + \sigma(x_i)\epsilon_i,$$

with $\text{E}[\epsilon_i] = 0$ and $\text{var}(\epsilon_i) = 1$. Since the variance must be positive, a natural model to consider is

$$Z_i = \log\left[(Y_i - f(x_i))^2\right] = \log\left[\sigma^2(x_i)\right] + \log(\epsilon_i^2)$$
$$= g(x_i) + \delta_i, \qquad (11.51)$$

where $\delta_i = \log(\epsilon_i^2)$. A simple approach to implementation is to first estimate $f(\cdot)$ under the assumption of constant variance, obtain fitted values, and then form residuals. One may then estimate $g(\cdot)$ using a nonparametric estimator to produce $\widehat{\sigma}(x)^2 = \exp[\widehat{g}(x)]$, for $i = 1, \ldots, n$. Subsequently, confidence intervals may be constructed based on $\widehat{\sigma}(x)$. For further details, see Yu and Jones (2004). A more statistically rigorous approach would simultaneously estimate $f(\cdot)$ and $g(\cdot)$.

Example: Light Detection and Ranging

Using the natural cubic spline fit, the variance estimate based on the residual sum of squares (11.46) is 0.080^2. The estimates based on the first differences (11.49) and second differences (11.50) are 0.082^2 and 0.083^2, respectively. In this example, therefore, the estimates are very similar though of course for these data, the variance of the error terms is clearly nonconstant. In Fig. 11.15(a), we plot the residuals (from a natural cubic spline fit) versus the range. To address the nonconstant error variance, we assume a model of the form (11.51). Figure 11.15(b) plots the log squared residuals z_i, as defined in (11.51), versus the range. Experimentation with smoothing models for z_i indicates that a simple linear model

$$\mathrm{E}[Z_i \mid x_i] = \alpha_0 + \alpha_1 x_i$$

is adequate, and this is added to the plot. Figure 11.15(c) plots the estimated standard deviation, $\widehat{\sigma}(x) = \sqrt{\exp(\widehat{\alpha}_0 + \widehat{\alpha}_1 x)}$, versus x, and Fig. 11.15(d) shows the standardized residuals:

$$\frac{y_i - \widehat{f}(x_i)}{\widehat{\sigma}(x_i)}$$

versus x_i. We see that the spread is constant across the range of x_i, suggesting that the error variance model is adequate.

11.5 Spline and Kernel Methods for Generalized Linear Models

So far we have considered models of the form, $Y = f(x) + \epsilon$, with independent and uncorrelated constant variance errors ϵ. We outline the extension to the situation in which generalized linear models (GLMs, Sect. 6.3) are appropriate in a parametric framework. To carry out flexible modeling, penalty terms or weighting may be applied to the log-likelihood and smoothing models (e.g., based on splines or kernels) may be used on the linear predictor scale.

Recall that, for a GLM, $\mathrm{E}[Y_i \mid \theta_i, \alpha] = b'(\theta_i) = \mu_i$, with a link function $g(\mu_i)$ and a variance function $\mathrm{var}(Y_i \mid \theta_i, \alpha) = \alpha b''(\theta_i) = \alpha V_i$. In a smoothing context, we may relax the linearity assumption and connect the mean to the smoother via $g(\mu_i) = f(x_i)$. The log-likelihood for a GLM is

$$l(\boldsymbol{\theta}) = \sum_{i=1}^n l_i(\boldsymbol{\theta}) = \sum_{i=1}^n \frac{y_i \theta_i - b(\theta_i)}{\alpha} + c(y_i, \alpha). \tag{11.52}$$

11.5.1 Generalized Linear Models with Penalized Regression Splines

Let $l(\boldsymbol{f})$ denote the log-likelihood corresponding to the smoother $f(x_i)$, $i = 1, \ldots, n$. Maximizing over all smooth functions $f(\cdot)$ is not useful since there are an infinite number of ways to interpolate the data. Consider a regression spline model on the scale of the canonical link:

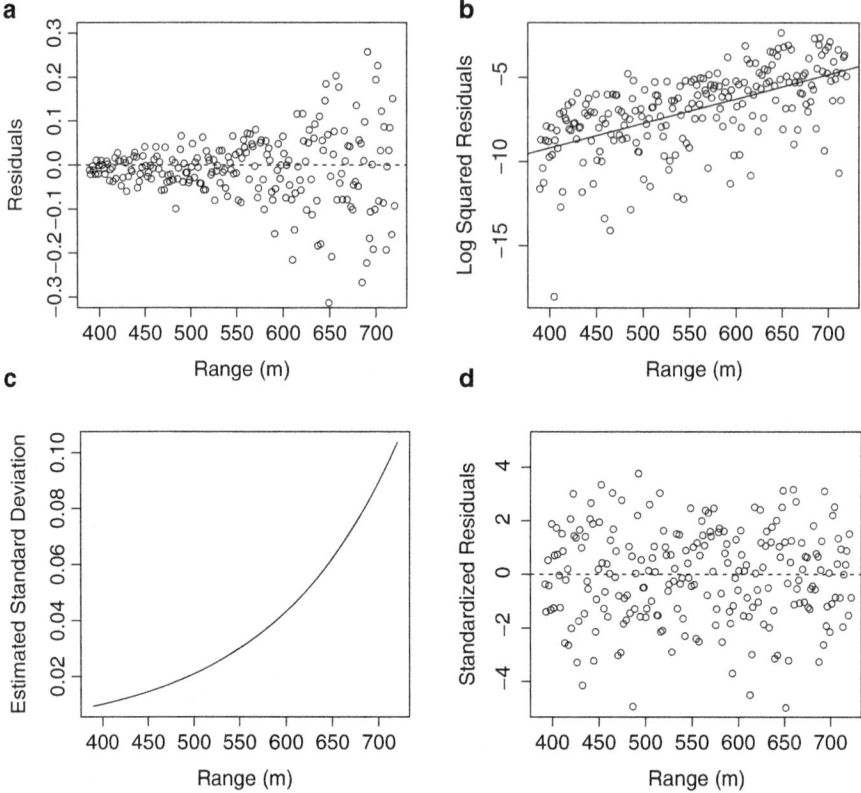

Fig. 11.15 Examination of heteroscedasticity for the LIDAR data. In all plots, the range is plotted on the x-axis, and on the y-axis we have (**a**) residuals from a natural cubic spline fit to the response data; (**b**) log squared residuals, with a linear fit; (**c**) the estimated standard deviation $\widehat{\sigma}(x)$; and (**d**) standardized residuals

$$\theta_i = f(x_i) = \beta_0 + \beta_1 x_i + \ldots + \beta_p x_i^p + \sum_{l=1}^{L} b_l (x_i - \xi_l)_+^p$$
$$= x_i \beta + z_i b,$$

with penalty term $\lambda b^\mathrm{T} D b$, where D denotes a known matrix that determines the nature of the penalization, as in Sect. 11.2.5. For example, an obvious form is $\lambda \int f''(t)^2 \, dt$. As in Sect. 11.2.8, we may write $f(x_i) = c\gamma$ with $c = [x, z]$ and $\gamma = [\beta, b]^\mathrm{T}$, and $D = \mathrm{diag}(0_{p+1}, 1_L)$ to give penalty $\lambda \gamma^\mathrm{T} D \gamma$. To extend the penalized sum of squares given by (11.21), consider the penalized log-likelihood which adds a penalty to (11.52) to give

11.5 Spline and Kernel Methods for Generalized Linear Models

$$l_p(\boldsymbol{\gamma}) = l(\boldsymbol{\gamma}) - \lambda \boldsymbol{\gamma}^{\mathrm{T}} \boldsymbol{D} \boldsymbol{\gamma}, \qquad (11.53)$$

where

$$l(\boldsymbol{\gamma}) = \sum_{i=1}^{n} \frac{y_i \theta_i - b(\theta_i)}{\alpha} + c(y_i, \alpha), \qquad (11.54)$$

and $\theta_i = \boldsymbol{c}\boldsymbol{\gamma}$.

For known λ, the parameters $\boldsymbol{\gamma}$ can be estimated as the solution to

$$\frac{\partial l_p}{\partial \gamma_j} = \sum_{i=1}^{n} \frac{\partial \mu_i}{\partial \gamma_j} \frac{y_i - \mu_i}{\alpha V_i} - 2\lambda \boldsymbol{D} \gamma_j = 0.$$

To find a solution, a hybrid of IRLS (as described in Sect. 6.5.2) termed the penalized IRLS (P-IRLS) algorithm can be used. At the tth iteration, we minimize a penalized version of (6.16):

$$(\boldsymbol{z}^{(t)} - \boldsymbol{x}\boldsymbol{\gamma})^{\mathrm{T}} \boldsymbol{W}^{(t)} (\boldsymbol{z}^{(t)} - \boldsymbol{x}\boldsymbol{\gamma}) + \lambda \boldsymbol{\gamma}^{\mathrm{T}} \boldsymbol{D} \boldsymbol{\gamma}, \qquad (11.55)$$

where, as in the original algorithm, $\boldsymbol{z}^{(t)}$ is the vector of pseudo-data with

$$z_i^{(t)} = \boldsymbol{x}_i \boldsymbol{\gamma}^{(t)} + (Y_i - \mu_i^{(t)}) \left. \frac{d\eta_i}{d\mu_i} \right|_{\boldsymbol{\gamma}^{(t)}}$$

and $\boldsymbol{W}^{(t)}$ is a diagonal matrix with elements:

$$w_i = \frac{\left(d\mu_i/d\eta_i|_{\boldsymbol{\gamma}^{(t)}}\right)^2}{\alpha V_i}.$$

The iterative strategy therefore solves (11.55) using the current versions of $\boldsymbol{z}$ and $\boldsymbol{W}$.

We define an influence matrix for the working penalized least squares problem at the final step of the algorithm as $\boldsymbol{S}^{(\lambda)} = \boldsymbol{x}(\boldsymbol{x}^{\mathrm{T}} \boldsymbol{W} \boldsymbol{x} + \lambda \boldsymbol{D})^{-1} \boldsymbol{x}^{\mathrm{T}} \boldsymbol{W}$. The effective degrees of freedom is then defined as $p^{(\lambda)} = \mathrm{tr}\left(\boldsymbol{S}^{(\lambda)}\right)$.

So far as inference is concerned, $\widehat{\boldsymbol{\gamma}}$ is asymptotically normal with mean $\mathrm{E}\left[\widehat{\boldsymbol{\gamma}}\right]$ and variance–covariance matrix:

$$\alpha (\boldsymbol{x}^{\mathrm{T}} \boldsymbol{W} \boldsymbol{x} + \lambda \boldsymbol{D})^{-1} \boldsymbol{x} \boldsymbol{W} \boldsymbol{x} (\boldsymbol{x}^{\mathrm{T}} \boldsymbol{W} \boldsymbol{x} + \lambda \boldsymbol{D})^{-1}.$$

For more details, see Wood (2006, Sect. 4.8).

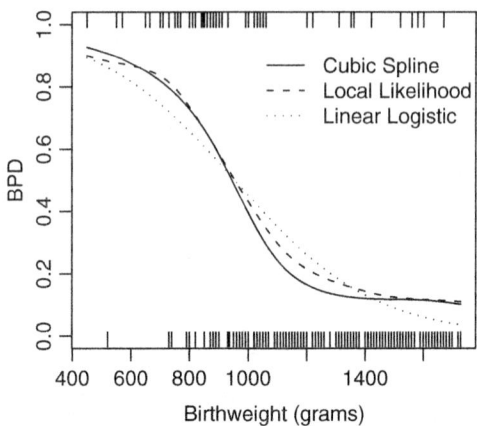

Fig. 11.16 Penalized cubic spline and local likelihood fits to the BPD/birthweight data, with linear logistic fit for comparison

Example: Bronchopulmonary Dysplasia

We illustrate GLM smoothing using the data introduced in Sect. 7.2.3, which consist of binary responses (BPD) Y_i along with birthweights x_i. We consider a logistic regression model:

$$Y_i \mid p(x_i) \sim_{ind} \text{Binomial}\,[\,n_i, p(x_i)\,], \qquad (11.56)$$

with

$$\log\left(\frac{p(x_i)}{1 - p(x_i)}\right) = f(x_i). \qquad (11.57)$$

The log-likelihood is

$$l\,(\boldsymbol{f}) = y_i f(x_i) - n_i \log\{\,1 + \exp\,[f(x_i)]\,\}.$$

A penalized spline model assumes

$$f(x_i) = \beta_0 + \beta_1 x_i + \ldots + \beta_p x_i^p + \sum_{l=1}^{L} b_l (x_i - \xi_l)_+^p$$
$$= \boldsymbol{x}_i \boldsymbol{\beta} + \boldsymbol{z}_i \boldsymbol{b}.$$

The predicted probabilities are therefore

$$p(x) = \frac{\exp(\boldsymbol{x}_i \boldsymbol{\beta} + \boldsymbol{z}_i \boldsymbol{b})}{1 + \exp(\boldsymbol{x}_i \boldsymbol{\beta} + \boldsymbol{z}_i \boldsymbol{b})}.$$

Figure 11.16 displays the data along with three fitted curves. The linear logistic model is symmetric in the tails, which appears overly restrictive for these data. We fit a penalized cubic spline model (11.53) with $L = 10$ knots using P-IRLS and pick the smoothing parameter using AIC. For this model,

$$\text{AIC}^{(\lambda)} = -2l\left(\boldsymbol{f}^{(\lambda)}\right) + 2p^{(\lambda)},$$

where we have now explicitly written $f(x)$ as a function of the smoothing parameter λ and $p^{(\lambda)}$ is the effective degrees of freedom. Figure 11.16 gives the resultant fit, which has an effective degrees of freedom of 3.0. It is difficult to determine the adequacy of the fit with binary data, but in terms of smoothness and monotonicity, the curve appears reasonable. Notice that the behavior for high birthweights is quite different from the linear logistic model.

11.5.2 A Generalized Linear Mixed Model Spline Representation

The regression spline model described in Sect. 11.5.1 has an equivalent specification as a *generalized linear mixed model* (Sect. 9.3) with the assumption that $b_l \mid \sigma_b^2 \sim_{iid} N(0, \sigma_b^2)$, $l = 1, \ldots, L$. The latter random effects distribution penalizes the truncated basis coefficients.

For a GLM with canonical link, maximization of the penalized log-likelihood (11.53) is then equivalent to maximization of

$$\frac{1}{\alpha}\left[\sum_{i=1}^{n}\{y_i(\boldsymbol{x}_i\boldsymbol{\beta} + \boldsymbol{z}_i\boldsymbol{b}) - b(\boldsymbol{x}_i\boldsymbol{\beta} + \boldsymbol{z}_i\boldsymbol{b}) + \alpha \times c(y_i, \alpha)\} - \frac{\alpha}{2\sigma_b^2}\boldsymbol{b}^{\mathsf{T}}\boldsymbol{b}\right] \quad (11.58)$$

with respect to $\boldsymbol{\beta}$ and $\boldsymbol{b}$, for fixed α, σ_b^2. In practice, estimates of α, σ_b^2 will also be required and will determine the level of smoothing. As discussed in Chap. 9, rather than maximize (11.58) as a function of both $\boldsymbol{\beta}$ and $\boldsymbol{b}$, an alternative is to integrate the random effects $\boldsymbol{b}$ from the model and then maximize the resultant likelihood. This approach is outlined for the case of a binomial model.

The likelihood as a function of $\boldsymbol{\beta}$ and σ_b^2 is calculated via an L-dimensional integral over the random effects $\boldsymbol{b}$:

$$L(\boldsymbol{\beta}, \sigma_b^2) = \prod_{i=1}^{n}\binom{n_i}{y_i}\int_{\boldsymbol{b}} \exp\{y_i(\boldsymbol{x}_i\boldsymbol{\beta} + \boldsymbol{z}_i\boldsymbol{b}) - n_i\log[1 + \exp(\boldsymbol{x}_i\boldsymbol{\beta} + \boldsymbol{z}_i\boldsymbol{b})]\}$$

$$\times (2\pi\sigma_b^2)^{-L/2}\exp\left(-\frac{\boldsymbol{b}^{\mathsf{T}}\boldsymbol{b}}{2\sigma_b^2}\right) d\boldsymbol{b}$$

and may be maximized to find $\boldsymbol{\beta}$ and σ_b^2. For implementation, some form of approximate integration strategy must be used; various approaches are described in Chap. 9. The latter also contains details on how the random effects b_l may be estimated (as required to produce the fitted curve), as well as Bayesian approaches to estimation. Under the mixed model formulation, smoothing parameter estimation is carried out via estimation of σ_b^2. Maximizing jointly for $\boldsymbol{\beta}$ and $\boldsymbol{b}$ is formally equivalent to penalized quasi-likelihood (Breslow and Clayton 1993); see Chaps. 10

and 11 of Ruppert et al. (2003) for the application of penalized quasi-likelihood to spline modeling.

Inference from a likelihood perspective may build on mixed model theory, as described in Chap. 9 (see also, Ruppert et al. 2003, Chap. 11). A Bayesian approach can be implemented using either INLA or MCMC, both of which are described in Chap. 3.

11.5.3 *Generalized Linear Models with Local Polynomials*

The extension of the local polynomial approach of Sect. 11.3.4 to GLMs is relatively straightforward with a locally weighted log-likelihood replacing the locally weighted sum of squares (11.42). Recall that for the ith data point, the canonical parameter is $\theta_i = x_i\beta$ (Sect. 6.3). The local polynomial replaces the linear model in θ_i so that we have a local polynomial on the linear predictor scale. We write the log-likelihood for β as

$$l(\beta) = \sum_{i=1}^{n} l\left[y_i, \theta_i(\beta)\right].$$

To obtain the fit at the point x under a local polynomial model, we maximize the locally weighted log-likelihood:

$$l_x(\beta) = \sum_{i=1}^{n} w_i(x)\, l_x\left[y_i, P_x(x_i; \beta)\right],$$

where $w_i(x) = K[(x_i - x)/\lambda]$ and $P_x(x_i; \beta)$ is the local polynomial with parameters β. Our notation also emphasizes that the likelihood is constructed for each point x at which a prediction is desired. The local likelihood score equations are therefore

$$\sum_{i=1}^{n} w_i(x) \frac{\partial}{\partial \beta} l_x\left[y_i, P_x(x_i; \beta)\right].$$

Once we have performed estimation, the estimate (on the transformed scale) for x is evaluated as $\widehat{\beta}_0$. This method is often referred to as *local likelihood*. The existence and uniqueness of estimates are discussed in Chap. 4 of Loader (1999). For a GLM, an iterative algorithm is required; Chap. 11 of Loader (1999) gives details based on the Newton–Raphson method. We stress that the equations are solved at all locations x for which we wish to obtain the fit. The smoothing parameter may again be chosen in a variety of ways, with cross-validation being an obvious approach.

Example: Bronchopulmonary Dysplasia

Returning to the BPD/birthweight example, local log-likelihood fitting at the point x is based on

$$l_x(\boldsymbol{\beta}) = \sum_{i=1}^{n} w_i(x) n_i \left(\frac{y_i}{n_i} P_x(\boldsymbol{x}_i; \boldsymbol{\beta}) - \log\{1 + \exp[P_x(\boldsymbol{x}_i; \boldsymbol{\beta})]\} \right), \quad (11.59)$$

with $w_i(x) = K[(x_i - x)/h]$. Writing the likelihood in this form emphasizes that $w_i(x)n_i$ is acting as a local weight.

Figure 11.16 shows the local linear likelihood fit with a tricube kernel and smoothing parameter chosen by minimizing the AIC. The latter produces a model with effective degrees of freedom of 4.1. The local likelihood cubic curve bears more resemblance to the penalized cubic spline curve than to the linear logistic model, but there are some differences between the former two approaches, particularly for birthweights in the 900–1,500-gram range.

11.6 Concluding Comments

In this chapter we have described smoothing methods for general data types based on spline models and kernel-based methods. A variety of spline models are available, but we emphasize that the choice of smoothing parameter will often be far more important than the specific model chosen. For simple scatterplot smoothing, the spline and kernel techniques of Sects. 11.2 and 11.3 will frequently produce very similar results. If inference is required, penalized regression splines are a class for which the theory is well developed and for which much practical experience has been gathered. To obtain confidence intervals for the complete curve, a Bayesian solution is recommended; see Marra and Wood (2012). For inference about a curve, including confidence bands, care must be taken in variance estimation, as described in Sect. 11.4. In terms of smoothing parameter choice, there will often be no clear optimal choice, and a visual examination of the resultant fit is always recommended.

Kernel-based methods are very convenient analytically, and we have seen that expressions for the bias and variance are available in closed form which allows insight into when they might preform well. Spline models are not so conducive to such analysis though penalized regression splines have the great advantage of having a mixed model representation which allows the incorporation of random effects and the estimation of smoothing parameters using conventional estimation techniques.

11.7 Bibliographic Notes

Book-length treatments on spline methods include Wabha (1990) and Gu (2002). A key early reference on spline smoothing is Reinsch (1967). The book of Wand and Jones (1995) is an excellent introduction to kernel methods. Local polynomial methods are described in detail in Fan and Gijbels (1996) and Loader (1999). Bowman and Azzalini (1997) provides a more applied slant. The work of Ruppert et al. (2003) is a readable account of smoothing methods, with an emphasis on the mixed model representation of penalized regression splines.

11.8 Exercises

11.1 Based on (11.7) and (11.8), write code, for example, within R, to produce plots of the B-spline basis functions of order $M = 1, 2, 3, 4$, with $L = 9$ knots and for $x \in [0, 1]$.

11.2 Prove that (11.22) and (11.23) are equivalent to (11.24).

11.3 Show that an alternative basis for the natural cubic spline given by (11.3), with constraints (11.4) and (11.5), is

$$h_1(x) = 1, \quad h_2(x) = x, \quad h_{l+2}(x) = d_l(x) - d_{L-1}(x),$$

where

$$d_l(x) = \frac{(x - \xi_l)_+^3 - (x - \xi_L)_+^3}{\xi_L - \xi_l}.$$

11.4 In this question, various models will be fit to the fossil data of Chaudhuri and Marron (1999). These data consist of 106 measurements of ratios of strontium isotopes found in fossil shells and their age. These data are available in the R package SemiPar and are named fossil. Fit the following models to these data:

(a) A natural cubic spline (this model has n knots), using ordinary cross-validation to select the smoothing parameter.
(b) A natural cubic spline (this model has n knots), using generalized cross-validation to select the smoothing parameter.
(c) A penalized cubic regression spline with $L = 20$ equally spaced knots, using ordinary cross-validation to select the smoothing parameter.
(d) A penalized cubic regression spline with $L = 20$ equally spaced knots, using generalized cross-validation to select the smoothing parameter.
(e) A penalized cubic regression spline with $L = 20$ equally spaced knots, using a mixed model representation to select the smoothing parameter.

In each case report $\widehat{f}(x)$, along with an asymptotic 95% confidence interval, for the (smoothed) function, at $x = 95$ and $x = 115$ years.

11.5 In this question a dataset that concerns cosmic microwave background (CMB) will be analyzed. These data are available at the book website; the first column is the wave number (the x variable), while the second column is the estimated spectrum (the y variable):

(a) Fit a penalized cubic regression spline using, for example, the R package mgcv.
(b) Fit a Nadaraya–Watson locally constant estimator.
(c) Fit a locally linear polynomial model.
(d) Which of the three models appears to give the best fit to these data?
(e) Obtain residuals from the fit in part (c) and form the log of the squared residuals. Model the latter as a function of x.

11.8 Exercises

(f) Compare the model for the fitted standard deviation with the estimated standard error (which is the third column of the data).

(g) Reestimate the linear polynomial model, weighting the observations by the reciprocal of the variance, where the latter is the square of the estimated standard errors (column three of the data). Repeat using your estimated variance function.

(h) Does the fit appear improved when compared with constant weighting?

At each stage provide a careful description of how the models were fitted. For example, in (a), how were the knots chosen, and in (b) and (c), what kernels and smoothing parameters were used and why?

11.6 For the locally linear polynomial fit described in Sect. 11.3.4, show that

$$\widehat{f}(x) = \frac{\sum_{i=1}^n w_i(x) Y_i}{\sum_{i=1}^n w_i(x)} + (x - \bar{x}_w) \frac{\sum_{i=1}^n w_i(x)(x_i - \bar{x}_w) Y_i}{\sum_{i=1}^n w_i(x)(x_i - \bar{x}_w)^2}$$

where $\bar{x}_w = \sum_{i=1}^n w_i(x) x_i / \sum_{i=1}^n w_i(x)$ and $w_i(x) = K[(x - x_i)/\lambda]$ is a kernel.

Chapter 12
Nonparametric Regression with Multiple Predictors

12.1 Introduction

In this chapter we describe how the methods described in Chaps. 10 and 11 may be extended to the situation in which there are multiple predictors. We also provide a description of methods for classification, concentrating on approaches that are more model, as opposed to algorithm based.

To motivate the ensuing description of modeling with multiple covariates, suppose that $x_{i1}, \ldots, x_{ik}$ are k covariates measured on individual i, with Y_i a univariate response. In Chap. 6 generalized linear models (GLMs) were considered in detail, and we begin this chapter by relaxing the linearity assumption via so-called generalized additive models. A GLM has Y_i independently distributed from an exponential family with $E[Y_i \mid \boldsymbol{x}_i] = \mu_i$. A link function $g(\mu_i)$ then connects the mean to a linear predictor

$$g(\mu_i) = \beta_0 + \beta_1 x_{i1} + \ldots + \beta_k x_{ik}. \tag{12.1}$$

This model is readily interpreted but has two serious restrictions. First, we are constrained to linearity on the link function scale. Transformations of x values or inclusion of polynomial terms may relax this assumption somewhat, but we may desire a more flexible form. Second, we are only modeling each covariate separately. We can add interactions but may prefer an automatic method for seeing the way in which the response is associated with two or more variables.

A general specification with k covariates is

$$g(\mu_i) = f(x_{i1}, x_{i2}, \ldots, x_{ik}). \tag{12.2}$$

Flexible modeling of the complete k-dimensional surface is extremely difficult to achieve due to the curse of dimensionality. To capture "local" behavior in high dimensions requires a large number of data points. To illustrate, suppose we wish to smooth a function at a point using covariates within a k-dimensional hypercube centered at that point, and suppose also that the covariates are uniformly distributed in the k-dimensional unit hypercube. To capture a proportion q of the unit volume requires the expected edge length to be $q^{1/k}$. For example, to capture 1% of the points in $k = 4$ dimensions requires $0.01^{1/4} = 0.32$ of the unit length of each variable to be covered. In other words, "local" has to extend a long way in higher dimensions, and so modeling a response as a function of multiple covariates using local smoothing becomes increasingly more difficult as the number of covariates grows.

The outline of this chapter is as follows. The modeling of multiple predictors via the popular class of generalized additive models is the subject of Sect. 12.2. Section 12.3 extends the spline models of Sect. 11.2 to the multiple covariate case, including descriptions of natural thin plate splines, thin plate regression splines, and tensor product splines. The kernel methods of Sect. 11.3 are described for multiple covariates in Sect. 12.4. Section 12.5 considers approaches to smoothing parameter estimation including the use of a mixed model formulation. Varying-coefficient models provide one approach to modeling interactions, and these are outlined in Sect. 12.6. Moving towards classification, regression tree methods are discussed in Sect. 12.7. Section 12.8 is dedicated to a brief description of methods for classification, including logistic modeling, linear and quadratic discriminant analysis, kernel density estimation, classification trees, bagging, and random forests. Concluding comments appear in Sect. 12.9. Section 12.10 gives references to additional approaches and more detailed descriptions of the approaches considered here.

12.2 Generalized Additive Models

12.2.1 Model Formulation

Generalized additive models (GAMs) are an extremely popular, simple and interpretable extension of GLMs (which were described in Sect. 6.3). The simplest GAM extends the linear predictor (12.1) of the GLM to the additive form

$$g(\mu_i) = \beta_0 + f_1(x_{i1}) + f_2(x_{i2}) + \ldots + f_k(x_{ik}) \qquad (12.3)$$

where β_0 is the intercept and $f_j(\cdot)$, $j = 1,\ldots,k$ are a set of smooth functions. Each of the functions $f_j(\cdot)$ may be modeled using different techniques, with splines and kernel local polynomials (as described in Chap. 11) being obvious choices. For reasons of identifiability, we impose $\sum_{i=1}^n f_j(x_{ij}) = 0$, for $j = 1,\ldots,k$. A GAM may also consist of smooth terms that are functions of pairs, or triples of variables, providing a compromise between the simplest model with k smoothers and the "full" model (12.2) which allows interactions between all variables. The multivariate spline models of Sect. 12.3 provide one approach to the modeling of more than a single variable. As a concrete example of a GAM, suppose that

12.2 Generalized Additive Models

univariate penalized regression splines (Sect. 11.5.1) are used for each of the covariates, with the spline for covariate j being of degree p_j and with knot locations $\xi_{jl}, l = 1, \ldots, L_j$. The GAM is

$$g(\mu) = \beta_0 + \sum_{j=1}^{k} \left[\sum_{d=1}^{p_j} \beta_{jd} x_j^d + \sum_{l=1}^{L_j} b_{jl}(x_j - \xi_{jl})_+^{p_j} \right],$$

with penalization applied to the coefficients $\boldsymbol{b}_j = [b_{j1}, \ldots, b_{jL_j}]^\mathrm{T}$, as described in Sect. 11.2.5. For example, penalty j may be of the form

$$\lambda_j \sum_{l=1}^{L_j} b_{jl}^2.$$

Model (12.3) is very simple to interpret since the smoother for element j of $\boldsymbol{x}$, $f_j(x_j)$, is the same regardless of the values of the other elements. Hence, each of the f_j terms may be plotted to visually examine the relationship between Y and x_j; Fig. 12.1 provides an example. Model (12.3) is also computationally convenient, as we shall see in Sect. 12.2.2.

A *semiparametric model* is one in which a subset of the covariates are modeled parametrically, with the remainder modeled nonparametrically. Specifically, let $\boldsymbol{z}_i = [z_{i1}, \ldots, z_{iq}]$ represent the sets of variables we wish to model parametrically and $\boldsymbol{\beta} = [\beta_1, \ldots, \beta_q]^\mathrm{T}$ the set of associated regression coefficients. Then (12.3) is simply extended to

$$g(\mu_i) = \beta_0 + \boldsymbol{z}_i \boldsymbol{\beta} + \sum_{j=1}^{k} f_j(x_{ij}).$$

We saw an example of this form in Sect. 11.2.9 in which spinal bone marrow density was modeled as a parametric function of ethnicity and as a nonparametric function of age.

12.2.2 Computation via Backfitting

The structure of an additive model suggests a simple and intuitive fitting algorithm. Consider first the linear link $g(\mu_i) = \mu_i$, to give the additive model

$$Y_i = \beta_0 + f_1(x_{i1}) + f_2(x_{i2}) + \ldots + f_k(x_{ik}) + \epsilon_i. \tag{12.4}$$

Define partial residuals

$$r_i^{(j)} = Y_i - \beta_0 - \sum_{l=1, l \neq j}^{k} f_l(x_{il}), \tag{12.5}$$

for $j = 1, \ldots, k$. For these residuals,

$$E[r_i^{(j)} \mid x_{ij}] = f_j(x_{ij}),$$

which suggests we can estimate f_j, using as response the residuals $r_1^{(j)}, \ldots, r_n^{(j)}$. Iterating across j produces the *backfitting* algorithm. Backfitting proceeds as follows:

1. Initialize: $\widehat{\beta}_0 = \frac{1}{n}\sum_{i=1}^n y_i$ and $\widehat{f}_j \equiv 0$ for $j = 1, \ldots, k$.
2. For a generic smoother S_j, cycle over j repeatedly:

$$\widehat{f}_j = S_j\left(r_1^{(j)}, \ldots, r_n^{(j)}\right)$$

with $r_i^{(j)}$ given by (12.5), until the functions $\widehat{f}_j$ change by less than some prespecified threshold.

Buja et al. (1989) describe the convergence properties of backfitting. For general responses beyond (12.4), the backfitting algorithm uses the "working" residuals, as defined with respect to the IRLS algorithm in Sect. 6.5.2. Wood (2006) contains details on how the P-IRLS algorithm (Sect. 11.5.1) may be extended to fit GAMs. An alternative method of computation for GAMs, based on a mixed model representation, is described in Sect. 12.5.2.

Example: Prostate Cancer

For illustration, we fit a GAM to the prostate cancer data (Sect. 1.3.1) in order to evaluate whether a parametric model is adequate. The response is log PSA, and we model each of log cancer volume, log weight, log age, log BPH, log capsular penetration, and PGS45 using smooth functions. The variable SVI is binary, and the Gleason score can take just 4 values. Hence, for these two variables, we assume a parametric linear model. The smooth functions are modeled as penalized regression cubic splines, with seven knots for each of the six variables. Generalized cross-validation (GCV, Sect. 10.6.3) was used for smoothing parameter estimation and produced effective degrees of freedom of 1, 1.1, 1.5, 1, 4.6, and 3.9 for the six smooth terms (with the variable order being as in Fig. 12.1). The resultant fitted smooths, with shaded bands indicating pointwise asymptotic 95% confidence intervals, are plotted in Fig. 12.1. Panels (e) and (f) indicate some nonlinearity, but the wide uncertainty bands and flatness of the curves indicate that little will be lost if linear terms are assumed for all variables. This figure illustrates the simple interpretation afforded by GAMs, since each smooth shows the modeled association with that variable, with all other variables held constant.

12.3 Spline Methods with Multiple Predictors

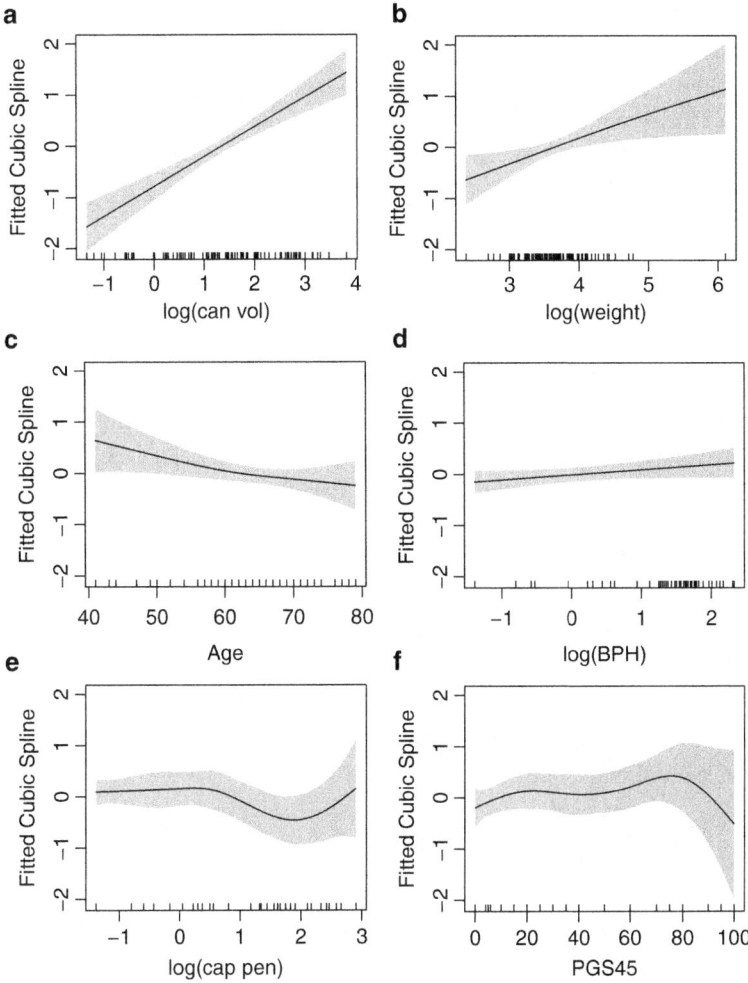

Fig. 12.1 GAM fits to the prostate cancer data. For each covariate, penalized cubic regression splines were fitted, with seven knots each. The *tick marks* on the x axis indicate the covariate values

12.3 Spline Methods with Multiple Predictors

In this section we describe how splines may be defined as a function of multivariate x. These models are of interest in their own right and may be used within GAM formulations alongside univariate specifications. For example, suppose associated with a response Y there are three variables temperature x_1, latitude x_2, and longitude x_3. In this situation we might specify a GAM with two smoothers, $f_1(x_1)$ for temperature and $f_2(x_2, x_3)$, a bivariate smoother for x_2, x_3 (since we might expect an interaction involving latitude and longitude).

12.3.1 Natural Thin Plate Splines

For simplicity, we concentrate on the two-dimensional case and begin by defining a measure of the smoothness of a function $f(x_1, x_2)$. In the one-dimensional case, the penalty was $P(f) = \int f''(x)^2 \, dx$. A natural penalty term to measure rapid variation in f in two dimensions is

$$P(f) = \int \int \left[\left(\frac{\partial^2 f}{\partial x_1^2} \right)^2 + 2 \left(\frac{\partial^2 f}{\partial x_1 \partial x_2} \right)^2 + \left(\frac{\partial^2 f}{\partial x_2^2} \right)^2 \right] dx_1 dx_2. \quad (12.6)$$

Changing the coordinates by rotation or translation in $\mathbb{R}^2$ does not affect the value of the penalty[1] which is an appealing property. The penalty is always nonnegative, and, as in the one-dimensional case, the penalty equals zero, if and only if $f(\boldsymbol{x})$ is linear in x_1 and x_2, as we now show. If $f(\boldsymbol{x})$ is linear, then it is clear that $P(f)$ is zero. Conversely, if $P(f) = 0$, all of the second derivatives are zero. Now, $\partial^2 f/\partial x_1^2 = 0$ implies $f(x_1, x_2) = a(x_2)x_1 + b(x_2)$ for functions $a(\cdot)$ and $b(\cdot)$. The condition $\partial^2 f/\partial x_1 \partial x_2 = 0$ gives $a'(x_2) = 0$ so that $a(x_2) = a$ for some constant a. Finally, $\partial^2 f/\partial x_2^2 = 0$ implies $b''(x_2) = 0$ so that $b'(x_2) = b$ and $b(x_2) = bx_2 + c$, for constants b and c. It follows that

$$f(x_1, x_2) = ax_1 + bx_2 + c$$

is linear.

We wish to minimize the penalized sum of squares

$$\sum_{i=1}^{n} [y_i - f(x_{i1}, x_{i2})]^2 + \lambda P(f) \quad (12.7)$$

with penalization term (12.6). As shown by Green and Silverman (1994, Chap. 7), the unique minimizer is provided by the *natural thin plate spline* with knots at the observed data, which is defined as

$$f(\boldsymbol{x}) = \beta_0 + \beta_1 x_1 + \beta_2 x_2 + \sum_{i=1}^{n} b_i \eta(||\boldsymbol{x} - \boldsymbol{x}_i||), \quad (12.8)$$

where

$$\eta(r) = \begin{cases} \frac{1}{8\pi} r^2 \log(r) & \text{for } r > 0 \\ 0 & \text{for } r = 0 \end{cases}$$

---

[1] This requirement is natural in a spatial context where the coordinate directions and the position of origin are arbitrary.

12.3 Spline Methods with Multiple Predictors

and the unknown b_i are constrained via

$$\sum_{i=1}^{n} b_i = \sum_{i=1}^{n} b_i x_{i1} = \sum_{i=1}^{n} b_i x_{i2} = 0,$$

that is, $\boldsymbol{x}^\mathrm{T} \boldsymbol{b} = \boldsymbol{0}$, where $\boldsymbol{x} = [\boldsymbol{x}_1, \ldots, \boldsymbol{x}_n]^\mathrm{T}$ is $n \times 3$ with $\boldsymbol{x}_i = [1, x_{i1}, x_{i2}]$ and $\boldsymbol{b} = [b_1, \ldots, b_n]^\mathrm{T}$.

Such a spline provides the unique minimizer of $P(f)$ among interpolating functions. Interested readers are referred to Theorems 7.2 and 7.3 of Green and Silverman (1994) and to Duchon (1977), who proved optimality and uniqueness properties for natural thin plate splines. Consequently, the one-dimensional result outlined in Sect. 11.2.3 holds in two dimensions also. If f is a natural thin plate spline, it can be shown that the penalty (12.6) is given by $P(f) = \boldsymbol{b}^\mathrm{T} \boldsymbol{E} \boldsymbol{b}$ where $\boldsymbol{E}$ is the $n \times n$ matrix with $E_{ij} = \eta(\|\boldsymbol{x}_i - \boldsymbol{x}_j\|)$, $i, j = 1, \ldots, n$ (Green and Silverman 1994, Theorem 7.1). The minimization (12.7) with penalty term (12.6) is

$$(\boldsymbol{y} - \boldsymbol{x}\boldsymbol{\beta} - \boldsymbol{E}\boldsymbol{b})^\mathrm{T} (\boldsymbol{y} - \boldsymbol{x}\boldsymbol{\beta} - \boldsymbol{E}\boldsymbol{b}) + \lambda \boldsymbol{b}^\mathrm{T} \boldsymbol{E} \boldsymbol{b} \quad (12.9)$$

subject again to $\boldsymbol{x}^\mathrm{T} \boldsymbol{b} = \boldsymbol{0}$ and where $\boldsymbol{\beta} = [\beta_0, \beta_1, \beta_2]^\mathrm{T}$. Green and Silverman (1994, p. 148) show that this system of equations has a unique solution.

In terms of a mechanical interpretation, suppose that an infinite elastic flat plate interpolates a set of points $[\boldsymbol{x}_i, y_i]$, $i = 1, \ldots, n$. Then the "bending energy" of the plate is proportional to the penalty term (12.6), and the minimum energy solution is the natural thin plate spline. Natural thin plate regression splines can be easily generalized to dimensions greater than two. Green and Silverman (1994, Sect. 7.9) contains details.

12.3.2 Thin Plate Regression Splines

Natural thin plate splines are very appealing since they remove the need to decide upon knot locations or basis functions; each is contained in (12.8). In practice, however, thin plate splines have too many parameters. A *thin plate regression spline* (TPRS) reduces the dimension of the space of the "wiggly" basis (the b_i's in (12.8)), while leaving $\boldsymbol{\beta}$ unchanged. Specifically, let $\boldsymbol{E} = \boldsymbol{U}\boldsymbol{D}\boldsymbol{U}^\mathrm{T}$ be the eigendecomposition of $\boldsymbol{E}$, so that $\boldsymbol{D}$ is a diagonal matrix containing the eigenvalues of $\boldsymbol{E}$ arranged so that $|D_{i,i}| \geq |D_{i-1,i-1}|$, $i = 2, \ldots, n$, and the columns of $\boldsymbol{U}$ are the corresponding eigenvectors. Now, let $\boldsymbol{U}_k$ denote the matrix containing the first k columns of $\boldsymbol{U}$ and $\boldsymbol{D}_k$ the top left $k \times k$ submatrix of $\boldsymbol{D}$. Finally, write $\boldsymbol{b} = \boldsymbol{U}_k \boldsymbol{b}_k$ so that $\boldsymbol{b}$ is restricted to the column space of $\boldsymbol{U}_k$. Then, under this reduced basis formulation, analogous to (12.9), we minimize with respect to $\boldsymbol{\beta}$ and $\boldsymbol{b}_k$

$$(\boldsymbol{y} - \boldsymbol{x}\boldsymbol{\beta} - \boldsymbol{U}_k \boldsymbol{D}_k \boldsymbol{b}_k)^\mathrm{T} (\boldsymbol{y} - \boldsymbol{x}\boldsymbol{\beta} - \boldsymbol{U}_k \boldsymbol{D}_k \boldsymbol{b}_k) + \lambda \boldsymbol{b}_k^\mathrm{T} \boldsymbol{D}_k \boldsymbol{b}_k$$

subject to $x^{\mathsf{T}} U_k b_k = 0$. See Wood (2006, Sect. 4.1.5) for further details, including the manner by which predictions are obtained and details on implementation. In addition, the optimality of thin plate regression splines as approximating thin plate splines using a basis of low rank is discussed. Thin plate regression splines retain both the advantage of avoiding the choice of knot locations and the rotational invariance of thin plate splines.

Example: Prostate Cancer

For illustration, we examine the association between the log of PSA and log cancer volume and log weight. Figure 12.2(a) shows the two-dimensional surface corresponding to a model that is linear in the two covariates (and in particular has no interaction term). We next fit a GAM with a TPRS smoother for log cancer volume and log weight, along with (univariate) cubic regression splines for age, log BPH, log capsular penetration, and PGS45, along with linear terms for SVI and the Gleason score. Figure 12.2(b) provides a perspective plot of the fitted bivariate surface. There are some differences between this plot and the linear model. In particular for high values of log cancer volume and low values of log weight, the linear no interaction model gives a lower prediction than the TPRS smoother. Overall, however, there is no strong evidence of an interaction.

12.3.3 Tensor Product Splines

As an alternative to thin plate splines, one may consider products of basis functions. Again, suppose that $x \in \mathbb{R}^2$ and that we have basis functions $h_{jl}(x_j)$, $l = 1, \ldots, M_j$, representing x_j, $j = 1, 2$. Then, the $M_1 \times M_2$ dimensional tensor product basis is defined by

$$g_{j_1 j_2}(x) = h_{1j_1}(x_1) h_{2j_2}(x_2), \quad j_1 = 1, \ldots, M_1; j_2 = 1, \ldots, M_2,$$

which leads to the two-dimensional predictive function:

$$f(x) = \sum_{j_1=1}^{M_1} \sum_{j_2=1}^{M_2} \beta_{j_1 j_2} g_{j_1 j_2}(x).$$

We illustrate this construction using spline bases. Suppose that we wish to specify linear splines with L_1 truncated lines for x_1 and L_2 for x_2. This model therefore has $L_1 + 2$ and $L_2 + 2$ bases in the two dimensions:

$$1, x_1, (x_1 - \xi_{11})_+, \ldots, (x_1 - \xi_{1L_1})_+, \tag{12.10}$$

$$1, x_2, (x_2 - \xi_{21})_+, \ldots, (x_2 - \xi_{2L_2})_+. \tag{12.11}$$

12.3 Spline Methods with Multiple Predictors

Fig. 12.2 Perspective plots of the fitted surfaces for the variables log cancer volume and log weight in the prostate cancer data: (**a**) linear model, (**b**) thin plate regression spline model, (**c**) tensor product spline model

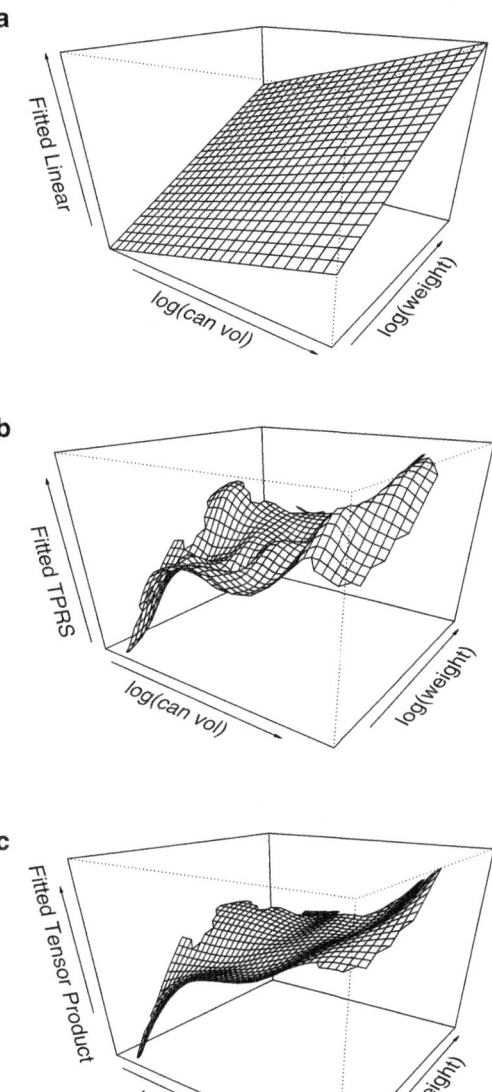

The tensor product model is

$$f(x_1, x_2) = \beta_0 + \beta_1 x_1 + \beta_2 x_2 + \beta_3 x_1 x_2$$
$$+ \sum_{l_1=1}^{L_1} b_{l_1}^{(1)}(x_1 - \xi_{1l_1})_+ + \sum_{l_2=1}^{L_2} b_{l_2}^{(2)}(x_2 - \xi_{2l_2})_+$$
$$+ \sum_{l_1=1}^{L_1} c_{l_1}^{(1)} x_2(x_1 - \xi_{1l_1})_+ + \sum_{l_2=1}^{L_2} c_{l_2}^{(2)} x_1(x_2 - \xi_{2l_2})_+$$
$$+ \sum_{l_1=1}^{L_1} \sum_{l_2=1}^{L_2} d_{l_1 l_2}^{(12)}(x_1 - \xi_{1l_1})_+(x_2 - \xi_{2l_2})_+. \qquad (12.12)$$

An additive model would correspond to the first two lines of this model only (with the $x_1 x_2$ term removed), illustrating that the last two lines are modeling interactions. The unknown parameters associated with this model are

$$\boldsymbol{\beta} = [\beta_0, \ldots, \beta_3]^{\mathrm{T}}$$
$$\boldsymbol{b}^{(1)} = [b_1^{(1)}, \ldots, b_{L_1}^{(1)}]^{\mathrm{T}}$$
$$\boldsymbol{b}^{(2)} = [b_1^{(2)}, \ldots, b_{L_2}^{(2)}]^{\mathrm{T}}$$
$$\boldsymbol{c}^{(1)} = [c_1^{(1)}, \ldots, c_{L_1}^{(1)}]^{\mathrm{T}}$$
$$\boldsymbol{c}^{(2)} = [c_1^{(2)}, \ldots, c_{L_2}^{(2)}]^{\mathrm{T}}$$
$$\boldsymbol{d}^{(12)} = [d_{11}^{(12)}, \ldots, d_{L_1 L_2}^{(12)}]^{\mathrm{T}}.$$

Consequently, there are

$$4 + L_1 + L_2 + L_1 + L_2 + L_1 L_2 = (L_1 + 2)(L_2 + 2)$$

parameters in the tensor product model. Clearly the dimensionality of the basis increases quickly with the dimensionality of the covariate space k. See Exercise 12.1 for an example of the construction and display of these bases. An example of a tensor product fit is given at the end of Section 12.5.

The fit from a tensor product basis is not invariant to the orientation of the coordinate axis. Radial invariance can be achieved with basis functions of the form $C(||\boldsymbol{x} - \boldsymbol{\xi}||)$, with $\boldsymbol{\xi} = [\xi_1, \xi_2]$ and for some univariate function $C(\cdot)$, see for example Ruppert et al. (2003). The value of the function at $\boldsymbol{x}$ only depends on the distance from this point to $\boldsymbol{\xi}$, and so the function is radially symmetric about this point.

12.4 Kernel Methods with Multiple Predictors

In principle, the extension of the kernel local polynomial smoothing methods of Sect. 11.3 is straightforward; one simply needs to choose a multivariate weight function (i.e., a kernel) and a multivariate local polynomial. For simplicity, we consider the case of two covariates and a continuous response with additive errors:

$$Y_i = f(x_{i1}, x_{i2}) + \epsilon_i$$

with $\mathrm{E}[\epsilon_i] = 0$, $\mathrm{var}(\epsilon_i) = \sigma^2$ and $\mathrm{cov}(\epsilon_i, \epsilon_j) = 0$ for $i \neq j$. A suitably smooth function may be approximated, for values $\boldsymbol{u} = [u_1, u_2]$ in a neighborhood of a point $\boldsymbol{x} = [x_1, x_2]$, by a second-order Taylor series approximation:

$$f(\boldsymbol{u}) \approx f(\boldsymbol{x}) + (u_1 - x_1)\frac{\partial f}{\partial x_1} + (u_2 - x_2)\frac{\partial f}{\partial x_2}$$
$$+ (u_1 - x_1)^2 \frac{1}{2}\frac{\partial^2 f}{\partial x_1^2} + (u_1 - x_1)(u_2 - x_2)\frac{\partial^2 f}{\partial x_1 \partial x_2} + (u_2 - x_2)^2 \frac{1}{2}\frac{\partial^2 f}{\partial x_2^2}.$$

We see that the model includes an interaction term $(u_1 - x_1)(u_2 - x_2)$, and the approximation suggests that for a prediction at the point $\boldsymbol{x}$, we can use the local polynomial:

$$P_{\boldsymbol{x}}(\boldsymbol{u}; \boldsymbol{\beta}_{\boldsymbol{x}}) = \beta_{0\boldsymbol{x}} + (u_1 - x_1)\beta_{1\boldsymbol{x}} + (u_2 - x_2)\beta_{2\boldsymbol{x}}$$
$$+ (u_1 - x_1)^2 \frac{\beta_{3\boldsymbol{x}}}{2} + (u_1 - x_1)(u_2 - x_2)\beta_{4\boldsymbol{x}} + (u_2 - x_2)^2 \frac{\beta_{5\boldsymbol{x}}}{2}.$$

Estimation proceeds exactly as in the one-dimensional case by choosing $\widehat{\boldsymbol{\beta}}_{\boldsymbol{x}}$ to minimize the locally weighted sum of squares

$$\sum_{i=1}^n w_i(\boldsymbol{x}) \left[Y_i - P_{\boldsymbol{x}}(\boldsymbol{x}_i; \boldsymbol{\beta}_{\boldsymbol{x}})\right]^2,$$

with the weights $w_i(\boldsymbol{x})$ depending on a two-dimensional kernel function. The simplest choice is the product of one-dimensional kernels, that is,

$$w_i(\boldsymbol{x}) = K_1\left(\frac{x_1 - x_{i1}}{\lambda_1}\right) \times K_2\left(\frac{x_2 - x_{i2}}{\lambda_2}\right).$$

The fitted value is

$$\widehat{f}(\boldsymbol{x}) = P_{\boldsymbol{x}}(\boldsymbol{x}; \widehat{\boldsymbol{\beta}}_{\boldsymbol{x}}) = \widehat{\beta}_{0\boldsymbol{x}}.$$

Embedding multivariate local polynomials within a generalized linear model framework is straightforward, by simple extension of the approach described in Sect. 11.5.3.

As with multivariate spline methods, the local polynomial approach becomes more difficult as the dimensionality increases, due to the sparsity of points in high dimensions.

12.5 Smoothing Parameter Estimation

12.5.1 Conventional Approaches

The simplest way to control the level of smoothing is to specify an effective degrees of freedom, df_j, for each of the $j = 1, \ldots, k$ smoothers (where we have assumed for simplicity that we are modeling k univariate smoothers).

As we saw in Chap. 10, there are two ways of estimating smoothing parameters. The first is to attempt to minimize prediction error which may be represented by AIC-like criteria or via cross-validation. Such procedures were described in Sect. 10.6. The second method is to embed the penalized smoothing within a mixed model framework and then use likelihood (ML or REML) or Bayesian estimation of the random effects variances. This approach is described in Sect. 12.5.2.

For GAMs the smoothing of multiple parameters may be estimated during the iterative cycle (e.g., within the P-IRLS iterates), which is known as *performance iteration*. As an alternative, fitting may be carried out multiple times for each set of smoothing parameters, which is known as *outer iteration*. The latter is more reliable but requires more work to implement. However, the methods for minimizing prediction error using outer iteration described in Wood (2008) are shown to be almost as computationally efficient as performance iteration.

12.5.2 Mixed Model Formulation

To illustrate the general technique, consider a linear additive model with penalized regression splines providing the smoothing for each of the k covariates. Further, assume a truncated polynomial representation with a degree p_j polynomial and L_j knots with locations ξ_{jl}, $l = 1, \ldots, L_j$, associated with the jth smooth, $j = 1, \ldots, k$. A mixed model representation is

$$y_i = \beta_0 + \sum_{j=1}^{k} \left[\sum_{d=1}^{p_j} \beta_{jd} x_j^d + \sum_{l=1}^{L_j} b_{jl}(x_j - \xi_{jl})_+^{p_j} \right] + \epsilon_i$$

$$= \beta_0 + \sum_{j=1}^{k} \boldsymbol{x}_{ij} \boldsymbol{\beta}_j + \sum_{j=1}^{k} \boldsymbol{z}_{ij} \boldsymbol{b}_j + \epsilon_i \quad (12.13)$$

with $\epsilon_i \mid \sigma_\epsilon^2 \sim N(0, \sigma_\epsilon^2)$,

$$\boldsymbol{x}_{ij} = [x_{ij}, \ldots, x_{ij}^{p_j}], \quad \boldsymbol{\beta}_j = \begin{bmatrix} \beta_{j1} \\ \vdots \\ \beta_{jp_j} \end{bmatrix},$$

and

$$\boldsymbol{z}_{ij} = [(x_{ij} - \xi_{j1})_+^{p_j}, \ldots, (x_{ij} - \xi_{jL_j})_+^{p_j}], \quad \boldsymbol{b}_j = \begin{bmatrix} b_{j1} \\ \vdots \\ b_{jp_j} \end{bmatrix}.$$

The parameters $\beta_1, \ldots, \beta_k$ are treated as fixed effects with $\boldsymbol{b}_1, \ldots, \boldsymbol{b}_k$ being a set of independent random effects. The penalization is incorporated through the introduction of k sets of random effects:

$$b_{jl} \mid \sigma_{bj}^2 \sim_{ind} N(0, \sigma_{bj}^2), \quad l = 1, \ldots, L_j$$

for $j = 1, \ldots, k$. Inference, from either a likelihood (Sect. 11.2.8) or Bayesian (Sect. 11.2.9) perspective, proceeds exactly as in the univariate covariate case. The extension of (12.13) to a tensor product spline model, such as (12.12), is straightforward. Comparison with (12.12) reveals the strong simplification of (12.13) (with $k = 2$ and $p_j = 1$), which includes no cross-product terms.

One can estimate the variance components (and hence the amount of smoothing) using a fully Bayesian approach or via ML/REML. With a likelihood-based approach, one requires the random effects to be integrated from the model. For non-Gaussian response models, these integrals cannot be evaluated analytically. Approaches to integration were reviewed in Sect. 3.7. One iterative strategy we mention briefly here linearizes the model, which allows linear methods of estimation to be applied. This strategy is known as penalized quasi-likelihood (PQL, Breslow and Clayton 1993) and is essentially equivalent to performance iteration. Using the more sophisticated Laplace approximation gives one approach to outer iteration. See Wood (2011) for details of a method that is almost as computationally efficient as performance iteration. Bayesian approaches typically use MCMC (Sect. 3.8) or INLA (Sect. 3.7.4).

Some theoretical work (Wabha 1985; Kauermann 2005) suggests that methods that minimize prediction error criteria give better prediction error asymptotically, but have slower convergence of smoothing parameters (Härdle et al. 1988). Reiss and Ogden (2009) show that the equations by which generalized cross-validation (GCV) and REML estimates are obtained have a similar form and use this to examine the properties of the estimates. They find that converging to a local, rather than a global, solution appears to happen more frequently for GCV than for REML. Hence, care is required in finding a solution, and Reiss and Ogden (2009) recommend plotting the criteria function over a wide range of values. Wood (2011) discusses how GCV can lead to "occasional severe under-smoothing," and this is backed up by Reiss and Ogden (2009) who argue, based on their theoretical derivations, that REML estimates will tend to be more stable than GCV estimates.

Example: Prostate Cancer

We return to the prostate cancer example and fit a GAM with a tensor product spline smoother for log cancer volume and log weight, along with (univariate) cubic regression splines for age, log BPH, log capsular penetration and PGS45, and with linear terms for SVI and the Gleason score. Each of the constituent smoothers in the tensor product is taken to be a cubic regression spline with bases of size 6 for each of the components. GCV is used for estimation of the smoothing parameters and results in an effective degrees of freedom of 12.4 for the tensor product term. Figure 12.2c provides a perspective plot of the fitted bivariate surface. It is reassuring that the fit is very similar to the thin plate regression spline in panel (b).

12.6 Varying-Coefficient Models

Varying-coefficient models (Cleveland et al. 1991; Hastie and Tibshirani 1993) provide another flexible model based on a linear form but with model coefficients that vary smoothly as a function of other variables. We begin our discussion by giving an example with two covariates, x and z. The model is

$$\mathrm{E}[Y \mid x, z] = \mu = \beta_0(z) + \beta_1(z)x \qquad (12.14)$$

so that we have a linear regression with both the intercept and the slope corresponding to x being smooth functions of z. The first thing to note is that the model is not symmetric in the two covariates. Rather, the linear association between Y and x is *modified* by z, and we have a specific form of interaction model.

The extension to a generalized linear/additive model setting is clear, on replacement of $\mathrm{E}[Y \mid \boldsymbol{x}, z]$ by $g(\mu)$. With covariates $\boldsymbol{x} = [x_1, \ldots, x_k]$ and z the model is

$$g(\mu) = \beta_0(z) + \sum_{j=1}^{k} \beta_j(z)x_j,$$

so that each of the slopes is modified by z. Computation and inference are straightforward for the varying-coefficient model.

12.6 Varying-Coefficient Models

We return to the case of just two variables, x and z, and consider penalized linear spline smoothers with L knots having locations ξ_k for each of the intercept and slope. Then, model (12.14) becomes:

$$E[Y \mid x, z] = \underbrace{\alpha_0^{(0)} + \alpha_1^{(0)} z + \sum_{l=1}^{L} b_l^{(0)} (z - \xi_l)_+}_{\beta_0(z)}$$

$$+ \underbrace{\left(\alpha_0^{(1)} + \alpha_1^{(1)} z + \sum_{l=1}^{L} b_l^{(1)} (z - \xi_l)_+ \right)}_{\beta_1(z)} x.$$

A mixed model representation (Ruppert et al. 2003, Sect. 12.4) assumes independent random effects with $b_l^{(0)} \mid \sigma_0^2 \sim_{iid} N(0, \sigma_0^2)$ and $b_l^{(1)} \mid \sigma_1^2 \sim_{iid} N(0, \sigma_1^2)$ for $l = 1, \ldots, L$.

An obvious application of varying-coefficient models is in the situation in which the modifying variables correspond to time. As a simple example, if a response and covariate x are collected over time, we might consider the model

$$Y_t = \alpha + \beta(t) x_t + \epsilon_t, \tag{12.15}$$

where we have chosen a simple model in which the slope, and not the intercept, is a function of time. We briefly digress to provide a link with Bayesian dynamic linear models, which were developed for the analysis of time series data and allow regression coefficients to vary according to an autoregressive model. The simplest dynamic linear model (see, for example, West and Harrison 1997) is defined by the equations

$$Y_t = \alpha + x_t \beta_t + \epsilon_t, \quad \epsilon_t \mid \sigma_\epsilon^2 \sim_{iid} N(0, \sigma_\epsilon^2)$$
$$\beta_t = \beta_{t-1} + \delta_t, \quad \delta_t \mid \sigma_\delta^2 \sim_{iid} N(0, \sigma_\delta^2).$$

Accordingly, we have a varying-coefficient model of the form of (12.15) with smoothing carried out via a particular flexible form: a first-order Markov model (the limiting form of an autoregressive model, see Sect. 8.4.2). This is also a mixed model but with the first differences ($\beta_t - \beta_{t-1}$) being modeled. A spatial form of this autoregressive model was considered in Sect. 9.7.

Example: Ethanol Data

We illustrate the use of a varying-coefficient model with the ethanol data described in Sect. 10.2.2. Figure 10.2 provides a three-dimensional plot of these data. An initial analysis, with NOx modeled as a linear function of C and a quadratic function

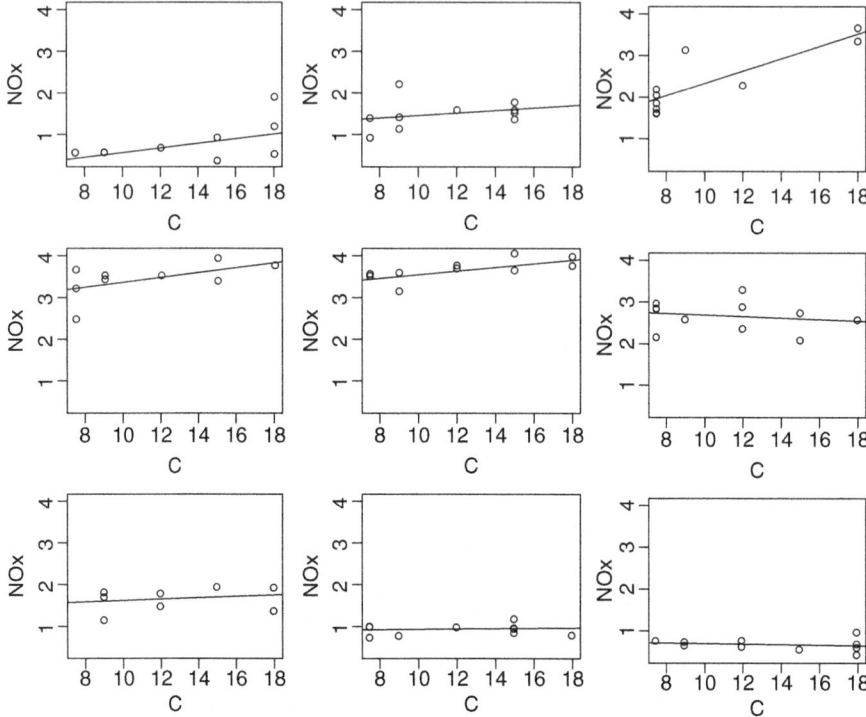

Fig. 12.3 NOx versus C for nine subsets of the ethanol data (defined via the quantiles of E), with linear model fits superimposed

of E, was found to provide a poor fit. Specifically, the association between NOx and E is far more complex than quadratic. To examine the association more closely and to motivate the varying-coefficient model, we split the E variable into nine bins, based on the quantiles of E, with an approximately equal number of pairs of [NOx, C] points within each bin. We then fit a linear model to each portion of the data. Figure 12.3 shows the resultant data and fitted lines. A linear model appears, at least visually, to provide a reasonable fit in each panel, though the intercepts and slopes vary across the quantiles of E.

Figures 12.4(a) and (b) plot these intercepts and slopes as a function of the midpoint of the bins for E, and we see that the coefficients vary in a non-monotonic fashion. Consequently, we fit the varying-coefficient model

$$E\,[\text{NOx} \mid C, E] = \beta_0(E) + \beta_1(E) \times C, \qquad (12.16)$$

with $\beta_0(E)$ and $\beta_1(E)$ both modeled as penalized cubic regression splines with 10 knots each. The smoothing parameters for the smoothers were chosen using GCV, which resulted in 6.4 and 4.7 effective degrees of freedom for the intercept and

12.6 Varying-Coefficient Models

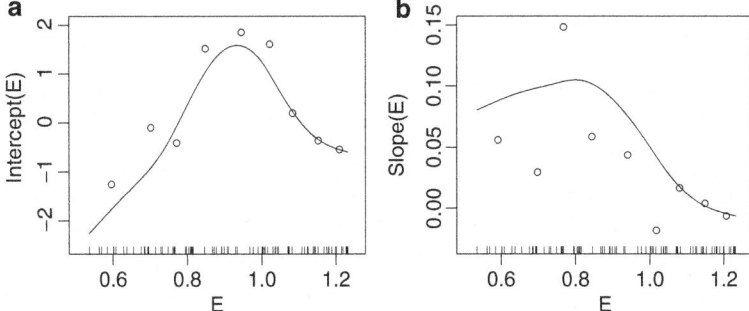

Fig. 12.4 (**a**) Intercepts and (**b**) slopes from linear models fitted to the ethanol data in which the response is NOx and the covariate is C, with the nine groups defined by quantiles of the E variable. The fitted curves are from a varying-coefficient model in which the intercepts and slopes are modeled as cubic regression splines in E

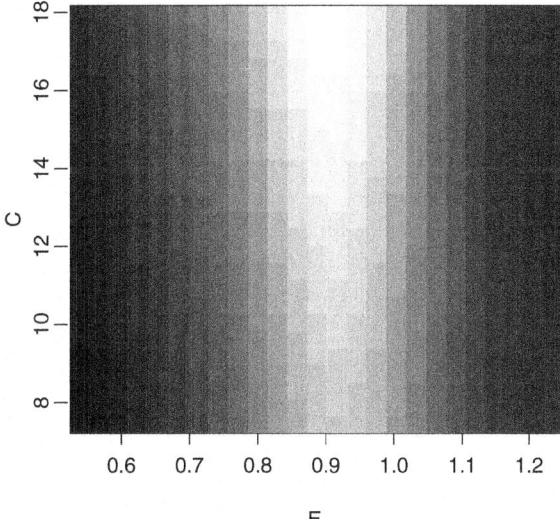

Fig. 12.5 Image plot of the predictive surface from the varying-coefficient model (12.16) fitted to the ethanol data. *Light* and *dark gray* values indicate, respectively, high and low values of expected NOx

slope, respectively. The fitted smooths are shown on Fig. 12.4, and we see that the intercept and slope rise then fall as a function of E.

Figure 12.5 gives the fitted surface. The inverted U-shape in E is evident. More subtly, as we saw in Fig. 12.4(b), the strength (and sign) of the linear association between NOx and C varies as a function of E.

12.7 Regression Trees

12.7.1 Hierarchical Partitioning

In this section we consider a quite different approach to modeling, in which the covariate space is partitioned into regions within which the response is relatively homogeneous. A key feature is that although a model for the data is produced, the approach is best described *algorithmically*. As we will see, a *tree-based* approach to the construction of partitions is both interpretable and amenable to computation. Our development follows similar lines to Hastie et al. (2009, Sect. 9.2).

In order to motivate tree-based models, we first take a step back and consider ways of constructing partitions; the aim is to produce regions of the covariate space within which the response is constant. An obvious statement is that, in practice, clearly the shapes and sizes of the partition will be dependent on the distribution of the covariates. There are clearly many possible ways (models) by which partitions might be defined, beginning with a completely unrestricted search in which there are no constraints on the shapes and sizes of the partition region. This is too complex a task to practically accomplish, however.[2] We examine a series of partitions for the case of two covariates, x_1 and x_2, leading to a particular mechanism for partitioning. Figure 12.6(a) shows partitions defined by straight lines in the covariate space, with the lines not constrained to be parallel to either axis (clearly we could start with partitions of even greater complexity). Explaining how the response varies as a function of x_1 and x_2 for the particular partition in Fig. 12.6(a) is not easy, however. In addition, searching for the best partitions defined with respect to lines of this form is very difficult, particularly when the covariate space is high dimensional. Figure 12.6(b) displays a partition in which the space is dissected with lines that are parallel to the axes, and, though simpler to describe than the previous case, the regions are still not straightforward to explain or compute.

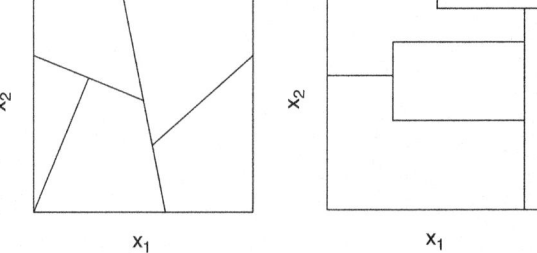

Fig. 12.6 Examples of flexible partitions of the $[x_1, x_2]$ space that use *straight lines* to define the partitions

---

[2]Methods aimed in this direction do exist, for example, in the spatial literature. Knorr-Held and Rasser (2000) and Denison and Holmes (2001) describe Bayesian partition models based on Voronoi tessellations. These models are computationally expensive to implement and have so far been restricted to two-dimensional covariate settings.

12.7 Regression Trees

Fig. 12.7 Hierarchical binary tree partition of the $[x_1, x_2]$ space

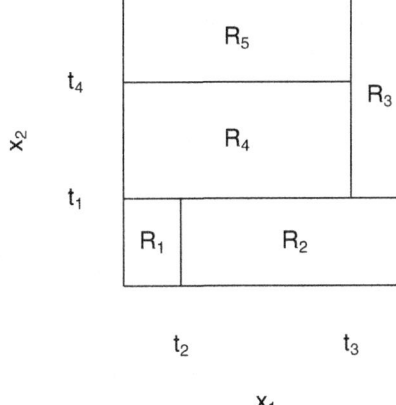

Fig. 12.8 Hypothetical regression tree corresponding to Fig. 12.7. The four splits lead to five terminal nodes (leaves), labeled $R_1, \ldots, R_5$

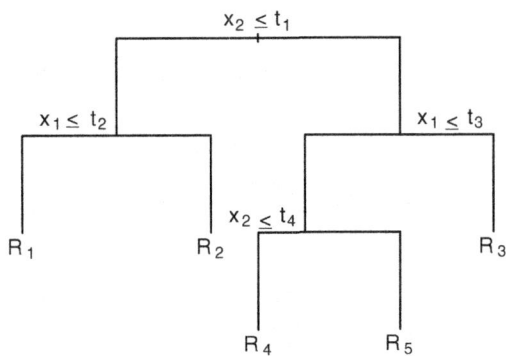

Figure 12.7 shows a *tree-based* partition that is based on successive binary partitions of the predictor space, again to produce subsets of the response which are relatively constant. Splits are only allowed within, and not between, partitions. Such a method has the advantage of producing models that are relatively easy to explain, since they follow simple rules, and may be computed without too much difficulty. The partition in Fig. 12.7 is generated by the algorithm illustrated in the form of a "tree" in Fig. 12.8 (notice that trees are usually shown as growing down the page). We describe in detail how this partition is constructed.

The terminology we use is graphical. Decisions are taken at *nodes*, and the *root* of the tree is the top node. The terminal nodes are the *leaves*, and covariate points x assigned to these nodes are assigned a constant fitted value (which is called a classification if the response is discrete). Attached to each nonterminal node is a question that determines a split of the data. Suppose a tree T_0 is grown. A *subtree* of T_0 is a tree with root a node of T_0; it is a *rooted subtree* if its root is the root of T_0. The *size* of a tree, denoted $|T|$, is the number of leaves.

In Fig. 12.8, the first split is according to $X_2 \leq t_1$. If this condition is true, then we follow the left branch and next split on $X_1 \leq t_2$, to give leaves with labels R_1

Fig. 12.9 Hypothetical surface corresponding to the partition of Fig. 12.7 and the tree of Fig. 12.8

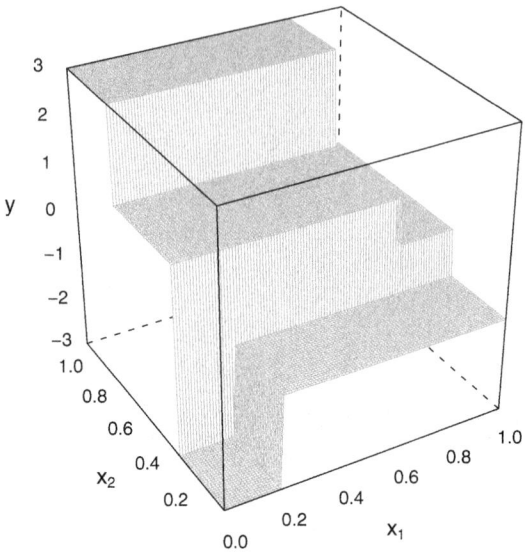

and R_2. If we follow the right hand branch and $X_1 > t_3$, we terminate at the R_3 leaf. If $X_1 \leq t_3$, we split again via $X_2 \leq t_4$ to give the leaves R_4 and R_5. The model resulting from these operations is

$$f(x_1, x_2) = \sum_{j=1}^{5} \beta_j I\left([x_1, x_2] \in R_j\right),$$

where the indicator $I\{[x_1, x_2] \in R_j\}$ is 1 if the point $[x_1, x_2]$ lies in region R_j and is equal to 0 otherwise. Figure 12.9 is a hypothetical surface corresponding to the tree shown in Fig. 12.8.

The five basis functions h_j, which correspond to R_j, $j = 1, \ldots, 5$, are:

$$h_1(x_1, x_2) = I(x_2 \leq t_1) \times I(x_1 \leq t_2)$$
$$h_2(x_1, x_2) = I(x_2 \leq t_1) \times I(x_1 > t_2)$$
$$h_3(x_1, x_2) = I(x_2 > t_1) \times I(x_1 > t_3)$$
$$h_4(x_1, x_2) = I(x_2 > t_1) \times I(x_1 \leq t_3) \times I(x_2 \leq t_4)$$
$$h_5(x_1, x_2) = I(x_2 > t_1) \times I(x_1 \leq t_3) \times I(x_2 > t_4).$$

Basis h_j corresponds to R_j, $j = 1, \ldots, 5$. These bases cover the covariate space and at any point x only one basis function is nonzero, so that we have a partition. We emphasize that these bases are not specified a priori, but selected on the basis of the observed data, so that they are locally *adaptive*. A regression tree provides a hierarchical method of describing the partitions (i.e., the partitioning is defined through a nested series of instructions), which aids greatly in describing the model. Tree models effectively perform variable selection, and discovering interactions is implicit in the process.

12.7 Regression Trees

There are many ways in which we could go about "growing" a tree. Clearly we could continue to split the data until each leaf contains a single unique set of x values, but this would lead to overfitting. Many approaches grow a large tree and then *prune* it back, to avoid such overfitting. There are different ways to both split nodes (e.g., only binary splits may be performed) and prune back the tree.

We now describe an approach to tree-building based on binary splits. Consider a simple situation in which we have a response Y and k continuous predictors x_l, $l = 1, \ldots, k$. A common implementation considers recursive binary partitions in which the x space is first split into two regions on the basis of one of $x_1, \ldots, x_k$, with the variable and split point being chosen to achieve the best fit (according to, say, the residual sum of squares, or more generally the deviance). There are a maximum of $k(n-1)$ partitions to consider. Next, one or both of the regions are split into two more regions. Only partitions within, and not between, current partitions are considered at each step of the algorithm. This process is continued until some stopping rule is satisfied. The final tree may then be pruned back.

For ordered categorical variables, the above procedure poses no ambiguity, but for unordered categorical variables with more than two levels, we may divide the levels into two groups; with L levels there are $2^{L-1} - 1$ pairs of groups. Note that, in general, monotonic transformations of quantitative covariates produce identical results.

When the algorithm terminates, we end up with a regression model having fitted values $\widehat{\beta}_j$ in region R_j, that is,

$$\widehat{f}(x) = \sum_{j=1}^{J} \widehat{\beta}_j I(x \in R_j). \tag{12.17}$$

An obvious estimator is

$$\widehat{\beta}_j = \frac{1}{n_j} \sum_{i:x_i \in R_j} y_i$$

where n_j is the number of observations in partition R_j, $j = 1, \ldots, J$ (so that there are J leaves). Inherent in the construction of this unweighted estimator is an assumption that the error terms are uncorrelated with constant variance (which is consistent with choosing the splits on the basis of minimizing the residual sum of squares).

We now give more detail on how regression trees are "grown." The algorithm automatically decides on both the variable on which to split and on the split points. To find the best tree, we start with all the data and proceed with a greedy algorithm.[3] Consider a particular variable x_l and a split point s and define

$$R_1(l, s) = \{ x : x_l \leq s \}$$
$$R_2(l, s) = \{ x : x_l > s \}.$$

---

[3] A greedy algorithm is one in which "locally" optimal choices are made at each stage.

We seek the splitting variable index l and split point s that solve

$$\min_{l,s} \left[\min_{\beta_1} \sum_{i:x_i \in R_1(l,s)} (y_i - \beta_1)^2 + \min_{\beta_2} \sum_{i:x_i \in R_2(l,s)} (y_i - \beta_2)^2 \right],$$

that is, that minimizes the residual sum of squares among models with two response levels, based on a split of one of the k variables. For any choice of l and s, the inner minimization is solved by

$$\widehat{\beta}_1 = \frac{1}{|R_1(l,s)|} \sum_{i:x_i \in R_1(l,s)} y_i$$

$$\widehat{\beta}_2 = \frac{1}{|R_2(l,s)|} \sum_{i:x_i \in R_2(l,s)} y_i.$$

Each of the covariates, $x_1, \ldots, x_k$, is scanned and, for each, the determination of the best split point s is found, which is fast. Having found the best split, we partition the data into the two resulting regions, and the splitting process is then repeated on each region to find the next split.

We now return to the key question: How large a tree should be grown? If the tree is too large, then we will overfit, and if too small, the tree will not capture important features. The tree size is therefore acting as a tuning parameter that determines complexity. By analogy with forward selection, growing a tree until (say) the sum of squares is not significantly reduced in size is shortsighted, since splits below the current tree may be highly beneficial. In practice, a common approach is to first grow a large tree, T_0, stopping when some minimum node size is reached (in the extreme case we could continue until each leaf contains a single observation); the tree is then pruned back. The space of trees becomes large very quickly, as k increases. Consequently, searching over all subtrees and using, for example, cross-validation or AIC to select the "best" is not feasible. We discuss an alternative way to penalize overfitting.

Let T be a subtree of T_0 that is obtained by weakest-link pruning T_0, that is, by collapsing any number of its internal (nonterminal) nodes. We let

$$S_j(T) = \frac{1}{n_j} \sum_{i:x_i \in R_j} (y_i - \widehat{\beta}_j)^2 \qquad (12.18)$$

denote the within-partition residual sum of squares and $|T|$ be the number of terminal nodes in T. With respect to (12.17), $J = |T|$. Following, Breiman et al. (1984) define the *cost complexity criterion* as the total residual sum of squares plus a penalty term that consists of a smoothing parameter λ multiplied by the size of the tree:

$$C_\lambda(T) = \sum_{j=1}^{|T|} n_j S_j(T) + \lambda |T|. \qquad (12.19)$$

Hence, we have a penalized sum of squares. For a given λ, we can find the subtree $T_\lambda \in T_0$ that minimizes $C_\lambda(T)$. The tuning parameter $\lambda \geq 0$ obviously balances the tree size and the goodness of fit of the tree to the data, with larger values giving smaller trees. As usual we are encountering the bias-variance trade-off. Large trees exhibit low bias and high variance, with complementary behavior being exhibited by small trees. With $\lambda = 0$, we obtain the full tree, T_0.

For each λ, it can be shown that there exists a unique smallest subtree, $T_{\hat{\lambda}}$ that minimizes $C_\lambda(T)$. See Breiman et al. (1984) and Ripley (1996) for details. This tree can be found using weakest-link pruning. The estimation of the smoothing parameter λ may be carried out via cross-validation to give a final tree $T_{\hat{\lambda}}$. Specifically, cross-validation splits are first formed, and then, for a given λ, the tree that minimizes (12.19) can be found. For this tree, the cross-validation sum of squares $(y - \hat{y})^2$ can be calculated over the left-out data y, where $\hat{y}$ is the prediction from the tree. This procedure is carried out for different values of λ, and one may pick the value that minimizes the sum of squares.

Before moving to an example, we make some general comments about regression trees. See Hastie et al. (2009, Sect. 9.2.4) and Berk (2008, Chap. 3) for more extensive discussions.

In applications there are often missing covariate values. One approach to accommodating such values that is applicable to categorical variables is to create a "missing" category; this may reveal that responses with some missing values behave differently to those without missing values. Another approach is to drop cases down the tree, as far as they will go, until a decision on a missing value is reached. At that point, the mean of y can be calculated from the other cases available at this node and can be used as the prediction. This can result in decisions being made based on little information, however. A general alternative strategy is to create surrogate variables that mimic the behavior of the missing variables. When considering a predictor for a split only, the non-missing values are used. Once the best predictor/split combination is selected, other predictor/split points are examined to see which best mimics the one selected. For example, suppose the optimal split based on the non-missing observations is based on x_1. The binary outcome defined by the split on x_1 is then taken as response, and we try to predict this variable using splits based on each of x_l, $l = 2, \ldots, k$. The classification rate is then examined with the best, second best, etc., surrogates being recorded. When training data with missing values on x_1 are encountered, one of the surrogates is then used instead, with the variable chosen being the one that is available with the best classification rate. The same strategy is used for new cases with missing values. The basic idea is to exploit correlation between the covariates. The best advice with regard to missing data is obvious: one should avoid having missing values as much as possible when the study is conducted, and if there are a large proportion of missing values predictions should be viewed with a fair amount of skepticism, whatever the correction method employed. A number of authors have pointed out that variables having more values are favored in the splitting procedure (e.g., Breiman et al. 1984, p. 42). Variables with more missing values are also favored (Kim and Loh 2001).

An undesirable aspect of the fitted surface is that, by construction, it is piecewise constant, which will often not be plausible a priori. Multiple adaptive regression splines (MARS, to be described in Sect. 12.7.2) constructs a basis function from linear segments, in order to alleviate this problem. The flexibility of trees can be a disadvantage since one cannot build in structure which one might think is present. For example, consider an additive model with two variables. Specifically, suppose the true model is $E[Y \mid x_1, x_2] = \beta_1 I(x_1 \leq t_1) + \beta_2 I(x_2 \leq t_2)$ and the first split is at $x_1 \approx t_1$. Then, two subsequent splits would be needed, one on each branch at $x_2 \approx t_2$.

Fundamentally, carrying out inference with regression trees is difficult, because one needs to consider the stepwise nature of the search algorithm (Sect. 4.8.1 discussed the inherent difficulties of such an approach). One solution, based on permutation methods, has been suggested by Hothorn et al. (2006). Gordon and Olshen (1978) and Gordon and Olshen (1984); Olshen (2007) (among others) have produced results on the conditions under which tree-based approaches are consistent.

A major problem with trees is that they can exhibit high variance, in the sense that a small change in the data can result in a very different tree being formed. The hierarchical nature of the algorithm is responsible for this behavior, since the effect of changes is propagated down the tree. Later in the chapter we will describe *bagging* (Sect. 12.8.5) and *random forest* (Sect. 12.8.6) approaches that consider *collections* of trees in order to alleviate this instability.

We now illustrate the use of regression trees with the prostate cancer data. In Sect. 12.8.4 we consider tree-based approaches to classification.

Example: Prostate Cancer

We fit a binary regression tree model treating log PSA as the response and with the splits based on the eight covariates. We grow the tree with a requirement that there must be at least three observations in each leaf. This specification leads to a regression tree with 27 splits.

We now choose the smoothing parameter λ based on cross-validation and minimizing (12.19), with weakest-link pruning being carried out for each candidate value of λ. Figure 12.10 plots the cross-validation score (along with an estimate of the standard error) as a function of "complexity" (on the bottom axis) and tree size (top axis). The complexity score here is the improvement in R^2 (Sect. 4.8.2) when the extra split is made. The tree that attains the minimum CV is displayed in Fig. 12.11 and has four splits and five leaves (terminal nodes). We saw in Fig. 10.7 that when the lasso was used, log cancer volume was the most important variable (in the sense of being the last to be removed from the model), followed by log weight and SVI. Consequently, it is no surprise that two of the splits are on log cancer volume, with one each for log weight and SVI.

12.7 Regression Trees

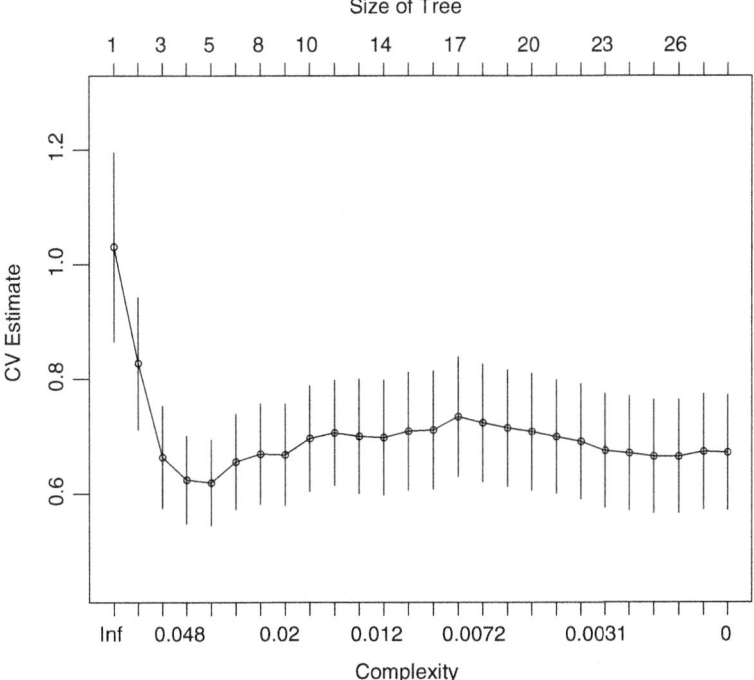

Fig. 12.10 Cross-validation score versus complexity, as measured by tree size (top axis) and improvement in R^2 (bottom axis), for the prostate cancer data

The final model is

$$\widehat{f}(\boldsymbol{x}) = \sum_{j=1}^{5} \widehat{\beta}_j h_j(\boldsymbol{x}),$$

where the numerical values of $\widehat{\beta}_j$ are given in Fig. 12.11 and

$h_1(\boldsymbol{x}) = I(\text{lcavol} < -0.4786)$

$h_2(\boldsymbol{x}) = I(\text{lcavol} \geq -0.4786) \times I(\text{lcavol} < 2.462) \times I(\text{lweight} < 3.689) \times I(\text{svi} < 0.5)$

$h_3(\boldsymbol{x}) = I(\text{lcavol} \geq -0.4786) \times I(\text{lcavol} < 2.462) \times I(\text{lweight} < 3.689) \times I(\text{svi} > 0.5)$

$h_4(\boldsymbol{x}) = I(\text{lcavol} \geq -0.4786) \times I(\text{lcavol} < 2.462) \times I(\text{lweight} \geq 3.689)$

$h_5(\boldsymbol{x}) = I(\text{lcavol} \geq 2.462).$

In terms of assigning a prediction to a new observation with covariates $\boldsymbol{x}$, we simply read down the tree in Fig. 12.11.

Fig. 12.11 Hierarchical regression tree for the prostate cancer data. For each leaf we give the estimated mean response and the number of observations

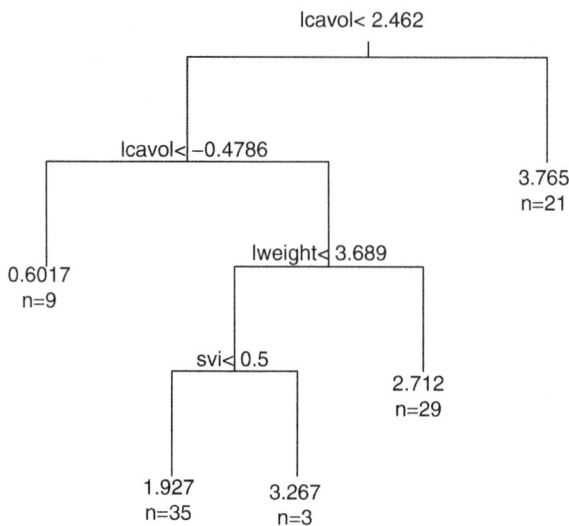

12.7.2 Multiple Adaptive Regression Splines

We briefly describe the multiple adaptive regression splines (MARS) algorithm that combines stepwise linear regression with a spline/tree model; MARS was introduced in Friedman (1991). MARS overcomes the discreteness of the regression trees fitted model by using piecewise linear basis functions of the form $(x_j - t)_+$ and $(t - x_j)_+$ for $j = 1, \ldots, k$; these are known as a *reflected pair*. Here, x_j refers to a generic covariate, and t to an observed value of that covariate. Hence, we have a pair of linear truncated line segments, which we have already seen used as building blocks for splines in Sect. 11.2.1. The collection of basis functions is

$$\{ (x_l - t)_+, (t - x_l)_+,\ t \in \{x_{1l}, \ldots, x_{nl}\},\ l = 1, \ldots, k \}. \tag{12.20}$$

If all of the covariates are distinct, there are $2nk$ basis functions in total.

The model is of the form

$$f(\boldsymbol{x}) = \beta_0 + \sum_{j=1}^{J} \beta_j h_j(\boldsymbol{x})$$

where each $h_j(\boldsymbol{x})$ is a particular reflected pair from the collection (12.20) or a product of two or more pairs. To select basis functions, forward selection is used (Sect. 4.8.1). At a particular step suppose we have functions $h_l(\boldsymbol{x}), l = 1, \ldots, L$ in the current model. We then add the term of the form

$$\widehat{\beta}_{L+1} h_l(\boldsymbol{x}) \times (x_{l'} - t)_+ + \widehat{\beta}_{L+2} h_l(\boldsymbol{x}) \times (t - x_{l'})_+$$

12.7 Regression Trees

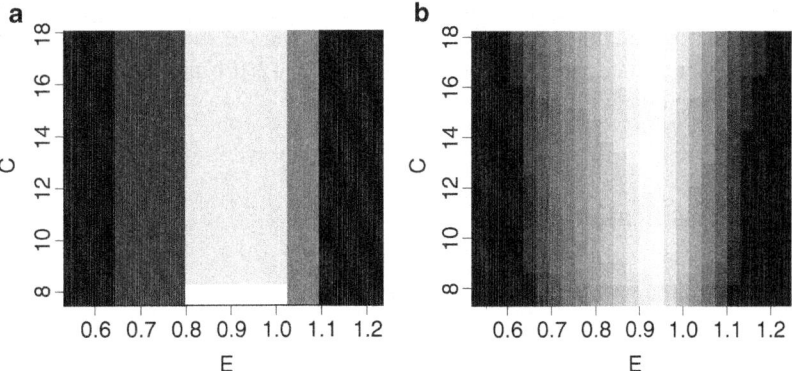

Fig. 12.12 Image plots of the fitted surfaces for the ethanol data obtained from (**a**) a pruned regression tree and (**b**) the MARS algorithm

that gives the largest decrease in the residual sum of squares. As with regression trees the process is continued until some preset maximum number of terms are present. This typically gives overfitting, and so backward elimination (Sect. 4.8.1) is used to reduce the size of the model by removing one by one the term that gives the smallest increase in the residuals sum of squares. Note that whereas in the forward direction the terms are added in pairs, in the backward direction, single terms can be removed. The balance between the size of the model and the closeness of the predictions to the observations is decided upon using generalized cross-validation (recall that generalized cross-validation requires less computation than ordinary cross-validation, Sect. 10.6.3).

Like regression trees, MARS is effectively performing variable selection. The model and parameter estimates produced by MARS are also quite interpretable (unlike boosting and random forests, which we will meet shortly) though inference, as with regression trees, is not straightforward.

MARS is particularly appealing as the dimensionality of the covariate space increases since, as we saw in Sect. 12.3.3, the use of tensor products of splines is prohibitive in higher dimensions, as the number of bases explodes. MARS avoids this problem by adaptively choosing bases and then carrying out a kind of pruning. The manner in which the terms are added has a flavor of following the hierarchy principle (Sect. 4.8) since interaction terms are added on top of the main effects. A more detailed discussion of MARS can be found in Hastie et al. (2009, Sect. 9.4).

Example: Ethanol Data

We applied both regression trees and the MARS approach to the ethanol data. The pruned regression tree had five splits with four involving the E variable. The resultant fitted surface is shown in Fig. 12.12(a) with the discreteness being apparent and undesirable. Applying the MARS method to the ethanol data results in six bases in the model (in addition to the intercept), with four of the six involving the E variable. The resultant fitted surface is shown in Fig. 12.12b and is far more visually appealing than the regression tree surface.

12.8 Classification

The classification problem is to predict the class of a response, given covariates x, from the set $\{0, 1, \ldots, K-1\}$. The true outcome is denoted Y and the classification $\widehat{Y} = g(x)$, where $g(\cdot)$ is the classifier. There are many approaches to classification, and we will only scratch the surface in this section. More extensive treatments are referenced in Sect. 12.10.

We distinguish two broad approaches. In the first, we fit (or *train*) a model $Y \mid x$, with, for example, $E[Y \mid x] = f(x)$ for a class of functions $f(\cdot)$. A classification is then made on the basis of the fitted model. The spline and kernel generalized linear model methods discussed in Sect. 11.5 are clearly applicable, if we model the data as multinomial. For example, logistic smoothers may be used in the binary case. The fitted values $\widehat{f}(x)$ can be simply converted to classifications $\widehat{g}(x)$, for example, using the Bayes classifier that assigns an observation to the class with the highest probability (Sect. 10.3.2).

In the second approach, we reverse the conditioning and model $X \mid y$. Suppose initially that for class k the distribution of x is known with the prior probabilities of class k being π_k, $k = 0, 1, \ldots, K-1$. Then the posterior probability that a case with covariates x is of class k is

$$\Pr(Y = k \mid x) = \frac{p_k(x) \times \pi_k}{\sum_{l=0}^{K-1} p_l(x) \times \pi_l}, \tag{12.21}$$

where $p_k(x)$ is the distribution of x for class k. Given class probabilities, we wish to decide upon a classification. Minimization of the expected prediction error (which is the expected loss with equal misclassification losses) gives the classifier that assigns the class that maximizes the posterior probability (Sect. 10.4.2).

We may draw an analogy with Bayes model selection, as described in Sect. 4.3.1. For simplicity, suppose we have to decide between just two actions: in the model selection context, two models, and in the classification context, two classes. In the former, model M_1 is preferred if

$$\frac{\Pr(M_1 \mid y)}{\Pr(M_0 \mid y)} = \frac{p(y \mid M_1)}{p(y \mid M_0)} \times \frac{\pi_1}{\pi_0} > \frac{L_{\text{I}}}{L_{\text{II}}},$$

12.8 Classification

where L_I and L_II are the losses associated with type I and type II errors and π_0 and π_1 are the prior probabilities of M_0 and M_1. In the classification context, we classify to class 1 if

$$\frac{\Pr(Y=1\mid x)}{\Pr(Y=0\mid x)} = \frac{p(x\mid Y=1)}{p(x\mid Y=0)} \times \frac{\pi_1}{\pi_0} > \frac{L(0,1)}{L(1,0)}, \qquad (12.22)$$

where $L(0,1)$ is the loss associated with assigning $g(x) = 1$ when the truth is $Y = 0$ and $L(1,0)$ is the loss associated with assigning $g(x) = 0$ when the truth is $Y = 1$ and π_0 and π_1 are the prior probabilities of $Y = 0$ and $Y = 1$ (Table 10.1).

Now that we have briefly described the two basic approaches, we outline the structure of this section. In Sect. 12.8.1, we briefly describe a multinomial version of the logistic model that may be used for more than $K = 2$ categories. We then proceed to describe two methods for modeling $p(X \mid y)$, linear and quadratic discriminant analysis in Sect. 12.8.2 and kernel density estimation in Sect. 12.8.3. The former is a parametric method based on normal distributions for the distribution of $X \mid y$, and the latter is nonparametric. Turning to the approach of directly modeling $Y \mid x$, we describe classification trees, bagging, and random forests in, respectively, Sects. 12.8.4–12.8.6.

12.8.1 Logistic Models with K Classes

We describe extensions to logistic regression modeling when $K > 2$, and the categories are nominal, that is, have no ordering. For K classes we may specify the model in terms of $K - 1$ odds where, for simplicity, we assume univariate x:

$$\frac{\Pr(Y=k\mid x)}{\Pr(Y=K-1\mid x)} = \exp(\beta_{0k} + \beta_{1k}x), \qquad (12.23)$$

to give

$$\Pr(Y=k\mid x) = \frac{\exp(\beta_{0k} + \beta_{1k}x)}{1 + \sum_{l=0}^{K-2} \exp(\beta_{0l} + \beta_{1l}x)}, \quad k = 0, \ldots, K-2, \qquad (12.24)$$

with $\Pr(Y = K-1 \mid x) = 1 - \sum_{k=0}^{K-2} \Pr(Y = k \mid x)$ (Exercise 12.3). The use of the last category as reference is arbitrary, and the particular category chosen makes no difference for inference. If we do wish to interpret the parameters, then $\exp(\beta_{0k})$ is the baseline probability of $Y = k$, relative to the probability of the final category, $Y = K - 1$, so that we have a specific generalization of odds. The parameter $\exp(\beta_{1k})$ is the odds ratio that gives the multiplicative change associated with a one-unit increase in x in the odds of response k relative to the odds of response

$K-1$. We emphasize that in a classification (prediction) setting, we will often have little interest in the model coefficients.

In terms of nonparametric modeling, one may model the collection of $K-1$ logits as smooth functions, along with a multinomial likelihood. For example, a simple model is of the form

$$\log\left[\frac{\Pr(Y=k \mid x)}{\Pr(Y=K-1 \mid x)}\right] = f_k(x)$$

with smoothers (such as splines or local polynomials) $f_k(\cdot)$, $k = 0, \ldots, K-2$. Yee and Wild (1996) describe how GAMs can be extended to this situation using penalized spline models and also describe the extension to ordered classes.

12.8.2 *Linear and Quadratic Discriminant Analysis*

If we wish to follow the approach summarized in (12.21), then a key element is clearly the specification of the distribution of the covariates for each of the different classes. In this section we assume these distributions are multivariate normal. In a slight change of notation from previous sections, we assume the dimensionality of x is p. We begin by assuming that $X \mid y = k \sim N_p(\boldsymbol{\mu}_k, \boldsymbol{\Sigma})$ so that the $p \times p$ covariance matrix is common to all classes. The within-class distribution of covariates is therefore

$$p_k(\boldsymbol{x}) = (2\pi)^{-p/2} |\boldsymbol{\Sigma}|^{-1/2} \exp\left[-\frac{1}{2}(\boldsymbol{x}-\boldsymbol{\mu}_k)^{\mathsf{T}} \boldsymbol{\Sigma}^{-1}(\boldsymbol{x}-\boldsymbol{\mu}_k)\right],$$

for $k = 0, 1, \ldots, K-1$. From (12.21) we see that maximizing $\Pr(Y=k \mid \boldsymbol{x})$ over k is equivalent to minimizing $-\log \Pr(Y=k \mid \boldsymbol{x})$, i.e., minimizing

$$(\boldsymbol{x}-\boldsymbol{\mu}_k)^{\mathsf{T}} \boldsymbol{\Sigma}^{-1} (\boldsymbol{x}-\boldsymbol{\mu}_k) - 2\log \pi_k, \qquad (12.25)$$

where the first term is the Mahalanobis distance (Malahanobis 1936) between $\boldsymbol{x}$ and $\boldsymbol{\mu}_k$. If the prior is uniform over $0, 1, \ldots, K-1$, then we pick the class that minimizes the within-class sum of squares. Expanding the square in (12.25), it is clear that the term $\boldsymbol{x}^{\mathsf{T}} \boldsymbol{\Sigma}^{-1} \boldsymbol{x}$, which depends on $\boldsymbol{\Sigma}$ and not k, can be ignored. Consequently, we see that the above rule is equivalent to picking, for fixed $\boldsymbol{x}$, the class k that minimizes

$$a_k + \boldsymbol{x}^{\mathsf{T}} \boldsymbol{b}_k \qquad (12.26)$$

where

$$a_k = \boldsymbol{\mu}_k^{\mathsf{T}} \boldsymbol{\Sigma}^{-1} \boldsymbol{\mu}_k - 2\log \pi_k$$
$$\boldsymbol{b}_k = -2\boldsymbol{\Sigma}^{-1} \boldsymbol{\mu}_k$$

12.8 Classification

for $k = 0, \ldots, K-1$. Hence, we have a set of K linear planes, and for any $\boldsymbol{x}$, we pick the class k whose plane at that $\boldsymbol{x}$ is a minimum. Said another way, we have a decision boundary that is linear in $\boldsymbol{x}$, and the method is therefore known as *linear discriminant analysis* (LDA). The decision boundary between classes k and l, that is, where $\Pr(Y = k \mid \boldsymbol{x}) = \Pr(Y = l \mid \boldsymbol{x})$ is linear in $\boldsymbol{x}$ and the regions in $\mathbb{R}^p$ that are classified according to the different classes, $0, 1, \ldots, K-1$, are separated by hyperplanes. An example of the linear boundaries, in the case of univariate x, is given in Fig. 12.15.

The parameters of the normal distributions are, of course, unknown and may be estimated from the training data via MLE:

$$\widehat{\pi}_k = \frac{n_k}{n} \tag{12.27}$$

$$\widehat{\boldsymbol{\mu}}_k = \frac{1}{n_k} \sum_{i:y_i=k} \boldsymbol{x}_i \tag{12.28}$$

$$\widehat{\boldsymbol{\Sigma}} = \frac{1}{n-K} \sum_{k=0}^{K-1} \sum_{i:y_i=k} (\boldsymbol{x}_i - \widehat{\boldsymbol{\mu}}_k)(\boldsymbol{x}_i - \widehat{\boldsymbol{\mu}}_k)^\mathsf{T} \tag{12.29}$$

where n_k is the number of observations with $Y = k$, $k = 0, 1, \ldots, K-1$, and $n = \sum_k n_k$. To estimate π_k from the data, as in (12.27), depends on a random sample of observations having been taken. Otherwise, we might use prior information to specify class probabilities.

We now relax the assumption that the covariance matrices are equal. In this case, we pick k that minimizes

$$\log |\boldsymbol{\Sigma}_k| + (\boldsymbol{x} - \boldsymbol{\mu}_k)^\mathsf{T} \boldsymbol{\Sigma}_k^{-1} (\boldsymbol{x} - \boldsymbol{\mu}_k) - 2 \log \pi_k,$$

as shown by Smith (1947). Expanding the quadratic form gives a term $\boldsymbol{x}^\mathsf{T} \boldsymbol{\Sigma}_k^{-1} \boldsymbol{x}$ which cannot be ignored since the variance–covariance matrix depends on k; hence, the method is known as *quadratic discriminant analysis* (QDA). Again, we need to estimate the parameters, with the estimators for π_k and $\boldsymbol{\mu}_k$ corresponding to (12.27) and (12.28) with

$$\widehat{\boldsymbol{\Sigma}}_k = \frac{1}{n_k} \sum_{i:y_i=k} (\boldsymbol{x}_i - \widehat{\boldsymbol{\mu}}_k)(\boldsymbol{x}_i - \widehat{\boldsymbol{\mu}}_k)^\mathsf{T}$$

for $k = 0, 1, \ldots, K-1$.

We now examine the connection between logistic regression and LDA in the case of two classes, that is, $K = 2$. Under LDA we define the log odds function

$$L(\boldsymbol{x}) = \log \left[\frac{\Pr(Y=1 \mid \boldsymbol{x})}{\Pr(Y=0 \mid \boldsymbol{x})} \right]$$

$$= \underbrace{\log\left(\frac{\pi_1}{\pi_0}\right) - \frac{1}{2}(\boldsymbol{\mu}_1 + \boldsymbol{\mu}_0)^\mathsf{T} \boldsymbol{\Sigma}^{-1}(\boldsymbol{\mu}_1 - \boldsymbol{\mu}_0)}_{\alpha_0} + \underbrace{\boldsymbol{\Sigma}^{-1}(\boldsymbol{\mu}_1 - \boldsymbol{\mu}_0)^\mathsf{T}}_{\alpha_1} \boldsymbol{x},$$

so that the function upon which classifications are based has the linear form $\alpha_0 + \alpha_1 x$. If the losses associated with the two types of errors are equal, then we assign a case with covariates x to class $\widehat{Y} = 1$ if $L(x) > 0$, see (12.22). Notice that x only enter through the term $\Sigma^{-1}(\mu_1 - \mu_0)^{\mathrm{T}}$. Under (linear) logistic regression,

$$\log\left[\frac{\Pr(Y = 1 \mid x)}{\Pr(Y = 0 \mid x)}\right] = \beta_0 + \beta_1 x.$$

Consequently, the rules are both linear in x, but differ in the manner by which the parameters are estimated. In general, we may factor the distribution of x_i, y_i in two ways, which correspond to the two approaches to classification that we have highlighted. Modeling the x distributions corresponds to the factorization

$$\prod_{i=1}^{n} p(x_i, y_i) = \prod_{i=1}^{n} p(x_i \mid y_i) \prod_{i=1}^{n} p(y_i).$$

For example, under LDA, it is assumed that $p(x_i \mid y_i)$ is normal, and then $\prod_{i=1}^{n} p(x_i, y_i)$ is maximized with respect to the parameters of the normals. In contrast, under linear logistic regression the factorization is

$$\prod_{i=1}^{n} p(x_i, y_i) = \prod_{i=1}^{n} p(y_i \mid x_i) \prod_{i=1}^{n} p(x_i),$$

and we maximize the first term, under the assumption of a linear logistic model, while ignoring the second term. Logistic regression therefore leaves the marginal distribution $p(x)$ unspecified, and so the method is more nonparametric than LDA, which is usually an advantage. Asymptotically, there is a 30% efficiency loss when the data are truly multivariate normal but are analyzed via the logistic regression formulation (Efron 1975).

The original derivation of LDA in Fisher (1936) was somewhat different to the presentation given above and was carried out for $K = 2$. Specifically, a linear combination $a^{\mathrm{T}} x$ was sought that separated (or discriminated between) the classes as much as possible to, "maximize the ratio of the difference between the specific means to the standard deviations", Fisher (1936, p. 466). This difference is maximized by taking $a \propto \Sigma^{-1}(\mu_1 - \mu_0)^{\mathrm{T}}$, an expression we have already seen, see Exercise 12.4 for further detail.

Example: Bronchopulmonary Dysplasia

We return to the BPD example and classify individuals on the basis of their birth weight using linear and quadratic logistic models, and linear and quadratic discriminant analysis. We emphasize that these rules are relevant to the sampled

12.8 Classification

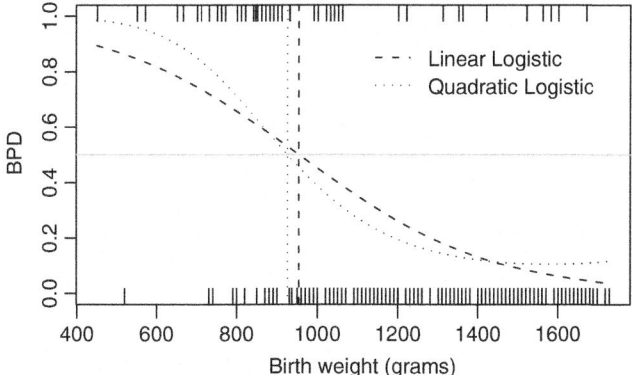

Fig. 12.13 Logistic linear and quadratic fits to the BPD and birth weight data. The *horizontal line* indicates $p(x) = 0.5$, and the two *vertical lines* correspond to the linear and quadratic logistic decision rules

population of children for whom data were collected and not to the general populations of newborn babies. This is important since this is far from a random sample, and so the estimate of the probability of BPD (the outcome of interest) is a serious overestimate. In general this example is illustrative of techniques rather than substantively of interest, not least because of the lack of other covariates that one would wish to base a classification rule upon (including medications used by the mother).

Figure 12.13 shows the linear and quadratic logistic regression fits as a function of x. The horizontal $p(x) = 0.5$ line is drawn in gray, and we see little difference between the classification rules based on the two logistic models. The birth weight thresholds below/above which individuals would be classified as BPD/not BPD, for the linear and quadratic models, are 954 g and 926 g, respectively. The fitted curves are quite different in the tails, however. In particular, we see that the quadratic curve seems to move toward a nonzero probability for higher birth weights, a feature we have seen in other analyses. For example, the penalized cubic splines and local likelihood fits shown in Fig. 11.16 display this behavior. The similarity between the classification rules, even though the models are quite different, illustrates that prediction is a different enterprise to conventional modeling.

Turning to a discriminant analysis approach, Figures 12.14(a) and (b) display normal QQ plots (Sect. 5.11.3) of the birth weights for the BPD = 0 and BPD = 1 groups, respectively. The babies with BPD in particular have birth weights which do not appear normal.

The parameter estimates for LDA and QDA are

$$\widehat{\pi}_0 = 0.66$$
$$\widehat{\mu}_0 = 1{,}287, \quad \widehat{\mu}_1 = 953$$
$$\widehat{\Sigma} = 76{,}309, \quad \widehat{\Sigma}_0 = 77{,}147, \quad \widehat{\Sigma}_1 = 74{,}677.$$

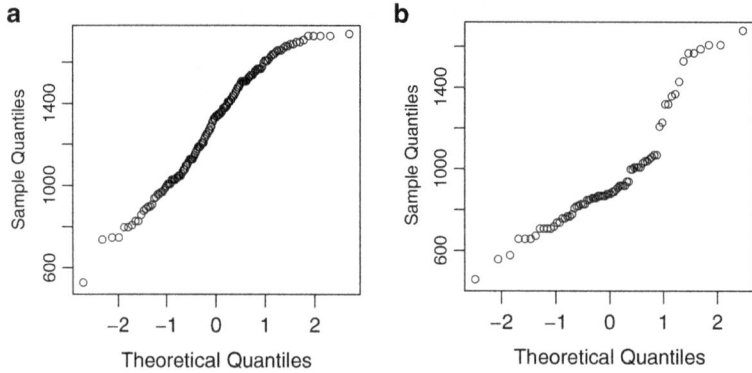

Fig. 12.14 (a) Normal QQ plot of birth weights for the BPD=0 group, (b) normal QQ plot of birth weights for the BPD=1 group

Crucially, we see that the variances within the two groups (not BPD/BPD) are very similar so that we would expect LDA and QDA to give very similar answers in this example. This is indeed the case as the linear and quadratic discriminant boundaries are at birth weights of 970 and 972 g, respectively.

Figure 12.15 gives the lines that are proportional to $-2 \log \Pr(Y = k \mid x)$ for $k = 0, 1$ (with x the birth weight here), that is the lines given by (12.26). The crossover point gives the birth weight at which we switch from a classification of $\widehat{Y} = 1$ to a classification of $\widehat{Y} = 0$. Figure 12.16 shows the fitted normals under the model with differing variances.

There are only small differences in this example, because the within-class birth weights are not too far from normal, the variances in each group are approximately equal, and the sample sizes are relatively large.

12.8.3 Kernel Density Estimation and Classification

We now describe a nonparametric method for classification based on kernel density estimation (Sect. 11.3.2). With estimated densities $\widehat{p}_k(\boldsymbol{x})$, the classification is

$$\Pr(Y = k \mid \boldsymbol{x}) = \frac{\widehat{p}_k(\boldsymbol{x}) \times \pi_k}{\sum_{l=0}^{K-1} \widehat{p}_l(\boldsymbol{x}) \times \pi_l}.$$

When classification is the goal, then effort should be concentrated estimating the class probabilities $\Pr(Y = k \mid \boldsymbol{x})$ accurately near the decision boundary. As we saw in Sect. 11.3.2, the crucial aspect of kernel density estimation is an appropriate choice of smoothing parameter with the form of the kernel being usually unimportant.

Kernel density estimation is hard when the dimensionality p of $\boldsymbol{x}$ is large. The *naive Bayes* method assumes that, given a class $Y = l$, the random variables $X_1, \ldots, X_p$ are independent to give joint distribution

12.8 Classification

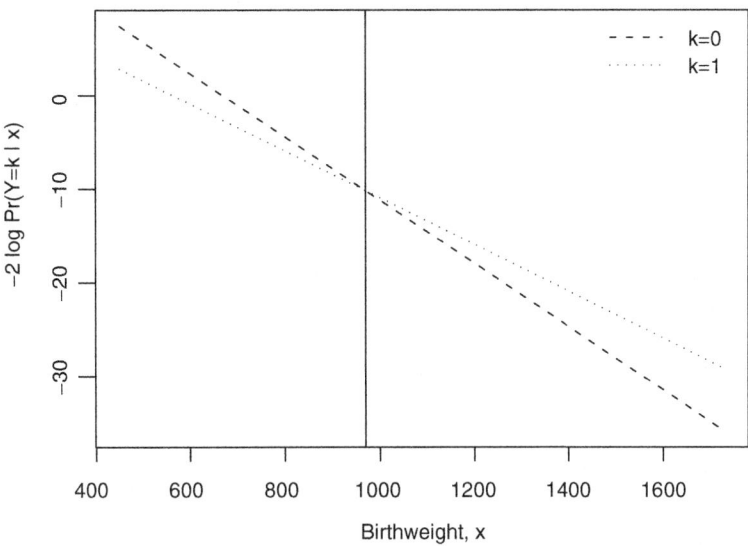

Fig. 12.15 Linear discriminant boundaries (12.26) for the two BPD groups with $k = 0/1$ representing no disease/disease, the rule is based on whichever of the two lines is lowest. The *vertical line* is the decision boundary so that to the left of this line the classification is to disease and to the right, to no disease

$$p_l(\boldsymbol{x}) = \prod_{j=1}^{p} p_{lj}(x_j). \tag{12.30}$$

The naive Bayes method is clearly based on heroic assumptions but is also clearly simple to apply since one need only compute p univariate kernel density estimates for each class. An additional advantage of the method is that elements of $\boldsymbol{x}$ that are discrete may be estimated using histograms, allowing the simple combination of continuous and discrete variables. Taking the logit transform of (12.30), as described in Sect. 12.8.1, we obtain

$$\log\left[\frac{\Pr(Y = l \mid \boldsymbol{x})}{\Pr(Y = K-1 \mid \boldsymbol{x})}\right] = \log\left[\frac{\pi_l p_l(\boldsymbol{x})}{\pi_{K-1} p_{K-1}(\boldsymbol{x})}\right]$$

$$= \log\left[\frac{\pi_l \prod_{j=1}^{p} p_{lj}(x_j)}{\pi_{K-1} \prod_{j=1}^{p} p_{K-1,j}(x_j)}\right]$$

$$= \log\left[\frac{\pi_l}{\pi_{K-1}}\right] + \sum_{j=1}^{p} \log\left[\frac{p_{lj}(x_j)}{p_{l,K-1}(x_j)}\right]$$

$$= \beta_l + \sum_{j=1}^{j} f_{lj}(x_j)$$

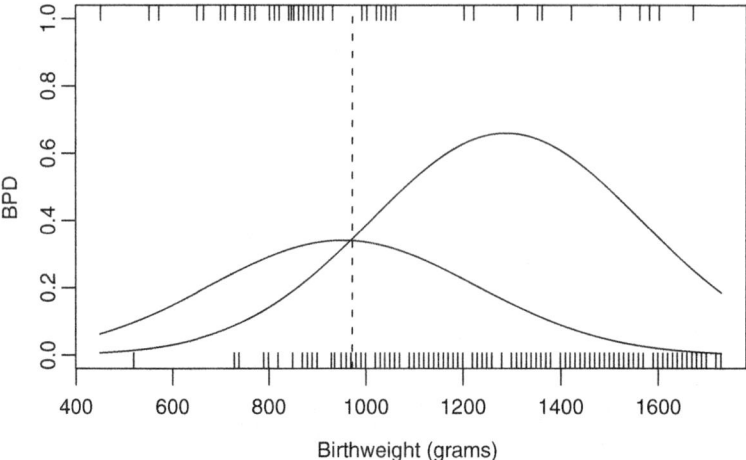

Fig. 12.16 For the BPD data, fitted normal distributions with different variances for each of the two classes (i.e., $\Sigma_0 \neq \Sigma_1$) and with areas proportional to π_1 and π_0 (for the left and right normals, respectively); the *dashed line* represents the quadratic discrimination rule and corresponds to the crossover point of the two densities. The *dashes* on the top and bottom axes represent the observed birth weights for those babies with and without BPD

which has the form of a GAM (Sect. 12.2) and provides an alternative method of estimation to kernel density estimation under the assumption of independence of elements of x in different classes. The same form of decision rule arising via two separate estimation approaches is similar to that seen when comparing LDA and logistic regression.

Example: Bronchopulmonary Dysplasia

We illustrate the use of kernel density estimation in a one-dimensional setting using the birth weight/BPD example. The choice of smoothing parameter λ is crucial, and we present three different analyses based on different methods. We let λ_k represent the smoothing parameter under classification k, $k = 0, 1$.

First, we use the optimal λ_k, given by (11.36), that arises under the assumption that each of the densities is normal. This leads to estimates of $\widehat{\lambda}_0 = 108$ and $\widehat{\lambda}_1 = 122$. Figure 12.17(a) shows the estimated densities for both classes. The non-normality of birth weights for the $k = 0$ class, that was previously seen in Fig. 12.14(a), is evident. The log ratio,

$$\log\left[\frac{\Pr(Y = 1 \mid x)}{\Pr(Y = 0 \mid x)}\right] = \log\left[\frac{p_1(x)}{p_0(x)}\right] + \log\left[\frac{\pi_1}{\pi_0}\right],$$

is shown in panel (b) and gives a decision threshold of $x = 1{,}162\,\mathrm{g}$.

12.8 Classification

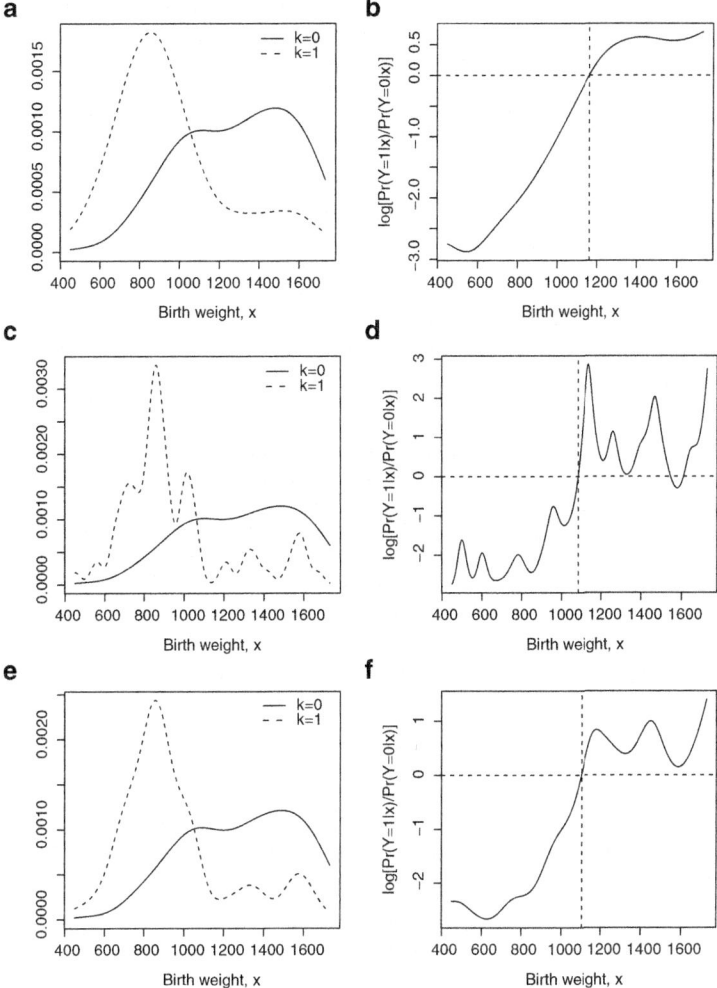

Fig. 12.17 The left column shows kernel density estimates for the birth weights under the two classes (with $k = 0/1$ corresponding to no BPD/BPD) and the right column the log of the ratio $\Pr(Y = 1 \mid x)/\Pr(Y = 0 \mid x)$ with the *vertical line* indicating the decision threshold. The three rows correspond to choosing the smoothing parameters based on normality of the underlying densities, cross-validation, and upon a plug-in method

We next use cross-validation to pick the smoothing parameters (as described in Sect. 11.3.2). The resultant estimates are $\widehat{\lambda}_0 = 103$ and $\widehat{\lambda}_1 = 28$ with the resultant density estimates plotted in Fig. 12.17c. The estimate for the disease group ($k = 1$) is very unsatisfactory, though the decision boundary (as shown in Fig. 12.17(d)) is very similar to the previous approach, with a threshold of $x = 1,083$ g.

Finally, we use the plug-in method of Sheather and Jones (1991) to pick the smoothing parameters, giving $\widehat{\lambda}_0 = 97$ and $\widehat{\lambda}_1 = 59$. The resultant density estimates are plotted in Fig. 12.17(e). The birth weight threshold is $x = 1{,}102$ g under these smoothing parameters with the log ratio, shown in Fig. 12.17(f), being more smooth than the cross-validation version but less smooth than the normal version.

12.8.4 Classification Trees

In this section we consider how the regression trees described in Sect. 12.7 can be used in a classification context. Classification and regression trees, or CART, has become a generic term to describe the use of regression trees and classification trees.

In the classification setting, the criteria for splitting nodes needs refinement. For regression, we used the residual sum of squares within each node as the impurity measure $S_j(T)$, defined in (12.18). This measure was then used within the cost complexity criterion, (12.19), to give a penalized sum of squares function to minimize. A sum of squares is not suitable for classification, however (for a variety of reasons, including the nonconstant variance aspect of discrete outcomes). In order to define an impurity measure, we need to specify, for each of the J terminal nodes (leaves), a probability distribution over the K outcomes. Node j represents a region R_j with n_j observations, and the obvious estimate of the probability of observing class k at node j is

$$\widehat{p}_{jk} = \frac{1}{n_j} \sum_{i:x_i \in R_j} I(y_i = k),$$

which is simply the proportion of class k observations in node j, for $k = 0, \ldots, K-1, j = 1, \ldots, J$. We may classify the observations in node j to class

$$k(j) = \arg\max_k \widehat{p}_{jk},$$

the majority class (Bayes rule) at node j. Given a set of classification probabilities, we turn to defining a measure of impurity. In a regression setting, we wished to find regions of the x space within which the response was relatively constant, and the impurity measure in this setting was the residual sum of squares about the mean of the terminal node in question. By analogy, we would like the leaves in a classification setting to contain observations of the same class. An impurity measure should therefore be 0 if all the probability at a node is concentrated on one class, that is, if $\widehat{p}_{jk} = 1$ for some k, and the measure should achieve a maximum if the probability is spread uniformly across the classes, that is, if $\widehat{p}_{jk} = 1/K$ for $k = 0, \ldots, K-1$. Three different impurity measures are discussed by Hastie et al. (2009, Sect. 9.2.3).

12.8 Classification

The *misclassification error* of node j is the proportions of observations at node j that are misclassified:

$$\frac{1}{n_j} \sum_{i:x_i \in R_j} I[y_i \neq k(j)] = 1 - \widehat{p}_{k(j),j}.$$

The *Gini index* associated with node j is

$$\sum_{k \neq k'} \widehat{p}_{jk}\widehat{p}_{jk'} = \sum_{k=0}^{K-1} \widehat{p}_{jk}(1 - \widehat{p}_{jk}).$$

The Gini index has an interesting interpretation. Instead of assigning observations to the majority class at a node, we could assign to class k with probability $\widehat{p}_{jk}$. With such an assignment, the training error of the rule at the node is

$$\sum_{k=0}^{K-1} \Pr(\text{Truth} = k) \times \Pr(\text{Classify} \neq k) = \sum_{k=0}^{K-1} \widehat{p}_{jk}(1 - \widehat{p}_{jk}),$$

which is the Gini index. It may be better to use this than the misclassification error because it "has an element of look ahead" Ripley (1996, p. 327), that is, it considers the error in a hypothetical training dataset. The final measure is the *deviance*, which is just the negative log-likelihood of a multinomial:

$$-\sum_{k=0}^{K-1} \widehat{p}_{jk} \log \widehat{p}_{jk}.$$

This measure is also known as the entropy.[4] The deviance and Gini index are differentiable and hence more amenable to numerical optimization.

For two classes, let p_j be the proportion in the second class at node j, for $j = 1, \ldots, J$. In this case, the misclassification error, Gini index, and deviance measures are, respectively,

$$1 - \max(\widehat{p}_j, 1 - \widehat{p}_j)$$

$$2\widehat{p}_j(1 - \widehat{p}_j)$$

$$-\widehat{p}_j \log \widehat{p}_j - (1 - \widehat{p}_j)\log(1 - \widehat{p}_j)$$

The worst scenario is $\widehat{p} = 0.5$ since we have a 50:50 split of the two classes in the partition (and hence the greatest impurity). Figure 12.18 graphically compares the three measures, with the deviance scaled to pass through the same apex point as the other two measures.

---

[4]In statistical thermodynamics, the entropy of a system is the amount of uncertainty in that system, with the maximum entropy being associated with a uniform distribution over the states.

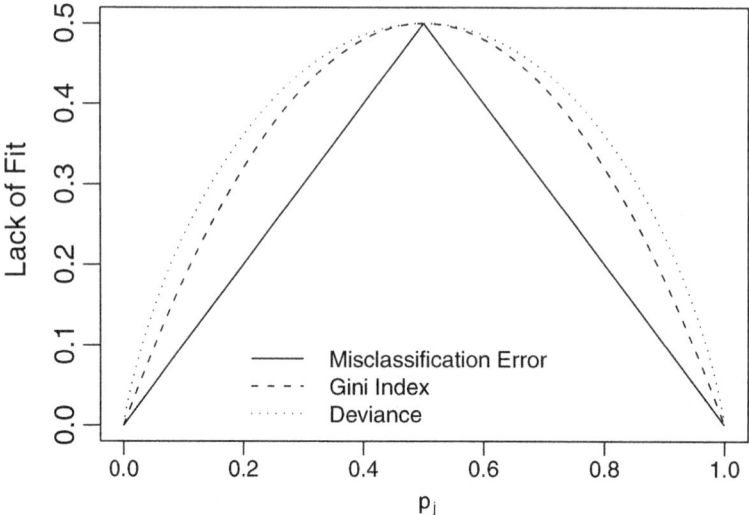

Fig. 12.18 Comparison of impurity measures for binary classification. The classes are labeled 0 and 1 and p_j is the proportion in the second class (i.e., $k = 1$) at node j

12.8.5 Bagging

We previously noted in Sect. 12.7.1 that regression trees can be unstable, in the sense that a small change in the learning data can induce a large change in the prediction/classification. Classification trees can also produce poor results when there exist heterogeneous terminal nodes, or highly correlated predictors.

Bootstrap aggregation or *bagging* (Breiman 1996) averages predictions over bootstrap samples (Sect. 2.7) in order to overcome the instability. The intuition is that idiosyncratic results produced by particular trees can be averaged away, resulting in more stable estimation. Although bagging is often implemented with regression or classification trees, it may be used with more general nonparametric techniques. To demonstrate the variability in tree construction; Figs. 12.19(a)–(c) show three pruned trees based on three bootstrap samples for the prostate cancer data. The first splits in (a) and (b) are on log cancer volume but at very different points, while in (c), the first split is on SVI. The variability across bootstrap samples is apparent.

As usual, let $[x_i, y_i]$, $i = 1, \ldots, n$ denote the data. The aim is to form a prediction, $f(x_0) = \mathrm{E}[Y \mid x_0]$ at a covariate value x_0. Bagging proceeds as follows:

1. Construct B bootstrap samples

$$[x_b^\star, y_b^\star] = \{x_{bi}^\star, y_{bi}^\star, i = 1, \ldots, n\},$$

12.8 Classification

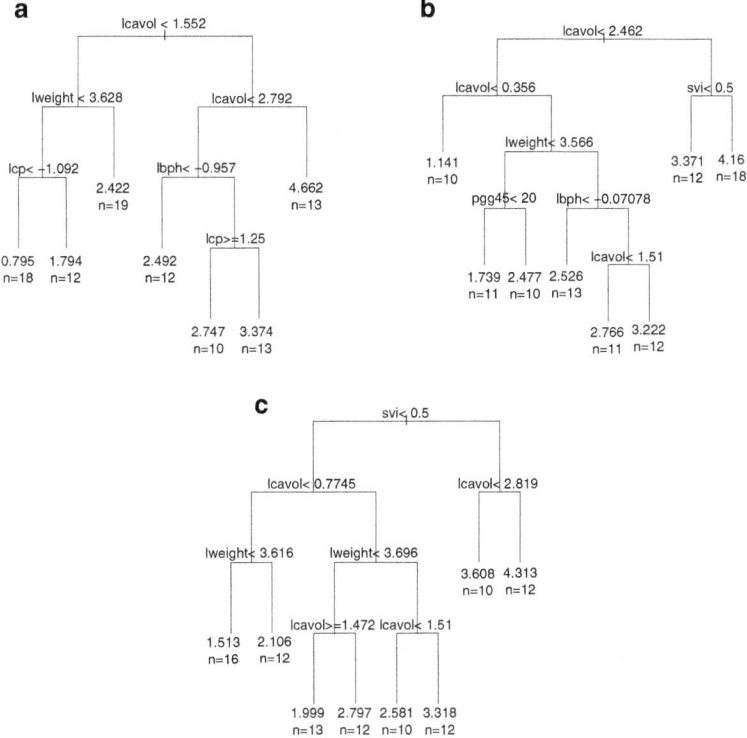

Fig. 12.19 Three pruned trees for the prostate cancer data, based on different bootstrap samples

for $b = 1, \ldots, B$. The bootstrap samples are formed by resampling cases (Sect. 2.7.2), that is, we sample with replacement from $[\boldsymbol{x}_i, y_i]$, $i = 1, \ldots, n$.

2. If the outcome is continuous (in which case, we might use regression trees for prediction), form the averaged prediction

$$\widehat{f}_{\text{B}}(\boldsymbol{x}_0) = \frac{1}{B} \sum_{b=1}^{B} \widehat{f}_b^{\star}(\boldsymbol{x}_0), \qquad (12.31)$$

where $\widehat{f}_b^{\star}(\boldsymbol{x})$ is the prediction constructed from the b-th bootstrap sample $[\boldsymbol{x}_b^{\star}, \boldsymbol{y}_b^{\star}]$, $b = 1, \ldots, B$. If classification is the aim and regression trees are constructed from the samples, one may take a majority vote over the B samples in order to assign a class label.

If tree methods are used, there is evidence that pruning should not be carried out (Bauer and Kohavi 1999). By not pruning, more complex models are fitted which reduces bias, and since bagging averages over many models the variance can be reduced also.

We examine the continuous case in greater detail. Expression (12.31) is a Monte Carlo estimate of the theoretical bagging estimate $E_{\widehat{P}}\left[\widehat{f}_B^\star(x_0)\right]$ where the expectation is with respect to sampling $[x^\star, y^\star]$ from $\widehat{P}$, which is the empirical distribution having probability $1/n$ at $[x_i, y_i]$, $i = 1, \ldots, n$.

We now examine the mean squared error in an idealized setting. Let P be the population from which $[y_i, x_i]$, $i = 1, \ldots, n$ are drawn. For analytical simplicity, suppose we can draw bootstrap samples from the population rather than the observed data. Let $\widehat{f}_\star(x_0)$ be a prediction at x_0, based on a sample from P, and

$$f_{\text{AGG}}(x_0) = E_P\left[\widehat{f}_\star(x_0)\right]$$

be the ideal bagging estimate which averages the estimator over samples from the population.

We consider a decomposition of the MSE of the prediction, in a regression setting, based on the single sample estimator, $\widehat{f}_\star(x_0)$, only:

$$E_P\left\{\left[Y - \widehat{f}_\star(x_0)\right]^2\right\} = E_P\left\{\left[Y - f_{\text{AGG}}(x_0) + f_{\text{AGG}}(x_0) - \widehat{f}_\star(x_0)\right]^2\right\}$$

$$= E_P\left\{[Y - f_{\text{AGG}}(x_0)]^2\right\} + E_P\left\{\left[\widehat{f}_\star(x_0) - f_{\text{AGG}}(x_0)\right]^2\right\}$$

$$+ 2E_P\left\{[Y - f_{\text{AGG}}(x_0)]\right\} E_P\left\{\left[\widehat{f}_\star(x_0) - f_{\text{AGG}}(x_0)\right]\right\}$$

$$= E_P\left\{[Y - f_{\text{AGG}}(x_0)]^2\right\} + E_P\left\{\left[\widehat{f}_\star(x_0) - f_{\text{AGG}}(x_0)\right]^2\right\}$$

(12.32)

$$\geq E_P\left\{[Y - f_{\text{AGG}}(x_0)]^2\right\}.$$

Hence, the MSE of idealized population averaging (which is the expression in the last line) never increases the MSE of an estimate from a single prediction. The second term in (12.33) is the variability of the estimator $\widehat{f}_\star(x_0)$ about its average. The above decomposition is relevant to a regression setting but is not valid for classification (0–1 loss), and bagging a bad classifier can make it even worse.

Bagging is an example of an *ensemble learning* method; another such method that we have already encountered is Bayesian model averaging (Sect. 3.6). The bagged estimate (12.31) will differ (in expectation) from the original estimate $\widehat{f}(x_0)$, only when $\widehat{f}(x_0)$ is a nonlinear function of the data. So, for example, bagged prediction estimates from spline and local polynomial models that produce linear smoothers will be the same as those from fitting a single model using the complete data.

The original motivation for bagging (Breiman 1996) was to reduce variance. However, bagging can also reduce (or increase!) bias (Bühlmann and Yu 2002). Bias may be reduced if the true function is smoothly varying and tree models are used (since the averaging of step functions will produce smoothing). If the true function is "jaggedy," bias can be introduced through averaging. As just noted, bagging was originally designed to reduce variance and works well in examples in which the data are "unstable," that is, in situations in which small changes in the data can cause large changes in the prediction. We give an example of a scenario in which bagging can increase the variance. Suppose there is an outlying point (in x space). This point may stabilize the fit when the model is fitted to the complete data, and if it is left out of a particular bootstrap sample, the fitted values may be much more variable for this sample.

To bag a tree-based classifier, we first grow a classification tree for each of the B bootstrap samples. Recall, we may have two different aims: reporting a classification or reporting a probability distribution over classes. Suppose we require a classification. The bagged estimate $\widehat{f}_B(x)$ is the K-vector: $[\widehat{p}_0(x), \ldots, \widehat{p}_{K-1}(x)]$ where $\widehat{p}_k(x)$ is the proportion of the B trees that predict class k, $k = 0, 1, \ldots, K-1$. The classification is the k that maximizes $p_k(x)$, that is, the Bayes rule. If we require the class-probability estimates, then we average the underlying functions that produce the classifications $g_b(x)$. We should not average the classifications. To illustrate why, consider a two class case. Each bootstrap sample may predict the 0 class with probability 0.51, and hence the classifier for each would be $\widehat{Y} = 0$, but we would not want to report the class probabilities as (1,0).

The simple interpretation of trees is lost through bagging, since a bagged tree is not a tree. For each tree, one may evaluate the test error on the "left-out" samples (i.e., those not selected in the bootstrap sample). On average, around 1/3 of the data do not appear in each bootstrap sample. These data are referred to as the "out-of-bag" (oob) estimate. These test estimates may be combined, removing the need for a test dataset.

Bagging takes the algorithmic approach to classification to another level beyond tree-based methods and was important historically as it was an intermittent step toward various other methods including random forests, which we describe next.

12.8.6 Random Forests

Random forests (Breiman 2001a) are a very popular and easily implemented technique that build on bagging by reducing the correlation between the multiple trees that are fitted to bootstrap samples of the data.

The random forest algorithm is as follows:

1. B bootstrap samples of size n are drawn, with replacement, from the original data.
2. Suppose there are p covariates, a number $m \ll p$ is specified, and at each node m variables are selected at random from the p available. The best split from these m is used to split the node.
3. Each tree is grown to be large, with no pruning. We emphasize that a different set of m covariates is selected at each split so the input variables are changing within each tree.

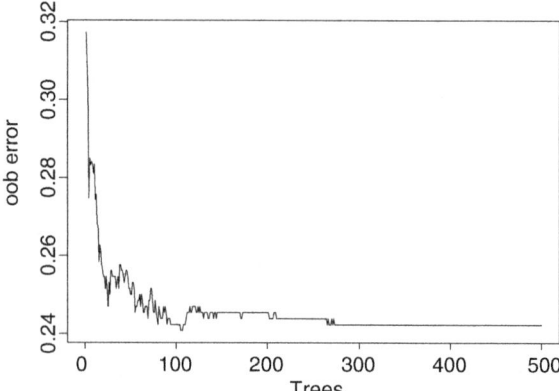

Fig. 12.20 Out-of-bag error rate as a function of the number of trees, for the outcome after head injury data

Once this process is completed, the output is a collection of B trees. As with bagging, in a regression context, the prediction may be taken to be the average of the fits, as in (12.31), while in a classification context the majority vote may be taken.

There are two conflicting aims when we consider the size of m. Increasing the correlation between any two trees in the forest increases the forest error rate, but the forest error rate decreases as the strength of each individual tree increases. Reducing m reduces both the correlation and the strength, while increasing m increases both. We heuristically explain why reducing the correlation between predictions leads to a lowering of the forest error rate. Suppose we wish to estimate a prediction at a value x_0 using an average of B predictions. Let $\widehat{f}_{\text{AVE}}(x_0) = \frac{1}{B}\sum_{b=1}^{B} \widehat{f}_b(x_0)$ and suppose that the predictions each have variance σ^2 and pairwise correlations ρ. Then it is straightforward to show that the variance of the average is

$$\text{var}(\widehat{f}_{\text{AVE}}(x_0)) = \frac{(1-\rho)}{B}\sigma^2 + \rho\sigma^2.$$

The first term decreases to zero as the number of predictor functions increases, while the second term is a function of the dependence between the functions. Hence, the closer the predictor functions are to independence, the lower the variance.

As with bagging, when the training set (i.e., the bootstrap sample) for the current tree is drawn by sampling with replacement, about 1/3 is left out of the sample, and these form the oob (Sect. 12.8.5). These set aside data are used to get a running unbiased estimate of the classification error, as trees are added to the forest. An example of such a plot is given in Fig. 12.20. A typical recommended value for m is the integer part of $\sqrt{p}$ in a classification setting, and the integer part of $p/3$ for regression (Hastie et al. 2009, Sect. 15.3) though these values should not be taken as written in stone, and some experimentation should be performed.

The concept of only taking a subset of variables seems totally alien statistically, since information is reduced, but the vital observation is that this produces classifiers that are close to being uncorrelated. This is a key difference with bagging, with

12.8 Classification

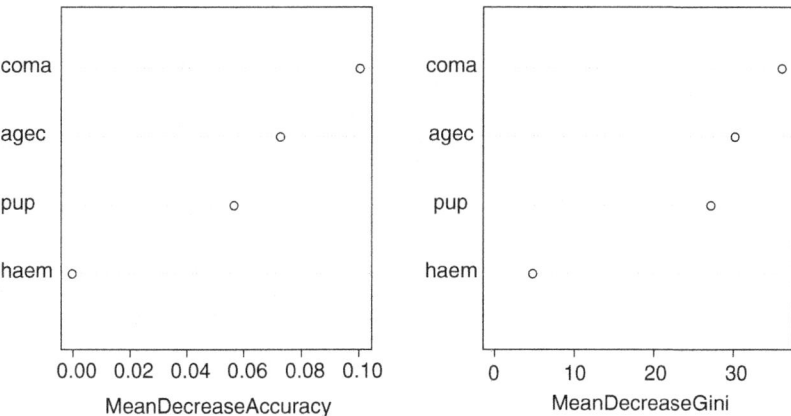

Fig. 12.21 Random forest variable importance for one split of the outcome after head injury data. The *left panel* shows the decrease in predictive ability (as measured by the misclassification error) when the variable is permuted and the *right panel* the decrease in the Gini index when the variable is not included in the classification

which the random forests method shares many similarities. By injecting randomness into the algorithm, through the random selection of covariates at each split, the constituent trees are more independent. Selecting a random set of m covariates also allows random forests to cope with the situation in which there are more covariates than observations (i.e., $n < p$).

As we have noted, random forests lose the relatively simple interpretation of tree-based methods. Although prediction is the objective of random forests, it may still often be of interest to see which of the variables are making contributions to the overall prediction (averaged over trees). If one is interested in gaining this insight into which predictors are performing well, then two measures of variable importance are popular. One approach is to obtain the decrease in the fitting measure each time a particular variable is used. The average of this decrease over all trees can then be calculated with important variables having large decreases. For regression the fitting measure is the residual sum of squares, and for classification it is often the Gini index (Sect. 12.7). The right panel of Fig. 12.21 shows this average. This measure seems intuitively reasonable, but, as discussed by Berk (2008, Sect. 5.6.1), it has a number of drawbacks. First, reductions in the fitting criteria do not immediately translate into improvements in prediction. Second, the decreases are calculated using the data that were used to build the model and not from test data. Finally, there is no absolute scale on which to judge reductions. As an alternative, one may, for each predictor, calculate the error rate using a random permutation of the predictor. The difference between the two is then averaged over trees. This second approach is more akin to setting a coefficient to zero in a regression model and then assessing the reduction in predictive power (say in a test dataset). The importance of each variable is assessed by creating trees using random permutations of the values of the variable, rather than the variable itself. The predictive power is then assessed using the "true" variable in the oob data compared to the permuted version. We give more detail in a classification setting and assuming we measure the predictive power in terms of the misclassification error rate. Suppose that, for the bth tree, this rate is v_b when using all of the true variables and is v_{bj} when the jth variable is shuffled. Then, the change in the predictive power is summarized as

$$\frac{1}{B}\sum_{b=1}^{B}(v_{bj} - v_b). \qquad (12.33)$$

Note that if the variable is not useful predictively, then this measure might by chance be negative. The left panel of Fig. 12.21 shows the decrease in predictive power for each of four variables.

Example: Outcome After Head Injury

We now compare classification methods on the head injury data described in Sect. 7.2.1. The binary response is outcome after head injury (dead/alive), and there are four discrete covariates: pupils (good/poor), coma score (depth of coma, low/high), hematoma present (no/yes), and age (categorized as 1–25, 26–54, ≥ 55). We found in Sect. 7.6.4 that these data are explained by relatively simple models. For example, a model with all main effects and three two-way interactions H.P, H.A, P.A had a deviance of 13.6 on 13 degrees of freedom which indicates a good fit. The main effects only model has a deviance of 34.1 on 18 degrees of freedom and an associated p-value of 1.2% so although not a good fit, it is not terrible either.

The approaches to prediction we compare are the null model, main effects only model, subset selection over all models using AIC and BIC, unrestricted subset selection using AIC and BIC, classification trees, bagging trees, and random forests. We looked at two versions of AIC and BIC with one enforcing the hierarchy principle and the other not. The random forest method used two variables to split on at each node. In this example there are just four covariates, and the discrete nature of these covariates (each with few levels) and the good fit of simple models indicates that we would not expect to see great advantages in using tree-based methods.

We split the data into training and test datasets consisting of 70% and 30% of the data, respectively. Each of the methods was ran 100 times for different splits of the data and then the misclassification rates were recorded, along with the standard deviations of these rates. The results are given in Table 12.1. The striking aspect of this table is the lack of a clear winner; apart from the null model, all methods perform essentially equally.

Figure 12.20 shows the oob error rate as a function of the number of trees. We see that the error rate stabilizes at around 300 trees. Figure 12.21 shows two measures of the variable importance from one particular split of the data (i.e., one out of 100). For this split, coma is the most important variable for classification, with hematoma the least important. These importance measures are in line with the summaries from

Table 12.1 Average test errors over 100 train/test splits of the outcome after head injury data, along with the standard deviation over these splits

	Null	Main	AIC 1	BIC 1	AIC 2	BIC 2	Tree	Bagging	Ran For
Mean	50.5	26.1	26.1	25.8	26.1	25.9	25.4	25.7	25.6
SD	2.8	2.2	2.1	2.3	2.0	2.4	2.2	2.3	2.2

AIC 1 and BIC 1 enforce the hierarchy principle, while AIC 2 and BIC 2 do not

Table 12.2 Parameter estimates, standard errors, and p-values for the main effects only model and one split of the outcome after head injury data

	Estimate	Std. Err.	p-value
Haem	0.169	0.194	0.386
Pup	1.26	0.192	<0.0001
Coma	−1.60	0.198	<0.0001
Age 1	0.724	0.209	0.00054
Age 2	2.36	0.303	<0.0001

the main effects only model presented in Table 12.2. The left-hand panel shows that replacing the coma score with a permuted version leads to, on average, an increase in the predictive error rate, as measured by (12.33), of around 10%. In contrast, replacing the hematoma variable with a permuted variable actually gives a slight decrease, indicating that this variable is not useful for forecasting the outcome status (dead/alive) of a child.

This example is not typical of classification problems since the number of predictors is so small, Exercise 12.2 describes a setting that is more usual.

12.9 Concluding Comments

In this chapter we have discussed various nonparametric methods for prediction and classification. For exploration and description it is clear that the GAM models described in Sect. 12.2 are very useful. Formal inference requires more care since, as we have seen repeatedly, the appropriateness of inference depends critically on smoothing parameter choice. The potential loss of efficiency as compared to a parametric approach should also be borne in mind.

Classification is a huge topic, and the surface has only been scratched here with a focus on model-based, as opposed to algorithm-based, techniques. Bagging and random forests have been included, however, to provide a hint of the algorithmic approaches that are available. Neural networks (Ripley 1996; Neal 1996), boosting (Freund and Schapire 1997; Friedman et al. 2000), and support vector machines (Vapnick 1996) are three popular classification techniques which have not been discussed. For a very interesting exposition on the algorithmic approach to regression, see Breiman (2001b) and the accompanying discussion. We have not considered large datasets in this chapter and in particular have only briefly discussed the situation in which the sample size is small relative to the number of available predictors (the "small n, large p problem"). If prediction is the sole aim, then ensemble methods such as bagging, random forests, and Bayesian model averaging have been shown to be very powerful.

This chapter has almost exclusively considered frequentist approaches to prediction and classification (apart from the mixed model approach to fitting GAMs). We briefly mention some Bayesian approaches. The book-length treatment of Denison et al. (2002) describes Bayesian analogs of a number of the techniques that we have discussed including spline and classification models. Bayesian CART models are described in Chipman et al. (1998). Gaussian process models are an important topic that are considered by Rasmussen and Williams (2006).

12.10 Bibliographic Notes

An influential early work on GAMs is the book-length treatment of Hastie and Tibshirani (1990). Wood (2006) is an excellent mix of the theory and practice of using GAMs, with an emphasis on thin plate regression splines. Ruppert et al. (2003) also consider GAMs from a mixed model standpoint. Natural thin plate splines are described in Wabha (1990) and Green and Silverman (1994, Chap. 7).

Early references to tree-based strategies include Morgan and Sonquist (1963), Morgan and Messenger (1973), and Friedman (1979). Approaches based on trees were expanded and popularized in Breiman et al. (1984). Ripley (1996), Izenman (2008), and Berk (2008) describe machine learning techniques from a statistical perspective. Hastie et al. (2009) is a broad and in-depth treatment.

12.11 Exercises

12.1 For model (12.12), form and graphically display (via perspective plots) the 16 tensor product bases functions with $L_1 = L_2 = 2$, $\xi_{11} = \xi_{21} = 1/3$, $\xi_{12} = \xi_{22} = 2/3$.

12.2 For the ethanol data in the R package SemiPar, fit a tensor product cubic spline model.

12.3 Show that (12.24) follows from (12.23).

12.4 The background to this question on discriminant analysis can be found in Sect. 12.8.2. Suppose that under the two classes, $X_0 \sim N_p(\mu_0, \Sigma)$ and, independently, $X_1 \sim N_p(\mu_1, \Sigma)$. Consider the statistic

$$\frac{\{E[a^\top X_0] - E[a^\top X_1]\}^2}{\text{var}(a^\top X_0 - a^\top X_1)}$$

12.11 Exercises

as a function of the $p \times 1$ vector a. Show that $a \propto \Sigma^{-1}(\mu_0 - \mu_1)$ maximizes the statistic, using a Lagrange multiplier approach. Explain why this result provides one justification for the use of linear discriminant analysis.

12.5 In this question, the famous iris data analyzed in Fisher (1936) will be considered. These data may be found at the book website and contain three classes of iris (Setosa, Versicolour, and Virginica) and four covariates (sepal length, sepal width, petal length, and petal width) all measured in cm.

(a) Based on the full data for Setosa and Versicolour only, build classifiers based on the approaches listed below. In each case, explain carefully how you implemented the approach, and provide graphical summaries of the output.

(1) Linear discriminant analysis.
(2) Quadratic discriminant analysis.
(3) Linear logistic regression.
(4) Classification trees.
(5) Bagging.
(6) Random forests.

(b) Repeat the previous part for the data on all three classes of iris.

12.6 At the book website of Hastie et al. (2009), you will find data that have been extensively used to test binary classification methods. The data concern 4601 emails, and the aim is to predict which are spam, in order to filter out such emails. There are 1813 spam messages and 57 potential predictors that concern the content of the emails. The data have been split into a training set of 3065 emails, with 1536 remaining for testing the models. Following Hastie et al. (2009), analyze these data using linear logistic regression, a GAM with splines having fixed degrees of freedom equal to 3 for each smoother, classification trees, bagging, and random forests. Summarize your findings based on the test error.

Part V
Appendices

Appendix A
Differentiation of Matrix Expressions

For univariate x and $f : \mathbb{R} \to \mathbb{R}$ we write the derivative as

$$\frac{df}{dx}.$$

We define

$$\frac{\partial}{\partial \boldsymbol{x}} = \begin{bmatrix} \frac{\partial}{\partial x_1} \\ \vdots \\ \frac{\partial}{\partial x_p} \end{bmatrix}$$

to be differentiation with respect to elements of a vector $\boldsymbol{x} = [x_1, \ldots, x_p]^{\mathrm{T}}$. Let $\boldsymbol{a}$ and $\boldsymbol{x}$ represent $p \times 1$ vectors, then

$$\frac{\partial}{\partial \boldsymbol{x}}(\boldsymbol{a}^{\mathrm{T}}\boldsymbol{x}) = \boldsymbol{a} = \frac{\partial}{\partial \boldsymbol{x}}(\boldsymbol{x}^{\mathrm{T}}\boldsymbol{a}), \tag{A.1}$$

the second equality arising because $\boldsymbol{a}^{\mathrm{T}}\boldsymbol{x} = \boldsymbol{x}^{\mathrm{T}}\boldsymbol{a}$. Also

$$\frac{\partial}{\partial \boldsymbol{x}^{\mathrm{T}}}(\boldsymbol{a}^{\mathrm{T}}\boldsymbol{x}) = \left[\frac{\partial}{\partial \boldsymbol{x}}(\boldsymbol{a}^{\mathrm{T}}\boldsymbol{x})\right]^{\mathrm{T}} = \boldsymbol{a}^{\mathrm{T}} = \frac{\partial}{\partial \boldsymbol{x}^{\mathrm{T}}}(\boldsymbol{x}^{\mathrm{T}}\boldsymbol{a}). \tag{A.2}$$

Suppose $\boldsymbol{u} = \boldsymbol{u}(\boldsymbol{x})$ is an $r \times 1$ vector and $\boldsymbol{x}$ is $p \times 1$. Then

$$\frac{\partial \boldsymbol{u}^{\mathrm{T}}}{\partial \boldsymbol{x}}$$

is a matrix of order $p \times r$ with (i,j)th element

$$\frac{\partial u_j}{\partial x_i}, \quad i = 1, \ldots, p, \ j = 1, \ldots, r.$$

The transpose
$$\left(\frac{\partial \boldsymbol{u}^{\mathrm{T}}}{\partial \boldsymbol{x}}\right)^{\mathrm{T}} = \frac{\partial \boldsymbol{u}}{\partial \boldsymbol{x}^{\mathrm{T}}}$$
is a matrix of order $r \times p$ with (j, k)th element
$$\frac{\partial u_j}{\partial x_k}, \quad j = 1, \ldots, r, \quad k = 1, \ldots, p.$$

For example,
$$\frac{\partial \boldsymbol{x}}{\partial \boldsymbol{x}^{\mathrm{T}}} = \frac{\partial \boldsymbol{x}^{\mathrm{T}}}{\partial \boldsymbol{x}} = \boldsymbol{I}_p,$$
the $p \times p$ identity matrix.

Consider the matrix $\boldsymbol{A}$ of dimension $p \times p$. If $\boldsymbol{A}$ is not a function of $\boldsymbol{x}$:
$$\frac{\partial}{\partial \boldsymbol{x}^{\mathrm{T}}}(\boldsymbol{A}\boldsymbol{x}) = \boldsymbol{A}\frac{\partial \boldsymbol{x}}{\partial \boldsymbol{x}^{\mathrm{T}}} = \boldsymbol{A}$$
and
$$\frac{\partial}{\partial \boldsymbol{x}^{\mathrm{T}}}(\boldsymbol{x}^{\mathrm{T}}\boldsymbol{A}) = \frac{\partial \boldsymbol{x}^{\mathrm{T}}}{\partial \boldsymbol{x}}\boldsymbol{A} = \boldsymbol{A}.$$

If $\boldsymbol{u} = \boldsymbol{u}(\boldsymbol{x})$ then
$$\frac{\partial}{\partial \boldsymbol{x}^{\mathrm{T}}}(\boldsymbol{A}\boldsymbol{u}) = \boldsymbol{A}\frac{\partial \boldsymbol{u}}{\partial \boldsymbol{x}^{\mathrm{T}}},$$
and
$$\frac{\partial}{\partial \boldsymbol{x}^{\mathrm{T}}}(\boldsymbol{u}^{\mathrm{T}}\boldsymbol{A}) = \frac{\partial \boldsymbol{u}^{\mathrm{T}}}{\partial \boldsymbol{x}}\boldsymbol{A}.$$

Let $\boldsymbol{u} = \boldsymbol{u}(\boldsymbol{x})$ and $\boldsymbol{v} = \boldsymbol{v}(\boldsymbol{x})$ be $p \times 1$ vectors. Then the derivative of the inner product $\boldsymbol{u}^{\mathrm{T}}\boldsymbol{v}$ is
$$\frac{\partial}{\partial \boldsymbol{x}}(\boldsymbol{u}^{\mathrm{T}}\boldsymbol{v}) = \frac{\partial \boldsymbol{u}^{\mathrm{T}}}{\partial \boldsymbol{x}}\boldsymbol{v} + \frac{\partial \boldsymbol{v}^{\mathrm{T}}}{\partial \boldsymbol{x}}\boldsymbol{u}.$$

If $\boldsymbol{A}$ is again a $p \times p$ matrix then
$$\frac{\partial}{\partial \boldsymbol{x}}(\boldsymbol{u}^{\mathrm{T}}\boldsymbol{A}\boldsymbol{u}) = \frac{\partial \boldsymbol{u}^{\mathrm{T}}}{\partial \boldsymbol{x}}\boldsymbol{A}\boldsymbol{u} + \frac{\partial \boldsymbol{v}^{\mathrm{T}}}{\partial \boldsymbol{x}}\boldsymbol{A}^{\mathrm{T}}\boldsymbol{u}.$$

If $\boldsymbol{A}$ is symmetric
$$\frac{\partial}{\partial \boldsymbol{x}}(\boldsymbol{u}^{\mathrm{T}}\boldsymbol{A}\boldsymbol{u}) = 2\frac{\partial \boldsymbol{u}^{\mathrm{T}}}{\partial \boldsymbol{x}}\boldsymbol{A}\boldsymbol{u}.$$

In particular, for a quadratic form
$$\frac{\partial}{\partial \boldsymbol{x}}(\boldsymbol{x}^{\mathrm{T}}\boldsymbol{A}\boldsymbol{x}) = 2\boldsymbol{A}\boldsymbol{x}. \tag{A.3}$$

A Differentiation of Matrix Expressions

Let $f : \mathbb{R} \to \mathbb{R}$ then
$$\frac{\partial f}{\partial \boldsymbol{x}}$$
is $p \times 1$ and
$$\frac{\partial}{\partial \boldsymbol{x}} \frac{\partial f}{\partial \boldsymbol{x}^\mathrm{T}} = \frac{\partial^2 f}{\partial \boldsymbol{x} \partial \boldsymbol{x}^\mathrm{T}}$$
is a $p \times p$ matrix with elements
$$\frac{\partial^2 f}{\partial x_i \partial x_j}, \quad i = 1, \ldots, p, \ j = 1, \ldots, p.$$

For example, with $p = 2$:

$$\frac{\partial}{\partial \boldsymbol{x}} \frac{\partial f}{\partial \boldsymbol{x}^\mathrm{T}} = \begin{bmatrix} \frac{\partial}{\partial x_1}\left(\frac{\partial f}{\partial x_1}\right) & \frac{\partial}{\partial x_1}\left(\frac{\partial f}{\partial x_2}\right) \\ \frac{\partial}{\partial x_2}\left(\frac{\partial f}{\partial x_1}\right) & \frac{\partial}{\partial x_2}\left(\frac{\partial f}{\partial x_2}\right) \end{bmatrix} = \begin{bmatrix} \frac{\partial^2 f}{\partial x_1^2} & \frac{\partial^2 f}{\partial x_1 \partial x_2} \\ \frac{\partial^2 f}{\partial x_2 \partial x_1} & \frac{\partial^2 f}{\partial x_2^2} \end{bmatrix}.$$

For a non-singular $p \times p$ matrix $\boldsymbol{A}$, whose elements are functions of x, we have

$$\frac{\partial \boldsymbol{A}^{-1}}{\partial x} = -\boldsymbol{A}^{-1} \frac{\partial \boldsymbol{A}}{\partial x} \boldsymbol{A}^{-1}.$$

Also,
$$\frac{\partial}{\partial x} \log |\boldsymbol{A}| = \mathrm{tr}\left(\boldsymbol{A}^{-1} \frac{\partial \boldsymbol{A}}{\partial x}\right).$$

The trace of a $p \times p$ square matrix is

$$\mathrm{tr}(\boldsymbol{A}) = \sum_{i=1}^{p} a_{ii}.$$

Appendix B
Matrix Results

We begin with two properties of determinants:

$$\det(A^\mathrm{T} A) = \det(A^\mathrm{T})\det(A) = \det(A)^2 \tag{B.1}$$

and

$$\left| \begin{matrix} T & U \\ V & W \end{matrix} \right| = |T| |W - VT^{-1}U|. \tag{B.2}$$

Let A be an $n \times n$ non-singular matrix, which we express as

$$\begin{bmatrix} A_{11} & A_{12} \\ A_{21} & A_{22} \end{bmatrix},$$

where A_{11} is $k \times k$ and A_{12} is $k \times (n-k)$. The inverse $B = A^{-1}$ has elements

$$B_{11} = (A_{11} - A_{12} A_{22}^{-1} A_{21})^{-1}$$
$$B_{22} = (A_{22} - A_{21} A_{11}^{-1} A_{12})^{-1}$$
$$B_{12} = -A_{11}^{-1} A_{12} B_{22}$$
$$B_{21} = -A_{22}^{-1} A_{21} B_{11}.$$

For matrices A, B and C of the appropriate dimensions:

$$(A + BCB^\mathrm{T})^{-1} = A^{-1} - A^{-1} B (B^\mathrm{T} A^{-1} B + C^{-1})^{-1} B^\mathrm{T} A^{-1}. \tag{B.3}$$

We now describe how the expectation, variance and covariance operators deal with vectors of random variables.

Suppose U is an $n \times 1$ vector of random variables, and A is an $m \times n$ matrix. Then

$$E[AU] = A\, E[U]$$
$$\text{var}(AU) = A\, \text{var}(U) A^\mathrm{T}.$$

Suppose V is an $m \times 1$ vector of random variables. Then $\text{cov}(U, V) = C$ is an $n \times m$ matrix with (i,j)th element $\text{cov}(U_i, V_j)$, $i = 1, \ldots, n$, $j = 1, \ldots, m$. Hence, $\text{cov}(V, U) = C^\mathrm{T}$. In addition,

$$\text{cov}(U, AU) = \text{cov}(U) A^\mathrm{T}$$
$$\text{cov}(AU, U) = A\, \text{cov}(U).$$

The iterated expectation and covariance formulas are given by:

$$E[Y] = E_X\left[E_{Y\mid X}(Y \mid X)\right]$$
$$\text{cov}(Y, Z) = E_X\left[\text{cov}_{Y,Z\mid X}(Y, Z \mid X)\right] + \text{cov}_X\left[E_{Y\mid X}(Y\mid X), E_{Z\mid X}(Z\mid X)\right].$$

Suppose Z is an $n \times 1$ random variable with $E[Z] = \mu$, $\text{var}(Z) = \Sigma$ and A is a symmetric $n \times n$ matrix. Then

$$E[Z^\mathrm{T} A Z] = \text{tr}(A\Sigma) + \mu^\mathrm{T} A \mu. \tag{B.4}$$

See Schott (1997, p. 391) for a proof.

Appendix C
Some Linear Algebra

Bases

Definition. Let S be a collection of $m \times 1$ vectors satisfying the following:

1. If $x_1 \in S$ and $x_2 \in S$, then $x_1 + x_2 \in S$.
2. If $x \in S$, and α is any real scalar, then $\alpha x \in S$.

Then S is called a *vector space* in m-dimensional space.

Definition. Let $\{x_1, \ldots, x_n\}$ be a set of $m \times 1$ vectors in the vector space S. If each vector in S can be expressed as a linear combination of $x_1, \ldots, x_n$ then the set $\{x_1, \ldots, x_n\}$ is said to *span S*.

Definition. Let $\{x_1, \ldots, x_n\}$ be a set of $m \times 1$ vectors in the vector space S. This set is called a *basis* if it spans S and if the vectors $x_1, \ldots, x_n$ are linearly independent.

Appendix D
Probability Distributions and Generating Functions

Continuous Distributions

Multivariate Normal Distribution

The p-dimensional random variable $X = [X_1, \ldots, X_p]^T$ has a normal distribution, denoted $N_p(\boldsymbol{\mu}, \boldsymbol{\Sigma})$, with mean $\boldsymbol{\mu} = [\mu_1, \ldots, \mu_p]^T$ and $p \times p$ variance–covariance matrix $\boldsymbol{\Sigma}$ if its density is of the form

$$p(\boldsymbol{x}) = (2\pi)^{-p/2} \mid \boldsymbol{\Sigma} \mid^{-1/2} \times \exp\left[-\frac{1}{2}(\boldsymbol{x} - \boldsymbol{\mu})^T \boldsymbol{\Sigma}^{-1} (\boldsymbol{x} - \boldsymbol{\mu})\right],$$

for $\boldsymbol{x} \in \mathbb{R}^p$, $\boldsymbol{\mu} \in \mathbb{R}^p$ and non-singular $\boldsymbol{\Sigma}$.

Summaries:

$$E[\boldsymbol{X}] = \boldsymbol{\mu}$$
$$\text{mode}(\boldsymbol{X}) = \boldsymbol{\mu}$$
$$\text{var}(\boldsymbol{X}) = \boldsymbol{\Sigma}.$$

Suppose

$$\begin{bmatrix} \boldsymbol{X}_1 \\ \boldsymbol{X}_2 \end{bmatrix} \sim N_p\left(\begin{bmatrix} \boldsymbol{\mu}_1 \\ \boldsymbol{\mu}_2 \end{bmatrix}, \begin{bmatrix} \boldsymbol{V}_{11} & \boldsymbol{V}_{12} \\ \boldsymbol{V}_{21} & \boldsymbol{V}_{22} \end{bmatrix} \right)$$

where:

- $\boldsymbol{X}_1$ and $\boldsymbol{\mu}_1$ are $r \times 1$,
- $\boldsymbol{X}_2$ and $\boldsymbol{\mu}_2$ are $(p-r) \times 1$,
- $\boldsymbol{V}_{11}$ is $r \times r$,
- $\boldsymbol{V}_{12}$ is $r \times (p-r)$, $\boldsymbol{V}_{21}$ is $(p-r) \times r$,
- $\boldsymbol{V}_{22}$ is $(p-r) \times (p-r)$.

Then the marginal distribution of X_1 is

$$X_1 \sim N_r(\mu_1, V_{11})$$

and the conditional distribution $X_1 \mid X_2 = x_2$ is

$$X_1 \mid X_2 = x_2 \sim N_r \left[\mu_1 + V_{12} V_{22}^{-1} (x_2 - \mu_2), W_{11} \right], \quad (D.1)$$

where $W_{11} = V_{11} - V_{12} V_{22}^{-1} V_{21}$.

Suppose

$$Y_j \mid \mu_j, \sigma_j^2 \sim N(\mu_j, \sigma_j^2),$$

for $j = 1, \ldots, J$, with $Y_1, \ldots, Y_J$ independent. Then, if $a_1, \ldots, a_J$ represent constants,

$$Z = \sum_{j=1}^{J} a_j Y_j \sim N \left(\sum_{j=1}^{J} a_j \mu_j, \sum_{j=1}^{J} a_j^2 \sigma_j^2 \right). \quad (D.2)$$

If Y is a $p \times 1$ vector of random variables whose distribution is $N(\mu, \Sigma)$ and A is an $r \times p$ matrix of constants, then

$$AY \sim N(A\mu, A\Sigma A^\mathsf{T}). \quad (D.3)$$

Beta Distribution

The random variable X follows a beta distribution, denoted $Be(a, b)$, if its density has the form:

$$p(x) = B(a, b)^{-1} x^{a-1} (1 - x)^{b-1},$$

for $0 < x < 1$ and $a, b > 0$ and where

$$B(a, b) = \frac{\Gamma(a) \Gamma(b)}{\Gamma(a + b)} = \int_0^1 z^{a-1} (1 - z)^{b-1} \, dz \quad (D.4)$$

is the beta function.

Summaries:

$$E[X] = \frac{a}{a + b}$$

$$\text{mode}(X) = \frac{a - 1}{a + b - 2} \quad \text{for} \quad a, b > 1$$

$$\text{var}(X) = \frac{ab}{(a + b)^2 (a + b + 1)}.$$

D Probability Distributions and Generating Functions

Gamma Distribution

The random variable X follows a gamma distribution, denoted $\mathrm{Ga}(a,b)$, if its density is of the form

$$p(x) = \frac{b^a}{\Gamma(a)} x^{a-1} \exp(-bx),$$

for $x > 0$ and $a, b > 0$.

Summaries:

$$\mathrm{E}[X] = \frac{a}{b}$$

$$\mathrm{mode}(X) = \frac{a-1}{b} \quad \text{for } a \geq 1$$

$$\mathrm{var}(X) = \frac{a}{b^2}.$$

A χ_k^2 random variable with degrees of freedom k corresponds to the $\mathrm{Ga}(k/2, 1/2)$ distribution.

Inverse Gamma Distribution

The random variable X follows an inverse gamma distribution, denoted $\mathrm{InvGa}(a,b)$, if its density is of the form

$$p(x) = \frac{b^a}{\Gamma(a)} x^{-(a+1)} \exp(-b/x),$$

for $x > 0$ and $a, b > 0$.

Summaries:

$$\mathrm{E}[X] = \frac{b}{a-1} \quad \text{for } a > 1$$

$$\mathrm{mode}(X) = \frac{b}{a+1}$$

$$\mathrm{var}(X) = \frac{b^2}{(a-1)^2(a-2)} \quad \text{for } a > 2.$$

If Y is $\mathrm{Ga}(a,b)$ then $X = Y^{-1}$ is $\mathrm{InvGa}(a,b)$.

Lognormal Distribution

The random variable X follows a (univariate) lognormal distribution, denoted LogNorm(μ, σ^2), if its density is of the form

$$p(x) = (2\pi\sigma^2)^{-1/2} \frac{1}{x} \exp\left[-\frac{1}{2\sigma^2}(\log x - \mu)^2\right],$$

for $x > 0$ and $\mu \in \mathbb{R}, \sigma > 0$.

Summaries:

$$E[X] = \exp(\mu + \sigma^2/2)$$
$$\text{mode}(X) = \exp(\mu - \sigma^2)$$
$$\text{var}(X) = E[X]^2 \left[\exp(\sigma^2) - 1\right].$$

If Y is $N(\mu, \sigma^2)$ then $X = \exp(Y)$ is LogNorm(μ, σ^2).

Laplacian Distribution

The random variable X follows a Laplacian distribution, denoted Lap(μ, ϕ), if its density is of the form

$$p(x) = \frac{1}{2\phi} \exp(-\mid x - \mu \mid /\phi),$$

for $x \in \mathbb{R}, \mu \in \mathbb{R}$ and $\phi > 0$.

Summaries:

$$E[X] = \mu$$
$$\text{mode}(X) = \mu$$
$$\text{var}(X) = 2\phi^2.$$

Multivariate t Distribution

The p-dimensional random variable $\boldsymbol{X} = [X_1, \ldots, X_p]^\text{T}$ has a (Student's) t distribution with d degrees of freedom, location $\boldsymbol{\mu} = [\mu_1, \ldots, \mu_p]^\text{T}$ and $p \times p$ scale matrix $\boldsymbol{\Sigma}$, denoted $\text{T}_p(\boldsymbol{\mu}, \boldsymbol{\Sigma}, d)$, if its density is of the form

D Probability Distributions and Generating Functions

$$p(x) = \frac{\Gamma[(d+p)/2]}{\Gamma(d/2)(d\pi)^{p/2}} |\Sigma|^{-1/2} \left[1 + \frac{(x-\mu)^{\mathrm{T}} \Sigma^{-1} (x-\mu)}{d} \right]^{-(d+p)/2},$$

for $x \in \mathbb{R}^p$, $\mu \in \mathbb{R}^p$, non-singular Σ and $d > 0$.

Summaries:

$$E[X] = \mu \quad \text{for} \quad d > 1$$

$$\text{mode}(X) = \mu$$

$$\text{var}(X) = \frac{d}{d-2} \times \Sigma \quad \text{for} \quad d > 2.$$

The margins of a multivariate t distribution also follow t distributions. For example, if $X = [X_1, X_2]^{\mathrm{T}}$ where X_1 is $r \times 1$ and X_2 is $(p-r) \times 1$, then the marginal distribution is

$$X_1 \sim T_r(\mu_1, V_{11}, d),$$

where μ_1 is $r \times 1$ and V_{11} is $r \times r$.

F Distribution

The random variable X follows an F distribution, denoted $F(a, b)$, if its density is of the form

$$p(x) = \frac{a^{a/2} b^{b/2}}{B(a/2, b/2)} \frac{x^{a/2-1}}{(b + ax)^{(a+b)/2}},$$

for $x > 0$, with degrees of freedom $a, b > 0$ and where $B(\cdot, \cdot)$ is the beta function, as defined in (D.4).

Summaries:

$$E[X] = \frac{b}{b-2} \quad \text{for} \quad b > 2$$

$$\text{mode}(X) = \frac{a-2}{a} \frac{b}{b+2} \quad \text{for} \quad a > 2$$

$$\text{var}(X) = \frac{2b^2(a+b-2)}{a(b-2)^2(b-4)} \quad \text{for} \quad b > 4.$$

Wishart Distribution

The $p \times p$ random matrix X follows a Wishart distribution, denoted $\text{Wish}_p(r, S)$, if its probability density function is of the form

$$p(x) = \frac{\mid x \mid^{(r-p-1)/2}}{2^{rp/2}\Gamma_p(r/2) \mid S \mid^{r/2}} \exp\left[-\frac{1}{2}\text{tr}(xS^{-1})\right],$$

for x positive definite, S positive definite and $r > p - 1$ and where

$$\Gamma_p(r/2) = \pi^{p(p-1)/4} \prod_{j=1}^{p} \Gamma[(r+1-j)/2]$$

is the generalized gamma function.

Summaries:

$$\text{E}[X] = rS$$
$$\text{mode}(X) = (r - p - 1)S \quad \text{for} \quad r > p + 1$$
$$\text{var}(X_{ij}) = r(S_{ij}^2 + S_{ii}S_{jj}) \quad \text{for} \quad i, j = 1, \ldots, p.$$

Marginally, the diagonal elements X_{ii} have distribution $\text{Ga}[r/2, 1/(2S_{ii})]$, $i = 1, \ldots, p$.

Taking $p = 1$ yields

$$p(x) = \frac{(2S)^{-r/2}}{\Gamma(r/2)} x^{r/2-1} \exp(-x/2S),$$

for $x > 0$ and $S, r > 0$, i.e. a $\text{Ga}[r/2, 1/(2S)]$ distribution, revealing that the Wishart distribution is a multivariate version of the gamma distribution.

Inverse Wishart Distribution

The $p \times p$ random matrix X follows an inverse Wishart distribution, denoted $\text{InvWish}_p(r, S)$, if its probability density function is of the form

$$p(x) = \frac{|x|^{-(r+p+1)/2}}{2^{rp/2}\Gamma_p(r/2) \mid S \mid^{r/2}} \exp\left[-\frac{1}{2}\text{tr}(x^{-1}S)\right],$$

for x positive definite, S positive definite and $r > p - 1$.

D Probability Distributions and Generating Functions

Summaries:

$$E[X] = \frac{S^{-1}}{r-p-1} \quad \text{for } r > p+1$$

$$\text{mode}(X) = \frac{S^{-1}}{r+p+1}$$

$$\text{var}(X_{ij}) = \frac{(r-p+1)S_{ij}^{-2} + (r-p-1)S_{ii}^{-1}S_{jj}^{-1}}{(r-p)(r-p-1)^2(r-p-3)} \quad \text{for } i,j = 1,\ldots,p.$$

If $p = 1$ we recover the inverse gamma distribution $\text{InvGa}[r/2, 1/(2S)]$ with

$$E[X] = \frac{1}{S(r-2)} \quad \text{for } r > 2$$

$$\text{mode}(X) = \frac{1}{S(r+2)}$$

$$\text{var}(X) = \frac{1}{S^2(r-2)(r-4)} \quad \text{for } r > 4.$$

If $Y \sim \text{Wish}_p(r, S)$, the distribution of $X = Y^{-1}$ is $\text{InvWish}_p(r, S)$.

Discrete Distributions

Binomial Distribution

The random variable X has a binomial distribution, denoted $\text{Binomial}(n, p)$, if its distribution is of the form

$$\Pr(X = x) = \binom{n}{x} p^x (1-p)^{n-x},$$

for $x = 0, 1, \ldots, n$ and $0 < p < 1$.

Summaries:

$$E[X] = np$$
$$\text{var}(X) = np(1-p).$$

Poisson Distribution

The random variable X has a Poisson distribution, denoted Poisson(μ), if its distribution is of the form

$$\Pr(X = x) = \frac{\exp(-\mu)\mu^x}{x!},$$

for $\mu > 0$ and $x = 0, 1, 2, \ldots$.

Summaries:

$$E[X] = \mu$$
$$\text{var}(X) = \mu.$$

Negative Binomial Distribution

The random variable X has a negative binomial distribution, denoted NegBin(μ, b), if its distribution is of the form

$$\Pr(X = x) = \frac{\Gamma(x+b)}{\Gamma(x+1)\Gamma(b)} \left(\frac{\mu}{\mu+b}\right)^x \left(\frac{b}{\mu+b}\right)^b,$$

for $\mu > 0, b > 0$ and $x = 0, 1, 2, \ldots$

Summaries:

$$E[X] = \mu$$
$$\text{var}(X) = \mu + \mu^2/b.$$

The negative binomial distribution arises as a gamma mixture of a Poisson random variable. Specifically, if $X \mid \mu, \delta \sim$ Poisson($\mu\delta$) and $\delta \mid b \sim$ Ga(b, b), then $X \mid \mu, b \sim$ NegBin(μ, b).

We link the above description, motivated by a random effects argument, with the more familiar derivation in which the negative binomial arises as the number of failures seen before we observe b successes from independent trials, each with success probability $p = \mu/(\mu + b)$. The probability distribution is

$$\Pr(X = x) = \binom{x+b-1}{x} p^x (1-p)^b,$$

for $0 < p < 1, b > 0$ an integer, and $x = 0, 1, 2, \ldots$

D Probability Distributions and Generating Functions

Summaries:

$$\mathrm{E}[X] = \frac{pb}{1-p}$$

$$\mathrm{var}(X) = \frac{pb}{(1-p)^2}.$$

Generating Functions

The *moment generating function* of a random variable Y is defined as $M_Y(t) = \mathrm{E}[\exp(tY)]$, for $t \in \mathbb{R}$, whenever this expectation exists. We state three important and useful properties of moment generating functions:

1. If two distributions have the same moment generating functions then they are identical at almost all points.
2. Using a series expansion:

$$\exp(tY) = 1 + tY + \frac{t^2 Y^2}{2!} + \frac{t^3 Y^3}{3!} + \cdots$$

so that

$$M_Y(t) = 1 + t m_1 + \frac{t^2 m_2}{2!} + \frac{t^3 m_3}{3!} + \cdots$$

where m_i is the ith moment. Hence,

$$\mathrm{E}[Y^i] = M_Y^{(i)}(0) = \left.\frac{d^i M_Y}{dt^n}\right|_{t=0}.$$

3. If $Y_1, \ldots, Y_n$ are a sequence of independent random variables and $S = \sum_{i=1}^{n} a_i Y_i$, with a_i constant, then the moment generating function of S is

$$M_S(t) = \prod_{i=1}^{n} M_{Y_i}(a_i t).$$

The *cumulant generating function* of a random variable Y is defined as

$$C_Y(t) = \log \mathrm{E}[\exp(tY)]$$

for $t \in \mathbb{R}$.

Appendix E
Functions of Normal Random Variables

If $Y_j \sim N(0,1)$, $j = 1, \ldots, J$, with $Y_1, \ldots, Y_J$ independent, then

$$Z = \sum_{j=1}^{J} Y_j^2 \sim \chi_J^2, \tag{E.1}$$

a chi-squared distribution with J degrees of freedom and $E[Z] = J$, $\text{var}(Z) = 2J$.

If $X \sim N(0,1)$, $Y \sim \chi_d^2$, with X and Y independent, then

$$\frac{X}{(Y/d)^{1/2}} \sim T(0, 1, d), \tag{E.2}$$

a Student's t distribution with d degrees of freedom.

If $U \sim \chi_J^2$ and $V \sim \chi_K^2$, with U and V independent, then

$$\frac{U/J}{V/K} \sim F(J, K), \tag{E.3}$$

the F distribution with J, K degrees of freedom.

Appendix F
Some Results from Classical Statistics

In this section we provide some definitions and state some theorems (without proof) from classical statistics. More details can be found in Schervish (1995). Let $\boldsymbol{y} = [y_1, \ldots, y_n]^\mathrm{T}$ be a random sample from $p(y \mid \theta)$.

Definition. The statistic $T(\boldsymbol{Y})$ is *sufficient* for θ within a family of probability distributions $p(y \mid \theta)$ if $p(\boldsymbol{y} \mid T(\boldsymbol{y}))$ does not depend upon θ.

Theorem. *The Fisher–Neyman factorization theorem states that $T(\boldsymbol{Y})$ is sufficient for θ if and only if*
$$p(\boldsymbol{y} \mid \theta) = g[T(\boldsymbol{y}) \mid \theta] \times h(\boldsymbol{y}).$$

Intuitively, all of the information in the sample with respect to θ is contained in $T(\boldsymbol{Y})$.

Definition. The statistic $T(\boldsymbol{Y})$ is *minimal sufficient* for θ within a family of probability distributions $p(y \mid \theta)$ if no further reduction from T is possible while retaining sufficiency.

Theorem. *The Lehmann–Scheffé theorem states that if $T(\boldsymbol{Y})$ satisfies the following property: for every pair of sample points $\boldsymbol{y}, \boldsymbol{z}$ the ratio $p(\boldsymbol{y} \mid \theta)/p(\boldsymbol{z} \mid \theta)$ is free of θ if and only if $T(\boldsymbol{y}) = T(\boldsymbol{z})$, then T is minimal sufficient.*

Example. Let $Y_1, \ldots, Y_n$ be independent and identically distributed from the one-parameter exponential family of distributions:
$$p(y \mid \theta) = \exp[\theta T(y) - b(\theta) + c(y)]$$

for functions $b(\cdot)$ and $c(\cdot)$. Then $\sum_{i=1}^n T(Y_i)$ is sufficient for θ by the factorization theorem and minimal sufficient by the Lehmann–Sheffé theorem.

Definition. A statistic $V = V(\boldsymbol{y})$ is *ancilliary* for θ within a family of probability distributions $p(y \mid \theta)$ if its distribution does not depend on θ.

Definition. If a minimal sufficient statistic is $T = [T_1, T_2]$ and T_2 is ancillary then T_1 is called *conditionally sufficient given* T_2.

Example. In a linear normal linear regression with covariate x, suppose that x has distribution $p(x)$ and

$$Y_i \mid X_i = x_i \sim \mathrm{N}(\beta_0 + \beta_1 x_i, \sigma^2), \qquad i = 1, \ldots, n.$$

Then, letting $\boldsymbol{x} = [x_1, \ldots, x_n]^\mathrm{T}$, $\boldsymbol{y} = [y_1, \ldots, y_n]^\mathrm{T}$ and $\boldsymbol{\beta} = [\beta_0, \beta_1]^\mathrm{T}$ the distribution for the data is

$$p(\boldsymbol{x}, \boldsymbol{y} \mid \boldsymbol{\beta}, \sigma^2) = p(\boldsymbol{x})(2\pi\sigma)^{-n/2} \exp\left[-\frac{1}{2\sigma^2} \sum_{i=1}^n (y_i - \beta_0 - \beta_1 x_i)^2\right].$$

The sufficient statistic for $[\boldsymbol{\beta}, \sigma^2]$ is

$$\boldsymbol{S} = \left[\widehat{\boldsymbol{\beta}}, \widehat{\sigma}^2, \sum_{i=1}^n x_i, \sum_{i=1}^n x_i^2\right],$$

with the last two components being an ancillary statistic.

Definition. A statistic T is *complete* if for every real-valued function $g(\cdot)$, $\mathrm{E}[g(T)] = 0$ for every θ implies $g(T) = 0$.

Definition. Suppose we wish to estimate $\phi = \phi(\theta)$ based on $Y \mid \theta \sim p(\cdot \mid \theta)$. An unbiased estimator $\widehat{\phi}$ of ϕ is a *uniformly minimum-variance unbiased estimator (UMVUE)* if, for all other unbiased estimators $\widetilde{\phi}$,

$$\mathrm{var}(\widehat{\phi}) \leq \mathrm{var}(\widetilde{\phi})$$

for all θ.

Lemma. *If T is complete then $\phi(\theta)$ admits at most one unbiased estimator $\widehat{\phi}(T)$ depending on T.*

Theorem (Rao–Blackwell–Lehmann–Scheffé). *Let $T = T(\boldsymbol{Y})$ be complete and sufficient for θ. If there exists at least one unbiased estimator $\widetilde{\phi} = \widetilde{\phi}(\boldsymbol{Y})$ for $\phi(\theta)$ then there exists a unique UMVUE $\widehat{\phi} = \widehat{\phi}(T)$ for $\phi(\theta)$, namely,*

$$\phi(\boldsymbol{Y}) = \mathrm{E}[\widetilde{\phi}(\boldsymbol{Y} \mid T].$$

Corollary. *Let $T = T(\boldsymbol{Y})$ be complete and sufficient for θ. Then any function $g(T)$ is the UMVUE of its expectation $\mathrm{E}[g(T)] = \phi(\theta)$.*

Theorem. *The Cramér–Rao lower bound for any unbiased estimator $\widehat{\phi}$ of a scalar function of interest $\phi = \phi(\theta)$ is*

F Some Results from Classical Statistics

$$\text{var}(\widehat{\phi}) \geq -\frac{[\phi'(\theta)]^2}{\text{E}\left[\frac{\partial^2 l}{\partial \theta^2}\right]},$$

where $l(\theta) = \sum_{i=1}^n \log p(y_i \mid \theta)$, is the log of the joint distribution, viewed as a function of θ. Equality holds if and only if $p(y \mid \theta)$ is a one-parameter exponential family.

Appendix G
Basic Large Sample Theory

We define various quantities, and state results, that are useful in various places in the book. The presentation is informal see, for example, van der Vaart (1998) for more rigour.

Modes of Convergence

Suppose that Y_n, $n \geq 1$, are all random variables defined on a probability space $(\Omega, \mathcal{A}, P)$ where Ω is a set (the sample space) $\mathcal{A}$ is a σ-algebra of subsets of Ω, and P is a probability measure.

Definition. We say that Y_n converges *almost surely* to Y, denoted $Y_n \to_{a.s.} Y$, if

$$Y_n(\omega) \to Y(\omega) \quad \text{for all } \omega \in A \text{ where } P(A^c) = 0 \tag{G.1}$$

or, equivalently, if, for every $\epsilon > 0$

$$P\left(\sup_{m \geq n} |Y_m - Y| > \epsilon\right) \to 0 \quad \text{as } n \to \infty. \tag{G.2}$$

Definition. We say that Y_n converges *in probability* to Y, denoted $Y_n \to_p Y$, if

$$P(|Y_m - Y| > \epsilon) \to 0 \quad \text{as } n \to \infty. \tag{G.3}$$

Definition. Define the distribution function of Y as $F(y) = \Pr(Y \leq y)$. We say that Y_n converges *in distribution* to Y, denoted $Y_n \to_d Y$, or $F_n \to F$, if

$$F_n(y) \to F(y) \quad \text{as } n \to \infty \quad \text{for each continuity point } y \text{ of } F. \tag{G.4}$$

Limit Theorems

Proposition (Weak Law of Large Numbers). *If $Y_1, Y_2, \ldots, Y_n, \ldots$ are independent and identically distributed (i.i.d.) with mean $\mu = \mathrm{E}[Y]$ (so $\mathrm{E}[|Y|] < \infty$) then $\overline{Y}_n \to_p \mu$.*

Proposition (Strong Law of Large Numbers). *If $Y_1, Y_2, \ldots, Y_n, \ldots$ are i.i.d. with mean $\mu = \mathrm{E}[Y]$ (so $\mathrm{E}[|Y|] < \infty$) then $\overline{Y}_n \to_{a.s.} \mu$.*

Proposition (Central Limit Theorem). *If $Y_1, Y_2, \ldots, Y_n$ are i.i.d. with mean $\mu = \mathrm{E}[Y]$ and variance σ^2 (so $\mathrm{E}[Y^2] < \infty$), then $\sqrt{n}(\overline{Y}_n - \mu) \to_d \mathrm{N}(0, \sigma^2)$.*

Proposition (Slutsky's Theorem). *Suppose that $A_n \to_p a$, $B_n \to_p b$, for constants a and b, and $Y_n \to_d Y$. Then $A_n Y_n + B_n \to_d aY + b$.*

Proposition (Delta Method). *Suppose $\sqrt{n}\,(\boldsymbol{Y}_n - \boldsymbol{\mu}) \to_d \boldsymbol{Z}$ and suppose that $\boldsymbol{g}: \mathbb{R}^p \to \mathbb{R}^k$ has a derivative $\boldsymbol{g}'$ at $\boldsymbol{\mu}$ (here $\boldsymbol{g}'$ is a $k \times p$ matrix of derivatives). Then the delta method gives the asymptotic distribution as*

$$\sqrt{n}\,[\boldsymbol{g}(\boldsymbol{Y}) - \boldsymbol{g}(\boldsymbol{\mu})] \to_d \boldsymbol{g}'(\boldsymbol{\mu})\boldsymbol{Z}.$$

If $\boldsymbol{Z} \sim \mathrm{N}_p(\boldsymbol{0}, \boldsymbol{\Sigma})$, then

$$\sqrt{n}\,[\boldsymbol{g}(\boldsymbol{Y}) - \boldsymbol{g}(\boldsymbol{\mu})] \to_d \mathrm{N}_k\left[\boldsymbol{0}, \boldsymbol{g}'(\boldsymbol{\mu})\boldsymbol{\Sigma}\boldsymbol{g}'(\boldsymbol{\mu})^\mathsf{T}\right].$$

References

Agresti, A. (1990). *Categorical data analysis*. New York: Wiley.
Akaike, H. (1973). Information theory and an extension of the maximum likelihood principle. In B.N. Petrov & F. Csaki (Eds.), *Second International Symposium on Information Theory* (pp. 267–281). Budapest: Akademia Kiado.
Allen, J., Zwerdling, R., Ehrenkranz, R., Gaultier, C., Geggel, R., Greenough, A., Kleinman, R., Klijanowicz, A., Martinez, F., Ozdemir, A., Panitch, H., Nickerson, B., Stein, M., Tomezsko, J., van der Anker, J., & American Thoracic Society. (2003). Statement of the care of the child with chronic lung disease of infancy and childhood. *American Journal of Respiratory and Critical Care Medicine, 168*, 356–396.
Altham, D. (1991). *Practical statistics for medical research*. Boca Raton: Chapman and Hall/CRC.
Altham, P. (1969). Exact Bayesian analysis of a 2×2 contingency table and Fisher's 'exact' significance test. *Journal of the Royal Statistical Society, Series B, 31*, 261–269.
Arcones, M., & E. Giné. (1992). On the bootstrap of M-estimators and other statistical functionals. In R. LePage & L. Billard (Eds.), *Exploring the limits of bootstrap*. New York: Wiley.
Armitage, P., & Berry, G. (1994). *Statistical methods in medical research, third edition*. Oxford: Blackwell Science.
Bachrach, L., Hastie, T., Wang, M.-C., Narasimhan, B., & Marcus, R. (1999). Bone mineral acquisition in healthy Asian, Hispanic, Black and Caucasian youth. A longitudinal study. *Journal of Clinical Endocrinology and Metabolism, 84*, 4702–4712.
Bahadur, R. (1961). A representation of the joint distribution of responses to n dichotomous items. In H. Solomon (Ed.), *Studies on item analysis and prediction* (pp. 158–168). Stanford: Stanford Mathematical Studies in the Social Sciences VI, Stanford University Press.
Barnett, V. (2009). *Comparative statistical inference* (3rd ed.). New York: Wiley.
Bartlett, M. (1957). A comment on D.V. Lindley's statistical paradox. *Biometrika, 44*, 533–534.
Bates, D. (2011). Computational methods for mixed models. Technical report, http://cran.r-project.org/web/packages/lme4/index.html.
Bates, D., & Watts, D. (1980). Curvature measures of nonlinearity (with discussion). *Journal of the Royal Statistical Society, Series B, 42*, 1–25.
Bates, D., & Watts, D. (1988). *Nonlinear regression analysis and its applications*. New York: Wiley.
Bauer, E., & Kohavi, R. (1999). An empirical comparison of voting classification algorithms: Bagging, boosting, and variants. *Machine Learning, 36*, 105–139.
Bayes, T. (1763). An essays towards solving a problem in the doctrine of chances. *Philosophical Transactions of the Royal Society of London, 53*, 370–418. Reprinted, with an introduction by George Barnard, in 1958 in *Biometrika, 45*, 293–315.

Beal, S., & Sheiner, L. (1982). Estimating population kinetics. *CRC Critical Reviews in Biomedical Engineering*, *8*, 195–222.

Beale, E. (1960). Confidence regions in non-linear estimation (with discussion). *Journal of the Royal Statistical Society, Series B*, *22*, 41–88.

Beaumont, M., Wenyang, Z., & Balding, D. (2002). Approximate Bayesian computation in population genetics. *Genetics*, *162*, 2025–2035.

Benjamini, Y., & Hochberg, Y. (1995). Controlling the false discovery rate: A practical and powerful approach to multiple testing. *Journal of the Royal Statistical Society, Series B*, *57*, 289–300.

Berger, J. (2003). Could Fisher, Jeffreys and Neyman have agreed on testing? (with discussion). *Statistical Science*, *18*, 1–32.

Berger, J. (2006). The case for objective Bayesian analysis. *Bayesian Analysis*, *1*, 385–402.

Berger, J., & Bernardo, J. (1992). On the development of reference priors (with discussion). In J. Bernardo, J. Berger, A. Dawid, & A. Smith (Eds.), *Bayesian statistics 4, Proceedings of the Fourth Valencia International Meeting* (pp. 35–60). Oxford: Oxford University Press.

Berger, J. & Wolpert, R. (1988). *The likelihood principle: A review, generalizations, and statistical implications*. Hayward: IMS Lecture Notes.

Berk, R. (2008). *Statistical learning from a regression perspective*. New York: Springer.

Bernardo, J. (1979). Reference posterior distributions for Bayesian inference (with discussion). *Journal of the Royal Statistical Society, Series B*, *41*, 113–147.

Bernardo, J., & Smith, A. (1994). *Bayesian theory*. New York: Wiley.

Bernstein, S. (1917). *Theory of probability (Russian)*. Moscow-Leningrad: Gostekhizdat.

Besag, J., & Kooperberg, C. (1995). On conditional and intrinsic auto-regressions. *Biometrika*, *82*, 733–746.

Besag, J., York, J., & Mollié, A. (1991). Bayesian image restoration with two applications in spatial statistics. *Annals of the Institute of Statistics and Mathematics*, *43*, 1–59.

Bickel, P., & Freedman, D. (1981). Some asymptotic theory for the bootstrap. *Annals of Statistics*, *9*, 1196–1217.

Bishop, Y., Feinberg, S., & Holland, P. (1975). *Discrete multivariate analysis: Theory and practice*. Cambridge: MIT.

Black, D. (1984). *Investigation of the possible increased incidence of cancer in West Cumbria*. London: Report of the Independent Advisory Group, HMSO.

Bliss, C. (1935). The calculation of the dosage-mortality curves. *Annals of Applied Biology*, *22*, 134–167.

de Boor, C. (1978). *A practical guide to splines*. New York: Springer.

Bowman, A., & Azzalini, A. (1997). *Applied smoothing techniques for data analysis*. Oxford: Oxford University Press.

Breiman, L. (1996). Bagging predictors. *Machine Learning*, *24*, 123–140.

Breiman, L. (2001a). Random forests. *Machine Learning*, *45*, 5–32.

Breiman, L. (2001b). Statistical modeling: The two cultures (with discussion). *Statistical Science*, *16*, 199–231.

Breiman, L., & Spector, P. (1992). Submodel selection and evaluation in regression. the x-random case. *International Statistical Review*, *60*, 291–319.

Breiman, L., Friedman, J., Olshen, R., & Stone, C. (1984). *Classification and regression trees*. Monterrey: Wadsworth.

Breslow, N. (2005). Whither PQL? In D. Lin & P. Heagerty (Eds.), *Proceedings of the Second Seattle Symposium* (pp. 1–22). New York: Springer.

Breslow, N. & Chatterjee, N. (1999). Design and analysis of two-phase studies with binary outcome applied to Wilms tumour prognosis. *Applied Statistics*, *48*, 457–468.

Breslow, N., & Clayton, D. (1993). Approximate inference in generalized linear mixed models. *Journal of the American Statistical Association*, *88*, 9–25.

Breslow, N., & Day, N. (1980). *Statistical methods in cancer research, Volume 1- The analysis of case-control studies*. Lyon: IARC Scientific Publications No. 32.

Brinkman, N. (1981). Ethanol fuel – a single-cylinder engine study of efficiency and exhaust emissions. *SAE Transcations*, *90*, 1410–1424.

Brooks, S., Gelman, A., Jones, G., & Meng, X.-L. (Eds.). (2011). *Handbook of Markov chain Monte Carlo*. Boca Raton: Chapman and Hall/CRC.

Bühlmann, P., & Yu, B. (2002). Analyzing bagging. *The Annals of Statistics*, *30*, 927–961.

Buja, A., Hastie, T., & Tibshirani, R. (1989). Linear smoothers and additive models (with discussion). *Annals of Statistics*, *17*, 453–555.

Buse, A. (1982). The likelihood ratio, Wald, and Lagrange multiplier tests: an expository note. *The American Statistician*, *36*, 153–157.

Cameron, A., & Trivedi, P. (1998). *Regression analysis of count data*. Cambridge: Cambridge University Press.

Carey, V., Zeger, S., & Diggle, P. (1993). Modeling multivariate binary data with alternating logistic regressions. *Biometrika*, *80*, 517–526.

Carlin, B., & Louis, T. (2009). *Bayesian methods for data analysis* (3rd ed.). Boca Raton: Chapman and Hall/CDC.

Carroll, R., & Ruppert, D. (1988). *Transformations and weighting in regression*. Boca Raton: Chapman and Hall/CRC.

Carroll, R., Ruppert, D., & Stefanski, L. (1995). *Measurement error in nonlinear models*. Boca Raton: Chapman and Hall/CRC.

Carroll, R., Rupert, D., Stefanski, L., & Crainiceanu, C. (2006). *Measurement error in nonlinear models: A modern perspective* (2nd ed.). Boca Raton: Chapman and Hall/CRC.

Casella, G., & Berger, R. (1987). Reconciling Bayesian evidence in the one-sided testing problem. *Journal of the American Statistical Association*, *82*, 106–111.

Casella, G., & Berger, R. (1990). *Statistical inference*. Pacific Grove: Wadsworth and Brooks.

Chaloner, K., & Brant, R. (1988). A Bayesian approach to outlier detection and residual analysis. *Biometrika*, *75*, 651–659.

Chambers, R., & Skinner, C. (2003). *Analysis of survey data*. New York: Wiley.

Chan, K., & Geyer, C. (1994). Discussion of "Markov chains for exploring posterior distributions". *The Annals of Statistics*, *22*, 1747–1758.

Chatfield, C. (1995). Model uncertainty, data mining and statistical inference (with discussion). *Journal of the Royal Statistical Society, Series A*, *158*, 419–466.

Chaudhuri, P., & Marron, J. (1999). SiZer for exploration of structures in curves. *Journal of the American Statistical Association*, *94*, 807–823.

Chen, S., Donoho, D., & Saunders, M. (1998). Atomic decomposition by basis pursuit. *SIAM Journal of Scientific Computing*, *20*, 33–61.

Chipman, H., George, E., & McCulloch, R. (1998). Bayesian cart model search (with discussion). *Journal of the American Statistical Association*, *93*, 935–960.

Clayton, D., & Hills, M. (1993). *Statistical models in epidemiology*. Oxford: Oxford University Press.

Clayton, D., & Kaldor, J. (1987). Empirical Bayes estimates of age-standardized relative risks for use in disease mapping. *Biometrics*, *43*, 671–682.

Cleveland, W., Grosse, E., & Shyu, W. (1991). Local regression models. In J. Chambers & T. Hastie (Eds.), *Statistical models in S* (pp. 309–376). Pacific Grove: Wadsworth and Brooks/Cole.

Cochran, W. (1977). *Sampling techniques*. New York: Wiley.

Cook, R., & Weisberg, S. (1982). *Residuals and influence in regression*. Boca Raton: Chapman and Hall/CRC.

Cox, D. (1972). The analysis of multivariate binary data. *Journal of the Royal Statistical Society, Series C*, *21*, 113–120.

Cox, D. (1983). Some remarks on overdispersion. *Biometrika*, *70*, 269–274.

Cox, D. (2006). *Principles of statistical inference*. Cambridge: Cambridge University Press.

Cox, D., & Hinkley, D. (1974). *Theoretical statistics*. Boca Raton: Chapman and Hall/CRC.

Cox, D., & Reid, N. (2000). *The theory of the design of experiments*. Boca Raton: Chapman and Hall/CRC.

Cox, D., & Snell, E. (1989). *The analysis of binary data* (2nd ed.). Boca Raton: Chapman and Hall/CRC.
Craig, P., Goldstein, M., Seheult, A., & Smith, J. (1998). Constructing partial prior specifications for models of complex physical systems. *Journal of the Royal Statistical Society, Series D, 47*, 37–53.
Crainiceanu, C., Ruppert, D., & Wand, M. (2005). Bayesian analysis for penalized spline regression using WinBUGS. *Journal of Statistical Software, 14*, 1–24.
Craven, P., & Wabha, G. (1979). Smoothing noisy data with spline functions. *Numerische Mathematik, 31*, 377–403.
Crowder, M. (1986). On consistency and inconsistency of estimating equations. *Econometric Theory, 2*, 305–330.
Crowder, M. (1987). On linear and quadratic estimating functions. *Biometrika, 74*, 591–597.
Crowder, M. (1995). On the use of a working correlation matrix in using generalized linear models for repeated measures. *Biometrika, 82*, 407–410.
Crowder, M., & Hand, D. (1990). *Analysis of repeated measures*. Boca Raton: Chapman and Hall/CRC.
Crowder, M., & Hand, D. (1996). *Practical longitudinal data analysis*. Boca Raton: Chapman and Hall/CRC.
Darby, S., Hill, D., & Doll, R. (2001). Radon: a likely carcinogen at all exposures. *Annals of Oncology, 12*, 1341–1351.
Darroch, J., Lauritzen, S., & Speed, T. (1980). Markov fields and log-linear interaction models for contingency tables. *The Annals of Statistics, 8*, 522–539.
Davidian, M., & Giltinan, D. (1995). *Nonlinear models for repeated measurement data*. Boca Raton: Chapman and Hall/CRC.
Davies, O. (1967). *Statistical methods in research and production* (3rd ed.). London: Olive and Boyd.
Davison, A. (2003). *Statistical models*. Cambridge: Cambridge University Press.
Davison, A., & Hinkley, D. (1997). *Bootstrap methods and their application*. Cambridge: Cambridge University Press.
De Finetti, B. (1974). *Theory of probability, volume 1*. New York: Wiley.
De Finetti, B. (1975). *Theory of probability, volume 2*. New York: Wiley.
Demidenko, E. (2004). *Mixed models. Theory and applications*. New York: Wiley.
Dempster, A. P., Laird, N. M., & Rubin, D. B. (1977). Maximum likelihood from incomplete data via the em algorithm. *Journal of the Royal Statistical Society, Series B, 39*(1), 1–38.
Denison, D., & Holmes, C. (2001). Bayesian partitioning for estimating disease risk. *Biometrics, 57*, 143–149.
Denison, D., Holmes, C., Mallick, B., & Smith, A. (2002). *Bayesian methods for nonlinear classification and regression*. New York: Wiley.
Dennis, J., Jr, & Schnabel, R. (1996). *Numerical methods for unconstrained optimization and nonlinear equations*. Englewood Cliffs: Siam.
Devroye, L. (1986). *Non-uniform random variate generation*. New York: Springer.
Diaconis, P., & Freedman, D. (1986). On the consistency of Bayes estimates. *Annals of Statistics, 14*, 1–26.
Diaconis, P., & Ylvisaker, D. (1980). Quantifying prior opinion (with discussion). In J. Bernardo, M. D. Groot, D. Lindley, & A. Smith (Eds.), *Bayesian statistics 2* (pp. 133–156). Amsterdam: North Holland.
DiCiccio, T., Kass, R., Raftery, A., & Wasserman, L. (1997). Computing Bayes factors by combining simulation and asymptotic approximations. *Journal of the American Statistical Association, 92*, 903–915.
Diggle, P., & Rowlingson, B. (1994). A conditional approach to point process modelling of raised incidence. *Journal of the Royal Statistical Society, Series A, 157*, 433–440.
Diggle, P., Morris, S., & Wakefield, J. (2000). Point source modelling using matched case-control data. *Biostatistics, 1*, 89–105.

Diggle, P., Heagerty, P., Liang, K.-Y., & Zeger, S. (2002). *Analysis of longitudinal data* (2nd ed.). Oxford: Oxford University Press.

Doob, J. (1948). *Le Calcul des Probabilités et ses Applications*, Chapter Application of the theory of martingales (pp. 22–28). Colloques Internationales du CNRS Paris.

Duchon, J. (1977). Splines minimizing rotation-invariant semi-norms in Solobev spaces. In W. Schemp & K. Zeller (Eds.), *Construction theory of functions of several variables* (pp. 85–100). New York: Springer.

Dwyer, J., Andrews, E., Berkey, C., Valadian, I., & Reed, R. (1983). Growth in "new" vegetarian preschool children using the Jenss-Bayley curve fitting technique. *American Journal of Clinical Nutrition, 37*, 815–827.

Efron, B. (1975). The efficiency of logistic regression compared to normal discriminant analysis. *Journal of the American Statistical Association, 70*, 892–898.

Efron, B. (1979). Bootstrap methods: Another look at the jacknife. *Annals of Statistics, 7*, 1–26.

Efron, B. (2008). Microarrays, empirical Bayes and the two groups model (with discussion). *Statistical Science, 23*, 1–47.

Efron, B., & Tibshirani, R. (1993). *An introduction to the bootstrap*. Boca Raton: Chapman and Hall/CRC.

Efroymson, M. (1960). Multiple regression analysis. In A. Ralston & H. Wilf (Eds.), *Mathematical methods for digital computers* (pp. 191–203). New YOrk: Wiley.

Eilers, P., & Marx, B. (1996). Flexible smoothing with B-splines and penalties. *Statistical Science, 11*, 89–102.

Essenberg, J. (1952). Cigarette smoke and the incidence of primary neoplasm of the lung in the albino mouse. *Science, 116*, 561–562.

Evans, M., & Swartz, T. (2000). *Approximating integrals via Monte Carlo and deterministic methods*. Oxford: Oxford University Press.

Fan, J. (1992). Design-adaptive nonparametric regression. *Journal of the American Statistical Association, 87*, 1273–1294.

Fan, J. (1993). Local linear regression smoothers and their minimax efficiencies. *Annals of Statistics, 21*, 196–215.

Fan, J. & I. Gijbels (1996). *Local polynomial modelling and its applications*. Boca Raton: Chapman and Hall/CRC.

Faraway, J. (2004). *Linear models with R*. Boca Raton: Chapman and Hall/CRC.

Fearnhead, P., & Prangle, D. (2012). Constructing summary statistics for approximate bayesian computation: semi-automatic approximate bayesian computation (with discussion). *Journal of the Royal Statistical Society, Series B, 74*, 419–474.

Ferguson, T. (1996). *A course in large sample theory*. Boca Raton: Chapman and Hall/CRC.

Feynman, R. (1951). The concept of probability in quantum mechanics. In J. Neyman (Ed.), *Proceedings of the Second Berkeley Symposium on Mathematical Statistics and Probability* (pp. 535–541). California: University of California Press.

Fine, P., Ponnighaus, J., Maine, N., Clarkson, J., & Bliss, L. (1986). Protective efficacy of BCG against leprosy in Northern Malawi. *The Lancet, 328*, 499–502.

Firth, D. (1987). On the efficiency of quasi-likelihood estimation. *Biometrika, 74*, 233–245.

Firth, D. (1993). Recent developments in quasi-likelihood methods. In *Bulletin of the international Statistical Institute, 55*, 341–358.

Fisher, R. (1922). On the mathematical foundations of theoretical statistics. *Philosophical Transactions of the Royal Society of London, Series A, 222*, 309–368.

Fisher, R. (1925a). *Statistical methods for research workers*. Edinburgh: Oliver and Boyd.

Fisher, R. (1925b). Theory of statistical estimation. *Proceedings of the Cambridge Philosophical Society, 22*, 700–725.

Fisher, R. (1935). The logic of inductive inference (with discussion). *Journal of the Royal Statistical Society, Series A, 98*, 39–82.

Fisher, R. (1936). The use of multiple measurements in taxonomic problems. *Annals of Eugenics, 7*, 179–188.

Fisher, R. (1990). *Statistical methods, experimental design and scientific inference.* Oxford: Oxford University Press.
Fitzmaurice, G., & Laird, N. (1993). A likelihood-based method for analyzing longitudinal binary responses. *Biometrika, 80,* 141–151.
Fitzmaurice, G., Laird, N., & Rotnitzky, A. (1993). Regression models for discrete longitudinal responses (with discussion). *Statistical Science, 8,* 248–309.
Fitzmaurice, G., Laird, N., & Ware, J. (2004). *Applied longitudinal analysis.* New York: Wiley.
Fong, Y., Rue, H., & Wakefield, J. (2010). Bayesian inference for generalized linear models. *Biostatistics, 11,* 397–412.
Freedman, D. (1997). From association to causation via regression. *Advances in Applied Mathematics, 18,* 59–110.
Freund, Y., & Schapire, R. (1997). Experiments with a new boosting algorithm. In *Machine Learning: Proceedings for the Thirteenth International Conference, San Fransisco* (pp. 148–156). Los Altos: Morgan Kaufmann.
Friedman, J. (1979). A tree-structured approach to nonparametric multiple regression. In T. Gasser & M. Rosenblatt (Eds.), *Smoothing techniques for curve estimation* (pp. 5–22). New York: Springer.
Friedman, J. (1991). Multivariate adaptive regression splines (with discussion). *Annals of Statistics, 19,* 1–141.
Friedman, J., Hastie, T., & Tibshirani, R. (2000). Additive logistic regression: A statistical view of boosting (with discussion). *Annals of Statistics, 28,* 337–407.
Gallant, A. (1987). *Nonlinear statistical models.* New York: Wiley.
Gamerman, D. and Lopes, H. F. (2006). *Markov chain Monte Carlo: Stochastic simulation for Bayesian inference* (2nd ed.). Boca Raton: Chapman and Hall/CRC.
Gasser, T., Stroka, L., & Jennen-Steinmetz, C. (1986). Residual variance and residual pattern in nonlinear regression. *Biometrika, 73,* 625–633.
Gelfand, A. E., Diggle, P. J., Fuentes, M., & Guttorp, P. (Eds.). (2010). *Handbook of spatial statistics.* Boca Raton: Chapman and Hall/CRC.
Gelman, A. (2006). Prior distributions for variance parameters in hierarchical models. *Bayesian Analysis, 1,* 515–534.
Gelman, A., & Hill, J. (2007). *Data analysis using regression and multilevel/hierarchical models.* Cambridge: Cambridge University Press.
Gelman, A., & Rubin, D. (1992). Inference from iterative simulation using multiple sequences. *Statistical Science, 7,* 457–511.
Gelman, A., Carlin, J., Stern, H., & Rubin, D. (2004). *Bayesian data analysis* (2nd ed.). Boca Raton: Chapman and Hall/CRC.
Gibaldi, M., & Perrier, D. (1982). *Pharmacokinetics* (2nd ed.). New York: Marcel Dekker.
Giné, E., Götze, F., & Mason, D. (1997). When is the Student t-statistic asymptotically normal? *The Annals of Probability, 25,* 1514–1531.
Glynn, P., & Iglehart, D. (1990). Simulation output using standardized time series. *Mathematics of Operations Research, 15,* 1–16.
Gneiting, T., & Raftery, A. (2007). Strictly proper scoring rules, prediction, and estimation. *Journal of the American Statistical Association, 102,* 359–378.
Godambe, V., & Heyde, C. (1987). Quasi-likelihood and optimal estimation. *International Statistical Review, 55,* 231–244.
Godfrey, K. (1983). *Compartmental models and their applications.* London: Academic.
Goldstein, M., & Wooff, D. (2007). *Bayes linear statistics, theory and methods.* New York: Wiley.
Golub, G., Heath, M. & Wabha, G. (1979). Generalized cross-validation as a method for choosing a good ridge parameter. *Technometrics, 21,* 215–223.
Goodman, S. (1993). p values, hypothesis tests and likelihood: Implications for epidemiology of a neglected historical debate. *American Journal of Epidemiology, 137,* 485–496.
Gordon, L., & Olshen, R. A. (1978). Asymptotically efficient solutions to the classification problems. *Annals of Statistics, 6,* 515–533.

References

Gordon, L., & Olshen, R. A. (1984). Almost surely consistent nonparametric regression from recursive partitioning schemes. *Journal of Multivariate Analysis, 15*, 147–163.

Gourieroux, C., Montfort, A., & Trognon, A. (1984). Pseudo-maximum likelihood methods: Theory. *Econometrica, 52*, 681–700.

Green, P., & Silverman, B. (1994). *Nonparametric regression and generalized linear models*. Boca Raton: Chapman and Hall/CRC.

Green, P. J. (1995). Reversible jump MCMC computation and Bayesian model determination. *Biometrika, 82*, 711–732.

Greenland, S., Robins, J., & Pearl, J. (1999). Confounding and collapsibility in causal inference. *Statistical Science, 14*, 29–46.

Gu, C. (2002). *Smoothing spline ANOVA models*. New York: Springer.

Haberman, S. (1977). Maximum likelihood estimates in exponential response models. *Annals of Statistics, 5*, 815–841.

Hand, D. and Crowder, M. (1991). *Practical longitudinal data analysis*. Boca Raton: Chapman and Hall/CRC Press.

Haldane, J. (1948). The precision of observed values of small frequencies. *Biometrika, 35*, 297–303.

Härdle, W., Hall, P., & Marron, J. (1988). How far are automatically chosen smoothing parameters from their optimum? *Journal of the American Statistical Association, 83*, 86–101.

Hastie, T., & Tibshirani, R. (1990). *Generalized additive models*. Boca Raton: Chapman and Hall/CRC.

Hastie, T., & Tibshirani, R. (1993). Varying-coefficient models. *Journal of the Royal Statistical Society, Series B, 55*, 757–796.

Hastie, T., Tibshirani, R., & Friedman, J. (2009). *The elements of statistical learning* (2nd ed.). New York: Springer.

Hastings, W. (1970). Monte Carlo sampling methods using Markov chains and their applications. *Biometrika, 57*, 97–109.

Haughton, D. (1988). On the choice of a model to fit data from an exponential family. *The Annals of Statistics, 16*, 342–355.

Haughton, D. (1989). Size of the error in the choice of a model to fit from an exponential family. *Sankhya: The Indian Journal of Statistics, Series A, 51*, 45–58.

Heagerty, P., Kurland, B. (2001). Misspecified maximum likelihood estimates and generalised linear mixed models. *Biometrika, 88*, 973–986.

Heyde, C. (1997). *Quasi-likelihood and its applications*. New York: Springer.

Hobert, J., & Casella, G. (1996). The effect of improper priors on Gibbs sampling in hierarchical linear mixed models. *Journal of the American Statistical Association, 91*, 1461–1473.

Hodges, J., & Reich, B. (2010). Adding spatially-correlated errors can mess up the fixed effect you love. *The American Statistician, 64*, 325–334.

Hoerl, A., & Kennard, R. (1970). Ridge regression: Biased estimation for non-orthogonal problems. *Technometrics, 12*, 55–67.

Hoff, P. (2009). *A first course in Bayesian statistical methods*. New York: Springer.

Holst, U., Hössjer, O., Björklund, C., Ragnarson, P., & Edner, H. (1996). Locally weighted least squares kernel regression and statistical evaluation of LIDAR measurements. *Environmetrics, 7*, 401–416.

Hothorn, T., Hornik, K., & Zeileis, A. (2006). Unbiased recursive partitioning: A conditional inference framework. *Journal of Computational and Graphical Statistics, 15*, 651–674.

Huber, P. (1967). The behavior of maximum likelihood estimators under non-standard conditions. In L. LeCam & J. Neyman (Eds.), *Proceedings of the Fifth Berkeley Symposium on Mathematical Statistics and Probability* (pp. 221–233). California: University of California Press.

Inoue, L., & Parmigiani, G. (2009). *Decision theory: Principles and approaches*. New York: Wiley.

Izenman, A. (2008). *Modern multivariate statistical techniques: Regression, classification, and manifold learning*. New York: Springer.

Jeffreys, H. (1961). *Theory of probability* (3rd ed.). Oxford: Oxford University Press.

Jenss, R., & Bayley, N. (1937). A mathematical method for studying the growth of a child. *Human Biology, 9*, 556–563.

Johnson, N., Kotz, S., & Balakrishnan, N. (1994). *Continuous univariate distributions, volume 1* (2nd ed.). New York: Wiley.

Johnson, N., Kotz, S., & Balakrishnan, N. (1995). *Continuous univariate distributions, volume 2* (2nd ed.). New York: Wiley.

Johnson, N., Kotz, S., & Balakrishnan, N. (1997). *Discrete multivariate distributions*. New York: Wiley.

Johnson, N., Kemp, A., & Kotz, S. (2005). *Univariate discrete distributions* (3rd ed.). New York: Wiley.

Johnson, V. (2008). Bayes factors based on test statistics. *Journal of the Royal Statistical Society, Series B, 67*, 689–701.

Jordan, M., Ghahramani, Z., Jaakkola, T., & Saul, L. (1999). An introduction to variational methods for graphical models. *Machine Learning, 37*, 183–233.

Kadane, J., & Wolfson, L. (1998). Experiences in elicitation. *Journal of the Royal Statistical Society, Series D, 47*, 3–19.

Kalbfleisch, J., & Prentice, R. (2002). *The statistical analysis of failure time data* (2nd ed.). New York: Wiley.

Kass, R., & Raftery, A. (1995). Bayes factors. *Journal of the American Statistical Association, 90*, 773–795.

Kass, R., & Vaidyanathan, S. (1992). Approximate Bayes factors and orthogonal parameters, with application to testing equality of two binomial proportions. *Journal of the Royal Statistical Society, Series B, 54*, 129–144.

Kass, R., Tierney, L., & Kadane, J. (1990). The validity of posterior expansions based on Laplace's method. In S. Geisser, J. Hodges, S. Press, & A. Zellner (Eds.), *Bayesian and likelihood methods in statistics and econometrics* (pp. 473–488). Amsterdam: North-Holland.

Kauermann, G. (2005). A note on smoothing parameter selection for penalized spline smoothing. *Journal of Statistical Planning and Inference, 127*, 53–69.

Kauermann, G., & Carroll, R. (2001). A note on the efficiency of sandwich covariance matrix estimation. *Journal of the American Statistical Association, 96*, 1387–1396.

Kemp, I., Boyle, P., Smans, M., & Muir, C. (1985). *Atlas of cancer in Scotland, 1975–1980: Incidence and epidemiologic perspective*. Lyon: IARC Scientific Publication No. 72.

Kerr, K. (2009). Comments on the analysis of unbalanced microarray data. *Bioinformatics, 25*, 2035–2041.

Kim, H., & Loh, W.-Y. (2001). Classification trees with unbiased multiway splits. *Journal of the American Statistical Association, 96*, 589–604.

Knafl, G., Sacks, J., & Ylvisaker, D. (1985). Confidence bands for regression functions. *Journal of the American Statistical Association, 80*, 683–691.

Knorr-Held, L., & Rasser, G. (2000). Bayesian detection of clusters and discontinuities in disease maps. *Biometrics, 56*, 13–21.

Korn, E., & Graubard, B. (1999). *Analysis of health surveys*. New York: Wiley.

Kosorok, M. (2008). *Introduction to empirical processes and semiparametric inference*. New York: Springer.

Kotz, S., Balakrishnan, N., & Johnson, N. (2000). *Continuous multivariate distributions, volume 1* (2nd ed.). New York: Wiley.

Laird, N., & Ware, J. (1982). Random-effects models for longitudinal data. *Biometrics, 38*, 963–974.

Lange, N., & Ryan, L. (1989). Assessing normality in random effects models. *Annals of Statistics, 17*, 624–642.

Lehmann, E. (1986). *Testing statistical hypotheses* (2nd ed.). New York: Wiley.

Lehmann, E., & Romano, J. (2005). Generalizations of the familywise error rate. *Annals of Statistics, 33*, 1138–1154.

van der Lende, R., Kok, T., Peset, R., Quanjer, P., Schouten, J., & Orie, N. G. (1981). Decreases in VC and FEV1 with time: Indicators for effects of smoking and air pollution. *Bulletin of European Physiopathology and Respiration, 17,* 775–792.
Liang, K., & Zeger, S. (1986). Longitudinal data analysis using generalized linear models. *Biometrika, 73,* 13–22.
Liang, K.-Y., & McCullagh, P. (1993). Case studies in binary dispersion. *Biometrics, 49,* 623–630.
Liang, K.-Y., Zeger, S., & Qaqish, B. (1992). Multivariate regression analyses for categorical data (with discussion). *Journal of the Royal Statistical Society, Series B, 54,* 3–40.
Lindley, D. (1957). A statistical paradox. *Biometrika, 44,* 187–192.
Lindley, D. (1968). The choice of variables in multiple regression (with discussion). *Journal of the Royal Statistical Society, Series B, 30,* 31–66.
Lindley, D. (1980). Approximate Bayesian methods. In J. Bernardo, M. D. Groot, D. Lindley, & A. Smith (Eds.), *Bayesian statistics* (pp. 223–237). Valencia: Valencia University Press.
Lindley, D., & Smith, A. (1972). Bayes estimates for the linear model (with discussion). *Journal of the Royal Statistical Society, Series B, 34,* 1–41.
Lindsey, J., Byrom, W., Wang, J., Jarvis, P., & Jones, B. (2000). Generalized nonlinear models for pharmacokinetic data. *Biometrics, 56,* 81–88.
Lindstrom, M., & Bates, D. (1990). Nonlinear mixed-effects models for repeated measures data. *Biometrics, 46,* 673–687.
Lipsitz, S., Laird, N., & Harrington, D. (1991). Generalized estimating equations for correlated binary data: Using the odds ratio as a measure of association. *Biometrika, 78,* 153–160.
Little, R., & Rubin, D. (2002). *Statistical analysis with missing data* (2nd ed.). New York: Wiley.
Loader, C. (1999). *Local regression and likelihood.* New York: Springer.
Lumley, T. (2010). *Complex surveys: A guide to analysis using R.* New York: Wiley.
Lumley, T., Diehr, P., Emerson, S., & Chen, L. (2002). The importance of the normality assumption in large public health data sets. *Annual Reviews of Public Health, 23,* 151–169.
Machin, D., Farley, T., Busca, B., Campbell, M., & d'Arcangues, C. (1988). Assessing changes in vaginal bleeding patterns in contracepting women. *Contraception, 38,* 165–179.
Malahanobis, P. (1936). On the generalised distance in statistics. *Proceedings of the National Institute of Sciences of India, 2,* 49–55.
Mallows, C. (1973). Some comments on C_p. *Technometrics, 15,* 661–667.
Marra, G., & Wood, S. (2012). Coverage properties of confidence intervals for generalized additive model components. *Scandinavian Journal of Statistics, 39,* 53–74.
van Marter, L., Leviton, A., Kuban, K., Pagano, M., & Allred, E. (1990). Maternal glucocorticoid therapy and reduced risk of bronchopulmonary dysplasia. *Pediatrics, 86,* 331–336.
Matheron, G. (1971). The theory of regionalized variables and its applications. Technical report, Les Cahiers du Centre de Morphologie Mathématique de Fontainebleau, Fascicule 5, Ecole des Mines de Paris.
McCullagh, P. (1983). Quasi-likelihood functions. *The Annals of Statistics, 11,* 59–67.
McCullagh, P., & Nelder, J. (1989). *Generalized linear models* (2nd ed.). Boca Raton: Chapman and Hall/CRC.
McCulloch, C., & Neuhaus, J. (2011). Prediction of random effects in linear and generalized linear models under model misspecification. *Biometrics, 67,* 270–279.
McDonald, B. (1993). Estimating logistic regression parameters for bivariate binary data. *Journal of the Royal Statistical Society, Series B, 55,* 391–397.
Meier, L., van de Geer, S., & Bühlmann, P. (2008). The group lasso for logistic regression. *Journal of the Royal Statistical Society, Series B, 70,* 53–71.
Meinshausen, N., & Yu, B. (2009). Lasso-type recovery of sparse representations for high-dimensional data. *The Annals of Statistics, 37,* 246–270.
Mendel, G. (1866). Versuche über Pflanzen-Hybriden. *Verhandl d Naturfsch Ver in Bünn, 4,* 3–47.
Mendel, G. (1901). Experiments in plant hybridization. *Journal of the Royal Horticultural Society, 26,* 1–32. Translation of Mendel (1866) by W. Bateson.
Meng, X., & Wong, W. (1996). Simulating ratios of normalizing constants via a simple identity. *Statistical Sinica, 6,* 831–860.

Metropolis, N., Rosenbluth, A., Teller, A., & Teller, E. (1953). Equations of state calculations by fast computing machines. *Journal of Chemical Physics, 21*, 1087–1091.

Miller, A. (1990). *Subset selection in regression*. Boca Raton: Chapman and Hall/CRC.

von Mises, R. (1931). *Wahrscheinlichkeitsrecheung*. Leipzig: Franz Deutiche.

Montgomery, D., & Peck, E. (1982). *Introduction to linear regression analysis*. New York: Wiley.

Morgan, J., & Messenger, R. (1973). Thaid: a sequential search program for the analysis of nominal scale dependent variables. Technical report, Ann Arbor: Institute for Social Research, University of Michigan.

Morgan, J., & Sonquist, J. (1963). Problems in the analysis of survey data, and a proposal. *Journal of the American Statistical Association, 58*, 415–434.

Nadaraya, E. (1964). On estimating regression. *Theory of Probability and its Applications, 9*, 141–142.

Naylor, J., & Smith, A. (1982). Applications of a method for the efficient computation of posterior distributions. *Applied Statistics, 31*, 214–225.

Neal, R. (1996). *Bayesian learning for neural networks*. New York: Springer.

Nelder, J. (1966). Inverse polynomials, a useful group of multi-factor response functions. *Biometrics, 22*, 128–141.

Nelder, J. A., & Wedderburn, R. W. M. (1972). Generalized linear models. *Journal of the Royal Statistical Society, Series A, 135*, 370–384.

Neyman, J., & Pearson, E. (1928). On the use and interpretation of certain test criteria for purposes of statistical inference. Part i. *Philosophical Transactions of the Royal Society of London, Series A, 20A*, 175–240.

Neyman, J., & Pearson, E. (1933). On the problem of the most efficient tests of statistical hypotheses. *Philosophical Transactions of the Royal Society of London, Series A, 231*, 289–337.

Neyman, J., & Scott, E. (1948). Consistent estimates based on partially consistent observations. *Econometrica, 16*, 1–32.

Nychka, D. (1988). Bayesian confidence intervals for smoothing splines. *Journal of the American Statistical Association, 83*, 1134–1143.

O'Hagan, A. (1994). *Kendall's advanced theory of statistics, volume 2B: Bayesian inference*. London: Arnold.

O'Hagan, A. (1998). Eliciting expert beliefs in substantial practical applications. *Journal of the Royal Statistical Society, Series D, 47*, 21–35.

O'Hagan, A., & Forster, J. (2004). *Kendall's advanced theory of statistics, volume 2B: Bayesian inference* (2nd ed.). London: Arnold.

Olshen, R. (2007). Tree-structured regression and the differentiation of integrals. *Annals of Statistics, 35*, 1–12.

Ormerod, J., & Wand, M. (2010). Explaining variational approximations. *The American Statistician, 64*, 140–153.

O'Sullivan, F. (1986). A statistical perspective on ill-posed problems. *Statistical Science, 1*, 502–518.

Pagano, M., & Gauvreau, K. (1993). *Principles of biostatistics*. Belmont: Duxbury Press.

Pearl, J. (2009). *Causality: Models, reasoning and inference* (2nd ed.). Cambridge: Cambridge University Press.

Pearson, E. (1953). Discussion of "Statistical inference" by D.V. Lindley. *Journal of the Royal Statistical Society, Series B, 15*, 68–69.

Peers, H. (1971). Likelihood ratio and associated test criteria. *Biometrika, 58*, 577–587.

Pepe, M. (2003). *The statistical evaluation of medical tests for classification and prediction*. Oxford: Oxford University Press.

Pérez, J. M., & Berger, J. O. (2002). Expected-posterior prior distributions for model selection. *Biometrika, 89*, 491–512.

Pinheiro, J., & Bates, D. (2000). *Mixed-effects models in S and splus*. New York: Springer.

Plummer, M. (2008). Penalized loss functions for Bayesian model comparison. *Biostatistics, 9*, 523–539.

Potthoff, R., & Roy, S. (1964). A generalized multivariate analysis of variance useful especially for growth curve problems. *Biometrika, 51*, 313–326.

Prentice, R. (1988). Correlated binary regression with covariates specific to each binary observation. *Biometrics, 44*, 1033–1048.

Prentice, R., & Pyke, R. (1979). Logistic disease incidence models and case-control studies. *Biometrika, 66*, 403–411.

Prentice, R., & Zhao, L. (1991). Estimating equations for parameters in means and covariances of multivariate discrete and continuous responses. *Biometrics, 47*, 825–839.

Qaqish, B., & Ivanova, A. (2006). Multivariate logistic models. *Biometrika, 93*, 1011–1017.

Radelet, M. (1981). Racial characteristics and the imposition of the death sentence. *American Sociological Review, 46*, 918–927.

Rao, C. (1948). Large sample tests of statistical hypotheses concerning several parameters with applications to problems of estimation. *Proceedings of the Cambridge Philosophical Society, 44*, 50–57.

Rao, C., & Wu, Y. (1989). A strongly consistent procedure for model selection in a regression problem. *Biometrika, 76*, 369–374.

Rasmussen, C., & Williams, C. (2006). *Gaussian processes for machine learning.* Cambridge: MIT.

Ravishanker, N., & Dey, D. (2002). *A first course in linear model theory.* Boca Raton: Chapman and Hall/CRC.

Reinsch, C. (1967). Smoothing by spline functions. *Numerische Mathematik, 10*, 177–183.

Reiss, P., & Ogden, R. (2009). Smoothing parameter selection for a class of semiparametric linear models. *Journal of the Royal Statistical Society, Series B, 71*, 505–523.

Rice, J. (1984). Bandwidth choice for nonparametric regression. *Annals of Statistics, 12*, 1215–1230.

Rice, K. (2008). Equivalence between conditional and random-effects likelihoods for pair-matched case-control studies. *Journal of the American Statistical Association, 103*, 385–396.

Ripley, B. (1987). *Stochastic simulation.* New York: Wiley.

Ripley, B. (1996). *Pattern recognition and neural networks.* Cambridge: Cambridge University Press.

Ripley, B. (2004). Selecting amongst large classes of models. In N. Adams, M. Crowder, D. Hand, & D. Stephens (Eds.), *Methods and models in statistics: In honor of Professor John Nelder, FRS* (pp. 155–170). London: Imperial College Press.

Robert, C. (2001). *The Bayesian choice* (2nd ed.). New York: Springer.

Roberts, G., & Sahu, S. (1997). Updating schemes, correlation structure, blocking and parameterization for the Gibbs sampler. *Journal of the Royal Statistical Society, Series B, 59*, 291–317.

Roberts, G., Gelman, A., & Gilks, W. (1997). Weak convergence and optimal scaling of random walk Metropolis algorithms. *The Annals of Applied Probability, 7*, 110–120.

Robinson, G. (1991). That BLUP is a good thing (with discussion). *Statistical Science, 6*, 15–51.

Robinson, L., & Jewell, N. (1991). Some surprising results about covariate adjustment in logistic regression models. *International Statistical Review, 59*, 227–240.

Rosenbaum, P. (2002). *Observational studies* (2nd ed.). New York: Springer.

Rothman, K., & Greenland, S. (1998). *Modern epidemiology* (2nd ed.). Philadelphia: Lipincott, Williams and Wilkins.

Royall, R. (1986). Model robust confidence intervals using maximum likelihood estimators. *International Statistical Review, 54*, 221–226.

Royall, R. (1997). *Statistical evidence – a likelihood paradigm.* Boca Raton: Chapman and Hall/CRC.

Rue, H., & Held, L. (2005). *Gaussian Markov random fields: Theory and application.* Boca Raton: Chapman and Hall/CRC.

Rue, H., Martino, S., & Chopin, N. (2009). Approximate Bayesian inference for latent Gaussian models using integrated nested Laplace approximations (with discussion). *Journal of the Royal Statistical Society, Series B, 71*, 319–392.

Ruppert, D., Wand, M., & Carroll, R. (2003). *Semiparametric regression*. Cambridge: Cambridge University Press.
Salway, R., & Wakefield, J. (2008). Gamma generalized linear models for pharmacokinetic data. *Biometrics, 64*, 620–626.
Savage, L. (1972). *The foundations of statistics* (2nd ed.). New York: Dover.
Scheffé, H. (1959). *The analysis of variance*. New York: Wiley.
Schervish, M. (1995). *Theory of statistics*. New York: Springer.
Schott, J. (1997). *Matrix analysis for statistics*. New York: Wiley.
Schwarz, G. (1978). Estimating the dimension of a model. *Annals of Statistics, 6*, 461–464.
Seaman, S., & Richardson, S. (2004). Equivalence of prospective and retrospective models in the Bayesian analysis of case-control studies. *Biometrika, 91*, 15–25.
Searle, S., Casella, G., & McCulloch, C. (1992). *Variance components*. New York: Wiley.
Seber, G., & Lee, S. (2003). *Linear regression analysis* (2nd ed.). New York: Wiley.
Seber, G., & Wild, C. (1989). *Nonlinear regression*. New York: Wiley.
Sellke, T., Bayarri, M., & Berger, J. (2001). Calibration of p values for testing precise null hypotheses. *The American Statistician, 55*, 62–71.
Sheather, S., & Jones, M. (1991). A reliable data-based bandwidth selection method for kernel density estimation. *Journal of the Royal Statistical Society, Series B, 53*, 683–690.
Sidák, Z. (1967). Rectangular confidence region for the means of multivariate normal distributions. *Journal of the American Statistical Association, 62*, 626–633.
Silverman, B. (1985). Some aspects of the spline smoothing approach to non-parametric regression curve fitting. *Journal of the Royal Statistical Society, Series B, 47*, 1–52.
Simonoff, J. (1997). *Smoothing methods in statistics*. New York: Springer.
Simpson, E. (1951). The interpretation of interaction in contingency tables. *Journal of the Royal Statistical Society, Series B, 13*, 238–241.
Singh, K. (1981). On the asymptotic accuracy of Efron's bootstrap. *Annals of Statistics, 9*, 1187–1195.
Smith, A., & Gelfand, A. (1992). Bayesian statistics without tears: A sampling-resampling perspective. *The American Statistician, 46*, 84–88.
Smith, C. (1947). Some examples of discrimination. *Annals of Eugenics, 13*, 272–282.
Smyth, G., & Verbyla, A. (1996). A conditional likelihood approach to residual maximum likelihood estimation in generalized linear models. *Journal of the Royal Statistical Society, Series B, 58*, 565–572.
Sommer, A. (1982). *Nutritional blindness*. Oxford: Oxford University Press.
Spiegelhalter, D., Best, N., Carlin, B., & van der Linde, A. (1998). Bayesian measures of model complexity and fit (with discussion). *Journal of the Royal Statistical Society, Series B, 64*, 583–639.
Stamey, T., Kabalin, J., McNeal, J., Johnstone, I., Freiha, F., Redwine, E., & Yang, N. (1989). Prostate specific antigen in the diagnosis and treatment of adenocarcinoma of the prostate, II Radical prostatectomy treated patients. *Journal of Urology, 141*, 1076–1083.
Stone, M. (1977). An asymptotic equivalence of choice of model by cross-validation and Akaike's criterion. *Journal of the Royal Statistical Society, Series B, 39*, 44–47.
Storey, J. (2002). A direct approach to false discovery rates. *Journal of the Royal Statistical Society, Series B, 64*, 479–498.
Storey, J. (2003). The positive false discovery rate: A Bayesian interpretation and the q-value. *The Annals of Statistics, 31*, 2013–2035.
Storey, J., Madeoy, J., Strout, J., Wurfel, M., Ronald, J., & Akey, J. (2007). Gene-expression variation within and among human populations. *American Journal of Human Genetics, 80*, 502–509.
Sun, J., & Loader, C. (1994). Confidence bands for linear regression and smoothing. *The Annals of Statistics, 22*, 1328–1345.
Szpiro, A., Rice, K., & Lumley, T. (2010). Model-robust regression and a Bayesian "sandwich" estimator. *Annals of Applied Statistics, 4*, 2099–2113.

Thall, P., & Vail, S. (1990). Some covariance models for longitudinal count data with overdispersion. *Biometrics*, *46*, 657–671.

Tibshirani, R. (1996). Regression shrinkage and selection via the lasso. *Journal of the Royal Statistical Society, Series B*, *58*, 267–288.

Tibshirani, R. (2011). Regression shrinkage and selection via the lasso: a retrospective (with discussion). *Journal of the Royal Statistical Society, Series B*, *73*, 273–282.

Tierney, L., & Kadane, J. (1986). Accurate approximations for posterior moments and marginal densities. *Journal of the American Statistical Association*, *81*, 82–86.

Titterington, D., Murray, G., Murray, L., Spiegelhalter, D., Skene, A., Habbema, J., & Gelpke, G. (1981). Comparison of discrimination techniques applied to a complex data set of head injured patients. *Journal of the Royal Statistical Society, Series A*, *144*, 145–175.

Upton, R., Thiercelin, J., Guentert, T., Wallace, S., Powell, J., Sansom, L., & Riegelman, S. (1982). Intraindividual variability in Theophylline pharmacokinetics: statistical verification in 39 of 60 healthy young adults. *Journal of Pharmacokinetics and Biopharmaceutics*, *10*, 123–134.

van der Vaart, A. (1998). *Asymptotic statistics*. Cambridge: Cambridge University Press.

Vapnick, V. (1996). *The nature of statistical learning theory*. New York: Springer.

Verbeeke, G., & Molenberghs, G. (2000). *Linear mixed models for longitudinal data*. New York: Springer.

Wabha, G. (1983). Bayesian 'confidence intervals' for the cross-validated smoothing spline. *Journal of the Royal Statistical Society, Series B*, *45*, 133–150.

Wabha, G. (1985). A comparison of GCV and GML for choosing the smoothing parameter in the generalized spline problem. *Annals of Statistics*, *13*, 1378–1402.

Wabha, G. (1990). *Spline models for observational data*. Philadelphia: SIAM.

Wakefield, J. (1996). Bayesian individualization via sampling-based methods. *Journal of Pharmacokinetics and Biopharmaceutics*, *24*, 103–131.

Wakefield, J. (2004). Non-linear regression modelling. In N. Adams, M. Crowder, D. Hand, & D. Stephens (Eds.), *Methods and models in statistics: In honor of Professor John Nelder, FRS* (pp. 119–153). London: Imperial College Press.

Wakefield, J. (2007a). A Bayesian measure of the probability of false discovery in genetic epidemiology studies. *American Journal of Human Genetics*, *81*, 208–227.

Wakefield, J. (2007b). Disease mapping and spatial regression with count data. *Biostatistics*, *8*, 158–183.

Wakefield, J. (2008). Ecologic studies revisited. *Annual Review of Public Health*, *29*, 75–90.

Wakefield, J. (2009a). Bayes factors for genome-wide association studies: Comparison with p-values. *Genetic Epidemiology*, *33*, 79–86.

Wakefield, J. (2009b). Multi-level modelling, the ecologic fallacy, and hybrid study designs. *International Journal of Epidemiology*, *38*, 330–336.

Wakefield, J., Smith, A., Racine-Poon, A., & Gelfand, A. (1994). Bayesian analysis of linear and non-linear population models using the Gibbs sampler. *Applied Statistics*, *43*, 201–221.

Wakefield, J., Aarons, L., & Racine-Poon, A. (1999). The Bayesian approach to population pharmacokinetic/pharmacodynamic modelling. In C. Gatsonis, R. E. Kass, B. P. Carlin, A. L. Carriquiry, A. Gelman, I. Verdinelli, & M. West (Eds.), *Case studies in Bayesian statistics, volume IV* (pp. 205–265). New York: Springer.

Wald, A. (1943). Tests of statistical hypotheses concerning several parameters when the number of observations is large. *Transactions of the American Mathematical Society*, *54*, 426–482.

Wand, M., & Jones, M. (1995). *Kernel smoothing*. Boca Raton: Chapman and Hall/CRC.

Wand, M., & Ormerod, J. (2008). On semiparametric regression with O'Sullivan penalised splines. *Australian and New Zealand Journal of Statistics*, *50*, 179–198.

Watson, G. (1964). Smooth regression analysis. *Sankhya*, *A26*, 359–372.

Wedderburn, R. (1974). Quasi-likelihood functions, generalized linear models, and the Gauss-Newton method. *Biometrika*, *61*, 439–447.

Wedderburn, R. (1976). On the existence and uniqueness of the maximum likelihood estimates for certain generalized linear models. *Biometrika*, *63*, 27–32.

West, M. (1993). Approximating posterior distributions by mixtures. *Journal of the Royal Statistical Society, Series B, 55*, 409–422.

West, M., & Harrison, J. (1997). *Bayesian forecasting and dynamic models* (2nd ed.). New York: Springer.

Westfall, P., Johnson, W., & Utts, J. (1995). A Bayesian perspective on the Bonferroni adjustment. *Biometrika, 84*, 419–427.

White, H. (1980). A heteroskedasticity-consistent covariance matrix estimator and a direct test for heteroskedasticity. *Econometrica, 48*, 1721–746.

White, J. (1982). A two stage design for the study of the relationship between a rare exposure and a rare disease. *American Journal of Epidemiology, 115*, 119–128.

Wood, S. (2006). *Generalized additive models: An introduction with R.* Boca Raton: Chapman and Hall/CRC.

Wood, S. (2008). Fast stable direct fitting and smoothness selection for generalized additive models. *Journal of the Royal Statistical Society, Series B, 70*, 495–518.

Wood, S. (2011). Fast stable restricted maximum likelihood and marginal likelihood estimation of semiparametric generalized linear models. *Journal of the Royal Statistical Society, Series B, 73*, 3–36.

Wu, T., & Lange, K. (2008). Coordinate descent algorithms for lasso penalized regression. *The Annals of Applied Statistics, 2*, 224–244.

Yates, F. (1984). Tests of significance for 2×2 contingency tables. *Journal of the Royal Statistical Society, Series B, 147*, 426–463.

Yee, T., & Wild, C. (1996). Vector generalized additive models. *Journal of the Royal Statistical Society, Series B, 58*, 481–493.

Yu, K., & Jones, M. (2004). Likelihood-based local linear estimation of the conditional variance function. *Journal of the American Statistical Association, 99*, 139–144.

Yuan, M., & Lin, Y. (2007). Model selection and estimation in regression with grouped variables. *Journal of the Royal Statistical Society, Series B, 68*, 49–67.

Zeger, S., & Liang, K. (1986). Longitudinal data analysis for discrete and continuous outcomes. *Biometrics, 42*, 121–130.

Zhao, L., & Prentice, R. (1990). Correlated binary regression using a generalized quadratic model. *Biometrika, 77*, 642–648.

Zhao, L., Prentice, R., & Self, S. (1992). Multivariate mean parameter estimation by using a partly exponential model. *Journal of the Royal Statistical Society, Series B, 54*, 805–811.

Zou, H., & Hastie, T. (2005). Regularization and variable selection via the elastic net. *Journal of the Royal Statistical Society, Series B, 67*, 301–320.

Index

Symbols
F distribution, 661, 667
R^2 measure, 183
p-value, 72, 154
    Bayesian view, 160
    misinterpretation, 155
q-value, 170

A
Adjusted R^2, 183
AIC, 184, 533–537
Aliasing, 202
    one-way ANOVA, 225
Almost sure convergence, 673
Alternating logistic regression, 473
Analysis of variance, 224–231
    crossed designs, 226–229
    nested designs, 229
    one-way, 224–226, 371–373, 380–381, 417–420
    random and mixed models, 229–231
Ancillarity, 669
    of covariates, 197
Autocorrelation function, 404
    empirical, 404
Autoregressive error model, 362
Average mean squared error, 513, 515, 527
Average predictive risk, 513, 515, 529

B
Bagging, 635–639
Basis, 655
Bayes classifier, 508
Bayes factors, 137–140, 157
    for nonlinear models, 295–297
    Jeffreys–Lindley paradox, 161
    prior choice for nonlinear models, 295–297
    with improper priors, 158
Bayes risk, 509
Bayesian inference
    asymptotic properties of Bayes estimators, 89–90
    comparison with frequentist inference, 23–24
    implementation, 101–121
    sequential arrival of data, 86
    summarization, 85–89
    using the sampling distribution, 140–143
Bayesian model averaging, 100–101, 186
    prediction example, 538–541
Bernoulli distribution, 308
Bernstein–von Mises theorem, 89
Best linear unbiased predictor, 379
Beta distribution, 658–659
    conjugate prior, 104, 147, 315
Beta-binomial distribution, 105
    as predictive distribution, 105
    for overdispersion, 315
Bias-variance trade-off, 231–236, 512, 514, 532, 576, 580
BIC, 140, 184
Binomial data
    assessment of assumptions, 331–332
    hypothesis testing, 319–321
Binomial distribution, 307–310, 663
    conjugate prior for, 104, 147
    convolution, 310
    exponential family, 104
    genesis, 308–309
    Poisson approximation, 309
    quasi-likelihood version, 51
    rare events, 309

Bonferroni method, 167
    dependent tests, 167
Bootstrap methods, 63–69
    bagging, 636
    for testing, 72
    non-parametric bootstrap, 64
    parametric bootstrap, 64
    percentile interval, 65
    pivotal interval, 65, 83
    random forests, 639
    resampling cases, 65–66
    resampling residuals, 65–66

C
CART, 634
Cartesian product rule, 109
Case-control study, 3, 4, 337–343, 347–348
    frequency matching, 341
    individual matching, 341
    matched pairs, 343
    matching, 3, 341–343
    selection bias, 338
Causal
    effect, 21
    interpretation, 2
    models, 2
Causation
    versus association, 199–201
Central limit theorem, 674
    importance sampling, 111
Chi-squared distribution, 659, 667
Choice based sampling, 337
Classification, 511, 624–643
    logistic models, 625–626
Classification trees, 634–635
Coefficient of variation, 255
Cohort data, 413–415
Cohort study, 337
Collapsibility, 333–337
Compartmental models, 13–15
Complete statistic, 670
Compound symmetry error structure, 360
Conditional modeling, 18
Conditional likelihood, 327–330
    binary mixed model, 465–467
    individual-matched case-control study, 342–343
    mixed model, 437–438
Confidence intervals, 29
    Bayesian, in a smoothing context, 566
    binomial model, 41
    linear smoothing context, 561
    simultaneous, linear smoothing context, 562

Confounding, 2, 3, 21, 201
    bias due to, 2
    by location, 449
    definition of, 233
Conjugacy, 102–105
    random effects models, 449–450
Consistency, 24, 31, 33, 51, 54
    exponential family, 262
    GLMs, 261
    nonlinear models, 285
    quadratic exponential model, 454, 470
    sandwich estimation, 57
    variance estimation, 58
Convergence in distribution, 673
Convergence in probability, 673
Corner-point constraint, 202, 203
    crossed design, 227
    one-way ANOVA, 225
Correlogram, 404
Cost complexity criterion, 618
Cramér-Rao lower bound, 30, 670
Credible interval, 88
Cross-sectional data, 357, 413–415
Cross-validation
    K-fold, 529–532
    estimate of expected prediction error, 533
    generalized, 513, 532–533
    GLMs, 537–538
    kernel density estimation, 633
    leave-one-out, 530–532, 578
    ordinary, 513

D
Data dredging, 4, 6
de Finetti's representation theorem, 135
Decision theory
    hypothesis testing, 138, 157
    multiple testing, 172
    prediction, 506
Delta method, 674
    GLM example, 264–265
Design bias, 579, 583
Deviance, 268
    classification, 635
    scaled, 268
DIC, 144
Dirichlet distribution
    conjugate prior, 149
Discriminant analysis
    linear, 626–628
    quadratic, 626–628
Dynamic linear models, 611

Index

E
Ecological bias, 10
Ecological study, 10
EDA, 6
Effective degrees of freedom, 520, 522, 524, 532, 555, 560, 589
 prior on, 569
 variance estimation, 585
Efficiency, 31
Efficiency-robustness trade-off, 70
 Poisson model, 62
Elastic net, 525
Empirical Bayes
 random effects inference, 376
Empirical logits, 332
Ensemble learning, 638
Entropy, 635
Errors-in-variables, 8, 198, 251–252
Estimating equations, 32
 asymptotic properties, 33
 for GEE, 392
 for variance estimating in GEE, 395–397
Estimating functions, 32–36
 choice of, 69–72
 nonlinear models, 289
 quadratic, 71
 quasi-likelihood, 50
Excess-Poisson variability, 12, 60
Exchangeability, 134–137
 hierarchical models, 381
Exchangeable error structure, 360
Exchangeable variance model
 GEE, 396
Exhaustive search
 deficiencies of, 539
Expected number of false discoveries, 168
Expected squared prediction error, 512, 513, 515
Experimental data, 2
Exponential family, 30
 binomial model, 311
 conjugacy, 102
 linear, 102
 natural parameter, 102
 one-parameter, 257, 669
 two-parameter, 257
Extended hypergeometric distribution, 328

F
Factor variable, 197
Factorial design, 227

False discovery rate, 168
False positive fraction, 516
Family-wise error rate, 166
 k incorrect rejections, 167, 191
Fisher scoring, 263
Fisher's exact test, 330, 343–344
Fisher's information
 conditional, 329
 expected, 37
 exponential model, 82
 GLM, 261–262
 negative binomial model, 81
 nonlinear model, 286
 observed, 39
 sandwich estimation, 58
Fisher-Neyman factorization theorem, 669
Fixed effects, 355
 ANOVA, 230
Fractional polynomial model, 255
Frequentist inference
 comparison with Bayesian inference, 23–24

G
Gamma distribution, 659
 conjugate prior, 148
Gauss rules, 108
Gauss–Hermite rule, 108
 adaptive, 109
 Hermite polynomials, 108
Gauss–Markov theorem, 70, 215–216
Generalization error, 512, 513, 515, 527
Generalized additive models, 598–600
 backfitting, 599–600
Generalized cross-validation, 609
 MARS, 623
Generalized estimating equations, 390–398
 assessment of assumptions, 400–407
 binary data, 467–468
 estimation of variance parameters, 395–397
 GEE2, connected estimating equations, 452–454
 GLMs, 450–452
 nonlinear models, 487–488
 working variance model, 391–394
Generalized linear mixed models, 109, 429–449
 Bayesian inference, 440–442
 binary data, 458–462
 conditional likelihood, 437–438
 likelihood inference, 432–433
 spatial dependence, 444–447
 spline smoothing, 591